Metal Sustainability

Metal Sustainability

Global Challenges, Consequences, and Prospects

Edited by

REED M. IZATT

IBC Advanced Technologies, Inc.,
American Fork, UT, USA
and
Department of Chemistry and Biochemistry
Brigham Young University
Provo, UT, USA

WILEY

This edition first published 2016

Registered Office
John Wiley & Sons, Ltd, The Atrium, Southern Gate, Chichester, West Sussex, PO19 8SQ, United Kingdom

For details of our global editorial offices, for customer services and for information about how to apply for permission to reuse the copyright material in this book please see our website at www.wiley.com.

Library of Congress Cataloging-in-Publication Data

Names: Izatt, Reed M., 1926– editor.
Title: Metal sustainability : global challenges, consequences, and prospects / edited by Reed M. Izatt, IBC Advanced Technologies, Inc., American Fork, Utah, and Department of Chemistry and Biochemistry, Brigham Young University, Provo, Utah.
Description: Chichester, West Sussex : John Wiley & Sons, Ltd., 2016. | Includes bibliographical references and index.
Identifiers: LCCN 2016014406 (print) | LCCN 2016016291 (ebook) | ISBN 9781119009108 (cloth) | ISBN 9781119009146 (pdf) | ISBN 9781119009122 (epub)
Subjects: LCSH: Metals. | Metals–Fatigue. | Metallurgy. | Nonferrous metals–Metallurgy. | Metals–Recycling. | Fracture mechanics
Classification: LCC QD171 .M4164 2016 (print) | LCC QD171 (ebook) | DDC 669/.042–dc23
LC record available at https://lccn.loc.gov/2016014406

A catalogue record for this book is available from the British Library.

Front Cover image: Gettyimages/JacobH

Set in 10/12pt Times by SPi Global, Pondicherry, India

1 2016

Contents

List of Contributors **xvii**
Preface **xxi**
Acknowledgments **xxiii**

1 Recycling and Sustainable Utilization of Precious and Specialty Metals **1**
Reed M. Izatt and Christian Hagelüken

1.1 Introduction 1
1.2 How did we come to this Situation? 4
1.3 Magnitude of the Waste Problem and Disposal of End-of-Life Products 7
1.4 Benefits Derived by the Global Community from Effective Recycling 8
1.5 Urban Mining 13
1.6 Technologies for Metal Separations and Recovery from EOL Wastes 16
1.6.1 Collection, Conditioning, and Pre-processing of Waste 16
1.6.2 Separation and Recovery Technologies 17
1.6.2.1 Integrated Smelter and Advanced Refining Technologies 17
1.6.2.2 Informal Recycling 18
1.7 Conclusions 19
References 21

2 Global Metal Reuse, and Formal and Informal Recycling from Electronic and Other High-Tech Wastes **23**
Ian D. Williams

2.1 Introduction 23
2.2 Metal Sources 24
2.3 E-waste 28
2.4 Responses to the E-waste Problem 29
2.5 Reuse of Metals from High-tech Sources 31
2.5.1 Reuse by Social Enterprises 33
2.5.2 Reuse in the Private Sector 35
2.5.3 Reuse Research 35
2.6 Recycling of Metals from High-tech Sources 36
2.6.1 Ferrous and Non-ferrous Metals 36
2.6.2 Speciality and Precious Metals 37
2.6.3 Formal Recycling 39
2.6.3.1 Collection and Sorting of Metals for Recycling 39
2.6.3.2 Role of the Third Sector 40
2.6.3.3 Technical Aspects of Formal Recycling 40

2.6.3.4 *Metal Extraction* 42
2.6.3.5 *Economics of Formal Recycling* 43
2.6.4 Informal Recycling 43
2.6.4.1 *Collection and Sorting of Metals for Informal Recycling* 44
2.6.4.2 *Informal Sorting Methods* 44
2.6.4.3 *Legal Issues* 45
2.6.4.4 *Health, Safety and Environmental Issues* 45
2.7 Conclusions 46
References 47

3 Global Management of Electronic Wastes: Challenges Facing Developing and Economy-in-Transition Countries 52
Oladele Osibanjo, Innocent C. Nnorom, Gilbert U. Adie, Mary B. Ogundiran, and Adebola A. Adeyi

3.1 Introduction 52
3.1.1 Electronic waste (E-waste): Definitions, Categories and Composition 52
3.1.2 Typology and Categories of E-waste 53
3.2 E-waste Composition 56
3.3 E-waste Generation 61
3.3.1 Estimated Global Quantities of E-waste Generated 61
3.4 Problems with E-waste 63
3.5 E-waste Management Challenges Facing Developing Countries 65
3.5.1 Introduction 65
3.5.2 Poor Feedstock Collection Strategies 67
3.5.3 Lack of State-of-the-Art Technologies to Recover Resources from E-Waste 68
3.5.4 Lack of Specific E-Waste Regulations and Enforcement in Developing Countries 68
3.6 Environmental and Health Impacts of E-Waste Management in Developing Countries 71
3.6.1 Environmental Impacts of E-Waste 71
3.6.2 Health Impacts of E-Waste 71
3.7 Solutions for Present and Future Challenges 73
3.7.1 Optimizing and Promoting E-Waste as a Resource 73
3.7.2 Role of Product Design in Defining Product EoL Scenario 73
3.7.3 Recovering EoL Products 74
3.7.4 E-Waste as a Resource for Socioeconomic Development 75
3.7.5 Urban Mining 76
3.8 Conclusions 77
References 78

4 Dynamics of Metal Reuse and Recycling in Informal Sector in Developing Countries 85
Mynepalli K. C. Sridhar and Taiwo B. Hammed

4.1 Introduction 85
4.2 Science of Metals 86

4.3 Technosphere, Demand and Mobility of Metals 89
4.4 Waste Dumpsites and Treasures of Heavy Metals 92
4.4.1 African Countries 92
4.4.2 Latin American Countries 94
4.4.3 Asian Countries 94
4.4.4 Metals and Global Business 94
4.5 Scrap Metal and Consumer Markets 96
4.6 Export of Metal Scrap 99
4.7 E-waste Scavenging and End-of-Life Management 102
4.8 Scrap Metal Theft 105
4.9 Conclusions 106
References 106

5 Metal Sustainability from Global E-waste Management 109
Jinhui Li and Qingbin Song

5.1 Introduction 109
5.2 E-Waste Issues 109
5.3 E-Waste Management in China 112
5.3.1 Generation and Flows 112
5.3.2 Policies 113
5.3.3 Formal and Informal Sectors 115
5.3.3.1 Formal Sectors 115
5.3.3.2 Informal Sectors 116
5.4 Recycling of Metals Found in E-waste 119
5.4.1 Base or Major Metals (Fe, Al, Cu, Pb, etc.) 119
5.4.2 Toxic Metals 120
5.4.2.1 Lead 120
5.4.2.2 Cadmium and Chromium(VI) 120
5.4.3 Precious Metals 123
5.4.4 Rare Earth Elements (REEs) 123
5.5 Challenges and Efforts in Metal Sustainability in China 124
5.5.1 Challenges 124
5.5.2 Efforts 124
5.6 Summary 127
5.7 Acknowledgment 130
References 131

6 E-waste Recycling in China: Status Quo in 2015 134
Martin Streicher-Porte, Xinwen Chi, and Jianxin Yang

6.1 Introduction 134
6.2 Formal E-waste Collection and Recycling System in China 135
6.2.1 The Policy Framework of E-waste Management 135
6.2.2 E-waste flow in China 137
6.2.3 The Mechanism and Practice of WEEE Recycling in China 137

6.3 Informal E-waste Collection and Recycling 139
6.3.1 Informal Sector and E-waste Management 139
6.3.2 Informal E-waste Collection and Recycling in China 140
6.3.2.1 Casual Waste Workers and Recycling Jobs 141
6.3.2.2 Organization of Manual Sorting and Dismantling 143
6.3.3 Interactions between the Formal and Informal Sectors 145
6.4 Conclusions 146
References 147

7 Metallurgical Recovery of Metals from Waste Electrical and Electronic Equipment (WEEE) in PRC 151
Xueyi Guo, Yongzhu Zhang, and Kaihua Xu

7.1 Introduction 151
7.2 Major Sources of E-Waste in China 152
7.3 Strategies and Regulations for WEEE Management and Treatment 153
7.3.1 Strategies for WEEE Management 153
7.3.2 European Regulations 154
7.3.3 Regulations for WEEE Management in China 154
7.3.4 Implementation of Regulations Related to E-Waste 156
7.3.5 Collection System of WEEE Materials 157
7.3.6 WEEE Materials Processing Companies 158
7.3.7 International Cooperation 158
7.4 Recycling and Processing of WEEE 159
7.4.1 Operational Strategies 159
7.4.2 General Processing Technology 160
7.4.3 Disassembly 161
7.4.4 Upgrading 161
7.4.4.1 Comminuting 161
7.4.4.2 Separation 162
7.4.5 Metal Refining 163
7.4.5.1 Copper Smelting Route 164
7.4.5.2 Lead Smelting Route 165
7.4.5.3 Industrial Practices for the Recovery of Metals from E-Waste 166
7.5 Current Issues in WEEE Treatment in China 167
7.6 Conclusions 167
References 168

8 Metal Pollution and Metal Sustainability in China 169
Xiaoyun Jiang, Shengpei Su, and Jianfei Song

8.1 Introduction 169
8.2 Heavy Metal Pollution in China 170
8.2.1 Heavy Metal Pollution Status 170
8.2.1.1 Heavy Metal Pollution in Water 171
8.2.1.2 Heavy Metal Pollution of Soil 174
8.2.1.3 Heavy Metal Pollution of Atmosphere 175

8.2.2 Heavy Metal Pollution in China: Prevention and Control 177
8.2.2.1 Laws and Regulations for Heavy Metal Pollution Prevention and Control 177
8.2.2.2 Policies for Heavy Metal Pollution Prevention and Control 181
8.3 Metal Sustainability in China 185
8.3.1 Metal Recycling in China 185
8.3.2 Metal Recycling from Wastewater, Solid Waste and Flue Gas 186
8.3.2.1 Metal Recycling from Wastewater 186
8.3.2.2 Metal Recycling from Solid Waste 187
8.3.2.3 Metal Recycling from Flue Gas 189
8.3.2.4 Metal Recycling from E-waste 191
8.4 Metal Sustainability in China: Future Prospects 192
References 193

9 Mercury Mining in China and its Environmental and Health Impacts **200**
Guangle Qiu, Ping Li, and Xinbin Feng

9.1 Introduction 200
9.2 Mercury Mines and Mining 201
9.2.1 Mercury Mines 201
9.2.2 Mercury Production 201
9.2.3 Mercury Usage 202
9.3 Mercury in the Environment 202
9.3.1 Air 203
9.3.1.1 Levels 203
9.3.1.2 Emission Sources 204
9.3.2 Mine-waste Tailings (Calcines) 204
9.3.3 Soil 205
9.3.3.1 Levels 205
9.3.3.2 Spatial Distribution 205
9.3.3.3 Bioavailability 208
9.3.4 Water 208
9.3.5 Biota 209
9.3.5.1 Fish 209
9.3.5.2 Rice 210
9.3.5.3 Other Crops 210
9.4 Human Exposure and Health Risk Assessment 211
9.4.1 Human Exposure 211
9.4.1.1 Hair 212
9.4.1.2 Blood 213
9.4.1.3 Urine 214
9.4.2 Health Risk Assessment 215
9.4.2.1 IHg Exposure 215
9.4.2.2 MeHg Exposure 215
9.5 Summary 216
References 216

10 Effects of Non-Essential Metal Releases on the Environment and Human Health 221
Peter G.C. Campbell and Jürgen Gailer

10.1 Introduction 221
10.2 Metal Biogeochemical Cycles 222
10.2.1 Natural and Anthropogenic Sources 222
10.2.2 Notions of Metal Speciation 223
10.2.3 Environmental Fate of Metals 224
10.3 Metal Environmental Toxicology 226
10.3.1 How Do Metals Interact with Aquatic Freshwater Organisms? 226
10.3.2 The Biotic Ligand Model (Chemical Equilibrium Approach) 227
10.3.3 The Dynamic Multi-Pathway Bioaccumulation Model (Chemical Kinetics Approach) 228
10.3.4 Metal Detoxification 229
10.4 Case Study: Cadmium 229
10.4.1 Bioaccumulation (BLM vs. DYM-BAM) 230
10.4.2 Subcellular Partitioning 231
10.4.3 Evidence for Cd-Induced Effects in Aquatic Organisms 232
10.5 Chronic Low-Level Exposure of Human Populations to Non-Essential Metals 232
10.5.1 Historical Perspective 233
10.5.2 Assessment of Human Exposure to Non-Essential Metals 235
10.5.3 Bioavailability of Non-Essential Metal Species 237
10.5.3.1 Respiratory System 237
10.5.3.2 Gastrointestinal System 238
10.5.3.3 Skin 239
10.5.4 Metabolism of Non-Essential Metals 239
10.5.4.1 Blood Circulation 239
10.5.4.2 Organs 240
10.5.5 Linking Non-Essential Metal Exposure to the Etiology of Human Diseases 241
10.5.6 Global Ecosystem Contamination by Arsenic, Cadmium, Lead and Mercury as an Underestimated Threat to Human and Ecosystem Health: A Summary 242
References 243

11 How Bacteria are Affected by Toxic Metal Release 253
Mathew L. Frankel, Sean C. Booth, and Raymond J. Turner

11.1 Introduction to Bacteria in the Environment 253
11.2 Bacterial Interactions with Metals 255
11.2.1 Essential Metals 255
11.2.2 Non-essential Metals 256
11.3 Bacterial Response to Toxic Metals 257
11.3.1 What Are the Toxicity Levels of Metals to Bacteria? 257
11.3.2 Resistance Mechanisms of Bacteria to Metals 258

11.4 How Are Metals Toxic to Bacteria? 261
11.4.1 Reactive Oxygen Species 261
11.4.1.1 Disruptive Reactions of ROS. 261
11.4.2 Thiol Chemistry 262
11.4.3 Replacement of Co-factor Metals in Metalloproteins 263
11.4.4 Mutagenic Effects 263
11.4.5 Other Mechanisms for Metal Toxicity 264
11.5 Conclusions 265
References 265

12 Application of Molecular Recognition Technology to Green Chemistry: Analytical Determinations of Metals in Metallurgical, Environmental, Waste, and Radiochemical Samples 271
Yoshiaki Furusho, Ismail M.M. Rahman, Hiroshi Hasegawa, and Neil E. Izatt

12.1 Introduction 271
12.2 Technologies Used for Green Chemistry Trace Element Analysis 272
12.3 Elemental Analysis Instrumentation 273
12.4 Arsenic Speciation in Food Analysis 275
12.5 Metal Separation Resins and Their Application to Elemental Analyses 275
12.5.1 Ion Exchange Resins 277
12.5.2 Chelating Resins 278
12.5.3 Molecular Recognition Technology Resins 279
12.6 Green Chemistry Analytical Applications of Metal Separation Resins 279
12.6.1 Analysis of Trace Levels of Rare Earth Elements in Rainwater in Suburban Tokyo, Japan 279
12.6.2 Analysis of Metal Pollutants in Aqueous Environmental Samples 279
12.6.3 Analysis of Trace Levels of Lead in High Matrix Plating Solutions 280
12.6.4 Analysis of Trace Levels of Precious Metals in Recycled Materials 282
12.6.5 Analysis of Radioactive Strontium and Other Radionuclides using MRT Rad Disks 286
12.7 Conclusions 288
References 290

13 Ionic Liquids for Sustainable Production of Actinides and Lanthanides 295
Paula Berton, Steven P. Kelley, and Robin D. Rogers

13.1 Introduction 296
13.2 f-Element Chemistry in Ionic Liquids 297
13.3 Applications of Ionic Liquids in f-Element Isolation 298
13.3.1 Liquid-Liquid Extractions 298
13.3.2 Processing of Ore, Spent Fuel, and Recycling 304
13.3.2.1 Use of ILs for Dissolution of Metals and Metal Salts 304
13.3.2.2 Strategies for Isolating f-Elements from Solid Resources 306

13.3.3 Uranium from Seawater: A Case Study 307
13.4 Summary 308
13.5 Acknowledgments 308
References 309

14 Selective Recovery of Platinum Group Metals and Rare Earth Metals from Complex Matrices Using a Green Chemistry/Molecular Recognition Technology Approach 317
Steven R. Izatt, James S. McKenzie, Ronald L. Bruening, Reed M. Izatt, Neil E. Izatt, and Krzysztof E. Krakowiak

14.1 Introduction 317
14.2 Molecular Recognition Technology 319
14.3 Strengths of Molecular Recognition Technology in Metal Separations 320
14.3.1 Significant Improvement in Process Conditions 320
14.3.2 Short Process Time 320
14.3.3 High Selectivity for Target Species 320
14.3.4 Availability of SuperLig® Products for a Wide Range of Species 321
14.3.5 Significant Operating Advantages 321
14.3.6 Environmentally and Ecologically Friendly Processes 322
14.3.7 Cost Effectiveness 322
14.4 Applications of Molecular Recognition Technology to Separations Involving Platinum Group Metals 322
14.5 Applications of Molecular Recognition Technology to Separations Involving Rare Earth Elements 327
14.6 Comparison of Opex and Capex Costs for Molecular Recognition Technology and Solvent Extraction in Separation and Recovery of Rare Earth Metals 330
14.7 Conclusions 331
References 331

15 Refining and Recycling Technologies for Precious Metals 333
Tetsuya Ueda, Satoshi Ichiishi, Akihiko Okuda, and Koichi Matsutani

15.1 Introduction 333
15.2 Precious Metals Supply and Demand 334
15.2.1 Supply 334
15.2.1.1 Platinum 334
15.2.1.2 Palladium 335
15.2.1.3 Gold 336
15.2.2 Demand 336
15.2.2.1 Platinum 336
15.2.2.2 Gold 337
15.2.3 Outlook for Supply and Demand 337
15.3 Autocatalysts (Pt, Pd, Rh) 337
15.3.1 Demand for Autocatalysts by Region 337

15.3.2 Recycling System for Autocatalysts 341
15.3.3 Extraction and Refining Technologies for End-of-Life Autocatalysts 342
15.3.4 Outlook for Recycling 343
15.4 Electronic Components 344
15.4.1 Demand for Electronic Components 344
15.4.2 Recycling System for Electronic Components 345
15.4.3 Extraction and Refining Technologies for Electronic Waste 347
15.4.4 Outlook for Recycling 348
15.5 Catalysts for Fuel Cell Application 349
15.5.1 Platinum, Platinum/Cobalt Alloy/Carbon and Platinum Ruthenium Alloy/Carbon Catalysts for Polymer Electrolyte Membrane Fuel Cells 349
15.5.1.1 Fuel Cells 349
15.5.1.2 Highly Active Platinum and Platinum Alloy Catalysts for Cathodes (Air Poles) 350
15.5.1.3 Highly Durable Platinum Catalysts and Platinum Alloy Catalysts for Cathodes (Air Poles) 351
15.5.1.4 Platinum/Ruthenium Alloy Catalysts 352
15.5.2 Outlook for Recycling 354
15.6 Extraction and Refining Technologies for Precious Metals 355
15.6.1 Extraction Technologies 355
15.6.1.1 Dissolving Precious Metals 356
15.6.1.2 Chemistry Behind Precious Metal Aqueous Solutions 356
15.6.1.3 Ion Exchange Resin and Activated Carbon 357
15.6.2 Refining Technologies 357
15.6.2.1 Precipitation Crystallization 357
15.6.2.2 Solvent Extraction 358
15.6.2.3 Molecular Recognition Technology (MRT) 359
15.6.2.4 Electrolytic Refining 359
15.7 Conclusions 359
References 360

16 The Precious Metals Industry: Global Challenges, Responses, and Prospects 361
Michael B. Mooiman, Kathryn C. Sole, and Nicholas Dinham

16.1 Introduction: The Precious Metals Industry 361
16.1.1 Structure of the Industry 362
16.1.2 Precious Metal Demand and Prices 364
16.2 Current and Emerging Challenges 365
16.2.1 Increased Demand 365
16.2.2 Increasing and Volatile Prices 366
16.2.3 Decreasing Grades and Increasingly Complex Mineralogy 368
16.2.4 Increasing Production Costs 369
16.2.5 Deleterious Byproducts 370

16.2.6 Geopolitics, Public Perception, and Regulations 371
16.2.6.1 Government–Mining Company Interactions 371
16.2.6.2 Safety in Mining and Processing 373
16.2.6.3 Environmental Impacts 373
16.2.6.4 Fungibility of Precious Metals 374
16.2.7 Labor Relations 374
16.2.8 Artisanal and Illegal Mining 375
16.2.9 Sustainability and Sustainable Development 376
16.2.10 Water and Energy Use 379
16.2.11 Technology Cycles 380
16.3 Responding to the Challenges: Mitigating Approaches and New Developments 380
16.3.1 Recycling of Precious Metals 381
16.3.1.1 Recycling of High-Grade Materials 381
16.3.1.2 Recycling of Low-Grade Materials 382
16.3.1.3 Trends and Efficiencies in Precious Metals Recycling 383
16.3.2 Thrifting and Substitution 384
16.3.3 Mining and Recovery from Lower-Grade Materials 385
16.3.4 Improved Mining, Recovery, and Separation Technologies 386
16.4 Concluding Remarks: A Long-Term View of the Precious Metals Industry 388
References 389

17 Metal Sustainability from a Manufacturing Perspective: Initiatives at ASARCO LLC Amarillo Copper Refinery 397
Luis G. Navarro, Tracy Morris, Weldon Read, and Krishna Parameswaran

17.1 Introduction 397
17.2 General Overview of Sustainability from the Copper Industry Perspective 398
17.3 A Brief History of ASARCO LLC 399
17.3.1 Asarco's Footprint in Amarillo, Texas 399
17.4 How Refined Copper Is Produced 400
17.5 Introduction to Physical Chemistry of Copper Electrorefining 402
17.6 Electrolyte Purification 404
17.6.1 Conventional Methods for Electrolyte Purification 404
17.6.2 Molecular Recognition Technology (MRT) 406
17.6.2.1 Use of MRT for Bismuth Removal at ACR 406
17.7 Recovery of Metals by Precipitation from Acidic Streams 409
17.7.1 Nickel Carbonate Recovery 410
17.7.1.1 Nickel Carbonate precipitation 410
17.7.2 Tellurium Recovery 413
17.7.2.1 Atmospheric Oxidizing Slimes Leaching Process 415
17.7.2.2 Pressurized Leaching Process of Anodic Copper Slimes 416
17.7.2.3 Detellurization Process 417
17.8 Other Sustainable Development Efforts at ACR 419
17.8.1 Implementation of Quality Management System 421
17.9 Conclusions 421
References 422

18 Sustainability Initiatives at ASARCO LLC: A Mining Company Perspective **424**
Dr. Krishna Parameswaran

18.1 Introduction 424
18.2 What is Sustainable Mining? 425
18.3 Exploration 427
18.3.1 Montana, USA 427
18.3.1.1 Troy Mine 427
18.3.1.2 Rock Creek 429
18.3.2 Camp Caiman Gold Exploration Project, French Guiana, South America 431
18.4 Innovative Reclamation Methods 436
18.4.1 Use of Biosolids 436
18.4.2 Use of Cattle 439
18.5 Reclamation of San Xavier Tailings Storage Facilities and Waste Rock Deposition Areas 441
18.6 Fostering Renewable Energy Projects on Disturbed Lands 442
18.7 Utilization of Mining Wastes 448
18.8 Conclusions 450
References 451

19 Recycling and Dissipation of Metals: Distribution of Elements in the Metal, Slag, and Gas Phases During Metallurgical Processing **453**
Kenichi Nakajima, Osamu Takeda, Takahiro Miki, Kazuyo Matsubae, and Tetsuya Nagasaka

19.1 Introduction: Background, Motivation, and Objectives 453
19.2 Method: Chemical Thermodynamic Analysis of the Distribution of Elements in the Smelting Process 454
19.3 Element Distribution Tendencies in Recycling Metals 456
19.3.1 Copper Smelting 456
19.3.2 Lead and Zinc Smelting 457
19.3.3 Aluminum Remelting 457
19.4 Metallurgical Knowledge for Recycling: Element Radar Chart for Metallurgical Processing 463
References 465

20 Economic Perspectives on Sustainability, Mineral Development, and Metal Life Cycles **467**
Roderick G. Eggert

20.1 Introduction 467
20.2 The Many Faces of Sustainability 468
20.3 Economic Concepts 469
20.3.1 Economic Efficiency and Equity 469
20.3.2 Discounting 470
20.3.3 Externalities 470
20.3.4 Capital 471

20.4 Implications for Mine Development 471
20.4.1 Sustainability of Production 471
20.4.2 Sustainability of Wellbeing Originating from Mining 472
20.5 Implications for Regional and National Mineral Development 473
20.5.1 Sustainability Funds 473
20.5.2 Green Accounting 474
20.6 Implications for Metal Life Cycles, Material Efficiency, and the Circular Economy 476
20.6.1 Nonrenewability of Mineral Resources and Metals 477
20.6.2 Environmental Damages and Wastes 480
20.7 What to Do? 481
Acknowledgments 482
References 483

21 Closing the Loop: Minerals Industry Issues 485
William J. Rankin and Nawshad Haque

21.1 Introduction 485
21.2 The Waste Hierarchy 486
21.3 Reducing and Eliminating Wastes 487
21.3.1 Cleaner Production 490
21.3.2 Wastes as Co-products 490
21.3.3 Process Re-engineering 491
21.3.4 Closing the Loop 492
21.3.5 Stewardship 494
21.4 Tools for Closing the Loop 497
21.4.1 A Case Study: Steelmaking Using Biomass 497
21.4.1.1 Economic Benefits 499
21.4.1.2 Environmental Benefits 501
21.4.1.3 Summary 501
21.5 Closing the Loop: Barriers and Drivers 503
References 505

Index **508**

List of Contributors

Adebola A. Adeyi, Department of Chemistry, University of Ibadan, Ibadan, Nigeria

Gilbert U. Adie, Department of Chemistry, University of Ibadan, Ibadan, Nigeria

Paula Berton, Department of Chemistry, University of Alabama, Tuscaloosa, AL, U.S.A; Department of Chemistry, McGill University, Montreal, Canada.

Sean C. Booth, Department of Biological Sciences, University of Calgary, Calgary, Alberta, Canada

Ronald L. Bruening, IBC Advanced Technologies, Inc., American Fork, UT, U.S.A.

Peter G.C. Campbell, Institut national de la Recherche scientifique, INRS-ETE, Centre Eau Terre Environnement, Québec, Canada

Xinwen Chi, School of Environmental Science & Engineering, South University of Science and Technology of China, Nanshan District, Shenzhen, Guangdong, China

Nicholas Dinham, Platinum Group Metals Consultant, Johannesburg, South Africa

Roderick G. Eggert, Division of Economics and Business, Colorado School of Mines, Golden, CO, U.S.A.

Xinbin Feng, State Key Laboratory of Environmental Geochemistry, Institute of Geochemistry, Chinese Academy of Sciences, Guiyang, China

Mathew L. Frankel, Department of Biological Sciences, University of Calgary, Calgary, Alberta, Canada

Yoshiaki Furusho, GL Sciences Inc., Shinjuku, Tokyo, Japan

Jürgen Gailer, University of Calgary, Department of Chemistry, Calgary, Alberta, Canada

Xueyi Guo, Research Institute for Resource Recycling, School of Metallurgy and Environment, Central South University, Changsha, Hunan, PRC

Christian Hagelüken, Umicore AG & Co, KG, Hanau, Germany

Taiwo B. Hammed, Department of Environmental Health Sciences, College of Medicine, University of Ibadan, Ibadan, Nigeria

Nawshad Haque, CSIRO Mineral Resources, Clayton, Australia

Hiroshi Hasegawa, Institute of Science and Engineering, Kanazawa University, Kakuma, Kanazawa, Japan

Satoshi Ichiishi, Chemical & Refining Company, Tanaka Kikinzoku Kogyo K.K, Nagatoro, Hiratsuka, Kanagawa, Japan

Neil E. Izatt, IBC Advanced Technologies, Inc., American Fork, UT, U.S.A.

Reed M. Izatt, IBC Advanced Technologies, Inc., American Fork, UT, U.S.A.; Department of Chemistry and Biochemistry, Brigham Young University, Provo, UT, U.S.A.

Steven R. Izatt, IBC Advanced Technologies, Inc., American Fork, UT, U.S.A.

Xiaoyun Jiang, Changsha Hasky Environmental Science and Technology Limited Co., Xinsheng Road, Changsha, Hunan, China

Steven P. Kelley, Department of Chemistry, University of Alabama, Tuscaloosa, AL, U.S.A; Department of Chemistry, McGill University, Montreal, Canada.

Krzysztof E. Krakowiak, IBC Advanced Technologies, Inc., American Fork, UT, U.S.A.

Jinhui Li, School of Environment, Tsinghua University, Beijing, China

Ping Li, State Key Laboratory of Environmental Geochemistry, Institute of Geochemistry, Chinese Academy of Sciences, Guiyang, China

Kenichi Nakajima, Center for Material Cycles and Waste Management, National Institute for Environmental Studies, Ibaraki, Japan

Kazuyo Matsubae, Graduate School of Engineering, Tohoku University, Miyagi, Japan

Koichi Matsutani, Shonan Plant, Chemical & Refining Products Division, Tanaka Kikinzoku Kogyo K.K., Nagatoro, Hiratsuka, Kanagawa, Japan

James S. McKenzie, Ucore Rare Metals, Inc., Bedford, Nova Scotia, Canada

Takahiro Miki, Graduate School of Engineering, Tohoku University, Miyagi, Japan

Michael B. Mooiman, Franklin Pierce University, Manchester, NH, U.S.A.

Tracy Morris, ASARCO LLC, Amarillo, TX, U.S.A.

Tetsuya Nagasaka, Graduate School of Engineering, Tohoku University, Miyagi, Japan

Luis G. Navarro, IBC Advanced Technologies, Inc., American Fork, UT, U.S.A.

Innocent C. Nnorom, Department of Industrial Chemistry, Abia State University, Uturu, Abia State, Nigeria

Mary B. Ogundiran, Department of Chemistry, University of Ibadan, Ibadan, Nigeria

Akihiko Okuda, Shonan Plant, Chemical & Refining Products Division, Tanaka Kikinzoku Kogyo K.K., Hiratsuka, Kanagawa, Japan

Oladele Osibanjo, Basel Convention Coordinating Centre For Training & Technology Transfer for the African Region, University of Ibadan, Ibadan, Nigeria & Department of Chemistry, University of Ibadan, Ibadan, Nigeria

Krishna Parameswaran, tfgMM Strategic Consulting, Scottsdale, AZ, U.S.A.

Guangle Qiu, State Key Laboratory of Environmental Geochemistry, Institute of Geochemistry, Chinese Academy of Sciences, Guiyang, China

Ismail M.M. Rahman, Institute of Environmental Radioactivity, Fukushima University, Fukushima City, Fukushima, Japan

William J. Rankin, CSIRO Mineral Resources, Clayton, Australia

Weldon Read, ASARCO LLC, Amarillo, TX, U.S.A.

Robin D. Rogers, Department of Chemistry, McGill University, Montreal, Canada; Department of Chemistry, University of Alabama, Tuscaloosa, AL, U.S.A.

Kathryn C. Sole, Consulting Hydrometallurgist, Johannesburg, South Africa

Jianfei Song, Changsha University of Science & Technology, Changsha, Hunan, China

Qingbin Song, School of Environment, Tsinghua University, Beijing, China

Mynepalli K. C. Sridhar, Department of Environmental Health Sciences, Faculty of Public Health, University of Ibadan, Ibadan, Nigeria

Martin Streicher-Porte, FHNW, University of Applied Sciences and Arts Northwestern Switzerland, Institute for Biomass and Resource Efficiency, Windisch, Switzerland

Shengpei Su, Hunan Normal University, Changsha, Hunan, China

Osamu Takeda, Graduate School of Engineering, Tohoku University, Miyagi, Japan

Raymond J. Turner, Department of Biological Sciences, University of Calgary, Calgary, Alberta, Canada

Tetsuya Ueda, Shonan Plant,Tanaka Kikinzoku K.K., Hiratsuka, Kanagawa, Japan

Ian D. Williams, Faculty of Engineering and the Environment, University of Southampton, Highfield, Southampton, U.K.

Kaihua Xu, GEM CO., Ltd, Marina Bay Center, South of Xinghua Rd., Bao'an Center Area, Shenzhen, PRC

Jianxin Yang, Research Center for Eco-Environmental Sciences, Chinese Academy of Sciences, Beijing, China

Yongzhu Zhang, School of Metallurgy and Environment, Central South University, Changsha, Hunan, PRC

Preface

Achievement of improved metal sustainability is a critical global goal for the 21st century. There is room for significant improvement in global metal sustainability throughout metal life cycles from mining ore to beneficiation processes to product manufacture to recovery from end-of-life materials. Serious global environmental and health issues resulting from unrecovered metals entering the commons exist for each of these life-cycle steps, especially in non-Organization for Economic Cooperation and Development (OECD) nations. Greater use of green chemistry principles is needed in these life-cycle steps to maximize metal conservation while minimizing metal loss to the commons. Maintenance of adequate global metal supplies requires greater use of formal recycling and increased urban mining. A particular challenge to metal sustainability is informal recycling, which is widespread, particularly in non-OECD nations, resulting in significant metal losses and severe environmental and health problems in populations least able to confront them. Informal recycling is considered by some to be the most pressing global environmental issue associated with e-waste. Despite these concerns, informal recycling is an important economic activity for large segments of the population in many non-OECD nations, presenting a 'catch-22' situation for government policy makers.

A few decades ago, about ten metals were in common use globally, mainly for infrastructure, transportation, and construction purposes. In 2016, as many as 40 metals are in use, most being essential, usually in small quantities per item, for optimal performance of high-technology products, which have become an essential part of our society. Many of these metals are used once, then discarded, with recycling rates $<1\%$. To the extent that metals are not recycled, the need to mine virgin ore to meet demand is increased, with attendant environmental damage and greater use of energy and water resources.

A unique feature of this book is its coverage in a single volume of many aspects of metal life cycles together with discussion of relevant environmental, health, political, economic, industrial, and societal issues. These issues are presented and discussed by individuals knowledgeable in various aspects of metal life cycles as given above. Special emphasis is given to precious, specialty, toxic, and radioactive metals. Economic considerations are presented, since these are the driving forces on the pathway to metal sustainability. Global societal effects related to metal sustainability are presented and discussed including those involving health, environmental, political, industrial, and other stakeholder issues. The increasing presence of toxic metals, such as Hg, Pb, As, and Cd, in the environment poses challenging questions to all stakeholders. Mercury, for example, can be released in China, but be a global health threat because it may remain airborne long enough to circle the globe. Arsenic is concentrated in rice in China, where it becomes a health issue, since rice is a food staple in that nation and may be exported. A broad perspective is important because the tendency is to look at metal sustainability from a specific stakeholder's standpoint at a particular location and not consider the global interdependence of the many aspects of metal life cycles.

There is a global distribution of chapter authors representing non-OECD as well as OECD nations in order to obtain first-hand information about metal sustainability issues worldwide. Major goals are to provide information that will make readers aware of the increasingly important role technology metals play in our high-tech society, the need to conserve our metal supply throughout the metal life cycle through application of green chemistry principles, the importance of improved metal recycling, and the dire effects that unhindered metal loss can have on the environment and on human health.

The material presented will be useful to scientists, engineers, and other researchers in the field; policy makers as they consider alternatives; companies as they make key decisions that impact how metals are used and how products and processes can be optimized to enhance recycling; press/media as they communicate with the public; and the public who ultimately, as they become aware of the issues, will demand of other stakeholders conservation of natural resources, including technology metals, that make their quality of life secure. The book will be successful if it creates a greater awareness among stakeholders of the adverse consequences of continuing on the present course and makes these stakeholders aware of alternatives that can lead to greater achievement of global high-technology metal sustainability.

Reed M. Izatt
Provo
April 2016

Acknowledgments

I appreciate and thank the authors of each chapter who have worked diligently to deliver high- quality content for this volume. I have enjoyed the constant guidance of a terrific set of Wiley editors who have provided help whenever needed by me and by the authors. My computer-literate daughter, Anne Marie Izatt, has been of immeasurable assistance throughout this editing experience. Knowing she was there when needed, which was often, brought me great comfort. Finally, I thank my wife, Janet, for her patience, understanding and support throughout the preparation of this book. Janet, who has a university degree in English, has been a valuable sounding-board, has made many helpful suggestions, and has come to know the names of and become familiar with many technology metals, like dysprosium. She is amazed that few people have heard of this technology metal, but that is what the book is about.

1

Recycling and Sustainable Utilization of Precious and Specialty Metals

Reed M. Izatt[1] *and Christian Hagelüken*[2]

[1]*Department of Chemistry and Biochemistry, Brigham Young University, Provo, Utah, 84602, U.S.A.*
[2]*Umicore AG & Co., KG, Hanau-Wolfgang, Germany*

1.1 Introduction

The need for increased and more effective recycling of our technology metal supply is urgent. This supply consists of both precious and specialty metals. Both sets of metals are essential to functioning of our high-technology products, but for economic reasons there is much more interest in recycling the former than the latter. Average recycling rates for precious metals are above 50% [1], but huge differences exist depending on their application. For example, from chemical and oil refining process catalysts used in "closed cycles" over 90% of the precious metals contained therein are recovered even in case of long lifecycles of over 10 years. Closed cycles prevail in industrial processes where precious metals are used to enable the manufacture of products or intermediates. Hence, a closed cycle is typically taking place in a business-to-business (B2B) environment with no private consumers involved in its different steps. In such systems, the user of the metal-containing product (e.g., the chemical plant) returns the spent product directly to a refiner who recovers the metals and returns them to the owner for a new product cycle. In most cases, the metals remain the property of the user for the entire cycle and the metal-refiner conducts recycling as a service (so called toll refining). Third parties are hardly involved, and, if so, only as other-service contractors (e.g., burning off carbon-contaminated oil refining catalysts), but not taking property of the material. With such a setup, the whole cycle flow becomes very transparent and professionally managed by industrial stakeholders, resulting in very small metal losses.

Metal Sustainability: Global Challenges, Consequences, and Prospects, First Edition. Edited by Reed M. Izatt.

Recycling rates are usually much lower in "open cycles" taking place in a business-to-consumers (B2C) environment. Typical examples are electronics and car catalysts. The owner of the spent product (e.g., an ELV or a PC), who might be number *x* in line after a number of preceding (second-hand) product owners, does not return the product directly to a metals refiner. Instead, the product goes through a usually, long, complex and sometimes opaque chain of collectors and scrap dealers until it reaches the real metal recyclers, in a consolidated way. In this process, ownership of the metal changes each time a transaction occurs, transparency is low, business transactions can be rather strange and special, and resulting metal losses are usually much higher than in B2B closed-loop systems. Important impact factors that determine the overall recycling rates of open cycles are intrinsic value, the ease or difficulty of accessing the relevant component or product, and legal or other boundary conditions that can help channel consumer products into appropriate recycling processes along the chain. An example on the high side (>95% recycling rate) is jewelry, where the high metal and emotional value of a gold ring, for example, prevents losses. Recovery rates of platinum group metals (PGM) can be 60–70%, in the case of automotive catalysts [2], which are quite successfully recycled (easy to disassemble from a car and high intrinsic value). However, metallurgical recovery rates for PGM are>95% with the gap being due to exports of end-of-life (EoL) cars and long and opaque chains before a spent catalyst reaches a precious metals refinery. On the low side with average precious metal recycling rates below 15% are EoL electronic wastes (e-wastes). This low recycling rate is caused by poor collection, often inappropriate pre-treatment, and a high share of precious metal-containing fractions that enter sub-standard or informal recycling processes. Such processes operate with untrained personnel using crude equipment and result in severe adverse environmental and health effects [3]. Recovery rates of precious metals from e-wastes, if treated in state-of-the-art integrated smelter operations, would be>95%, but the waste materials need to get there. The concept of open versus closed cycles has been described [4]. Summarizing, in open cycles metal losses are significantly higher than those that would be found in metallurgical refining. The net effect is that highly efficient state-of-the-art technology [2] is used for only a small portion of waste products containing these precious and specialty metals. Products that are recycled properly are mainly those of high economic value and/or those from closed industrial loops. Recycling of specialty metals from such products is even more challenging. Metals in these products face the same limits of open cycles, but in addition with a lower economic value their recovery is far less attractive, and in some cases there are also thermodynamic limits. As has been elaborated [2,3,5] and is discussed later in this chapter, advanced metallurgical processes can co-recover a number of specialty metals if they fit chemically into a specific extraction system, e.g., in addition to the precious metals, Se, Te, Sb, Sn and In, partially, can be extracted pyrometallurgically by the collector metals Cu, Pb or Ni. However, others like Ta, Ga, and rare earth metals do not extract well. This situation leads overall to very low recycling rates for many specialty metals. Although of high strategic importance in our society, many specialty metals are not recycled but are usually discarded to the commons after one, often brief, use.

The subject of recycling is central to the thrust of this book. Most chapters have sections dealing with the status of metal recycling. For example, Ueda *et al.* [6] describe Pt metal recovery at Tanaka Kikinzoku Kogyo K.K. in Japan. From these accounts, one can obtain an appreciation for the successes, inadequacies, and challenges associated with metal recycling throughout the world. The amount of e-waste generated globally is enormous, estimated by several chapter authors as being 30–50 million tons yearly [7,8] with an estimated growth rate of 4–5% [8]. These numbers are startling and provide evidence for why it is incumbent on involved stakeholders to find technical and practical ways to improve global recycling processes [9,10]. However, it needs to be understood that only a fraction of this global waste is relevant for the recycling of precious and specialty metals. This fraction comprises of EoL information and communications technology (ICT) devices encompassing cellular phones, computer and network hardware, etc., and of audio-video devices (radio, television, etc.). White goods as well as electric household devices such as vacuum cleaners, toasters or electric tools are of importance for the recycling of steel, base metals (e.g., Cu) and plastics but contain very small amounts of precious and specialty metals. In addition, especially for electronic devices, miniaturization and new types of products lead to a reduction of weight although sales numbers are still on the rise. Examples are TVs (CRT-TV > 30 kg; LCD-TV ≈ 16 kg, LED-TV ≈ 14 kg) and computers (desktop PC ≈ 12 kg, notebook 2–3 kg, tablet 0.3 kg) [11]. Continuing on the current course has dire consequences for Earth's metal supply as well as negative consequences for the global environment and health of Earth's inhabitants, human and otherwise [3].

Recycling of metals from modern high-technology products, including waste electronics, EoL vehicles, and automotive catalytic devices is a complex procedure. Current recycling procedures from collection of EoL products to disassembling them into component parts to recovering target metals have been presented and discussed [9]. Important global benefits are derived from effective recycling, including the possibility of 'mining' target metals at a fraction of the economic and environmental costs associated with mining virgin ore [2,3]. However, there is a fundamental difference between a geological and an urban mine deposit. In general, a geological deposit is characterized by the composition and grade of its ore and by the total volume of the ore body leading to an estimation of the tonnage of target metals to be extracted. In a mining deposit, the ore body is concentrated in a specific location. It might be difficult to access and to mine the ore, but it exists in a defined space and it stays there. Hence, if total ore volume and metal prices justify, the necessary infrastructure will be built up and mining will start. The high investments and capital costs of operating a mine, consequently, force many operators to keep the mine running even at depressed prices as long as at least the variable operating costs can be covered.

In these respects, the challenge for secondary deposits, such as are found in an urban mine, is much greater. Although the "ore grade" might be significantly higher than in natural deposits, the urban mining activities are scattered over a vast area. In the case of consumer products, this area comprises millions of individual households. To make a real urban mine, it is first necessary to bring or pull the millions of devices — think about mobile phones or computers — towards the recycling facilities. Once there is a big pile of EoL devices at the gate of a recycling facility, it forms a real deposit, but not before. High metal prices and metal content in an EoL device (i.e., a high intrinsic value) can **push** these devices towards recycling, as it is the case with jewellery scrap or catalysts. However, if the intrinsic value is not sufficiently attractive, then **pull** mechanisms like waste legislation or business models such as leasing or deposit systems will be needed to generate an economically viable urban mine. Also, other than in primary mines, the system is much more vulnerable to price fluctuations. Decreasing metal prices can immediately stop the push mechanism, as the logistical costs involved are mainly variable. So, metallurgical recycling operations down the chain, which usually have high capital costs to bear, might be "overnight" faced with decreasing feeds. Hence, in the urban mine not only can the logistics be more challenging than in primary mines, but the economic drivers and feedback effects are often more complex. This is the reason that societal and legislative frame conditions are crucial for harvesting the urban mine.

Of equal importance to the technical and economic aspects of recycling is the involvement of stakeholders in decisions and actions involving recycling and sustainable utilization of precious and specialty metals [3,9]. Stakeholders include the public; media personnel; local, regional and national decision makers; industry executives; scientists and engineers; and others. Issues of importance might include: wastage of a critical recoverable resource, depletion of a non-renewable resource; environmental damage associated with inefficient mining and recycling, and target metal recovery technologies; and/or irresponsible mining of virgin ore. It is a major purpose of this chapter to supply information intended to make stakeholders aware of these issues and of the advantages of overcoming them by involvement of an informed public and media, passage and enforcement of appropriate legislation, and working together to conserve our valuable metal resources. It has been observed [12] that an effective way of promoting innovation on environmental matters is to pass and enforce legislation requiring compliance. This action often prompts companies involved to consider and develop new technologies to achieve compliance with the legislation. Another way to express the principle is that 'necessity is the mother of invention.'

The magnitude of the global waste problem is large and is expected to continue to increase into the indefinite future. Since technology metals are integral parts of electronic and other high-tech wastes, it is desirable to recycle these metals to aid in conserving our metal supply. With present usage, there is a constant drain on the global technology metal resource. In this chapter, several aspects of recycling are presented and discussed with the aim of improving metal sustainability. Global benefits of recycling specialty and precious metals are explored. Urban or above ground mining of metals is presented as a viable but little used means of conserving our metal supply. Reasons for this situation are complex, but urban mining holds great promise of providing a means of conserving the technology metal supply. Technologies currently used are evaluated in terms of the needs in formal and informal recycling. The need for innovation in all areas of recycling is stressed. Important roles are suggested that interested stakeholders can fill in ensuring that responsible and sustainable utilization of precious and specialty metals occurs. The chapter material makes clear that there are negative consequences to society and to the planet of inadequately maintaining metal sustainability.

1.2 How did we come to this Situation?

Our world faces major societal challenges that are unique to our time. These challenges include a rapidly growing global population, increasing global demand for high-technology consumer products, growing global affluence as the standard of living increases steadily in non-Organization for Economic Cooperation and Development (OECD) nations, management of our global energy supply amid demands for greater clean chemistry and non-greenhouse-gas-emitting energy sources, and improving stewardship over Earth's critical resources. An important common component of these challenges and a major critical resource is our global metal supply, particularly our technology metals. These metals are the group of specialty and precious metals that are essential for the functioning of high-technology products due to the specific and often unique chemical and physical properties of the metals. In general, these properties cannot be duplicated by other metallic or non-metallic substances without significant loss of function, making their replacement difficult [13]. Sustainability of our high-tech society at its present level depends on continued availability of these technology metals. An increasingly important part of metal use is recognized to be the need to recycle these important resources when the products containing them reach their EoL state. Generally, the global track record of society in recycling technology metals is poor [3,14].

The use of technology metals as essential components of high-tech products is a recent one. It is instructive to learn how this role developed and how the need for recycling has become critical. Metal recycling is as old as man's use of metals, extending into antiquity. The dozen or so metals available for use in earlier times were valuable for many reasons, including weapons, structural material, infrastructure, transportation, trade items, currency, and ornament. In general, these metals were used in bulk where they were familiar to those who used them. The impact of metal use on developing civilizations over the ages has been enormous [15] and has accelerated during the past half century [3]. In principle, metals are indestructible and can be reused indefinitely without loss of function, provided there is an economically viable means to recover them from EoL products and to purify them. Recovering metals for reuse was a simpler task prior to the 20th century. Fewer metals were involved, generally major metals used in large-scale operations, such as Fe, Co, Ni, Cu, Zn, Sn, and Pb, and the precious metals Au and Ag used for jewelry, coins and (religious) artifacts. Environmental laws were largely non-existent and recycling usually involved working with large quantities of waste metals. Environmental damage caused by metals was mainly limited to specific areas of mining and refining and was largely accepted as part of the price to be paid. Serious health effects from metal poisoning were present but were poorly understood, and few safety features were available to workers. Environmental effects were largely ignored.

This situation had changed significantly by the mid-20th century. Blank spaces in the periodic table up to U were completed with the identification of Pm in 1947. Intensive studies of the chemical and physical properties of the newly discovered elements paid rich dividends in providing a base for those trained in material sciences to design and construct products never before envisaged that could accomplish tasks formerly relegated to science fiction. The metals essential to these products were used in small quantities and were usually hidden behind casings, so they were little known to the public who used the products. This lack of knowledge can be ascertained quickly by asking literate users of high-tech devices, such as an iPhone, if they are familiar with, say, Dy, Tb, Nd, or In. The answer is usually no with the added comment that chemistry was not one of their favorite subjects.

During the last half of the 20th century and continuing to the present, a remarkable change has occurred with the appearance of new products containing these technology metals which literally transformed global society, including our personal lives, and how we do business, communicate, transmit information, entertain ourselves, produce clean energy, catalyse reactions to produce new products and control unwanted emissions, diagnose and treat a myriad of medical conditions, and many more. These products are used with little or no knowledge by users of the essential role of the technology metals in their function or of the consequences of depleting our technology metal supply. It is desirable to change this situation and make every effort to make the public and other stakeholders aware that in discarding e-waste and other wastes containing technology metals, as is done at present, we are sending enormous amounts of valuable and critical metal resources to locations in the commons where they are unrecoverable with current technologies.

Changes in global society are slowly bringing a realization that resources are finite and that mankind has a responsibility to manage them for the good of present and future generations.

This recognition of responsible stewardship is a positive development and needs to be encouraged. Metals are among the most important resources on the planet and their very nature makes it possible to recycle them repeatedly without loss of function, unlike other resources such as those based on fossil fuels or synthesized from organic compounds. The unique and remarkable physical and chemical properties of metals make possible the advanced society we all enjoy. However, without proper control, metals can harm the environment and have devastating effects on human health. Environmental and human health disasters such as Minamata Bay and Toyama Bay in Japan in the mid-20th century [16–18], involving uncontrolled release of Hg and Cd, respectively, into the environment, contributed immeasurably to the early movement in OECD nations of establishing and enforcing legislative controls on metal emissions. Most non-OECD nations still lag far behind in metal sustainability efforts. Rare earth mining in China [19] and essentially uncontrolled use of Hg in artisanal Au mining in China [20] and Peru [21] illustrate the externality effects of improper use of metals.

Beginning in the 1960s, new products began to appear that were characterized by the incorporation in them of a wide range of technology metals. These metals imparted novel properties and functions unlike any seen previously. One of the earliest of these products was color television, which was made possible by the unique phosphor properties of Eu that produced the red color on television screens. Europium is still the metal of choice for production of this color on screens, illustrating the uniqueness of this property to Eu. The unique electronic properties of rare earth metals made them favorite targets for use in novel product development. The resulting spectacular growth in rare earth mining and production is shown in Figure 1. Neodymium and Dy together with Fe and B were fabricated to form miniature magnets of superior strength, making miniaturization of high-tech products and other devices possible.

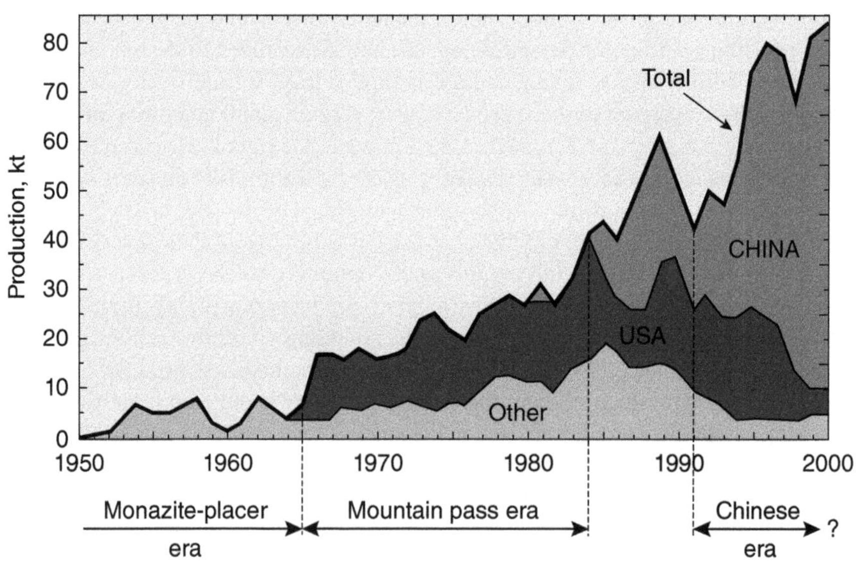

Figure 1 *Global rare earth oxide production trends, 1950–2000. The Mountain Pass deposit is in California, U.S.A. Reproduced from Wikipedia: <https://en.wikipedia.org/wiki/Rare_earth_element> Accessed from website March 10, 2016. kt=kilotonnes. (See insert for color representation of the figure.)*

Another significant discovery was that In had the nearly unique ability, when coupled with Sn, to form the transparent conductive coating now used in all flat screen devices [3]. The list could go on. In mid-20th century, a few metals were used largely for bulk purposes. By 1980, 25 were in use, but in 2015 over 40 are needed for products used in our high-technology global society [2]. Recycling rates for many of these technology metals in electronic products remain <1% [1].

1.3 Magnitude of the Waste Problem and Disposal of End-of-Life Products

The magnitude of the global waste problem is enormous and is steadily growing. Williams [7] indicates that e-waste is one of the chief sources of metal and organic pollutants in solid waste and the fastest growing waste stream. Williams cites one estimate that 41.8 million metric tonnes of electrical and electronic products were discarded globally in 2014, with a rise to 50 million metric tonnes predicted by 2018. The number beyond that date is expected to continue to rise due to global population increase, greater consumer affluence, increased production of new products to replace old ones, and, inevitably, the introduction of additional new products containing technology metals. In other words, there is no leveling off of the generation of e-waste. Osibanjo et al. [8] report that the amount of generated e-waste increases annually by about 4–5%. Williams [7] points out that a 1987 Nokia Cityman mobile phone weighed 770 g and contained few elements, whereas a modern Nokia smartphone weighs about 100 g and contains more than 40 elements. The increased number of elements is closely connected to the greatly increased number of functions the modern phone can perform. What is the fate of the technology metals in these discarded products? The simple answer is that they are to a great part lost to the commons [1].

Main options for disposal of EoL products include storage or stockpile, landfill, incineration, use as second-hand units, and formal/informal recycling of constituents [22]. Estimates are that nearly half of the 1.8 billion new cell phones purchased in 2014 will end up 'hibernating' in drawers within a few years [23]. This statistic is a reflection on the lack of effective collection systems for high-tech products. The effectiveness of collection systems is a bit higher in the European Union and Japan where more attention is focused on this activity [7]. In the U.S., no national program exists and individual states have programs ranging from non-existent to fair. In non-OECD nations like China [7] and Nigeria [24], few formal collection systems exist, but informal systems are fairly successful. The number of electronic products discarded in 2010 in the U.S.A., 400 million [25], was matched by an equal number discarded in China [3]. These numbers increase globally each year. Adding the expected number of waste electronic products from Europe and other nations, one can appreciate the growing magnitude of the global waste management problem. It has been estimated that of the 27% of generated e-waste collected for recycling in the U.S. in 2010, up to 80% was shipped illegally [25] to non-OECD nations, such as China, India, and Nigeria, where some may be reused. However, the ultimate fate of most of these exported waste electronic products is that they are manually dismantled or burned and have a few metals of value recovered by the process known as informal recycling [3].

Landfills are the most common means for disposal of waste worldwide [25]. Amounts of e-waste deposited in landfills are large. For example, a large fraction of e-waste products generated in the U.S. was probably stored in drawers, consigned to landfills, or incinerated.

In the latter two cases, contained metals are usually unrecoverable by present technology and, hence are lost to the potential metal supply. Potential dangers in landfills include leakage due to improper sealing, weather events, population increases in the area with subsequent human exposure to the contents, and inevitable deterioration of the site with time. Effectiveness of landfills varies with global location. In OECD nations, legislative action has resulted in environmental laws mandating requirements for landfills. In non-OECD nations, such laws, if they exist, are usually less stringent and enforcement is spotty. Sridhar and Hammed [24] point out that in some African nations, there is simply no money for maintenance of landfills; hence they become open dumps. Landfills require space, which is often at a premium and which often is the source of ongoing environmental problems. Furthermore, the amount of e-waste discarded in landfill grows steadily each year.

Incineration is becoming the method of choice worldwide for waste disposal [3]. Less space is required and potential environmental dangers of landfills are avoided. However, incineration is not without problems. It is a thermal process and has the drawbacks associated with high-temperature operations. The high energy input required for high-temperature incineration is usually obtained from electricity, which is produced from coal combustion with attendant emissions of carbon dioxide and metals, such as Hg [3]. Incineration results in ash which contains residual metals. Disposal of the ash poses environmental problems similar to those associated with landfills. Space is conserved in incineration, but dispersion of toxic and other metals into the commons remains a concern. Effectiveness of regulatory and environmental controls on incineration processes and atmospheric emissions varies significantly from location to location, especially between OECD and non-OECD nations. Metals in the ash resulting from incineration are concentrated and could be recovered, although at present this is usually not done in the U.S. In Europe, an increasing amount of bottom ashes is treated for recovery of the base metals, steel, Cu, and Al, but recovering precious and special metals from such ashes faces physical limits.

1.4 Benefits Derived by the Global Community from Effective Recycling

Recycling of technology metals, where practiced, has important beneficial features. These have been summarized by Hagelüken [2] and are presented in Table 1. Reduction of the environmental burden of not recycling (Table 1, 1) refers to the fact that, otherwise, EoL products would either be discarded, in an uncontrolled and visually unpleasant way, into the environment with often severe impacts on the commons due to emissions of hazardous

Table 1 *Global Benefits Derived from Effective Metal Recycling Programs*

1. Reduce environmental burden of not recycling
2. Mitigate environmental impact of mining
3. Extend lifetime of and preservation of primary geological resources
4. Reduce geopolitical dependencies involving critical metals
5. Contribute to supply security of technology metals
6. Support ethical sourcing of technology metals
7. Dampen technology metal price fluctuations
8. Create significant employment potential

substances into soil, water and air, or these products would go to controlled landfills or incineration plants. In the latter cases, negative environmental impacts can be minimized by proper management of such landfills or waste incineration plants, but in many countries landfills are far from being well controlled and incineration often takes place in open space or in crude plants without appropriate off-gas treatment. Moreover, especially in populated regions, landfills need land, which is consequently then not available for housing or agriculture. However, it is often the case that such land is later used for these purposes with dire environmental and health effects.

Mining is inherently a process that generates large amounts of waste. This waste derives from the fact that most ore bodies contain a few percent of the target metal, necessitating the removal of a large overburden and separation of a few percent, at most, of the desired metal. As a result, the environmental burden of the mining process is large and grows as the demand for metals increases. As ore is mined to meet the increased raw material demands of society, grades of available ore decrease prompting the need to find new deposits, develop improved technologies to mine, economically, deposits having lower grades, and/or go to greater depths in existing deposits. The gradual depletion of deposits in main parts of Europe and the exploration of new deposits in South America and Africa as well as increasingly stringent environmental regulations in OECD nations have resulted in recent decades in the movement of much mining and ore beneficiation processing to non-OECD nations, which often have less strict regulations or lax enforcement of existing regulations.

Recycling of metals, if effectively done, has the potential to reduce significantly the need to mine new ore to fill the continuing need for metals to replace those discarded in EoL products. This effort has had success in the case of recycling PGM from process catalysts and autocatalytic converters. As elaborated in Section 1.1, the value of PGM and/or the closed industrial cycle has made possible their recovery in significant amounts [6,9]. On the other hand, in open cycles, such as for e-waste, valuable precious metals are recovered at low rates while low-value technology metals, such as indium and rare earth metals, are not recovered at all. Since the need for the products containing these metals is large and increasing, new virgin ore must be mined to meet this demand.

Mining has high energy and water requirements [26,27]. Metals recovered through sound recycling result in significant reduction in the amount of mining required, with associated reduction in coal combustion, carbon dioxide emissions, land and water use, and impacts on the biosphere, e.g. in rain forests, Arctic regions, ocean floors, and so forth (Table 1, 2). A further important benefit of recycling is the reduction in the discard of solid, liquid, and/or gaseous waste into surrounding land, streams, and atmosphere, which is common in many processes, where environmental regulations either do not exist or are poorly enforced. Examples of this waste generation are abundant and are found associated with present mining activities in many countries, such as China [19] and Peru [21].

An additional benefit of recycling is that fewer toxic metals inherent in mining activities as "unwanted companions" enter the technosphere. Examples are As, Hg, Cd, Tl, U, and Th. The first four metals are present, usually as by-products, in sulfide ores from which many of the technology metals are obtained. Thorium and U and their decomposition products are commonly found in rare earth metal deposits. Unless great care is taken, these toxic metals enter the environment through discard to tailings, emissions to the atmosphere, and inefficient beneficiation processes. The effect of these toxic metals on the environment and on human health have been documented [3,16,19–21,28].

Continued use of Earth's metal supply without replenishing it will lead to eventual, serious depletion of this resource. Recycling can delay this scenario and provide time to develop improved mining and processing techniques that might facilitate extraction of metals from lower grades or other undeveloped ore bodies, and to search for alternatives (Table 1, 3). The status of individual technology metals differs. However, a major effect will be that the cost of metals will increase as supply diminishes. Another effect is that, in some cases, such as for certain technology metals deemed critical, demand may exceed supply, resulting in major price fluctuations. Indium is an example of a metal that, in the form of indium-tin-oxide (ITO), is critical to the function of all flat screen devices, which, in turn, are essential to the efficiency of a whole host of modern high-tech products. Indium is one of the rarest of metals in the Earth's crust (Figure 2). Its sole commercial source is sphalerite, ZnS, from which it is recovered as a by-product. Eighty percent of global Zn is mined and processed in China [28], making this Nation a major source of In.

Forty percent of the global supply of In originates in China. Indium is of interest because it is a critical metal, but of low economic value. Hence, there is little economic incentive to recycle In. Some forecast consumption growth rates for LCDs and solar cells project that supplies of In from primary mined Zn sources will be severely affected by the early 2020s [3]. Estimated global reserves of In range from 2,800 (2006) to 11,000 (2007) tons with an annual consumption of 510 tons [3]. Consumption is expected to grow to 1,900 tons annually by 2030. Indium is a scarce commodity, yet its recycle rate is <1% [1]. Active programs are underway to find material that can replace InSnO in flat screens, but so far none can equal the effectiveness of InSnO. An effective process for economically recycling In would be valuable to provide an alternate supply of this metal.

Iridium is an example of a valuable metal that has a critical use but may be in danger of not being sufficiently accessible. Iridium is one of the scarcest of metals (Figure 3), but is the metal with the greatest corrosion resistance and very specific catalytic properties. Iridium finds multiple uses in commerce where corrosion resistance is required. Iridium alloys enable jet engines to operate at very high temperatures. The longevity of spark plugs results from the use of Ir in their construction. The availability of Ir depends on the mining of Pt ore, where it is found as a by-product. Gordon, Bertram, and Graedel [29] list Pt as an element at risk for supply. Platinum group metals are unusual in that they are found in commercial grades at only a few locations: South Africa, Russia, and Canada [30]. In South Africa, use over the past century has resulted in the need to go to lower grades and deeper operations, both of which have environmental, energy, and water consequences. Thus, Ir is an excellent candidate for recycling as a means of preserving and extending a critical natural resource. The recycling rate for Ir is <25% [2].

Several technology metals considered critical to domestic, commercial, and military uses in the U.S. are concentrated in a few countries, not all of which are politically stable or friendly (Table 1, 4) [2,13]. Effective recycling programs would reduce concern because they would provide alternate reliable sources of these metals. The PGM mining in South Africa has recently been affected by labor strikes in the PGM mines, although so far this has not had negative impacts on the PGM supply side. In the past few years, China used its monopoly on rare earth metal production and supply as an economic weapon against Japan during conflict over islands in the seas bordering the two nations [13]. This cutback in supply was of great concern to Japan because these metals are essential to the production of high-technology products produced and marketed by Japan. A fallout from this experience

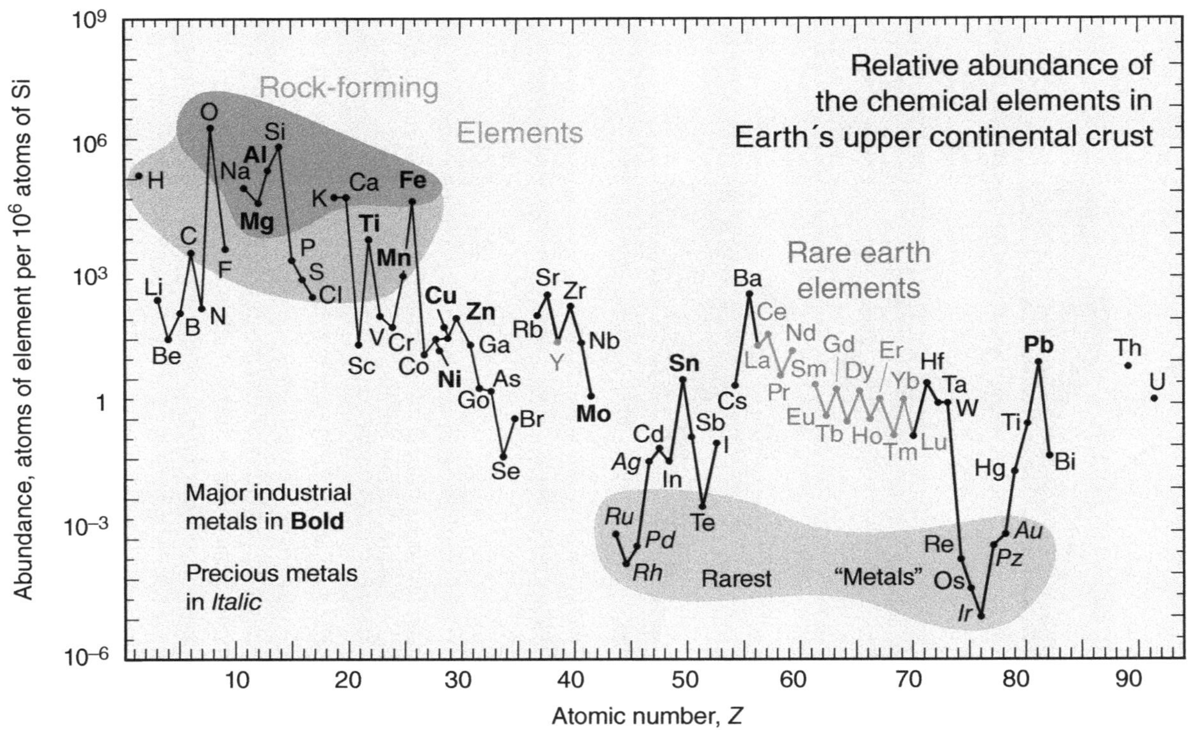

Figure 2 *Abundance (expressed as atoms of element per 10^6 atoms of Si) of the chemical elements in Earth's upper continental crust as a function of atomic number. Many of the elements are classified into (partially overlapping) categories: (1) rock-forming elements (major elements in green field and minor elements in light green field); (2) rare earth elements (lanthanides, La–Lu, and Y; labeled in blue); (3) major industrial metals (global production $> \sim 3 \times 10^7$ kg/year; labeled in bold); (4) precious metals (italic); and (5) the nine rarest "metals" — the six platinum group elements plus Au, Re, and Te (a metalloid). Source: U.S. Geological Survey. http://pubs.usgs.gov/fs/2002/fs087-02/*
(See insert for color representation of the figure.)

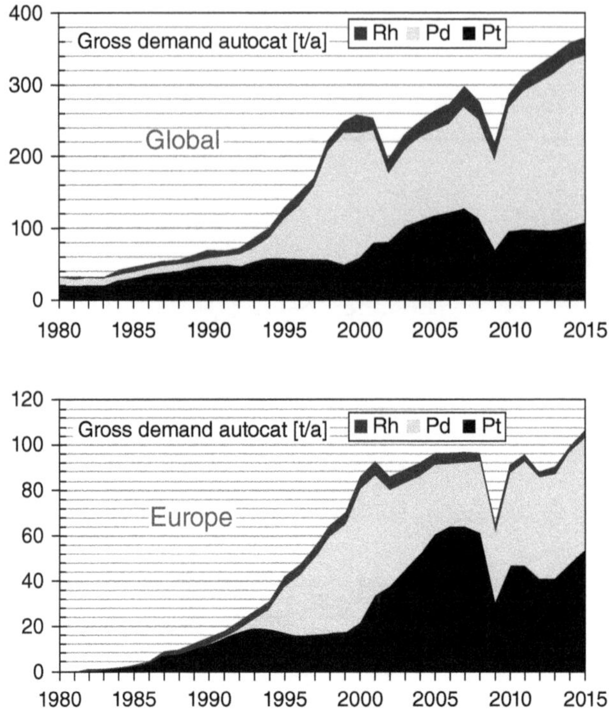

Figure 3 *Gross global and European demand for platinum, palladium, and rhodium for automotive catalysts from 1980 to 2015. (See insert for color representation of the figure.)*

is that greater effort has been expended by Japan to develop recycling capacity, as in recycling Dy from magnets. There has been political unrest in several nations in central Africa for decades [13]. Yet some of these countries are rich in metal resources of value to the world's economy.

A major benefit of recycling is that it could partially decouple the production of the technology metal from the major metal source from which it is derived as a by-product (Table 1, 5). The main objective of producers of commercial metals is production of the major metals, not of by-product metals. Production of technology metals is nearly always tied to the production of a major metal. Examples include (major metal in parentheses) In (Zn); Mo, Re, and Te (Cu); Ga (Zn, Al); and Ru, Ir, and Rh (Pt and Pd). Thus, any success in decoupling this link through recycling is advantageous to the security of the technology metal supply.

Effective metal recycling has the geopolitical advantage that it can make possible a more transparent supply chain reducing or avoiding sourcing of metals from regions involved in wartime activities or in the use of child labor, etc. (Table 1, 6) [13]. An alternative, reliable source of such metals could supply a degree of independence in such cases.

Effective recycling would make possible the dampening of price fluctuations by improving the supply-demand balance and limiting speculation by broadening the supply base (Table 1, 7). The rare earth market would benefit greatly by reliable sources of recycled metals. The rare earth metal market from mining to production and sale of >99% pure products has been and continues to be dominated by China [13]. The only other producing rare earth facility, MolyCorp, operating in California, filed for bankruptcy protection in 2015. Prices for individual rare earth elements have fluctuated widely during the past five years. The critical rare earth metals represent an interesting example of this problem. Reasons for the growth of markets which require Nd have been presented earlier. Despite the critical need for Nd in high-technology products, the sources of supply are limited primarily to China, except for any stored metal. This situation has attracted the attention of several groups who are developing recycling processes for recovery of Nd from high performance magnets. Dupont and Binnemanns [31] are among the leaders in this effort, as is the Critical Materials Institute in the U.S. At least one group is well on the way to the production of Nd and other rare earth metals from pregnant leach solutions derived from U.S. ore deposits using a molecular recognition technology (MRT) process [32]. The MRT procedure could also be used to recycle these metals. Hopefully, recycling efforts underway will bring some stability to the critical rare earth market by improving the supply-demand balance for these important metals, which presently have a recycling rate of <1%.

Development of recycling facilities and infrastructure would create significant employment potential (Table 1, 8), especially in collection and pre-processing. Informal recycling operations are major employers of individuals of all ages in non-OECD nations [3]. Diverting EoL products from landfills or uncontrolled discards requires a significant workforce. Beyond labor-intensive collection activities, in countries with low labor costs manual dismantling of e-waste and other wastes offers a large job potential and can be a viable alternative to mechanical treatment. Supervised manual dismantling of e-waste can be a way gradually to formalize the informal sector and, together with appropriate training on sound recycling, avoid the negative environmental and health effects of crude "backyard-recycling". Mechanized preprocessing in OECD countries as well as the metallurgical recovery of metals at the end of the recycling chain are less labor-intensive [4,9], but provide job potentials for skilled labor and for suppliers of technical recycling equipment.

1.5 Urban Mining

Metals differ from energy raw materials in an important way. Metals can be kept in an 'eternal' life cycle by proper management. The use of technology metals in personal, consumer, industrial, and military products has grown rapidly in recent decades. More than 80% of global mine production of PGM, rare earth metals, indium, and gallium since 1900 has occurred in the past three decades [2,14]. Many of these metals are still bound in the 'technosphere' or 'anthrosphere.' As a result, products such as automobiles, electronics, batteries, automotive and industrial catalysts, and other high-tech devices that reach their EoL state represent a potential 'renewable' metal resource of great value. This resource has been termed an 'urban' or 'above-ground' mine and proposed as an alternative source of these metals. There is good reason for this proposition. Concentrations of technology metals in many of these products are relatively high. For example, a typical primary Au mine will yield five grams of Au per tonne (g/t). In electronic scrap, this figure was, 10 years ago, as high as 200–250 g/t for computer circuit boards and 300–350 g/t for mobile phone handsets. High metal prices together with progress in materials development and product design then triggered both a significant miniaturization of devices and components and a thrifting of precious metal contents. Since then, Au and Ag in PC motherboards has declined by 40% and Pd by 60%. Similar trends can be recognized for mobile phones [11]. But still, even on today's lower level, grades this high are very uncommon in natural Au deposits. An autocatalytic converter contains approximately 2,000 g/t of PGM in the ceramic block, compared to average PGM concentrations of <10 g/t in most PGM mines. Considering the high environmental impact of primary production of precious metals arising from low ore concentrations, difficult mining conditions, high energy and water use, high chemical consumption, large waste generation, and other factors, recovery of metals from EoL products is appealing. Further environmental and economic benefits are realized if state-of-the-art technologies are used for the metal recoveries. However, use of these advanced metallurgical smelting and refining technologies usually requires collection and transport of the resource-relevant components from EoL products, partly over long distances. This is neither from an economic nor from an ecological point of view a problem as long as not the entire device but only the components/fractions with higher metal concentrations (i.e., circuit boards, catalysts, batteries) are shipped (comparable to shipping primary Cu concentrates or anode slimes to specialized large scale smelters around the world). To achieve this end result, it is required that EoL products be dismantled and pre-processed locally or regionally in an appropriate way. Subsequently, the resource-relevant complex metal fractions need to find their way to sophisticated integrated smelters/refineries. For this last step in the recycling chain, sufficient economies of scale are crucial; hence it does not make sense to install such operations in "every country". This matter is discussed further in Section 1.6.

With the positive scenario presented, one must wonder why urban mining is not employed extensively as a means to augment our metal supply. Major reasons are that most consumer products are widely distributed, difficult to trace around the planet, and the (precious) metal content in any single device is very low. Accordingly, economic exploitation of these urban mines requires collecting sufficient quantities of the dispersed product, such as cell phones or automotive catalysts, to create a true above-ground deposit from which the metal(s) can be mined. This collecting of e-waste fulfills the second basic criterion for an economically viable 'ore body', i.e. sufficient volume. The first criterion, that the 'ore body' have sufficient concentration, is amply fulfilled, as indicated earlier. Fulfilling the second criterion presents challenges which are addressed in several chapters in this book. The challenge of collection of e-waste has been met in informal recycling, but not adequately in formal recycling. The informal recycling situation is discussed in Section 1.6.

In Table 2, the total value of several metals in large quantities of collected mobile phones and laptop computers is given for the year 2010 [2]. Mobile phones contain over 40 different chemical elements including base metals, such as Cu, Ni, and Sn; specialty metals including Co, In, and Sb; and the precious metals Ag, Au, and Pd. Metals, mostly Cu, make up about one quarter of the metal content in each phone. One tonne of scrap mobile phones (equivalent to about 13,000 units without batteries), contained in 2010 an average of 3.5 kg Ag, 340 g Au, 130 g Pd, and as much as 130 kg Cu [2]. Today, these amounts have dropped to 1.3 kg Ag, 300 g Au, 40 g Pd and 125 kg Cu, on average. The value of these metals can approximate up to $10,000/t (although less at current low metal prices), with 80% or more of the total being due to the precious metals present. By contrast, a single unit contains mg of precious metals and ~9 g of Cu. Thus, the net value of one unit is below $1U.S., which

***Table 2** Average content of precious metals, copper, and cobalt in mobile phones and computers, and resulting metals demand from global sales in 2010, compared with world mine production. Reproduced with permission from* [2]

	a) Mobile phones		b) PCs and laptop computers			
	1,600 million units/year, each with a lithium-ion battery		350 million units/year, of which ~180 million have a lithium-ion battery		a + b = Urban mine	
Metal	Unit metal content	Total metal content	Unit metal content	Total metal content	Global mine production (2010)	Share a + b of global mine production
Silver	250 mg	400 t	1,000 mg	380 t	22,900 t	3%
Gold	24 mg	38 t	220 mg	77 t	2,650 t	4%
Palladium	9 mg	14 t	80 mg	28 t	225 t	19%
Copper	9 g	14,000 t	500 g	175,000 t	18Mt	<1%
Cobalt	3.8 g	6,100 t	65 g	11,700 t	88,000 t	20%

t, tonnes; Mt, million tonnes; g, grams; mg, milligrams

does not provide an economic incentive for recycling. It is the sheer number of mobile phones in use that attracts attention for possible metal recovery. About 1.6 billion of these phones were sold worldwide in 2010 alone. In 2014, this number rose to 1.9 billion, including 1.2 billion smart phones. The number of phones produced is increasing yearly to meet demand. The active lifetime of each phone is 2–3 years, after which it is out of use in drawers or landfills, or sent to non-OECD nations for reuse or informal recycling [13,22,33]. The estimated ten billion units produced in total by 2010 would contain a total of 2,400 tonnes of Ag, 230 tonnes of Au, and 90 tonnes of Pd. As seen in Table 2, the Au and Ag contents of the combined 2010 sales volumes of mobile phones and computers are equivalent to ~4 % of the global mine production for Ag and Au and ~20% of that of Pd and Co.

The modern automobile is rapidly becoming a 'computer on wheels.' The proliferation of electronic devices to control an increasing number of operations requires technology metals, making the automobile an urban mine by itself. As global affluence increases, the number of automobiles grows proportionately, especially in non-OECD countries. One example is the growth in the use of PGM as automotive catalysts. In Figure 31.3, annual demand figures are shown for PGM in automotive catalysts, both globally and for Europe. Cumulative gross demands (6,280 t PGM globally) and recycling volumes (1,250 t) up to 2015 are also shown. In Europe, the large increase in demand for PGM for automotive catalysts began in the mid-1990s. Because of the relatively long lifetime of automobiles (10–15 years), some 4,000 t of PGM are still in use in cars globally. Most of the car catalysts that are recycled today are from the early 2000s, so the potential for future recycling of PGM from cars on the road today is huge [4,9].

However, major deficiencies in recycling of automotive catalysts, mainly related to failure to collect vehicles at the end of their lives, could result in inability to close this cycle, and, thus, seriously undermine the potential of this urban mine. Hagelüken [2] estimated that the cumulative wastage from car catalysts already adds up to 900 t of PGM globally, with Europe contributing about 200 t. Updated for 2015, it can be assumed that about 1,000 t of PGM have been lost from automotive catalysts globally. PGM losses occur at all stages of the life cycle including some dissipative driving losses 'through the exhaust pipe' if the road is rough and the car not well maintained, losses from cars that were not recycled, losses from catalysts that were not removed before the car was shredded, losses during inappropriate handling and mechanical pre-treatment of car catalysts, and losses during metallurgical catalyst processing. PGM from these inadvertent losses remain in the commons and are not recoverable. It is expected that wastage of PGM by these losses will be particularly large in the near term in China, India, Brazil, Nigeria, Indonesia, and other non-OECD nations, where car ownership is increasing rapidly, but roads and cars are not as well maintained as in OECD nations and recycling structures are not adequately established.

Recycling PGM from automotive catalysts is relatively easily accomplished, since the catalyst can be rapidly removed from the automobile and the PGM recovered with high yields in state-of-the-art precious metal refineries [34]. Precious and specialty metals that are vital in the operation of many other features of modern automobiles, mainly in car electronics, are not so easily recovered. These metals are located in small amounts throughout the automobile. Removal and collection of the products containing them in EoL vehicles is usually not economic presenting the potential for significant losses of technology metals as their use in automobiles rapidly increases.

1.6 Technologies for Metal Separations and Recovery from EOL Wastes

It is apparent that several 'urban mines' for precious metals exist, i.e., automobiles, e-waste, scrap, petrochemical catalysts, to name a few. Effectively recovering metals from these sources requires well set-up recycling chains, starting with collection and product sorting to dismantling and pre-processing to metallurgical separation and purification of metals as the final step of the chain. Technologies applied at these steps range from small to large scale, from manual to automated pre-processing and from highly efficient integrated smelter-refineries to crude recovery processes with low yields and high pollution used in informal recycling. Metal recycling is done efficiently and to near completion in the case of closed cycles, but it is often done poorly in the case of open cycles, especially when final processing is conducted by informal recycling. Globally, most electronic waste is still consigned to landfill, incinerated, or treated in poor-performing "backyard" operations. Extension of sound recycling to this source is desirable, since large quantities of precious and critical metals are otherwise lost while hazardous substances are emitted. In this section, several technologies used and proposed for metal recycling are presented and discussed. Greater detail is provided in several references [3–5,9].

1.6.1 Collection, Conditioning, and Pre-processing of Waste

Before the actual recovery process for precious and specialty metals can take place, collected material to be recycled needs to be conditioned, in most cases. Conditioning may involve, for example, dismantling and/or mechanical pre-processing (e.g., by shredding and sorting) electronic scrap, decanning car catalysts (extracting the catalyst monolith from the steel case) and burning oil refining catalysts contaminated with carbon. Whatever kind of pre-processing is employed, it must always be conducted in a way that the output fractions provide an optimal fit to the subsequent metallurgical recovery processes and that losses of valuable substances during pre-processing are minimized. The layout of any specific recycling chain and its interfaces depends on the type of material to be treated and is crucial for the overall recycling success. Pre-processing is particularly difficult for technology metals in many high-tech products. For example, precious metals contained in circuit boards are associated with other metals in contacts, connectors, solders, hard disk drives, etc.; with ceramics in multilayer capacitors, integrated circuits (IC), hybrid ceramics, etc.; and with plastics in circuit board tracks, interboard layers, etc. Small-size material connections, coatings, and alloys cannot be separated by shredding. Hence, incomplete liberation and subsequent incorrect sorting result in losses of technology metals to side streams, including dust, from which they cannot be recovered by metallurgical treatment [9]. An industrial test with mixed information technology (IT) electronic scrap treated in a modern shredder without prior dismantling of circuit boards revealed that the percentages of Ag, Au, and Pd ending up in fractions from which they could be recovered (circuit board and Cu fractions) were only 12%, 26%, and 26%, respectively [35]. Thus, automotive catalysts, batteries, high-grade circuit boards, and mobile phones or MP3 players need to be removed or sorted out prior to mechanical pre-processing to prevent irrecoverable losses. These components/devices can be fed into a smelter-refinery process directly, with the effect of recovering most of the metals with high efficiency (over 90%) [9]. For larger items and for low-grade electronic scrap, direct feeding to a smelter is usually not applicable and some degree of mechanical pre-processing is required. Instead of intensely shredding the devices, a coarse size reduction, followed by manual or automated removal of circuit board fractions, can be a viable alternative. Trained workers can often remove certain complex target components more selectively than can be done by automated sorting wherever trained manual labor is available and affordable; in many developing and transition countries, for instance, this labor can be used as a valid alternative to dismantle, sort, and remove critical fractions, such as circuit boards or batteries by hand. It would be desirable to combine this manual recovery of metal-containing parts with state-of-the-art industrial metal recovery processes. Unfortunately, such processes are usually located far from the location where the parts are found [2]. Hence, manual dismantling/sorting and informal work organization are not negative *per se*. Such activities can be the adequate solution in some countries as long as training is provided and no crude backyard recovery of metals is taking place as the final step. Unfortunately, use of untrained manual labor coupled with corrosive chemicals in informal recycling processes in non-OECD countries is extensive and has resulted in extensive negative environmental and health conditions, as described in several chapters in this book. One conclusion that can be drawn is that there is no single optimal way to go from the EoL product to recycling of the metals contained therein.

1.6.2 Separation and Recovery Technologies

1.6.2.1 Integrated Smelter and Advanced Refining Technologies

Hagelüken [9] has described the process of recovering a number of metals from a variety of waste products by means of the Umicore integrated smelter. This process is highly efficient and allows the recovery of many metals at rates near 100%. The process, also, produces much less CO_2 than would have been the case with mining an equivalent amount of ore. In one case, working with Germany's ecological research group Öko-Institut, the potential CO_2 benefit was calculated based on real 2007 Umicore Hoboken data and the EMPA/ETH Zurich ecoinvent data base.The 2007 CO_2 emissions of the Umicore process when used to recover 75,000t of metals from 300,000t of feed materials was 0.28 Mt, compared to 1.28 Mt CO_2 that would have been generated if the metals were obtained through primary production. The generation of CO_2 is reduced nearly five-fold in this recycling process compared to an equivalent mining process. Recycling metals just in this single facility can thus prevent the emission of 1Mt CO_2 per annum and the environmental footprint of the recovered metals is reduced substantially. This calculation was based on a metal output of 1,100t Ag, 32t Au, 32t PGM, 70,000t Cu/Pb/Ni and 4,100t Sn/Se/Te/In/Sb/Bi/As from a mixture of low-grade industrial by-products and higher-grade materials such as circuit boards or catalysts. The savings in emissions would be even larger if only higher grade materials were taken into account [9].

The combination of valuable metals as well as toxic and organic substances with halogens in many EoL products requires special equipment and considerable investments for off-gas and effluent management to ensure environmentally sound operations, *i.e.*, to prevent heavy metal and dioxin emissions, etc. Many plants, in particular in Asian transition countries, are not adequately equipped for metal recovery from such EoL fractions. In such plants, electronic scrap is often 'industrially' treated in noncompliant smelters or leached with strong acids in hydrometallurgical plants with questionable effluent management, with a primary focus on recovering Au and Cu, only.

The chapter by Ueda *et al.* [6] describes the recovery of PGM from catalytic converters and other waste at the Tanaka Kikinzoku Kogyo K.K., refinery in Japan. The prediction is made that recycling volumes of PGM from end-of-life automotive catalysts will double from their current volumes in the next few years, so it is possible that the recycled PGM supply will reach approximately 30% of total platinum group metal supply (total of mining and recycling volumes.) These authors [6] state that "Predictions such as this are likely to also impact mine development and management going forward. This and increasing processing capabilities among platinum group metal recovery and refining businesses will become important issues going forward."

1.6.2.2 Informal Recycling

Taizhou has been a prominent e-waste recycling center in China for the past twenty-five years. Its e-waste industry employs around 40,000 people, with an annual dismantling capacity of over 2.2 million tons [36]. Streicher-Porte *et al.* [36] have pointed out some advantages of the manual processing of e-waste. "Informal manual processing has been criticized for its pollution and health impacts, but manual disassembly itself, if well organized and properly protected, can serve as an efficient way of separating reusable, hazardous and valuable components from e-waste, which greatly simplifies the subsequent recovery of materials and improves reusability."

Unfortunately, a major part of electronic scrap and electronic waste is handled in the informal sector in thousands of backyard informal recycling facilities [3,9]. Pre-processing of the EoL products includes open-sky incineration to remove plastics, 'cooking' of circuit boards over a torch for de-soldering, cyanide leaching, and mercury amalgamation. Over and above their disastrous effects on health and environment, the efficiency of such activities is very low, as well. An investigation in Bangalore, India revealed that only 25% of the Au contained in circuit boards was recovered, compared to over 95% in integrated smelters [9]. A UNEP report cited by Hagelüken and Grehl [9] provides a comprehensive overview on the situation in developing countries.

Informal recycling is successful because employment is provided to a large number of individuals who are willing to work for small wages. Average wages per day for e-waste recycling jobs in China are in the range of 50-100 RMB [36]. In the economies involved, this wage is considered good. There is no shortage of workers. The manual disassembly of electronic products is done with little regard for the safety of the worker or protection of the environment. Recovery rates of precious metals from the disassembled products are low with much of the value lost to the commons due to inefficiencies in the process. One of the important challenges in the global recycling industry is how to establish an effective formal recycling industry, perhaps in cooperation with the informal recycling industry, in which these EoL electronic products are collected, disassembled, and the value retrieved by a safe and efficient recovery process. The dynamic relationships between the formal and informal sectors in China have been presented [36].

It has been suggested [33] that efforts be made to introduce clean chemistry methods [37] into the informal recycling system. Whatever is done, at present, the amount of value retrieved from e-waste by informal recycling exceeds that recovered by formal recycling procedures. Those engaged in informal recycling have coupled the availability of large quantities of e-waste from OECD nations, usually shipped illegally, and increasing amounts from non-OECD nations, with the large labor force available in non-OECD nations to produce a thriving business in the recovery of precious metals. However, this form of urban mining is inefficient and has large negative environmental and health consequences. Primitive recovery methods are used by untrained persons, often children, with unknown, but appreciable, loss of precious and other metals to the commons. Despite these limitations, informal recycling is a major contributor to urban mining and its use is expanding. If combined with state-of-the-art metallurgical recovery in integrated smelters, this "best of two worlds" approach can be a valid solution for many developing and transition countries [38].

1.7 Conclusions

Concerned stakeholders need to be made aware of the importance of the technology metal supply to global society. These metals are essential to the continuation of personal, commercial, industrial, and military activities world-wide, yet few people know of them or of their essential role in the high-technology products that have transformed our world. Every effort should be made to educate stakeholders who have much to gain by being aware of these metals. Awareness may be the key word and education should start with the public, from whom, in the long term, come all other stakeholders. Appropriate individuals must be aware and become involved before action can be taken and major changes made to the present global recycling systems. To the extent possible, the long-term vision is to turn the open loop for e-waste, cars, and other resource-relevant consumer products, as it exists today, into a closed one all over the world. Key aspects of this global vision include the following.

1. A paradigm shift is needed to convince industry of the high recycling potentials that exist in urban mines. Successful urban mining operations need to be acknowledged with publicity and the economic and environmental benefits of such mining promoted whenever possible.
2. Attitudes need to change from a waste-management to a resource-management perspective, reflecting the potential that our waste scrap and EoL products have for society. This attitude change will promote collection, appropriate treatment, enactment of appropriate legislation, and enforcement of this legislation. Specifically, for critical technology metals, such as the rare earth elements, Li, Co, and In, measures need to be put in place to advance their recycling even in the absence of (current) value, volume, and environmental drivers. Such action would represent a paradigm shift from present actions.
3. Goals need to be adjusted where appropriate. The current focus on mass is insufficient. Instead, much more emphasis should be placed on the quality and efficiency of recycling and on recovery of critical and precious metals. This goes hand in hand with a certain prioritization. It is indeed reasonable to recover less material mass (e.g., of plastics or steel) if this leads to tangible improvements in the recovery of technology metals. Adjustment of goals would be helped by an informed public, media, and other stakeholders who are aware of what the technology metals are, why they are important, and why their recovery should be prioritized.
4. Recycling practices need to reflect the increasing global emphasis on stewardship and use of green chemistry procedures. Since the recycling industry plays a key role for our future needs, traditional structures of the scrap and metal recovery business with its rather poor prestige do not fit any more. Recycling of high-technology metals plays in the same league as clean technology manufacturing and renewable energy generation. Economic structures; public esteem; new technologies, where needed; and stakeholder cooperation in recycling efforts should reflect this, with more emphasis on transparency and business ethics than is the case today. Opportunities abound for the development of new procedures and technologies to improve formal and informal recycling operations as well as the active pursuit of means to bring the two together to produce more effective and environmentally responsible recycling.
5. Education of the public; media; policy makers at local, regional, and national levels; and industrial organizations concerning the need to responsibly control Earth's technology metal supply must be more effective. Media should be encouraged, wherever possible, to provide balanced coverage of the importance of technology metals in our global society. Media coverage could be significant in educating the public and policy makers on significant issues concerning recycling.
6. Greater efforts must be made to interest young people in fields that are important to recycling and metal sustainability, such as separation science, metallurgy, waste management, and economics. Individuals trained in these and related subjects are needed to generate the new ideas and create the new technologies required to solve many of the pressing recycling problems of the present and future.
7. Policy makers at all levels and in all nations need to be informed about the need to preserve Earth's technology metal supply and the consequences of not doing so. These individuals have responsibility for enacting and enforcing legislation. Many factors are

at play here. Economic considerations, environmental and health concerns, resource depletion, National security, and global warming are but a few of these factors. Recycling today is spotty around the world. Informed and committed policy makers are needed to improve this situation.

8. Finally, the vision from a manufacturer's perspective needs to change. Today, producer responsibility and recycling is often seen as a burden, imposed by law. In fact, responsible recycling is an opportunity for manufacturers to sustainably and with greater reliability get access to the raw materials they need for future production. This reliable access can be a valuable asset. To fulfil this vision, creative business models to close the loop are essential. Recycling EOL products efficiently today provides insurance for the future [2].

Effective recycling systems can make a significant contribution to conserving natural resources of scarce metals, securing sufficient supplies of technology metals for coming generations, contributing to national security, and reducing the environmental burden connected to the mining of metals. Moreover, such systems would dampen metal price volatility and decrease the climatic impact of metal production, which is energy intensive, especially in the case of low-concentration ores mined for precious metals.

References

1. Reck, B.K., Graedel, T.E., 2012, Challenges in Metal Recycling, *Science*, 337, 690–695.
2. Hagelüken, C., 2014, Recycling of (Critical) Metals, in *Critical Metals Handbook,* Gunn, G. (Ed.), Wiley, Oxford, 41–69.
3. Izatt, R.M., Izatt, S.R., Bruening, R.L. *et al.*, 2014, Challenges to Achievement of Metal Sustainability in Our High-Tech Society, *Chemical Society Reviews*, 43, 2451–2475.
4. Hagelüken, C., Buchert, M., Ryan, P., 2009, Materials Flow of Platinum Group Metals in Germany, *International Journal of Sustainable Manufacturing*, 1, 330–346.
5. Reuter, M., Hudson, C., Hagelüken, C., et al., 2013, Metal Recycling – Opportunities, Limits, Infrastructure — A Report of the Working Group on the Global Metal Flows to the International Resource Panel, United Nations Environment Program, Paris, France.
6. Ueda, T., Ichiishi, S., Okuda, A., Matsutani, K., 2016, Refining and Recycling Technologies for Precious Metals, this volume.
7. Williams, I.D., Global Metal Reuse, and Formal and Informal Recycling from Electronic and Other High-Tech Wastes, this volume.
8. Osibanjo, O., Nnorom, C.I., Sdie, G.U. *et al.*, Global Management of Electronic Wastes: Challenges Facing Developing and Economy-in-Transition Countries, this volume.
9. Hagelűken, C., Grehl, M., 2012, Recycling and Loop Concept for a Sustainable Usage, In *Precious Materials Handbook*, Sehrt, U. (Ed), Umicore D et Co, KG, Hanau-Wolfgang, Germany.
10. Mooiman, M.B., Sole, K.C., Dinham, N., 2016, The Precious Metals Industry: Global Challenges, Responses, and Prospects, this volume.
11. Van Kerckhoven, T., 2014, Optimizing (Precious) Metals Recovery out of Electronic Scrap, presentation, *Electronics Recycling Asia Conference, Singapore*, 13.11.
12. Taylor, M.B., Rubin, E.S., Hounshell, D.A., 2005, Regulation as the Mother of Innovation: The Case of SO_2 Control, *Law and Policy*, 27, 348–378.
13. Abraham, D.S., 2015, *The Elements of Power: Gadgets, Guns, and the Struggle for a Sustainable Future in the Rare Metal Age,* Yale University Press, New Haven, CT.
14. Hagelüken, C., Meskers, C.E.M., 2010, Complex Lifecycles of Precious and Special Metals, in *Linkages of Sustainability*, Graedel, T., Van der Voet, T.E. (Eds), MIT Press, Boston, 165–197.
15. Wilson, A. J., 1994, *The Living Rock*, Woodhead, Cambridge, UK.

16. Pan, J., Chon, H.-S., Cave, M.R., *et al.*, 2012, Toxic Trace Elements, in *Pollutants, Human Health and the Environment: a Risk Based Approach*, Plant, J.A., Voulvoulis, N., and Ragnarsdottir, K.V. (Eds.), Wiley-Blackwell, Oxford, pp. 87–114.
17. Normile, D., 2013, In Minamata, Mercury Still Divides, *Science*, 341, 1446–1447.
18. Lubick, N., Malakoff, D., 2013, With Pact's Completion, the Real Work Begins, *Science*, 341, 1443–1445.
19. Yang. X. J., Lin, A., Li, X.-L., *et al.*, (2013), China's Ion-adsorption Rare Earth Resources, Mining Consequences and Preservation, *Environmental Development*, 8, 131–136.
20. Qiu, G., Ping Li, P., Feng, X., 2016, Mercury Mining in China and Its Environmental and Health Impacts, this volume.
21. Wade, L., 2013, Gold's Dark Side, *Science,* 341, 1448–1449.
22. Izatt, R.M., 2015, Recovering Metals from Electronic Wastes, *Chemical Engineering Progress,* January, 48–54.
23. Scott, A. 2014, Dialing Back on Cell Phone Waste, *Chemical & Engineering News*, September 4, 30–33.
24. Sridhar, M.K.C., and Hammed, T.B., 2016, Dynamics of Metal Reuse and Recycling in Informal Sector in Developing Countries, this volume.
25. Ritter, S.K., 2013, Toward Sustainable Electronics, *Chemical & Engineering News*, April 1, 41–43.
26. Norgate, T.E., 2010, Deteriorating Ore Resources, in *Linkages of Sustainability*, Graedel, T.E., van der Voet, E. (Eds.), The MIT Press, Cambridge, MA, pp. 131–148.
27. Mudd, G.M., 2012, Sustainability Reporting and the Platinum Group Metals: A Global Mining Industry Leader, *Platinum Metals Reviews*, 56, 2–19.
28. Jiang, X., Su, S., Song, J., 2016, Metal Pollution and Metal Sustainability in China, in *Metal Sustainability: Global Challenges, Consequences, and Prospects,* this volume.
29. Gordon, R.B., Bertram, M., Graedel, T.E., 2006, Metal Stocks and Sustainability, *Proceedings of the National Academy of Sciences (U. S. A).* 103, 1209–1214.
30. Wilburn, D.R., 2012, Global Exploration and Production Capacity for Platinum-Group Metals from 1995 Through 2015 (Version 1.1), U.S. Geological Survey Scientific Investigations Report 2012–5164, 26 pp., http://pubs.usgs.gov/sir/2012/5164/, accessed 10 December 2015.
31. Dupont, D., Binnemans, K., 2015, Recycling of Rare Earths from NdFeB Magnets Using a Combined Leaching/Extraction System Based on the Acidity and Thermomorphism of the Ionic Liquid [Hbet][Tf_2N], *Green Chemistry*, 17, 2150–2163.
32. Izatt, S.R., McKenzie, J.S., Bruening, R.L., et al., 2016, Selective Recovery of Platinum Group Metals and Rare Earth Metals from Complex Matrices Using a Green Chemistry-Molecular Recognition Technology Approach, this volume.
33. Izatt, R.M., Izatt, S.R., Izatt, N.E., *et al.*, 2015, Industrial Applications of Molecular Recognition Technology to Green Chemistry Separations of Platinum Group Metals and Selective Removal of Metal Impurities from Process Streams, *Green Chemistry*, 17, 2236–2245.
34. Hagelüken, C. (2012): Recycling the Platinum Group Metals: A European Perspective, *Platinum Metals Reviews,* 56, 29–35.
35. Chancerel, P., Meskers, C.E.M., Hagelüken, C., Rotter, S., 2009, Assessment of Precious Metal Flows During Preprocessing of Waste Electrical and Electronic Equipment, *Journal of Industrial Ecology*, 13, 791–810.
36. Streicher-Porte, M., Chi, X., Yang, J., 2016, E-waste Recycling in China: Status Quo in 2015, this volume.
37. Anastas, P., Eghbali, N., 2010, Green Chemistry: Principles and Practice, *Chemical Society Reviews,* 39, 301–312.
38. Wang, F., Huisman, J., Meskers, C.E.M., *et al.*, 2012, The Best-of-2-Worlds philosophy: Developing local dismantling and global infrastructure network for sustainable e-waste treatment in emerging economies, *Waste Management* 32, 2134–2146.

2

Global Metal Reuse, and Formal and Informal Recycling from Electronic and Other High-Tech Wastes

Ian D. Williams

International Centre for Environmental Science, University of Southampton, UK

2.1 Introduction

Metals have been used for structural and decorative purposes since prehistoric times. Of the 118 elements listed in the Periodic Table, about 75% (91) are metals. Their chemical properties and hence their reactivity are principally governed by the extent to which they form cations. They have characteristic physical properties that give them their renowned wide range of utility: many elemental metals are good conductors of heat and electricity; are hard, reusable and recyclable, malleable and/or ductile; have an attractive sheen and lustre; are non-biodegradable; can be sonorous; and can be blended with other materials to form alloys. Some metals are radioactive, e.g. actinium, plutonium, promethium, protactinium, technetium, thorium, uranium. The form in which metals occur (elemental, ionic or organo-metallic compounds, alloys) provides their physical and chemical properties. Sixteen metals can occur in nature in the elemental form (free or native metals): bismuth, cobalt, copper, gold, iridium, iron, lead, mercury, nickel, osmium, palladium, platinum, ruthenium, rhodium, tin and silver. However, most metals occur naturally as compounds and mineral ores, mainly in combination with O (as oxides) or S (as sulfides). The science of extracting metals from mineral ores and preparing them for use is called extractive metallurgy.

Metal Sustainability: Global Challenges, Consequences, and Prospects, First Edition. Edited by Reed M. Izatt.

As the population of the Earth has grown, so has the demand for water, food, energy, goods, services and infrastructure. Prevailing market-economy models have led to a so-called "throwaway society" that is founded on economies of scale, planned obsolescence and a growing demand for new products [1]. Accordingly, sustainable consumption has been identified as one of the most important societal challenges for the 21st century [2]. In the case of metal consumption, this is an exceptional challenge; although modern-day lifestyles depend upon the availability of an abundant stock of a wide range of metals, the public is generally unaware of its share of metals in infrastructures (buildings, structures, power generation and distribution systems, communications and transport networks), as well as in their personal homes, gardens and vehicles. The average more-developed country citizen's in-use metal stock has been estimated as 10-15 metric tons, with five metals — iron, aluminium, copper, zinc, and manganese — responsible for >98% [3]. Although the Earth's crust contains significant quantities of exploitable natural resources, many of these vital assets are non-renewable and are depleting rapidly. The total amounts of geogenic stocks of metals — those deposited by geological processes in concentrations suitable for being extracted and processed, now and in the future — are difficult to quantify accurately [3], but some global estimates are publicly reported (see Table 1).

2.2 Metal Sources

Mining for primary raw materials is capital-, water- and energy-intensive, requires significant quantities of land and has a high environment impact. Concerns about issues such as health and safety, land-use conflicts, mineral rights, social and commercial exploitation of miners, and ecological collapse in certain areas have dogged the mining industry in recent years. Furthermore, some countries have imposed trade restrictions on the export of high-value metals and minerals in order to protect their own demand. China currently produces >90% of the world's rare earth elements (REEs), with an estimated 39% of the proven reserve [4], and has imposed trade barriers for some metals. The REEs are a group of 17 chemically similar metallic elements (15 lanthanides, plus scandium and yttrium) that may be subdivided into "light" (Sc, La, Ce, Pr, Nd, Pm, Sm, Eu, Gd) and "heavy" (Tb, Dy, Ho, Er, Tm, Yb, Lu, Y) groups. The REEs are becoming critical as society moves towards a more sustainable, low-carbon economy, they play an essential role in facilitating the functionality of desirable modern technologies, including permanent magnets in automobiles, computers, wind turbines and mobile phones; rechargeable nickel metal hydride batteries; catalysts; phosphors in lamps and plasma screens; and other applications [5]. Demand for certain REEs (e.g. Dy, Nd) is expected to increase substantially over the next 25 years [6].

As demand for scarce metals — especially rare earth and precious metals — has increased, political concerns about heavy dependence on supplies from China, tightening restrictions on exports and higher prices have triggered searches for alternative sources of REE around the world [7]. Consequently, there has been a recent focus on the second, increasingly important stock of metals available to us. Anthropogenic stocks are those already in society, having been previously extracted, processed and put into use; currently providing service; or discarded or dissipated over time [3]. There is relatively little information available for reliable estimates of anthropogenic stocks (also known as urban mines), with sufficient quality only available for a few metals and just for more-developed

Table 1 *Metal sources and estimated metal stocks*

Metal	Main minerals	Main occurrence and production	Main uses in EEE (electrical and electronic equipment)	Reserve base in geogenic stock 2009 (Tg) [9]	Per-capita anthropogenic stock [3]*
Precious metals					
Gold (Au)	Native	South Africa, USA, Russia, China, Australia, Canada	Corrosion-free electrical connectors, switch contacts	0.1	35–90 g
Iridium (Ir)	Native, natural alloys (osmiridium, iridosmium)	South Africa, Russia, Canada, USA	Indium-tin-oxide in flat panels, thin-film-photovoltaic cells, semiconductors, InGaN in LEDs	0.03	No data
Osmium (Os)			Electrical contacts	0.14	
Platinum (Pt)	Native	Canada, South Africa	Capacitors, connectors, contacts,	2.9	1–3 g
Palladium (Pd)	Native	Canada	diodes, transistors, Ag-Cu-Pd-	0.48	1–4 g
Rhodium (Rh)	Mixed ores	South Africa, Russia, Canada	soldering	0.053	0.2 g
Ruthenium (Ru)			Electrical contacts, thick-film chip resistors	0.076	No data
Silver (Ag)	Native, Argentite	Mexico, USA, Peru, Canada, Russia	Contacts, lead-free soldering, batteries, capacitors, RFID-chips, photovoltaic cells	0.57	13 g
Speciality metals					
Antimony (Sb)	Native, Stibnite	China	Semiconductors, batteries, flame retardants for plastics	0.23	1 kg
Arsenic (As)	Fe, Ni, Co-minerals, realgar	China, Chile, Peru, Morocco	Semiconductors, batteries, LEDs	1.7	No data
Barium (Ba)	Barite, Witherite	USA	Electrode coatings in fluorescent lamps	880	No data
Beryllium (Be)	Beryl	Brazil, Colombia, Siberia, USA	Beryllium-copper-alloys, beryllium oxide-ceramics, metallic beryllium	3.4	No data
Boron (B)	Borax	USA, Turkey, China, Chile, Italy	Semiconductors, magnets	410	No data
Cadmium (Cd)	Greenockite	China, Japan, South Korea	Batteries, semiconductors	1.2	80 g
Cerium (Ce)	Allanite, Monazite and Bastnäsite minerals	India, Madagascar, South Africa, Hungary, Greece, USA	Phosphors**	58	No data
Cesium (Cs)	Pollucite, Petalite	Canada, Zimbabwe, Namibia	Photoelectric cells	0.11	No data

(Continued)

Table 1 *(continued)*

Metal	Main minerals	Main occurrence and production	Main uses in EEE (electrical and electronic equipment)	Reserve base in geogenic stock 2009 (Tg) [9]	Per-capita anthropogenic stock [3]*
Dysprosium (Dy)	Ion-adsorption clays	China	Magnets, compact fluorescent bulbs, phosphors, fuel cells, batteries	3.1	No data
Erbium (Er)			Fiber amplifier in optical communications	2.0	No data
Europium (Eu)	Bastnäsite, Loparite, Xenotime, and Monazite minerals		Phosphors, fluorescent glass	0.8	No data
Gadolinium (Gd)	Bastnäsite and Monazite minerals	China, Australia, USA, Brazil, Sri Lanka, India	Magnetic resonance imaging. phosphors	3.4	No data
Gallium (Ga)	Zinc ores, bauxite	Guinea, Australia, Vietnam, Jamaica, Brazil, India, China	Laser diodes, light- emitting diodes, solar cells, photodetectors, integrated circuits, switches	19	No data
Germanium (Ge)	Zinc-copper-lead ore bodies, coal deposits	China, Russia, USA	Semi-conductive chips, photovoltaic cells, glass fibre, optical glasses glass fibre	1.8	No data
Hafnium (Hf)	Heavy mineral sands, carbonatite intrusions	Brazil, Malawi, Australia	Microprocessors	1.1	No data
Holmium (Ho)	Ion-adsorption clays, Monazite and Gadolinite minerals	Australia, China, USA, Brazil, India, Sri Lanka	Magnets, microwave equipment	0.7	No data
Indium (In)	Slag and dust from zinc production	China, Canada, Japan, South Korea	Flat display panels, thin-film photovoltaic cells, LEDs, semiconductors	0.57	No data
Lanthanum (La)	Bastnäsite and Monazite minerals	China, Australia, USA, Brazil, Sri Lanka, India	Magnets, compact fluorescent bulbs, phosphors, fuel cells, batteries	27	No data
Lutetium (Lu)	Ion-adsorption clays, Monazite minerals		Phosphors	0.3	No data
Mercury (Mg)	Cinnabar	USA, Spain, Peru, Italy, Slovenia	Switches and relays, fluorescent lamps	80	10 g
Neodymium (Nd)	Bastnäsite and Monazite minerals	China, Australia, USA, Brazil, Sri Lanka, India	Magnets, compact fluorescent bulbs, phosphors, fuel cells, batteries	22	No data
Praseodymium (Pr)			Magnets, phosphors, optical amplifiers	6.3	No data

Rhenium (Rh)	Molybdenite minerals	Chile, Peru, USA, Poland	Thermocouples, electrical contacts, filaments	0.053	No data
Samarium (Sm)	Bastnäsite and Monazite minerals	China, Australia, USA, Brazil, Sri Lanka, India	Magnets	4.0	No data
Scandium (Sc)	Thortveitite	Scandinavia	Magnets, compact fluorescent bulbs, phosphors, fuel cells, batteries	16	No data
Selenium (Se)	Selenide in sulfide ores, including Cu, Ni and Pb	China, USA	Solar cells, photocells, photocopiers, light meters, LEDs	57	No data
Strontium (Sr)	Celestite, Strontianite	Scotland, Mexico, Spain, Turkey, Iran, China, USA, Canada	Cathode ray tubes	0.4	No data
Tantalum (Ta)	Tantalite, Microlite, Wodginite, Euxenite minerals	Australia, China, Ethiopia, Mozambique, Brazil, Canada, Thailand, Malysia	Capacitors, surgical applications, cemented carbides, aerospace, automobile	0.9	No data
Tellurium (Te)	Anode sludges from copper production, Tellurite minerals	USA, Canada, Japan, Peru	Thin-film photovoltaic cells, photoelectrical devices, semiconductors	1.1	No data
Terbium (Tb)	Ion-adsorption clays, Xenotime, Euxenite minerals	China	Phosphors	0.6	No data
Thallium (Tl)	Sulfide ores, including Cu, Zn and Pb	China, USA	IR optical materials, photoresistors, metal halide lamps	0.00065	No data
Thulium (Tm)	Ion-adsorption clays, Monazite minerals	China, Australia, Brazil, USA, Greenland, India	High-temperature superconductors, lasers	0.3	No data
Tungsten (W)	Wolframite, Scheelite, Ferberite and Hübnerite minerals	China, Russia, Canada	W carbide, electrodes, cables, electrical components, additives in cathode ray tube glass	6.3	1 kg
Ytterbium (Yb)	Monazite, Xenotime, Euxenite minerals	China, USA, Brazil, India, Australia, Sri Lanka	Doping materials in lasers and amplifiers	2.0	No data
Yttrium (Y)	Ion-adsorption clays, Bastnäsite, Xenotime and Monazite minerals	China, USA, Canada, India, Brazil	Magnets, compact fluorescent bulbs, phosphors, fuel cells, batteries	25	No data
Zirconium (Zr)	Zircon	Australia, South Africa, USA, Russia, India, Brazil	Light filaments	77 (ZrO_2)	No data

*The anthropogenic stock reported here is from more-developed countries only: Australia, Canada, the European Union EU15, Norway, Switzerland, Japan, New Zealand, and the United States of America (altogether about 860 million people in 2005).
**A phosphor is a substance that exhibits the phenomenon of luminescence, the emission of light not resulting from heat.

countries (Table 1). In addition, little or no information is available on metals stocks in public, commercial and industrial infrastructure; governmental stockpiles; landfills; mine tailings; manufacturing, processing and recycling facilities; and scrap industries. A further significant gap in knowledge is the lack of data on the recycling performance of existing systems and current and potential individual metal recycling rates [8]. As anthropogenic stocks increase, towns, cities and industrial sites will become the metal mines of the future.

2.3 E-waste

Since the 1990s in particular, there has been a massive global explosion of new electronic goods and services. Today, electrical and electronic products are an essential part of many people's personal and professional lives. The production of modern electronics is energy-intensive and requires the use of scarce resources: many are valuable, some are hazardous and some are both. The average computer chip today contains 61 different elements compared to the 12 elements that were used in the 1980s [10]. However, increasingly short product lifecycles and rapidly advancing technology have led to vast quantities of relatively new electrical and electronic equipment (EEE) being discarded [11]. EEE is defined as equipment that is dependent on electric currents or electromagnetic fields in order to work properly and equipment for the generation, transfer and measurement of such currents. EEE that has come to its end-of-life (EoL) either by ceasing to function or ceasing to be of any value to its owners is commonly referred to as e-waste, or waste electrical and electronic equipment (WEEE), as it is usually called in the European Union (EU). With a rapidly growing and more demanding global population, it is essential to improve the environmental management of e-waste in order to contribute to a circular economy and enhance resource efficiency.

Whilst the key factors that affect the generation, collection and disposal of e-waste, and the spatial and temporal interactions of these factors, have been identified [12], it is proving enormously challenging to devise strategies to manage e-waste effectively on a global scale. E-waste is one of the chief sources of metals and organic pollutants in solid waste and the fastest growing waste stream [13]. There are varied calculations on the amount of e-waste generated annually, with estimates of between 30 and 50 million tonnes [13,14]. The most recent monitoring report [84] estimated that 41.8 million metric tonnes (megatonnes, Mt) of electrical and electronic products were discarded globally in 2014, with a rise to 50 million metric tonnes predicted by 2018. Information and communications technology (ICT) and large domestic appliances (commonly known as 'white goods') dominate the e-waste fractions by weight.

Four key global issues make e-waste a priority waste stream: global quantities of e-waste; resource impacts; potential health and environmental impacts; and ethical concerns [12]. E-waste is non-homogeneous and forms a complex mixture of materials and components, often containing hundreds of different substances, many of which are potentially highly toxic. It also contains high residual value metals and hence represents a huge loss of key metals (if discarded) or a potential reservoir (if stockpiled). Many modern electrical and electronic products contain metals that have been classified as Critical Raw Materials (CRMs) by the European Commission. CRMs are those for which supply and economic impact in the EU (and elsewhere) is at risk and higher than for other raw materials. Most

CRMs are virtually unrecovered from e-waste (estimated rate <1% [5]). CRMs are found in smaller electronic and electrical products that have a higher propensity to leak from the official e-waste collection channels.

Generally e-waste contains about 40–60% metal, with iron/steel, copper and aluminium being the main metallic components [15]. High-tech e-waste also often contains metals such as antimony, gold, hafnium, indium, iron, lead, nickel, palladium, rare earth metals, silver, tantalum, tin and zinc in combination with valuable reusable plastics, including polycarbonates, polyethylene, polyesters, polypropylene and phenolformaldehyde [16]. Material recovery from such secondary sources is dynamically related to fluctuating material prices due to resource scarcity, which has a direct effect on recycling practices [17].

Major categories of hazardous materials and components of e-waste include asbestos, batteries, cathode ray tubes, gas discharge lamps, components containing mercury and chlorinated and fluorinated organics, plastics containing halogenated fire retardants, polychlorinated biphenyl (PCB) containing capacitors, printed circuit boards, toner cartridges and toner [8]. Consequently, when e-waste is disposed of or recycled without any controls, there are potentially negative impacts on human health and the environment. Ethical concerns centre on shipments of e-waste (both legal and illegal) from affluent countries to poorer developing countries that lack the facilities to properly treat such wastes [18]; incidences of children working in informal e-waste industries [19,20]; and a lack of safety standards when wastes are handled and processed [12].

Each year around 12 million tonnes of e-waste are disposed of in the EU, but only 30% of e-waste generated is reported as properly collected and recycled, with significant proportions either leaking from the official system or being hoarded by householders and businesses. In practice, there is no such thing as a typical e-waste material/metal; it depends on the type, origin, age and manufacturer of the equipment [21]. For example, a 1987 Nokia Cityman mobile phone weighed 770 g and contained few elements whereas a modern Nokia smart phone weighs about 100 g and contains >40 elements (Source: Nokia).

2.4 Responses to the E-waste Problem

The EU has led the way globally in terms of drafting and implementing legislation to improve the reuse, recycling and other forms of recovery from e-wastes by adopting the "producer responsibility" principle[1]. The WEEE Directive (Directive 2002/96/EC) requires producers, and sometimes importers and distributors, in EU member states to fund the separate 'take-back', collection and recycling of their products and ensure that they are disposed of using environmentally sound methods. These companies have to register with

[1] The Producer Responsibility Principle as a policy principle can be summarized as "the concept that manufacturers and importers of products bear a degree of responsibility for the environmental impacts of their products throughout the products' life-cycles, including upstream impacts inherent in the selection of materials for the products, impacts from manufacturers' production process itself, and downstream impacts from the use and disposal of the products. Producers accept their responsibility when they design their products to minimize the lifecycle environmental impacts and when they accept legal, physical or economic responsibility for the environmental impacts that cannot be eliminated by design" [22]. There are two types of producer responsibility. Extended Producer Responsibility is a policy principle to promote life-cycle environmental improvements of product systems by extending the responsibility of the producers beyond the production phase and particularly to the end-of-life management of products. Individual Producer Responsibility is an operational approach in which producers take physical and/or financial responsibility in the post-consumer (or end-of-life management) stage of their own products.

Table 2 *Categories of electrical and electronic equipment under the two European Union Waste Electrical and Electronic Equipment (WEEE) Directives*

Categories of EEE according to the WEEE Directive (2002/96/EC)	Categories of EEE according to the Recast WEEE Directive (2012/19/EU), to apply from 15 August 2018
1. Large household appliances 2. Small household appliances 3. IT and telecommunications equipment 4. Consumer equipment 5. Lighting equipment 6. Electrical and electronic tools 7. Toys, leisure and sports equipment 8. Medical devices 9. Monitoring and control instruments 10. Automatic dispensers	1. Temperature exchange equipment 2. Screens, monitors, and equipment containing screens having a surface greater than 100 cm^2 3. Lamps 4. Large equipment (any external dimension more than 50 cm) 5. Small equipment (no external dimension more than 50 cm) 6. Small IT and telecommunication equipment (no external dimension more than 50 cm)

national schemes, often called "compliance schemes." A compliance scheme is a non-profit organisation that manages the collection, treatment and recycling of e-waste, batteries and accumulators from authorised collection points on behalf of its members. The scheme's members pay fees according to the share of product put onto the market in the previous compliance year. The regulations also allocate responsibility for "historic" waste made before they came into force, based on producers' current market shares.

There are a series of rising targets to recycle e-waste set at a European level and treatment standards that must be met. The regulations state that manufacturers must label equipment in a standard way and consumers must be able to return their equipment free of charge. Retailers are obliged to provide information to their customers about how they can do this, and are responsible for establishing take-back schemes. Businesses must ensure that the EEE that they use is collected properly at the end of its life. There are ten categories covered by the rules — see Table 2.

The first WEEE Directive (Directive 2002/96/EC) came into force in February 2003 but had to be recast in order to tackle the fast increasing waste stream. The new WEEE Directive (2012/19/EU) came into force in August 2012 and became effective in February 2014. The Recast affects many areas of the original Directive including but not limited to: collection and recovery targets; scope and categorisation of e-waste; distributor take-back provisions; proper treatment; tougher restrictions on the illegal export of e-waste. For example, under the new regulations, a collection target of 45% of EEE placed on the market will apply from 2016, and then 65% of equipment placed on the market or 85% of electronic waste generated must be collected by 2019, with Member States able to choose which 2019 measure they wish to implement.

The WEEE Directive is supported by two related directives:

- The restriction of the use of certain hazardous substances in EEE (RoHS Directive 2002/95/EC) prohibits the placing on the EU market of new EEE containing more than agreed levels of selected metals and flame retardants. This Directive also had to be recast in response to the fast increasing waste stream. The RoHS recast Directive (2011/65/EU),

which became effective in January 2013, aims to increase the amount of e-waste that is appropriately treated, to reduce the volume that goes to disposal and to reduce administrative burdens and ensure coherency with newer policies and legislation.
- The energy-using products directive (EuP Directive 2005/32/EC) establishes a framework for the setting of EU eco-design requirements for energy-using products with the aims of ensuring the free movement of those products within the internal EU market, increasing energy efficiency and the level of protection of the environment.

Regulatory systems for e-waste management have been or are being established in countries such as Australia, China, India, Japan, Kenya, Nigeria, the Philippines, South Africa, South Korea, Taiwan, Vietnam and some parts of Canada and the U.S. [13]. Despite several attempts, there is no Federal obligation to recycle e-waste in the U.S., although some individual states have established mandatory electronics recovery programs.

2.5 Reuse of Metals from High-tech Sources

The enormous resource potential of using EoL products, e-waste and stockpiled W/EEE as secondary sources of materials has focused attention on:

- The variety of metals and other materials they contain;
- How these materials are handled, generated, and dispersed around the globe, often from developed to developing countries;
- Ways in which they can be brought safely, sustainably and cost-effectively back into the value chain, particularly via reuse and recycling.

The EU Directive on waste (Waste Framework Directive 2008/98/EC) sets the basic concepts and definitions related to waste management. Under this Directive:

- Reuse is defined as "any operation to by which products or components that are not waste are used again for the same purpose for which they were conceived"
- Preparing for reuse is defined as "checking, cleaning or repairing recovery operations, by which products or components of products that have become waste are prepared so that they can be re-used without any other pre-processing."

Reuse is an important waste-prevention measure as it extends product life, delaying or even negating the extraction of virgin raw materials to manufacture new products. In developing countries and economies in transition, used goods are regarded as valuable materials that can be repaired or have components recovered for reuse [23]. The goods and components are then traded via intermediates, roadside stands and markets [23,24]. Whilst the methods used to recover components for reuse would usually be regarded as inappropriate in developed countries, they contribute to poverty alleviation by providing an income for significant numbers of poor and marginalized workers in the informal sector and facilitate greater material recovery [23,25].

In the developed world, reuse has been usually overshadowed by recycling. The EU has recently recognized the importance of reuse in terms of materials recovery by making legislative changes within the recast of Directive 2012/19/EE on WEEE that oblige EU member states to prioritize reuse at the earliest stages of e-waste take-back, separate it for reuse and enable access for refurbishment centres. Revised collection reporting now enables reuse to count towards collection targets both within the business-to-business (B2B) and business-to-consumer (B2C) markets.

However, even though reuse is — apart from waste prevention — designated within the waste hierarchy as the prime waste-management process, barriers and questions remain outstanding. In terms of barriers, reuse is dependent on goods being collected and channeled to this activity [26]. In regions/countries with dispersed and relatively small populations, the economics of reuse is highly influenced by low material quantities, irregular collections, uncertain supply, vulnerability to inconsistent market prices, and absent or low technology infrastructure. Regions/countries with larger and more concentrated populations are able to collect and bulk materials for reuse in a more economically viable manner because comprehensive collection networks with regular and more predictable supplies, higher collected volumes and ready access to markets are easier to establish. In terms of questions, the debate about when the potential benefits of purchasing a new, more energy-efficient appliance outweigh the benefits of reuse is significant and ongoing [27].

Nevertheless, refurbishment, repair and reuse of EEE have become increasingly common globally, although the systems and processes that support these activities tend to possess distinctive local characteristics. Due to the short innovation cycles and resultant arisings of e-waste, there is a need to extend the lifetime of these products. Lifetime extension can be achieved through designing easily replicable and reusable sub-assemblies or devices that reduce the risks and costs associated with disassembly using the D4R concept (D4R = design for recycling, repair, refurbishment and reuse). Unlike recycling, which requires the breaking down of technical components, re-use keeps units or components in their original state, which means it is possible to capture both the materials contained within the product and the value added during design and manufacturing. Although a number of standards and labels relating to reuse of EEE have been developed (see Table 3), no recognized reuse standard is really globally used.

The literature on reuse is often not distinguished from recycling because the two processes are interconnected, especially in developing countries where there is a strong interdependency between the formal and informal sectors. There is recognition that the formality or informality of waste management systems varies considerably between countries, although many possess both formal and informal attributes and therefore could be categorized as intermediate [25]. It is worth highlighting that the drivers for e-waste flows, reuse and recycling in China are particularly distinctive [25] for several reasons:

- Whilst China produces, exports and consumes the largest portion of global EEE, it also imports massive quantities of e-waste, mostly via illegal channels [28].
- The large disparity between rural and urban China in terms of income and living standards facilitates the collection and movement of second-hand EEE from urban to rural communities [29,30].
- Almost 60% of the e-waste generated in China is handled by the informal sector [31] because Chinese households prefer to sell obsolete EEE to individual collectors [32].
- Recyclers in China prioritize device or component reuse because their residual functional economic value is usually higher than the intrinsic recoverable material value [25].
- The reuse and remanufacturing markets in China are generally unregulated [25].

Table 3 *Reuse standards relevant to electrical and electronic equipment*

Standard	Application
ISO 9001:2008 Quality Management Systems	Sets out the requirements of a quality management system e.g. to certify repair centres for ICT and imaging inspection systems.
British Standards Institution (BSI) Publicly Available Specification (PAS) BSI-PAS 141	A process management specification for the reuse of used and waste electrical and electronic equipment.
WEEELABEX ('WEEE LABel of EXcellence')	A set of European standards on collection, handling, storage, recycling, preparation for reuse and disposal of WEEE and monitoring the processing companies through audits conducted by trained auditors. Auditors use the same audit process documents, apply the same set of standards and report their findings to the WEEELABEX Office, which records those that have successfully verified.
Revisie	A Belgian (Flanders) quality label that provides a guarantee for up to 6 months for second-hand electric and electronic equipment.
Microsoft Authorised Refurbisher (MAR)	A programme for large refurbishers that provides a reduced royalty licence for refurbished personal computers with a previous Window operating system software licence.
Electro/REV	A Belgian (Brussels and Wallonia) quality label that provides a guarantee for second-hand electrical and electronic appliances in terms of technical work, after-sale service and personnel training. The label is recognized by Recupel, the official Belgian organization responsible for the collection and recycling of e-waste.
The e-Stewards Standard for Responsible Recycling and Reuse of Electronic Equipment®	A North American market-based voluntary standard. It aims to identify and promote recyclers that ensure that used electronics are managed with high standards of environmental and social responsibility. Supported by the U.S. EPA.
Responsible Recycling (R2) Standard – R2:2013 Standard	This voluntary standard establishes so-called responsible recycling ("R2") practices for the recycling of e-waste globally. It aims to certify that e-waste is managed in an environmentally responsible manner, protects the health and safety of workers and the public, and ensures that all data on all media devices is secure until destroyed. Supported by the US EPA.

2.5.1 Reuse by Social Enterprises

Social enterprises such as voluntary organizations, charities and not-for-profit companies are involved in the repair, refurbishment and reuse of various EEE. These organizations collect EEE, especially white goods, for treatment prior to sale or redistribution to disadvantaged people both inside and outside of their own country. Some social enterprises operate very professional reuse and repair operations that utilize detailed systems for monitoring and reporting and guarantees for environmental and health and safety compliance, standardized treatment procedures, reused products' liability, reporting and traceability, insurance, technical expertise, training, license requirements and after-sales service. However, a large number of barriers to the reuse operations of socio-economic enterprises' have been identified

Table 4 *Barriers to socioeconomic enterprises' reuse operations [33]. Reprinted from [33] with permission of Elsevier*

Issue	Barriers
Marketing	• Consumer non-confidence in refurbished equipment • Promotion and marketing reuse as the preferred option when considering purchasing • Lack of shop premises to increase sales • Publicity regarding fraud and identity theft leading to a circumspect approach to the donation of equipment • Social enterprises seen as 'tree huggers' rather than professionals • Finding customers for bulk volume sales
Legislation	• Manufacturers' obligated WEEE takeback services channelling potentially reusable WEEE to disposal/recycling • IT equipment held in lease agreements with large suppliers means companies cannot donate without breaching contracts • Legislation aimed at curtailing rogue reuse firms drives up licensing fees; extra expense for legitimate refurbishers • Consignment note regulations (extra administration charge to customer) • Companies are reluctant to pass products on for reuse because of concerns about data security
Collection and transport of WEEE	• High fuel costs • Practical and logistical factors, including distance travelled to collect WEEE • Limited collection capacity • Companies not wanting to pay for collection services offered
Supply	• Access to large quantities of quality equipment • Organisations upgrading equipment less regularly • Equipment stripped of vital components before donation • Too few laptops in good condition • Difficulties in obtaining operating system disks to go with the computers
Costs	• Time/cost of refurbishment versus final item cost • Low value of some used equipment • Company's perception of item value • Lack of awareness of residual value
Processing of WEEE	• Time-intensive processing • Vast range of technical knowledge required in order to process the variety of appliances received
Storage	• Limited storage space for equipment • Poor storage practices can lead to damage and breakage of equipment
Staffing	• Low staff numbers

(Table 4). There have also been criticisms of the way in which some social enterprises operate, such as allowing computers that are too old to function to be collected by charities for reuse in schools in developing parts of the world and/or are supplied to schools that have no reliable electricity supply or internet connection. Further, the charities subsequently take no responsibility for the EoL management of the equipment and computer recycling, effectively resulting in a transfer of hazardous waste to countries with no infrastructure for dealing with it. Such inappropriate actions are naïve (at best) or criminal (at worst) and they have undoubtedly adversely impacted all parties concerned as well as creating a negative public perception of charitable IT reuse schemes.

There are numerous examples of successful IT reuse by socioeconomic enterprises, although very few studies have addressed this subject [34,35]. One recent investigation [33] focused on reuse of Information and Communication Technology (ICT) EEE by UK-based socio-economic enterprises. The study reported that the most common ICT products handled are computers and related equipment and that the most common barriers to reuse operations are difficulties in marketing products and in understanding and complying with numerous legislative requirements. Successful and trusted networks for IT reuse have developed in the UK, Ireland, Netherlands, Belgium, Austria and Germany. The next step for organizations involved in reuse of ICT could be to cluster operations into "reuse parks" (e.g. the Alelyckan reuse park operated and owned by the City of Gothenburg, Sweden) and industrial networking to exchange components and by-products resulting from their activities.

2.5.2 Reuse in the Private Sector

Trade (reuse) of items when they are no longer needed/wanted has traditionally been via second-hand shops and market stalls, car boot sales, bring-and-buy sales and swap shops. The particular disadvantage of these trading methods for EEE is that there is a risk of buying items that may not be electrically tested and safe and with limited or no guarantee or after-sales support. Whilst such trading methods remain popular, the Internet has transformed people's ability to trade used items. The American multinational corporation eBay Inc., founded in 1995, has been at the forefront of e-commerce. The company manages eBay.com, a global online auction and shopping website in which people and businesses buy and sell a broad variety of goods and services, including EEE; phones and accessories and computers are amongst the most traded items on eBay. Most second-hand shops also have online trading to supplement their physical presence.

The Internet has also enabled reuse to thrive in a relatively non-commercial way as people and organisations give away or swap goods via organizations such as Freecycle (www.freecycle.org), Swapit (www.swapit.co.uk), Gumtree (www.gumtree.com) and WarpIT (www.warp-it.co.uk).

One organization that has drawn worldwide attention to the repair and reuse of EEE is iFixit (http://ifixit.org/). The company aims to reduce e-waste by teaching people to repair their own equipment using free wiki-like online repair guides for consumer electronics and offering tools, parts, and a forum to discuss repairs.

2.5.3 Reuse Research

Demonstration research projects have shown that IT products can be successfully made by combining the D4R concept with used components and recycled materials. The EU-funded Project Zerowin — 'Towards zero waste in industrial networks' (see www.zerowin.eu) — examined approaches to eliminate the wasteful consumption of resources in selected industrial sectors in Europe, primarily via the formation of industrial networks [36]. The project's plan was to show that the approach could be commercially viable whilst meeting at least two of the following stringent targets:

- 30% reduction of greenhouse gas emissions;
- 70% overall reuse and recycling of waste;
- 75% reduction of fresh water use.

The project focused on two waste types in four industry sectors, including high-tech waste from the EEE and photovoltaic (PV) sectors.

In two demonstration case studies [37,38], holistic designs for recycling, repair, refurbishment and reuse (D4R) products were investigated and successfully delivered. A D4R laptop was designed to create a new life for used IT components via an industrial network established by a laptop manufacturer with end-of-life IT asset management firms, IT refurbishers, component manufacturers and local industries [37]. A prototype D4R PV system was designed to operate more efficiently, whilst higher resource efficiency was developed via an industrial network [38].

In addition, an existing reuse network for ICT (ReUse ICT) was substantially advanced by involving new partners and organizations that offered data about used ICT equipment, scrap paper/cardboard, scrap plastics, wood offcuts and scrap metals [39]. The network developed a quality control scheme, its own label and an internet portal — the resource exchange platform (RXP) — offering additional business opportunities through clustering of reuse firms. The study facilitated regional transfer of the concept, the inclusion of further product categories and opportunities for additional service offers (for example, transportation of reuse or end-of-life products, take-back, service instead of sales) [39].

2.6 Recycling of Metals from High-tech Sources

The EU Directive on waste (2008/98/EC) defines recycling as "any recovery operation by which waste materials are reprocessed into products, materials or substances whether for the original or other purposes. It includes the reprocessing of organic material but does not include energy recovery and the reprocessing into materials that are to be used as fuels or for backfilling operations."

Metals cannot be destroyed, only dissipated, and theoretically many can be recycled almost indefinitely, with little loss in their technical qualities irrespective of how many times they are recovered. However, in practice recycling is often inefficient because of the limits imposed by product design, social behavior, recycling technologies and the thermodynamics of separation [40]. Metal recycling reduces the requirement for primary production with all the associated adverse impacts of mining and refining, including the extra processing required to exploit low-grade ores. In addition, secondary metal production, including collection and transport, is 2–20 times less energy-intensive per kg produced metal than primary production [8]. However, whilst theoretically it should be possible to appreciably increase metal recycling by deliberately designing systems with this goal in mind, it will only be possible to reduce primary production when secondary production approaches market demand.

2.6.1 Ferrous and Non-ferrous Metals

The most commonly recycled metals (by volume) are ferrous and non-ferrous metals, typically iron and scrap steel, aluminium, copper, lead, zinc and stainless steel. Ferrous scrap includes construction materials, cars, ships, railroads and railcars, food cans, mill wastes from primary processing, appliances, machinery, etc. Non-ferrous scrap tends to come from industrial sources: cables and pipes (copper), automotive die-cast alloys (zinc), battery plates (lead), beverage cans and building materials (aluminium). In developing countries and economies in transition, ferrous and non-ferrous metals find their way from the informal sector to industry via long-term cooperative agreements between collectors/

Table 5 *Typical processes involved in the recycling of ferrous and non-ferrous metals*

Process	Process purpose and overview
Collection, transport, loading and unloading	Conveying scrap to the recycling plant. Sorting materials into broad groups for subsequent processing.
Breaking and separating	Size reduction to break up large or complex items into dimensions suitable for further processing. Ball-drop breaking often used on large, solid pieces of metal. Separating can also occur after shredding. Physical separation technologies include: magnetic, eddy-current, air, sink-float and hot crush separation; spectrographic techniques; colour sorting.
Cutting	Size reduction using torches and/or mechanical/hydraulic shears. Gas torches often used to cut steel scrap. Heat resistant scrap cut using non-thermal methods (e.g. plasma or powder cutting torches).
Baling, compacting, and shredding	Hydraulic balers compact pre-sorted metal to promote more efficient furnace melting. Shredding is a generic term for crushing, grinding and shearing. Automated shredders are used for volume reduction to simplify material handling, minimize storage space and reduce transportation and disposal costs. Specialist shredders used for different product types and materials.
Melting and baking	High-temperature heat treatment is used to separate different metal components, increase purity, burn off contaminants, bake out non-metals, etc.
Chemical treatment	Used to: separate scrap into component metals, remove contaminants, clean metal prior to physical processing, extract selected metals. Processes include dissolution, reduction, chemical separation, plating, leaching, galvanizing, electrolytic refining, high-temperature chlorination.

traders and industrial workers who have special authorization to procure them [23]. The global recycling industry for these metals is sizeable and long-standing, and uses well-established sequential processing techniques (see Table 5).

Note that care must be taken when interpreting recycling rates for ferrous and non-ferrous metals as the definitions of recovery, recycling and scrap used vary from organisation to organisation, obscuring real recovery and recycling performance.

2.6.2 Speciality and Precious Metals

As anthropogenic mining is a relatively recent phenomenon, the recycling industry for the speciality and precious metals found in e-waste is less well established and rapidly emerging. The major economic driver for recycling of e-waste has typically been from the recovery of precious metals [41], although this may change as technology for recovering REE improves. Myriad factors are liable to impact the scale of recycling of such metals both now and in the future, including:

- The increasing worldwide use of metals, which is set to continue as the global population expands, demand for goods rises and the anticipated shift to renewable energy systems escalates the metal intensity of energy production.
- The significant energy demands of primary metal production (currently 7–8% of total global energy consumption [8] and consequent emissions of greenhouse gases, predominantly carbon dioxide.

- The prevailing recycling infrastructure and technologies in a country/region.
- The prevailing legislation and enforcement regimes in a country/region.
- Accelerated demand for and potential scarcity of some metals, especially REEs. This may be partially mitigated if: (i) replacement of critical REEs by less critical metals can be achieved and (ii) mining companies are successful in finding/developing new exploitable rare-earth deposits.
- Declining ore grades in geogenic metal stocks.
- Increasing anthropogenic metal stocks in EEE and WEEE (although such sources may not be available or accessible in sufficient quantities in some parts of the world, due to lack of infrastructure in many developing countries).
- Lack of data on anthropogenic metals stocks in EEE and WEEE, including the thermodynamic and physical properties of the material mix [42]. Note that thermodynamic considerations will almost certainly prevent full closure of the materials cycle, meaning that primary production of selected metals may always be necessary.
- The historical and ongoing use of hazardous substances in EEE, giving rise to concerns about potential environmental, health and safety hazards in handling and processing, as well as cross-contamination into recovered secondary resources.
- The economics of recycling and volatile metal markets.
- Relatively low recycling rates for many metals and the decreasing use of some (e.g. As, Cd).
- The increased and rapidly increasing complexity of modern EEE alongside factors that impact swift replacement (fashion, low durability and built-in obsolescence); electronic product design and complexity; and an expected move to dematerialization of EEE (using less metals in the product to achieve the same functions).
- The complex, unpredictable and dynamic behavior of EEE and WEEE flows, including factors such as time delays in the release of material stock due to residence time in households [43] and the impacts of planned large-scale technological changes, such as the so-called "digital switchover" in the UK [44].
- A widespread lack of data about the distribution and concentration of metals in materials and products.
- Lack of scientific and engineering knowledge about how to recycle speciality and precious metals efficiently and cost-effectively from modern materials and products. Despite considerable recent research efforts, technological difficulties are a significant factor in the exceedingly low commercial recycling rate of REEs and the so-called "balance problem" — the requirement that the demand and supply of individual REEs has to be equal at any time or there will be shortages/excesses of some elements — has not yet been solved [5].
- The ongoing development of new electronic products, recycling technologies and more efficient processing methods.
- Lack of political and industrial leadership and will to embrace the strategic and policy mechanisms required to facilitate a change towards the development of regional or global circular economies.

The sheer number and complexity of these factors, and the likely linkages and feedback among them, signify that it is extremely challenging to predict the extent of future increased secondary metal production and substitution. Nevertheless, Binnemans et al. [5] warn that "A drastic improvement in End-of-Life recycling rates for REEs is a strategic necessity, even more so in countries possessing no or few rare-earth deposits."

A 2013 UNEP report [8] has outlined in detail the opportunities and limiting factors in critical metal recycling and their consequences, together with suggestions for infrastructure to optimize metal recycling, tools to aid decision making, and policy recommendations. The authors make a powerful case for a move away from the traditional Material (& Metal)-Centric approach that is driven by weight-based recycling targets towards a new and more sophisticated Product-Centric approach. This is because electrical and electronic products have become progressively more complex and consequently there is a need to focus on the recycling of high-value, low-volume metals that are essential elements of high-tech products. These critical elements are usually present in very low concentrations in high-tech products and hence are consistently lost in current recycling systems and processes. The new economic, strategic and political importance of selected REEs, base and precious metals means that recycling systems will need to prioritize them in future, and weight-based recycling targets may hinder the extraction of metals that are present in small quantities in products. The authors argue that in order to increase resource efficiency via maximizing recycling rates, a Product-Centric viewpoint that is based upon a good understanding of separation physics, thermodynamics, metallurgy and the associated economics should be adopted [8].

2.6.3 Formal Recycling

2.6.3.1 Collection and Sorting of Metals for Recycling

The sequence for recycling e-waste has three fundamental steps: (i) collection, (ii) sorting/dismantling and pre-processing (including sorting, dismantling, mechanical treatment) and (iii) end processing (including refining and disposal) [45]. The second step can be further divided into: a) disassembly — selecting hazardous or valuable components for specific treatment and b) upgrading — targeted processing to upgrade the content of desirable materials [46]. Specialized operators, plant and machinery has evolved to support activities in each step.

Carefully designed collection schemes have to be created to minimize losses and leakage of materials from the system. A considerable literature on household collection systems for recyclables is available and a detailed overview has recently been published [47]. Aspects to be optimized for economic and environmental efficiency include collection infrastructure and frequency, energy/fuel usage during collection, convenience, communication and data-monitoring systems, health & safety, public participation and sorting accuracy. Detailed overviews of the formal networks, systems, compliance and take-back schemes used for the collection of e-waste from households can be found for various regions and countries [18,26,48–54].

Industrial and business e-waste is typically collected by private contractors and compliance schemes before being transferred to metal recycling companies that are attracted to significant quantities of high-value recyclables associated with high metal and material prices. Business-to-business (B2B) EEE includes equipment used by public sector organizations, small- to medium-sized enterprises (SME) and large businesses [55]. Coordination of collection and treatment following business use is normally independent of municipal authorities and may be performed by a range of organizations, including private companies and those that are manufacturer-owned/supported. In the EU, B2B e-waste systems are not reported under current regulations and thus there is no accountability for substandard treatment practices [55].

Some technological innovations for e-waste collections have been developed or are being investigated. Reverse vending machines for small appliances such as mobile phones, tablets and MP3 music players operate in many parts of the world (e.g. ecoATM; www.ecoatm.com). An EU-funded research project (IDENTIS WEEE; http://identisweee.net/en/) has been developing and testing containers for the separate collection of e-waste in Italy. The containers are equipped with devices for user identification and waste recognition interfacing with a data-processing centre capable of monitoring a container's fill level and supervising its operation.

2.6.3.2 Role of the Third Sector

Third-sector organizations (TSOs) are social enterprises in the voluntary or community sectors, also known as the "not-for-profit" sector. TSOs make a valuable but often hidden and underappreciated contribution to resource recovery in many areas across Europe. TSOs are motivated by the need to supply to low-cost household items to people in hardship and provide opportunities for volunteering and for long-term unemployed and other socially excluded individuals to gain practical skills and re-engage with society [56].

TSOs have gradually become more involved in the reuse and recycling of EEE such as mobile phones [57]. There are potentially huge reservoirs of products to be collected; in 2009, it was estimated that students in higher education alone stockpiled 29.3 and 28.1 million mobile phones for Europe and USA, respectively [57]. A survey of takeback schemes in the UK found more than 100 voluntary schemes offering online takeback of mobile phone handsets. The majority (>80%) of schemes were either for-profit/commercial-oriented and/or operated to raise funds for charities. They used various methods to advertise for and collect mobile phones, including pre-paid postal services; courier and in-store (see Figure 1). Most returned phones were of low quality and were therefore probably destined for recycling. There is no doubt that such schemes have the potential to divert EoL mobile phones (and other small EEE) from hoarding and landfills towards reuse and recycling.

2.6.3.3 Technical Aspects of Formal Recycling

The post-collection treatment of e-waste varies by category and the technology used. Treatment facilities can vary from large-scale automated shredding technologies to disassembly processes, which can be manual or automated. Facilities using shredders may not need to remove these materials, depending upon the type and size of technologies used, although certain hazardous substances/components must still be removed to avoid health and safety risks and potential damage to equipment. In Europe, disassembly operations must comply with the minimum requirements set out in Annex II of the WEEE Directive to remove certain substances and components. Guidance is available on the best available treatment, recovery and recycling techniques for use at authorized treatment facilities [e.g. 58–60].

Some facilities only part-treat e-waste and send it onto other facilities for further treatment, e.g. refrigerators sent onto a specialist treatment plant. Treatment facilities try to maximize the quality of the materials they derive from e-waste in order to maximize economic value and minimize residual waste that requires further treatment or disposal.

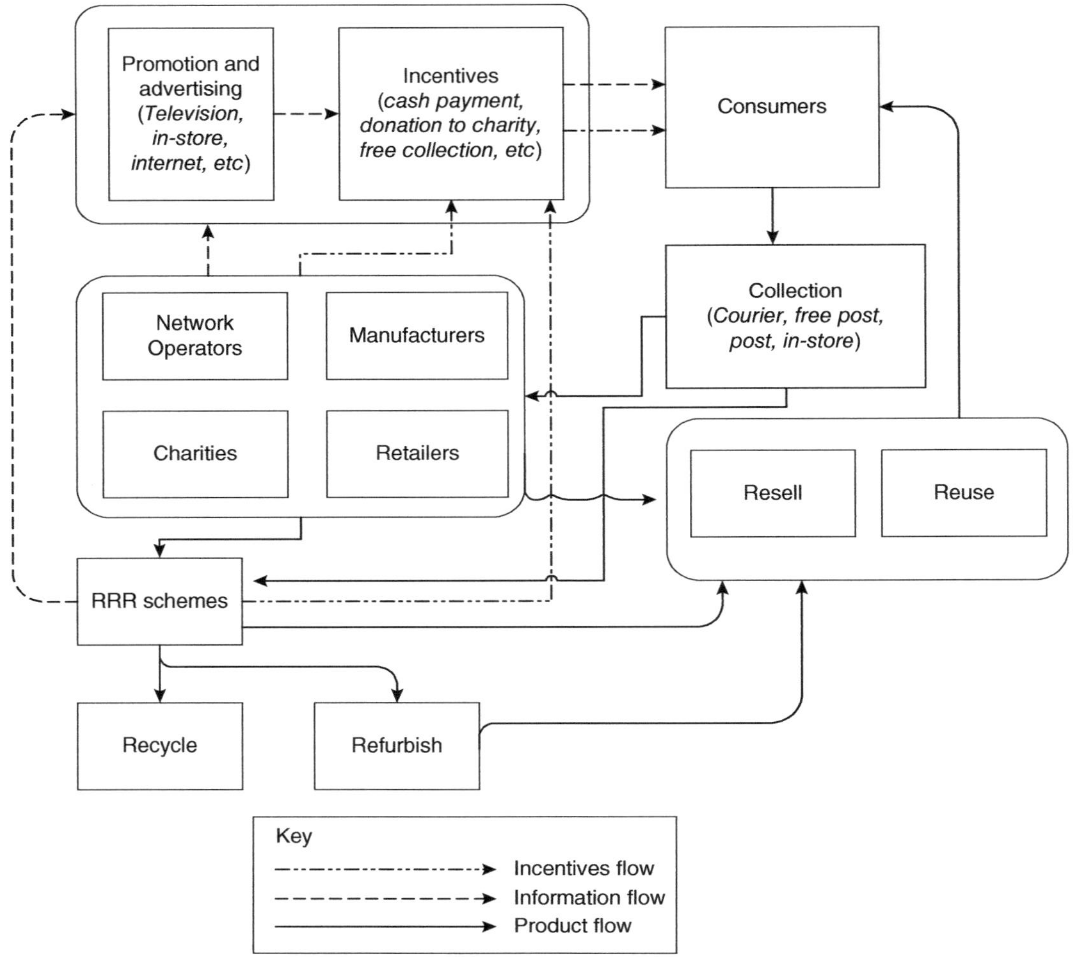

Figure 1 *FC Information, product and incentives flows in the voluntary UK mobile phone takeback network [57]. Reprinted from [57] with permission from Elsevier*

Organisations need to be extremely careful when disposing of IT equipment that contains sensitive data. Deleting files, overwriting or reformatting a hard drive does not remove the information, as this can be restored using readily available or forensic software. The only way to guarantee the secure removal of the information stored on any form of data media is by destruction via shredding.

2.6.3.4 Metal Extraction

Precious and speciality metals may be extracted from pre-processed e-waste by pyrometallurgical, hydrometallurgical and biometallurgical processing.

Hydrometallurgy is a process that uses aqueous chemistry to separate metal from ores, ore concentrates, recycled or residual materials by liquid processes. The core steps in hydrometallurgical processing involve [41]:

i. A series of acid or caustic leaches of solid material.
ii. Concentration of the desired metals by subjecting the resulting solutions to separation and purification procedures (e.g. solvent extraction, precipitation of impurities, adsorption on activated carbon, ion exchange).
iii. Metal recovery by treating the solutions via electrorefining, crystallization or chemical reduction.

Detailed technical summaries of hydrometallurgical methods for recovery of non-ferrous, speciality and precious metals from e-waste have recently been published [5,41,61] and can be found elsewhere in this book.

Pyrometallurgy is a branch of metallurgy in which ores and other materials are heated in order to extract metals. Pyrometallurgic techniques include drying, roasting, incineration, smelting, sintering, refining, calcination and alloying, among others. Various methods of heating are employed and the high temperatures cause chemical and exothermic reactions in materials, causing them to separate or bond. In a typical process, crushed scrap is burned in a furnace or molten bath to remove plastics and the refractory oxides form a slag phase together with some metal oxides. Comprehensive technical summaries of pyrometallurgical methods for recovery of metals from e-waste have recently been published [5,41,62] and can be found elsewhere in this book.

In recent years, recovery of metals via biometallurgy has been increasingly investigated. There are two main areas of biometallurgy for recovery of metals, bioleaching and biosorption. Bioleaching is the extraction of metals from their ores through the use of ferrous iron and sulfur oxidizing bacteria, including *Acidithiobacillus ferrooxidans* and *Acidithiobacillus thiooxidans* [63]. Biosorption is regarded as a very promising and potentially inexpensive future technology for the industrial recovery of precious metals from e-waste, although it has a long way to go before commercialization [64,65]. Biosorption utilises the properties of selected types of dead/inactive microbial biomass materials that permit them to bind and concentrate metal ions from aqueous solutions and industrial effluents [66].

Smelter and refinery production remains located mainly in developed countries, although this balance has started to change with the quick growth of Chinese production capability. The separation and processing of REEs from e-waste in particular is carried out by very few companies, including Boliden Ltd. (Sweden), Umicore (Belgium), AMR Technologies Inc.(Canada), Baotou HEFA Rare Earth (China), Rhodia Electronics and

Catalysts (France), Grace Davison (USA), Santoku America Inc (USA) and Dowa Holdings (Japan) [97].

2.6.3.5 Economics of Formal Recycling

The UNEP's International Resource Panel has unambiguously stated that "Recycling is thus driven by the value of the recovered metal (and material). In any case, all metallurgical plants always try to recover all valuable elements. If there is an economic incentive, recovery will happen." [8]. In fact, the efficiency — and hence the economic viability — of the e-waste recycling chain depends on the efficiency of each step and on how well the boundaries between interdependent steps are managed. If, for example, for a certain device/metal the efficiency of collection is 60%, the combined dismantling pre-processing efficiency is 80% and the materials recovery efficiency 90% (all optimistic assumptions), the resulting net metal yield along the chain would be only 43%.

2.6.4 Informal Recycling

In terms of waste management, the informal sector refers to "individuals or enterprises who are .... not sponsored, financed, recognised or allowed by the formal waste authorities, or who operate in violation of or in competition with formal authorities" [67]. Informal economic activities can be categorized as self-sufficient economy (legal) and shadow economy (illegal), with the latter further subdivided into an irregular sector that produces illegally but has legal output (e.g. goods, services), and a criminal sector that has illegal outputs (e.g. drugs) [68].

The informal recycling sector has evolved to provide an innovative, cost-effective service to market needs for goods and materials [69] by focusing its activities on the primary collection and processing of reusable and recyclable materials from mixed municipal solid waste. Their work is usually low-paid, labour-intensive, low-technology, small-scale, principally unregulated and unregistered, and often completed by individuals or family groups [70,71] that may have migrated to cities from rural areas [23], have no trading license, pay no taxes and are not involved in government insurance or social welfare schemes [72]. Amalgamations of up to 10 (micro) or 20 (small) individuals can form more complicated organisations that operate with low capital investment [71].

The informal sector operates globally in regions and countries with different levels of development, albeit at diverse scales and in various ways. Detailed overviews of the characteristics and activities of the informal recycling sector are available in the literature for a range of countries, including Bangladesh [73], Brazil [74], China [25], Egypt [75], India [76], Indonesia [23], Latin America [77,78], Uganda [79], and worldwide [80].

In developing countries and those in transition, the informal sector dominates activities along the WEEE recycling chain, with collection, manual dismantling, open burning to recover metals, and widespread dumping of residual fractions being standard practice [45]. The reasons for this are outlined in detail in a recent review [25] but basically boil down to a lack of finance; infrastructure; organization, management and technical skills; absence of or lax legislation and compliance activities; and lack of controls on imports of e-wastes.

Informal recycling activities are typically undertaken by individuals as quantities are too small to motivate large-scale intervention by the organized informal sector. Informal recyclers often belong to minorities or marginalized social groups such as the Roma people across Europe; Harijans in India; Zabbaleen in Egypt; Buscabotes, Catraneros and Pepenadores in Mexico; Chamberos in Ecuador; Cirujas in Argentina; Buzos in Costa Rica; and Basuriegos, Cartoneros, Chatarreros and Traperos in Columbia. However, large organized informal sectors dominate e-waste recycling in rapidly developing economies such as China, Nigeria and India. E-waste has become a particular problem in Asian nations because they produce immense quantities of domestic e-waste due to their rapid consumption rates of EEE and importation of vast amounts of used information technology (IT) devices from overseas [81]. China manufactures, exports and consumes the largest portion of global EEE [25] and its swift industrialization and urbanization have rapidly amplified local demand for refurbished appliances and second-hand components, the most common outputs from informal recyclers. Consequently, the informal W/EEE sector is especially important in China, notably in some coastal regions [28,82], and also in countries such as Nigeria and Ghana that are subject to concerted trade of second-hand equipment and illegal waste shipments. Recent literature provides fascinating, detailed histories and overviews of the development of the informal electronic waste recycling sectors in China [25], Kenya [51], India [83], Indonesia [23], Nigeria [18,24].

2.6.4.1 Collection and Sorting of Metals for Informal Recycling

Several collection methods for secondary raw materials, including EEE and E-waste, are employed by the informal recycling sector. The most basic method involves waste pickers/scavengers picking materials from dumps after wastes have been deposited and before they are covered. This practice is widespread in communities that live in shacks near dump sites in urban areas through the developing world. Some waste pickers take secondary materials directly from vehicles that are transporting MSW to transfer stations, dumps or other intermediate points. Street-waste scavenging involves the recovery of materials from mixed waste deposited in bins before formal collection teams arrive or simply dropped on the streets. Itinerant collectors go from door to door in urban areas in order to buy sorted dry recyclable materials from householders or domestic servants, although middle- or higher-income householders may donate used goods free of charge [23]. They typically barter for the best price and transport their materials via a handcart or bi-/tricycle to a merchant who buys the recyclables from them at a markup.

In some countries, the informal sector runs a waste collection service in communities where no formal municipal system is in place. Whilst a fee is charged for providing the service, the principal economic incentive is often the income that can be made from sorting and recycling the collected waste [71].

Informal waste sectors are habitually organized in a network of individuals and small businesses of collectors, traders, intermediates, dealers, brokers and recyclers, each adding value, and creating jobs at every point in the recycling chain [49]. Since many poor people rely on small incomes generated in this chain, impressive collection rates of up to 95% of waste generated are achieved [9], which is far above what can be achieved by contemporary take-back schemes in the formal sector [50].

2.6.4.2 Informal Sorting Methods

Sorting and dismantling are key processes for the informal sector as they increase the value of the materials recovered: the higher the source separation, the higher the material value obtained. For instance, if EEE is grouped into one major category, its value is lower than when it is further separated into subcategories. Sorting usually occurs according to the end-users' specifications and may be by size, shape, colour and potential use or reuse of the materials. Cleaned materials typically increase the selling value by 40–50% from base or dirty prices. Irreparable used EEE are usually disassembled for valuable components that can be sold to dealers, such as copper wires, screws, circuit boards, plastic hard covers, etc.

Sorting and dismantling methods employed by the informal sector usually involve a non-skilled workforce using crude, low-technology or manual approaches that contrast sharply with the highly automated processes utilised by well developed formal-sector recyclers. There are numerous examples of these crude techniques, including:

- Physical separation of materials via dismantling by hand or using tools such as hammers, screw-drivers and chisels [20,85].
- Burning plastic insulation off cables to recover copper [86].
- Heating printed circuit boards to remove components [20].
- Recovery of gold using open-pit acid baths to strip metals [86].
- Open dumping of unsalvageable materials on land and rivers/riverbanks [87].

Wilson *et al.* [71] provide a concise overview of the methods and approaches used in the informal sector to extract and add value to recovered waste materials.

2.6.4.3 Legal Issues

In some countries, the informal recycling sector's activities are not considered illegal. For example, in Indonesia the informal sector has provided work and economic activity during the global economic crisis when the formal sector could not provide adequate job opportunities [23]. Consequently, informal sector activities may be performed openly, although law enforcement officers sometimes become involved if dealings disrupt civic order (e.g. by blocking pavements during business) or occupy prohibited spaces, such as city parks.

2.6.4.4 Health, Safety and Environmental Issues

When e-waste is disposed of or recycled without any controls, there are potentially negative impacts on human health. Since they contain more than 1000 different substances, many of which are potentially highly toxic, there are potentially serious health impacts if e-waste is not disposed of properly. For instance, the open burning of plastics, widespread general dumping, exposure to toxic solders and other malpractices associated with improper dismantling and treatment of WEEE as observed in various countries can have serious health consequences [20,51,88].

Metals do not degrade in the environment and hence can accumulate in soils, sediments, fresh and marine waters, biota, and in urban and industrial settings. Certain metals have historically been associated with environmental problems e.g. mercury, lead, cadmium, arsenic. The use of platinum-group metals in automotive catalytic converters and subsequent deposition in roadside dust is well established [89]. The metals found in e-waste can contaminate drinking water by leaching into groundwater from sources such as landfills. It is reported that ~70% of the so-called heavy metals in U.S. landfills come from e-waste [20]. Although the use of selected brominated flame retardants has been restricted in the manufacture of EEE by the EU, there are still legacy issues: older e-waste containing such chemicals still pose health concerns [88]. Environmental issues also arise from the manufacturing, transport and energy use of EEE.

The issues highlighted in this section demonstrate that it is imperative to understand the factors that precipitate the generation of WEEE, variables that influence its stockpiling, collection (takeback) and decisions on how it is disposed in order to help avoid the negative impacts associated with WEEE. Even more important is an understanding of how these factors interact to affect decisions about EoL EEE. Such an understanding is vital for decision making on the design of strategies and systems to manage WEEE that maximise the recovery of resources with minimal ethical malpractices, health and environmental impacts.

Detailed overviews of the general health, safety and environmental risks and hazards associated with informal recycling can be found in the literature [71,88]. In China in particular, very large numbers of informal-sector workers routinely apply crude, backyard methods for separation of reusable components and recyclable materials, especially metals. These practices frequently occur in primitive conditions, exposing workers to wide-ranging health hazards [25,78,86,90–92] and environmental impacts on soil and sediment [93–95], air [96] and water bodies.

2.7 Conclusions

With 9 billion people expected on our planet by 2050, the amount of high-tech equipment in our homes, offices and industries is going to proliferate. Whilst continuing miniaturization of EEE is likely, overall volumes and mass flows are expected to rise steadily in the future, as appliances become cheaper and consequently more accessible, especially in the non-saturated markets of developing countries. This in turn will lead to increased quantities of e-waste that will require safe and efficient processing and treatment.

Concerns about the negative impacts of e-waste upon biological health, ecosystems and the environment and adequate future supplies of metals mean that anthropogenic stocks of metals, especially rare earth and precious metals, will become the mines of the future. Repair, refurbishment and reuse are important waste-prevention measures but they have usually been overshadowed by recycling. Demonstration research projects have shown that high-tech products can be successfully made with used components and recycled materials, but the systems for manufacturing and selling these products will take time to become established and accepted in society. Recycling of metals from anthropogenic stocks will therefore take centre stage over the next 20 years, probably via structured global systems that combine the best features of both the public/private and formal/informal sectors. However, myriad technical and nontechnical factors, including political will, public education and behavior-change programmes, regulatory interventions and the development of a global circular economy, are liable to impact the speed and scale of metal recycling.

Sceptics claim that the anticipated future global circular economic system for materials is based upon a naïve argument and that the barriers to successfully delivering effective and efficient metal recycling via this system are huge. However, case studies and ambitious joint research projects among commerce, industry and academics are starting to show that this type of system can simultaneously deliver environmental benefits whilst retaining commercial credibility. An ongoing supply of key materials such as metals will be sustainably secured only if it is commercially viable and technically possible. Future metal management must become more about lifestyle choices and developing a sound circular system for the e-wastes we generate if our society and global economy are to develop in a safe and sustainable fashion.

References

1. Mont, O., 2008, Innovative Approaches to Optimizing Design and Use of Durable Consumer Goods, *International Journal of Product Development*, 6, 227–50.
2. Wils, A., 1998, End-use or Extraction Efficiency in Natural Resource Utilisation: Which Is Better? *System Dynamics Review*, 14, 163–88.
3. United Nations Environment Programme (UNEP), 2010, *Metal Stocks in Society: Scientific Synthesis*. UNEP, Nairobi, Kenya. ISBN: 978-92-807-3082-1.
4. United States Geological Survey (USGS) (2014). *Mineral Commodity Summaries, February 2014*, http://minerals.usgs.gov/minerals/pubs/commodity/rare_earths/mcs-2014-raree.pdf (accessed 25 March 2015).
5. Binnemans, K., Jones, P.T., Blanpain, B., *et al.*, 2013, Recycling of Rare Earths: a Critical Review, *Journal of Cleaner Production*, 51, 1–22.
6. Alonso, E., Sherman, A.M., Wallington, T.J., *et al.*, 2012, Evaluating Rare Earth Element Availability: A Case With Revolutionary Demand from Clean Technologies, *Environmental Science & Technology*, 46, 3406–3414.
7. Chen, Z. 2011, Global Rare Earth Resources and Scenarios of Future Rare Earth Industry, *Journal of Rare Earths*, 29, 1–6.
8. UNEP, 2013, *Environmental Risks and Challenges of Anthropogenic Metal Flows and Cycles*, United Nations Environment Programme (UNEP), Paris, France.
9. Graedel, T.E., Barr, R., Cordier, D., *et al.*, 2011, *Estimating Long-Run Geological Stocks of Metals*, UNEP International Panel on Sustainable Resource Management, Working Group on Geological Stocks of Metals, Working Paper April 2011, http://www.unep.org/resourcepanel/Portals/24102/PDFs/GeolResourcesWorkingpaperfinal040711.pdf (accessed 15 April 2015).
10. Dent, P., Walmer, M. (2010). Supply chain sustainability — rare earth materials,. In *Proceedings of Vacuum Electronics Conference* (IVEC), 2010 IEEE International, May 18-20, 2010, Monterey, California, USA, 57–58.
11. Goosey, M., 2004, End-of-life Electronics Legislation: an Industry Perspective, *Circuit World*, 30(2), 41–45.
12. Ongondo, F., Williams, I.D., 2011, Are WEEE in control? Rethinking Strategies for Managing Waste Electrical and Electronic Equipment. In *Waste Management* Book 2 (Kumar, S., Ed.), Chapter 18. Intech Open Access Publisher, Rijeka, Croatia.
13. Ongondo, F., Williams, I.D., Cherrett, T.J., 2011, How Are WEEE Doing? A Global Review of the Management of Electrical and Electronic Wastes, *Waste Management*, 31(4), 714–730.
14. STEP (2011). *Annual Report 2011*, http://www.step-initiative.org/tl_files/step/_documents/StEP%20Annual%20Report%202011_lowres.pd.f (accessed 23 March 2015).
15. European Commission (EC) (2010). Critical Raw Materials for the EU. *Report of the Ad-hoc Working Group on Defining Critical Raw Materials*, http://ec.europa.eu/enterprise/policies/raw-materials/files/docs/report-b_en.pdf (accessed 12 January 2015).
16. Dimitrakakis, E., Janz, A., Biletewski, B., Gidarakos, E, 2009, Determination of Heavy Metals and Halogens in Plastics from Electric and Electronic Waste, *Waste Management*, 29, 2700–06.
17. Wang, F., Huisman, J., Meskers, C.E.M., *et al.*, 2012, The Best-of-2-Worlds philosophy: Developing Local Dismantling and Global Infrastructure Network for Sustainable E-Waste Treatment in Emerging Economies, *Waste Management*, 32(11), 2134–2146.

18. Nnorom, I.C., Osibanjo, O., 2008. Electronic Waste (E-waste): Material Flows and Management Practices in Nigeria, *Waste Management*, 28(8), 1472–1479.
19. Shinkuma, T., Huong, N.T.M., 2009, The Flow of E-waste Material in the Asian Region and a Reconsideration of International Trade Policies on E-waste, *Environmental Impact Assessment Review*, 29(1), 25–31.
20. Puckett, J., Byster, L., Westervelt, S., *et al.*, 2002, *Exporting Harm: The High-Tech Trashing of Asia. Basel Action Network and Silicon Valley Toxics Coalition,* http://www.ban.org/E-waste/technotrashfinalcomp.pdf (accessed 12 January 2015).
21. Cui, J., Zhang, L., 2008, Metallurgical Recovery of Metals from Electronic Waste: a Review, *Journal of Hazardous Materials,* 158, 228–256.
22. Wilt, C.A., Davis, G., 1994, Extended Producer Responsibility: A New Principle for a New Generation of Pollution Prevention. In *Proceedings of the Extended Producer Responsibility Symposium,* 14-15 November 1994, Washington D.C., USA, 1-14.
23. Damanhuri, E., Padmi, T., 2012. The Role of Informal Collectors of Recyclable Waste and Used Goods in Indonesia. In *Post-Consumer Waste Recycling and Optimal Production* (ed. E. Damanhuri), Chapter 2. Intech Open Access Publisher, Rijeka, Croatia.
24. Öko-Institut e.V. (2011). *Informal E-Waste Management in Lagos, Nigeria — Socio-Economic Impacts and Feasibility of International Recycling Co-Operations*, final report of component 3 of the UNEP SBC E-waste Africa Project. Öko-Institut e.V., Freiburg, Germany.
25. Chi, X., Streicher-Porte, M., Wang, M.Y.L., Reuter, M.A., 2011, Informal Electronic Waste Recycling: a Sector Review with Special Focus on China. *Waste Management*, 31, 731–742.
26. Ongondo, F., Williams, I.D., 2012, A Critical Review of the UK Household WEEE Collection Network In *ICE Proceedings, Waste and Resource Management*, 165(WR1), 13–23.
27. Geyer, R., Blass, V.D., 2010, The Economics of Cell Phone Reuse and Recycling., *The International Journal of Advanced Manufacturing Technology*, 47, 515–25.
28. Yang, J.X., Lu, B., Xu, C., 2008, WEEE Flow and Mitigating Measures in China, *Waste Management*, 28(9), 1589–1597.
29. He, K.T., Li, L., Ding, W.Y., 2008, Research on Recovery Logistics Network of Waste Electronic and Electrical Equipment in China. In *Proceedings of the 3rd IEEE Conference on Industrial Electronics and Applications (ICIEA 2008)*, Singapore, 3–5 June 2008, 1797–1802.
30. He, W.Z., Li, G.M., Ma, X.F., *et al.*, 2006, WEEE Recovery Strategies and the WEEE Treatment Status in China, *Journal of Hazardous Materials*, 136(3), 502–512.
31. Liu, X.B., Tanaka, M., Matsui, Y., 2006, Electrical and Electronic Waste Management In China: Progress and the Barriers to Overcome, *Waste Management and Research*, 24(1), 92–101.
32. Streicher-Porte, M., Geering, A.C., 2010, Opportunities and Threats of Current E-Waste Collection System in China: A Case Study From Taizhou with a Focus on Refrigerators, Washing Machines, and Televisions, *Environmental Engineering Science*, 1(27), 29–36.
33. Ongondo, F.O., Williams, I.D., Dietrich, J., Carroll, C., 2013, ICT Reuse in Socio-Economic Enterprises, *Waste Management*, 33, 2600–2606.
34. Alexander, C., Smaje, C., 2008, Evaluating Third Sector Reuse Organisations in the UK: Case-Studies and Analysis of Furniture Reuse Schemes. *Resources, Conservation and Recycling*, 52, 719–730.
35. Curran, A. and Williams, I.D., 2010, The Role of Furniture and Appliance Re-Use Organisations in England and Wales, *Resources, Conservation and Recycling*, 54, 692–703.
36. Curran, A., Williams, I.D., 2012, A Zero Waste Vision for Industrial Networks in Europe, *Journal of Hazardous Materials*, 207–208, 3–7.
37. Hickey, S., Fitzpatrick, C., Maher, P., *et al.*, 2014, A Case Study of the D4R Laptop, *Waste and Resource Management*, 167(WR3), 101–108.
38. Arranz, P., Anzizu, M., Pineau, A., *et al.*, 2014, The Development of a Resource-Efficient Photovoltaic System, *Waste and Resource Management*, 167(WR3), 109–122.

39. Dietrich, J., Becker, F., Nittka, T., *et al.* 2014, Extending Product Lifetimes: A Reuse Network for ICT Hardware, *Waste and Resource Management*, 167(WR3), 123–135.
40. Reck, B., Graedel, T., 2012, Challenges in Metal Recycling. *Science*, 337, 690–695.
41. Cui, J., Zhang, L., 2008. Metallurgical Recovery of Metals from Electronic Waste: a Review, *Journal of Hazardous Materials*, 158, 228–256.
42. Schluep, M., Müller, E., Hilty, L.M., *et al.,* 2013, Insights from a Decade Of Development Cooperation in E-Waste Management. In *Proceedings of the First International Conference on Information and Communication Technologies for Sustainability* (Hilty et al., Eds.) ETH Zurich, February 14-16, 2013.
43. Lin, P., Huang, Y., 2012, The Influence Factors on Choice Behavior Regarding Green Products Based on the Theory of Consumption Values, *Journal of Cleaner Production*, 22, 11–18.
44. Ongondo, F., Williams, I.D., Keynes, S., 2011, Estimating the Impact of the "Digital Switchover" on Disposal of WEEE at Household Waste Recycling Centres in England., *Waste Management*, 31(4), 743–753.
45. Schluep, M., Hagelueken, C., Kuehr, R., *et al.,* 2009, *Recycling —from E-waste to Resources, Sustainable Innovation and Technology Transfer Industrial Sector Studies*, United Nations Environment Programme (UNEP), Paris, France.
46. Cui, J., Forssberg, E., 2003, Mechanical Recycling of Waste Electric and Electronic Equipment: a Review, *Journal of Hazardous Materials*, B99, 243–263.
47. Williams, I.D., 2015, A Change of Emphasis: Waste to Resource Management. In *Issues in Environmental Science and Technology, Still Only One Earth: A 40-Year Perspective on the State of the Planet*, Vol 40 (eds. R.E. Hester and R.M. Harrison), RSC Publishing, Cambridge.
48. Kahhat, R., Kim, J., Xu, M., *et al.,* 2008, Exploring E-waste Management Systems in the United States, *Resources, Conservation and Recycling*, 52, 955–964.
49. Chi, X., Wang, M.Y.L., Reuter, M.A., 2014, E-waste Collection Channels and Household Recycling Behaviors in Taizhou of China, *Journal of Cleaner Production*, 80, 87–95.
50. Friege, H., Oberdörfer, M., Günther, M., 2015, Optimising Waste From Electric and Electronic Equipment Collection Systems: A Comparison of Approaches in European Countries, *Waste Management Research*, 33(3), 223–231.
51. Mureithi, M., Waema, T., 2008, E-waste Management in Kenya. Kenya ICT Action Network (KICTANet), Kenya, http://ewasteguide.info/Waema_2008_KICTANet (accessed 31 March 2015).
52. Pérez-Belis, V., Bovea, M.D., Ibáñez-Forés, V., 2015, An In-Depth Literature Review of the Waste Electrical and Electronic Equipment Context: Trends and Evolution, *Waste Management Research*, 33(1), 3–29.
53. Lee, J., Song, H.T., Yoo, J.M., 2007. Present Status of the Recycling of Waste Electrical and Electronic Equipment in Korea, *Resources, Conservation and Recycling,* 50, 380–397.
54. Suja, F., Abdul Rahman, R., Yusof, A., Shahbudin Masdar, M., 2014, e-Waste Management Scenarios in Malaysia, *Journal of Waste Management*, Article ID 609169, 7 pages, doi:10.1155/2014/609169.
55. Peagram, R., Williams, I.D., Curran, T., *et al.,* 2014, Business to Business End-of-Life IT Industrial Networks, *Waste and Resource Management*, 167(WR4), 178–192.
56. Williams, I.D., Curran, T., Schneider, F., 2012, Editorial: The Role and Contribution of the Third Sector in Terms of Waste Management and Resource Recovery, *Waste Management*, 32, 1739–1741.
57. Ongondo, F. and Williams, I.D., 2011, Mobile Phone Collection, Reuse and Recycling in the UK, *Waste Management*, 31(6), 1307–1315.
58. Department for Environment, Food and Rural Affairs (Defra) (2006). *Guidance on Best Available Treatment Recovery and Recycling Techniques (BATRRT) and treatment of Waste Electrical and Electronic Equipment (WEEE)*. Defra, London, UK.

59. United States Environmental Protection Agency (USEPA) (2013). eCycling Regulations/Standards, www.epa.gov/epawaste/conserve/materials/ecycling/rules.htm (accessed 16 April 2015).
60. WEEELABEX ('WEEE label of excellence') 2015. WEEELABEX Standards, www.weeelabex.org/#!standards/component_41229 (accessed 16 April 2015).
61. Tuncuk, A., Stazi, V., Akcil, A., *et al.*, 2012), Aqueous Metal Recovery Techniques from E-Scrap: Hydrometallurgy in Recycling, *Minerals Engineering*, 25, 28–37.
62. Shuey, S.A., Taylor, P., 2005, Review of Pyrometallurgical Treatment of Electronic Scrap, *Mining Engineering*, 57(4), 67–70.
63. Rohwerder, T., Gehrke, T., Kinzler, K., Sand, W., 2003, Bioleaching Review Part A, *Applied Microbiology and Biotechnology*, 63(3), 239–48.
64. Mack, C., Wilhelmi, B., Duncan, J.R., Burgess, J.E., 2007, Biosorption of Precious Metals, *Biotechnology Advances*, 25, 264–271.
65. Das, N., 2010, Recovery of Precious Metals Through Biosorption — A review, *Hydrometallurgy*, 103, 180–189.
66. Dobson, R.S., Burgess, J.E., 2007, Biological Treatment of Precious Metal Refinery Wastewater: A Review. *Minerals Engineering*, 20, 519–532.
67. Obersteiner, G., Linzner, R., Pertl, A., *et al.*, 2011, Dimension of Informal Waste Collection Activities in Central Europe — an Austrian-Hungarian Case Study. In *Proceedings of the Thirteenth International Waste Management and Landfill Symposium*. S. Margherita di Pula, Cagliari, Sardinia, Italy, October 3/7. Session C4.
68. Schneider, F., Enste, D.H., 2003, *The Shadow Economy: An International Survey*, Cambridge University Press, Cambridge, UK.
69. Ahmed, A. S., Ali, M., 2004, Partnerships for Solid Waste Management in Developing Countries: Linking Theories to Realities, *Habitat International*, 28(3), 467–479.
70. Wilson, D., Whiteman, A., Tormin, A., 2001, *Strategic Planning Guide for Municipal Solid Waste Management*. Washington, DC: World Bank, http://www.worldbank.org/urban/solid_wm/erm/start_up.pdf (accessed 22 February 2015).
71. Wilson, D., Velis, C., Cheeseman, C., 2006, Role of Informal Sector in Waste Management in Developing Countries. *Habitat International*, 30, 797–808.
72. Haan, H. C., Coad, A. and Lardinois, I. (1998). *Municipal Waste Management: Involving Micro- and Small Enterprises. Guidelines for Municipal Managers*, Turin, Italy: International Training Centre of the ILO, SKAT, WASTE.
73. Matter, A., Dietschi, M., Zurbrügg, C., 2013, Improving the Informal Recycling Sector Through Segregation of Waste in the Household: the Case of Dhaka Bangladesh, *Habitat International*, 38, 150–156.
74. Gutberlet, J., 2015, More Inclusive and Cleaner Cities with Waste Management Co-Production: Insights from Participatory Epistemologies and Methods, *Habitat International*, 46, 234–243.
75. Fahmi, W.S., Sutton, K., 2006, Cairo's Zabaleen Garbage Recyclers: Multi-Nationals' Takeover and State Relocation Plans, *Habitat International*, 30, 809–837.
76. Asim, M., Batool, S., Chaudhry, M., 2012. Scavengers and Their Role in the Recycling of Waste in South Western Lahore. *Resources, Conservation and Recycling*, 58, 152–162.
77. Medina, M., 2000, Scavenger Cooperatives in Asia and Latin America. *Resources, Conservation and Recycling*, 31, 51–69.
78. Gutberlet, J. and Baeder, A. (2008). Informal recycling and occupational health in Santo Andre, Brazil. *International Journal of Environmental Health Research*, 18, 1–15.
79. Katusiimeh, M.W., Burger, K., Mol, A.P.J., 2013, Informal Waste Collection and its Coexistence with the Formal Waste Sector: The Case of Kampala, *Uganda, Habitat International*, 38, 1–9.
80. Ezeah, C., Fazakerley, J.A., Roberts, C.L., 2013, Emerging Trends in Informal Sector Recycling in Developing and Transition Countries, *Waste Management*, 33, 2509–2519.

81. Brigden, K., Labunska, I., Santillo, D., Johnston, P., 2008. *Chemical Contamination at E-Waste Recycling and Disposal Sites in Accra and Korforidua, Ghana.* Greenpeace Research Laboratories Technical Note. Greenpeace International, Amsterdam, The Netherlands, www.greenpeace.org/international/press/reports/chemical-contamination-at-e-wa (accessed 8 January 2015).
82. Terazono, A., Murakami, S., Abe, N., *et al.,* 2006, Current Status and Research on E-Waste Issues in Asia. *Journal of Material Cycles and Waste Management*, 8(1), 1–12.
83. Raghupathy, L., Chaturvedi, A., Arora, R., Mehta, V., 2010. *E-Waste Recycling in India — Bridging The Gap Between the Informal and Formal Sector*, http://www.iswa.org/uploads/tx_iswaknowledgebase/Krueger.pdf (accessed 22 February 2015).
84. Baldé, C.P., Wang, F., Kuehr, R., Huisman, J., 2015, *The global e-waste monitor — 2014*, United Nations University, IAS – SCYCLE, Bonn, Germany.
85. Wen, X.F., Li, J.H., Hao, L., Yin, F.F., Hu, L.X., Liu, H.P., Liu, Z.Y., 2006, An Agenda to Move Forward E-Waste Recycling and Challenges in China. In *Proceedings of the 2006 IEEE International Symposium on Electronics and the Environment*, 8–11 May 2006. pp. 315–320.
86. Wong, M.H., Wu, S.C., Deng, W.J., *et al.,* 2007, Export of Toxic Chemicals — A Review of the Case of Uncontrolled Electronic-Waste Recycling, *Environmental Pollution*, 149(2), 131–140.
87. Huo, X., Peng, L., Xu, X.J., *et al.*, 2007, Elevated Blood Lead Levels of Children in Guiyu, an Electronic Waste Recycling Town in China, *Environmental Health Perspectives*, 115(7), 1113–1117.
88. Sepúlveda, A., Schluep, M., Renaud, F.G. *et al.* 2010, A Review of the Environmental Fate and Effects of Hazardous Substances Released from Electrical and Electronic Equipments During Recycling: Examples from China and India, *Environmental Impact Assessment Review*, 30(1), 28–41.
89. Williams, I.D., McCrae, I.S., Ward, N.I., Richmond, N.A., 1998, Roadside particulate and Trace Element Concentrations at Four London Sites, *International Journal of Vehicle Design*, 20(1–4), 21–29.
90. Bi, X.H., Thomas, G.O., Jones, K.C., *et al.* 2007, Exposure of Electronics Dismantling Workers to Polybrominated Diphenyl Ethers, Polychlorinated Biphenyls, and Organochlorine Pesticides in South China, *Environmental Science and Technology*, 41(16), 5647–5653.
91. Li, Y., Xu, X.J., Wu, K.S., *et al.,* 2008, Monitoring of Lead Load and its Effect on Neonatal Behavioral Neurological Assessment Scores in Guiyu, an Electronic Waste Recycling Town in China, *Journal of Environmental Monitoring*, 10(10), 1233–1238.
92. Zhao, G.F., Wang, Z.J., Zhou, H.D., Zhao, Q., 2009, Burdens of PBBs, PBDEs, and PCBs in Tissues of The Cancer Patients in the E-Waste Disassembly Sites in Zhejiang, China, *Science of the Total Environment*, 407(17), 4831–4837.
93. Leung, A., Cai, Z.W., Wong, M.H., 2006, Environmental Contamination from Electronic Waste Recycling at Guiyu, Southeast China, *Journal of Material Cycles and Waste Management*, 8(2), 21–33.
94. Luo, Y., Luo, X.J., Lin, Z., *et al* (2009). Polybrominated Diphenyl Ethers in Road and Farmland Soils from an E-Waste Recycling Region in Southern China: Concentrations, Source Profiles, and Potential Dispersion and Deposition, *Science of the Total Environment*, 407(3), 1105–1113.
95. Zhang, J.H., Min, H., 2009. Eco-toxicity and Metal Contamination of Paddy Soil in an E-Wastes Recycling Area, *Journal of Hazardous Materials*, 165(1–3), 744–750.
96. Deng, W.J., Zheng, J.S., Bi, X.H., *et al.,* 2007, Distribution of PBDEs in Air Particles from an Electronic Waste Recycling Site Compared with Guangzhou and Hong Kong, South China, *Environment International*, 33(8), 1063–1069.
97. Massari, S., Ruberti, M., 2013, Rare Earth Metals as Critical Raw Materials: Focus on International Markets and Future Strategies, *Resources Policy*, 38, 36–43.

3

Global Management of Electronic Wastes: Challenges Facing Developing and Economy-in-Transition Countries

Oladele Osibanjo[1,2], *Innocent C. Nnorom*[2,3], *Gilbert U. Adie*[1,2], *Mary B. Ogundiran*[1,2], *and Adebola A. Adeyi*[1,2]

[1]*Department of Chemistry, University of Ibadan, Ibadan, Nigeria.*
[2]*Basel Convention Coordinating Centre for Training & Technology Transfer for the African Region, University of Ibadan, Nigeria.*
[3]*Department of Industrial Chemistry, Abia State University, Uturu, Abia State, Nigeria.*

3.1 Introduction

3.1.1 Electronic waste (E-waste): Definitions, Categories and Composition

Waste electrical and electronic equipment (WEEE), electronic waste or e-waste, electronic scrap or e-scrap, and end-of-life (EoL) electronic devices are synonyms used to describe discarded electrical or electronic devices. E-waste is composed of a variety of disposed electrical and electronic devices that can be categorized on the basis of their size (large or small), areas of application/uses (toys, consumer or household goods), and reusability/recyclability (Baldé et al., 2014). Attempts have also been made to differentiate e-waste from WEEE thus: while e-waste describes electronic goods, such as computers, television and radio sets, and mobile phones, WEEE refers to all equipment powered by electricity and, therefore, includes non-electronic goods such as refrigerators, air conditioners, and washing machines (Robinson, 2009). Finlay (2005) gave a standard definition of e-waste to include all end-of-life (EoL) electronic products, components and peripherals, such as computers, cell phones, fax machines, photocopiers, radio sets and TV. Widmer et al.

Metal Sustainability: Global Challenges, Consequences, and Prospects, First Edition. Edited by Reed M. Izatt.

Table 1 *Selected definitions of electronic waste or E-waste*

Definitions	References
"Electrical or electronic equipment which is waste including all components, sub-assemblies and consumables, which are part of the product at the time of discarding." "Directive 75/442/EEC, Article 1(a) defines waste as any substance or object which the holder disposes of or is required to dispose of pursuant to the provisions of national law in force."	(EU, 2003)
"E-waste encompasses a broad and growing range of electronic devices ranging from large household devices such as refrigerators, air conditioners, cell phones, personal stereos, and consumer electronics to computers which have been discarded by their owners."	BAN/SVTC, 2002
"Any appliances using an electric power supply that has reached its end-of-life"	OECD, 2001
"An electrically powered appliance that no longer satisfies the current owner for its original purpose."	Widmer et al., 2005
E-waste or WEEE: "Electrical and electronic equipment that is no longer suitable for use or that the last owner has discarded".	SBC, 2011
"E-waste is a term used to cover all items of electrical and electronic equipment (EEE) and its parts that have been discarded by its owner as waste without the intent of reuse"	StEP, 2012
E-waste means: "waste electrical electronic equipment (WEEE) including old, end-of-life (EoL) or discarded electrical/electronic appliances that use electricity".	Osibanjo, 2009a

(2005), however, claim that there is, yet, no standard definition or general agreement as to whether the term e-waste applies to resale, reuse, and refurbishing industries or to specific products that cannot be used for their intended purpose. Some of the different definitions of e-waste in literature are summarized in Table 1. A common understanding can be inferred from these definitions that e-waste and WEEE refer to electrical and electronic gadgets that are at their end of life, obsolete or discarded by their original owners.

E-waste is considered hazardous waste under the List A1180 of Annex VIII of the Basel Convention on the control of trans-boundary movements of hazardous wastes and their disposal (SBC, 2011). A decision tree to determine whether used electronic equipment is waste to be controlled under the Basel Convention or not is shown in Figure 1. The fact that the electrical and electronic equipment (EEE) imported into developing countries is usually an admixture of EEE, used EEE, near end-of-life EEE and WEEE make such devices difficult to control under the Basel Convention Control (Osibanjo, 2009b).

3.1.2 Typology and Categories of E-waste

According to the European Union WEEE directive *2002/96/EC* (EU, 2003), which is widely adopted internationally, the term electronic waste (e-waste) or waste electrical and electronic equipment (WEEE) refers to unwanted EEE that are obsolete, at the end of their lives or have been discarded by their original users. The typology of the 10 WEEE categories according to the EU WEEE Directive is listed in Table 2.

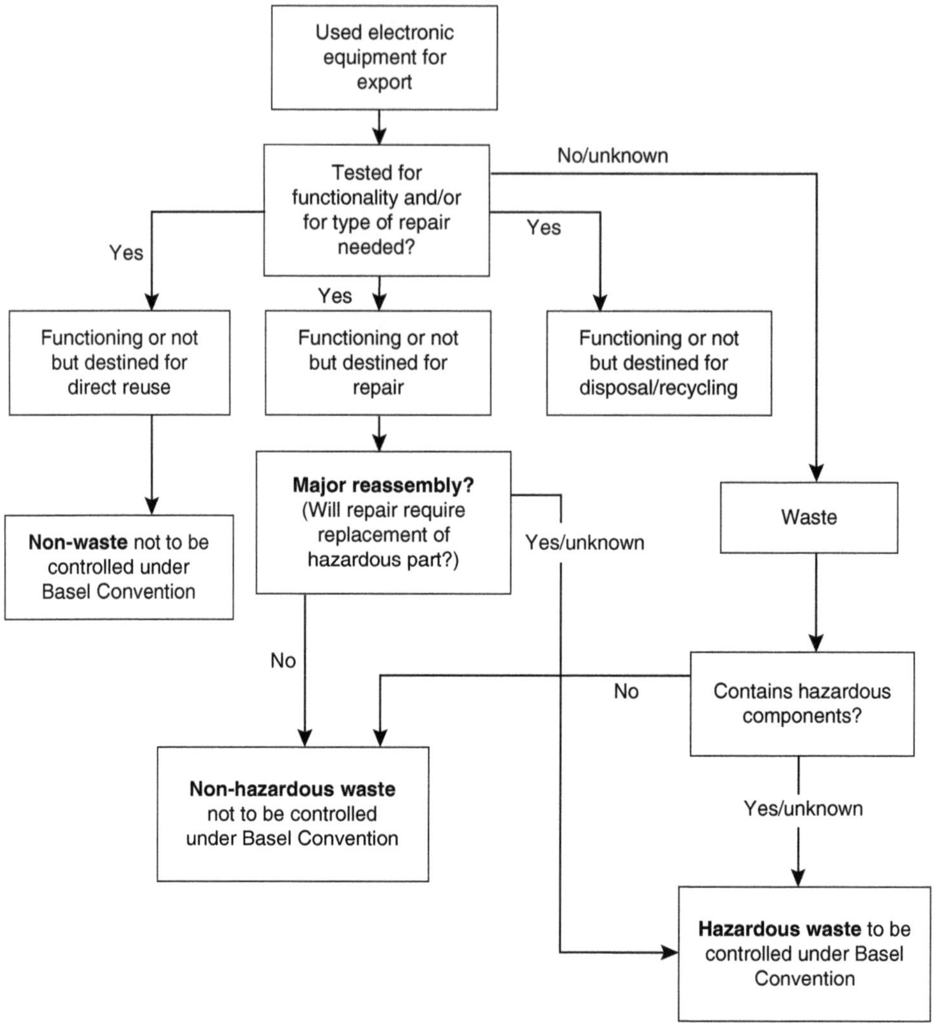

Figure 1 *Decision tree to determine whether used electronic equipment is a waste to be controlled under the Basel Convention or not. Source: Puckett et al. (2005)*

E-waste has been reported to comprise large household electrical appliances, such as fridges, air conditioners, washing machines, and microwave ovens as the largest proportion (about 50%) of e-waste, followed by information and communications technology (ICT) equipment (about 30%), and consumer electronics such as fluorescent light bulbs, electronic products, television sets and stereo equipment (about 10%) (ILO, 2012). Balde *et al.* (2014) recently classified six new categories of EEE and WEEE (Table 3) based on original function, weight, size, and material composition of each category of device, by streamlining the 10 categories in EU WEEE directive (Table 2) into six categories. Table 4 provides a list of common items of e-waste (Robinson, 2009).

Table 2 *Ten Categories of E-waste according to the EU WEEE Directive 2002/96/EC. Source: (UNEP 2007)*

	Category	Typical Examples
1	Large Household Appliances	Refrigerators, freezers, washing machines, clothes dryers, microwaves, heating appliances, radiators, fanning/exhaust ventilation/conditioning equipment
2	Small Household Appliances	Vacuum cleaners, other cleaners, sewing/knitting/weaving textile appliances, toasters, fryers, pressing irons, grinders, opening/sealing/packaging appliances, knives, hair cutting/drying/shaving devices, clocks, watches
3	IT and Telecommunication Equipment	Mainframes, microcomputers, printers, PCs (desktop, notebooks, laptops), photocopiers, typewriters, fax/telex equipment, telephones
4	Consumer Equipment	Radio and TV sets, video cameras/decoders, hi-fi recorder, audio amplifiers, musical instruments
5	Lighting Equipment	Bulbs for fluorescent lamps, low-pressure sodium lamps
6	Electrical and Electronic Tools (excluding large-scale industrial tools)	Drills, saws, sewing machines, turning/milling/sanding/sawing/cutting/shearing/drilling/punching/folding/bending equipment, riveting/nailing/screwing tools, welding/soldering tools, spraying/spreading/dispersing tools,
7	Toys, Leisure and Sports Equipment	Electric trains, car racing sets, video games, sports equipment, coin slot machines, biking/diving/running/rowing computers
8	Medical Devices	Devices for radiotherapy/cardiology/dialysis, ventilators, analyzers, freezers, fertilization tests, detecting/preventing/monitoring/treating/alleviating illness, injury or disability
9	Monitoring and Control Instruments	Smoke detectors, heating regulators, thermostats, measuring/weighing/adjusting appliances for household or laboratory use, other industrial monitoring and control instruments
10	Automatic Dispensers	For hot drinks, hot or cold bottles/cans, solid products, money, and all kinds of products

Table 3 *Six categories of E-waste according to Balde et al. (2014). Source: Balde et.al (2014)*

	Category	Typical Examples
1	Temperature exchange cooling and freezing equipment	Refrigerators, freezers, air conditioners, heat pumps
2	Screens, monitors.	Televisions, monitors, laptops, notebooks, tablets
3	Lamps	Straight lamps, LED lamps
4	Large Equipment	Washing machines, clothes dryers, dish washing machines, electric stoves, large printing machines, copying equipment and photovoltaic panels.
5	Small Equipment	Vacuum cleaners, microwaves, ventilation equipment, toasters, electric kettles, electric shavers, scales, calculators, radio sets
6	Small IT and telecommunication equipment	Mobile phones, GPS, pocket calculators, routers, personal computers, printers, telephones

Table 4 *Some toxic and hazardous components of e-waste*

Item	Hazardous Components
Cathode ray tube	Lead, antimony, mercury, phosphorus
Liquid crystal display	Mercury
Circuit board	Lead, beryllium, antimony, brominated flame retardants (BFR)
Fluorescent lamp	Mercury, phosphorus, flame retardants
Cooling systems	Ozone-depleting substance (ODS)
Plastic	BFRs, phthalate plasticizer
Insulation	Ozone-depleting substances in foam, asbestos, refractory ceramic fibre
Rubber	Phthalate plasticizer, BFRs, lead
Electrical Wiring	Phthalate plasticizer, BFRs
Batteries	Lead, lithium, cadmium, mercury

Source: UNEP, 2007; MoEF, 2008.

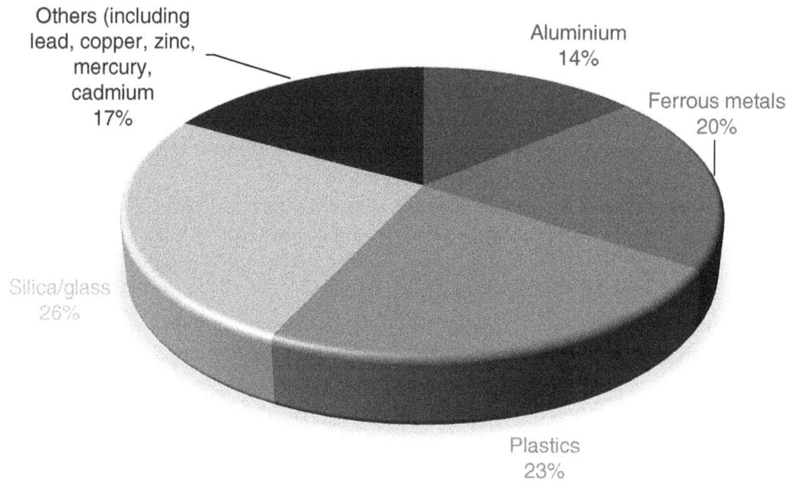

Figure 2 *Material composition of waste personal computer*

3.2 E-waste Composition

Electrical and electronic equipment (EEE) and WEEE are made of components/materials that are highly valuable as well as those that are toxic and hazardous to both man and the environment. Generally electronic products and the e-waste arising from them consist of ferrous and non-ferrous metals, plastics, glass, wood and plywood, printed circuit boards, concrete and ceramics, rubber and other items. The relative composition of the components depends on product type and production technology. Iron and steel constitute about 50% of e-waste, followed by plastics (21%), non-ferrous metals (13%), and other constituents (UNEP, 2007). The material composition of a personal computer (PC) shown in Figure 2 indicates that ferrous metal and aluminum constitute 20% and 14%, respectively. In addition, there are other metals

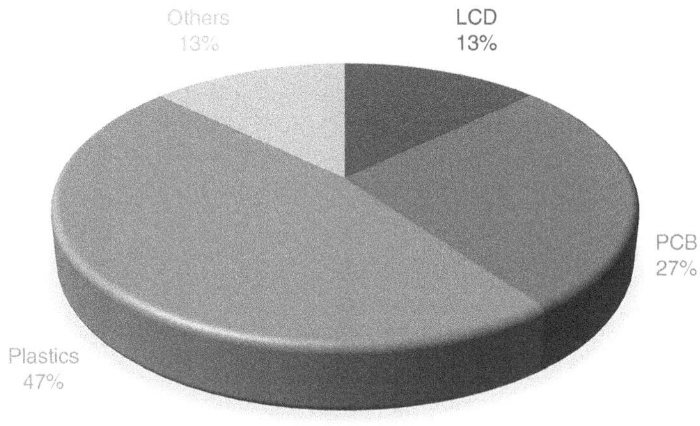

Figure 3 Material composition of waste mobile phone. (See insert for color representation of the figure.)

such as copper, lead, cadmium, mercury and zinc, which together make up 17% of the PC. Figure 3 indicates the material composition of a mobile phone as comprising plastics (47%), printed circuit board (27%), liquid crystal display (LCD) (13%) and others including metals (13%) (Babayemi et al., 2014).

In developing countries like Nigeria, erratic electricity power supply is a major challenge to sustainable socioeconomic and industrial development. Rechargeable and non-rechargeable electric torches and rechargeable lamps serve as alternatives that have been used extensively for domestic lighting sources, as they are readily available, affordable and portable, compared to diesel or gasoline-powered generator engines. Hitherto, local and international attention in scientific literature to have focused mainly on e-wastes from information and communications equipment and from consumer electronics (categories 3 and 4 of EU WEEE directive), such as mobile phones, computers, TVs and large household appliances, as the main contributors to electronic waste. But domestically generated WEEE also constitute a significant portion of domestic e-waste and should not be ignored, especially in developing countries. Figure 4 shows percentage composition of spent rechargeable lamps with the following composition: battery (50%); plastic (40%); glass (5%), PCBs (4%) and metal (<1%) respectively, while Figures 5 and 6 indicate pictorial samples of waste rechargeable electric lamps (WRELs) and samples of waste rechargeable electric torches (WRETs) and their batteries imported into Nigeria, mainly from China (Ogundiran et al., 2014a, 2014b, 2015).

There is a paucity of information about the toxicity and environmental fate of some of the over 1,000 assorted chemicals identified in e-waste streams. This is largely a result of rapidly changing manufacturing processes, the chemicals used, and increasing public awareness of the harm to humans and the environment from environmental releases of these chemicals which enhance regulatory control in some regions and countries.

Electrical and electronic products and e-waste often contain several persistent, bioaccumulative and toxic substances, including heavy metals such as lead, nickel, chromium and mercury (UNEP, 2011), and persistent organic pollutants (POPs), such as polychlorinated

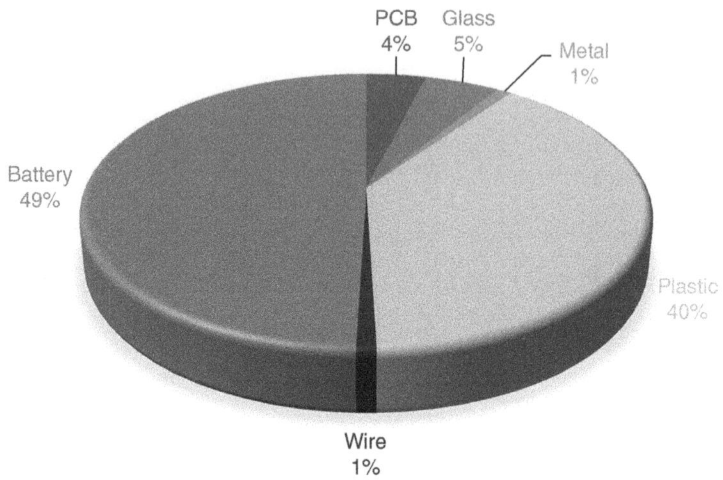

Figure 4 *Mean percentage composition of spent rechargeable lamps*

biphenyls (PCBs) and brominated flame retardants (BFRs) (Sindiku et al., 2014). Presented in Table 4 are some components of WEEE and their potentially hazardous and toxic contents. For instance, the cathode ray tube (CRT) of a TV or computer monitor could contain as much as 4 kg of lead (Nnorom *et al.*, 2011).

CRTs also contain antimony, phosphorus, etc. in some proportions, while circuit boards in different electrical and electronic products contain lead, beryllium, antimony and brominated flame retardants (BFRs). Other toxic substances contained in various electronic items include selenium, antimony trioxide, cadmium, cobalt, manganese, bromine and barium, amongst many others. Table 5 indicates that valuable metals are also present in the printed circuit boards of selected electronic products.

There is also increasing international concern about the large amounts of epoxy resins, fibre glass, PVC (polyvinyl chlorides), thermosetting plastics, lead, tin, copper, silicon, beryllium, carbon, iron and aluminium, and the trace amounts of germanium, tantalum, vanadium, terbium, gold, titanium, ruthenium, palladium, manganese, bismuth, niobium, rhodium, platinum, carbon, cerium, antimony, arsenic, barium, boron, cobalt, europium, gallium, indium, lithium, manganese, nickel, palladium, ruthenium, selenium, silver, tantalum, molybdenum, thorium and yttrium present in e-waste (Chi, 2011).

While functional and in use, these potentially toxic and hazardous components of EEE and e-waste portend little or no danger to users. However, the negative environmental effects from the toxic materials content of e-wastes are most visible in the end-of-life (EoL) stage (Nnorom and Osibanjo, 2009a, b; Nnorom et al., 2010). The exponential growth in consumer waste in recent years has started to threaten the environment and is posing significant challenge to waste management experts. The magnitude of the problem is exemplified by the fact that in the decade between 1994 and 2003, about 500 million personal computers containing approximately 718,000 tonnes of lead, 1,363 tonnes of cadmium and 287 tonnes of mercury, reached their end-of-life (Smith et al., 2006) and the heavy metals content will be released into the environment if not properly managed.

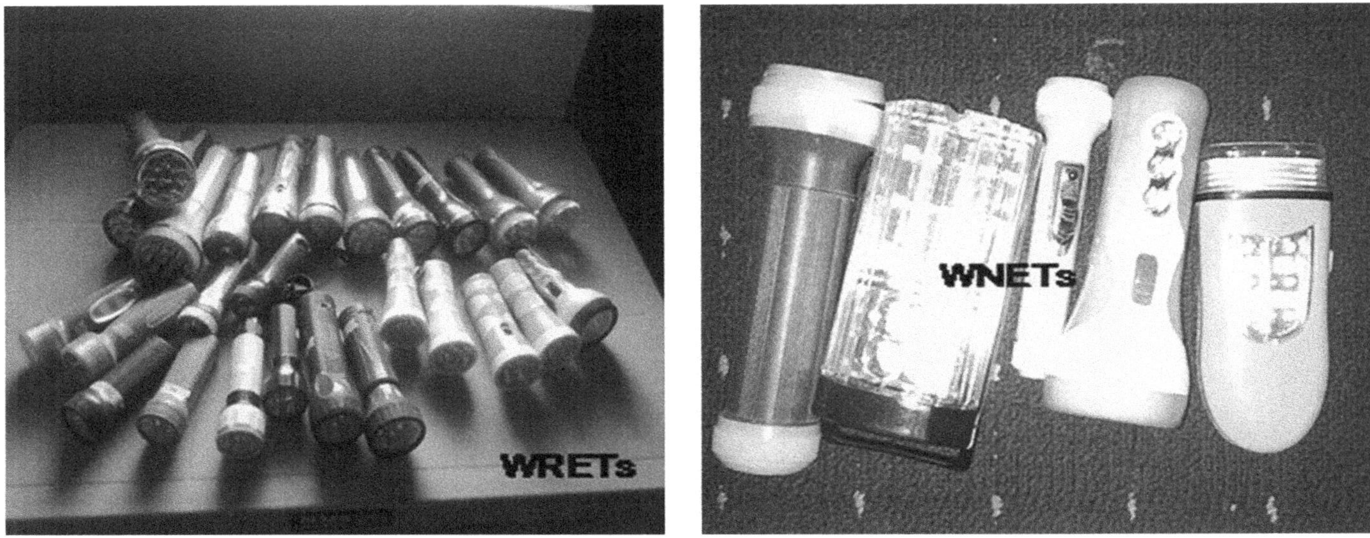

Figure 5 *Samples of waste rechargeable electric torches (WRETs) and waste non-rechargeable electric torches (WNETs)*

Figure 6 *Samples of waste rechargeable electric lamps (WRELs) and their batteries*

Table 5 *Contents of selected valuable metals in the printed circuit boards of selected electronic products*

Product	Valuable metals contained in products
TV (CRT monitor)	gold, silver, copper, platinum, antimony, nickel, yttrium, neodymium, iron, and aluminium
Washing machine, air conditioner, refrigerator	gold, silver, copper, platinum, antimony, iron, and aluminium
TV (LCD, plasma)	gold, silver, platinum, antimony, indium, yttrium, iron, aluminium

3.3 E-waste Generation

Over the years, the rapid growth in ICT (an umbrella term that includes any communication device or application: radio, TV, mobile phones, computer and network hardware and software, satellite systems, etc.) has led to phenomenal improvement in the capacity and features of electronics but simultaneously to a drastic decrease in the product's lifetime. In fact, the growth of the electronics sector and the rapid changes in technology mean that more consumers are replacing more equipment more often than ever before. One reason for the increasing generation of e-waste is the constant availability of newer technologies and designs and an increasingly early obsolescence. For example, the average lifespan of a new-model computer has decreased from 4.5 years in 1992 to an estimated 2 years in 2005 and is continuing to decrease (Widmer et al., 2005). Electronic waste is generated by three major sectors: individuals and small businesses; large businesses, institutions, and governments; and, original equipment manufacturers (OEMs). There are a number of reasons for a product reaching its end-of-life. These reasons include technical obsolescence (the product itself is worn out and no longer functions properly); economic obsolescence (new products in the market are more economical in terms of cost); feature obsolescence (new products have come onto the market that offer more or better features), and aesthetic obsolescence (new products in the market have a nicer look or more fashionable design from the point of view of the consumer) (Osibanjo and Nnorom, 2008; Nnorom et al., 2009).

3.3.1 Estimated Global Quantities of E-waste Generated

The last 2–3 decades have witnessed a tremendous upsurge in the global generation of e-waste, with estimated global annual quantities in 2006 reaching 40 to 50 million metric tons (UNEP, 2006; Schluep et al., 2009; Sepulveda et al., 2010; StEP, 2012). Estimates by Cobbing (2008) also revealed that by 2015, personal computers, mobile phones and television sets would have contributed an additional 9.8 million tons to the global e-waste stream. Estimates of e-waste generation in different countries and regions across the globe based on available literature/data have been reported and are indicated in Table 6 though measurement of actual quantities in a reliable manner remains a challenge. The order of e-waste generation based on Table 6 is Asia > North America > Europe > Africa > Latin America > Oceania. Nonetheless, Nigeria generates the highest e-waste annually in Africa of about 1.2 million tons (Table 6; SBC, 2011).

Table 6 Global quantities of e-waste generated by continent

Continent	Country	Quantities (1000 tons/yr)	Per capita (Kg/person)	Year of generation
Europe	Germany	1,100	13.3	2005[a]
	UK	940	15.8	2003[a]
	Switzerland	66.04	9.0	2003[a]
Asia	China	3, 620[b]	2.6[c]	2011[b,c]
	India	439	0.4	2007[a]
	Japan	860	6.7	2005[a]
Africa	Nigeria	1,200	7.1	2011[d,e]
	South Africa	59.65	1.2	2007[a]
	Ghana	179	41.0	2011[d,e]
	Cote d'Ivoire	15	4.8	2011[d,e]
	Benin	9.7	6.32	2011[d,e]
	Kenya	7.35	0.2	2007[a]
North America	USA	2,250	7.5	2007[a]
	Canada	86	2.7	2007[a]
Latin America	Brazil	709	3.77	2008[f]
	Mexico	47.5	0.44	2006[f]
	Argentina	20	0.49	2007[f]
	Colombia	7.4	0.17	2007[f]
	Peru	7.3	0.26	2007[f]
	Chile	7.0	0.42	2008[f]
Oceania	Australia	130	1.4	2008[g]

Agamuthu and Herat, 2012[a]; CHEARI, 2012[b]; Wang et al., 2013[c]; Ogungbuyi et al., 2011[d]; SBC 2011[e]; Araujo et al., 2012[f]; Davis and Herat, 2010[g]

A more recent estimate of global e-waste generation has been put at about 41.8 million tons (Mt) (Balde et al., 2014). Asia generated most of the e-waste, to the tune of 16 Mt in 2014, with 3.7 kg per capita. The highest per-capita e-waste quantity (15.6 kg/per capita) was generated in Europe. The whole region (including Russia) generated 11.6 Mt. The lowest quantity of e-waste was generated in Oceania and was 0.6 Mt, but the per-capita generation was nearly as high as Europe's (15.2 kg per capita). The lowest amount of e-waste per capita was generated in Africa, where only 1.7 kg/capita was generated in 2014. The whole of African continent generated 1.9 Mt of e-waste. The Americas generated 11.7 Mt of e-waste (7.9 Mt for North America, 1.1 Mt for Central America, and 2.7 Mt for South America), which represented 12.2 kg per capita, and comprised 1.0 Mt of lamps, 3.0 Mt of small IT, 6.3 Mt of screens and monitors, 7.0 Mt of temperature-exchange equipment (cooling and freezing equipment), 11.8 Mt of large equipment, e.g. washing machines, large printing machines, etc, and 12.8 Mt of small equipment, e.g. vacuum cleaners, microwaves etc. Thus small equipment (30.6%) and large equipment (28.2%) together represent about 60% of total e-waste generated globally, while small IT (7.2%), e.g. mobile phones, personal computers, and screens and monitors (15.1%), e.g. TVs and monitors, together represent about (22.3%) of global e-waste generated in 2014.

The amount of e-waste generated is expected to grow to 49.8 Mt in 2018, with an annual growth rate of 4 to 5 per cent, making it the fastest-growing waste stream in the world. Table 7 shows common WEEE items and their typical life span. There is speculation

Table 7 *Characteristics of some common waste electrical and electronic equipment (WEEE) items. Source: Robinson, (2009)*

Item	Weight of item (kg)	Typical lifespan (years)	References
Computer	25	3	Betts, 2008b
Facsimile machine	3	5	Robinson, 2009
Mobile telephone	0.1	2	Cobbing, 2008
Electronic games	3	5	Cobbing, 2008
Television	30	5	Li et al., 2009
Radio	2	10	Cobbing, 2008
Photocopier	60	8	Robinson, 2009
Video and DVD player	5	5	Cobbing, 2008
High-fidelity system	10	10	Cobbing, 2008

that developing and economy-in-transition countries may generate more e-waste than developed countries by 2020. Specifically, it is foreseen that in 2030 developing countries will be discarding 400–700 million obsolete personal computers per year, compared to 200 million–300 million in developed countries.

It can be affirmed that the exponential increase and near-tsunami global generation of e-waste is due to a number of factors, such as increasing market penetration of e-products in developing countries and improving economies in transition countries (CEIT), implementation of product "take-back" schemes in developed countries, and a pervading high product obsolescence rate (UNEP, 2007; Nnorom and Osibanjo, 2008), as well as a decrease in prices enhancing affordability by the relatively poor and the growth in internet use.

3.4 Problems with E-waste

E-waste represents the dark side of the information communication technology (ICT) revolution that has transformed modern living, international business, global governance, communication, entertainment, transport, education, and health care with fast communication gadgets, such as personal computers and mobile phones (Schluep et al., 2012). E-waste contains about 60 elements in the periodic table, including hazardous substances as well as scarce and valuable resource materials. E-waste generation and its environmentally sound management represent no doubt one of the foremost environmental challenges of the 21st century. It is a globalized problem affecting both developed and developing countries. The developed countries generate most of the e-waste in uncontrollable quantities and externalize the problem by shipping electrical and electronic devices as second hand or end-of-life electronic equipment into developing and economy-in-transition countries (CEIT), which lack the infrastructure and resources for environmentally sound management of e-waste, with risk to human health (Table 8) and the environment, under the guise of bridging the so-called digital divide (Osibanjo and Nnorom, 2007). In appreciating the relevance of ICT to the achievement of sustainable development and millennium development goals (MDGs), as well as bridging the digital divide, UN Secretary-General Ban Ki-Moon encouraged member states to bridge the digital divide by turning "the digital divide into digital opportunity" (Jhin, 2007).

Table 8 *Potential adverse health effects of toxic components of e-waste on humans*

Toxin	Typical Sources	Effects on Humans
Mercury	Fluorescent lamps, LCD monitor, switches, flat panel screens	Impairment of neurological development in foetuses and small children, tremors, changes in emotions, cognition, motor function, insomnia, headaches, changes in nervous response, kidney effects, respiratory failures, death
Lead	CRT of TV, computer monitor, circuit boards	Probable human carcinogen, damage to brain and nervous systems, slows growth in children, hearing problems, blindness, diarrhoea, cognition, behavioural changes (e.g. delinquency), and physical disorder.
Chromium	Untreated and galvanized steel plates, decoration or hardener for steel housings	Asthmatic bronchitis, skin irritation, ulceration, respiratory irritation, perforated eardrums, kidney damage, liver damage, pulmonary congestion, oedema, epigastric pain, erosion and discolouration of the teeth, motor function
BFR	Plastic casings, circuit boards	May increase cancer risk to digestive and lymph systems, endocrine disorder
Cadmium	Light-sensitive resistors, as corrosion retardant, Ni-Cd battery	Inhalation due to proximity to hazardous dump can cause severe damage to the lungs, kidney, cognition

Source: UNEP, 2007; MoEF, 2008; Pinto, 2008; Osuagwu, 2010; Chen et al., 2007.

In the past two decades e-wastes have garnered significant interest among policymakers and waste-management experts as a problem of crisis proportions in virtually all countries because they are a waste stream with the following unique combination of problematic characteristics (Balde et al., 2014):

- High volumes – High volumes are generated due to the rapid obsolescence of gadgets combined with the high demand for new technology (BAN, 2011).
- Toxic design – E-waste is classified as hazardous waste (SBC, 2011; Tsydenova and Bengtsson, 2011) having adverse health and environmental implications. Approximately 40 per cent of the heavy metals found in landfills come from electronic waste (Montrose, 2011).
- Poor design and complexity – E-waste imposes many challenges to the recycling industry (Smith et al., 2006) as it contains many different materials that are mixed, bolted, screwed, snapped, glued or soldered together. Toxic materials are attached to non-toxic materials, which makes separation of materials for reclamation difficult. Hence, responsible recycling requires intensive labour and/or sophisticated and costly technologies that safely separate materials (BAN, 2011).
- Recycled materials compete unfavourably in some circumstances with virgin materials due to variations in composition and contamination. For instance, effective reuse of recycled cathode ray-tube glass (CRTs used in TVs and monitors) is hampered by uncertainty on the composition of recycled glass as well as possible contamination with lead.
- They contain valuable scarce materials such as gold and palladium that are not easily recoverably by simple techniques, as well as specific products of concern: CRTs, flat screen, batteries, CFCs/fridges

- Labour issues: These include occupational exposure, informal sector domination causing health and environmental problems, lack of labour standards and rights.
- Financial incentives: There is a high cost to reverse logistics; hence in general, there is not enough value in most e-waste to cover the costs of managing it in a responsible way. However, in line with extended producer responsibility (EPR) policies, new opportunities can be realized with the rise in the price of many of the materials in electronics, such as gold and copper (Widmer et al., 2005). Furthermore, with rising e-waste quantities, formal recyclers are increasingly entering the e-waste recycling sector (Raghupathy et al., 2010).
- Environmentally unsound recycling practices are adopted in developing countries.
- Lack of regulation: Many nations either lack adequate regulations applying to this relatively new waste stream, or lack effective enforcement of new e-waste regulations (BAN, 2011).

3.5 E-waste Management Challenges Facing Developing Countries

3.5.1 Introduction

Increasing consumer demand, arising from population explosion, for electrical and electronic devices in developing countries is fuelling the exponential and sometimes uncontrollable generation of e-waste globally. Developing countries have increased their share of the world's total number of internet users from 44% in 2006, to 62% in 2011. Today, internet users in China represent almost 25% of the world's total internet users (ITU, 2012). Thus it is foreseen that by 2030, there will be a reversal of the present trend and developing countries would generate significantly more e-waste than developed countries.

The major global issues facing e-waste management in the 21st century hinge on rising e-waste quantities as a result of short lifespan of electrical and electronic equipment (EEE), poor feedstock collection of e-waste, high cost of/crude resource recovery technologies, poor product design, poor regulatory and enforcement frameworks and ethical issues involving externalisation of risks. Most often, illegal, trans-boundary movement of e-waste from the developed countries to the poorer developing countries occur regularly (Figure 7). The regulatory framework and the capacity for the prevention and control of trans-boundary movement of used and end-of-life electronic products are weak and grossly inadequate in developing countries. Although most of the latter have ratified the Basel Convention, which forbids the dumping of e-waste from developed countries in developing countries and CIET, it is generally yet to be reflected in national laws.

E-waste poses a sweet-sour situation in developing countries. The hazardous constituents in e-waste are harmful to human health and the environment, which raises human exposure concerns in developing and economy-in-transition countries because of crude processing/treatment technologies employed, lack of human health safeguards, and environmental control regulations. Yet the precious metals in e-waste, such as gold and silver, provide entrepreneurship, employment, and poverty alleviation opportunities for the informal sector that dominates the e-waste recycling business. For example, experts estimate that recycling 1 million mobile phones can recover about 24 kg (50 lb) of gold, 250 kg (550 lb) of silver, 9 kg (20 lb) of palladium, and more than 9,000 kg (20,000 lb) of copper.

Thus there is a compelling need to adopt innovative policies and approaches in e-waste management. We need to modernize the 20th-century thinking about waste management as

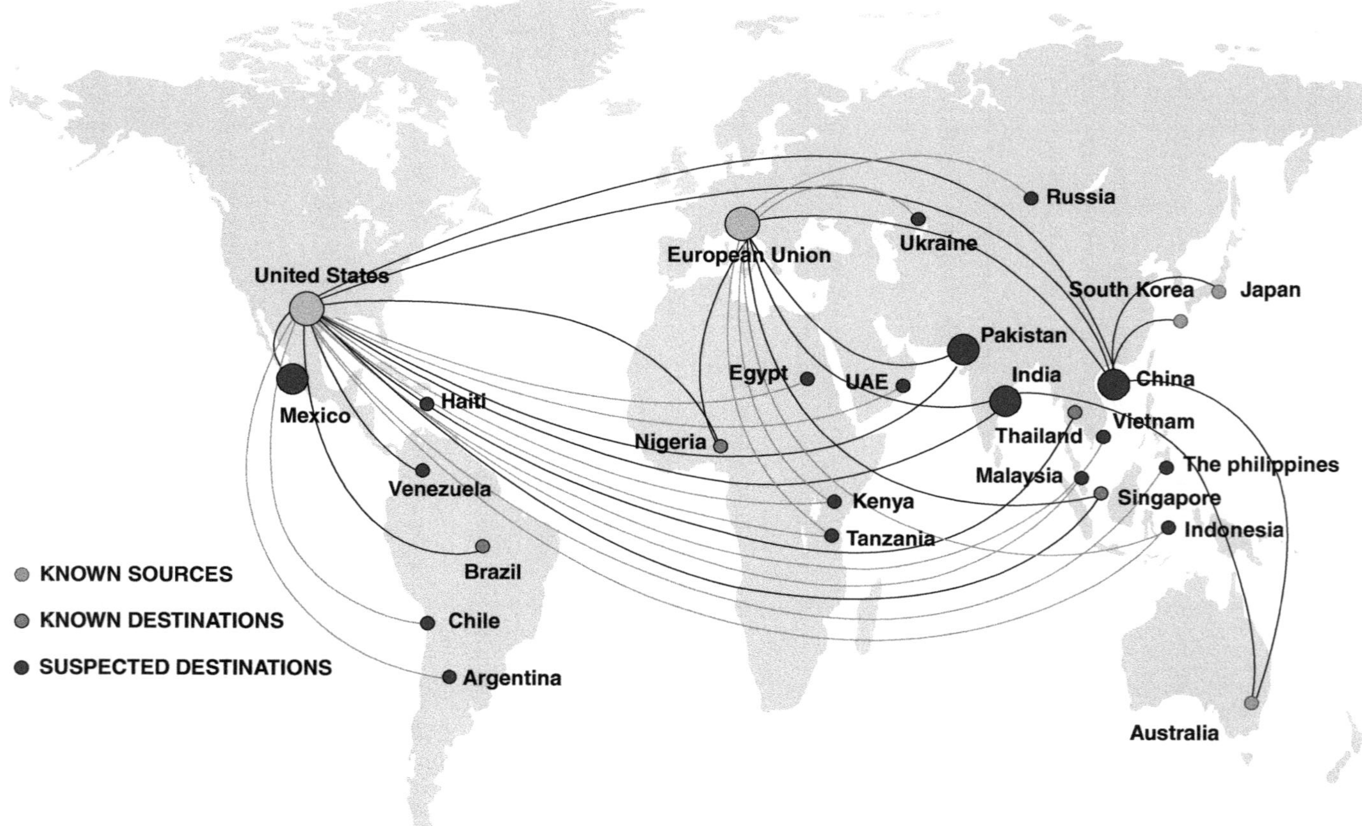

Figure 7 E-waste flows to developing countries. (See insert for color representation of the figure.)

"*How do we get rid of our waste efficiently with minimum damage to public health and the environment?*" to 21st-century thinking of resource management, "*How do we handle our discarded resources in ways which do not deprive future generations of some, if not all, of their value?*" This is consistent with the Basel Convention 10th conference of the parties (COP 10) Cartagena Declaration in 2010, which called for waste to be recognized as a secondary resource rather than material to be thrown away; as well as the Rio+20 outcome document "The Future We Want", which promotes transition to a "Green Economy" including waste minimization and waste utilization.

Waste generation occurs along the supply chain during various processing activities in the life cycle of electronic products (Figure 8). The major challenges associated with e-waste management in developing countries are discussed below.

3.5.2 Poor Feedstock Collection Strategies

Collection of e-waste in the developed countries is organized, though the strategies vary from country to country. In many cases there are collection points either provided by government or original equipment manufacturers (OEMs) where end-of-life (EoL) EEE or e-waste is dropped for further treatment or disposal. In some countries like Japan, consumers pay some disposal fees prior to dropping (Ogushu and Kandlikar, 2007). In most developing countries e-waste is still co-disposed with other municipal wastes, making sorting practically impossible. In a typical developing country like Nigeria, most e-wastes are collected from either refurbishers'/repairers' workshops by the informal sector or mined by scavengers (otherwise called urban miners) from dumpsites after co-disposal with other municipal wastes. This attitude in developing countries could be born primarily from lack of awareness and ignorance on the harmful effects of improper disposal of e-waste, absence of "government will" on "take-back" incentives and lack of collection centres. These deficiencies present a huge socioeconomic

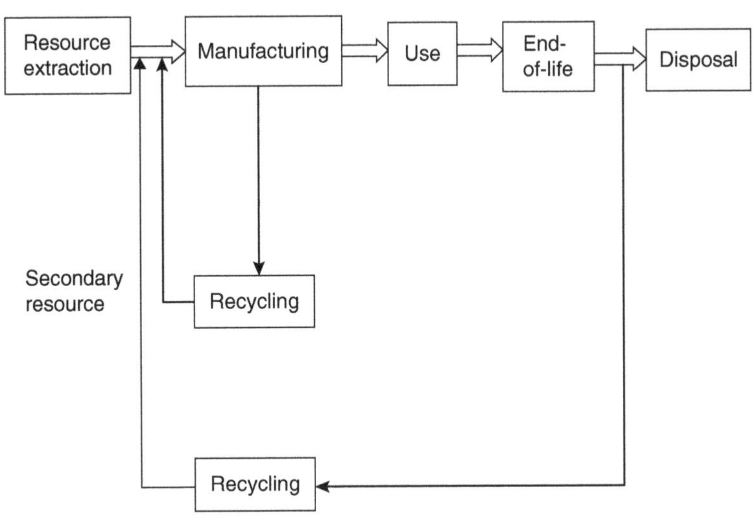

Figure 8 *Life cycle of electronic equipment. Source: Adapted from Hoang, (2009)*

challenge involving gender and child labour issues (Osibanjo, 2015), as women and children, who are more vulnerable to the deleterious effects of exposure to hazardous substances from improper e-waste management, feature prominently in e-waste collection from dumpsites.

3.5.3 Lack of State-of-the-Art Technologies to Recover Resources from E-Waste

Conventionally, e-waste after collection is sorted into different components like CRT glass, printed wiring board (PWB), plastics and others (including metals and ceramic components, etc). The flow diagram in Figure 9 depicts the major components that are disassembled from typical WEEEs and the available management methods.

The e-waste component shown in Figure 3 that is of greatest economic importance to the recycler is the PWB because of a variety of precious metals that are found glued, bolted and smeared within the board. CRT glass is the most challenging component of e-waste to manage globally, and a great percentage of this waste category is still landfilled (Kang and Schoenung, 2005; Manhart et al., 2013). The two most popular methods employed by the formal recyclers (mainly in developed countries) to recover precious metals from PWBs are pyrometallurgy and hydrometallurgy. Debates are ongoing on which method is better. Other methods that are used in the poor developing countries where the expensive methods are grossly lacking are uncontrolled open burning and backyard acid leaching. Table 9 describes each method.

Table 9 shows that the so-called high-tech methods, pyrometallurgy and hydrometallurgy have environmental and health concerns, but there are remedies in place to reduce these issues to the barest minimum. Open burning and backyard acid leaching methods are carried out by poor and uneducated people in the developing countries. These activities have deleterious effects on both the environment and human health.

3.5.4 Lack of Specific E-Waste Regulations and Enforcement in Developing Countries

E-waste-specific regulations in most developing countries are generally lacking and where they exist they are still in draft forms (in countries such as Ghana and Kenya). Where regulations exist, they are weakly enforced. Nigeria is the only African country with a National E-waste Regulation, enacted in 2011. The regulation is anchored on the 5R ('Reduce, Repair, Recover, Recycle and Re-use') principle as the primary drivers including all the categories and lists of WEEE (Ogungbuyi et al., 2012). In South Africa and many other developing countries, only generalized waste management regulations exist and there are still no e-waste-specific regulations (ATE, 2012). In China, where specific e-waste regulations have come into force, there is still a huge gap between the estimated generated quantities and the quantities actually collected by government-approved vendors (Wang et al., 2013). Because of this laxity in regulations and enforcement, the informal sector, which lacks the capacity to handle e-waste in an environmentally sound manner, is still having a field day. Therefore, the management of the increasing volumes of e-waste effectively and efficiently, in terms of resource recovery and minimal environmental impact, is still a very difficult challenge (Sinha-Khetriwal et al., 2005).

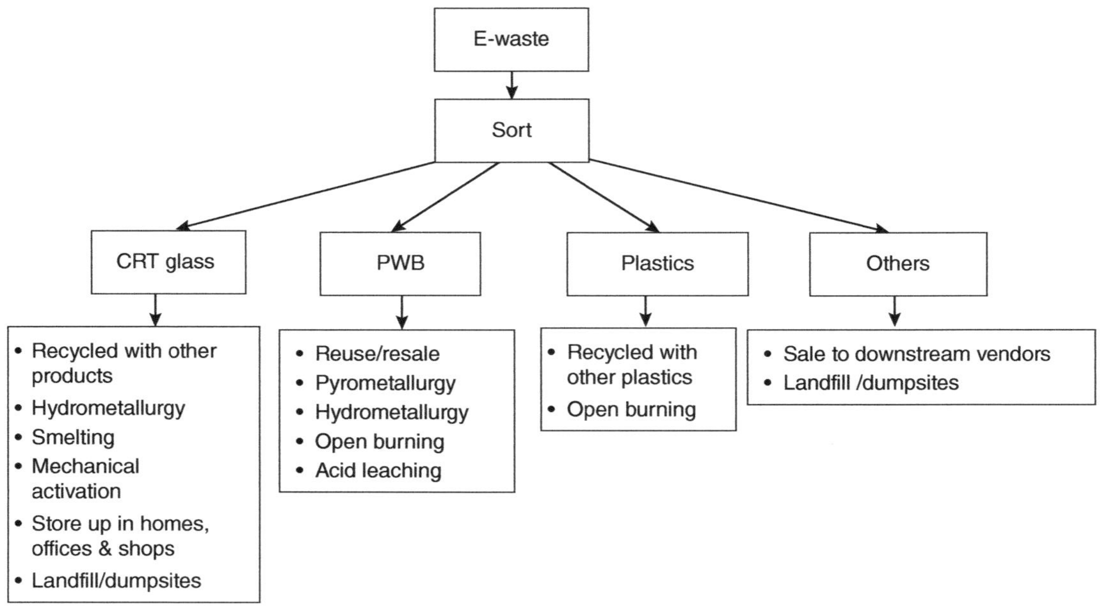

Figure 9 *Flow diagram of major components of E-waste and available management methods*

Table 9 *Major available technologies used to recover metals from e-waste*

Parameter	Pyrometallurgy	Hydrometallurgy	Open burning	Backyard acid leaching
Principle	Smelting of e-waste in furnace at high temperature to recover precious metals	Leaching of e-waste with different chemicals in controlled environment to recover precious metals	Open burning of e-waste cable to recover Cu	Soaking of e-waste in uncontrolled environment in different acid solutions to recover precious metals
Highlights of operation	• High cost of energy required to generate heat • Full capacity of furnace before processing • High cost of scrubbing system • Achieves reduced volumes through burning in controlled environment • High recovery efficiency of precious metals • No recovery of plastics and other useful non-metals	• Less energy as no heat is required • Any quantity can be processed • Uses mainly particulate and chemical scrubbing system • High volumes must be dealt with • Recovery of precious metals not as efficient as in pyro. • Plastics could be reused or recycled	• Open burning of cables to recover Cu • Burning of plastics to reduce volumes • No protection of nose • Practiced in developing and transitioning countries	• Backyard leaching with substandard tools • Low recovery and highly impure precious metals obtained • No protection of nose and hands • Practiced in developing and transitory countries
Environmental/ occupational &. health concerns	• High particulate and gaseous emissions • Loss of volatile toxic metals • Release of dioxins and furans • Generates toxic ash/slag	• Release of toxic HCN, HCl and Cl_2 gases • Generates wastewater and toxic sludge • Generates dust from crushing	• Release of dioxins and furans through burning to air, water and soil • Threat to occupational health	• Release of acid fumes • Release of acid water loaded with toxic metals like Pb on soil and water bodies • Occupational.exposure to acid fumes and toxic metals
Remedies	• Installation of particulate and wet electrostatic precipitators and chemical scrubbing system • Use of safety clothing and equipment • Installation of other pollution control equipment	• Installation of particulate and chemical scrubbing system • Use of safety clothing and equipment • Installation of other pollution control equipment	None in place	None in place

3.6 Environmental and Health Impacts of E-Waste Management in Developing Countries

E-waste has profound potential to cause damage both to the environment and to human health as the hazardous substances in e-waste may be released or leached into the environment in landfills, with potential human exposure to these pollutants during e-waste processing. This occurs during treatment processes for recyclable materials comprising plastics, glass and precious metals like gold, palladium, platinum, silver and copper. There are emerging environmental and health issues arising from e-waste management, especially in the developing countries.

3.6.1 Environmental Impacts of E-Waste

The commonest e-waste management methods in the developing countries are undertaken in the informal sector and include crude dismantling with hammers, uncontrolled open burning of cables to recover copper wire, residues of e-waste repairs and refurbishment activities, dumping on open land/spaces, and co-disposal with municipal waste on dumpsites. These methods may result in emission of persistent organic pollutants such as polychlorinated dibenzo-p-dioxins (PCDDs), polychlorinated dibenzofurans (PCDFs) (Sindiku et al., 2014), polychlorinated biphenyls (PCBs), polycyclic aromatic hydrocarbons (PAHs), polybrominated diphenyl ethers (PBDEs) (Sindiku et al., 2014) and polybrominated biphenyls (PBBs), as well as hazardous heavy metals such as Pb, Cd, Cr, Hg, and As (Babayemi et al., 2014). The methods of e-waste treatments in the developing countries encourage contamination of environmental media (air, soil, water), with consequent bioaccumulation in plants and biological organisms. The environmental impacts of e-wastes as reported from some developing countries are presented in Table 10.

3.6.2 Health Impacts of E-Waste

Resolution II/4 on emerging policy issues adopted by the International Conference on Chemicals Management of the Strategic Approach to International Chemicals management at its second session, held in Geneva in May 2009, included the adoption of hazardous substances within the lifecycle of electrical and electronic products as an emerging environmental policy issue. The contaminants released into the environment through e-wastes have been linked with arrays of health problems including endocrine disruption, cancer, liver and DNA damage, behavioural changes and developmental problems. Humans are exposed to health effects of e-waste through dermal contact, inhalation of burning smoke/dust and dietary intake through contaminated water and food (Song and Li, 2015).

Children and adults working or living near e-waste recycling sites are susceptible to health problems related to e-waste. However, children are more affected due to high gastrointestinal uptake of heavy metals (Song and Li, 2015). Song and Li (2015) documented increases in spontaneous abortions, stillbirths, premature births, reduced birth weights and infant lengths in pregnant women exposed to heavy metals from a e-waste sites etc. Changes in thyroid function, changes in cellular expression and function, adverse neonatal outcomes, changes in temperament and behaviour, and decreased lung function have been linked with exposure to e-waste (Grant et al., 2013). Cancer incidence in the e-waste disassembly sites was related to higher burdens of PBBs, PBDEs, and PCBs in kidney,

Table 10 *Environmental impact of e-waste recycling*

Country	Environmental media	Contaminants and concentrations	References
China	Dumpsite soil, sediment	Soil PAHs (593), sediment PAHs (514) and PCBs (743) μg/kg	Leung et al., 2006
China	Dust	Workshop dust: Pb (110 000), Cu(8360), Zn (4420), Ni (1500) mg/kg; Adjacent road dust: Pb (22600), Cu (6170), Zn (2370), Ni (304) mg/kg	Leung et al., 2008
China	Soil and vegetables	Soils: PAHs (0.13–10.6); vegetable PAHs (0.20–2.42) μg/g	Wang et al., 2012
India	Soil	V (24–37), Cr (46–160), Mn (286-849), Co (5.2–42), Cu (61.7–4790), Zn (126–2530), Mo (0.84–11.0), Ag (2.2–320) μg/g	Ha et al., 2009
Hong Kong	Soil	Total 16 PAHs 107–2300 ng/g, total PCBs (6.7–143.7) ng/g, PBDEs (27.5–32340 ng/g), Cr(7.05–1717), Cu(13.8–756), Cd (0.28–11.0), Pb (132–3254), Zn (123–1717) mg/kg	Lopez et al., 2011
Nigeria	Soil and plants	Soil: PAHs (116 mg/kg), PBDEs (37.7), PCBs (4.06) mg/kg; Pb (1535), Cu(4308) Cr (13.9), Ni (32.0), Cd (7.69), Mn (270) mg/kg; Plants: PAHs (15.0), PBDEs (31.1), PCBs (0.21) mg/kg; Pb (34.34), Cu (69.55) Cr (2.0) Ni (1.24) Cd (1.29) Mn (24.65) – Plants	Alabi et al., 2012
China	Biota (snails, prawns, fish, water snakes)	PBDEs (52.7–1702), PCBs (20.2–25958) ng/g	Wu et al., 2008

Table 11 *Indicators of health impacts of e-waste in developing countries*

Country	Indicators	Contaminants and concentrations	References
China	Blood	Pb (4.40–32.7) μg/Dl	Huo et al., 2007
China	Placenta	Pb (0.007–3.47), Ni (0.001–1.11) μg/g	Wang et al., 2012
Vietnam	Breast milk	PBDEs (20–250) ng/g),	Tue et al., 2010
China	Hair	Cu (39.8), Pb (49.5) μg/g	Wang et al., 2009
China	Hair	PBBs (57.8), PBDEs (29.6), PCBs (180) ng/g	Zhao et al., 2008

liver and lung (Zhao et al., 2009). Elevated levels of heavy metals and persistent organic pollutants in blood, placenta of babies, breast milk and hair of people living close to e-waste treatment sites have been documented, as shown in Table 11. Most of the reports are from China.

3.7 Solutions for Present and Future Challenges

3.7.1 Optimizing and Promoting E-Waste as a Resource

Considering the environmental and health risks of adopting inappropriate management options for e-waste, as well as the loss of valuable resources, it is necessary to seek eco-friendly and sustainable sound options throughout the life cycle of electronics. An analysis of the electronics life cycle indicates that material extraction and manufacturing steps, energy requirements in production and use, and the negative environmental effects of EoL management are of most concern (Nnorom, 2012a,b). The ever-increasing technological complexity of EEE and the ever- shortening product life expectancy compounds this. However, the consumer electronic industry in general has accomplished a greater deal in reducing its impact on the environment by focusing on efficient use of its products, reducing products energy consumption and implementing environmental management systems to make the manufacturing processes increasingly resource efficient (Haugen, 2002). However, much still needs to be done in finding short- to long-term solutions to WEEE management, especially in the developing countries.

3.7.2 Role of Product Design in Defining Product EoL Scenario

End-of-life strategies are particularly important for EEE and innovative measures are needed to manage this, since EEE product life depends more on technological obsolescence than on wear-out life. However, the EoL strategy chosen depends on the characteristics of the product. A good EoL strategy for any product is to choose the alternative that causes minimal environmental damage while maximizing reusability of the products and components.

As the sustainability debate progresses, there is an urgent need to control consumer issues and increasing waste generation in the EEE sector by extending the lifetime of electronic products (Boks et al., 1998). Product design and development are essential in ensuring that products pose minimum challenges to health and the environment. Consequently, within the sustainability framework, it is important that manufacturers set the right design priorities, taking into consideration the entire life cycle of the products. For instance, the design should take into consideration possible reuse options as well as the appropriate EoL scenario of the product, such as:

a. will a product be reused?
b. will a product be disassembled for components reuse?
c. will a product be shredded for material recovery (recycling)?
d. will a product be incinerated (with energy recovery)?
e. will the product be dumped to landfill? (Nnorom, 2012a,b).

These issues should be considered at the design stage in order to have products that meet the tenets of sustainable development. In dealing with e-waste, it is essential that reuse options be integrated into the design to reduce the volume generated. To achieve this, it is essential that designers adhere to the tenets of design for environment (DfE) or design for recycling (DfR) to ensure that products are built for reuse, repair, and/or upgradeability. Emphasis should be placed on the use of less toxic, easily recoverable, and recyclable materials that could be taken back for refurbishment, remanufacturing, disassembly, and reuse.

3.7.3 Recovering EoL Products

Today, material disposition is driven primarily by design and economics, and products or components that lack effective reverse logistics networks or are designed so that they cannot be economically re-manufactured or recycled are disposed of into the solid waste stream. Efforts to develop products with improved environmental performance should not be seen as a threat by electronics manufacturers, but rather as an opportunity to increase business ventures and sales and to create awareness in consumers that the enterprise is 'environmentally conscious'. In fact, a 2009 University of Illinois study observed that the present global e-waste management system is generally not sustainable because mechanisms for collecting, sorting, reuse, refurbishing, repairing, and remanufacturing are not well developed and/or implemented. The adoption of sustainable management strategies for end-of-life electronics is critical in averting the loss of precious scarce resources and the environmental consequences of inappropriate management practices. Product recovery, the broad set of activities designed to reclaim value from a product at end-of-life, is used to describe activities designed to reuse, recover, refurbish, remanufacture, demanufacture, or recycle durable product assets at the end of a product life-cycle.

Product recovery plays three main roles:

- it lessens the environmental and economic costs of waste disposal
- it reduces the economic cost of purchasing and processing of new materials. This is because reusing components from used products ensures that their embodied value is retained
- it can be used as an environmental marketing tool by companies to differentiate their products and services

Several authors have reviewed the complexity of developing an integrated recovery process for WEEE (Ferrer, 1997; Ahluwalia and Nema, 2006). End-of-life costs are dependent on reverse logistics costs, product disassembly costs, the net value of materials to be recycled or processed, and the likelihood and revenue from component reuse or remanufacturing (Spicer and Johnson, 2004).

The five common options for material recovery from EoL products are repair, refurbish, remanufacturing, cannibalization and recycling (Nnorom et al., 2007). Reuse and recycling of EoL electronics are very demanding but advantageous alternatives to incineration or landfill of electronic scrap (Knoth et al., 2002). Factors such as cost, labour availability, return flow volume, and optimal disassembly level determine what recovery processes are feasible (Ritchey et al, 2001). Recovery of EoL products is constrained by the large variety of product models available in the market, size changes, and compatibility issues (Kumar et al., 2005). Meanwhile, the decision to remanufacture, disassemble and then recycle, recycle without prior disassembly, or simply dispose of an EoL product is based on product durability, rate of technological obsolescence, product complexity, duration of a design cycle, and reason for redesign, among other factors.

A recent study (Sindiku et al., 2014) has underscored the importance of screening plastics from e-waste for hazardous substances such as brominated flame retardants prior to recycling; otherwise products from such recycled plastics may become future sources of contamination and human exposure to these chemicals with health risks. The screening study suggests that average PBDE levels (of c-OctaBDE + DecaBDE) in Nigerian-stockpiled CRT casings were

1.1% for TV and 0.13% for PC CRTs. These are above the Restriction of Hazardous Substances (RoHS) limit and should therefore be separated for RoHS-compliant recycling. The Nigerian e-waste inventory of 237,000 t of CRT plastic (Ogungbuyi et al., 2011) would therefore contain approximately 594 t of c-OctaBDE and 1,880 t of DecaBDE. In Nigeria, as in most developing countries, there is currently no adequate e-waste management, plastics separation or destruction capacity. The data highlighted the urgent need to develop environmentally sound management strategy for this large plastic material flow. It further raises the question: What can developing countries and CEIT do with WEEE plastics/polymer-containing BFRs that the original equipment manufacturers (OEMs) and electronics recycling industry cannot take back? What support can the global OEMs give that Africa can address and solve the WEEE polymer recycling and end-of-life management challenge?

Product recovery will be required to control the various inappropriate management practices of EoL electronics in the developing countries (such as disposal with solid waste, into surface water bodies and crude backyard recycling practices), save resources and ensure environmental protection. Remanufacturing is important in achieving a green economy and saving scarce resources (Nnorom, 2012a, b; 2013). Designing products for remanufacture is required to assure their adaptability to remanufacturing operations. For example, designing products with high levels of modularity will be required if the products are to be remanufactured at their EoL. Xerox has been cited frequently in literature as a leader in the remanufacturing of their copiers. Xerox concurrently designs manufacturing and remanufacturing facilities for new models of their copiers and, in steady state, most of their products are actually "newly remanufactured" copiers (Ishii, 1998). Xerox has saved hundreds of millions of dollars through asset recovery and remanufacturing programs, while having a significant positive effect on the environmental bottom line (Kerr and Ryan, 2001).

Similarly, manufacturers of EEE have advanced research into design for environment (DfE), and significant progress had been made in the past two decades. For instance, IBM has established a research arm called Design for Environment and has established a worldwide asset-recovery organization that has been providing global remanufacturing and refurbishment focus for corporate and institutional accounts. Remanufacturing of EEE is becoming increasingly necessary and important in ensuring that future economic and manufacturing growth is sustainable.

To avoid negative environmental impacts by today's practice in demand markets with slack environmental regulations, clean remanufacturing activities must be initiated at the returned product's origin (Kernbaum et al., 2006). The reuse of EoL EEE conserves resources and feedstock that supply steel, glass, plastics and precious metals. Such reuse activities also avoid air and water pollution as well as greenhouse gas emissions associated with material production and manufacturing. The reused resources in a remanufacturing operation are the material in the product, energy, machine time, labour and other costs that have accumulated in the new production process (Östlin et al., 2009).

3.7.4 E-Waste as a Resource for Socioeconomic Development

The semi-formal and informal take-back system for e-waste management and the recovery of valuable materials from these in developing countries are contributing to the socioeconomic development of these countries. Typical examples are Ghana and Nigeria in

Africa and China and India in Asia (Grant and Oteng-Ababio, 2012; Manhart et al., 2011; Ogungbuyi et al., 2012). Manhart et al., (2011) assessed the socioeconomic impacts of the second-hand EEE and e-waste recycling formal and informal sector in Nigeria as part of the Secretariat of Basel Convention (SBC) E-waste Africa Project (SBC, 2011). This study observed that both formal and informal operators and individuals in the (second-hand) EEE sector are partly organized in associations to protect their business interests. The E-waste Africa Project observed that no formal education is required for collecting and sorting of e-waste, or in recovering of valuables from the wastes. Despite the limited formal education, all the waste collectors and recyclers interviewed in the study had very good knowledge of the kind of wastes they were interested in collecting and recycling. A sizeable number of graduates are also in the business of repairing and refurbishing EoL EEE in Nigeria. The sector is male dominated (>70 %).

Ogungbuyi et al. (2012) observed that wages in the e-waste sector are structured according to the waste volumes collected or treated, and hence the motivation for most individuals is the economic returns/benefits rather than concern for the environment. Between 144 and 1985 kg/week of e-waste mixed with other metal scrap are collected by a waste picker. The collected e-waste is co-mingled with other metal scrap. Up to 80000 persons in Nigeria are involved in this sector. The main sources of collected materials are homes/dump sites, refurbishers, streets, and importers. The waste picker sells to scrap dealers or vendors.

A typical vendor gets N3000–N5000/week (approximately US$20–30) when he sells his scrap. Mobile phone repairers often require the customers to register by paying a certain fee before the phones are investigated for faults, a charge that is usually not part of the repair charge. Revenue per refurbished EEE is between N1000–N3000 depending on the nature of the fault; when faulty components or modules are to be replaced the charge may be higher. An estimated 52000 persons are engaged in the refurbishing business in Nigeria (Ogungbuyi et al., 2012).

Approximately 66–68% of EEE brought to repairers and refurbishers shops are effectively repaired. Un-repairable EEE abandoned in the repairers' shops are disposed of or sold between 6 months to 3 years (mean: 1.5 years) of storage in the repairer's shop. 12–25% of the refurbishers dispose of all e-waste generated in their operations with general waste. Others (estimated at 66%) store and sell the waste to collectors and dispose of the useless wastes with general waste. The metal/steel and the plastic sectors of the country have been the main beneficiary of the informal collection of scraps and recyclable items. The e-waste aspect is also becoming a profitable venture for those who export printed circuit boards through various informal channels of downstream vendors across political borders.

3.7.5 Urban Mining

The life cycle of EEE begins with development and production, followed by use and maintenance, and leads right up to the reuse and recycling of the product in whole or in part. If reuse is not possible, recycling should be given preference; only as a last resort should a product be incinerated or dumped in a landfill. Urban mining is increasingly being recognised as an important component of resource strategies of public authorities, not only because it contributes to environmental protection but also because it is a source of valuable recyclable materials.

Simoni et al. (2015) used the rare earth element (REE) group to illustrate an overview of information and knowledge gaps concerning urban mining. The analysis shows that rare earth element recycling can be more environmentally friendly than primary production, particularly if the latter comes from countries with weak enforcement of environmental legislation. On the other hand, REE recycling often cannot compete with large-scale primary production because market prices do not reflect the social and environmental impacts of production, and because the avoided impacts of waste decontamination and waste production are not considered. The analysis of urban mining potential can be used to support decision making and the setting of priorities for future research and public action. The findings of the study and expert opinions based thereon contribute to the selection of measures and the formulation of public waste management and resource strategies in general. Urban mining especially using crude technologies is presently ongoing in some cities in developing countries, in particular Chennai in India and Guiyu in China. In these cities, large quantities of discarded internally generated computers, phones and television sets as well as imported e-wastes are sorted, disassembled, crushed and eventually chemically treated to recover the precious metals and traces of rare earth elements.

Estimates show that metal deposits in e-waste are up to 40–50 times richer than ore extracted from mines. For example, one ton of gold ore yields about 5 grams of gold, but one ton of phone circuitry yields about 150 grams, 30 times as much (Harvey, 2013). Unfortunately, only about 15–20% of the world's e-waste (estimated at about 50 million tons/year globally by the United Nations Environment Programme) is recycled annually. It is unfortunate and ironic too that, even with all efforts at achieving resource conservation and sustainability globally, only about 15% of the estimated $21 billion worth of gold and silver used in electronics is recovered from e-waste worldwide (Harvey, 2013). Consequently, e-waste is a promising reserve of valuable resources for any urban miner. Urban mining presents an opportunity to reclaim and recycle precious metals and REEs from e-waste and this requires the use of state-of-the-art facilities to ensure high recovery rates and high purity of recovered material while ensuring that environmental standards are maintained.

3.8 Conclusions

Information and communication technology (ICT), driven by electrical and electronic equipment, especially computers and mobile phones, has in the last two to three decades transformed the world beyond imagination. It has become a critical factor in achieving sustainable development for developing countries and fostering productivity and innovation, as well as helping to achieve the Millennium Development Goals (MDGs). These countries have achieved rapid advances in ICT in recent years to bridge the digital divide with developed countries. The ICT explosion is facilitated by the import of second-hand or used computers and mobile phones from developed countries, especially Europe and North America, as most of the population in developing countries can not afford the price of new electronic gadgets.

However, the near-tsunami generation of e-waste, classified as hazardous waste under the Basel Convention, from unsustainable production and consumption of electronic products; and the export of e-waste from developed to developing countries that lack the infrastructure and resources for their environmentally sound management, with the externalization of the adverse effects, are the dark sides of the ICT revolution and the resultant globalisation of the e-waste challenge. E-waste contains hazardous substances such as heavy metals, cadmium, lead, and mercury, as well as persistent organic pollutants (POPs) such as polybrominated biphenyl ethers (PBDEs). The crude e-waste management methods prevalent in developing countries are environmentally unsound and have potential risks to human health and the environment. It has been predicted that by 2020 developing and CEIT countries will generate more e-waste than developed countries. E-waste has therefore become a global crisis, not only from its quantity, as the fastest growing waste stream in the world, but also from various hazardous contents such as heavy metals and endocrine-disrupting substances, e.g. brominated flame retardants. E-waste has thus become an important risk to health and the environment, especially in developing countries and CEITs.

E-waste is somehow a paradox as it is both a problem and an opportunity, since it also contains valuable ferrous (e.g. iron), non-ferrous (e.g. copper), precious (e.g. gold and silver) and strategic metals (e.g. indium, gallium) that are scarce and may be lost if e-waste is landfilled or improperly processed; including the uncontrolled open burning currently practised in developing countries and CEITs. It is noteworthy that approximately 40% of the heavy metals found in landfills come from electronic waste (Montrose, 2011).

The simultaneous depletion of key metals and minerals, and the continuous production of e-waste streams, are certainly risky situations. Hence, there is a need for a paradigm shift from a perception of e-waste as a waste-disposal problem to a resource-management challenge, in line with the Rio+20 outcome document, "The Future We Want", which also promotes a transition to a green economy. This would mean "mining" the e-waste streams for raw materials such as precious metals (e.g. gold, silver, copper,) and strategic minerals such as rare earth metals. This will slow down the extraction and depletion of minerals from the earth, reduce their waste, and lessen the environmental, human health, and other impacts associated with electronic gadgets production cycle, including the reduction of greenhouse gases emission. Thus, e-waste is of special interest because most of it contains strategic minerals whose recovery and recycling could be cost-effective, create employment, and help alleviate poverty. However, recycling of WEEE plastic containing brominated flame retardants poses a special challenge for developing countries, as the BFRs must be removed from plastic before recycling and reuse. Thus, e-waste is one of the environmental challenges of the 21st century. Developing countries require international support from the United Nations agencies and donors to solve the monumental challenges in acquiring the infrastructure, resources, and capacity necessary for environmentally sound management of e-waste for sustainable development.

References

Advanced Tropical Environment (ATE). (2012) Identification of the Magnitude of the Electrical and Electronic (E-Waste) Situation in South Africa: A Strategic Approach to International Chemicals Management (SAICM) E-Waste as an Emerging Policy Issue, Department of Environmental Affairs, People's Republic of South Africa.

Agamuthu, P., Herat, S. (2012) E-waste: A Problem or an Opportunity? Review of Issues, Challenges and Solutions in Asian Countries. *Waste Management and Research* 30(11): 1113–1129.

Ahluwalia, P.K., Nema, A.K. (2006) Multi-objective Reverse Logistics Model for Integrated Computer Waste Management, *Waste Management and Research*, 24: 514–27.

Alabi, O.A., Bakare, A.A., Xu, X., Li, B., Zhang, X. Huo, Y. (2012) Comparative Evaluation of Environmental Contamination and DNA Damage Induced by Electronic-Waste in Nigeria and China, *Science of the Total Environment*, 423: 62–72.

Araujo, M.G., Magrini, A., Mahler, C.F., Bilitewski, B. (2012) A Model for Estimation of Potential Generation of Waste Electrical and Electronic Equipment in Brazil, *Waste Management*, 32: 335–342.

Babayemi, J.O., Osibanjo, O., Weber, R. (2014) Assessment of Use, Reuse and End-of- Life Disposal and X-Ray Fluorescence Analysis Screening of Waste Mobile Phones in Nigeria, *Environmental Quality Management*, 23(4): 1–12.

Baldé, C. P., J. Huisman, F. Wang and L. Herreras (2014) WEEE Market Assessment, Deliverable 4.2 of Countering WEEE Illegal Trade Project (Internal document), United Nations University.

BAN/SVTC (Basel Action Network and Silicon Valley Toxics Coalition). (2002) Exporting Harm: The High-Tech Trashing of Asia, www.ban.org, www.svtc.org

Basel Action Network (BAN). (2011) Toxic Trade News: Cochin Port a Safe Conduit for Imported E-waste. Available at http://www.ban.org/January. 2012.

Betts K. (2008) Producing Usable Materials from E-waste, *Environmental Science and Technology*, 42: 6782–3.

Boks, C., Nilsson, J., Masui, K., Suzuki, K., Rose, C., Lee, B.H. (1998) An International Comparison of Product End-of-Life Scenarios and Legislation for Consumer Electronics, in *Proceedings of the International Symposium on Electronics and the Environment*, 1998.

Chan, J.K.Y., Xing, G.H., Xu, Y., Liang, Y., Chen, L.X., Wu, S.C., et al. (2007) Body Loadings and Health Risk Assessment of Polychlorinated Dibenzo-p-dioxins and Dibenzofurans at an Intensive Electronic Waste Recycling Site in China, *Environmental Science and Technology*, 41: 7668–74.

CHEARI. (2013) *White Paper on WEEE Recycling Industry in China 2012*, 2013, China Household Electric Appliance Research Institute.

Chi, X. (2011) Informal Electronic Waste Recycling: A Sector Review with Special Focus on China, *Waste Management*, 31(4): 731–742.

Cobbing, M. (2008) Toxic Tech: Not in Our Backyard. Uncovering The Hidden Flows of E-Waste. Report from Greenpeace International. www.greenpeace.org/seasia/th/Global/seasia/.../2/e waste-report-full.pdf. *Accessed September 1, 2013.*

Davis, G., Herat, S. (2010) Opportunities and Constraints for Developing a Sustainable E-Waste Management System at Local Government Level in Australia, *Waste Management and Research*, 28, 705–713.

EU (2003) European Union WEEE Directive 2002/96/EC on Waste Electrical and Electronic Equipment (WEEE); European Commission. *Official Journal of the European Union: Luxembourg*, 2003; pp 24–38.

Ferrer, G. (1997) The Economics of Personal Computer Remanufacturing, *Resources Conservation and Recycling*, 21, 79–108.

Finlay, A., 2005, *E-Waste Challenges in Developing Countries: South Africa Case Study*. APC Issue Papers. Association for Progressive Communications. www.apc.org.

Grant, K., Goldizen, F.C., Sly, P.D., Brune, M., Neira, M., Berg, M. and Norman, R.E. (2013) Health Consequences of Exposure to E-Waste: A Systematic Review, *Lancet Global Health*, 1: 350–61.

Grant, R., Oteng-Ababio, M. (2012) Mapping the Invisible and Real "African" Economy: Urban E-Waste Circuitry, *Urban Geography*, 33: 1–21.

Ha, N.N., Agusa, T., Ramu, K., Phu, N., Tu, C., Murata, S., Bulbule, K.A., Parthasaraty, P., Takahashi, S., Subramanian, A., Tanabe S. (2009) Contamination by Trace Elements at E-Waste Recycling Sites in Bangalore, India, *Chemosphere*, 76: 9–15.

Harvey, G. (2013) The Potential of Urban Mining, *The Diplomat*, November 19, 2013, Accessed July 27, 2015 http://thediplomat.com/2013/11/the-potential-of-urban-mining/.

Haugen, H.M. (2002) Economic Forces Driving the Electr(On)Ic Industry Towards Sustainability. In *Proceedings of CARE INNOVATION*, Vienna. http://aix.meng.auth.gr/helcare/CARE02/Ordnerstruktur/Paper%20Presentation/0.10.pdf.

Hoang, A., Tseng, W., Viswanathan, S. and Evans, H. 2009. Life cycle assessment of a laptop computer and its contribution to greenhouse gas emission. Paper presented at the 8th World Congress of Chemical Engineering, Quebec, Canada, 23–7 Aug. 2009.Retrieved July 7, 2014 from www.docstoc.com/docs/75386698/Life-Cycle-Assessment-of-a-Laptop-Computer-and-its-Contribution-to greenhouse gas emission.

Huo, X., Peng, L., Xu, X., Zheng, L., Qiu, B., Qi, Z., Zhang, B., Han, D. and Piao, Z. (2007) Elevated Blood Lead Levels of Children in Guiyu, an Electronic Waste Recycling Town in China, *Environmental Health Perspectives*, 115: 1113–1117.

Ishii, K. (1998) Design for Environment and Recycling: Overview of Research in the United States. *CIRP 5th International Seminar on Lifecycle Engineering*, September 16–18, 1998, 1–10.

ITU (International Telecommunication Union). (2012) Measuring the Information Society. http://www.itu.int/en/ITU-D/Statistics/Documents/publications/mis2012/MIS2012_without_Annex_4.pdf. Accessed April 14, 2016.

Jhin, P. (2007) Collaboration Towards Information and Communications Technology for Development (ICT4D). Accessed on December 12, 2007 www.gkpcms.com/GK3/documents/07.12.12-GK3-SS4-Paul%20Jhin.pdf

Kang, H.-Y., Schoenung, J.M. (2005) Electronic Waste Recycling: a Review of U.S. Infrastructure and Technology Options, *Resources, Conservation and Recycling*, 45: 368–400.

Kernbaum, S., Franke, C., Seligar, G. (2006) Flat Screen Monitor Disassembly and Testing for Remanufacturing. In *Proceedings of LCE 2006, 13th CIRP International Conference in Life Cycle Engineering*, 435–440.

Kerr, W., Ryan, C. (2001) Eco-efficiency Gains from Remanufacturing: a Case Study of Photocopier Remanufacturing at Fuji Xerox Australia, *Journal of Cleaner Production*, 9: 75–81.

Knoth, R., Brandstotter, M., Kopacek, B., Kopacek, P. (2002) Case Study: Getting Most Value Out Of Used IT Equipment. In *Proceeding of the International Symposium on Electronics and the Environment* 2002 IEEE, May 6–9 2002. San Francisco, 280–283.

Kumar, V., Bee, D.J., Shirodkar, P.S., Tumkor, S., Bettig, B.P., Sutherland, J.W. (2005) Towards Sustainable Product and Material Flow Cycles: Identifying Barriers to Achieving Product Multi-Use and Zero Waste. In *Proceedings of IMECE 2005. 2005 ASME International Mechanical Engineering Congress and Exposition*, November 5–11, 2005 Orlando, Florida.

Leung, A., Cai, Z.W. and Wong, M.H. (2006) Environmental Contamination from Electronic Waste Recycling at Guiyu, Southeast China, *Journal of Material Cycles and Waste Management*, 8: 21–33.

Leung, A.O.W., Duzgoren-Aydin, N.S., Cheung, K.C.,Wong, M.H. (2008) Heavy Metals Concentrations of Surface Dust from e-Waste Recycling and Its Human Health Implications in Southeast China, *Environmental Science and Technology*, 42: 2674–2680.

Li, J.H., Gao, S., Duan, H.B., Liu, L.L. (2009) Recovery of Valuable Materials from Waste Liquid Crystal Display Panel, *Waste Management* 29:2033–2039.

Lopez, B.N., Man, Y.B., Zhao, Y.G., Zheng, J.S., Leung, A.O.W., Yao, J., Wong, M.H. (2011) Major Pollutants in Soils of Abandoned Agricultural Land Contaminated by e-Waste Activities in Hong Kong, *Archives of Environmental Contamination and Toxicology*, 61: 101–114.

Manhart, A., Amera, T., Belay, M. (2013) E-waste Country Study Ethiopia, StEP Green Paper, United Nations University/StEP Initiatives.

Manhart, A., Osibanjo, O., Aderinto, A., Prakash, S. (2011) Informal E-waste Management in Lagos, Nigeria - Socio-Economic Impacts and Feasibility of International Recycling Co-operations. Report of Component 3 of the UNEP SBC E-waste Africa Project for Nigeria. Institute for Applied Ecology and Basel Convention Coordinating Centre for Africa (BCCC-Nigeria), Freiburg/Germany & Ibadan/Nigeria.

MoEF. (2008) Guidelines for Environmentally Sound Management of E-waste. (As approved by Ministry of Environment and Forestry (MoEF), Letter No. 23–23/2007-HSMD, Delhi, India. March 12, 2008.

Montrose, P. (2011) Think before you Shuffle, News to Use (Wellcorps International). Available: http://www.wellcorps.com/Think-Before-You-Shuffle.html [5 January. 2012].

Ni, H.-G., Zeng, E.Y. (2009) Law Enforcement and Global Collaboration are the Keys to Containing E-Waste Tsunami in China, *Environmental Science and Technology*, 43: 3991–3994.

Nnorom, I.C, Ohakwe, J., Osibanjo, O. (2009) Survey of Willingness of Residents to Participate in Electronic Waste Recycling in Nigeria — A Case Study of Mobile Phone Recycling, *Journal of Cleaner Production*, 17(8): 1629–1637.

Nnorom, I.C. (2012a) Bridging the Digital Divide and Creating an ICT Dump: An Overview of the Unsustainability if Exporting Used and End-if-Life ICT-Wares to Developing Countries. In *E-Waste: Management, Types and Challenges*, ed. Y.C. Li and B.L. Wang. Nova Science Publishers, New York, 67–88.

Nnorom, I.C. (2012b) Nigeria: Remanufacturing in the Information and Communication Technologies (ICT) Sector, for Green Economy and Trade Opportunities. United Nations Environment Program, International Trade Centre. International Centre for Trade and Sustainable Development. http://www.unep.org/greeneconomy/Portals/88/documents/research_products/Tradeopportunities/Nigeria%20case%20study.pdf

Nnorom, I.C. (2013) Manufacturing: Case Study: Nigeria. Contribution to United Nations Environment Programme Report on: Green Economy and Trade Opportunities. UNEP; International Trade Centre, International Centre for Trade and Sustainable Development (ICTSD), October 9, 2012. http://www.unep.org/greeneconomy/ResearchProducts/GreenEconomyandTradeOpportunities/tabid/104305/language/en-US/Default.aspx Published in: UNEP (2013) Green Economy and Trade: Trends, Challenges and Opportunities, http://www.unep.org/greeneconomy/GreenEconomyandTrade. http://www.unep.org/greeneconomy/portals/88/getreport/pdf/fullreport.pdf

Nnorom, I.C., Osibanjo, O. (2008) Overview of Electronic Waste (E-Waste) Management Practices and Legislations, and Their Poor Applications in the Developing Countries, *Resources Conservation Recycling*, 52: 843–858.

Nnorom, I.C., Osibanjo, O. (2009a) Toxicity Characterization of Waste Mobile Phone Plastic, *Journal of Hazardous Materials*, 161: 183–188.

Nnorom, I.C., Osibanjo, O. (2009b) Heavy Metal Characterization of Waste Portable Rechargeable Batteries Used in Mobile Phones, *International Journal of Environmental Science and Technology*, 6(4): 641–650.

Nnorom, I.C., Osibanjo, O., Kanno, O., Onwughara, N, Chukwuma, R.C. (2010) Evaluation of Heavy Metal Release from the Disposal of Waste Computer Monitors at an Open Dump, *International Journal of Environmental Science and Development*, 1(3): 227–233.

Nnorom, I.C., Osibanjo, O., Nnorom, S. (2007) Achieving Resource Conservation in Electronic Waste Management: a Review of Options Available to Developing Countries, *Journal of Applied Science*, 7(20): 2918–33.

Nnorom, I.C., Osibanjo, O., Ogwuegbu, M.O.C. (2011) Global Disposal Strategies for Waste Cathode Ray Tubes, *Resources Conservation Recycling*, 55(3): 275–290.

Ogundiran M.B., Olujobi, T., Osibanjo O. (2014a) Composition and Management of Rechargable Electric Torch Wastes in Ibadan, Nigeria, *Journal of Material Cycles and Waste Management*, 16(1): 115–123.

Ogundiran, M.B., Oyetade, A., Babayemi, J.O., Osibanjo, O. (2014b) Potential Environmental Hazards of Non-rechargable Electric Torch Wastes in Nigeria, *International Journal of Environment and Waste Management*, 13(2): 115–130.

Ogundiran, M.B., Tordue, G.B., Babayemi, J.O., Osibanjo, O. (2015) Waste Rechargeable Electric Lamps: Characterisation and Recovery of Lead from their Lead-acid Batteries, *Journal of Material Cycles and Waste Management,* May 2015. DOI: 10.1007/s10163-015-0394-7.

Ogungbuyi, O., Nnorom, I.C., Osibanjo, O., Schluep, M. (2012) E-waste Country Assessment Nigeria. Report of Components 1 & 2 of the UNEP SBC E-waste Africa Project for Nigeria. Basel Convention Coordinating Centre for Africa (BCCC), EMPA. Available at http://www.ewasteguide.info/Ogungbuyi_2012_BCCC-Empa.

Ogungbuyi, O., Nnorom, I.C., Osibanjo, O., Schluep, M. (2012) E-waste Country Assessment Nigeria, e-waste Africa Project of the Secretariat of the Basel Convention. http://www.basel.int/Implementation/TechnicalAssistance/EWaste/EwasteAfricaProject/tabi 2546/Default.aspx.

Ogushu, Y., Kandlikar, M. (2007) Assessing Extending Producer Responsibility Laws in Japan, *Environmental Science and Technology*, 41(13): 4502–4508.

Organisation for Economic Cooperation and Development (OECD). (2001) Extended Producer Responsibility: A Guidance Manual for Governments. Paris OECD; 2001. Avaialable at http://www.oecd-ilibrary.org/environment/extended-producer-responsibility_9789264189867-en

Osibanjo O. (2009b) Electronic Waste: a Major Challenge to Sustainable Development in Africa. Paper presented at *R'09 Twin World Congress and World Resources Forum*, Davos, Switzerland. September 14–16, 2009.

Osibanjo, O. (2009a) E-Waste Control Along the Supply Chain: Challenges and Prospects. Paper presented at *National Environmental Standard Regulation and Enforcement Agency (NESREA) International Conference on E-Waste Control,* Abuja-Nigeria, July 20–21, 2009.

Osibanjo, O. (2015) Gender and E-waste Management in Africa, in *Gender Heroes: From Grassroots to Global Action: a Collection of Stories Featuring Gender Perspectives on the Management of Hazardous Wastes.* Basel Rotterdam and Stockholm Conventions Secretariat (BRS), April 2015.

Osibanjo, O., Nnorom, I.C. (2007) The Challenge of Electronic Waste (E-waste) Management in Developing Countries, *Waste Management Research*, 25: 489–501.

Osibanjo, O., Nnorom, I.C. (2008) Material Flows of Mobile Phone and Accessories in Nigeria: Environmental Implications and Sound End-of-Life Management Options, *Environmental Impact Assessment Review*, 28: 198–213.

Östlin, J., Sundin, E., Björkman, M. (2009) Product Life-Cycle Implications for Remanufacturing Strategies, *Journal of Cleaner Production*, 17: 999–1009.

Osuagwu, O.E., Ikerionwu, C. (2010) E-cycling E-waste: The Way Forward for Nigeria IT and Electro-Mechanical Industry, *International Journal of Academic Research*, 2(1): 142–149.

Pinto, V.N., Patil, D.Y. (2008) E-waste Hazards: The Impending Challenge, *Indian Journal of Occupational and Environmental Medicine*, 12(2): 65–70.

Puckett, J. and T. Smith (eds), (2002) Exporting Harm: the High Tech Trashing of Asia. Basel Action Network and Silicon Valley Toxics Coalition (Online). Available at http://www.ban.org/E-waste/technotrashfinalcomp.pdf (accessed 25 April 2013).

Puckett, J., Westervelt, S., Gutierrez, R., Takamiya, Y. (2005) The Digital Dump: Exporting Reuse and Abuse to Africa. BAN media release version, Seattle.www.ban.org.

Raghupathy, L., Krüger, C., Chaturvedi, A., Arora, R., Henzler, M.P. (2010) E-waste Recycling in India: Bridging the Gap between the Informal and Formal Sector. Accessed January 5, 2012, http://www.iswa.org/fileadmin/galleries/General%20Assembly%20and%20WC%202010%2011%20Hamburg/Presentations/Krueger.pdf

Ritchey, J., Mahmoodi, F., Frascatore M, Zander A (2001). Assessing the Technical and Economic Feasibility of Remanufacturing: Environmental Issues, In *Proceedings of the Twelfth Annual Conference of the Production and Operations Management Society* POM-2001, March 30-April 2, 2001, Orlando, FL.

Robinson, B.H. (2009) E-waste: An Assessment of Global Production and Environmental Impacts, *Science of the Total Environment* 408: 183–191.

Rochat, D., Hagelüken, C., Keller, M., Widmer, R. (2007) Optimal Recycling for Printed Wiring Boards (PWBs) in India. In *R'07 Recovery of Materials and Energy for Resource Efficiency*, 12, Davos, Switzerland.

SBC (2011) *Where Are WEee in Africa? Findings from the Basel Convention e-Waste Africa Programme*. Geneva/Switzerland. Secretariat of the Basel Convention.

Schluep, M., Hagelueken, C., Kuehr, R., Magalini, F., Maurer, C., Meskers, C., Mueller, E., Wang, F. (2009) *Recycling —from E-waste to Resources, Sustainable Innovation and Technology Transfer Industrial Sector Studies*. Paris, France: United Nations Environment Programme (UNEP).

Schluep, M., Terekhova, T., Manhart, A., Muller, E., Rochat, D., Osibanjo, O. (2012). Where Are WEEE in Africa? 2012 Electronics Goes Green 2012+ (EGG).

Schmidt, C.W. (2006) Unfair Trade — E-waste in Africa, *Environmental Health Perspectives*, 114: 232–235.

Sepúlveda, A., Schluep, M., Renaud, F.G., Streicher, M., Kuehr, R., Hagelüken, C., Gerecke, A.C. (2010) A Review of the Environmental Fate and Effects of Hazardous Substances Released from Electrical and Electronic Equipments During Recycling: Examples from China and India, *Environmental Impact Assessment Review*, 30: 28–41.

Simoni, M., Kuhn, E.P., Morf, L.S., Kuendig, R., Adam, F. (2015) Urban Mining as a Contribution to the Resource Strategy of the Canton of Zurich., *Waste Management*, doi:10.1016/j.wasman.2015.06.045

Sindiku, O., Babayemi, J., Osibanjo, O., Schlummer, M., Schluep, M., Weber, R., (2014) Polybrominated Diphenyl Ethers Listed as Stockholm Convention Pops and Other Brominated Flame Retardants and Heavy Metals in E-Waste Polymers in Nigeria, *Environmental Science and Pollution Research*, 22(19): 14489–14501.

Sinha-Khetriwal, D., Kraeuchi, P., Schwaninger, M. (2005) A Comparison of Electronic Waste Recycling in Switzerland and in India, *Environmental Impact Assessment*, 25: 492– 504.

Smith, T., Sonnenfeld, D.A., Naguib-, Pellow, D. (2006) Challenging the Chip: Labour Rights and Environmental Justice in the Global Electronics Industry (Philadelphia, PA, Temple University Press).

Song, Q., Li, J. (2015) A Review on Human Health Consequences of Metals Exposure to E-Waste in China, *Environmental Pollution*, 196: 450–461.

Spicer, A.J., Johnson, M.R. (2004) Third-party Demanufacturing as a Solution for Extended Producer Responsibility, *Journal of Cleaner Production*, 12: 37–45.

StEP — Solving the E-Waste Problem. (2012). What is E-waste? Accessed July 7, 2013, http://www.step-initiative.org/index.php/Initiative_WhatIsEwaste.html.

Tsydenova, O., Bengtsson M. (2009) Environmental and Human Health Risks Associated with the End of-Life Treatment of Electrical and Electronic Equipment, *Institute for Global Environmental Strategies (IGES), Integrated Waste Management and Resource Efficiency Project*, Hayama, Kanagawa, Japan: 2009, pp. 1–49. (www.iges.or.jp)

Tue, N.M., Sudaryanto, A., Minh, T.B., Isobe, T., Takahashi, S., Viet, P.H., Tanabe, S. (2010) Accumulation of Polychlorinated Biphenyls and Brominated Flame Retardants in Breast Milk from Women Living in Vietnamese E-waste Recycling Sites, *Science of the Total Environment* 408, 2155–2162.

UNEP (2006) E-waste: The Hidden Side of IT Equipment's Manufacturing and Use". *Early Warnings on Emerging Environmental Threats No. 5*, United Nations Environment Programme (UNEP), Geneva.

UNEP (2011) Study on the Possible Effects on Human Health and the Environment in Asia and the Pacific of the Trade of Products Containing Lead, Cadmium and Mercury: Draft Final. Accessed August 5, 2013, http://www.unep.org/hazardoussubstances/Portals/9/Lead_Cadmium/docs/Trade_Reports/AP/Rev_Draft%20Asia%20and%20Pacific%20Pb%20Cd%20Hg.pdf.

UNEP (2007) E-waste Volume I: Inventory Assessment Manual. United Nations Environmental Programme Division of Technology, Industry and Economics, International Environmental Technology Centre Osaka/Shiga. http://www.unep.org/ietc/Portals/136/Publications/Waste%20Management/EWasteManual_Vol1.pdf.

Wang, F., Kuehr, R., Ahlquist, D., Li, J. (2013) E-waste in China: A Country Report. StEP Green Paper Series. Accessed October 30, 2013, http://www.ehs.unu.edu/file/get/11082.pdf.

Wang, T., Fu, J., Wang, Y., Liao, C., Tao, Y., Jiang, G. (2009) Use of Scalp Hair As Indicator of Human Exposure to Heavy Metals in an Electronic Waste Recycling Area, *Environmental Pollution*, 157: 2445–2451.

Wang, Y., Tian, Z., Zhu, H., Cheng, Z., Kang, M., Luo, C., Li, J., Zhang, G. (2012) Polycyclic Aromatic Hydrocarbons (PAHs) in Soils and Vegetation Near an E-waste Recycling Site in South China: Concentration, Distribution, Source, and Risk Assessment, *Science of the Total Environment*, 439: 187–193.

Wath, Sushant B., Dutt, P.S., Chakrabarti, T. (2011) E-waste Scenario in India, *Its Management and Implications, Environmental Monitoring and Assessment* 172: 249–262.

Widmer, R., Oswald-Krapf, H., Sinha-Khetriwal, A., Scnellmann, M. and Boni, H. (2005) Global Perspectives on the E-waste, *Environmental Impact Assessment Review*, 25: 436–458.

Wu J.-P., Luo, X.-J., Zhang, Y., Luo, Y., Chen, S.-J., Mai, B.-X., Yang, Z.-Y. (2008) Bioaccumulation of Polybrominated Diphenyl Ethers (PBDEs) and Polychlorinated Biphenyls (PCBs) in Wild Aquatic Species from an Electronic Waste (E-Waste) Recycling Site in South China, *Environment International*, 34: 1109–1113.

Zhao, G., Wang, Z., Dong, M.H., Rao, K., Luo, J., Wang, D., Zha, J., Huang, S., Xu, Y., Ma, M. (2008) PBBs, PBDEs, and PCBs levels in Hair of Residents Around E-Waste Disassembly Sites in Zhejiang Province, China, and Their Potential Sources, *Science of the Total Environment*, 397: 46–47.

Zhao, G., Wang, Z., Zhou, H., Zhao, Q. (2009) Burdens of PBBs, PBDEs, and PCBs in Tissues of the Cancer Patients in the E-Waste Disassembly Sites in Zhejiang, China, *Science of the Total Environment*, 407: 4831–4837.

4

Dynamics of Metal Reuse and Recycling in Informal Sector in Developing Countries

Mynepalli K. C. Sridhar and Taiwo B. Hammed

Department of Environmental Health Sciences, Faculty of Public Health, University of Ibadan, Ibadan, Nigeria

4.1 Introduction

Metals are precious natural resources of the Earth. Metals fall naturally into two groups: light metals (densities <4) and heavy metals (densities >7). Metals play a very important role in industry, in technological and economic development, and in health and disease. The importance of iron in hemoglobin and myoglobin, cobalt and copper in coenzymes, and zinc, selenium and chromium (Cr^{3+}) in enzymes are well known. In traditional practices, wearing copper ornaments is said to reduce arthritis pain and inflammation, magnets supposedly increase circulation, relieving stiffness and pain, gold and silver supposedly enhance cognitive function and the body's ability to repair itself, and titanium, though unknown in ancient times, is said to help stabilize the body's flow of energy (many major-league baseball players have been draping themselves with titanium-coated necklaces in official team colors).

It is extremely unlikely that metals will be exhausted but it is more likely that as shortages develop, prices will rise. The Bronze Age, some 2,000 years ago, caused copper supplies to run dry. In response, the early humans recycled bronze, looked for new mines, and spent 200 years optimizing a more available but less ideal alternative — iron — to do the same job. Modern societies do not wait such long times but mine and metabolize large quantities of

Metal Sustainability: Global Challenges, Consequences, and Prospects, First Edition. Edited by Reed M. Izatt.

materials. For sustainable living, modern societies depend on material flows. The material flow has several inputs and outputs. The inputs are grouped into six classes: energy, construction minerals, industrial minerals, metals, ecosystem products and agriculture. The outputs are grouped into five categories: domestic stock, atmospheric emissions, dissipation, other wastes, and recycled materials. Metals form a major group that is environmentally sensitive, and heavy metals in particular have played significant roles in the dynamics of biological and ecosystems. Use of national material flows gives a good index for development.

Americans are known to consume materials at the rate of 50 kg per head, according to a 1990 survey. An estimate in the same period in the U.S. showed that the chemical industry generated about 350 million metric tonnes (MMT) of hazardous waste. Some of these heavy metals get into municipal wastes and also into biosolids generated from the nation's 16,024 wastewater treatment plants (WWTPs as of 1996) that provided service to 190 million people, representing 73% of the total population (258 million); this number was expected to increase to 90% of the U.S. population by 2016 as communities install sewers and as population increases in urban areas (Paul et al., 2015). The total recycled products consumed in the U.S. were about 5% of the total material flow. Recycled metals constitute over half of metals consumed in the U.S.

Metals constitute some 2% in terms of mass and they played an important role in human development due to their tensile strength, toughness, thermal and electrical conductivity properties (Wernck and Ausubel 1995). Recycling metals contributed 76.9 metric tonnes of metal valued at $14.2 billion, or 58% of apparent U.S. metal supply. But still information is scanty on beryllium, magnesium, manganese, nickel, molybdenum, and tungsten (Papp, 2001).

The world's biggest 'ship graveyard' is located at Gadani, 65 km SW of Karachi, Pakistan; here 100 ships a year are reduced into sheets and angles of metal, pipes and working machines and produce about a million tonnes of steel, fulfilling most of Pakistan's demand for metal from the construction sector (Daniel 2013).

Metal production today represents about 8 per cent of the total global energy consumption, and a similar percentage of fossil-fuel-related CO_2 emissions. Therefore, metal scrap with a density of 244 kg/m^3, recycling will help decreasing this "footprint," as it usually requires less energy to remelt 'end-of-life (EoL)' products. World War II brought out more metal into recycling (Box 1).

4.2 Science of Metals

The periodic table represents the earth's known chemical elements, including 94 naturally occurring elements and 24 synthetic elements artificially produced in extreme conditions such as in particle accelerators. The periodic table also comprises over 3,000 isotopes of 114 elements. Recently, four more — berkelium Bk (atomic number 97) and neptunium (Np, 93) and two isotopes of the element americium (Am, 95) — were discovered.

The periodic table is divided into nine families of elements with similar properties: alkali metals, alkaline metals, transition metals, other metals, metalloids, non-metals, halogens, noble gases and rare-earth metals. The rare-earth metals, along with a selection of base and precious metals, provide today's investors and active traders with a variety of trading opportunities.

Metals are defined in various ways. Substances that are crystalline when solid and characterized by opacity, ductility, conductivity, and a unique luster when freshly fractured

Box 1 World War II and Scrap Metal Demand

World War II was an industrial war, and metals were needed to build the weapons that won the war. The shortage of rubber was the most serious impediment to the Allied war effort, but metals of all kinds were needed in huge quantities. Building tanks, ships, planes, and other weapons required massive amounts of metals, more than any other war in history. An Army Sherman M-4 tank required more than 20 tons of metal. A Navy battleship needed more than 900 tons. And building the world's largest air force meant that aluminum would be needed in unprecedented quantities. Thus the Government after Pearl Harbor either cut off the supply of metal to the consumer economy or strictly rationed it. Everything from barbed wire to farm equipment was rationed. Kids were unable to get bicycles, tricycles, and pedal cars because of both the metal and the rubber. Expanding mine production took time. And increasing imports meant that ships had to be built, which also took time. There was metal that was immediately available. One estimate suggests that 1.5 million tons of scrap lay useless on U.S. farms. And there were also large quantities in the cities as well. The government urged Americans to turn in scrap metal for recycling, and schools and community groups like the Scouts across the country held scrap metal drives. Celebrities pitched in to help promote these drives. The metals that could be obtained through scrap drives included aluminum, copper, iron, nickel, steel, and tin. Given the need for aluminum for aircraft production, drives were launched for old pots. Cans and even tinfoil were collected. People saved tinfoil from gum wrappers. Often they made tin foil balls which were taken to the collection sites. A shortage of nickel developed. The Victory Key campaign was launched as some keys (especially Yale and Corbin) contained large quantities of nickel needed by the Navy. Americans got caught up in the patriotic feeling and sometimes brought historic cannons or even monuments to the collection sites that after the War they wished they had saved.

(Source: http://histclo.com/essay/war/ww2/cou/us/aod/aod.scrap.html)

are referred to as metals. Any of the large group of chemical elements, namely iron, gold, copper, lead, and magnesium, which readily become cations and form ionic bonds having relatively free valence electrons (electrons in the outer shells) are called metals. Metals are generally good conductors of electricity. Metals also generally conduct heat well, and in solid form are malleable and ductile compared to other solids. All metals except mercury are solid at room temperature. The toxicity or extractability of metals depends on the partitioning properties such as water soluble, ion exchangeable, associated with carbonates, manganese oxides, amorphous iron oxides, crystalline iron oxides, sulfide minerals, and occurring as or in silicate minerals.

Metals are grouped under various categories depending on their properties (Table 1). They are grouped under the nutritional use basis and otherwise. Nutritionally essential

Table 1 *Metals in various groups and their nature*

Metal Type	Properties	Element
(a) **Base metals**	Those that oxidize or corrode easily; widely used in industry, such as copper plumbing, aluminum cans and the steel used in automobile production.	Cu, Ni, Al, Zn, Pb, Sn, Fe and steel (an alloy of Fe and C).
(b) **Alkali metals**	Found in Group 1 of the periodic table, are very reactive and do not occur freely in nature. The alkali metals are malleable, ductile, and good conductors of heat and electricity. They are softer than most other metals. Ce and Fr are the most reactive elements in this group. Alkali metals can explode when exposed to water	Li, Na, K, Rb, Ce and Fr.
(c) **Alkaline earth elements**	Metallic elements found in the second group of the periodic table. All alkaline earth elements have an oxidation number of +2 and are very reactive.	Be, Mg, Ca, Sr, Ba, and Ra.
(d) **Metalloids**	Elements found along the stair-step line that distinguishes metals from non-metals (between B and Al to the border between Po and Astatine). At, however is classified under "other metals". Metalloids have properties of both metals and non-metals. Some of the metalloids, such as Si and Ge, are semiconductors that can carry an electrical charge under special conditions. This property makes metalloids useful in computers and calculators.	B, Si, Ge, As, Sb, Te, and Po.
(e) Other metals	The 7 elements classified as "other metals" are located in groups 13, 14, and 15. These elements are ductile and malleable. They do not exhibit variable oxidation states, and their valence electrons are present only in their outer shell. All of these elements are solid and have a relatively high density.	Al, Ga, In, Sn, Tl, Pb, and Bi.
(f) Rare earth elements	These are 30 in number and are composed of the lanthanide and actinide series. These may be synthetic or man-made and are found in group 3 of the periodic table and the 6th and 7th periods. Rare earth metals are becoming increasingly important to the electronics industry and technological advancements and in electronics like cell phones, computer memory chips, cameras and e-readers. Rare earths are also a critical component of several military and defense applications, such as night-vision goggles, precision-guided weapons and stealth technology. Although none of the rare-earth metals is especially rare, extracting and processing them is challenging due to their widespread geographic distribution and the environmental concerns with processing. Rare earth metals are valuable because they are hard to get, and they have special properties of luminescence, magnetism and electronics that could be used to develop many new materials. Europium provides red phosphor for TVs and computer monitors and cerium is widely used in the glass-polishing industry.	Lanthanum Series: La, Ce, Pr, Nd (Light Rare-earths), Sm, Eu, Gd (Medium rare-earths), Dy, Ho, Er, Yb (Heavy rare-earths), Pr, Tb,Tm, Lu, Actinide Series: Ac, Th, Pr, U, Np, Pu, Am, Cm, Bk, Cf, Es, Fm, Md, No, Lr

Table 1 *(continued)*

Metal Type	Properties	Element
(g) Precious metals	These are naturally occurring metallic chemical elements that have a high luster and melting point, are softer and more ductile than other metals and are less reactive; they are used for jewelry, art, coins, dental work, medical devices, and electronics and for investment/holding purposes.	Ag, Au, Pt, and Pd

metals include cobalt, chromium III, copper, iron, manganese, molybdenum, selenium, and zinc. Metals with possible other beneficial effects are boron, nickel, silicon, and vanadium. Metals with no known beneficial effects in nutrition or human health are aluminum, antimony, arsenic, barium, beryllium, cadmium, lead, mercury, silver, strontium, and thallium.

4.3 Technosphere, Demand and Mobility of Metals

The word 'technosphere', borrowed from the European Union (2002) publication 'Heavy Metals in Waste,' implies those metals widely used in industry and technology. Figure 1 gives the dynamics of metals flow in the environment.

A study on U.S. recycling rates of about 20 products from plastic to metal showed that the one with the highest rate of recovery is lead-acid batteries, used primarily in cars. Their recovery rate is 98% compared to about 50% for aluminum cans. Recycling is considered the best route forward for elements for which demand is expected to level off in the long run. Demand for terbium and europium, for example, will likely fade as fluorescent bulbs are eventually replaced by much smaller LEDs. But for other elements, like neodymium, this cannot be the only solution. For the earbuds of modern smart phones a tiny amount of neodymium is sufficient. But a high-performance wind turbine needs about two tonnes. Recent estimates suggest that 32 tonnes of gold is locked up in all the world's cell phones and there is a "huge goldmine in our urban landfills" (Nicola 2013). It is difficult to develop economies of scale when dealing with materials only used in tiny amounts. Global demand for tellurium in 2009, for example, was just 200 metric tonnes. All of that came as a by-product of copper or gold mining. Tellurium is extremely valuable, at $145 per kilogram.

Metal and waste recycling expect to process in excess of 100,000 tonnes of non-ferrous metals annually, producing a high-grade secondary raw material that meets national and international specifications. A full range of non-ferrous grades is produced and can be loaded into containers, or delivered on pallets, in stillages or loose as required including: aluminum (turnings, radiators, extrusion, wheels, cuttings, litho, old rolled, shredded old rolled, cans), brass (rod swarf, mixed), copper (radiators, wire, tanks, heavy), cast aluminum, electric motors, miscellaneous cable, zorba, shredded armatures and shredded cable.

A shortage of rare earth metals, used in everything from electric car batteries to solar panels to wind turbines, is hampering the growth of renewable energy technologies. Researchers are now working to find alternatives to these critical elements or better ways to recycle them. Present-day thin, cheap solar panels need tellurium, which makes up a scant 0.0000001 per cent

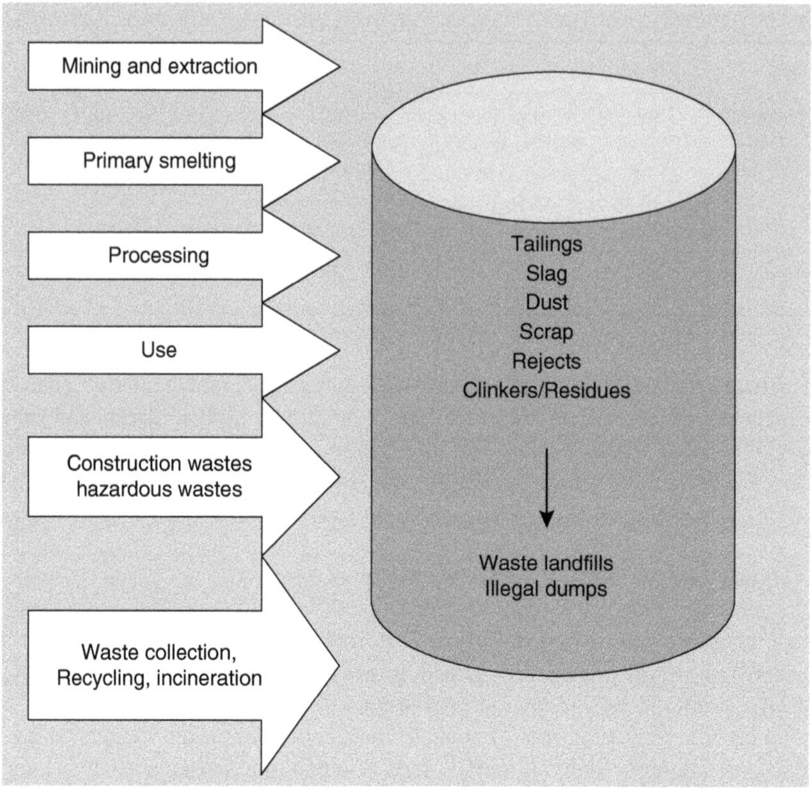

Figure 1 *Dynamics of metals flow in the environment*

of the earth's crust, making it three times rarer than gold. High-performance batteries need lithium, which is only easily extracted from briny pools in the Andes. Platinum, needed as a catalyst in fuel cells that turn hydrogen into energy, comes mainly from South Africa. Future threats exist for rare earth elements or 'critical elements' (Nicola 2013).

Sub-Saharan Africa is described as a 'technological desert' (Odedra et al. 1993), though a large variety of resources exist unexplored or underexplored. The major metal resources are shown in Figure 2. Between 1976 and 1978, Nigeria commenced the construction of two integrated iron and steel plants located at Ajaokuta (Ajaokuta Steel Company, ASC) and Aladja (Delta Steel Company, DSC) and three rolling mills at Oshogbo, Jos and Katsina. The 1.3 MTpa (metric tonnes per annum) ASC is based on blast furnace/basic oxygen furnace (BF/BOF) technology with rolling product capacity of 5.2 MTpa. DSC has a 1.0 MTpa steel melting plant for the production of 0.96 MTpa of billets and 0.32 MTpa of rolled products, while supplying 210,000 tonnes of billets each to Oshogbo, Jos and Katsina rolling mills. These projects were expected to kick-start a vibrant iron and steel sector in Nigeria. However, due to political, technical, logistical and managerial challenges, all these publicly owned iron and steel companies folded. The privately owned iron and steel companies, which are mostly rolling mills that depend on the integrated mills for billets, are now threatened due to lack of raw materials (Ohimain 2013).

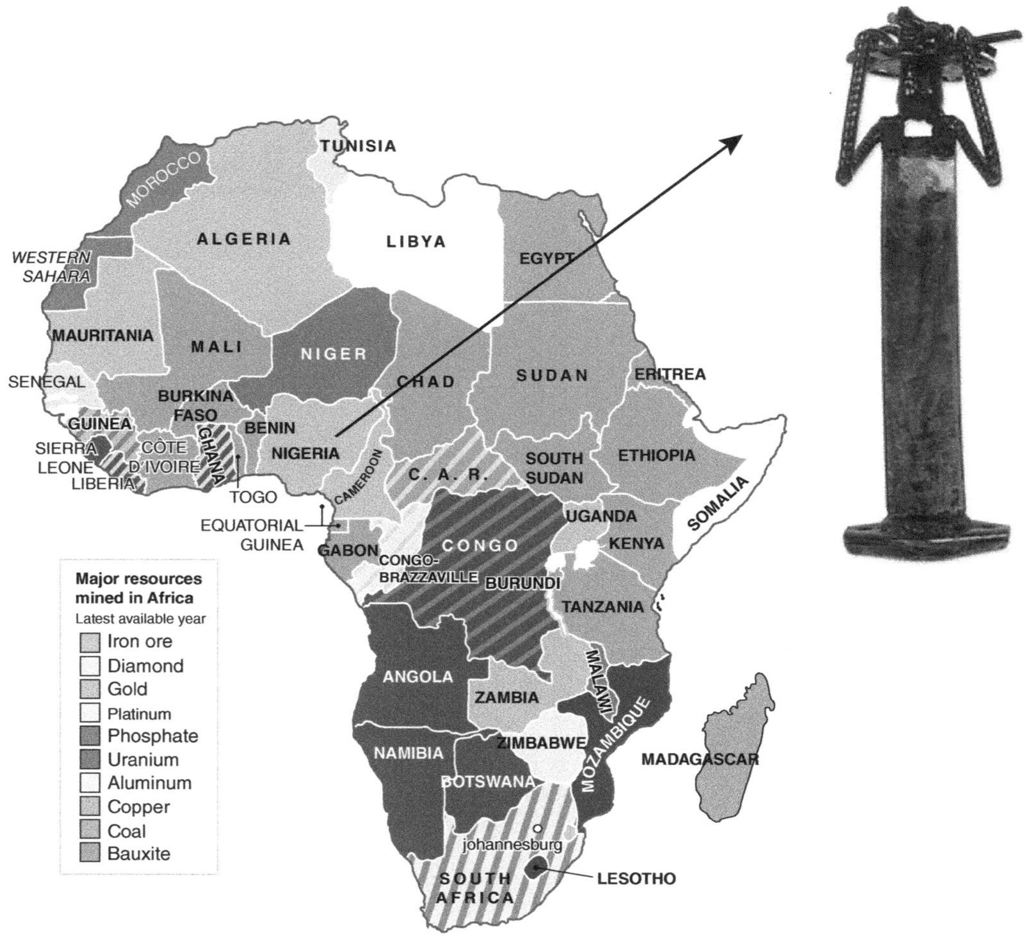

Figure 2 *The African continent showing the major resources; on the right is an art piece made from scrap metal. (Source: Raw Material Group & Ernst Y (2009). (See insert for color representation of the figure.)*

4.4 Waste Dumpsites and Treasures of Heavy Metals

Waste dumps are common in developing countries. Some of the less industrialized countries, though developing, are referred to as 'emerging markets,' countries in the process of rapid growth and development. They include the BRICs (Brazil, Russia, India, and China) and are distinguished from 'developing countries'.

4.4.1 African Countries

In African countries, waste is not segregated and very few engineered landfills exist (mainly in South Africa, Botswana and Zimbabwe), while most countries use the open dumping method. Waste from some of the East African countries with high biodegradable component show bulk density of the waste in the range 180–310 kg/m^3. Metal content of waste in some East African cities showed (%): Dar Es Salaam 3, Moshi 2, Kampala 0.3, Jinja 0.5, Lira 2.2 and Nairobi (James 2012). Some landfills have been downgraded and now function as open dumps because municipalities do not have the financial capacity to maintain them. The environment around the dumping areas is degraded in a number of ways through contamination by solid and hazardous wastes and leachates. In a study on a dumpsite in Kariba, Zimbabwe, trace metal concentrations were determined in soil samples collected from the area during 1996 and 1997. Similarly, trace elements were found concentrated by water lettuce growing in polluted waters (Sridhar 1988).

An overview of landfill practices was compiled by Johannessen and Boyer (1999) from visits and observations made during 1997–1998 on over 50 landfills in Africa, Asia, and Latin America. The presence of waste pickers on disposal sites can have a major impact on how the site is operated. Waste pickers pose a safety hazard to themselves and to landfill employees, reducing productivity by interfering with operations at the tipping face and starting fires, which cause serious air pollution. The negative impacts of scavenging have been reduced in some places by formalizing this work, either by employing waste pickers directly or by engaging contractors to do their work.

On-site scavenging disrupts landfill operations in many parts of Africa. In Accra, Ghana, waste pickers sorted through waste from incoming garbage trucks before and immediately after unloading. Waste pickers often prevented the compactor from leveling and compressing the newly disposed waste. Elsewhere in Ghana, scavenging was uncommon, as the cost of transporting recyclable materials to recycling industries in Accra and Côte d'Ivoire exceeded the value of the recyclables.

Uncontrolled scavenging at controlled and semi-controlled dumps also took place in South Africa. At the Krugersdorp landfill in the North West Province, an entire village of waste pickers had sprung up close to the dump site. More than 600 waste pickers subsisted on income generated from sorting waste, interfering with daily operations by starting fires in order to access metals and glass. Their actions prevented landfill operators from making optimal use of compactors at the tipping face. At the Bisasar Road landfill in Durban, a more controlled form of scavenging took place. Registered waste pickers living in a squatter community immediately adjacent to the landfill were allowed into the site after regular hours. This is also common in San Mateo, Philippines. Part of the tipping face remained open for the waste pickers at the end of each working day. During regular working hours, armed guards kept waste pickers out of the landfill. Scavenging at Bisasar Road generated approximately US$15,500 to support close to 200 families, equivalent to approximately US$77 per family per month.

In Gaborone, Botswana, scavengers are given permits to scavenge some specific recyclables. These permits bear code numbers: for aluminum cans 16, scrap metal 1, waste paper 10, and tyres 3. These are legal. Estimated metal collection per day were copper wire 30 kg, aluminum cans 40 kg, and scrap metal 30 kg. Battery waste is exported to neighbouring countries (Gwebu 2003).

The densities of various components of waste generated at Wa central market in Ghana gave the following (kg/m^3): organic 468, paper/cardboard 142, plastics, metals/cans 79, metals/cans 244, textiles/fabrics 152, and others 312. Thus the bulk density was found to be 261. The composition showed the following percentages: organic waste 47%, other waste 29%, paper/card board 13%, plastics 5%, metals/cans 3% and textiles/fabrics 3% (Bowan and Tierobaar, 2014).

One of the major sources of metal scrap is from importation of 5- to 15-year-old automobiles in African countries including Nigeria. Typically an automobile is composed of about 75% iron and steel, 2% of nonferrous metals, 15% of rubber and plastics and 3% glass as structural materials. These cars end up as scrap. Mazda Motor Corporation developed the world's first recycling technology for recycling bumpers from vehicles whose useful life has ended into raw plastic resin for use in new vehicle bumpers (Agbo 2011).

Metal content of wastes from selected West African countries was Ghana (Accra 2%, Kefa 5%), Ivory Coast (Abidjan 2%), Nigeria (Lagos 5%), and Senegal (Dakar 4%). In Accra, a state-owned agency is actively involved in recycling scrap metal by buying it from informal collectors and recyclers. At least six steel-melting companies process scrap steel in Accra; in Dakar, at least five companies recycle metal and non-metal scrap. Dealers located on the informal scrap yards tend to export most metal scrap to China. Many informal aluminum smelting activities are found in Ghana and Senegal. Other countries have many informal copper smelting operations, mostly for jewellery making. High prices of the metals in foreign countries have driven scrap collectors to export at the cost of local recycling. Côte d'Ivoire presently exports an estimated 100,000 tonnes of scrap metal each year, employing 226,000 people and earning around 24 million dollars in taxes for the government. The sector contributes 3.5% of the country's gross domestic product.

Accumulation of copper (Cu), lead (Pb), iron (Fe), and zinc (Zn) were found within the disposal sites (Chifamba 2007). Concentrations of Zn, Pb, and Cu were in surface soil samples up to 75 meters away from the disposal site. Leachates collected from Ibadan and Lagos dumpsites had appreciable levels of dissolved solids, chloride, ammonia, chemical oxygen demand (COD), lead, iron, copper, and manganese. This was most likely a result of rampant dumping of lead acid car batteries and metal scraps (Sridhar et al. 1992, Ikem et al. 2002). In a study at Dandora dumpsite, 42% of soil samples had ten times greater lead levels than normal (Sridhar & Bammeke, 1986, Sridhar et al., 2000, Oyaro, 2003).

Our group in the Department of Environmental Health Sciences, University of Ibadan is actively working on e-waste management (Sridhar, Taiwo, Hannah 2014 unpublished data). Here, the group has developed chemical and biological leaching techniques to release industrially important heavy metals and purify them using electrochemical/electrolytic methods. Thus extraction of copper and lead were found to be promising. These techniques will find application in preventing exposure of informal waste handlers to crude methods of tearing and burning of e-wastes.

4.4.2 Latin American Countries

Waste pickers are prevalent in the less prosperous parts of Latin America and many depend on waste recovery for their livelihood. Unorganized scavenging does not take place at any of the well-managed landfills visited. In Mexico City, 300 to 500 waste pickers are organized at the Bordo Xochiaca dump, with little opposition from landfill operators. Waste pickers occasionally set fires to recover non-combustible materials (primarily metals).

At the Zapalla landfill in Peru, two organized waste-picker cooperatives shared operations. The cooperatives selected valuable materials from the waste at the tipping front and sorted it off-site. According to municipal authorities, the waste pickers had to wear blue uniforms during operations to be allowed on the landfill premises. The municipality of Belo Horizonte, Brazil, organized waste pickers through a city programme, providing opportunities to recover reusable and recyclable materials before waste collection. Waste pickers in Rio de Janeiro were removed from the landfill's tipping front and given a designated space near the reception area of the landfill. The municipality allowed waste pickers to select trucks from high-income areas and pick from the waste to recover aluminum cans, plastic bottles, metal scrap, cardboard, and paper. The landfill operator then removed the container with the residuals from the picking belt. The waste pickers were organized into a cooperative — everyone was paid an equal amount of money at the end of each week, depending on income from the sale of recyclable materials. The average income for a waste picker was R$400 in June 1997, well above Brazil's minimum wage of R$106.

4.4.3 Asian Countries

In East Asia and Pacific Region, the presence of waste pickers was observed only at some of the not so well managed landfills. These activities presented safety issues to the waste pickers themselves and to landfill employees. In addition, waste pickers reduced productivity by hampering operations at the tipping face and sometimes set fires to separate the metals from the rest of the waste. At the Jakarta landfill, Bantar Gebang, approximately 640 registered waste pickers operate and scavenge the recyclables. As reported by Johannessen and Boyer (1999), a landfill in San Mateo, the Philippines, employs some of the inhabitants of the nearest squatter community to work at the site. Scavenging was not observed at any of the well-operated landfills visited.

4.4.4 Metals and Global Business

The earth's 118 known elements appear on the periodic table. Among these elements are base metals, precious metals and rare-earth metals, which provide a variety of opportunities for investors and traders. While gold has been revered for thousands of years and is likely to remain a popular investment, other elements like palladium and neodymium are attracting attention through a variety of trading vehicles including stocks, futures, options and exchange-traded funds. The metal stocks used in more developed countries equate to between 10 and 15 metric tonnes per citizen. Of this amount, five metals — iron, aluminum, copper, zinc, and manganese — make up more than 98% (UNEP 2013). The geology of urban mines is unpredictable and therefore realistic estimates are difficult to compute. An interesting trend in scrap processing and recycling in the U.S. is that year 2010 recorded a 26% increase in iron and steel, 41% increase in aluminum, 53% increase in copper, 4%

increase in lead and 49% increase in zinc. Sims Metal Management is the world's largest listed metals and electronics recycler. Sims Metal Management in the U.S. are global leaders in the secure and sustainable management of resources for industry, organizations and the public at large. Today, Sims Metal Management has over 250 locations on five continents, and more than 6,000 employees. In recent years, most of the processes in industrially advanced countries are going into automation because of increasing labour costs and stringent quality requirements at the global level.

The SA Metal Group, one of the largest metal recyclers in South Africa, recently commissioned the second of the three Terex Fuchs material handling units supplied by Terex Africa, Cape Town. The company collects and processes all forms of ferrous and non-ferrous scrap metal, some of which is consumed by local smelting, alloying and casting operations, while the balance is exported throughout the world (Treding Stories, 2008).

The state-owned minerals research organization Mintek and the e-Waste Association of South Africa (eWasa) started test work on the Urban Mining Project, after signing a five-year memorandum of understanding (MoU) to jointly tackle the growing challenge of metal-bearing waste in South Africa. This work will entail demonstrating the relevant high-temperature beneficiation technologies to be used in treating metal-bearing waste on a pilot scale in Mintek's pilot furnace facilities, which can treat 1 t/d of material. The Urban Mining Project, a type of mining beneficiation involving the extraction of value from any and all types of metal-bearing waste streams, will focus on the recovery and beneficiation of all sources of metal-bearing waste, including e-waste and industrial waste, as well as mine tailings and residues in South Africa, to recover a greater portion of the metal value from metal-bearing waste and residues (Chantelle 2014).

Industrial metals and bulk commodities are struggling in the face of significant demand and supply challenges. A continued weakening emerging -market growth, particularly in China, has become a "cyclical demand" problem. Some of the opportunities for investment in metal business include the following (Jean 2011):

a. **Base metals.** Investments can be made in individual companies specializing in particular base metals production, including the steel company U.S. Steel (NYSE: X) or the aluminum company Alcoa (NYSE: AA). Futures and options contracts of individual metals can be traded, such as copper futures (product symbol HG) and options (product symbol HX) on CME Globex. In addition, a wide variety of base metal exchange-traded funds (ETFs) exist, such as PowerShares DB Base Metal Fund (NYSE: DBB), which is composed of futures contracts on aluminum, zinc and copper; SPDR Metals & Mining ETF (NYSE: XME), which is made up of companies involved in the metals and mining industries; and iShares Dow Jones U.S. Basic Materials (NYSE:IYM), which consists of companies involved in the production of basic materials.
b. **Precious metals.** Investors can trade stocks, futures, options, mutual funds and exchange-traded funds based on precious metal holdings. Examples of stocks include Eldorado Gold Corporation (NYSE: EGO) and Agnico-Eagle Mines Limited (NYSE: AEM), both Canadian-based gold producers. CME Group offers investors a choice of gold futures and options contracts. The full-sized contract is based on 100 troy ounces (GC), while the "miNY" gold contract is 50 troy ounces (QO), and the e-micro gold contract is 10 troy ounces (MGC). Gold options are also available in a contract size of 100 troy ounces (OG). Futures and options contracts are also available for silver,

platinum and palladium. Platinum group metals (PGM), essentially platinum and palladium, have recently outperformed gold and silver after underperforming relative to gold since the credit crisis as the auto industry collapsed and was bailed out by Washington. Weak demand and volatility in the auto sector caused the industrial side of the PGM market to be in surplus.

c. **Rare earth metals.** Investors can gain exposure to rare-earth metals through exploration and processing companies such as Freeport-McMoRan (NYSE: FCX) and Thompson Creek Metals Company (NYSE: TC). Market Vectors Rare Earth Strategic Metals (NYSE: REMX) is a popular exchange-traded fund with holdings in rare-earth metal companies.

The noble metals are resistant to oxidation and corrosion. The following metals are considered noble metals (in order of increasing atomic number): ruthenium, rhodium, palladium, silver, osmium, iridium, platinum, and gold. In recent years tungsten, tantalum and rare earth elements are emerging as global critical metals and China is leading.

4.5 Scrap Metal and Consumer Markets

Open dumping will continue to be the way of disposing waste by municipalities in Africa due to their inability to construct engineered landfills. Scavenging in developing countries provides livelihoods for the poor. It is a survival strategy for low-income communities to cope with scarcity. The dynamics of consumer market and consumer behavior must be understood in evaluating the success. The consumer buys goods and services for personal consumption, whereas the consumer market addresses all individuals and households who buy and acquire goods. Nigeria imports large amounts of used vehicles and refrigerators completely knocked down spare parts, and as a result, large quantities of scrap waste are generated and are mixed with normal waste. Nigerian studies indicate that the amount of metal in waste varies depending on the region: southwest 18.0% (Olanrewaju & Ilemobade 2009), southeast 10.8% (Ayotamuno & Gabo 2004), and northwest 3-20% (Sridhar & Bammeke 1986, Nabegu 2010). The metals identified are iron and steel, copper, brass and aluminum.

Scavengers, mostly youth aged 18-40 years, collect scrap from 'dumpsites, auto-mechanic workshops, scrap vehicle markets, dustbins and public places'. The largest source of these is auto-mechanic workshops and refrigerator repair shops. Some amounts also come from metal fabricators (Sridhar et al., 1992, Adebowale & Peter 2014). The tools and equipment the scavengers use include shovel, digger, sickle-like iron rods and two-wheel carts or wheel barrows. In the waste management recycling value chain, some specialization has emerged. For instance, Adebola (2006) reported the role of five different groups in the waste recycling value chain in Nigeria including (i) cart builders (who construct cart for sales or hire), (ii) cart pushers (who collect wastes from the street and household), (iii) scavengers (who recover recyclable material from waste dumps), (iv) waste merchants (i.e. middlemen who purchase recyclable material from scavengers) and (v) recyclers (enterprises that convert recovered materials into useful products) (Figure 3). The collectors (Iii, iv and v) may be grouped into three categories:

a. **Primary**: they usually go everywhere to collect any available scrap and are exposed to environmental hazards; they are paid a small amount of money based on weight of the metal.

Figure 3 Cooking stoves or braziers in scrap metal for wood or charcoal. (Source: https://commons.wikimedia.org/wiki/File:Cooking_stoves_or_braziers_in_scrap_metal_for_wood_or_charcoal.jpg)

b. **Secondary**: they partly finance the primary collectors in the form of hiring tools, transport vehicles, and they supervise 4-6 primary collectors; they also own small depots of capacity less than 10,000 tonnes of metal.
c. **Tertiary**: these are merchants and are accredited suppliers to the needy industries, which are usually small and medium entrepreneurs; they may have about 8-12 workforce sorting the scrap into various categories — iron, copper, aluminum, tin and brass. Some of them also own trucks to transport bulk waste.

Adebola (2006) further reported that the informal private sector is involved in virtually all aspects of integrated solid waste management including waste collection, transportation, recovery, recycling and merchandising of both recovered and recycled materials. The informal solid waste recycling has provided employment for men, women, youth and even children. Among the secondary users, itinerant artisans buy the scavenged metals and use for secondary processing such as stoves, lamps, pans, boxes, fence poles, etc. Some women's groups also use it for handicrafts like toys, decorative pieces, etc. In this hierarchy, the value chain increases and is governed by demand and supply and quality and specifications required by the end user.

A method was described by a South African group (Potgieter 2004) to selectively recover noble metals, silver, gold and platinum from jewelery and laboratory wastes and produce technical grade material suitable for training purposes. Nitric acid was employed to dissolve and separate silver and other impurities from gold and platinum, which were then dissolved in aqua regia and precipitated selectively. An increased liquid-to-solid ratio enhanced the recovery of the silver. Recovery yields of 97.1%, 99.9% and 99.4% for silver, gold and platinum, respectively, were obtained. The recovered silver, gold and platinum had purities of 98.3%, 99.1% and 99.0%, respectively.

In Yaoundé, Cameroun, scrap metal scavenging is a roaring business. In this country, with its 3.5 million population, the generation of waste increased from 500 tons in 1999 to 2000 tons in 2014. Metals are picked up from streets, dug-up soils and electronic wastes such as home electronics, car batteries, surplus materials, cardboard, plastics, glass, metals, zinc, building supplies and engine spare parts. Young waste pickers sift through the city's dumping grounds, companies' backyards and workshops for metal that they sell to smelters and dealers. The income generated per day ranges between 2000 CFC (USD $4) and 20,000 CFC (USD $40), depending on the type and quality of metal. The separation and sale of recyclables from mixed waste provides a livelihood to, and affordable goods for, marginalized urban populations. Officials of the Ministry of Mines advocate recycling and they believe that each ton of steel that is recycled can save 1.5 tons of iron ore, 5 tons of coal, 40 percent of the water used in production, and 75 percent of the energy needed to make steel from raw material. It also saves 1.28 tons of solid waste and reduces air emissions by 86% and water pollution by 76% compared to making steel from raw material. "Recycling 1 kg of aluminum saves up to 6 kg of bauxite, 4 kg of chemical products and 14 kWh of electricity, enough energy to light many homes for several hours" (Monde 2014).

Across Africa, local scrap metal workshops recycle all sorts of thin metal scrap into charcoal or wood-burning stoves of various shapes, styles and sizes (Figure 3). In West Africa, the small size is used for small teapots used for the popular mint tea. Lightweight and with a ring, it can be easily carried, even to and from any daily job to provide a handy means for a tea fire on breaks. The large size is used for grilling fish or meats. Even city families with modern ovens keep one handy for recipes that call for grilling over a fire and one for mint tea. They wear out somewhat quickly, so there is always a steady demand for replacements.

South Africa has a well articulated Second-Hand Goods Act of 1955. Fine Trading, the scrap metal industry in South Africa, is the main source of income for nearly 500,000 South Africans. They must abide by the Act, which stipulates:

- Any person selling scrap must be the legal owner of that scrap;
- Every person intending to sell scrap must present proof of identity;
- They will take copies of that identification, as well as vehicle registrations and contact details;
- They will keep the scrap with a unique identifier separately in a secure demarcated area for 7 days, during which time any relevant authority may inspect the goods;
- If suspected that the supplier is not the rightful owner they will immediately contact the police;
- They will not buy burned cable (e.g. copper) unless the supplier can provide a valid reason for having burned cable in his or her possession;
- Scrap vehicles must be accompanied by a scrapping certificate;
- All the scrap procured is made available to local consumers first, with the excess exported.
- Scrap generated by any legitimate business and sold on a company Invoice is generally exempt from any of these provisions.

In the West African scenario, scrap collectors and scrap recycling are found to be as follows:

- Scrap metal workers are typically not registered and do not pay any taxes (informal sector)
- Scrap metal workers are typically poor migrants from rural areas of the Sahel region
- Refurbishing and scrap metal sector provide many jobs informally
- Environmental hot spots are also in the scrap metal sector
- Environmentally sound recycling cannot compete with crude recycling
- Prices for non-reparable equipment in Lagos 2010 were: CRT-monitor 50 Naira (US$ 0.34), refrigerator 100 Naira (US$ 0.67), desktop PC 100 Naira (US$ 0.67); informal collectors offer a convenient pick-up, serve and pay money for e-waste.

Ghana is the world champion in e-waste collection, achieving collection rates of 95%.

Nassar et al (2015) made a comprehensive "companionality" evaluation for 62 metals, indicating for each the global percentage mined as a companion (Figures 4 and 5). The dynamic nature of companionality (the degree to which a metal is obtained largely or entirely as a by-product of one or more host metals from geologic ores), risk factors for companion metals, and an economic evaluation of companionality, were evaluated based on mineral supply and availability. The dependence of companion metal availability on the production of the host metals introduces a new facet of supply risk to modern technology. The evaluation for 62 different metals and metalloids showed that 61% (38 of 62) have companionality greater than 50%. Eighteen of the 38 — including such technologically essential elements as germanium, terbium, and dysprosium — are further characterized as having geopolitically concentrated production and extremely low rates of end-of-life recycling. It is this subset of companion metals — vital in current technologies such as electronics, solar energy, medical imaging, energy-efficient lighting, and other state-of-the-art products — that may be at the greatest risk of supply constraints in the coming decades.

The degree of companionality varies greatly among different metals. In the case of indium, for instance, typical concentrations in the zinc host ore are only a few parts per million, too low to mine for indium itself. Thus, nearly all indium production occurs as a by-product of zinc (with much smaller amounts from tin and copper ores, also as a by-product). In contrast, silver sometimes has deposits that are sufficiently rich to permit direct silver mining. More often, however, silver is a minor constituent in the ores of host metals, especially zinc, lead, copper, and gold, and it is from those ores that most silver (about 71% of total global production) is derived. Companionality is intertwined with the complexities of supply and demand of metals. The supply of a companion metal is often not significantly influenced by changes in its demand bur the production of companion metals is dependent on their host(s)' supply(ies), rendering companion metal supply relatively price inelastic.

4.6 Export of Metal Scrap

Many countries are exporting scrap metals. Germany, USA, Russia, Ukraine, France, UK, Japan and the Netherlands are the leading scrap iron exporting countries, while Turkey, South Korea, Spain, Belgium, Luxemburg, China and Italy are the major importers (International Iron and Steel Institute, 2001; Fenton, 2001). In Detroit, USA, in the past, scrap dealers mostly dealt with customers fairly informally. They had to record the identity

Table (UNEP, 2011b).

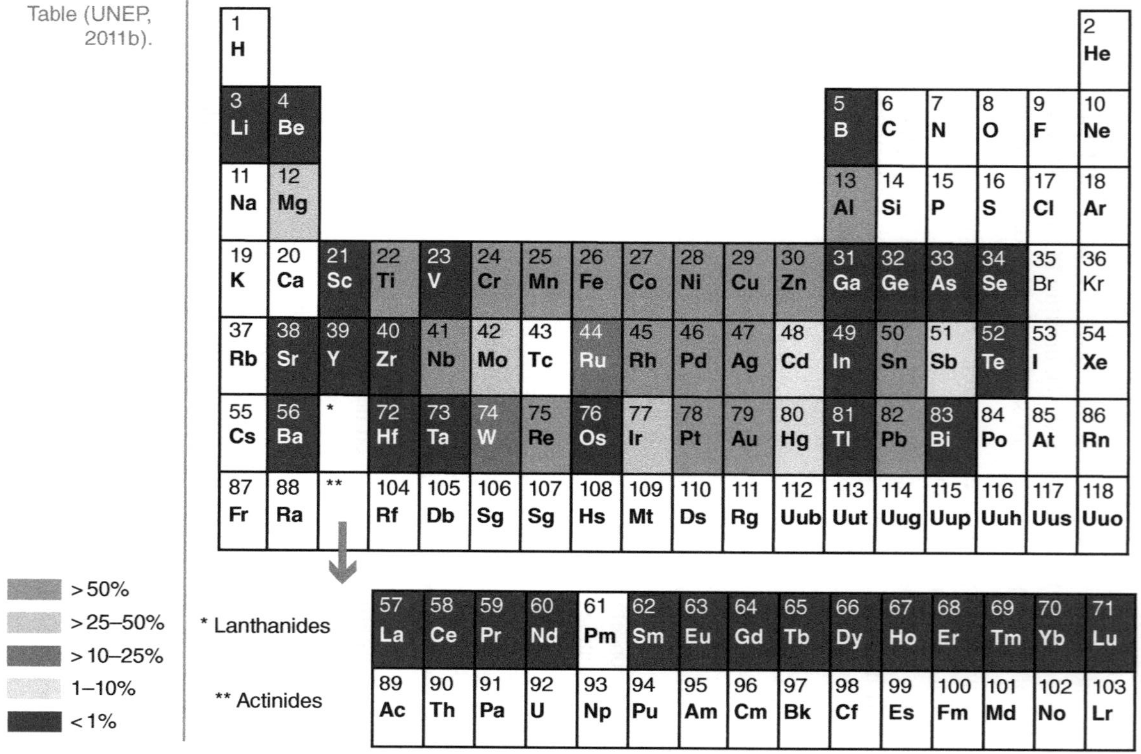

Figure 4 *Recycling rates of various metals at global level. (Source: http://www.unep.org/resourcepanel/Portals/24102/PDFs/Metal_Recycling_Full_Report.pdf) (See insert for color representation of the figure.)*

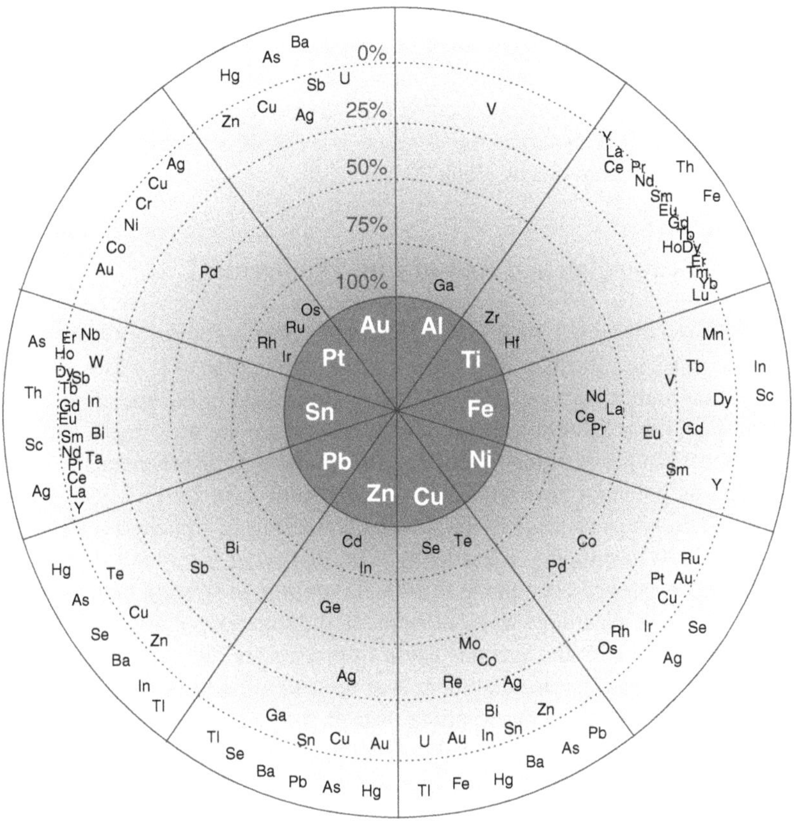

Figure 5 Companionality evaluation of selected metals. (Source: Nassar 2015. http://advances.sciencemag.org/content/1/3/e1400180)

and fingerprints of most of the people bringing them scrap, but they could still pay in cash instantly. Under a new law, scrappers are required to make any payment of \$25 or more by post, sending either a cheque or a bank card which can be used only at an on-site machine fitted with a camera (Anon 2015). Due to high demand for metal, several countries in West and Sub-Saharan Africa, e.g. Botswana, Nigeria and others, have introduced stringent procedures to export any metal waste out of the country. Scrap metal trade bans are rarely employed as a strategy for the prevention of scrap metal theft. However, the economic context of developing countries may warrant this stringent approach. Limited infrastructure and limited resources imply, in relative terms, larger economic and social consequences to immoderate levels of scrap metal theft than obtains in wealthier countries. In Botswana, by law, all scrap dealers seeking to export ferrous and non-ferrous metal waste and scrap are advised to register and obtain an application form from the Department of Industrial Affairs in the Ministry of Trade and Industry; this instrument took off from April 1, 2015 (BOPA 2015). Between 2000 and 2010, the EU has imported considerable quantities of metalliferous ores and metal scrap from Guinea and Sierra Leone (64.9% and 32.2%) and gold and monetary ores from Burkina-Faso (62.9%) (Nicolas 2012).

Ghana in 2010 lost an amount of GH¢ 60,512,100 through the illegal export of 110,022 metric tonnes of ferrous scraps, despite a ban placed on it by the Ministry of Trade and Industry in 2004. The shipping lines in the country succeeded in exporting 5,001 containers of the scrap in 2010 alone. Local steel companies are currently producing under capacity, due to the inadequacy of raw materials, which is a result of the increase in the illegal exportation of ferrous scrap metals (John 2012).

4.7 E-waste Scavenging and End-of-Life Management

E-waste comprises discarded electronic appliances, mostly computers and mobile telephones which have short lifespan. The current global production of E-waste is estimated to be 20–25 million tonnes per year. Most E-waste is produced in Europe, the United States and Australasia. China, Eastern Europe and Latin America are growing to become major E-waste producers in the next ten years. E-waste contains valuable metals (Cu, platinum group) as well as potential environmental contaminants, especially Pb, Sb, Hg, Cd, Ni, polybrominated diphenyl ethers (PBDEs), and polychlorinated biphenyls (PCBs). It is estimated that by 2030 developing countries will be disposing of around 400–700 million outdated computers per year, compared to 200–300 million in Western nations. Despite the international treaty on hazardous waste movement, the shipment of e-waste especially into Africa is still very much on the agenda. These shipments are often disguised as donations or used equipment. Usually this equipment is at the end of its lifespan and only a few are used or refurbished for further consumption; instead, they are dismantled for the valuables, often by children who are exposed to the dangerous chemicals in these items (Finland Team 2014). Of the E-waste, only 5% is recycled formally whereas 95% goes through the informal sector.

Cathode ray tubes (CRTs) used as computer and TV displays contain large quantities of lead, estimated at between 0.5 and 4 kg depending on the size of the CRT, and have been identified as the most polluting of all electronic waste components (Nnoroma et al. 2011). The five West Africa countries Benin, Côte d'Ivoire, Ghana, Liberia, and Nigeria generate between 650,000 and 1,000,000 tons of domestic E-waste each year, which must be managed to protect human health and the environment in the region (Figure 6, Box 2). Domestic consumption makes up the majority (up to 85%) of waste electronic and electrical equipment (WEEE) produced in the region, and used equipment is also imported from industrialized countries, significant volumes of which prove unsuitable for re-use (EMPA 2012).

The CO_2 emission potential of the primary metals from E-waste is (metric tons): copper 15.3, cobalt 0.08, tin 1.45, indium 0.05, silver 0.86, gold 5.10, palladium 0.30, platinum 0.18 and ruthenium 0.08). It is recommended that e-waste materials containing lead should be recovered and separated for hand-dismantling and mechanical extraction of old solder (lead) by use of soldering iron. The lead-extract should be treated in plastic (PVC) reactors to get lead (II) oxide for recycling to the industry (Onyenekenwa and Jonah 2011).

Jirang and Lifeng (2008) compared various extraction methods for metal recovery. Pyrometallurgical processing is a traditional technology for recovery of precious metals from waste electronic equipment. However, state-of-the-art smelters are highly dependent on investments. Recent research on recovery of energy from PC waste gives an example of using plastics in this waste stream: it indicates that thermal processing is a feasible approach

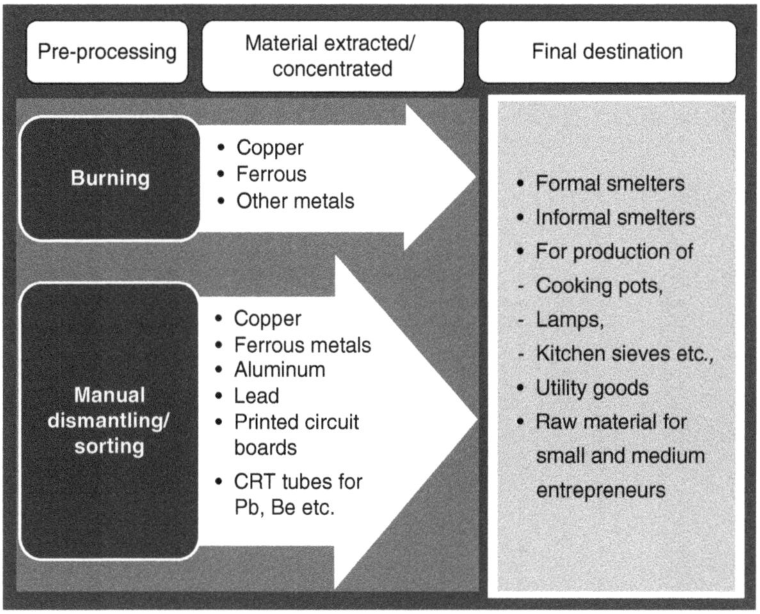

Figure 6 *E-waste scavenging and processing procedures for metals*

Box 2 Narrative by a High-Tech Trash Collector, Ghana

Each day scrap sellers bring loads of old electronics from.... They break copper yokes off picture tubes, littering the ground with shards containing lead, a neurotoxin, and cadmium, a carcinogen that damages lungs and kidneys. They rip resalable parts such as drives and memory chips. Then they rip out wiring and burn the plastic. He sells copper stripped from one scrap load to another. The key to making money is speed, not safety. "The gas goes to your nose and you feel something in your head"... "Then you get sick in your head and your chest. Nearby, hulls of broken monitors float in the lagoon. According to USEPA, an estimated 30 to 40 million PCs will be ready for "end-of-life management" in the next few years.

(Source: *Chris Carroll, 2008.* http://ngm.nationalgeographic.com/2008/01/high-tech-trash/carroll-text)

for recovery of energy from electronic waste if a comprehensive emission-control system is installed. In the last decade, attention has been diverted from pyrometallurgical processing to hydrometallurgical processing for metal recovery from electronic waste. The flowchart practiced in India is shown in Figure 7.

The Basel Convention is the only global treaty controlling transboundary movements and requiring the environmentally sound management of hazardous and other wastes. E-waste is listed in Annex VIII as A1180 (hazardous waste) and Annex IX as B1110

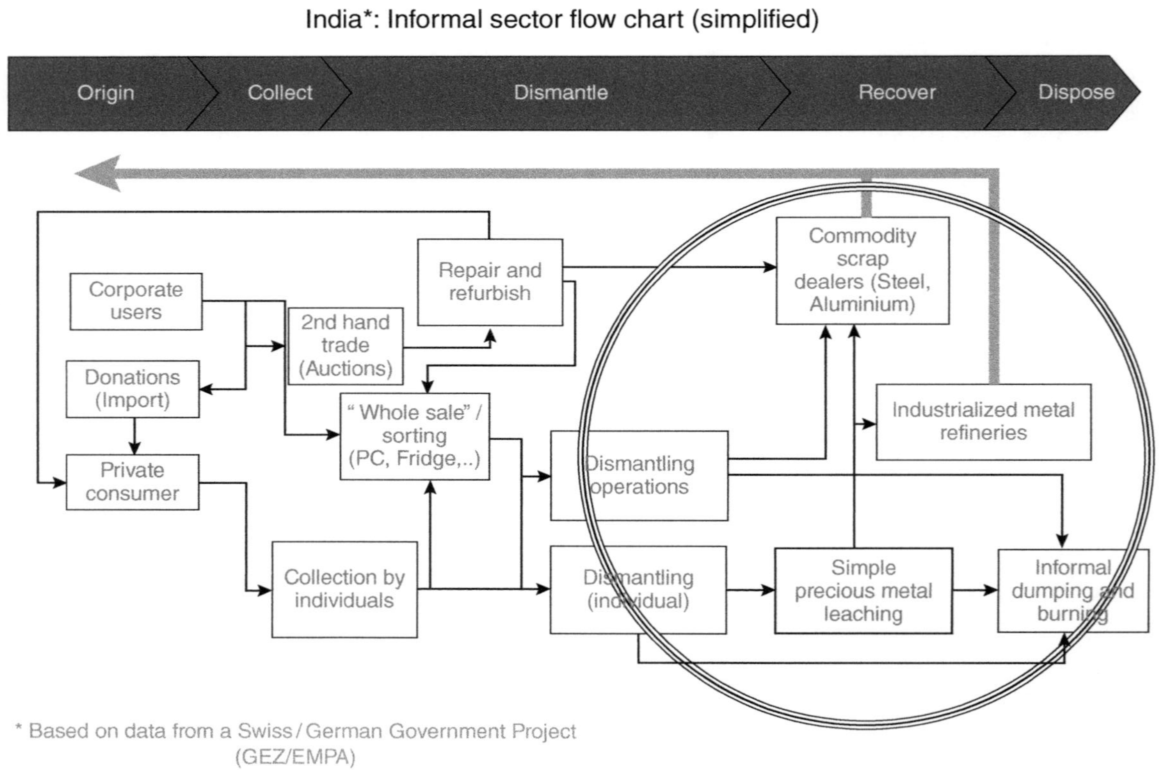

Figure 7 *Flow chart of E-waste management in informal sector in India. (Source: Klaus Hieronymi 2015. http://www.c3p.org/Workshop%20 2009/Presentations/WEEE%20and%20E-Waste%20Recovery/Klaus%20Hieronymi_Domestic%20EWaste%20in%20Africa%20Recycling.pdf)*

(non-hazardous waste) of the Convention. Specifically, e-wastes are characterized as hazardous wastes under the Convention when they contain components such as accumulators and other batteries, mercury switches, glass from cathode-ray tubes (CRTs) and other activated glass, PCB-containing capacitors or when contaminated with cadmium, mercury, lead or PCBs (BCCC-Nigeria & EMPA 2011).

4.8 Scrap Metal Theft

Scrap metal theft is a global problem. Metal theft is "the theft of items for the value of their constituent metals". It usually increases when worldwide prices for scrap metal rise, as has happened dramatically due to rapid industrialization in India and China. Apart from precious metals like gold and silver, the metals most commonly stolen are non-ferrous metals such as copper, aluminum, brass, and bronze. However, even cast iron and steel are now being taken owing to high prices as scrap. The metals most sought are copper, aluminum, brass, zinc, nickel, platinum, and bronze. These metals have value only when sold to a scrap metal dealer who arranges for the metal to be melted and reshaped for other uses. By contrast, gold and silver commonly have intrinsic value, either to the thief or to someone else who values the metal in its original shape. Theft occurs through: bicycle theft, burglary, shoplifting, stolen goods markets, theft from construction sites, theft from cafés and bars, theft of and from vehicles, and theft of outdoor household items. Vehicle parts, especially catalytic converters, are worth anywhere from $50 to $200 apiece from local scrap metal dealers due to the amounts of platinum, palladium, and radium present in them. Catalytic converters have been installed on vehicle undercarriages since 1975 and convert harmful pollutants into less-harmful emissions before they leave the vehicle's exhaust system. Replacing a catalytic converter can run anywhere from $200 to over $1,000. SUVs and jeeps provide the best opportunity for catalytic converter theft because their ground clearance is so high and often the converter is held on by four bolts, easily detached with a socket wrench.

There is an increasing number of incidents of platinum theft (in 2002 platinum traded for about $608 per troy ounce and but by 2008 it had risen to $2,083 per troy ounce on the New York Mercantile Exchange). Beer kegs have become targets because many scrap-metal dealers value them at $15 to $30 apiece, often more than the liquor store deposit. Estimates in 2007 reported a loss of $50 million to the beer keg industry due to theft. In addition, aluminum siding, gutters and roofs are targeted for their copper flashing. Bronze plaques and statues in cemeteries and museums, manhole cover and sewer grate theft has risen sharply throughout the industrialized world, and thus cities are finding innovative ways to secure the covers (Brandon 2010). Iraqi freedom looters removed "contaminated *scrap metal* and dozens of 50-gallon barrels … to yellowcake, sealed radioactive sources, and activated *metals" (Science News,* www.esciencenews.com *2015).*

Nigeria turns out to be a region of metallurgical activity dating back to ancient times (8th century BC) and demonstrating a remarkable continuity: today, ironworkers still play an active role in the country's economic, social and cultural life. The sites are spread across every region of the country. In the savanna region of northern Nigeria and indeed most parts of Nigeria and West Africa in general, iron technology has been widespread and abundant for over 25 centuries, in other words well before the coming of the Europeans (Hamady 2004).

During the colonial era, spherical cast-iron pots from different countries were introduced in different parts of Africa. These pots rapidly became popular in urban areas because of their durability, quick heating ability and retention of heat. They had handles on the top and legs at the base, making them popular for easy handling and use for cooking. They have been around for centuries.

4.9 Conclusions

The current global trend is a circular economy, which has a serious impact on the waste sector. Globally, most businesses are moving waste up the hierarchy as integrated waste management by focusing on increasing recycling rates, better waste prevention, greater waste reuse, sending zero waste to landfill sites and energy recovery. Waste recycling and recovery have become the focus of integrated waste management. Metals are an important component of waste and world's economies are revolving on effective management of this important resource with its limited lifespan.

As a UNEP report pointed out, sustainable metals management requires improving recycling rates of selected materials, and also changing the public's mindset on recycling itself. There is a need to move away from a materials-centric approach to a product-centric approach, keeping in view the changing pattern of technology and complexity of products and their interactions within recycling systems. Further, modern technology systems require effective sorting after collection, and then improved physical separation and adoption of metallurgical technologies for economically viable recovery of metals.

References

Adebola, O.O., 2006, The Role of Informal Private Sector in Integrated Solid Waste Management (ISWM) in Lagos, Nigeria, a Developing Country. *Proceedings of the 21st International Conference of Solid Waste Technology and Management.* March 26th-29th 2006, Philadelphia, PA.

Adebowale, B.A., Peter, O.O., 2014, Scrap Metal Scavenging: Consumer Market, Consumer Buying Behavior and the Environment, *Academic Research International*, 5, 270–283.

Agbo, C.O.A., 2011, Recycle Materials Potential of Imported Used Vehicles in Nigeria, *Nigerian Journal of Technology*, 30, 1–12.

Anon., 2015, Crime in Detroit: The Great Scrap-Metal Scrap, *The Economist*, Feb. 21, 2015.

Ayotamuno, J.M., Gobo, A.E., 2004, Municipal Solid Waste Management in Port Harcourt, Nigeria, *Management of Environmental Quality, An International Journal*, 15 389–398.

BCCC-Nigeria and EMPA 2011, UNEP SBC E-waste Africa Project: Building Local Capacity to Address the Flow of E-wastes and Electrical and Electronic Products Destined for Reuse in Selected African Countries and Augment the Sustainable Management of Resources Through the Recovery of Materials in E-Wastes. Contribution to components 1 and 2: Nigeria e-Waste Country Assessment. Ibadan/Nigeria and St. Gallen/Switzerland. Cited in *Where Are WEE in Africa? Findings from the Basel Convention E-waste aFriCa Programme*, Secretariat of the Basel Convention, 1-52.

BOPA, 2015, Botswana: Ministry Moves to Control Export of Metal Waste, Scrap, *Botswana Daily News*, www.allafric.com/misc, accessed May 3, 2015.

Bowan, P.A., Tierobaar, M.T., 2014, Characteristics and Management of Solid Waste In Ghanaian Markets: A Study of WA Municipality, *Civil and Environmental Research* www.iiste.org ISSN 2224-5790 (Paper) ISSN 2225-0514 (online), 6, 114–119.

Brandon, R.K., 2010, *Theft of Scrap Metal*, Guide No. 58, Center for Problem-Oriented Policies, Center for Problem-Oriented Policing, Inc. U.S. Department of Justice, 1-70.

Chifamba, P., 2007, Trace Metal Contamination of Water at a Solid Waste Disposal Site at Kariba, Zimbabwe, *African Journal of Aquatic Science*, 32, 71–78.

Chris, C., 2008, High Tech Trash, *National Geographic Magazine*, January 2008, www.nationalgeographic.com, accessed May 15, 2015.

Daniel, M., 2013, World's Biggest Ship Graveyard: Where Huge Tankers and Cruise Liners are Scrapped on the Shorefront and Workers Toil for £2 a Day, *Mailonline*, May 14, 2013, accessed May 7, 2015.

EMPA (Swiss Federal Laboratories for Materials Science and Technology) 2012, Domestic Consumption Main Contributor to Africa's Growing E-Waste Problem, http://www.ewasteguide.info/UNEP_2009_eW2R, accessed June 15, 2015.

European Commission, 2002, *Heavy Metals in Waste*, European Union DG ENV. E3, Final Report, COWI AS, Denmark, 1-86.

Fenton, M.D., 2001, Recycling Metals: Iron and Steel. In *USGS Minerals Yearbook* 2001. United States Geological Survey, United States Government Printing Office, Washington, pp. 1–195.

Finland Team, 2014, E-Waste Evolution in Africa, Embassy of Finland, *Pretoria News*, 3/10/2014.

Gwebu, T.D., 2003, Population, Development and Waste Management in Botswana: Conceptual and Policy Implications for Climate Change, *Environmental Management*, 31, 348–354.

Hamady, B., 2004, The Origins of Iron Metallurgy in Africa: New Light on Its Antiquity: West and Central Africa, Memory of Peoples, United Nations Educational, Scientific and Cultural Organization, 7, place de Fontenoy, 75732 Paris 07 SP, 1-230.

Ikem, A., Osibanjo, O., Sridhar, M.K.C., Sobande, A. 2002, Evaluation of groundwater quality characteristics near two waste sites in Ibadan and Lagos. *Water, Air and Soil Pollution*, 140,.1-4, 307–333.

International Iron and Steel Institute, 2001, *Steel Statistical Yearbook* 2001. Brussels, Belgium, 1–111.

IWWA, 2013, *Solid Waste Management Practices in Western Africa: A Report*, project funded by the European Commission within the Seventh Framework Programme (2007–2013), 1-24, project website: http://www.iwwa.eu, accessed May 10, 2015.

James, O., 2012, Solid Waste Management in African Cities – East Africa 10.5772/50241, accessed June 10, 2015.

Jean, F., 2011, *Investing in the metals market*, www.investopedia.com/corp, accessed June 15, 2015.

Jirang, C., Lifeng, Z., 2008, Metallurgical Recovery of Metals fom Electronic Waste: a Review, *Journal of Hazardous Materials*, 158, 228–256.

John, B., 2012, Local Steel Industry Collapsing as Illegal Export of Ferrous Scrap Booms and New Legislation to Stem Tide Goes AWOL, *Ghanaian Chronicle*, July 30, 2012.

Johannessen, L.M., Boyer, G., 1999, Observations of Solid Waste Landfills in Developing Countries: Africa, Asia, and Latin America, Urban Development Division, Waste Management Anchor Team, World Bank, pp. 1–47.

Klaus, H., 2015, Domestic E-waste in Africa: Recycling for a Profit? Powerpoint presentation, Chairman of the Environmental Board, Hewlett Packard EMEA, pdf, accessed May 1, 2015.

Monde, N., 2014, A Look Inside Yaounde's Scrap Metal Economy, Informality, Waste Management, Cameroon, Yaounde. Www.Urbananafrica.Net, 29 September 2014, accessed May 8, 2014.

Nabegu, A.B., 2010, An Analysis of Municipal Solid Waste in Kano Metropolis, *Nigeria, Journal of Human Ecology*, 31, 111–119.

Nassar, N.T., Graedel, T.E., Harper, E.M., 2015, By-product Metals Are Technologically Essential but Have Problematic Supply, *Science Advances*, 1, e1400180–, DOI: 10.1126/sciadv.

Nicola, J., 2013, A Scarcity of Rare Metals Is Hindering Green Technologies, Yale Environment 360 Report, Yale School of Forestry & Environmental Studies, www.environment.yale.edu, accessed May 2, 2015.

Nicolas, M., 2012, Trade Between EU and West Africa, 2000-2010 EU-27 Surplus with ECOWAS Countries for Both Goods and Services. General and Regional Statistics in Focus, External Trade Economy and Finance, *Eurostat Statistics in Focus*, 1-8.

Nnoroma, I.C., Osibanjo, O., Ogwuegbua, M.O.C., 2011, Global Disposal Strategies for Waste Cathode Ray Tubes, *Resources, Conservation and Recycling*, 55, 3, 275–290.

Odedra, M., Lawrie, M., Bennett, M., Goodman, S., 1993, Sub-Saharan Africa: A Technological Desert, *Communications of the CACM, International Perspective*, 36, 25–29.

Ohimain, E.I., 2013, The Challenge of Domestic Iron and Steel Production in Nigeria, Greener. *Journal of Business and Management Studies*, 3, 231–240.

Olanrewaju, O.O., Ilemobade, A.A., 2009, Waste to Wealth: A Case Study of Ondo State Integrated Wastes Recycling and Treatment Projects, Nigeria, *European Journal of Social Sciences*, 8, 1–9.

Onyenekenwa, C.E., Jonah, C.A., 2011, Managing Hazardous Wastes in Africa: Recyclability of Lead from E-waste Materials. *Journal of Applied Sciences*, 11, 3215–3220.

Oyaro, K., 2003, Month after Dump Scare, Problems Persist. InterPress Service News Agency (Johannesburg), 13 May 2008. www.unep.org/cpi/briefs/2008May14.doc, accessed April 20, 2014.

Papp, J.F., 2001, Recycling Metal, *US Geological Survey Minerals Yearbook*, USGS.

Paul, W., Sungyun, L.Y.Y., Gwyneth, W.G., Kiril, H., Rolf, U.H., Pierre, H. 2015, Characterization, Recovery Opportunities, and Valuation of Metals in Municipal Sludges from U.S. Wastewater Treatment Plants Nationwide, *Environmental Science & Technology*, 49, 9479–9488.

Potgieter, J.H., Potgieter, S.S., Mbaya, R.K.K., Teodorovic, A., 2004 Small-scale Recovery of Noble Metals from Jewellery Wastes, *The Journal ofThe South African Institute of Mining and Metallurgy*, 104, 563–571.

Rankokwane, B., Gwebu, T.D., 2006, Characteristics, *Threats and Opportunities of Landfill Scavenging: The Case of Gaborone-Botswana, GeoJournal*, 65, 151–163.

Raw Material Group and Ernst Y 2009, US Geological Survey, *The Economist*, www.usgeologicalsurvey,org, accessed October 10, 2012; also in African Memories, 2009, Edinburgh University Press., 7, 373–386.

Sridhar, M.K.C., 1988, Uptake of Trace Elements by Water Lettuce (*Pistia stratiotes* L.), *Acta Hydrochimica et Hydrobiologica*, DDR, 16, 293–297.

Sridhar, M.K.C., Bammeke, A.O., 1986, Heavy Metal Contents of Some Solid Wastes in Ibadan, Nigeria, *Water, Air and Soil Pollution*, USA, 29, 51–56.

Sridhar, M.K. C., Ajayi, A.A., Arinola, A.M., 1992, Collecting Recyclables in Nigeria, *Biocycle*, USA, 33, 46–47.

Sridhar, M.K.C., Olawuyi, J.F., Adogame, L.A., Okekearu, I.R., Osagie, C.O., Aborkar, L., 2000, Lead in the Nigerian Environment: Problems and Prospects, in *11th Annual International Conference on Heavy Metals in the Environment* (J. Nriagu, editor), No. 1019, University of Michigan, School of Public Health, Ann Arbor, MI (CD-Rom), 1-4.

Trending Stories 2008, Handling Scrap Metal, www.miningreview.com, September 29, accessed May 7, 2015.

UNEP 2013, Metal Recycling: Opportunities, Limits, Infrastructure, A Report of the Working Group on the Global Metal Flows to the International Resource Panel. Reuter, M.A., Hudson, C., van Schaik, A., Heiskanen, K., Meskers, C., Hagelüken, C., 1–320.

Wernck, I.K., Ausubel, J.H., 1995, National Materials Flows and the Environment, *Annual Review of Energy Environment*, 20, 463–492.

Wikimedia 2012, Cooking Stoves or Braziers in Scrap Metal for Wood or Charcoal, www.cookingstoves.jpg, accessed June 15, 2015.

5

Metal Sustainability from Global E-waste Management

Jinhui Li and Qingbin Song

School of Environment, Tsinghua University, Beijing, 100084, China

5.1 Introduction

Electrical and electronic waste (e-waste) is one of the fastest growing global waste streams. In the last decade, e-waste has become well known due to its serious threats to the environment and human health. However, e-waste is also valuable as a source for secondary raw materials, particularly metals. E-waste includes many of the metals in the periodic table, such as Fe, Al, Cu, Pb, Hg, As, Cd, Au, Ag, Pt, and Pd. E-waste has become an important source of several of these metals. As rapid economic development and technological innovation continue, the volume of e-waste will increase. Therefore, recycling of metal components of e-waste that have value will become more and more important in the attempt to increase global metal sustainability.

5.2 E-Waste Issues

E-waste is a term used to cover all types of electrical and electronic equipment (EEE) that has or could enter the waste stream. Globally, there is no standard or generally accepted definition of e-waste (Widmer et al., 2005). A number of countries have their own definitions. The most widely accepted definition is given in a European Union (EU) Directive on Waste Electrical and Electronic Equipment (WEEE Directive) (EPC, 2003), and this definition, including: (1) Large household appliances; (2) Small household appliances;

Metal Sustainability: Global Challenges, Consequences, and Prospects, First Edition. Edited by Reed M. Izatt.

(3) IT and telecommunications equipment; (4) Consumer equipment; (5) Lighting equipment; (6) Electrical and electronic tools; (7) Toys, leisure and sports equipment; (8) Medical devices; (9) Monitoring and control instruments; (10) Automatic dispensers, is followed in a number of member countries of the EU and other countries of Europe (Li et al. 2006; Yang et al. 2008). In 2011, it was estimated that global e-waste generation was growing by about 40 million tonnes a year (about 0.63 kg/year/capita). The amount generated in the EU was estimated to be 16–18 kg/year/capita) (Song et al. 2012a). The amount of e-waste generated in selected countries is shown in Table 1.

The majority of electronic equipment is sold in developed countries, such as the United States, Japan, Australia, and several European nations. It is estimated that 50% to 80% of the e-waste collected for recycling in the United States is shipped to less developed countries (Kahhat et al. 2008; Namias 2013). The UK is the dominant European exporting country, followed by France and Germany (Torretta et al. 2013). Certain regions have become centers for informal electronic waste recycling in which e-waste products, such as cell phones and computer motherboards, are dismantled manually by untrained personnel and metals of value extracted, often using hazardous chemicals and procedures. The most prominent of these regions are located in Asia and Africa, as shown in Figure 1. Informal e-waste recycling has been reported in Qingyuan, Guiyu, Fengjiang, Zhejiang, and Guangdong Provinces, China; Manila, Philippines; Bangalore and Chennai, India; Seoul, South Korea; New Territories, Hong Kong; and HaiPhong City, Dong Mai, and Bui Dau, Vietnam. Generally, the e-waste export market is quite diverse, ranging from small family-based networks to large and well organized trading firms.

Electronic devices are complex and are made of a wide variety of materials. If these devices are improperly recycled, toxic substances, such as polychlorinated dibenzo-p-dioxins and dibenzofurans, flame retardants, and toxic metals, e.g., Pb, Hg, As, Cd, and Cr(VI), can be released and cause serious environmental pollution (Duan et al. 2011; Lundgren 2012). Workers and local residents are exposed to toxic chemicals through inhalation, dust ingestion, dermal exposure and oral intake. Inhalation and dust ingestion impose a range of potential occupational hazards including silicosis. Overall, human health risks from e-waste include breathing difficulties, respiratory irritation, coughing, choking, pneumonia, tremors, neuropsychiatric problems, convulsions, coma and even death (Samarasekera 2005; Wang et al. 2011).

Table 1 *E-waste generation in selected countries*

Country	Total E-waste Generated (tonnes/year)	Year
Global	40,000,000	2011
Switzerland	66,042	2003
Germany	1,100,000	2005
United Kingdom	1,200,000	2010
USA	3,160,000	2008
Taiwan	14,036	2003
Thailand	60,000	2003
Denmark	118,000	1997
Canada	67,000	2005
India	380,000	2011
Philippines	58,000	2010
Nigeria	60,000	–

Figure 1 *Potential Global E-waste Flows. (Source: Song et al., 2014. Reproduced with permission from Elsevier)*

5.3 E-Waste Management in China

As one of the largest exporters of EEE and importers of e-waste worldwide, China plays a key role in the social, economic, environmental and material life cycles of much of the world's electrical and electronic equipment. As a result of increased Chinese and worldwide consumption and turnover of EEE, China is now facing serious e-waste problems from both growing domestic generation and foreign imports.

5.3.1 Generation and Flows

In China, e-waste comes from three major sources: households; business offices, institutions, and government offices; and original equipment manufacturers. Household appliances are the source of most household e-waste. Waste computers and other office equipment (e.g., copy machines and printers) are the major e-waste flows generated by business offices, institutions, and government offices. The upgrade rate of computers from these offices is higher than that for households. As seen in Table 2, domestic generation for five kinds of e-waste in China reached 148.97 million units, which was about 4.48 million tonnes, in 2014. In China, both domestic generation and illegal importation were important e-waste sources. From estimates by an environmental group (BAN &SVTC, 2002), around 80% of the e-waste collected for recycling in the United States was exported into Asia, and around 90% of this e-waste went to China. Attempts of the Chinese government to reduce the quantity of illegal e-waste imported have been only partially successful.

According to the traditional economic custom, Chinese people seldom discharge their used electrical and electronic products, even if these products are out of date or broken

Table 2 *Domestic e-waste generation in China*

Year		2014	2015	2016	2017	2018	2019	2020
Refrigerators	Number (million units)	12.32	13.75	18.73	23.29	35.77	42.45	54.47
	Mass (million tonnes)	0.55	0.62	0.84	1.05	1.61	1.91	2.45
Washing machines	Number (million units)	19.04	20.86	24.17	26.59	29.71	35.94	44.68
	Mass (million tonnes)	0.48	0.52	0.6	0.67	0.74	0.9	1.12
Air containers	Number (million units)	35.15	37.52	35.08	40.57	42.9	57.07	71.92
	Mass (million tonnes)	1.79	1.91	1.79	2.07	2.19	2.91	3.67
TV sets	Number (million units)	28.33	38.11	42.95	44.55	52.05	56.96	66.7
	Mass (million tonnes)	0.85	1.14	1.29	1.34	1.56	1.71	2.00
PCs	Number (million units)	54.13	57.1	59.7	61.97	63.94	65.63	67.08
	Mass (million tonnes)	0.81	0.86	0.9	0.93	0.96	0.98	1.01

Table 3 Disposal methods for unwanted electronic and electrical products in China

Categories	Ningbo/2003	Taizhou/2007	Beijing/2006	Beijing/2005	Beijing/2005
Sell to peddlers	43	34	30	56.8	49.7
Sell to second use	12	8		13.3	13.6
Return to retail or manufacturers	–	6	–	3.9	–
Direct discarding	16	5	–	16.9	3.5
Donation	8	13			26.1
Storage at home	14	26	70	9.1	7.1
Others	8	8	–	–	–

(Li et al. 2006). The owners hope that their used or broken electrical and electronic products may be useful in the future or may be sold to collectors. Table 3 shows methods for disposal of unwanted electronic and electrical products from households in China (Huang et al. 2006; Geering 2007; Yang et al. 2008; Liu et al. 2006a; Liu et al. 2006b). It is common for people in households and small businesses to sell their e-waste to peddlers, accounting for 30%–56.8% of the items discarded. It is seen in Table 3 that "storage at home" was an important disposal method. Such storage time may vary from several months to several years or more. Key factors that determine storage time are the existence or non-existence of collection systems and collection costs. Only 3.9%–6% of e-waste is directly returned to retail merchants or manufacturers. In addition, unwanted products are donated to others, especially poor relatives and friends, by the owners.

5.3.2 Policies

In an attempt to solve the serious e-waste issues in China, the Chinese government has drafted a comprehensive body of law on environmental protection. In the initial stage, the government enacted several general environmental laws (Figure 2) for e-waste recycling management, including General Environmental Law, Solid Waste Pollution Control Law, and Circular Economy Promotion Law. Although these laws gave an overall strategy of environmental protection and resources recycling, they were not feasible in practice.

As seen in the titles of the regulations and laws listed in Figure 2, the purpose of the legislation is to standardize e-waste recycling and renewable resources reuse. The Regulation for Management of Recycling and Treatment of WEEE (2009), a counterpart to the EU WEEE directive, is the basic law for processing e-waste in China. Its purpose is to reduce the volume of e-waste and to establish modern recycling systems. It stipulates that e-waste can be collected separately but must be recycled together in licensed enterprises. In addition, it requires that the stakeholders (i.e., producers and/or importers of e-waste products) must pay the costs for e-waste recycling. According to this law, a special treatment fund is set up by the government to assist in the e-waste recycling system, as shown in Table 4.

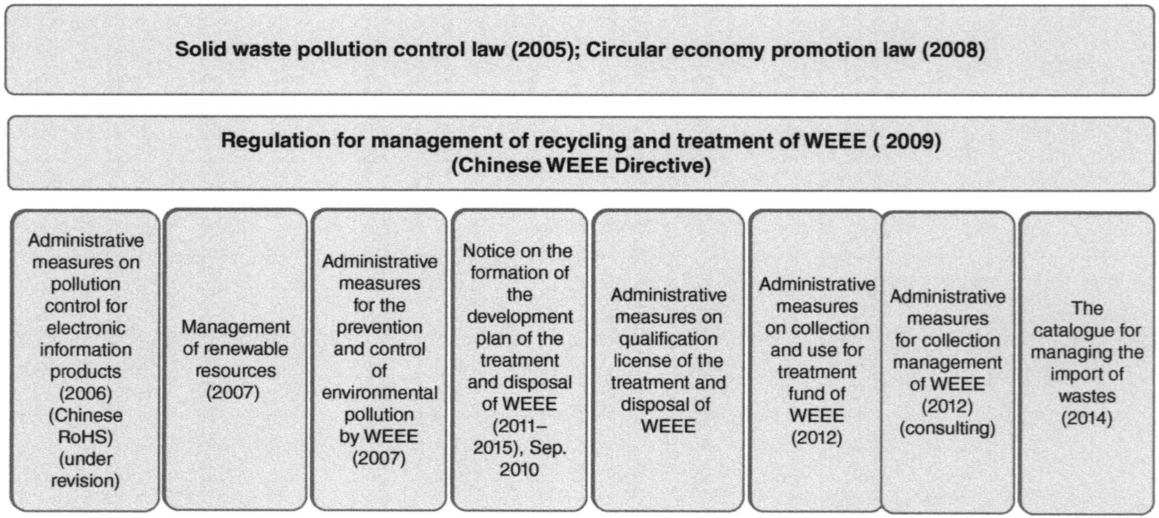

Figure 2 *E-waste Laws and Regulations in China*

Table 4 National specialized treatment fund for e-waste in China (RMB/per unit)

Categories	TV set	Refrigerator	Washing machine	Air conditioner	PC
Producer fee	13	12	7	7	10
Treatment subsidy	85	80	35	35	85

In addition to the above regulations, there are additional laws that establish the legal framework of e-waste disposition in China. For example, Administrative Measures on Pollution Control for Electronic Information Products (2006) is recognized as a counterpart to the EU's Restriction of the use of certain Hazardous Substances directive (RoHS Directive 2002/95/EC), which restricts six substances in electrical information products, i.e., Pb, Hg, Cd, Cr(VI), polybrominated biphenyl (PBB), and polybrominated diphenyl ethers (PBDE) but does not set a fixed deadline for the prohibition. Compared with RoHS, the Chinese Ordinance requires that manufacturers provide information about hazardous substances present both in components and in the whole product, as well as the product lifespan and related environmental protection information, such as product name, product content, and whether or not the product can be recycled. This Ordinance reflects the evolving trend of Chinese environmental policy regarding disposal of end-of-life (EOL) products. The Catalogue for Managing the Import of Wastes (2014), which closely resembles the Basel Convention (Basel Convention, 1989) in nature, focuses on the control of illegal e-waste shipments through customs administration.

The promulgation of these e-waste laws reflects the evolving trend of Chinese environmental policy from "End-to-end management" to "Source management", meaning that the Chinese government has changed the focus on e-waste management from "waste disposal" to "pollution source control and resource recycling". The essential objectives of current Chinese policy are to reduce, minimize, recycle, and reuse e-wastes. Current laws not only encourage environmentally friendly processing for e-waste recycling, but also provide a special fund to support the e-waste recycling system, and employ a licensed scheme for the e-waste recycling industry.

5.3.3 Formal and Informal Sectors

Formal and informal recyclers co-exist within China's current e-waste treatment system. The informal collection and recycling of e-waste emerged as a way of extracting value from e-waste through refurbishment and re-use, as well as through scrap recovery and recycling. However, e-waste recycling in the informal sector is characterized by product dismantling and metal recovery practices that pose significant risks to both workers' health and the environment (Izatt et al. 2014). In recent years, the formal recycling sectors have grown steadily under governmental regulation and financing, which has enabled the establishment of e-waste recycling processes that are safer and more environmentally and economically sound. Nevertheless, large volumes of e-waste continue to flow to and through the informal sector.

5.3.3.1 Formal Sectors

According to "Regulation for Management of Recycling and Treatment of WEEE" (2009), enforced since January 2011, enterprises that handle any or all of the five primary types of e-waste, i.e., TV sets, refrigerators, washing machines, air conditioners and personal computers (PC), must first receive a treatment license signifying that they have received qualification permission. Formal e-waste recyclers in China can thus be defined as designated recycling enterprises that are included on the e-waste Dismantling Enterprise List and have received a treatment license. Recycling licenses for e-waste are issued by provincial Environmental Protection Bureaus.

In China, development of the formal recycling system can be divided into three stages: (1) Embryonic stage (before 2009). In this stage, some resource-recycling enterprises began to pay increased attention to e-waste recycling. Only eight large-scale facilities with formal and advanced technical processes either had been built or were being planned for construction in China during the period 2003–2005. All these facilities were located along the eastern coast of China, where economic development was concentrated. The most important reason for the small number of recycling facilities is that those engaged in formal recycling were required to pay the cost of collecting the e-waste, because of the relatively inefficient collection network. (2) Development stage (2009–2011). By the end of 2010, about 81 formal enterprises had been established, due to the "Old for New" policy for home appliances. Under this policy, formal recyclers could obtain e-waste at lower cost. In 2011, 61.3 million (m) home appliances were collected and treated by the formal sector. This number included: TV sets, 51.5 m; refrigerators, 2.2 m; washing machines, 4.7 m; air conditioners, 0.2 m; and PCs, 2.6 m. (3) Mature (2012+). As the Regulation for Management of Recycling and Treatment of WEEE (2009) was enforced, beginning in 2011, the extended producer responsibility (EPR) system of e-waste treatment began to come into force, and the formal recycler could get a treatment subsidy for e-waste recycling (see Table 4). In order to ensure that the operating capacities of the local e-waste recycling industries match local needs, provincial governments have developed five-year plans for the e-waste recycling industry in their administrative territories and awarded a corresponding number of licenses to local recyclers. Up to December 2014, 106 e-waste recycling enterprises in 29 provinces and municipalities had obtained the official license for the treatment and disposal of e-waste, and were included in the Treatment Fund of E-waste list, as shown in Figure 3. In 2013, about 39.87 million units were recycled through the formal recyclers, and about 3.3 billion RMB in subsidies were provided for the formal recyclers.

5.3.3.2 Informal Sectors

According to "Administrative Measures for the Prevention and Control of Environmental Pollution by WEEE (2007)" (enforced since February 2008), any individuals or enterprises who were engaged in recycling activities but were not included in the e-waste dismantling enterprise list were performing these activities illegally. In other words, informal e-waste recyclers are not registered with the State and their activities are therefore illegal.

The informal recycling sector in China did not develop overnight. Rather, it has developed logically along with the improved economy and rising standard of living, both nationally and locally. In the 1990s, as imported electronic waste began to enter China as a cheap source for second-hand goods and secondary resources, individual recyclers began harvesting valuable materials from this imported waste. Legal restraints did not exist in China at the time and the enterprises flourished (Izatt et al. 2014). It is noteworthy that in 1994 the Basel Convention, representing most developed nations at the time, voted by consensus for a full ban on the export of hazardous waste materials from Organization of

Figure 3 *Formal recycling enterprises in China as of December 2013. (See insert for color representation of the figure.)*

Economic and Commercial Development (OECD) nations to non-OECD nations (Izatt et al. 2014). Informal recycling of e-waste is prevalent in China in 2015, particularly in some coastal regions. Informal recyclers generally use substandard processes and lack appropriate facilities to safeguard human health and the environment. The most common outputs of informal recyclers are second-hand components, refurbished appliances, and valuable materials, such as precious metals. In areas with sizeable electronics manufacturing enterprises, informal e-waste recycling plays an important role by providing useful components and materials for production. By trading with various business partners, including e-waste importers, private collectors, and dealers in secondary materials and reusable components, informal recyclers have been able to incrementally expand their trading networks and supply chains from the bottom up.

Duan and Eugster (2007) estimated that around 250,000 people in China work in the informal e-waste recycling industry, engaging in manual dismantling and material recovery. While informal e-waste recycling occurs in several locations in China, the two largest centers are Guiyu, Guangdong Province, and Taizhou, Zhejiang Province. With a population of 150,000, including 100,000 migrants, Guiyu is home to more than 300 companies and 3,000 individual workshops that are engaged in informal e-waste recycling. Of Guiyu's 28 villages, 20 are engaged in e-waste recycling (Bi et al., 2010). Most of the recycling laborers are rural migrants from outlying agrarian regions, such as Hunan and Anhui provinces, who benefit from higher wages for dismantling and processing e-waste products informally. Unfortunately, their training for this dangerous enterprise is meager, resulting in many accidents and health problems (Izatt et al. 2014; Song & Li 2014). Taizhou, a city famous for secondary material production, has been involved in informal e-waste recycling for nearly 25 years (Tang et al. 2010). In the early 1990s, workers in Taizhou began to process imported wastes such as scrap metals, obsolete electric capacitors, household appliances, electric generators, and cable wires, with an annual volume of dismantled e-waste exceeding 2.2 million tonnes.

In recent years, we believe that less regional informal e-waste recycling is occurring in China, largely as a result of more formal e-waste recyclers being available and of stricter environmental regulations related to e-waste recycling. In terms of e-waste categories, most personal computers (PC), TV sets, washing machines, refrigerators, and air conditioners are recycled in the formal sector. However, other e-waste, e.g., mobile phones, printers, and other small high-tech products, is still recycled in the informal sector.

Much of the research on e-waste issues in China was triggered by the recent environmental calamity in the Guiyu area. In 2002, Greenpeace and Basel Action Network first exposed the pollution and ecological damage resulting from informal recycling in Guiyu in their report titled "Exporting Harm" (Puckett et al. 2002). This report focused global attention on China's backyard e-waste recyclers and sparked a great deal of research seeking to assess the presence of heavy metals, persistent organics and other hazardous substances in the water, air, soil and even human organs and tissues in e-waste recycling sites like Guiyu and Taizhou. However, most research related to the informal sector has been limited to investigations of toxicity. More research is needed by individuals trained in the social sciences, geography and economics in order to understand the underlying social and economic aspects of the issue, which can in turn help address the root causes that fuel the informal recycling sector. An important consideration is that informal recycling is widespread among many non-OECD nations and has become vital to the economies of many nations, both those which provide e-waste and those which process it (Izatt et al. 2014). Disrupting the informal recycling situation as it exists would, in many cases, have severe economic repercussions, especially in non-OECD nations.

5.4 Recycling of Metals Found in E-waste

E-waste components include most of the metals in the periodic table, such as Fe, Al, Cu, Pb, Hg, As, Cd, Au, Ag, Pt, Pd, rare earth metals, etc. E-wastes represent an important and largely untapped secondary source of these metals. Reck and Graedel (2012) have reviewed the status and challenges involved in metal recycling. They summarize findings of an earlier 2011 United Nations study that defined and quantified global EOL recycling rates (EOLRR) for 60 elements, which reported that EOLRR values for common base (Fe, Ni, Co, Cu, Zn) and precious (Au, Ag, Pt, Pd) metals are a little above 50%, while that for Pb is about 80%.

However, the recycling situation becomes direr when recovery of metals from EOL e-wastes is considered (Izatt et al. 2014). EOLRR values for electronic products are low (<20%), but those for many of the metals they contain are even lower, 0–15%. For example, the EOLRR value for Au, an important constituent of cell and smart phones, is 10–15%, that for Ru, an essential component of computer hard disks is 0–5%, and that for Pd, another important component of cell and smart phones, is 0–5%. Izatt et al. (2014) list the possible barriers to effective recycling of e-waste shown in Table 5. It is apparent that improvement of metal recycling rates from e-waste will be challenging.

5.4.1 Base or Major Metals (Fe, Al, Cu, Pb, etc.)

Base metals such as Fe, Cu, Al, and Pb are present in large amounts in e-waste, primarily as structural components. Table 6 gives the percentage of these metals in five kinds of e-waste (Song et al. 2012b). Steel is the most abundant of these metals in e-waste in nearly every case, accounting for more than 50% by weight, except for the cathode ray tube (CRT) and liquid crystal display (LCD) monitors. Of the metals listed in Table 6, copper brings the most economic return from e-waste recycling. Lead (as the base or major metal) is found only in the CRT monitor (16%), and can be recycled only through complicated

Table 5 Barriers to effective recycling of e-waste (Izatt et al. 2014). Reproduced from Izatt, et al. 2014 with permission of The Royal Society of Chemistry

Item	Potential barriers
1	Economic incentive to recycle due to low economic benefits per unit is lacking
2	Traditional commercial recycling technologies are not adequate for recovering the metals present in modern products
3	Products contain a complex mixture of metals, and the composition changes over time
4	Collection procedures are scarce or do not exist
5	Cost of collection and transportation of units to smelters are large
6	Recycling is not part of a collection that includes smelters
7	Product design makes disassembly and material separation difficult
8	Public awareness of impending loss of crucial resources is low
9	Public is not attuned to local recycling procedures, if any

Table 6 Percentages by weight of base or major metals in e-waste

Categories	Steel	Copper	Aluminum	Lead	Total
PC mainframe	55%	8%	14%	-	77%
CRT monitor(PC and TV set)	12%	3%	1%	13%	29%
LCD monitor(PC and TV set)	39%	7%	1%	-	47%
Refrigerator	49%	4%	1%	-	54%
Washing machine	52%	2%	4%	-	58%
Air conditioner	54%	18%	9%	-	81%

technologies and procedures, resulting in its relatively lower recycling rate. Overall, the base or major metals constitute the main part of e-waste, accounting for 29–81 wt % for the five kinds of e-waste. These metals can be recycled easily through manual dismantling, mechanical crushing and sorting, and, finally, smelting. Metals of this class showed the highest recyclability. In 2013, according to our estimation, about 96,300 tonnes of steel, 19,800 tonnes of Cu and 5,200 tonnes of Al were recycled in formal e-waste recycling enterprises in China.

5.4.2 Toxic Metals

Toxic metals are those that are toxic at low concentrations to living organisms. E-waste contains multiple toxic metals, such as Pb, Cd, and Cr(VI). The toxic nature of these metals will now be discussed.

5.4.2.1 Lead

The negative effects of lead are well established and recognized (Huo et al., 2007; Julander et al., 2014). In humans, Pb causes damage to the central and peripheral nervous systems, blood systems, kidneys, and the reproductive system. Main sources of Pb in e-waste are: glass panels in CRT monitors, and solder in printed circuit boards. As shown in Table 7 (Yuan et al. 2012), lead levels in CRT monitors were highest, about 1.128 kg per unit (funnel glass + panel glass + frit glass + welding glass). For the CRT monitors, lead content of welding glass was highest, reaching 70–80%, followed by funnel glass, frit glass, and pannel glass. However, due to the large amount of funnel glass present, it contained more lead, about 0.91 kg. Technology innovation has resulted in the replacement of CRT monitors by LCD monitors. At the same time, most e-waste contains PCBs, which complicate Pb recycling. It is expected that the Pb-PCB issues will become more important in the future.

5.4.2.2 Cadmium and Chromium(VI)

Cadmium compounds are toxic and bioaccumulate in human tissues and organs, with a risk of irreversible effects on human health (Yang et al., 2013). Cadmium occurs in certain high-tech products, such as surface mount devices, chip resistors, infrared detectors, and semiconductor chips. Chromium(VI) is still used for corrosion protection of untreated and galvanized steel plates and as a decorative or hardener for steel housings (Rao et al., 2002). It easily passes through cell membranes and is then absorbed, producing various toxic effects in contaminated cells.

Table 7 *Lead levels in cathode ray tube monitors and printed circuit boards (PCB)*

Components	Percentage (%)	Mass (kg)
Funnel glass	22~28	0.91
Panel glass	0~4	0.18
Frit glass	26~32	0.012
Welding glass	70~80	0.026
PCBs	–	0.051

Table 8 *Environmental pollution of toxic metals in China*

Location	Sampling time	Sampling site	Concentrations	References
Toxic metals in air				
Guiyu	2007.9	PCB recycling workshop	TSP: Cd: 0.08; Pb: 4.4 μg/m^3	Bi et al. 2010)
Jiangsu (Formal)	2011.3	PCB recycling workshop, 1.7 m above ground level	TSP: Cr: 0.17; Cd: 0.028; Pb: 1.40 μg/m^3 PM_{10}: Cr: 0.17; Cd: 0.016; Pb: 0.56 μg/m^3	Xue et al. 2012)
Toxic metals in dust				
Guiyu	2004.12	Workshop dust; adjacent roads	Workshop dust: Pb 110,000 mg/kg; Adjacent roads: Pb 22,600 mg/kg	Leung et al 2008)
Guiyu	2010	Family-run workshops	Cd: 1.9; Cr: 61; Pb: 892 mg/kg	Zhu et al. 2012)
Jiangsu (Formal)	2011.3	PCB recycling workshop	Cr: 43; Cd: 14; Pb: 4,320 mg/kg	Xue et al. 2012)
Toxic metals in soils				
Guiyu	2003.8	Burnt plastic dump site, printer roller dump site	Cd 1.7; 3.1mg/kg; Cr 28.6; 74.9 mg/kg; Pb 104; 190 mg/kg	Leung et al. 2006)
Taizhou	2005.11	Paddy soil	Cd: 1.19, Cr:6.12, Pb: 55.81 mg/kg	Fu et al. 2008)
Luqiao in Taizhou	2008.8–9	Agricultural soils	Pb: 115.1; Cd: 1.77; Cr: 61.4 mg/kg	Tang et al. 2010)
Longtang	2005.6	Open burning sites	Pb: 1,714.5; Cd: 10.3; Cr:63.3 mg/kg	Luo et al. 2008)

In Tables 8 and 9, potential environmental and human health effects of toxic metals in China are presented. Levels of toxic metals other than Pb in e-waste are very low. Toxic metals are also dispersive in the components and parts of e-waste. At present, there are no effective recycling technologies for these metals. In the formal sectors, due to their low recyclability and low economic value, most toxic metals are treated as hazardous waste. In the informal sector, these metals often flow into the surrounding air, soil or water, resulting in serious threats to the environment and human health. Metals of this class showed lower recyclability.

Table 9 *Human body burden of heavy metals from e-waste in China*

Pollutants	Sampling time	Exposure groups	Sampling sites	Human body burden	References
Placenta					
Pb, Cd, Cr	2008.10–2009.5	220 mother–infant pairs (101:119)	Guiyu vs. Chaonan	Pb: 301.43 vs. 165.82 ng/g, p=0.010; Cd: 108.75 vs. 104.15 ng/g; Cr: 234.31 vs. 228.40 ng/g;	Guo et al., 2010)
Cd	2006	423 mother–infant pairs (289:134)	Guiyu vs. Chaonan	170 ±48 0 vs. 100 ±110 ng/L, p<0.01	Li et al., 2011)
Cd, Pb	2008.10–2009.6	105 mother–infant pairs (55:50)	Guiyu vs. Shantou	Cd: 83.99 vs. 51.59 ng/g, P<0.001; Pb: (521.01 vs. 273.24 ng/g, P=0.299)	Zhang et al., 2011)
Umbilical cord blood					
Pb, Cd, Cr	2012.3–2013.1	201 mother–infant pairs (126:75)	Guiyu vs. Shantou	Pb: 110.45 vs. 57.31μg/L, p<0.001; Cd: 2.50 vs. 0.33μg/L, p<0.001; Cr: 27.52 vs. 26.42 μg/L	Ni et al., 2013)
Pb	2006	152 mother–infant pairs	Guiyu vs. control area	113.28 vs 60.43 μg/L, p=0.00	Li et al., 2008)
Pb	2001–2008; 2004–2009	531 mother-infant pairs	Guiyu vs. Xiamen	Pb: 107.8 vs. 22.5 μg/L, p<0.01	Xu et al., 2012)
Blood					
Pb and Cd	–	246 children (aged 3–8)	Only Guiyu	Pb: 73.0 (43.3–154.3) (μg/L); Cd: 0.69 (0.36–1.45) (μg/L); No statistical analysis.	Yang et al., 2013)
Pb	2004	226 children (165:61)	Guiyu vs. Chendian	153.0 ± 57.9 vs. 99.4 ± 40.5 μg/L, p<0.01	Huo et al., 2007)
Pb	2006	278 children (aged 1–7) (154:124)	Guiyu vs. Chendian	131.7±59.8 vs. 100.4±48.5 μg/L, p<0.01	Zheng et al., 2008)
Pb	2008.1–2	303 children (aged 3–7) (153:150)	Guiyu vs. Chendian	144.3±69.3 vs. 87.2 ±43.4 μg/L, p<0.01	Liu et al., 2011)
Pb	2010.6	178 children (aged 11–12) (108:70)	Luqiao vs. Chun'an	6.97 vs. 27.8 μg/L, p<0.001	Wang et al., 2012)

5.4.3 Precious Metals

The demand for precious metals by manufacturers of e-products has increased significantly over the past ten years. In the USA in 2007, 6% of mined gold was used for manufacturing e-products (Tay et al., 2013; Zhang et al., 2012). The great advantage of precious metals is their high resistance to oxidation and corrosion and high conductivity, which gives them excellent contact properties. Although precious metal concentrations in e-products are very low, these metals have high economic value and significant environmental effects compared to other substances present in much higher quantity (e.g., iron, copper, plastics) (Chancerel et al., 2009). Mobile phones contain less than 0.01% gold, but ~50% of the economic value and environmental impacts of the phone result from this gold content (Geyer and Doctori Blass, 2009).

Precious metals mainly occur in printed circuit boards (PCB) as components of complex material mixtures, as they are mixed with or connected to other metals in contacts, connectors, cables, solders, hard disk drives, etc. Table 10 compares precious metal concentrations in PCBs, as determined by different researchers. It is seen that precious metal concentrations varied in the different studies, e.g., Au (80–800g/t); Ag (110–3300 g/t); Pt (4.6–30 g/t); Pd (10–294 g/t). Reasons for this variance are likely related to differences in ages, origins and/or manufacturers of the products.

Though concentrations of precious metals in e-waste products are appreciable, recovery rates for these metals are low (<5 %). As mentioned earlier, informal recycling has been much more effective than formal recycling in recovering precious metals from e-waste in China. However, informal recycling is also a major contributor to China's serious environmental and health problems.

5.4.4 Rare Earth Elements (REEs)

REEs are considered to be in the most critical materials group, with the highest supply risk. The global demand for REEs is growing at a rate of 3.7–8.6% annually. REEs are used in the widest range of consumer products of any element group because of their unique physical and chemical properties (Castor and Hedrick, 2006). Here, we mainly consider the recovery of REEs from the fluorescent materials in e-waste. Fluorescent material is used mainly in the production of lighting such as fluorescent lamps (FLs), the backlight of flat panel displays, and in CRT monitors and display devices. Yttrium, Eu, Ce, La and Tb are used as phosphors for FLs.

Ratios of the amounts of REEs used in phosphors to the total usage amounts of REEs have been calculated from the application proportions of REEs used in various industries

Table 10 *Precious metal concentrations in several different printed circuit boards (g/tonne)*

Metals	Hao et al., 2008	Yu et al., 2011	Sum, 1991	Zhao et al., 2004	Zhang and Forssberg, 1997	Iji and Yokoyama, 1997
Au	800	570	80	280	420	350
Ag	800	3301	3300	110	1240	1300
Pt	–	30	–	–	–	4.6
Pd	210	294	–	–	10	250

Table 11 *Flows into use and in-use stock of REEs in phosphors*

Element	La	Ce	Eu	Gd	Tb	Y
Flows into use (tonnes/yr)	350.4	558.0	300.0	385.0	255.6	9434.1
In-use stock (tonnes)	1376.0	2880.0	400.0	630.0	596.4	5292.3

(Tan et al. 2014). It was found that all of the Eu, 85.2% of the Tb, and 76.7% of the Y were used for the production of phosphors, while the proportions of La and Ce were negligible, only 1.6% and 2.0%, respectively. Flows into use and in-use stock of REEs in phosphors have been estimated and combined with studies on global REE in-use stock estimates (Table 11; Du and Graedel, 2011). Fluorescent lamp production is one of the main areas requiring use of phosphors. Due to the phasing out of CRT display devices, the percentage of total phosphor use for FLs has risen gradually to about 90% in 2011 (Tan et al. 2014). Consequently, considerable quantities of future rare earth resources, especially of Eu, Tb, and Y, will be used in the manufacture of phosphors. However, due to low REE concentrations in FL products and the lack of effective REE separation and recovery technologies, no REEs are being recycled from FL products.

5.5 Challenges and Efforts in Metal Sustainability in China

5.5.1 Challenges

As illustrated in Figure 4, China's metal production and consumption since the 1990s has accelerated because of the rapid increase of domestic industrialization and urbanization, improved infrastructure, social wealth accumulation, and the steady improvement of overall living standards. It is expected that the production and consumption of metals will continue to increase in the foreseeable future. However, compared with the large metal consumption envisioned, it is important to note that the basic reserves of metals are finite. As extraction of present ore supplies continues, ore grades decline and more energy and water are required to remove and process the ore. The search for new minable ore deposits is always ongoing, but the results are not certain. The data in Figure 5 indicate that metal reserves in China will decrease in the future. Thus, as the growing demand for metals increases, a supply crisis has emerged in China. This is illustrated in Figure 6, where the increasing demand for iron, as the infrastructure expands, has resulted in a high import ratio, 56.76% in 2012.

Metals are infinitely recyclable in principle, but in practice recycling is often inefficient or essentially nonexistent because of limits imposed by social behavior, product design, recycling technologies, and the thermodynamics of separation processes. The metal supply in China would be greatly augmented by more effective metal recycling.

5.5.2 Efforts

The average annual GDP growth in China is 8% (Zhang et al., 2014), making it one of the fastest developing countries in the world. Resource shortages have become a key constraint for China's sustainable development. Recycling is considered an effective means of waste

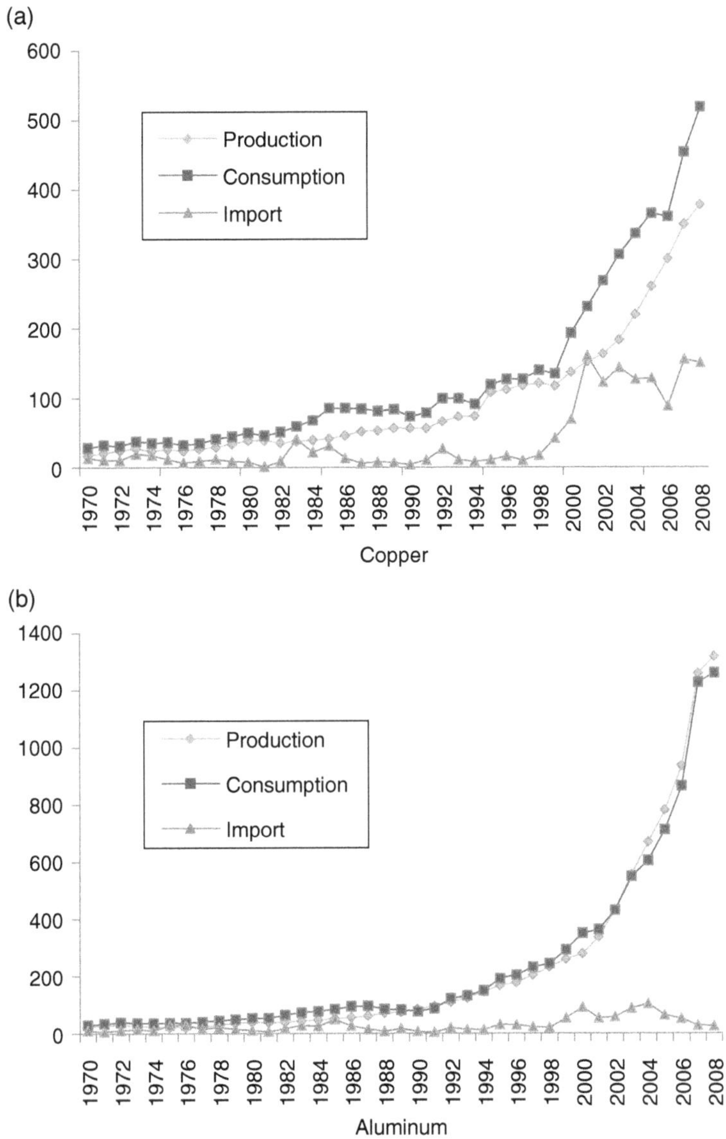

Figure 4 *Production, Import and Consumption Trends of Copper, Aluminum and Lead in China*

management aimed at partially alleviating this shortage. Benefits of recycling include resource conservation and indirect energy and water savings from reducing raw material extraction. Since around 2005, the Chinese government has made great efforts to improve resource recycling, especially of metals. Governmental policy regarding recycling has undergone a dramatic transformation in the last decade. Previously, the main concern was pollution prevention, but the new position gives increased attention to recovering the recyclable resources in EOL products. The center of this policy shift is the Circular Economy

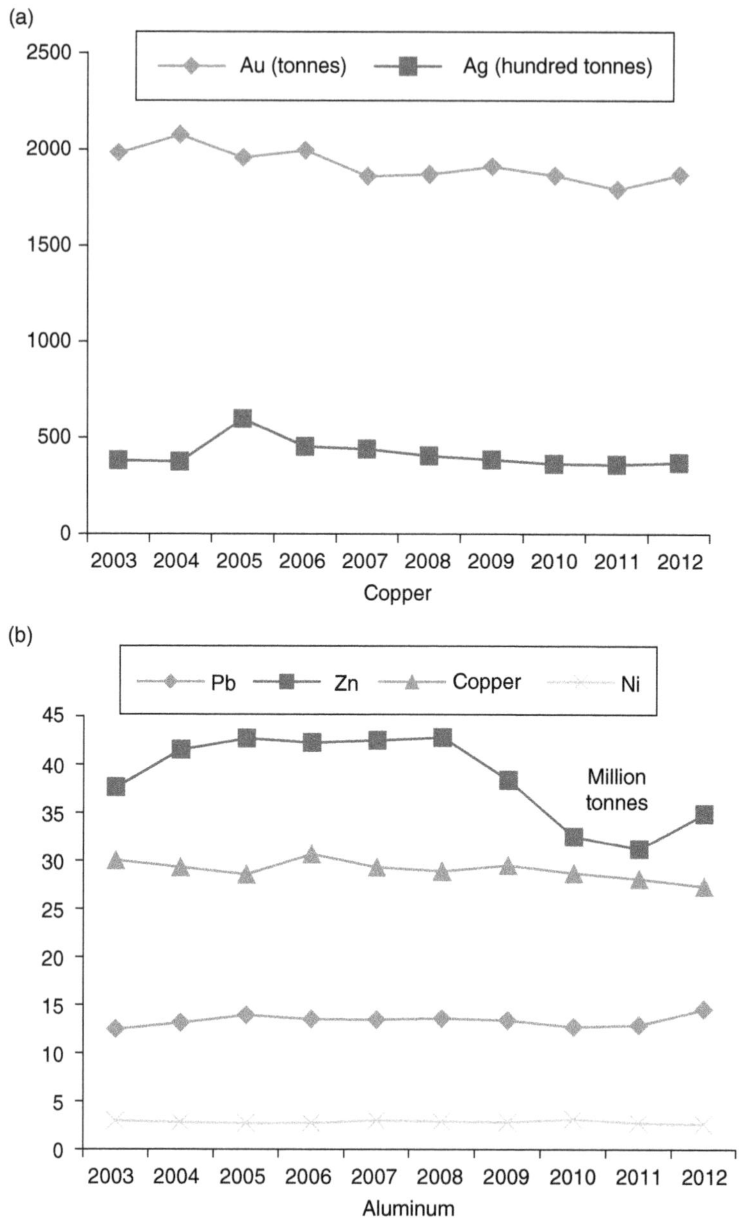

Figure 5 *Basic Reserves of Metals in China*

Promotion Law passed in 2008 (NPC, 2008), which aims to restructure the country's linear resource-consumption economy into a circular one in which all waste products (heat, water, gas and materials) are reduced in magnitude and recycled to the maximum extent possible. This dual emphasis on pollution prevention and resource recovery by recycling is intended to become the basis of an entirely new economic system (Mo et al. 2009). Along with this

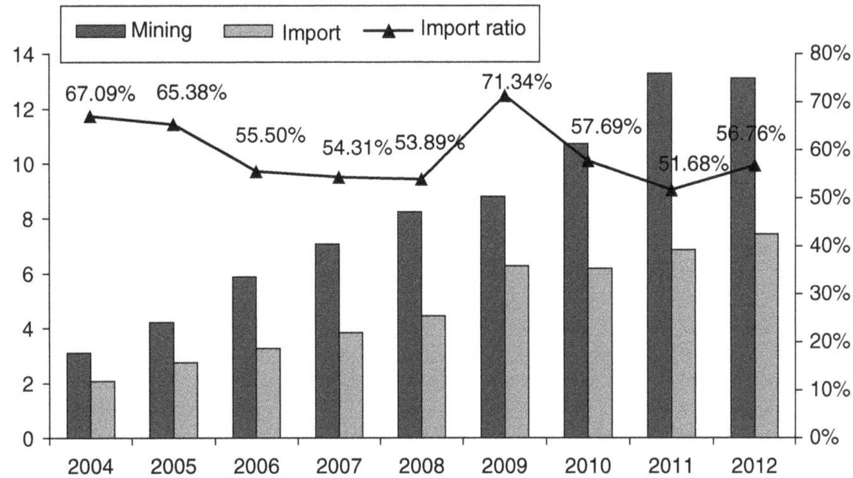

Figure 6 *Iron Ore Domestic Extraction and Import in China (100 Million Tonnes)*

key law, a recycling supervision policy system has been built. The whole recycling policy system now encompasses about 90 related laws and regulations, 26 of which are established or revised by legislative action or by departments of the central government. Regional governments also establish their own practical regulations for recycling systems in line with the central policy system.

Under the recycling supervision policy, China is establishing effective systems for various recyclable resources to achieve effective metal recycling. The recycling system has the following parts: recycling network in communities, recycling network in enterprises and institutions, waste metals recycling network, and renewable resources terminal market. Recycling processes for waste iron and copper are given in Figure 7. By the end of 2013, approximately 100,000 recycling enterprises of renewable resources, 0.23 million registered recycling outlets, 0.6 million unregistered recycling outlets, and 18 million employees had been involved in the recycling systems. In addition, some regional recycling centers have been built in China, as shown in Figure 8.

With effective policies and a working recycling system, metal recycling in China has made good progress. Large amounts of certain metals are recovered and reused in China, as shown by the examples in Table 12 and Figure 9. However, it should be noted that the ratio of secondary metals to metal production is still lower than that of the developed countries. More efforts are required in the future to further improve metal recycling in China.

5.6 Summary

As rapid global economic development and technological innovation occur, e-waste recycling will become increasingly important for maintaining metal sustainability. Achievement of effective recycling of metals from spent products is an important goal in this endeavor.

(a)

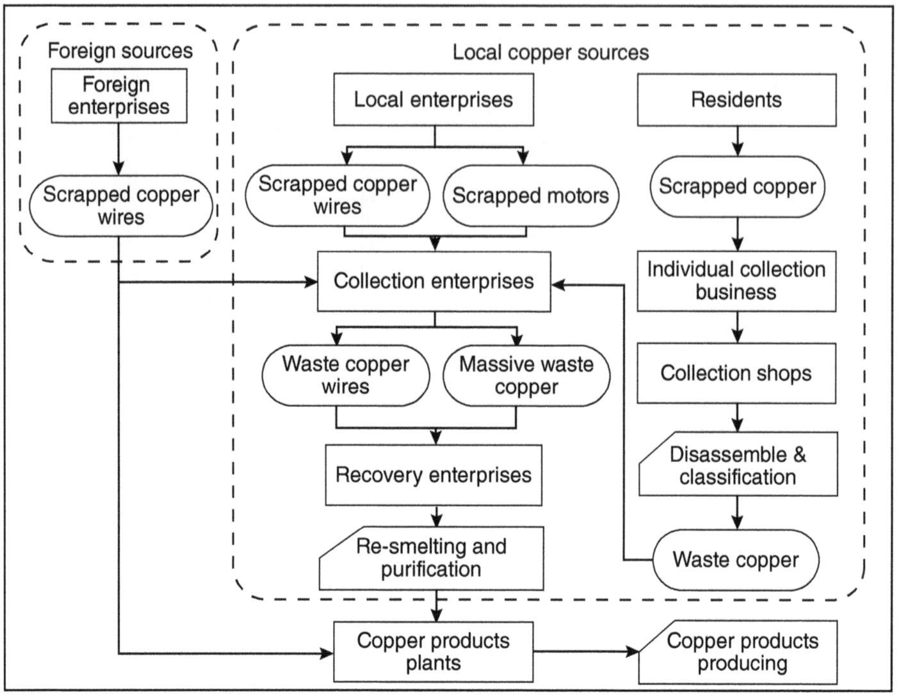

Iron

(b)

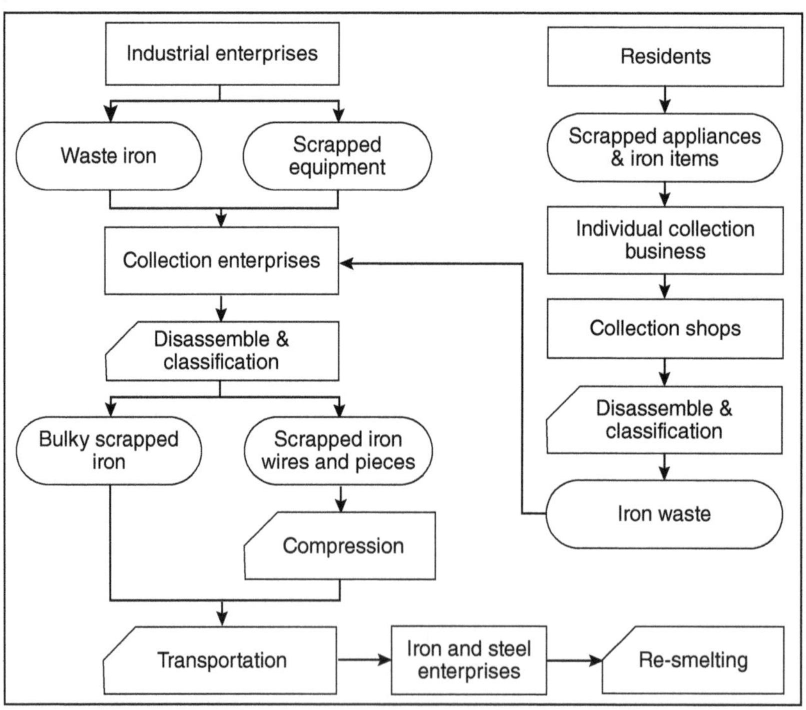

Copper

Figure 7 *Recycling Processes of Waste Iron and Copper in China*

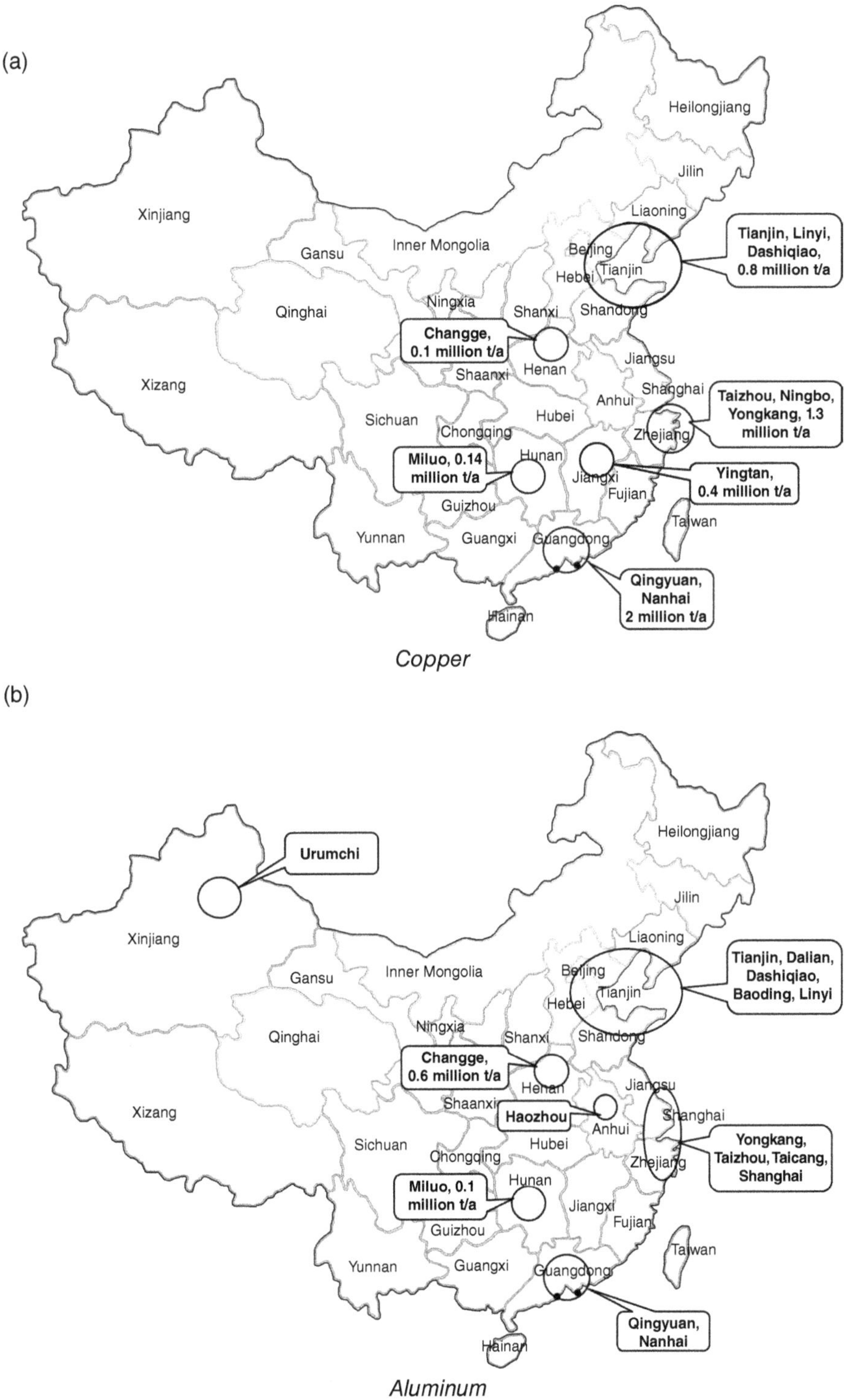

Figure 8 *Regional Recycling Centers of Waste Copper and Aluminum in China*

Table 12 Recovery and output value of metals

Categories	2011		2012	
	Mass	Output value	Mass	Output value
Iron and steel (million tons)	91	274	84	223
Nonferrous metals (	455	89	530	103
E-waste (million units)	3.7	12	1.9	6
Scrapted vehicles (million units)	2.9	8	2.5	6
Scrapted ships (million units)	2.0	6	2.3	6

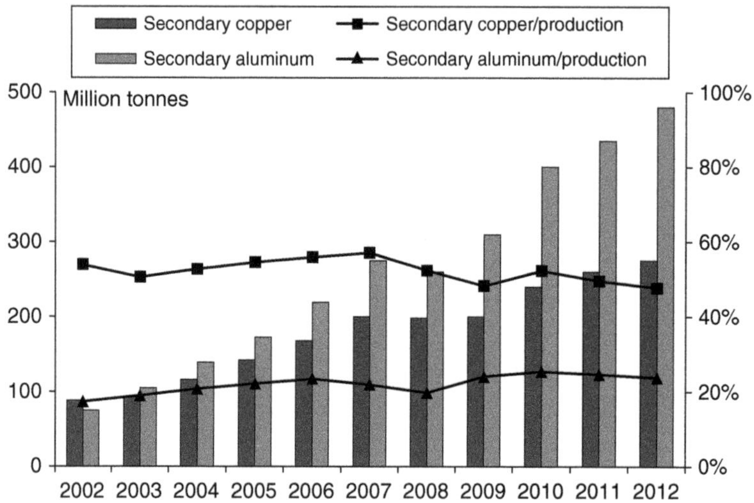

Figure 9 *Secondary metals recycling in China*

In this chapter, we first introduced global e-waste issues and then, using China as an example, analyzed the management status (generation, flows, policies, and formal and informal recycling) of e-waste. We next discussed recycling of metals (base or major, toxic, precious, and rare earth) from spent e-waste products. Finally, the challenges and efforts involved in maintaining metal sustainability in China were presented and discussed. Metal recycling in China has made good progress and, with legislative laws in place, should be an important contributor to the Nation's resources in the future.

5.7 Acknowledgment

Assistance from Prof. Reed M. Izatt of Brigham Young University on this effort is greatly appreciated.

References

Basel Action Network (BAN) & Silicon Valley Toxics Coalition (SVTC) (2002) Exporting Harm: the High-Tech Trashing of Asia. [Online], http://www.ban.org/E-waste technotrashfinalcomp.pdf [10 May 2015].

Basel Convention (1989) Basel Convention on the Control of Transboundary Movements of Hazardous Wastes and Their Disposal, http://www.basel.int/TheConvention/Overview/Textofthe Convention/tabid/1275/Default.aspx [10 May 2015].

Bi, X., Simoneit, B. R. T., Wang, Z., Wang, X., Sheng, G., Fu, J., 2010, The Major Components of Particles Emitted During Recycling of Waste Printed Circuit Boards in a Typical E-Waste Workshop of South China, *Atmos. Environ.* 44, 4440–4445.

Chancerel, P., Meskers, C. E. M., Hagelüken, C., Rotter, V. S., 2009, Assessment of Precious Metal Flows During Preprocessing of Waste Electrical and Electronic Equipment. *J. Ind. Ecol.* 13, 791–810.

Duan, H., Li, J., Liu, Y., Yamazaki, N., Jiang, W., 2011, Characterization and Inventory of PCDD/Fs and PBDD/Fs Emissions from the Incineration of Waste Printed Circuit Board. *Environ. Sci. Technol.* 45, 6322–6328.

The European Parliament and Council (EPC) (2003) Directive 2002/95/EC of the European Parliament and of the Council of 27 January 2003 on Waste Electrical and Electronic Equipment (WEEE). *Official Journal of the European Union.*

Fu, J., Zhou, Q., Liu, J., Liu, W., Wang, T., Zhang, Q., Jiang, G., 2008, High Levels of Heavy Metals in Rice (Oryza Sativa L.) from a Typical E-Waste Recycling Area in Southeast China and its Potential Risk to Human Health, *Chemosphere* 71, 1269–1275.

Geyer, R., Blass, V.D., 2009, The Economics of Cell Phone Reuse and Recycling, *International J. Adv. Manufac. Technol.* 47, 515–525.

Guo, Y., Huo, X., Li, Y., Wu, K., Liu, J., Huang, J., Zheng, G., Xiao, Q., Yang, H., Wang, Y., Chen, A., Xu, X., 2010, Monitoring of Lead, Cadmium, Chromium and Nickel in Placenta from an E-Waste Recycling Town in China, *Sci. Total Environ.*, 408, 3113–3117.

Huo, X., Peng, L., Xu, X., Zheng, L., Qiu, B., Qi, Z., Zhang, B., Han, D., Piao, Z., 2007, Elevated Blood Lead Levels of Children in Guiyu, an Electronic Waste Recycling Town in China, *Environ. Health Persp.*, 115, 1113–1117.

Izatt, R. M., Izatt, S. R., Bruening, R. L., Izatt, N. E., Moyer, B. A., 2014, Challenges to Achievement of Metal Sustainability in our High-Tech Society, *Chem. Soc. Rev.*, 43, 2451–2475.

Julander, A., Lundgren, L., Skare, L., Grandér, M., Palm, B., Vahter, M., Lidén, C., 2014, Formal Recycling of E-Waste Leads to Increased Exposure to Toxic Metals: an Occupational Exposure Study from Sweden, *Environ Int*, 73, 243–251.

Kahhat, R., Kim, J., Xu, M., Allenby, B., Williams, E., Zhang, P., 2008, Exploring E-Waste Management Systems in the United States, *Resour. Conserv. Recy.*, 52, 955–964.

Leung, A., Cai, Z.W., Wong, M.H., 2006, Environmental Contamination from Electronic Waste Recycling at Guiyu, Southeast China, *J. Mater. Cycl. Waste Manage.*, 8, 21–33.

Leung, A. O., Duzgoren-Aydin, N. S., Cheung, K. C., Wong, M.H., 2008, Heavy Metals Concentrations of Surface Dust from E-Waste Recycling and its Human Health Implications in Southeast China, *Environ. Sci. Technol.*, 42, 2674–2680.

Li, J., Tian, B., Liu, T., Liu, H., Wen, X., Honda, S.I., 2006, Status Quo of E-waste Management in Mainland China. *J. Mater. Cycl. Waste Manage.*, 8, 13–20.

Li, Y., Huo, X., Liu, J., Peng, L., Li, W., Xu, X., 2011, Assessment of Cadmium Exposure for Neonates in Guiyu, an Electronic Waste Pollution Site of China, *Environ. Monit. Assess.*, 177, 343–351.

Li, Y., Xu, X., Wu, K., Chen, G., Liu, J., Chen, S., Gu, C., Zhang, B., Zheng, L., Zheng, M., Huo, X., 2008, Monitoring of Lead Load and its Effect on Neonatal Behavioral Neurological Assessment Scores in Guiyu, an Electronic Waste Recycling Town in China, *J. Environ. Monitor.*, 10, 1233.

Liu, J., Xu, X., Wu, K., Piao, Z., Huang, J., Guo, Y., Li, W., Zhang, Y., Chen, A., Huo, X., 2011, Association Between Lead Exposure from Electronic Waste Recycling and Child Temperament Alterations. *Neurotoxicology*, 32, 458–464.

Lundgren, K., 2012, *The Global Impact of E-Waste: Addressing the Challenge*. International Labour Organization, Geneva.

Luo, Y., Yu, X., Yang, Z., Yuan, J., Mai, B. (2008) Studies on Heavy Metal Contamination by Improper Handling of E-Waste and its Environmental Risk Evaluation 1: Heavy Metal Contamination in E-Waste Open Burning Sites. *Asian J. Ecotoxicol.*, 3, 34–41.

Mo, H., Wen, Z., Chen, J., 2009, China's Recyclable Resources Recycling System and Policy: a Case Study in Suzhou, *Resour. Conserv. Recy.*, 53, 409–419.

Namias, J., 2013, *The Future of Electronic Waste Recycling in the United States: Obstacles and Domestic Solutions*, Doctoral dissertation, Columbia University.

Ni, W., Huang, Y., Wang, X., Zhang, J., Wu, K., 2013, Associations of Neonatal Lead, Cadmium, Chromium and Nickel Co-Exposure with DNA Oxidative Damage in an Electronic Waste Recycling Town, *Sci. Total Environ.*, 472C, 354–362.

National People's Congress of the PRC (NPC), 2008, *The Circular Economy Promotion Law*, http://www.gov.cn/flfg/2008-08/29/content_1084355.htm [10 May 2015].

Rao, J.R., Thanikaivelan, P., Sreeram, K.J., Nair, B.U., 2002, Green Route for the Utilization of Chrome Shavings (Chromium-Containing Solid Waste) in the Tanning Industry, *Environ. Sci. Technol.*, 36, 1372–1376.

Reck, B. K., Graedel, T. E., 2012, Challenges in Metal Recycling, *Science*, 337, 690–695.

Samarasekera, U., 2005, Electrical Recycling Exposes Workers to Toxic Chemicals, *Lancet*, 366, 1913–1914.

Song, Q., Wang, Z., Li, J., 2012a, Residents' Behaviors, Attitudes, and Willingness to Pay for Recycling E-Waste in Macau. *J. Environ. Manage.*, 106, 8–16.

Song, Q., Wang, Z., Li, J., Duan, H., 2012b, Sustainability Evaluation of An E-Waste Treatment Enterprise Based on Energy Analysis in China. *Ecol. Eng.* 42:223–231.

Song, Q., Li, J., Zeng, X., 2014, Minimizing the Increasing Solid Waste Through Zero Waste Strategy, *Journal of Cleaner Production*, http://dx.doi.org/10.1016/j.jclepro.2014.08.027

Song, Q., Li, J., 2014,.A Systematic Review of the Human Body Burden of E-Waste Exposure in China, *Environ Int*, 68, 82–93.

Tan, Q., Li, J., Zeng, X., 2014, Rare Earth Elements Recovery from Waste Fluorescent Lamps: a Review, *Critical Review in Environ. Sci. Technol.* 45, 749–776.

Tang, X., Shen, C., Chen, L., Xiao, X., Wu, J., Khan, M. I., Dou, C., Chen, Y., 2010, Inorganic and Organic Pollution in Agricultural Soil from an Emerging E-Waste Recycling Town in Taizhou Area, China, *J. Soils Sediments* 10, 895–906.

Tay, S.B., Natarajan, G., Rahim, M.N., Tan, H.T., Chung, M.C., Ting, Y.P., Yew, W.S., 2013, Enhancing Gold Recovery from Electronic Waste via Lixiviant Metabolic Engineering in Chromobacterium Violaceum, *Scientific Reports* 3, 2236.

Torretta, V., Ragazzi, M., Istrate, I.A., Rada, E.C., 2013, Management of Waste Electrical and Electronic Equipment in Two EU Countries: a Comparison, *Waste Manage.*, 33, 117–122.

Wang, H., Han, M., Yang, S., Chen, Y., Liu, Q., Ke, S., 2011, Urinary Heavy Metal Levels and Relevant Factors Among People Exposed to E-Waste Dismantling. *Environ. Int.*, 37, 80–85.

Wang, X., Miller, G., Ding, G., Lou, X., Cai, D., Chen, Z., Meng, J., Tang, J., Chu, C., Mo, Z., Han, J., 2012, Health Risk Assessment of Lead for Children in Tinfoil Manufacturing and E-Waste Recycling Areas of Zhejiang Province, China, *Sci. Total Environ.*, 426, 106–112.

Widmer, R., Oswald-Krapf, H., Sinha-Khetriwal, D., Schnellmann, M., Böni, H., 2005, Global Perspectives on E-waste, *Environ. Impact Assess. Rev.*, 25, 436–458.

Xu, X., Yang, H., Chen, A, Zhou, Y., Wu, K., Liu, J., Zhang, Y., Huo, X. 2012, Birth Outcomes Related to Informal E-Waste Recycling in Guiyu, China, *Reprod. Toxicol.*, 33, 94–98.

Xue, M., Yang, Y., Ruan, J., Xu, Z., 2012, Assessment of Noise and Heavy Metals (Cr, Cu, Cd, Pb) in the Ambience of the Production of Li Ne for Recycling Waste Printed Circuit Boards, *Environ. Sci. Technol.*, 46, 494–499.

Yang, H., Huo, X., Yekeen, T.A., Zheng, Q., Zheng, M., Xu, X., 2013, Effects of Lead and Cadmium Exposure from Electronic Waste on Child Physical Growth, *Environ. Sci. Pollut. Res. Internat.*, 20, 4441–4447.

Yang, J., Lu, B., Xu, C., 2008, WEEE Flow and Mitigating Measures in China, *Waste Manage.*, 28, 1589–1597.

Yuan, W., Li, J., Zhang, Q., Saito, F., 2012, Innovated Application of Mechanical Activation to Separate Lead from Scrap Cathode Ray Tube Funnel Glass, *Environ. Sci. Technol.*, 46, 4109–4114.

Zhang, Q., Zhou, T., Xu, X., Guo, Y., Zhao, Z., Zhu, M., Li, W., Yi, D., Huo, X., 2011, Downregulation of Placental S100P Is Associated with Cadmium Exposure in Guiyu, an E-Waste Recycling Town in China, *Sci. Total Environ.* 410–411, 53–58.

Zhang, Y., Liu, S., Xie, H., Zeng, X., Li, J., 2012, Current Status on Leaching Precious Metals from Waste Printed Circuit Boards, *Procedia Environmental Sciences*, 16, 560–568.

Zhang, H., Wen, Z.G., 2014, The Consumption and Recycling Collection System of PET Bottles: a Case Study of Beijing, China, *Waste Manag.*, 34, 987–998.

Zheng, L., Wu, K., Li, Y., Qi, Z., Han, D., Zhang, B., Gu, C., Chen, G., Liu, J., Chen, S., Xu, X., Huo, X., 2008, Blood Lead and Cadmium Levels and Relevant Factors Among Children from an E-Waste Recycling Town in China, *Environ. Res.*, 108, 15–20.

Zhu, Z., Han, Z., Bi, X., Yang, W., 2012, The Relationship Between Magnetic Parameters and Heavy Metal Contents of Indoor Dust in E-Waste Recycling Impacted Area, Southeast China, *Sci. Total Environ.*, 433, 302–308.

6

E-waste Recycling in China: Status Quo in 2015

Martin Streicher-Porte[1], *Xinwen Chi*[2], *and Jianxin Yang*[3]

[1]*FHNW, University of Applied Sciences and Arts Northwestern Switzerland, Institute for Biomass and Resource Efficiency, 5210, Windisch, Switzerland*

[2]*School of Environmental Science & Engineering, South University of Science and Technology of China, Shenzhen, Guangdong Province, 518055, PRC*

[3]*Research Center for Eco-Environmental Sciences, Chinese Academy of Sciences, Beijing, 100085, PRC*

6.1 Introduction

Waste from electrical and electronic equipment has become a waste stream that epitomizes the complexity of human living and consumption patterns. Electrical and electronic equipment are indispensable for everyday activities, being cheaper, smaller, more specific in their application, and are growing in number and uses. Being connected to digital networks, they need to be updated or replaced. The resulting waste stream is of considerable financial worth compared to other waste types. This is mainly due to the metals it contains and is particularly apparent when compared to the worth of natural ores, which have a considerably lower concentration of metals. Because of this, e-waste is often called an urban mine.

Digital network connectivity demands that these items often have an Internet Protocol number (IP) and connect, receive, send and store data. These factors add an additional dimension to the handling of e-waste, as the information to be gained from the data needs to be protected and safely destroyed. Also, the intellectual property of data producers and warranty issues are becoming increasingly important when treating obsolete equipment.

Metal Sustainability: Global Challenges, Consequences, and Prospects, First Edition. Edited by Reed M. Izatt.

Many business opportunities are created by dealing with obsolete electrical and electronic equipment. Not only can appliances be refurbished and reused but also components can be dismantled and used as spare parts. In addition, the valuable materials contained in the e-waste can of course be exploited. However, many hazardous substances and elements are often contained in e-waste or are created during handling processes.

Designing e-waste systems that are comprehensive, effective in recovering resources, cost-efficient and environmentally safe has been a key activity in many industrialised countries. But today all countries produce e-waste, even the least developed economies. Rapidly developing economies are already the largest producers of electrical and electronic equipment as well as e-waste. Due to the complexity of the waste stream and its origin, e-waste management systems need to be regulated under a legal framework and supported by the producing industry. In order to reach sufficient coverage, collect the e-waste stream within national borders and comply with minimal environmental and health standards, a management and financing mechanism needs to be installed.

Although the production process of electrical and electronic equipment has emerged as a globally optimised production chain, China is the gravitational centre of many key production processes and is also known to be one of the countries with a large informal e-waste recycling industry. For some years, China has been installing a formal e-waste management system that differs from many of the systems in industrialised countries. This chapter contrasts the formal e-waste recycling system with the grass-roots informal e-waste recycling system in China. Results of a case study from the Taizhou city in Eastern China are presented.

6.2 Formal E-waste Collection and Recycling System in China

6.2.1 The Policy Framework of E-waste Management

In 2009, the State Council issued the *Administration Regulation for the Recovery and Treatment of Waste Electric and Electronic Products* (China WEEE regulation), which was a milestone for Chinese sustainable management of e-waste. Prior to the China WEEE regulation, certain ministries had already put a series of regulations, policies and implementation measures in place that paved the way for the China WEEE regulation [1,2]. Two of the most prominent are

- The Ordinance on Management of Prevention and Control of Pollution from Electronic and Information Products by the Ministry of Industry and Information Technology (MIIT), which has been in force since 2007and is also called the China RoHs, named after the corresponding EU regulation.
- Administrative Measure on Pollution Prevention of Waste Electrical and Electronic Equipment by the Ministry of Environmental Protection (MEP), adopted in 2008.

These laws and regulations constitute a policy system for promoting the sustainable management of e-waste in China (Figure 1) and cover the entire life cycles of e-products, from design, production and use, to recycling and disposal. The National Standard Institute also drafted the relevant technical codes for e-waste management. Currently, the Ministry of Environment Protection (MEP) plays a key role in implementing the China WEEE regulation.

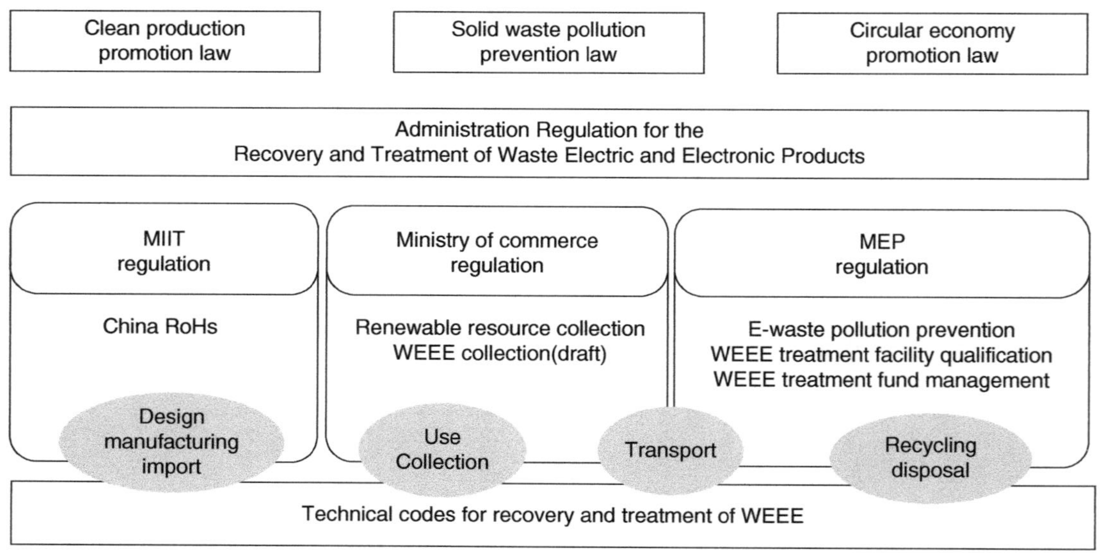

Figure 1 *The Administrative Regulation for the Recovery and Treatment of Waste Electrical and Electronic Products combines serveral e-waste-related laws and regulations. The main impact of this regulation to introduce a standardised financing mechanism*

6.2.2 E-waste flow in China

Today, China is the country with the largest production and consumption of electrical and electronic equipment. Consequently, China is advancing rapidly to become the biggest generator of e-waste. The China Household Appliance Research Institute published an "E-waste White Paper" [3] in 2013 that estimated the theoretical quantity of e-waste generation in the past decade in accordance with the peak of the normal retirement theoretical model (Table 1). The forecast of waste volumes calculated from past sales data shows the number of theoretically scrapped appliances in 2013: 32.04 million TVs, 12.78 million refrigerators, 12.62 million washing machines, 15.30 million air-conditioners, and 37.06 million PCs. The total volume of e-waste within these five categories reached over 100 million, out of which 70% were TV sets and PCs. As statistical data on e-waste in China is still scarce, it can be assumed that these volumes represent only a small portion of the entire Chinese e-waste.

6.2.3 The Mechanism and Practice of WEEE Recycling in China

The China WEEE regulations as shown in Figure 1 unite the concept of life cycle with the principle of extended producer responsibility (EPR). It presents such characteristics as multi-channel collection, centralized treatment, development planning, qualified licensing and treatment subsidies.

The collection system is still mainly dominated by informal sectors and e-waste trade is made both in formal and informal recycling companies. But recently the formal collection system has undergone considerable construction within the area of industry. An increasing number of authorized e-waste recyclers have started to build their own collection systems by installing collection sites in the communities or with the aid of online-to-offline strategies, and even web-based collection systems are now emerging. Four focal measures were implemented in the past few years: planning, e-waste lists, authorized recyclers and e-waste funds. E-waste management planning is implemented by local government on the mainland China at the provincial level. This planning helps to identify local e-waste flow and determine recycling capacity.

The first e-waste list, released in 2010, included a variety of end-of-life products such as

- Television
- Refrigerator
- Washing machine
- Air conditioner and
- Personal computer.

Table 1 *Estimation of e-waste generation in China (1,000 units) [3]. Source: Adpated from CHEARI [3]*

Year	TV set	Refrigerator	Washing machine	Air-conditioner	Personal Computers	Total
2001	6,547.5	379.4	2,985.5	–	–	9,912.4
2006	15,660.3	3,379.7	7,311.8	–	3,934.1	30,286.0
2009	21,980.4	5,462.2	9,813.6	963.8	13,264.0	51,484.0
2010	23,750.2	6,542.5	10,501.5	1,220.6	16,525.5	58,540.3
2011	25,480.4	7,441.7	11,305.2	981.7	21,498.2	66,707.2
2012	27,729.1	8,679.0	12,635.8	1,508.1	25,298.1	75,850.1
2013	32,037.2	12,785.7	12,616.6	15,299.5	37,062.8	109,801.8

Currently e-waste formal recycling practices are confined to these five categories. An expanded e-waste list was released at the end of 2014 that covers nine additional categories:

- Printer
- Fax machine
- Copying machine
- Mobile phone
- Telephone
- Electric water heater
- Gas water heater
- Cooker hood extractor
- Monitor.

The current e-waste fund only covers the five e-product categories. The funding measures for extended categories have not been clarified. In addition, the funds only allocate to authorized recyclers. This does not include collectors.

Some of the principles and approaches (e.g., EPR) of e-waste management that have been implemented in the European Union [4] and other countries have been used as a blueprint for establishing the Chinese e-waste system. In the EU, member states must design a system that ensures that producers are responsible for their end-of-life products. This is often done by public-private or entirely private business partners — so-called "producer responsible organisations".

The China e-waste fund established July, 2012 was also based on the principle of EPR. But one of the key elements of the Chinese system — the e-waste fund itself — differs considerably from other e-waste systems [5]. The core organization in the e-waste fund is the Ministry of Finance (MOF), which is in charge not only of defining the charging rate per electrical or electronic equipment, but also of allocating the collected subsidies to recyclers. The e-waste fund is partly fed by money from producers and importers of e-products based on production volume or on importation volume. The payment rate of products varies for the first five product categories. These are fixed at 13 Chinese yuan renminbi (CNY) per television set, 12 CNY per refrigerator, 7 CNY per washing machine, 7 CNY per air conditioner and 10 RMB per PC (the exchange rate as of July 2015 was US$1 = 6.21 CNY). Producers and importers are charged quarterly a special fee by the national taxation or custom offices. The fee is not included in the value-added tax but is collected by the same means and is listed as a new item in the central government budget.

Authorized recyclers can acquire subsidies from the e-waste fund. The reimbursement is based on the disassembling volume of each recycler. In order to become authorised, recyclers must be assessed and certified by the MEP with the help of the local Environmental Protection Bureau (EBP). Currently, the allocation rate for recycling five e-waste categories is defined as 85 CNY per television, 80 CNY per refrigerator, 35 CNY per washing machine and air conditioner and 85 CNY per PC in the end-of-life phase.

At this point in time, the e-waste fund balance lacks transparency. From the above numbers, it is clear that the fixed rates collected per item do not cover the expenses of recyclers, unless 5–8 times more items are sold than are being recycled. It can be assumed that direct government subsidies are used in order to match the fund's earnings with the reimbursement to recyclers. Figure 2 illustrates the e-waste fund flow and responsibilities of stakeholders. Only recyclers are reimbursed; collectors, transporters or providers of

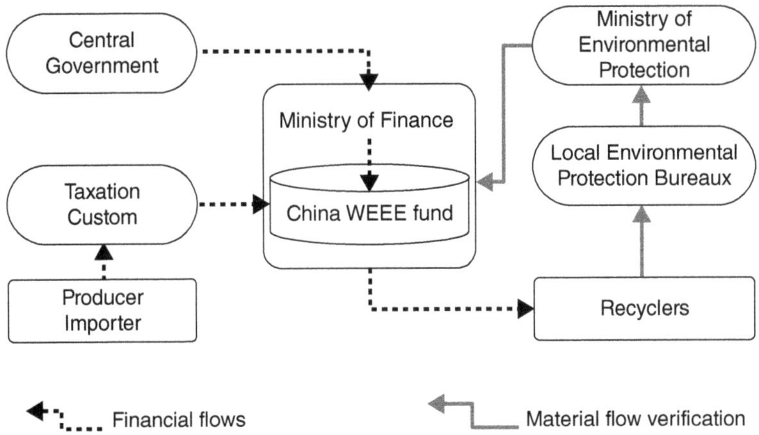

Figure 2 Financial and material flows under the China WEEE fund system

environmental facilities as yet receive no funding. From a life-cycle perspective and the principle of EPR, China's e-waste fund is still in an infant stage. In the future all stakeholders of e-waste flow should be coordinated and supported by the fund. Added to this, current practices only cover the five categories of e-products. The funding measures for extended categories have yet to be clarified.

According to the China WEEE regulation, only authorized recyclers can treat e-waste and receive subsides from the e-waste fund. Implementation of the China WEEE regulation in 2011 triggered rapid growth within the WEEE recycling industry throughout China. At the end of 2012, 43 processing enterprises had been authorized. By July 2014, 106 authorised enterprises offered processing capacity of 112 million units per year. According to MEP, 129 enterprises are due to be authorized, with recycling capacity of 127 million units, by the end of 2015 [6].

The end of 2013 saw about 40 million units of e-waste, which is only 6% of the estimated e-waste generation, recycled by authorized enterprises. The recycled units weighed about 810,000 tons, 94% of which stemmed from television sets. In return, authorised recyclers received 3.3 billion CNY of subsidies from the China e-waste fund.

In general, the formal e-waste recycling system in China can be said to be on the fast track. The central and local government have declared that they will make more efforts to promote more industries to enter recycling business in an environmentally sound way. New business models, e.g. for collection, new processing technologies and recycling equipment, are being installed and tested. Further professionalization of the formal e-waste recycling sector is predicted.

6.3 Informal E-waste Collection and Recycling

6.3.1 Informal Sector and E-waste Management

Informal economies can constitute a considerable amount of the gross national product (GNP) of developing or transitional countries [7]. Recycling e-waste in countries without a stringent e-waste policy often leads to a daisy-chain of processes that are carried out in the informal economy. The informal e-waste sector is characterised by a range of legal, unregistered and publicly accepted businesses, with little concern for illegal and clandestinely executed processes. This has considerable consequences for the environment and human health. The businesses collect, sort and manually separate electrical and electronic equipment, applying crude methods to segregate substances or materials of interest from the electrical/electronic equipment.

The physical characteristics of electrical and electronic equipment have imposed significant and unprecedented challenges for recycling. E-waste consists of a large number of components of various sizes and shapes. Complex connections between substances are often difficult to break up and separate, due to the limitations of separation techniques as well as incompatible thermodynamics [8]. Liberation and separation of the compound materials during shredding or dismantling determines the quality of recycling streams [9]. Consequently, increasing focus is now placed on optimizing interfaces among dismantling, shredding/sorting and integrated metals smelting [10]. Due to the physical and particle properties of e-waste and the diversity of the e-waste stream, selective disassembly has emerged as an indispensable process in e-waste recycling practice [11]. Physical and mechanical processing techniques are already well developed in the industry, but the adoptability of Western recycling technologies in developing countries remains questionable, considering the special material and financial flows of recycling models in these countries.

The literature on manual disassembly is voluminous, though disagreement still exists regarding the profitability and efficiency of this approach [12–14]. Inside China, much research has focused on physical, chemical and biological methods for recycling metallic and non-metallic fractions from waste printed circuit boards [15–19], whilst studies on the applications and possible improvements of manual sorting and dismantling are far less common. The advantage of manual disassembly over automatic or mechanical separation in China is discussed by Wen [20], Li [21] and Wang [22]. The study by Liu [23] on five types of obsolete home appliances in Beijing points out that labour accounts for only about 10% of the total costs, despite the fact that dismantlers account for 70–90% of the total labour in a recycling facility. Low labour costs considerably increase the adoptability of manual disassembly in developing countries.

Numerous studies have described crude e-waste recycling techniques that pose high risks for recyclers themselves both in the immediate vicinity of the recycling locations and in areas further away, especially when the emissions are carried by air or water [24–27]. These techniques include the open burning of printed circuit boards and cables for metal recovery [8,22,24,26–29], toner sweeping, plastic chipping and melting, burning wires to recover copper, heating and acid leaching of printed circuit boards [28–30], gold recovery from printed circuit boards with cyanide salt leaching or nitric acid and mercury amalgamation [31–33], and manual dismantling of cathode ray tubes and open burning of plastics [17–21]. Very often women and children are involved in collecting and processing e-waste with the methods listed above. The persons involved are not aware of the toxic compounds contained in e-waste and do not wear even minimal protective gear such as goggles, gloves, masks or protective clothes [30,34–36].

6.3.2 Informal E-waste Collection and Recycling in China

For China, managing e-waste effectively and efficiently is a continuous challenge, considering its quantum of domestically generated and illegally imported e-wastes [37]. Because of low costs, labour-intensive manual sorting and dismantling of e-waste has been extensively used in the informal recycling sector. Informal manual processing has been criticized for its pollution and health impacts, but manual disassembly itself, if well organized and properly protected, can be an efficient way of separating reusable, hazardous and valuable components from e-waste, greatly simplifying the subsequent recovery of materials and improving reusability [27].

Based on the study by Chi [38], this section introduces how manual sorting and dismantling of e-waste is conducted by informally employed casual workers in Taizhou city, Zhejiang Province, China. Located at the central coastal section of Zhejiang Province, Taizhou is divided into three urban districts (Jiaojiang, Huangyan, and Luqiao), two county-level cities (Linhai and Wenling), and four counties (Yuhuan, Tiantai, Xianju, and Sanmen). The job types and employment situations of casual waste workers are assessed. An illustration of the execution of manual separation of e-waste is then given, highlighting the process flexibility and market adaptability of such an approach. Finally a proposal is put forward for the improved application of manual sorting and dismantling in order to achieve better separation efficiency in the pre-processing stage and to protect the safety and rights of migrant workers.

6.3.2.1 Casual Waste Workers and Recycling Jobs

As a prominent waste recycling region, Taizhou, has been involved in e-waste recycling for nearly 25 years. By the end of 2013, Taizhou had a total population of 5.96 million of which the urban population accounted for 18.3% [39]. Its e-waste industry employs around 40,000 people, with an annual dismantling capacity of over 2.2 million tons [40]. Taizhou processes various waste materials such as scrap metals, waste home appliances, obsolete electric motors, transformers, and cable wires, most of which are imported and have considerable diversity in composition and pre-treatment status.

An investigation in Jiaojiang, Huangyan, and Luqiao in Taizhou in 2010 found that the recycling workforce mainly consists of mobile casual waste workers who are hired on a temporary basis without contract and frequently float between firms [38]. Informal or flexible employment is common in both the formal and informal recycling sectors. In the Fengjiang subdistrict of Luqiao District in Taizhou, a centralized area for processing imported wastes, 62,000 migrant workers from over 20 provinces of China performing waste sorting and dismantling tasks [41]. Casual waste workers are either paid by hour or by weight of processed materials. Table 2 lists twelve casual sorting and dismantling jobs identified during interviews in the "Fengjiang Secondary Metal Processing and Utilization Base" (the largest recycling centre in Taizhou, with area 1.1 km^2, referred as "Fengjiang Base" henceforth). These are:

1. Preliminary sorter: roughly groups materials into five major categories, i.e. iron, aluminium, motors, copper, and box cases, for downstream sorting and dismantling;
2. Big box case dismantler: dismantles large-size electronic devices (e.g. photocopies and printers, etc.) and separates iron, aluminium, motors, copper, plastics, and printed circuit boards;
3. Small box case dismantler: dismantles small-size electronic devices (e.g. DVD players, mainframe boxes of computers, and radios) and separates iron, aluminium, motors, copper, plastics, and printed circuit boards;
4. Big motor dismantler: dismantles motors (unit weight more than 6 kg) and removes aluminium cores, silicon steel, iron cases, and copper;

5. Copper dismantler: separates different types of copper (e.g. red copper, brass, bronze, etc.) and picks out intermixed iron, aluminium, motors, and stainless steel;
6. Small motor dismantler: dismantles motors (unit weight below 6 kg) and takes out copper wires, silicon steel, etc.;
7. Aluminium dismantler: separates cast aluminium, wrought aluminium, copper, iron, etc.;
8. Stainless steel dismantler: separates iron, stainless steel, and stainless iron;
9. Cutting operator: cuts and separates the iron, stainless steel, stainless iron parts and other large interconnected parts by using a oxy-fuel cutting torch;
10. Compressor dismantler: separates the iron and copper parts from compressors by oxy-fuel cutting;
11. Junk sorter: carries out in-depth sorting of fragments from leftovers of previous sorting and disassembly steps;
12. Plastic sorter: sorts different kinds of plastics.

Casual waste workers are knowledgeable in the physical properties of various metals and the interior structures of different electrical and electronic appliances. Using tools like hammers, screwdrivers, pliers, scissors and magnets, they successfully classify highly complex waste mixtures by judging physical characteristics such as color, weight, size,

Table 2 *Overview of casual waste recycling jobs in Taizhou of China*

Job type	Male	Female	Percentage of male	Job type	Average wage (CNY)
Junk sorter	0	16	0%	Hourly paid	50–70 per day
Plastic sorter	2	4	33.3%	Hourly paid	60–80 per day
Aluminium dismantler	4	5	44.4%	Hourly paid	90 per day
Small motor dismantler	3	3	50%	Contract	20 per 500g dismantled copper
Preliminary sorter	10	5	66.7%	Hourly paid	80–90 per day
Box case dismantler	12	2	85.7%	Contract	50–75 per one forklift load of box cases
Big motor dismantler	10	0	100%	Contract	90 per day
Copper dismantler	3	0	100%	Hourly paid	75–85 per day
Stainless steel dismantler	5	0	100%	Hourly paid	90 per day
Cutting operator	2	0	100%	Hourly paid	100–120 per day
Compressor dismantler	4	0	100%	Hourly paid	90–120 per day
Rubbish washer	3	0	100%	Hourly paid	80 per day
Total	**58**	**35**	**62.4%**		

Source: Chi [38].

shape, flexibility, magnetism and smell. Such know-how in recognizing and separating materials, usually accumulated through years of practice and on-the-job learning, is the core element sustaining the wide-scale execution of manual disassembly in Taizhou.

The primary employers of casual waste workers in Taizhou are recycling firms, waste material owners, and foremen (i.e. leaders of waste workers). Casual waste workers seek employment opportunities through individual job search, referral by fellow villagers, and offers from foremen. Wages of sorting and dismantling jobs fluctuate, influenced by the labour supply situation and metal market trends. During apprenticeship the wage is extremely low, for example 20 CNY per day for junior junk sorters and 40 CNY per day for junior box case dismantlers. In addition to their limited access to formal jobs in recycling firms, the mobility of casual jobs encourages migrant workers to remain in the informal employment sector, despite being vulnerable to wage change and arbitrary dismissal in casual employment.

Because of the precarious nature of their employment and the loose relationships with employers, casual workers are not enrolled in any social welfare programs, but are exposed to poor working conditions, low safety and health standards and environmental hazards. Direct exposure to the toxics in e-wastes impairs their health, general well-being and quality of life and those of their families. Dust, noise, smoke are the common types of pollution and dangers that workers must face. Many workers interviewed reported chronic stomach ache and throat pain [38]. In fact, human exposure to toxics by inhalation of ambient air, dust ingestion, dermal contact and dietary intake has already been signalled in e-waste dismantling sites [42–44]: for instance, Xing et al. [42] detected high levels of polychlorinated biphenyls in the transformer recycling workshops in Fengjiang. Poor working conditions in closed warehouses with no ventilator/fan or dust collection facility have increased the exposure to toxics and hazardous substances. The lack of basic protection (e.g. masks and gloves) makes the chances of injury even higher. It is thus imperative to provide employers and workers with information and guidance regarding how to carry out manual dismantling in ecologically and environmentally harmless ways. Minimum investment in simple and inexpensive protective measures can improve the working condition for these waste workers.

6.3.2.2 *Organization of Manual Sorting and Dismantling*

Depending on the inputs, manual sorting and disassembly can have diverse arrangements of sequence and depth. Figure 3 illustrates the manual sorting and dismantling processes conducted by casual workers under supervision of foremen, as observed in the Fengjiang Base. The processes aim at achieving quick separation of mixed wastes and dismantling of major parts like plastic frames, cables, and printed circuit boards from electrical and electronic devices, containing three key stages:

Stage 1: preliminary sorting: coarse classification of original materials into five groups, i.e. iron, aluminium, copper, motors, and box cases;

Stage 2: categorized dismantling: in-depth separation and upgrading of materials in several parallel dismantling workshops. Both small appliances like radios and big appliances like photocopiers are taken apart by hand. Various categories of materials including ferrous metals, non-ferrous metals, cables, plastics, rubber, and the composite materials

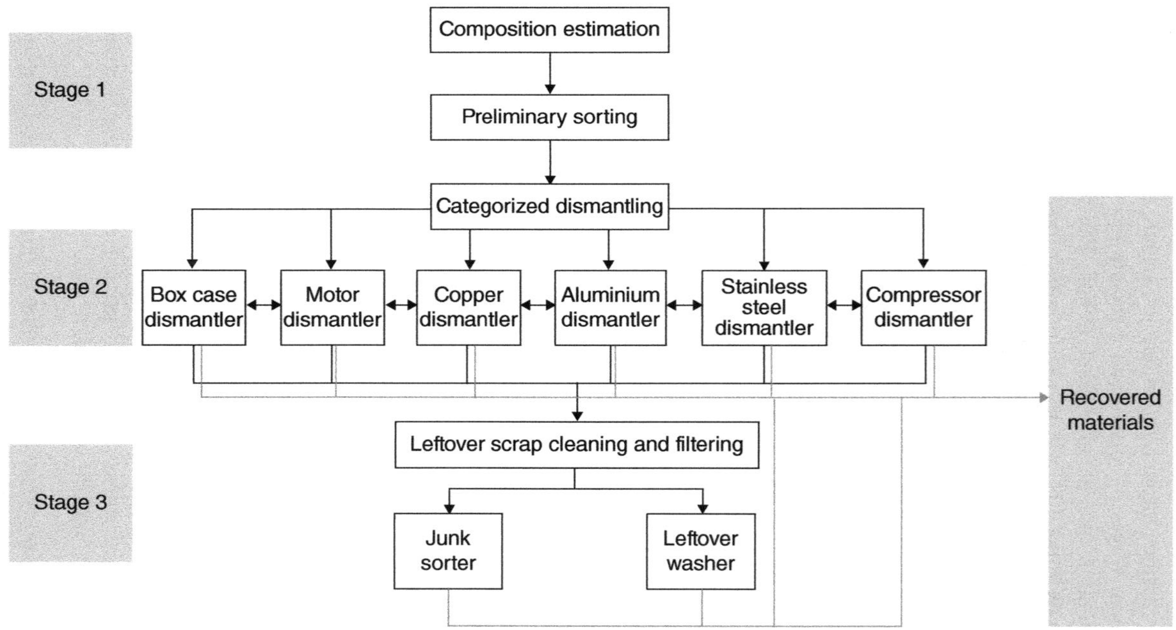

Figure 3 *General flowchart of manual sorting and dismantling procedures in Fengjiang Subdistrict of Taizhou, China. Source: Chi [38]*

are separated by respective dismantlers. Recyclates are also circulated among different dismantling workshops to improve the purity of outputs.

Stage 3: leftover scrap cleaning and filtering: picking out remaining recyclable particles such as chips and metal oddments from the consolidated leftovers of previous dismantling sessions. Junk sorters and rubbish washers will carry out meticulous final sorting to reclaim tiny valuable materials.

Such sorting and dismantling model lead to a flexible and predictable recycling path that reacts quickly to changes in demand. This generates highly marketable products that flow to diverse downstream buyers: e-waste components such as printed circuit boards, sound and media cards, and compact drives are often handed over to informal workshops or second-hand markets; metal fractions are supplied to metal and steel firms for final recovery; and waste plastics are usually recycled locally [38].

The diversity and heterogeneity of imported wastes in Taizhou leads to variations in labour requirements of the pre-processing stage, and encourages recyclers to shift from formal wage workers to casual workers. Casual employment results in immediate and discretionary adjustments of labour arrangement according to the properties of incoming materials. Meanwhile, growing international waste trading and the extension of global recycling chains have led to regional specialization, in which materials are first sorted by low-cost labourers in developing countries and then sent to advanced refining facilities located in developed countries.

6.3.3 Interactions between the Formal and Informal Sectors

Informal e-waste recycling has not only generated income opportunities for numerous low-income people, but has also led to the emergence of dynamic entities with intense links between the formal and informal economy [46]. The similarity of the formal and informal sectors, in terms of processing methods and efficiency, seems most obvious at the pre-processing stage where manual separation takes place. On the other hand, the financial and technical requirements of setting up end-processing facilities have prevented most informal stakeholders from entering the formal sector. Also, the existing e-waste collection network is diverted into formal collection (mainly businesses and industry) and informal collection (mainly households). Households generally tend to sell e-waste to informal collectors. The convenient services and buyout prices provided by informal collectors discourage households from returning their e-wastes to the formal sector [38,47,48].

Meanwhile, though regulations have included legal definitions, differentiating formal e-waste recyclers from informal ones in the actual recycling chain presents challenges, since there are not always clear splits between the two sectors. The co-dependency between the formal and informal sectors is reflected by the frequent material interflows between the two. Segmentation of the recycling industry and the variety of semi- and secondary products result in active movements of materials between treatment stages. The existence of unregulated second-hand markets, trading profits and the space of upgrading recyclate purity through repeating manual separation also encourage interactions among stakeholders from different sectors [47].

Currently there are overlaps between the formal and informal sectors in the e-waste cycle that result in unnecessary competition and waste of resources. Logically, the informal sector is most efficient during the collection and disassembly stages, whilst the formal sector should control most of the end solutions for metal recovery. However, in China informal recyclers also participate in rudimentary metal extraction, and formal recyclers start developing their own collection networks and installing pre-processing operations. In order to achieve an optimal division of labour in the recycling chain, regulations and policies need to provide enough incentives and the right controls. For instance, the practical skills of casual waste workers provide alternatives for more formal organization with improved safety and working conditions. Cooperation and partnership between the formal and informal sectors would help not only reduce the pollution and harm caused by improper e-waste management, but also significantly increase overall reuse and recycling efficiency.

Informal recycling activities with diverse applications should not be simply defined as "a sector". The connections between the formal and informal sectors at different recycling stages should be signalled. The choice of recycling technique is a complicated outcome of various interlinked factors, including the complexities of waste materials, market demands on reusable and recyclable resources, existent actors and their skill levels, legislative frameworks and degree of government control, cultural and social norms, global e-waste movements, and the labour cost advantage in developing countries. Despite some negative impacts, the informal sector has also made its own contributions to the local recycling chain, such as extending product life cycles through device and component reuse, supplying sorted materials to formal refineries, and creating job opportunities for migrant workers. Government interventions need to take into account market conditions and economic stimuli influencing recycling behaviours. The interactions between the formal and informal sectors indicate not only the increasing links between these two sectors but also the importance of more supportive policies for building cooperative and complementary relationships between them.

6.4 Conclusions

Current situation e-waste management in China offers opportunities but also bears certain risks. The opportunity to build a formal, national and government-controlled e-waste management system is globally unique. Such a situation would bring forth strong financial resources, creating freedom of operation and opportunities for guiding and steering the system. However, it also risks lack of transparency, distorting the recycling market and preventing competition as well as innovation — not to mention the danger of corruption.

Notably, the case study showed the large variety of jobs created in the informal e-waste industry employing casual waste workers. Many of the jobs are created in a very competitive business environment without any subsidies or additional funding. The strength of manual sorting and dismantling models lies in three aspects: its capacity of handling diversified inputs, its great cost-efficiency and its adaptability to changing market demands. E-waste streams frequently need specific adaptations — e.g., when toxic elements or valuable components must be taken out before mechanical treatment. Such selective separation can best be done by manual or semi-manual sorting.

The drawback of many of the e-waste recycling jobs is their relatively low pay, which mainly attracts casual or migrant workers. This results in a high cost -effectiveness for the informal recycling industry. Jobs directly involved in material recovery activities are often carried out without any protective measures for human or environmental health. The flexibility to adjust to changing market conditions is due to the nature of informal market activities of daily paid workforces.

An improved application of informal sorting and dismantling procedures in formal recycling operations deserves closer attention. China should develop its own recycling procedures, matching the mass flow of fractions in e-wastes, and fully consider the application of labour-intensive manual separation, which can both achieve efficient separation and create jobs. Comparison of the wages paid in the informal sectors and the reimbursement offered by the e-waste fund points to a large discrepancy between the formal and informal sector. The reimbursement from the e-waste fund for one PC is equal to the daily wage of four junior junk sorters; reimbursement for one refrigerator is equal to the cost of a junior dismantler. This cost comparison does not take into account the profits gained from the recovered material. Some reimbursement funding could be used to upgrade informal e-waste recycling jobs.

Currently, the formal system fails to reach satisfactory collected e-waste volumes from households. Neither does it manage to investigate adequately all means of improving collection by integration of the informal (collection) sector. The informal e-waste collection domain could offer opportunities for the upgrading of collectors as associated participants in a formal e-waste management system. By assuring minimal reimbursement for the collection and delivery of entire electrical and electronic equipment (without taking out valuable parts before delivery), a formalised e-waste collection industry sector might be created.

A process of legalization to bring casual waste workers within the legal framework, so that they are registered, recognized and well protected, is essential. The introduction of an educational and certifying program to recognise and respect recycling as a profession might help in such a process.

Experience in Europe shows that management systems of e-waste need to be designed in a multi-stakeholder dialogue. Producers, retailers, importers, consumers, collectors and recyclers have to participate when setting up such a system. Today, advanced mechanical recycling is combined with educational programmes for recycling personnel on academic and non-academic levels.

China now has a system installed that creates financial resources for e-waste treatment. If parts of the informal jobs can be granted benefits from the e-waste fund and integrated into the formal system, a truly sustainable e-waste recycling system could be created: environmentally safe, efficient and profitable in resource recovery and upgrading health and societal status of the workers involved.

References

1. Streicher-Porte, M., Kummer, K., Chi, X., Denzler, S., Wang, X., 2010, Chinese E-Waste Legislation, Current Status and Future Development. *elni Rewiew*, ISSN 1618-2502.
2. Streicher-Porte, M., Yang, J., 2007, WEEE Recycling in China: Present Situation and Main Obstacles for Improvement, in: *Proceedings of the 2007 IEEE International Symposium on Electronics and Environment*, Orlando.
3. CHEARI, 2014, China Home Electric Appliance Research Institute (CHEARI), 2013 White paper on WEEE recycling industry in China, May 2014.

4. EU, 2012/19/EU, 2012, DIRECTIVE 2012/19/EU of the European Parliament and of the Council of 4 July 2012 on Waste Electrical And Electronic Equipment (WEEE) (recast).
5. Zeng, X., Li, J., Stevels, A.L.N., Liu, L., 2013, Perspective of Electronic Waste Management in China Based on a Legislation Comparison Between China and the EU, *Journal of Cleaner Production,* 51, 80–87. doi:10.1016/j.jclepro.2012.09.030
6. Ministry of Environmental Protection, 2015, Solid Waste and Chemicals Management Center, Ministry of Environmental Protection. Progress of E-waste Regulations in China, presentation on the Seminar on EPR of WEEE in China, 14 May 2015.
7. Schneider, F., Enste, D.H., 2003, *The Shadow Economy, an International Survey*, Cambridge University Press, Cambridge, UK.
8. Reuter, M.A., Verhoef, E.V., 2004, A Dynamic Model for the Assessment of the Replacement of Lead in Solders, *Journal of Electronic Materials*, 33, 1567–1580. doi:10.1007/s11664-004-0100-3
9. Van Schaik, A., Reuter, M.A., 2010, Dynamic Modelling of E-Waste Recycling System Performance Based on Product Design, *Minerals Engineering*, 23, 192–210. doi:10.1016/j.mineng.2009.09.004
10. Sepúlveda, A., Schluep, M., Renaud, F.G., Streicher, M., Kuehr, R., Hagelüken, C., Gerecke, A.C., 2010, A Review of the Environmental Fate and Effects of Hazardous Substances Released from Electrical and Electronic Equipment During Recycling: Examples from China and India, *Environmental Impact Assessment Review*, 30, 28–41.
11. Cui, J., Forssberg, E., 2003, Mechanical Recycling of Waste Electric and Electronic Equipment: a Review, *Journal of Hazardous Materials,* 99, 243–263. doi:10.1016/S0304-3894(03)00061-X
12. Stevels, A.L.N., Ram, A.A.P., Deckers, E., 1999. Take-Back of Discarded Consumer Electronic Products from The Perspective of the Producer: Conditions for Success, *Journal of Cleaner Production*, 7, 383–389. doi:10.1016/S0959-6526(99)00152-3
13. Grit Walther, T.S., 2005, Impact of WEEE-Directive on Reverse Logistics in Germany, *International Journal of Physical Distribution & Logistics Management*, 35, 337–361. doi:10.1108/09600030510607337
14. Prakash, S., Manhart, A., 2010, E-Waste Africa Project — Introduction to the Study on the Feasibility of International E-Waste Recycling Co-Operations Between Ghana and Europe [WWW Document]. URL http://www.resourcefever.com/detail/items/e-waste-africa-project—introduction-to-the-study-on-the-feasib.html (accessed 3/3/15).
15. Mou, P., Wa, L., Xiang, D., Jiangang, G., Duan, G., 2004, A Physical Process for Recycling and Reusing Waste Printed Circuit Boards, in *2004 IEEE International Symposium on Electronics and the Environment,* 2004. Conference Record, 237–242. doi:10.1109/ISEE.2004.1299722
16. Wen, X., Li, J., Hao, L., Yin, F., Hu, L., Liu, H., Liu, Z., 2006, An Agenda to Move Forward E-waste Recycling and Challenges in China, in: *Proceedings of the 2006 IEEE International Symposium on Electronics and the Environment*, 2006, 315–320. doi:10.1109/ISEE.2006.1650083
17. Zhao, Y., Wen, X., Li, B., Tao, D., 2004, Recovery of Copper from Waste Printed Circuit Boards. *Minerals & Metallurgical Processing*, 21(2), 99–102.
18. Guo, J., Guo, J., Xu, Z., 2009, Recycling of Non-Metallic Fractions from Waste Printed Circuit Boards: A Review, *Journal of Hazardous Materials* 168, 567–590. doi:10.1016/j.jhazmat.2009.02.104
19. Huang, K., Guo, J., Xu, Z., 2009, Recycling of Waste Printed Circuit Boards: a Review of Current Technologies and Treatment Status in China, *Journal of Hazardous Materials* 164, 399–408. doi:10.1016/j.jhazmat.2008.08.051
20. Wen, X., Zhao, Y., Duan, C., Zhou, X., Jiao, H., Song, S., 2005, Study on Metals Recovery from Discarded Printed Circuit Boards by Physical Methods, in *Proceedings of the 2005 IEEE International Symposium on Electronics and the Environment*, 2005. pp. 121–128. doi:10.1109/ISEE.2005.1437005

21. Li, J.H., 2007, E-waste Management and Fecycling in China. Presented at Thailand's Electrical and Electronic Green Society International Conference, TEEGS, Thailand, 14-15 September 2007.
22. Wang, F., Huisman, J., Meskers, C.E.M., Schluep, M., Stevels, A., Hagelüken, C., 2012, The Best-of-2-Worlds Philosophy: Developing Local Dismantling and Global Infrastructure Network for Sustainable E-Waste Treatment in Emerging Economies, *Waste Management* 32, 2134–2146. doi:10.1016/j.wasman.2012.03.029
23. Liu, X., Tanaka, M., Matsui, Y., 2009, Economic Evaluation of Optional Recycling Processes for Waste Electronic Home Appliances. *Journal of Cleaner Production* 17, 53–60. doi:10.1016/j.jclepro.2008.03.005
24. Steiner, S., 2004, Risk Assessment of e-Waste Burning in Delhi, India (Master's Thesis). Swiss Federal Institute of Technology (ETH), Materials Science and Technology Research Institute, Zurich and St. Gallen, Switzerland.
25. Toxics Link, 2004, E-waste in Chennai: Time Is Running Out. Chennai.
26. Brigden, K., Labunska, I., Dadntillo, D., Allsopp, M., 2005, *Recycling of Electronic Wastes in China and India: Workplace and Environmental Contamination*. Greenpeace International.
27. Chi, X., Streicher-Porte, M., Wang, M.Y.L., Reuter, M.A., 2011, Informal Electronic Waste Recycling: a Sector Review with Special Focus on China, *Waste Management* 31, 731–742. doi:10.1016/j.wasman.2010.11.006
28. Hicks, C., Dietmar, R., Eugster, M., 2005, The Recycling and Disposal of Electrical and Electronic Waste in China — Legislative and Market Responses, *Environmental Impact Assessment Review* 25, 459–471.
29. Leung, A., Wei Cai, Z., Hung Wong, M., 2006, Environmental Contamination from Electronic Waste Recycling at Guiyu, Southeast China, *Journal of Material Cycles & Waste Management*, 8, 21–33.
30. Leung, A.O.W., Duzgoren-Aydin, N.S., Cheung, K.C., Wong, M.H., 2008, Heavy Metals Concentrations of Surface Dust from e-Waste Recycling and Its Human Health Implications in Southeast China, *Environmental Science & Technology*, 42, 2674–2680. doi:10.1021/es071873x
31. Keller, M., 2006, Assessment of Gold Recovery From Printed Wiring Boards in Bangalore, India and Evaluation of an Alternative Recycling Path, Master's thesis, Department of Environmental Sciences, Federal Institute of Technology.
32. Rochat, D., Hagelüken, C., Keller, M., Widmer, R., 2007, Optimal Recycling for Printed Wiring Boards (PWBs) in India, in *Proceedings of the R'07 World Congress, Recovery of Materials and Energy for Resource Efficiency*, Davos, Switzerland.
33. M. Torre, D.B., 2006, Cyanide Recovery from Gold Extraction Process Waste Effluents by Ion Exchange I. Equilibrium and Kinetics, *Solvent Extraction and Ion Exchange*, 24, 99–117. doi:10.1080/07366290500388657
34. Lundgren, K., 2012, *The Global Impact of E-Waste: Addressing The Challenge*, International Labour Organization.
35. Robinson, B.H., 2009, E-waste: An Assessment of Global Production and Environmental Impacts, *Science of The Total Environment,* 408, 183–191.
36. Li, Y., Xu, X., Liu, J., Wu, K., Gu, C., Shao, G., Chen, S., Chen, G., Huo, X., 2008, The Hazard of Chromium Exposure to Neonates in Guiyu of China, *Science of The Total Environment*, 403, 99–104. doi:10.1016/j.scitotenv.2008.05.033
37. Qu, Y., Zhu, Q., Sarkis, J., Geng, Y., Zhong, Y., 2013, A Review of Developing an E-Wastes Collection System in Dalian, China. *Journal of Cleaner Production*, 52, 176–184. doi:10.1016/j.jclepro.2013.02.013
38. Chi, X., 2012, *Between Formality and Informality: Recycling of Waste Electrical and Electronic Equipment in China* (Dissertation), Melbourne School of Land and Environment - Resource Management and Geography, Melbourne.

39. Taizhou Statistical Yearbook, 2013, http://www.zjtz.gov.cn/zwgk/xxgk/028/05/0504/201409/t20140910_292135.shtml (in Chinese).
40. Chan, Xing, G.H., Xu, Y., Liang, Y., Chen, L.X., Wu, S.C., Wong, C.K.C., Leung, C.K.M., Wong, M.H., 2007, Body Loadings and Health Risk Assessment of Polychlorinated Dibenzo-p-dioxins and Dibenzofurans at an Intensive Electronic Waste Recycling Site in China, *Environmental Science & Technology*, 41, 7668–7674. doi:10.1021/es071492j
41. Shen, Y.B., 2008, Fengjiang xiao zhen shang de nong min gong (Migrant workers in Fengjiang Town). *Law & Life*, 10 (in Chinese).
42. Ma, J., Kannan, K., Cheng, J., Horii, Y., Wu, Q., Wang, W., 2008, Concentrations, Profiles, and Estimated Human Exposures for Polychlorinated Dibenzo-p-Dioxins and Dibenzofurans from Electronic Waste Recycling Facilities and a Chemical Industrial Complex in Eastern China, *Environmental Science & Technology*, 42, 8252–8259. doi:10.1021/es8017573
43. Wang, J., Ma, Y.-J., Chen, S.-J., Tian, M., Luo, X.-J., Mai, B.-X., 2010, Brominated Flame Retardants in House Dust from E-Waste Recycling and Urban Areas in South China: Implications on Human Exposure, *Environment International*, 36, 535–541. doi:10.1016/j.envint.2010.04.005
44. Zhao, X.-R., Qin, Z.-F., Yang, Z.-Z., Zhao, Q., Zhao, Y.-X., Qin, X.-F., Zhang, Y.-C., Ruan, X.-L., Zhang, Y.-F., Xu, X.-B., 2010, Dual Body Burdens of Polychlorinated Biphenyls and Polybrominated Diphenyl Ethers Among Local Residents in an E-Waste Recycling Region in Southeast China, *Chemosphere*, 78, 659–666. doi:10.1016/j.chemosphere.2009.12.013
45. Xing, G.H., Liang, Y., Chen, L.X., Wu, S.C., Wong, M.H., 2011, Exposure to PCBs Through Inhalation, Dermal Contact and Dust Ingestion at Taizhou, China — a Major Site for Recycling Transformers. *Chemosphere*, 83, 605–611. doi:10.1016/j.chemosphere.2010.12.018
46. Oteng-Ababio, M., 2012, When Necessity Begets Ingenuity: E-Waste Scavenging as a Livelihood Strategy in Accra, Ghana, *African Studies Quarterly*, 13(1/2), 1–21.
47. Chi, X., Wang, M.Y.L., Reuter, M.A., 2014, E-waste Collection Channels and Household Recycling Behaviors in Taizhou of China, *Journal of Cleaner Production*, 80, 87–95. doi:10.1016/j.jclepro.2014.05.056
48. Streicher-Porte, M., Geering, A.-C., 2010, Opportunities and Threats of Current E-Waste Collection System in China: A Case Study from Taizhou with a Focus on Refrigerators, Washing Machines, and Televisions, *Environmental Engineering Science*, 27, 29–36. doi:10.1089/ees.2009.0134

7

Metallurgical Recovery of Metals from Waste Electrical and Electronic Equipment (WEEE) in PRC

Xueyi Guo[1], Yongzhu Zhang[1], and Kaihua Xu[2]

[1]*Research Institute for Resource Recycling, School of Metallurgy and Environment, Central South Univeristy, Changsha, Hunan, PRC*

[2]*GEM CO., LTD, Marina Bay Center, South of Xinghua Rd., Bao'an Center Area Shenzhen, 518101, PRC*

7.1 Introduction

The production of electric and electronic equipment (EEE) is one of the fastest-growing businesses in the world. In the meantime, both technological innovation and market expansion of EEE are accelerating the replacement of outdated EEE, leading to a significant increase in waste EEE (WEEE) that induces a new environmental challenge.

WEEE encompasses a broad and growing range of EEE from home appliances to computers at the end of their lives. Waste electrical and electronic equipment (WEEE) requires an individual description due to its particular characteristics. Also called electronic waste, e-waste or e-scrap, WEEE is a fast-growing waste stream that contains loosely discarded, surplus, obsolete, or broken electrical or electronic devices. WEEE materials are often considered hazardous waste. WEEE comprises of a range of electrical and electronic items such as: refrigerators, IT and telecommunications equipment, freezers, washing machines, medical equipment, televisions, etc. Therefore, sources of WEEE are from all users of electrical and electronic equipment from all sectors.

Metal Sustainability: Global Challenges, Consequences, and Prospects, First Edition. Edited by Reed M. Izatt.

The global production of e-waste/WEEE is increasing rapidly and is expected to accelerate in the near future. Currently, 20 to 25 million tons per year of e-waste are being generated globally, with major shares in Europe, China, USA and Australasia. However, Eastern Europe and Latin America are expected to become significant e-waste producers in the next decade.

WEEE is non-homogeneous and complex in terms of the materials and components. Many of the materials are highly toxic, such as chlorinated and brominated substances, toxic metals, photoactive and biologically active materials, acids, plastics and plastic additives. Major categories of hazardous materials and components of WEEE are

- Batteries: heavy metals such as lead, mercury and cadmium are present in batteries;
- Printed circuit boards (PCBs): cadmium occurs in certain components; and
- Cathode ray tubes (CRTs): lead in the cone glass and fluorescent coating cover the inside of panel glass.

With these hazardous elements, WEEE can cause serious environmental problems during disposal if not properly pre-treated. For example, the cadmium from one mobile phone battery is sufficient to pollute 600,000 liters of water. Growing attention is being given to the impacts of the hazardous components in WEEE on the environment. Due to the fast-growing nature of the WEEE and its potentially hazardous qualities it has qualified as a priority waste stream by the European Union. Disposal of WEEE is often inadequate and causes serious health and pollution problems.

WEEE also has high residual value. The main components of WEEE, in terms of weight, are iron and steel, followed by plastics [1]:

1. The relative composition is as follows: 38% ferrous, 28% non-ferrous, 19% plastics, 4% glass, 1% wood, and 10% other.
2. In general, printed circuit board (PCB) scrap contains approximately 40% metals, 30% plastics, and 30% ceramics. The typical metal scrap in PCB consists of copper (20%), iron (8%), tin (4%), nickel (2%), lead (2%), zinc (1%), silver (0.2%), gold (0.1%), and palladium (0.005%).
3. Polyethylene, polypropylene, polyesters, polycarbonates and phenolformaldehyde are the typical plastic components.

These materials are valuable and could be recycled by proper technologies. With the recognition of the deleterious environmental effects of dumping WEEE, the decreasing capacity of landfills, and the values of the reusable materials, legislation in many countries has refocused on the management of WEEE, and many technologies are being developed and used for WEEE recovery.

7.2 Major Sources of E-Waste in China

China is the largest emerging market region in the world with both global manufacturing and consumer market. Rapid economic growth, coupled with urbanization and a growing middle class, is expected to increase the consumption of EEE and thus increase the WEEE rapidly in this region, posing a severe threat to the environment and sustainable economic growth.

TV sets, refrigerators and washing machines are three kinds of widely used domestic electric equipment in China. Most of them were made in the middle or late 1980s and could have been replaced by new types after 2000, so that millions of them have been entering into the waste stream per year. The sales volumes of air conditioners and PCs, two more kinds of typical EEE, have been rapidly increasing since the 1990s.

The e-waste recycled in China comes mainly from three sources: consumption, importation and production. According to a recent study, China generated 1.7 million tons of e-waste in 2006, equal to 1.3 kg of e-waste per capita. By 2015, the figure is expected to increase to approximately 400 million units, or 5.4 million tons (Chi 2011).

1. The constant increase of domestic EEE consumption leads to a corresponding, time-delayed increase in e-waste; meanwhile we should also note that today there is still a significant rural–urban difference in the possession of five EE products in China. Between 2008 and 2012, there was a predicted sharp rise in obsolete personal computers (PCs), televisions (TVs) and air-conditioners, reaching by 2012 93.36 million units, 74.31 million units and 63.9 million units, respectively. While refrigerators and washing machines are estimated to have more moderate obsolescence rates, the numbers of discarded refrigerators and washing machines also almost doubled and trebled within the eleven years from 2001 to 2012.
2. At the same time, illegal imports of used EE products or WEEE from overseas have added to the volume of e-waste being treated in China. China now appears to be the largest dumping site of e-waste in the world, receiving continuous e-waste shipments from the U.S., Europe and neighboring Asian countries, including South Korea and Japan. Trans-boundary movement of e-waste is primarily profit-driven. Recyclers and waste brokers are taking advantages of lower recycling costs and higher revenues accruing from machine reuse in China, while also trying to evade their disposal responsibilities in their home countries. The legislation gaps in e-waste management between countries and relatively weak customs control in China also provide opportunities for illegal entry of e-waste.
3. The third source of the huge amount of e-waste is the electronics industry, which is a major economic driver of China and one of the fastest-growing sectors since the 1980s. Exports from the EEE sector earned China US$227.46 billion in 2003, accounting for 51.9% of the country's total exports. Today, China is the world's largest producer, consumer and exporter of EEEs. Consequently, scrap generated during EE manufacturing processes cannot be ignored in the domestic e-waste streams, though the total volume of production e-waste is much smaller and easier to control than that coming from domestic EEE consumptions and illegal imports.

Waste substances are secondary resources in the wrong place. Their effective recovery is important not only for environmental development, but also for rational utilization of natural resources. In view of the serious environmental problems caused by WEEE and the valuable materials they contain, great attention has been paid to them, and some corresponding projects have been put into action in China.

7.3 Strategies and Regulations for WEEE Management and Treatment

7.3.1 Strategies for WEEE Management

In view of the environmental problems and high residual value of WEEE, a WEEE management system should be established to extend the life cycle of EEE. This management system comprises collection, classification, pre-treatment, etc., and conventional end-of-life treatment strategies. In accordance with the potential economic and environmental efficiency, strategies can be categorized as follows:

1. reuse: the recovery and trade of used products or their components as originally designed;
2. servicing: a strategy aimed at extending the usage stage of a product by repair or maintenance;
3. remanufacturing: the process of removing specific parts of the waste product for further reuse in new products;
4. recycling (with or without disassembly): the treatment, recovery, and reprocessing of materials contained in the used products or components in order to replace the virgin materials in the production of new goods;
5. disposal: the processes of incineration (with or without energy recovery) or landfill.

7.3.2 European Regulations

The European Union (EU) has established comprehensive environmental legislation that has proven innovative and successful in many aspects. The current end-of-life model of EEE implemented in EU was established on the basis of EU Directives on WEEE and Restriction of Hazardous Substances (RoHS) approved in October, 2002.

The EU Directive on WEEE is based on the producer-pays principle. The scope of this Directive includes producers, distributors, consumers and all parties involved in the treatment of EEE. Producers are required to finance the collection, treatment, recovery, and environmentally sound disposal of WEEE. The directive imposes a high recycling rate for all targeted products that varies from 50% to over 80% depending on the WEEE type. The strategy includes collection, sorting, refurbishment, dismantling, shredding, treatment of recyclable components/materials, treatment of hazardous components-substances, and landfill or incineration (with heat recovery) of remaining wastes. The objective of the Directive is to reduce the amount of such waste going to landfill and to improve the overall environmental performance of EEE products during their life cycle.

The RoHS (Restriction of the use of certain Hazardous Substances) directive bans the presence of certain hazardous substances in EEE being traded in EU market. RoHS ensures that new EEE does not contain mercury, lead, cadmium, hexavalent chromium, polybrominated biphenyls, or polybrominated phenyl ethers. It aims to control the environmental impact of E waste by implementing clean production.

7.3.3 Regulations for WEEE Management in China

In China, the first legal document related to WEEE management is the List of Wastes Prohibited for Import, which was constituted in 1996 based on the Basel Convention. Waste batteries, air-conditioners, refrigerators, microwave ovens, phones, and TV sets, etc. are all on the list. Entering the 21st century, legislation, management strategies and recycling technologies on WEEE arouse great attention from government agencies and researchers. Legislation, research and international co-operation on WEEE treatment have been initiated.

In September 2003, an announcement on enhancing the environmental management of WEEE was made by the State Environmental Protection Administration of China (SEPA). On 29 December 2004, the revised Law of the People's Republic of China on the Prevention and Control of Environmental Pollution by Solid Waste was approved by National People's Congress and has been effect since 1 April 2005. This legislation has joined the regulation on dismantling, recycling and disposal of WEEE. Based on the above announcement and revised state law, a "Management Method on Prevention and Cure of Environmental Pollution in WEEE" was established by SEPA in early 2005 and is being discussed. The main contents of this legislation follow:

1. information on WEEE generation, transportation, storage, and disposal should be reported to the local environmental protection department at or above county level;
2. waste lead/acid batteries, nickel/cadmium batteries, mercury switches and relays, cathode-ray tubes, and PCB-bearing capacitors shall be collected, stored and disposed of. Duplicate forms designed for the transfer of hazardous wastes must be filled for WEEE transfer, and reported to the competent administrative department of environmental protection;
3. outdated recycling technology for WEEE that may induce pollution is prohibited. Open burning, simple incineration and rough extraction of metals from WEEE using acid are banned;
4. enterprises engaged in WEEE collection, storage, disposal shall apply to the competent administrative department of environmental protection at or above the provincial level for an operation license. Any illegal activities will be punished in accordance with the "Law of the PRC on the Prevention and Control of Environmental Pollution by Solid Waste";
5. local EPBs should take measures to encourage clean production in EEE manufacturing. Pollutants such as lead, mercury, cadmium, chromium (VI), PBB, PBDE, etc., should be gradually phased out in EEE manufacturing. Environmental sound design and packaging should be employed.

The present e-waste-related laws and regulations in China and their key contents have been summarized by Chi [2]. Present regulations set control on EEE production and WEEE management mainly from four aspects:

1. restriction of the usage of six toxic and hazardous substances in electronic information products ('Measures for the Administration of Prevention and Treatment of Pollution by Electronic Information Products');
2. pollution prevention and control on dismantling, utilizing and disposal of e-waste ('Technical Policies for Controlling Pollution of WEEE');
3. certification requirements for e-waste recycling operation ('Administrative Measures for the Prevention and Control of Environmental Pollution by WEEE');
4. responsibility of producers and importers for their products and for the e-waste collection and treatment fund ('Regulation on the Administration of the Recovery and Disposal of WEEE').

These regulations have successfully provided timely industrial guidance and basic legal constraint on e-waste management, and have responded the international trend of cleaner EEE production (with less toxic use) and better WEEE disposal (with more environmental concern). On the other side, it can also be observed that in these regulations some key elements and principles are still defined in very general terms which undoubtedly restrict their legal power, and the lack of subsequent implementation rules and measures makes enforcement difficult. Aside from national legislation development, a number of local administrative measures were introduced in major regions such as Beijing, Shanghai, Nanjing and Guangdong province to regulate the behavior of enterprises, dealers and customers, and to build the local system of e-waste take-back, reuse and recycling.

7.3.4 Implementation of Regulations Related to E-Waste

The sheer magnitude of e-waste generation in China and its linked environmental and health consequences call for clear policies, regulations, systems and institutions if effective management and control are to be established. In light of the potential impact on its EE industry and the current state of domestic recycling, the Chinese government has taken a gradual and cautious approach toward regulating e-waste [3].

In view of the successful WEEE strategies in current China, a "Management Regulation on the Recycling and Treatment of Disposed Appliances and Electronics Products (MRRTDAEP)" was drafted jointly by six ministries, including Ministry of Information Industry (MoII), Ministry of Commerce (MoC) and SEPA in September 2004. The revised version of the Regulation was approved and will soon be promulgated by State Development and Reform Commission (SDRC). This new regulation covers all types of EEE on the market, regardless of whether is produced in or imported to China. It takes source recycling and environmental protection as its goal, and puts into practice an extended producer responsibility system. Its main contents are:

1. local governments shall be responsible for the collection and treatment of WEEE and must establish corresponding detailed implementing rules;
2. multiple collections and centralized treatment of WEEE should be implemented. WEEE recycling enterprises should be approved and authorized by SEPA;
3. the state shall create a special fund for WEEE recycling;
4. WEEE recycling enterprises shall operate according to market rules;
5. The government promotes and supports WEEE enterprises by formulating favorable policies and regulations;
6. manufacturers of PCs, TV sets, mobile phones, DVD players, refrigerators and air-conditioners are required to select environmentally friendly raw materials for their products and to recycle the discarded ones;
7. manufacturers will be severely punished if they do not obey the new rules. They can either be fined and/or have their business licenses revoked.

Competition in the domestic electric appliance business is extremely intense in China. The profit per piece for domestic electric equipment is usually trivial. It is hardly oossible for producers entirely to finance the collection, treatment, recovery and disposal of WEEE like EU. With the reference of the successful experience in other countries, rules for charging for WEEE management that suit China's current situation are being studied by Chinese Household Electrical Appliances Association.

In view of the influence of EU directives of WEEE and RoHS on global supply chains (also including China's electronics industry) and on the basis of the state legislation on "Enhancing Clean Production" and the state law of "Prevention and Cure of Environmental Pollution by Solid Waste", the legal document titled "Pollution Prevention and Control Method in Electronics Production" was constituted by the MoII in 2004 with the aim of implementing green design and clean production concepts within the electronics industry so as to comply with upcoming EU and Chinese legislation in a proactive manner.

7.3.5 Collection System of WEEE Materials

In the 21st century, with the aggravation of the environmental impacts induced by WEEE, strategies of WEEE treatment began to attract wide attention in China. Many articles introduced legislation, management and recycling technologies of WEEE implemented in developed countries. Studies have been made on WEEE recycling. In July 2003, the Ministry of Science and Technology of China (MoST) initiated the National High Tech Research and Development Program (NHTRDP) on WEEE recycling technology with the aim of promoting research and development on technological aspects of WEEE management [3].

The system entails coordinated participation of several stakeholders including manufacturers, importers, retailers, consumers, and governmental offices. Manufacturers and importers are obliged to take back their products at designated collection sites and to recycle them according to governmental regulations. Retailers are required to take back used home appliances that they sold and to transfer them to the corresponding manufacturers or importers.

Governmental offices can also collect discarded appliances to transfer them to manufacturers or "independent bodies" for recycling. Consumers are obliged to cooperate in transferring used appliances to retailers or municipalities, and to pay the necessary fees for collection, transportation and recycling. In recycling plants, WEEE are sorted, treated, dismantled and crushed. Valuable materials — glass, aluminum, certain types of plastics, copper and iron — are recovered by electromagnetic, centrifugal and gravity separation techniques. Hazardous substances are recovered and destroyed via thermal or chemical processes.

In addition to lab investigations into WEEE recovery, recycling practices for WEEE are also promoted in China. Since December 2003, the SDRC designated Qingdao City, Zhejiang Province, Beijing City and Tianjin City as trial demonstration administrations to establish WEEE recycling systems and management methods. The four corresponding enterprises are: Qingdao Haier Group Company, Hangzhou Dadi Environmental Protection Company, China Huaxing Group Company and Tianjin Datong Copper Industry Company.

Zhejiang Province established a "Temporary Trial Method for Collection and Treatment of WEEE in Zhejiang Province" which came into effect at the beginning of 2005. This method follows the principle of fixed-point collection, centralizing treatment and extended producer responsibility, proposes to establish a formal collection network for WEEE covering the entire province, and stipulates the corresponding responsibilities of the government, production enterprises, sellers and consumers. Hangzhou Dadi Environmental Protection Company has constructed a recycling center WEEE and has set up 36 collection spots and four collection channels: (1) dealers, (2) communities, (3) institutions such as government agencies, universities and colleges, and (4) manufacturers of EEE for collection of products. The collected WEEE is delivered to the recycling center, where the products are repaired or maintained for use; when reuse is impossible, products are disassembled, comminuted and then separated for recovery of the valuable materials.

Qingdao Haier Group Company, China Huaxing Group Company and Tianjin Datong Copper Industry Company also began build workshops for WEEE recycling. Corresponding technologies are being developed.

7.3.6 WEEE Materials Processing Companies

Besides the above four trial demonstration models, some WEEE recycling plants have also been established. For example, in 2004 a WEEE processing center owned by the Fortune Group was completed and put into action for recovering precious metals from WEEE. In 2003, the Tianjin Loyalty Glass Material Company began to specialize in recycling glass materials from CRT by-products provided by the Tian-Samsung SDI Company. In 2004, the company along with Tianjin University successfully developed a waste CRT recovery technology that includes dismantling, separating and recycling processes (He 2006).

Shanghai WEEE Recycling Center Co., Ltd. has begun to construct workshops, and in collaboration with Tongji University is developing WEEE recycling technologies and setting up corresponding WEEE collection network system.

Guangzhou SE-MBA New Plastics Technology Co., Ltd. is a Sino-US joint venture established by American MBA Polymers and Guangzhou Iron and Steel Enterprises Group with total investment exceeding $12 million. The joint-venture contract was signed in January 2004, workshop construction and equipment installation were completed in March and August 2005, respectively, and the company started operation in November 2005. By utilizing an entirely mechanical separating process, the company recycles mixed scrap plastics of high-quality ABS, HIPS and PP in an environmentally safe way. Currently the company mainly imports mixed plastics for processing, but it aims at enormous quantities of end-of-life appliances, cars, plastic-rich consumer goods and WEEE.

Major companies for WEEE collection, pre-treatment (disassembly, upgrading), recycling and processing treatment include:

1. CRRC: China Resources Reuse Corp
2. DJE: Dong Jiang Environment Ltd
3. Dowa Corp
4. GEM: Green Eco-Manufacture Technology Corp
5. Gree: Gree Electric Ltd
6. SE: Sound Environment Co Ltd
7. TCL: TCL Corp

7.3.7 International Cooperation

Resolving the WEEE problem and thus improving sustainable development of the electronics industry in China require extensive international cooperation, not only in technologies but also in management, knowledge exchange, and education. An application to set up an Asia-Pacific WEEE Database and Management Center has just been approved by UNEP, and the center will soon be established in the UNEP-Tongji Institute of Environment for Sustainable Development. Its main activities are as follows (He 2006):

1. collecting information regarding WEEE in Asia and the Pacific and establishing a corresponding database;
2. organizing an annual international conference of WEEE to exchange related information;
3. analyzing the performance and logistic aspects of WEEE management systems currently implemented in the EU, the U.S. Japan and so forth, and then investigating regulations, policy and bylaws on WEEE management fit for local status to service-related office;

4. organizing an international forum on WEEE recycling technology in order to introduce successful, cost-effective and environmentally friendly technologies to the Asia-Pacific region;
5. inspecting the operation of new technologies and organizing researchers to make innovations in solving problems.

7.4 Recycling and Processing of WEEE

Recycling and processing of WEEE is an important step in WEEE treatment.

7.4.1 Operational Strategies

The important first step is reducing amounts of WEEE and emissions from WEEE treatment. To move from a situation in which a large portion of WEEE is either landfilled or incinerated, several measures can be taken:

- Tools for dismantling should be developed and improved so as to automate the dismantling process and increase the segregation of materials in this first step.
- Information for the recyclers about the location of parts containing dangerous substances and how to recognize them should be provided.
- The shredder process is the most problematic step in the pre-treatment chain, particularly when the input is not dismantled. Although this process is designed to facilitate material separation (ferrous, non-ferrous and plastic material), no pure fractions are obtained so that significant quantities of dangerous substances are dispersed in all fractions. This causes problems in the subsequent recycling facilities. Therefore efforts should be made to improve the process technology and to develop alternatives.
- The shredder residues are not recyclable, so this fraction should be incinerated in well controlled plants or used after further treatment for energy recovery in plants with high standard flue-gas cleaning systems.
- In order to reduce the amount of shredder residues, easily accessible parts should be removed before shredding.

For waste management purposes it is important to collect and separate different types of WEEE according to the subsequent waste treatment and recycling processes. According to the European Environment Agency (EEA), different categories of WEEE should be created (EEA, 2000). Recycling companies in China classified and ranked the WEEE materials based on their recyclable metals (refer to Table 1) and values:

1. Auto wreck;
2. Air conditioner;
3. TV sets, monitors to enable special treatment of circuit boards and parts containing flame retardant;
4. Refrigerators/freezers to enable separate treatment of CFC;
5. PC computers;
6. Laundry washer and dryer;
7. Lighting equipment for mercury recovery;

8. Large white goods; after removal of capacitor that might contain PCB the ferrous, non-ferrous and plastic fraction can be recycled directly;
9. Spent catalysts (NiCo, platinum-group metals (PGM), W, Mo, rare earth metals (REE)); and
10. Various types of battery (lead, Cd, Ni, Co, lithium, REE, etc.)

7.4.2 General Processing Technology

Disassembly and upgrading are two key processes in the mechanical recycling of WEEE. Maximizing valuable material recovery and consequent minimizing disposal rely on the technologies used in the process. With the steady decrease of precious metal content of EEE, the precious-metal-oriented recovery techniques such as hydrometallurgy and pyrometallurgy are facing great challenges. On the other hand, the mechanical/physical recycling of WEEE, due to its better environmental properties and easier operability, is drawing more attention. Compared with hydrometallurgy and pyrometallurgy, mechanical/physical processes can achieve full material recovery including plastics.

Mechanical recycling of WEEE can be broadly divided into three major stages (Table 2):

- Disassembly (dismantling): targeting on singling out hazardous or valuable components
- Upgrading: using mechanical/physical processing to upgrade desirable materials content, i.e. preparing materials for refining
- Refining: in the last stage, recovered materials return to their life cycle

Table 1 *Various Types of Metals for Different Kinds of WEEE Materials*

Waste Materials	Steel	Aluminum	Copper	Lead	Zinc	Ni & Co	Rare Earth	W & Mo
Auto	✓	✓	✓	✓	✓			
Appliances	✓	✓	✓	✓	✓			
Electronics	✓		✓	✓	✓			
PCB			✓	✓	✓			
Battery (Lead)				✓				
Battery (NiCo)						✓		
Battery (REE)							✓	
Spent Catalyst		□	□	□	□	✓	✓	✓

Table 2 *Various Types of WEEE Materials and Processing Companies*

Waste Materials	Disassembly and Upgrading	Refinery Copper	Refinery Lead	Refinery Aluminum
Auto	GEM			
Appliances	Gree	Fang Yuan Copper		Chalco Alum
Electronics	GEM; TCL	Fang Yuan Copper		
PCB	GEM	Daye Copper		
Battery (Lead)	Yuguang Gold	Yuguang Gold & Lead	Yuguang Gold	

7.4.3 Disassembly

Disassembly is a systematic process that removes a component, a part, a group of parts or a subassembly from a product (i.e., partial disassembly), or splits a product into all its parts (i.e., complete disassembly) for a given purpose. In WEEE recycling practice, selective disassembly (dismantling) is an indispensable process based on the following considerations:

1. reuse of components has highest priority;
2. dismantling the hazardous components is essential
3. it is important to dismantle highly valuable components and high- grade materials such as PCBs, cables, and engineering plastics in order to simplify the subsequent recovery of materials.

The implementation of disassembly needs highly efficient and flexible tools. A variety of tools are involved in the dismantling process. The disassembled cables, PCBs and metal/plastics mixture, being a mixture of various materials, should be further treated to upgrade their materials contents. A good example is utilized in Ragn-Sells Elektronikatervinning AB in Sweden [5].

7.4.4 Upgrading

WEEE can be regarded as a resource for metals, such as copper, aluminum and gold, and non-metals. Their effective separation, based on the differences in their physical characteristics, is another crucial process in WEEE recycling. Upgrading usually includes two stages: comminuting and separating.

7.4.4.1 Comminuting

Comminuting is the first step of the physical upgrading process. Only when the disassembled WEEE is shredded to proper granularity can its materials be liberated from one another, and then be separated effectively.

Basically, the materials present in EEE are attached by fastening, inserting, welding, binding, wrapping and so forth. Therefore, unlocking the associated materials like ceramics, glass, and metals having distinctive mechanical properties requires little intensive energy. The optimized comminuting result is that every comminuted particle is made of one material. The influence of granularities of the comminuted particles on the liberation of materials in PCBs was studied carefully.

Characterization of PC scrap and PCB scrap showed, after shredding by a laboratory-scale hammer mill, that the main metals present were in the –5 mm fraction for both PC and PCB scrap and showed excellent liberation (ca. 99%). In –3 mm fractions, the metals achieved approximately complete liberation; in coarser fractions, copper had poorer liberation due to the presence of copper pins associated with plastics and copper wire segments encapsulated with plastics; also, the ferromagnetics obtained good liberation for particles of sizes less than 16 mm, and aluminum in the +7 mm fraction for both PC and PCB scrap accomplished an excellent liberation. Industry scale tests showed that after two stages of comminution, the liberation of –5 mm fraction was between 96.5% and 99.5%. Figure 1 shows a similar flowchart of this process used in China [3].

7.4.4.2 Separation

Figure 1 also shows that, after liberation of the materials in the disassembled WEEE through comminuting, they can be separated by mechanical/physical methods. The differences in the physical characteristics of materials in non-homogeneous compounds, such as magnetism, electric conductivity and density, etc., are the bases of the mechanical/physical separation. Mechanical/physical separation processes include electronic magnetic separation, electronic-conductivity-based separation, density-based separation, and so forth. All of them have application instances in the WEEE recycling field.

1. Magnetic separation is widely used for the recovery of ferromagnetic metals from non-ferrous metals and other non-magnetic wastes. Over the past decade, advances in the design and operation of high-intensity magnetic separators have also made it possible to separate copper alloys from the waste matrix.
2. Electric conductivity-based separation is used to separate materials of different electric conductivity (or resistivity). There are three typical electric conductivity-based separation techniques: (1) eddy current separation, (2) corona electrostatic separation, and (3) triboelectric separation.

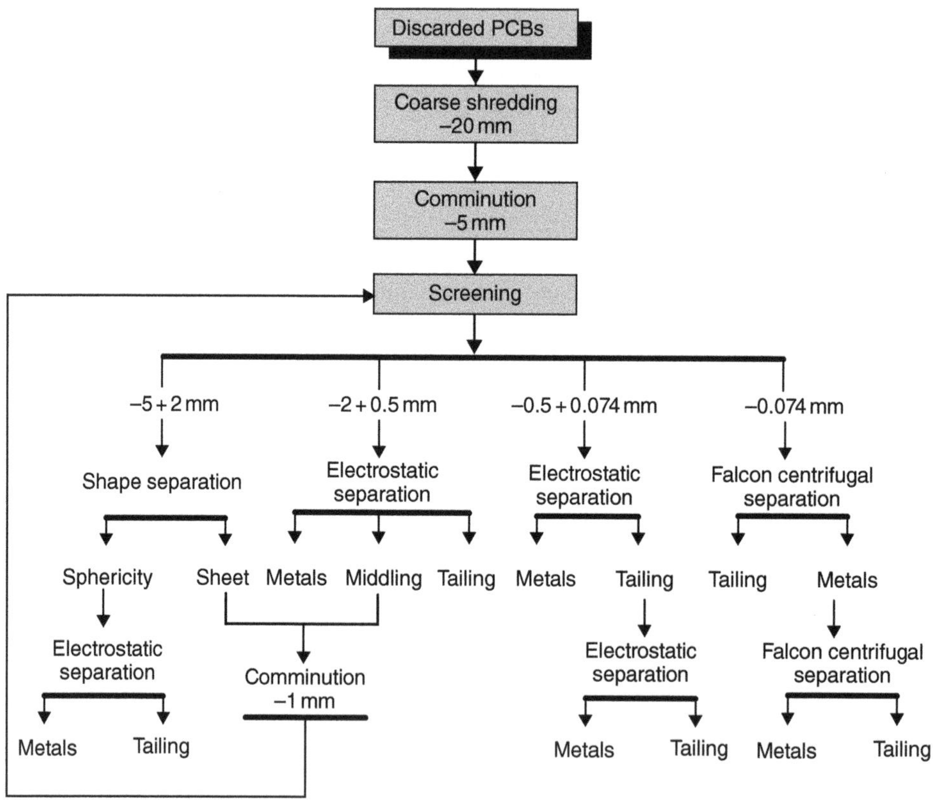

Figure 1 Flowchart for metals recovery from discarded PCBs by physical methods (He 2006). Reproduced from [3] with permission of Elsevier

3. With the marked density difference between metals and non-metals in WEEE powders, the heavier metal materials can be effectively separated from non-metal materials by density-based separation methods.
4. In the practice of recycling WEEE, according to the requirements of the task, some of the above methods can be combined. For example, in Germany, the Kamet Recycling Gmbh company has successfully reclaimed iron, aluminum, precious metals and plastic from waste PCBs through shredding, gravity separation, electromagnetic separation and eddy-current separation. About 90% of metal and plastic materials can be reclaimed by these processes.

7.4.5 Metal Refining

Table 3 summarizes a list of metal refinery companies in China for secondary metal processing.

The metal fractions separated during the preprocessing of e-waste are composed of Fe, Al, Cu, Pb and PMs. After Fe and Al, Cu and Pb are the main constituents of typical e-waste. Therefore, it is logical to send e-waste to smelters that accept copper/lead/aluminum scrap.

Currently, copper and lead smelters work as e-waste recyclers for the recovery of Pb, Cu and PMs. In these pyrometallurgical processes, e-waste/copper/lead scrap is fed into a furnace, wherein metals are collected in a molten bath and oxides form a slag phase. More details about the lead and copper smelting routes are given in following sections.

The metal fractions separated from e-waste during preprocessing can be further processed using hydrometallurgical, pyrometallurgical, electrometallurgical and biometallurgical processes and their combinations. Each route has advantages and disadvantages that must be considered in selecting an appropriate recycling process [5]:

1. Pyrometallurgical processes for recovering metals from various waste materials have been used during the last two decades. Smelting in furnaces, incineration, combustion and pyrolysis are typical e-waste recycling processes. Currently, e-waste recycling is dominated by pyrometallurgical routes, wherein the steel industry embraces the ferrous fractions for the recovery of iron, and the secondary aluminum industry takes the aluminum fractions.

Table 3 *Metal refinery companies for different types of metal recovery*

Company	Steel	Aluminum	Copper	Lead	Zinc	Ni & Co	REE	W & Mo
Wuhan Steel	✓							
Chalco Aluminum		✓						
Daye Copper Smelter			✓					
Yuguang Gold & Lead			✓	✓	✓			
Fang Yuan Copper			✓					
MinMetals SKS Smelter			✓	✓	✓			
Tongling Smelter			✓		✓			
Zhuzhou Smelter				✓	✓			
Zhuzhou Carbide						✓		✓
Baoto REE Smelter						✓	✓	
GEM						✓	□	□

2. Pyrometallurgical processes have the steps of liberation, separation/upgrading and purification that are fundamentally similar to those of mechanical or hydrometallurgical routes. However, the liberation of valuable metals is not achieved by leaching, crushing or grinding but by smelting in furnaces at high temperatures. In these processes, metals are sorted by exploiting their chemical and metallurgical properties, e.g., PMs are segregated into a solvent metal phase (copper or lead).
3. Hydrometallurgical routes have been successfully used to recover PMs from e-waste. However, these processes have certain disadvantages that limit their application on the industrial scale.

7.4.5.1 Copper Smelting Route

Primary and secondary copper smelting routes are adopted to recycle and extract PMs from e-waste. Copper smelting routes are reported to be more environmentally friendly than lead smelters, which generate toxic fumes. Copper-smelting facilities near population centers will minimize the cost of e-waste transportation so that recycling economy is improved. These advantages allow copper smelters to be installed near cities where e-waste is generated. In these processes, PMs are recovered via a conventional electrorefining process in which they are segregated in slimes.

Depending on copper grade, different types of technologies are used for copper scrap treatment:

1. High-grade copper scrap (Cu>90%): anode refinery furnaces are used for making anode;
2. Medium-grade copper scrap (Cu 70 to 90%): Kaldo furnaces are used in Hoboken smelter, and NGL furnace I is used in Jiang Xin Copper Smelter;
3. Medium-grade copper scrap (Cu <70%): ISA, Ausmelt, Noranda and SKS furnaces have been commonly used; ISA furnaces are used is Umicore and MinMetals Yunan Copper Smelter. an Ausmelt furnace is used in Daye smelter, a SKS furnace is used in Fangyuan copper smelter, etc.

Commonly, for low-grade copper scrap treatment, copper smelting routes including (1) matte; and (2) black copper are used for e-waste recycling:

1. In the sulfur-based route (primary copper smelting), copper matte (40%) and blister copper (98.5%) are produced. Finally, blister copper is refined by fire refining to produce pure copper.
2. In the black copper route (secondary copper smelting), crude copper is produced during a reduction process and is refined by oxidation in a converter. The black copper route is attractive because it can handle high levels of impurities including Fe, Zn, Pb and Sn, which are removed by oxidation [5]. The black copper smelting process entails reduction and oxidation cycles. Impurities are mostly segregated into the vapor phase and are off-gassed.

For the disposal of copper waste from WEEE in China (Table 4), the conventional blast furnace process has been replaced by the "SKS-BBS Oxygen Bottom Blowing Technology" invented by SKS and Fangyuan Copper Smelter [5]. The process has been implemented since 2008, with the advantages of low SO_2 emission (<100 ppm).

Table 4 *List of SKS Furnace Installations for Copper Smelting*

Case No	Smelter	Location	Furnace Size (m)	No of Lances	Year Built	Actual Cu(t/yr)
1	SKS	Hunan			1994	
2	Vietnam		Φ3.1×11.0	4	2007	10,000
3	Fangyuan I	Sandong	Φ4.4×16.5	9	2008	100,000
4	Hengbang	Yantai	Φ4.4×16.5	5	2010	100,000
5	Huading	Baotou	Φ3.8×13.5	9	2011	80,000
6	Yuanqu	Shanxi	Φ4.8×20.0	12	2012	200,000
7	West Mining	Qinghai	Φ4.8×20.0	12	2013	200,000
8	Yuguang	Sanxin	Φ4.4×18.0	10	2014	200,000
9	Fangyuan II	Sandong	Φ5.5×27.5	23	2015	200,000

7.4.5.2 Lead Smelting Route

Historically, primary lead is produced from sulfide ores containing iron, zinc, copper, PMs, and other trace elements. The concentrated ore and e-waste is treated for the extraction of lead and PMs.

1. The process consists of sintering (ores), reduction and refining stages. Sintering is carried out to reduce the sulfur content of feed material that is composed of pyrite, lime rock, silica and highly concentrated lead. The reduction process is conducted in blast furnaces with the help of coke where molten lead with about 85% purity is tapped from the bottom of the furnace. Use of sintering furnaces resulted in serious emission pollution;
2. The plastics fraction of e-waste can partially replace coke as a reducing agent during the reduction stage, and the metals fraction ends up in the metal phase. In the refining stage, copper dross is skimmed off from the lead dross and is treated in a reverberatory furnace. The lead dross is processed by adding wood chips, fine coke and sulfur. The sulfur dross produced is separated and transferred to the reverberatory furnace. The heating of lead dross in the reverberatory furnace separates the lead bullion (rich in lead), matte (copper and other metals sulfides) and speiss (high in arsenic and antimony contents). Matte and speiss are treated in copper smelters to extract copper and other associated metals such as silver and gold; and
3. In the last stage of e-waste processing via the lead-smelting route, PMs and other elements are separated from the lead bullion. PMs are separated by the Parkes process, in which zinc forms an insoluble intermetallic compound with gold and silver. Other impurities including antimony, tin, arsenic, bismuth, and trace elements are also separated during the refining stage. Final products of refining stage include metallurgical-grade lead (99.99%), PMs and other elements.

To reduce emission pollution in China, all sintering furnaces have been replaced by either Kivcet or most likely by SKS bottom oxygen-blowing technology since 2008.

For the disposal of lead waste from WEEE in China (Table 5), the conventional sintering process has been replaced by the "Direct Reduction of Molten Lead-Rich Slag" invented by SKS and Yuguang Lead Smelter (YGL) [6]. At the Yuguan lead smelter, waste lead batteries are pre-treated in an Evngitec CX system to separate lead from plastic, etc. After that, the lead is blended with lead concentrate at a 50:50 ratio and processed at a SKS furnace [7]. The process has been implemented since 2010, with the advantages of low SO_2 and lead emission ($SO_2 < 100$ ppm).

7.4.5.3 Industrial Practices for the Recovery of Metals from E-Waste

Pyrometallurgical routes are beneficial in segregating and upgrading valuable metals in the BMs from e-wastes that are further treated by hydrometallurgical routes to recover metals. Industrially, Ausmelt, IsaSmelt, Kaldo, rotary, plasma arc furnaces and SKS oxygen bottom-blowing technology are used for collecting valuable metals worldwide [4].

Industrial processes for recovering metals from e-waste are based on combined pyrometallurgical, hydrometallurgical and electrometallurgical routes. In pyrometallurgical processes, e-waste is blended with other materials and incorporated into primary/secondary smelting processes (e.g., into copper or lead smelters).

1. Copper smelting is the dominant route for e-waste recycling where PMs are collected in copper matte or black copper. In the final stage of copper production, i.e., the electrorefining process, pure copper metal is produced and the PMs are separated into slimes where they are recovered using hydrometallurgical routes;
2. Currently, various industrial processes are used globally for extracting metals from e-waste, including the Umicore integrate smelting and refining facility;
3. The Noranda process in Quebec, Rönnskär smelters in Sweden, Kosaka's recycling plant in Japan, the Kayser recycling system in Austria and the Metallo-Chimique N.V plants operate in Belgium and Spain.

At Umicore [4], PMs are extracted from e-waste using the secondary copper route. Umicore has extensively investigated building the world's largest waste recycling facility at Hoboken, Belgium. It is a specialty- material group that focuses on recovering gold and silver, PGMs (palladium, platinum, rhodium, iridium and ruthenium), special metals (selenium, tellurium and indium), secondary metals (antimony, tin, arsenic and bismuth), and the BMs (copper, lead and nickel) from various residues, including e-waste. Other by-products of the Umicore plant are sulfuric acid (from off-gas purification), and slag (consisting of silicon, aluminum and iron oxides) that are used as construction material.

Table 5 *Installations of SKS Lead Smelters in China*

Case No	Smelter	Location	Furnace Size (m)	Year Built	Status	Actual Pb(t/yr)
1	YGL	Henan	$\Phi 3.8 \times 11.5$	2002	operational	80,000
2	SKS	Hunan	$\Phi 4.1 \times 14.0$	2003	operational	100,000
3	YGL	Henan	$\Phi 3.8 \times 11.5$	2004	operational	80,000
4	Chizhou	Anhui	$\Phi 4.4 \times 16.5$	2005	operational	100,000
5	Xiangyun	Yunan	$\Phi 4.4 \times 16.5$	2006	operational	100,000
6	YGL	Henan	$\Phi 3.8 \times 13.5$	2007	operational	100,000
7	Yuanqu	Shanxi	$\Phi 4.8 \times 16.0$	2008	operational	100,000
8	West mining	Qinghai	$\Phi 4.8 \times 16.0$	2009	operational	100,000
9	Yuguang	Sanxin	$\Phi 4.4 \times 16.0$	2010	operational	100,000
10	Hengban	Sandong	$\Phi 4.8 \times 16.5$	2011	operational	100,000

The feed material for the Umicore smelting and refining process is composed of wastes from the non-ferrous industry, PM residues, and PCBs. On average 250,000 tons of feed material are treated annually for metals recovery. It is the world's largest recycling facility for the extraction of PMs and has an annual production of over 50 tons of PGMs, 100 tons of gold and 2400 tons of silver. The Umicore process flow sheet is composed of several pyrometallurgical, hydrometallurgical, and electrochemical processes for recovering PMs.

Most integrated e-waste recycling plants use PCBs directly in the smelting furnaces, with many advantages:

1. It maximizes the overall segregation of PMs in the copper fraction, so that the final recovery will be higher;
2. It partially replaces coke with plastics as a source of energy during the smelting process;
3. It provides a source for recycling of e-waste; and
4. It is efficient in terms of resource management by closing the loop of metals.

7.5 Current Issues in WEEE Treatment in China

As described above, as we enter the 21st century, with the constant increase of WEEE generated in China and the aggravation of corresponding environmental problems, some projects on WEEE management are being performed. However, systematic work on WEEE is just beginning. There are still some problems that restrict the operation of qualified WEEE recycling enterprises in China, and China must move quickly on the following:

1. Prevention of illegal importing of WEEE materials;
2. Enforcing implementation of regulations;
3. Improving collection system efficiency;
4. Improving metal recovery and reducing energy consumption in refining processes;
5. Minimizing emission and footprint; and
6. Minimizing reprocessing costs

7.6 Conclusions

Electric and electronic equipment has been developed, applied, and consumed world-wide at a very high speed. Subsequently, the huge amount of waste electric and electronic equipment (WEEE) has become a common global problem. In view of the deleterious effects of WEEE on the environment and the valuable materials that can be reused in them, legislation in many countries has focused on the management of WEEE and new techniques have been developed for its recovery.

In China, rapid economic growth, coupled with urbanization and growing demand for consumer goods, has increased the consumption of EEE in large quantity. WEEE has increased concomitantly, posing a severe threat to the environment and to sustainable economic growth as well.

This chapter reviewed implementation strategies for WEEE treatment and recovery technologies for WEEE. It presented the current status of WEEE and corresponding responses so far in China. The concept and implementation of scientific development is critical in the electronics sector, an important industrial sector in China's economy.

To achieve this objective, it is significant to recycle WEEE sufficiently to comply with regulations regarding WEEE management, and to implement green design and cleaner production concepts within the electronics industry to comply with upcoming legislation in a proactive manner.

Significant progress has been achieved in China since 2010:

1. Public awareness and social system-program development for WEEE sorting, collection, recycle and processing;
2. 3R (reuse, recycle, reprocess) strategies for WEEE management;
3. Regulation of the entire process from manufacturing, waste collection, disposal and environmental protection;
4. Development of processing technology such as physical separation, metal refining, non-metal materials recovery and landfill; and
5. Protection of environment by minimizing the air-soil-water pollution using European standards and world-class processing technology.

References

1. European Commission, 2000, Draft Proposal for a European Parliament and Council Directive on Waste Electric and Electronic Equipment, Brussels, Belgium. http://www.eia.org/download/eic/21/www Final Proposal June 2000.htm, 2000-07-31.
2. Chi, X., Streicher-Porte M, Wang, M., Reuter, M., 2011, Informal Electronic Waste Recycling, *Waste Management*, 31, 731–742.
3. He, W., Li, G., Ma, X., 2006, WEEE Recovery Strategies and the WEEE Treatment Status in China, *Journal of Hazardous Materials*, B136, 502–512.
4. Hagelüken, C., 2006, Recycling of Electronic Scrap at Umicore. Precious Metals Refining. *Acta Metallurgica Slovaca*, 12, 111–120.
5. Khaliq, A., Rhamdhanj, M., Brooks, G., Masood, S., 2014, Metal Extraction Processes for Electronic Waste and Existing Industrial Routes: a Review and Australian Perspective. *Resources*, 3, 152–179.
6. Guo, X., Zhang, Y., 2014, Direct Reduction of Molten Lead-Rich Slag Using Eco-Green YGL Technology, MetSoc Com2104, Vancouver, BC, Canada, 15: 8558.
7. Guo, X., Zhang, Y., 2014, Review of Eco Clean Fangyuan Copper Production Using SKS-BBS Oxygen Bottom Blowing Smelting Technology, MetSoc Com 2104, Vancouver, BC, Canada, 15: 8557.

8

Metal Pollution and Metal Sustainability in China

Xiaoyun Jiang[1], Shengpei Su[2], and Jianfei Song[3]

[1] *Changsha Hasky Environmental Protection Sci-Tech Development Co. Ltd., PRC*
[2] *College of Chemical Engineering, Hunan Normal University, PRC*
[3] *College of Chemistry, Changsha University of Science and Technology, PRC*

8.1 Introduction

During recent decades of reform and opening up, China's industrial base and economy have developed rapidly and the consumption of the nation's mineral resources has grown in parallel. In 2014, China's industrial production accounted for more than 22% of the world's output, and for 221 kinds of industrial products, it ranked first. As the factory of the world, China's rapid industrial development has inevitably led to huge demand for and consumption of mineral resources. China is the largest global user of coal, iron ore, aluminium oxide, copper and cement, and ranks second in oil consumption [1].

China's industry has developed rapidly, but there are problems, such as lack of industrial layout planning, unreasonable manufacturing structure, repetition of construction at a lower level and backward production technology. Extensive economic growth mode causes high consumption of mineral resources and has done tremendous damage to the environment [1]. Heavy metal concentrations in the atmosphere around enterprises of nonferrous metallurgy, lead acid battery, secondary lead, coal-fired power plant, cement, and ferrous metallurgy operations are high. Varying degrees of heavy-metal pollution of underground water and soil in these areas affect crop growth, the security of drinking water, and the health of people living in the vicinity. Environmental consciousness of people in China is also growing with economic development, causing them to pay more attention to their

Metal Sustainability: Global Challenges, Consequences, and Prospects, First Edition. Edited by Reed M. Izatt.

living environment. They desire more urgently to make our mountains greener, our water cleaner and our sky bluer.

To guarantee sustainable development of resources and environment, and to protect the health of ordinary people, the Chinese government devotes increasing efforts to controlling heavy metal pollution and promote the development of a circular economy for specific metals. First, China has accelerated efforts to change the economic growth mode by closing down backward production facilities and promoting cleaner production. Secondly, a national effort is underway to strengthen the supervision of key polluted areas and enterprises, intensify law enforcement, and severely punish illegal polluters. Thirdly, at the national and regional levels there is greater focus on improving laws and standards regarding heavy metal pollution control.

Faced with increasingly strict environmental protection standards and higher costs, industrial enterprises have taken measures to implement technology innovation, such as adopting advanced technologies to carry out production with greater efficiency and comply with low carbon goals. To achieve energy conservation, reduce emissions, and minimize adverse environmental effects, companies also design and implement strict technological regulations and standards, strengthen technological transformation, and develop low- or non-polluting technologies.

Under the concentrated and coordinated efforts of the Chinese government, commercial enterprises, and the general public, steady improvement is expected in the future in the wise use of available mineral resources, improvement of metal sustainability, and prevention and control of heavy metal pollution.

8.2 Heavy Metal Pollution in China

In environmental pollution, heavy metal mainly refers to mercury, cadmium, chromium, lead, and arsenic, which have significant biological toxicity. Other metals that can have considerable toxicity are zinc, copper, cobalt, nickel, tin, and antimony. Usually, the natural background concentrations of heavy metals do not reach harmful levels. Poisonous and harmful metals get into the atmosphere, water and soil, and cause serious environmental pollution, because of human activities such as heavy metal mining, smelting and processing of ores, large-scale industrial production, emissions during ore treatment and processing, and the widespread use of pesticides and chemical fertilizers that contain these metals,

8.2.1 Heavy Metal Pollution Status

Heavy metal pollution is a serious threat to human survival. Every year in China, heavy metal pollution affects more than 12 million tons of grain, causing a direct economic loss of more than 20 billion yuan. In 2008–2009, several heavy metal pollution events happened in Guizhou, Hunan, Guangxi, Yunnan, Henan and Shanxi. In October 2008, arsenic pollution in Hechi city, Guangxi Province, led 450 villagers to have high urine arsenic levels, and four villagers to have mild poisoning [2]. In 2009, 32 mass incidents were caused by heavy metal pollution. In these incidents, not only were the surrounding farms and woodlands polluted, but two persons died and more than 500 persons in Liuyang, Hunan province were found to have cadmium levels in their body exceeding the allowable standard [3].

In 2014, an unlicensed electrolytic factory in Shunde city, Guangdong province, discharged effluent during the night through the toilet every day [4]. This action made the heavy metal concentration in water greatly exceed the standard. Chromium, nickel, and copper each exceeded the standard by ten or even a hundred times, which was a big threat to the drinking water safety of local residents. Several recent pollution events are illustrated in Figure 1.

8.2.1.1 Heavy Metal Pollution in Water

With the speeding up of economic development and urbanization in China, increasing amounts of untreated urban garbage, industrial wastewater and domestic sewage are discharged into the water. The result is that heavy metal concentrations in sediments and suspended solids in water have increased sharply and major water pollution problems have occurred in most Chinese cities. Heavy metal pollution is a particularly serious national problem [13–15].

Many factors affect the heavy metal content in water. Industrial wastewater is the main source of heavy metal pollution in China. The papermaking, chemical, textile, iron, and steel industries are the main sources for emission of metals in industrial wastewater, accounting for ~48% of total heavy metal emissions. According to *The First National Pollution Source Census* [16] in 2010, the generation of heavy metals (cadmium, chromium, arsenic, mercury, lead and others) into industrial wastewater was 24,300 tons. Of this total, heavy metal pollutants discharged from plant outfall was 2100 tons. Following treatment by municipal treatment sewage plants and centralized treatment facilities, the actual heavy metal pollutants discharged into the aquatic environment was 900 tons. The main river basin emissions (Haihe River, Huaihe River, Liaohe River, Chaohu Lake, Dianchi Lake, and others) were 100 tons.

As mining and smelting of mineral resources has increased in China, heavy metal pollution of mining-area groundwater has become increasingly serious and has expanded in scope year by year. Heavy metal pollutants in mining-area groundwater include Pb, Cr, As, Ni, Cd, Hg, Cu, and Zn. Due to the development of the mining industry in Hunan, Shanxi, Qinghai and other provinces, the heavy metal levels in some areas of these provinces exceed standard limits. Taking the Xiangjiang River basin as an example (see Table 1), heavy metal pollution, which has a long history in the area, has increased dramatically. The extensive development of the non-ferrous mining and smelting industry is the main reason for the heavy metal pollution in this basin. According to Liu, *et al.* [17], the average recovery rate of metal from Hunan nonferrous mining operations is only 50% and the comprehensive recycling rate of associated ore is only 25%. As a result, a large amount of low-grade ore and associated minerals is abandoned as scrap slag. Hence emissions of industrial waste water, waste residue, waste gas and other waste, directly or indirectly, has caused the pollution of Xiangjiang River. According to statistics from Hunan Province Environmental Protection Authority, there are 146 key sources of pollution in industrial enterprises in the basin, and the total emission amount is as high as 8.39×10^8 m^3/year, in which the content of metal can reach 5,013 tons/year.

The national environmental quality data announced by the Ministry of Environmental Protection showed [18] that in the first half of 2013 heavy metal concentrations of the 12 surface water control section points exceeded the standard by 22 times. Mercury pollution was the most prevalent, accounting for 50% of the total exceeded number; this was

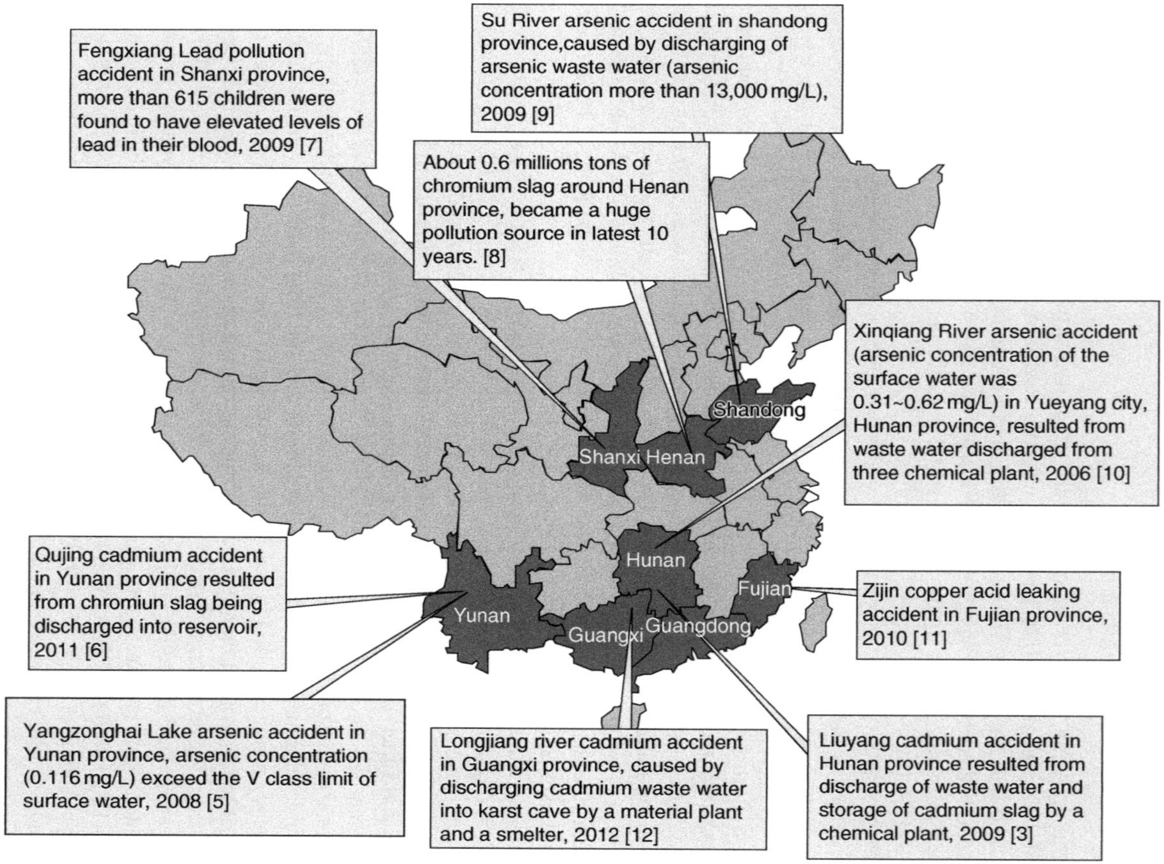

Figure 1 *Examples of Heavy Metal Pollution Accidents in China in 2008–2012*

Table 1 *Pollution statistics of key sources of industrial enterprises in Xiangjiang River Basin [17]. Source: Liu Y C, 2010 [17]. http://en.cnki.com.cn/Article_en/CJFDTOTAL-HJBH201004010.htm*

Industry types	Pollution sources	Wastewater emission/ $10^8 m^3 \cdot year^{-1}$	Heavy metal emission/tons·year^{-1}					
			Hg	Pb	Cr(VI)	Cd	Cu	Zn
Non-ferrous	31	2.044	2.478	29.664	498.447	132.105	86.307	132.105
Chemistry	65	2.519	1.302	1.374	104.558	1.08	53.229	1.08
Light industry	46	0.52	0.062	0.033	0.264	2.943	—	2.943
Electricity	4	3.307	—	—	6.345	0.081	1.254	0.081
Total	146	8.39	3.842	31.071	609.614	136.209	140.781	136.209

followed by arsenic, accounting for 36.4%. The sections (points) where standards were exceeded were distributed in Hongze Lake, the Weihe River, Paihe River and other regions; the sections (points) where standards were exceeded were distributed as follows: four in Tianjin city, two in Yunnan province and two in Inner Mongolia province. In Anhui, Shanxi and Jiangsu, each province had one point where the standard was exceeded. Different heavy metals exceeded the standard to different degrees: mercury exceeded the standard 0.2 to 9.0 times, arsenic 0.2 to 0.9 times, selenium 0.2 to 1.0 times and cadmium 0.8 to 3.0 times.

In China, especially in rural areas, it is urgent to reduce the level of arsenic in drinking water. Guifan Sun [19] examined samples from 39,514 wells in areas where the drinking water was seriously poisoned by arsenic. If the number 50 $\mu g \cdot L^{-1}$ is taken as the safety standard for maximum allowable arsenic concentration in drinking water, arsenic concentrations in Shanxi Province were over the limit by 50%. This province was followed by Jilin province (12.21%), Inner Mongolia province (11.30%), Qinghai province (8.33%), Xinjiang province (4.77%), and Ningxia province (1.06%), The number of wells that exceed World Health Organization (WHO) standards for arsenic in drinking water (10 $\mu g \cdot L^{-1}$) will be much greater., which is of great concern to people who live in the area.

According to data from 26 lakes and reservoirs and seven water systems, cadmium pollution in water is extensive in China. In the Yangtze River watershed, cadmium is the second most prevalent heavy metal pollutant, exceeded only by mercury. Concentrations of cadmium exceeded the standard by 16% in Yellow River, Huaihe River, and Luanhe River, all of which are seriously polluted [20].

Antimony production in China accounts for about 87% of the world output, making China the largest global producer. With the widespread use of antimony in all walks of life and the large-scale mining of antimony ore, the mass of antimony that gets into the atmosphere, surface water, groundwater, and soil can result in serious pollution to the environment. The natural background value of antimony in water is less than 1 μg/L [21]. Antimony is unlikely to be in relatively high concentrations in natural water bodies except for some hot springs or geothermal water in which the concentration of antimony can reach more than 500 μg/L. But due to the impact of human activities, the concentration of antimony can reach 100 μg/L, about 100 times the natural background value. The content of antimony in mining wastewater, river water, and well water of Xikuang mountain (Loudi city, Hunan Province) was 1.33–21.79, 0.063–0.037, and 24.02–42.03 mg/L [21], respectively.

The concentration of antimony in ponds around the Weishan antimony mining area (Yunnan Province) were 1,209–1,724 μg/L, well beyond the Chinese environment quality standard (GB3838-2002) of ≤5 μg/L [22].

8.2.1.2 Heavy Metal Pollution of Soil

The 2014 *National Soil Pollution Survey Bulletin* [23] showed that the Chinese soil environmental condition was not optimistic on the whole, and soil pollution in some areas was serious. The soil environmental quality of farmland was worrisome and the environmental problems of industry and mining wasteland were alarming. Heavy metal concentrations exceeded the standards at soil-monitoring sites nationwide by 16.1%, with severe pollution concentrations found in 1.1% of the cases. Heavy metal concentrations exceeded the standard set for cultivated land by up to 19.4%, with the major pollutants being Cd, Ni, Cu, As, Hg, and Pb. Heavy metal pollution of soil in China is mainly produced by 'three wastes' emissions, i.e., heavy metal mining processes, use of sewage for irrigation, and unreasonable application of inferior fertilizers and pesticides [24]. Heavy metals in water and the atmosphere will eventually recycle to the soil and cause soil pollution. Data from environmental monitoring stations showed that, in China, the most serious heavy metal pollutants in soils are Cd, Hg, Pb and As [25].

At present, in some provinces of South China, such as Hunan, Guangdong, Guizhou and Yunnan, arsenic pollution in soil due to mining operations is very serious. In Hunan Chenzhou city, a large area of soil had arsenic content as high as 1,217 $mg \cdot kg^{-1}$, caused by metal mining and smelting production [26]. In an abandoned tungsten mine in Shantou city, the concentration of arsenic in surrounding soil was found to be 935 mg/kg, while in the groundwater it reached 325 μg/L [27].

According to *The National Soil Pollution Soil Survey* [11] of China, the concentration of chromium exceeds the standard rate of the total soil monitoring sites by 1.1%. Chromium is the major pollutant in the investigated industrial wasteland. The improper disposal of chromium slag has been a major contributor to chromium pollution in the area. In cities like Jinzhou and Tianjin and in other chromium industrial areas, proper treatment and disposal of chromium slag has been ignored for a long time, with the result that this slag is piled up 'mountain-high.' The nearby soil suffers serious chromium pollution and the growth of trees, plants and crops in the region is adversely affected [28].

The average content of total mercury in China's urban soil was 0.2–0.7 mg/kg. In the agricultural soil, the mercury content was around 0.1~0.5 mg/kg, higher than in distant areas where levels were 0.02–0.2 mg/kg, but generally lower than the Grade II soil environmental quality standard in which the limit is 1 mg/kg, the maximum allowable limit of mercury content for agricultural land [29]. Serious mercury pollution of soil occurs mainly in areas where this metal is mined and smelted. In these areas, the total mercury content in soil not only exceeds the standard, it also impacts a much wider area. For example, total soil mercury content exceeds the standard 20 kilometers away from the mercury smelting operation in the Guizhou wanshan mercury area [30].

In arid and semiarid regions of northern China, the use of sewage in irrigation is very common. Due to inefficient wastewater treatment and lack of environmental consciousness, the use of wastewater of complex compositions has brought a series of problems to the water and soil environment and ecological security. Survey data from the Department of Agriculture showed that the sewage irrigated area was about 140 million hectares in China. Heavy metal pollution was found in 64.8% of the total polluted area, with seriously polluted, moderately polluted, and lightly polluted sub-areas accounting for 8.4%, 9.7%, and 46.7% of the total, respectively [31]. Mercury and cadmium were the major heavy metal contributors to the pollution.

Unreasonably large applications of chemical fertilizers, pesticides, and plastic films are major reasons for the presence of excess heavy metals in soil. At present, China's annual fertilizer application amount is about 5,800 million tons, accounting for a third of the world's total. China is the world leader in fertilizer consumption. However, excessive use of fertilizer in agricultural production in China is common. The efficiency of fertilizer use is only 35%, with most fertilizer being wasted. Heavy metals, along with the fertilizer itself, are leached into water or absorbed by soil. In 2011, the Ministry of Agriculture investigated 0.158 million hectares of rice paddy fields in 88 heavily polluted sites in Hubei, Hunan, Jiangxi, and Sichuan provinces. The excessive area amounted 0.107 million hectares, and the excessive rate is 67.8%. Among these pollutants, Cd pollution is most serious, followed by As, Ni, Cu, Hg and Cr. The excessive rates of other heavy metals are lower.

In Chinese soil, the natural concentration of antimony is 0.38~2.98 mg/kg. The maximum permissible content of antimony in soil is 3.5 mg/kg or 5 mg/kg [32]. The average background values of antimony are relatively high in Guangxi, Yunnan, Guizhou and Hunan provinces, being respectively 2.93 μg/g, 2.44 μg/g, 2.21 μg/g and 1.87 μg/g. The average antimony concentration in the surrounding soil of the largest antimony mine in the world, Hunan Xikuang Mountain Mine, reached 5,949 mg/kg [33]. The enrichment content of antimony in the Banpo antimony mining area in Guizhou province reached 51~7,369 mg/kg [34]. Due to long-lasting irrational development of mining sources and utilization, farmland along the DiaoRiver in Guangxi has been seriously polluted by heavy metals [35], with the highest antimony concentration reaching 1,300 mg/kg. The soil in lead and antimony smelting areas in Hechi city, Guangxi province, has been polluted by high antimony concentration. The antimony content has reached 155–30,439 mg/kg in these areas [36].

8.2.1.3 Heavy Metal Pollution of Atmosphere

Atmospheric pollutants are usually colorless and are easily overlooked compared with heavy metals in water. According to the results of the first national pollution census in 2007, lead, mercury, cadmium, chromium, and arsenic pollutant emissions in the atmosphere annually reached about 9,500 tons [37]. These heavy metal pollutants may move to water or soil and enter into the human body through respiration and/or the food chain.

Heavy metal pollution as atmospheric fine particulates is common. The Chinese Ministry of Environmental Protection officially released ambient air quality standards on February 29, 2012. Pollutant particles of 2.5 microns or lower (PM 2.5) were formally incorporated into the new air quality standards [38].

Heavy metal pollution in the atmosphere comes mainly from a large number of harmful gases and dust containing heavy metals from industrial production (such as metal smelting plants, thermal power plants, and chemical industries) and other sources including fuel burning, mining, automobile exhaust and automobile tire abrasion. Industrial production, such as metal smelters, coal-fired power plants and various chemical industries, results in discharge of many particles containing heavy metals during manufacture. Most of the heavy metals are transformed during pneumatic transportation, in which secondary pollutants are generated, often with increased toxicity. Heavy metal pollution from road traffic has a zonal distribution, mainly resulting from Zn, Cu and Fe dust produced by exhaust gas and tire abrasion. With the rapid increase of the number of vehicles producing emissions, the contents of heavy metal pollutants represented by Pb, Cd, Cu and Zn have risen sharply in the urban atmosphere. Lead pollution mainly comes from battery, smelting, metal, machinery, coating and electroplating industries and automobile exhaust. Cadmium pollution originates mainly from electroplating, dyestuffs, mining, smelting, chemicals, plastics, metallurgy and light sensor manufacturing. Nickel pollution mainly originates from industrial pollution and mining [39, 40].

In China, research on metal pollution in the PM 2.5 range is mainly concentrated in urban areas and involves primarily Cu, Sb, Zn, Pb, Cd, Cr, and As. The results in Table 2 [41] show that, except for Pb, the main pollution elements in PM2.5 in the urban areas of China indicated include As, Cu, Zn, Cd, and Cr. Arsenic pollution is seen to be the largest, with the highest concentration being in Xi'an city and the second highest in Changchun: arsenic concentrations reached 0.11 $g{\cdot}m^{-3}$ and 0.09 $g{\cdot}m^{-3}$, respectively, in these cities. Arsenic pollution levels in southern cities are generally lower than those in northern cities. Changsha city has the lowest arsenic concentration, 0.012 $g{\cdot}m^{-3}$. The levels of Cu, Zn, Cd and Cr differed significantly among different cities. Standard limits for environmental air quality of heavy metals in China are given in Table 3.

Production of leaded gasoline in China has been prohibited since January 1, 2000 and use since July 1, 2000. Use of unleaded gasoline eliminates lead pollution in the

Table 2 *Metal concentration levels in atmospheric PM 2.5 in typical regions of China ($\mu g{\cdot}m^{-3}$) [41]. Source: Tan J H, 2013 [41]. http://en.cnki.com.cn/Article_en/CJFDTOTAL-ZKYB201302002.htm*

Areas	PM2.5	Pb	Cd	Cr	As	Cu	Zn
Beijing	131.6	0.137	—	—	0.028	0.042	0.0239
Tianjin	101.7	0.63	—	—	0.06	—	1.34
Changchun	51.0	0.46	—	—	0.09	—	1.33
Jinan	139.3	0.30	0.01	0.02	0.02	0.03	0.59
Qingdao	27.3	0.27	—	—	0.02	—	0.43
Shanghai	33.7	0.013	—	0.18	—	0.006	0.024
Nanjing	139.5	0.490	—	—	0.03	—	0.0830
Hangzhou	80.0	0.75	—	—	0.04	—	1.35
Wuhan	70.7	0.157	0.004	0.0368	0.016	—	0.424
Changsha	92.3	0.054	—	0.020	0.012	0.028	0.299
Nanchang	99.1	0.124	—	—	—	0.029	0.278
Guangzhou	190.0	0.207	0.005	—	—	0.056	0.538
Xiamen[10]	N.A	0.22	—	—	0.02	—	0.71
Hongkong[10]	30.4	0.19	—	—	0.01	—	0.35
Urumqi[18]	—	0.081	0.003	—	—	0.136	—
Chongqi[10]	106.9	0.72	—	—	0.08	—	0.92
Xi'an[10]	102.8	1.68	—	—	0.11	—	2.41
Jinchang[10]	49.7	0.16	—	—	0.05	—	0.11
Hebei [19]	6.5	0.057	0.001	0.079	0.012	0.043	0.14

Table 3 *Standard limit of environmental air quality of heavy metals in China (annual average) (μg·m*$^{-3}$*)*

Standard name	Pb	Cd	Cr(VI)	As	Hg
GB3095-2012	0.5	0.005	0.000025	0.006	0.05

environment from vehicle exhaust and plays an important role in reducing lead concentrations in the urban atmosphere. Coal combustion is the major source of aerosol lead pollution. The pollution area from coal combustion is larger than that in vehicle exhaust and the exposure brings greater harm to human health. Zhu [42] determined the lead isotope content of dust and aerosols in the southeast coastal areas (especially in the Pearl River delta region). He concludes that air pollution of this area comes mainly from industrial lead, while automobile exhaust pollution is a relatively low proportion. The concentration of lead in aerosols in Guangzhou and Foshan city was about 1%, indicating that atmospheric environmental lead pollution is serious there.

In atmospheric mercury pollution, China is one of the most heavily contaminated areas in the world. Average atmospheric deposition values are >70 g/m^3 [41]. Atmospheric mercury emissions have increased rapidly in recent years (paralleling economic development in China) and have reached about 500–700 tons per year. The level of urban gaseous mercury is 1.5 to 5 times that in North America and Europe [43]. Color metal smelting, coal combustion, and waste incineration are the main sources of atmospheric mercury pollution in China. Joint research by Tsinghua University and the U.S. Department of Energy's Argonne National Laboratory pointed out that in 2003 China released 650 tons of mercury from various emission sources into the atmosphere. Emissions from non-ferrous metal smelting and coal combustion are the highest, accounting for 52% and 39% respectively of the total [44]. The iron and steel industry is a major mercury source. In China, the steel industry emitted about 8.89 tons of mercury into the atmosphere in 2003, accounting for about 1.4% of the total national emissions [45]. However, in 2013 steel production was nearly four times that in 2003. Since the electric power industry, which is the largest pollution source, began to control mercury emissions, the emission proportion from the steel industry will increase further. Thus, the iron and steel industry seems to be a key to the control of mercury pollution in the future.

The growth of heavy metal emissions in China is much lower than the growth rate of coal consumption. This stems, at least in part, from the wide promotion of dust removal by incorporating desulfurization and denitrification processes in the power sectors. Reduction of heavy metal atmospheric pollution should be further enhanced by legislated monitoring of the environment and of emission sources, the establishment of heavy metal emission standards for atmospheric pollution sources, and the modification of limit values of heavy metals in the Ambient Air Quality Standards [13].

8.2.2 Heavy Metal Pollution in China: Prevention and Control

8.2.2.1 Laws and Regulations for Heavy Metal Pollution Prevention and Control

Laws and regulations for environmental protection in China are mainly composed of constitutional provisions on environmental protection, fundamental laws on environmental protection, administrative regulations, department regulations, local regulations, and environmental standards. Existing laws and regulations related to environmental pollution number more than 50. Provisions on heavy metal pollution prevention and control mainly consist of control of pollutant discharges, exploitation and protection of natural resources such as minerals, prevention of heavy metal pollution to the environment and food from pesticides and sewage irrigation, prevention and control of electronic wastes and hazardous chemicals, establishment of standards and monitoring regulations of heavy metals, and regulatory documents for specialized affairs about heavy metal pollution prevention.

These legal provisions, administrative regulations, department regulations and local regulations have played a positive role in the prevention and control of heavy metal pollution, but much remains to be done. The environmental protection law in China established January 1, 2015, incorporated new regulations on basic systems such as environmental planning, environmental standards, environmental monitoring, environmental assessment, environmental economic policy, total amount control, ecological compensation, effluent charges, and discharge permits. This law also took public opinion into account, and stipulated the environmental public interest litigation system. Daily penalties, attachment, arrest and other punishment systems are designed to solve the problem of low illegal costs and high compliance costs. Based on scheduling conditions from January to April in 2015, nationwide implementation of consecutive daily penalties came to 160 cases. Among these cases, the highest fine reached 15.8 million yuan. But there are still many shortcomings in environmental legislation for heavy metal pollution control in China, including dispersed legislation, subordinate legislation, legislation that is too principled, low legislative level, uncompleted basic legal system, and others.

a. Urgent Demand for Special Legislation of Heavy Metal Pollution Prevention and Control In China, individual laws for heavy metal pollution prevention and control at the national level have still not been issued; only some relatively developed provinces and cities have put forward measures for implementation and have played a part in a management role. Although the Law of the People's Republic of China on Prevention and Control of Water Pollution, Regulations on the Control of Safety of Dangerous Chemicals, and the regulation of environmental standards and monitoring have some clauses related to heavy metal pollution prevention and control, these measures are ineffective for taking action, because they are merely principled legislation. The absence of legislation not only limits the industrialization process but also greatly weakens China's ability to attract potential social capital investment in the environmental remediation market. Therefore, the demand for special legislation has been very urgent under the current situation of severe heavy metal pollution.

b. Urgent Demand for Special Legislation of Soil Environmental Protection and Pollution Control There is a clear and definite plan for system construction and implementation in wastewater treatment and waste disposal that goes beyond technical research planning. This plan for wastewater clearly stipulates the technical route, investment proportion, and liability subject divisions, and formulates technical standards and specifications. This practical planning and execution is lacking in the process of soil contamination remediation.

It is very urgent to enact laws specifically for soil protection under the current severe soil pollution. Nowadays, more than 50 regulations on environmental pollution have been enacted in China, but China still lacks laws and regulations for soil environmental protection and pollution control. Current laws and regulations lack provisions for soil pollution monitoring, planning of systems, functional zoning systems, special soil conservation action, soil improvement areas planning, repair systems, and reuse of remediated polluted soil

c. Heavy Metal Pollution Standard Systems Remain to be Improved The history of the environmental standard system in China can be divided into three periods: a local development period, the period of industrial 'three wastes' treatment, and a comprehensive prevention and control period.

At present, China's environmental standard system has three levels: a national environmental standard system, local environmental standard systems, and environmental protection industrial standard system (Figure 2). After more than 40 years of development, China has established a preliminary environmental standard system in which *National Environmental Quality Standard (NEQS)* and *National Emission Standard (NES)* form the main body, form a complete set with environmental methods standard, standardized sample standard, and basic standard and is supplemented by the regional and local environmental standard system and environmental protection industrial standard.

The state standard on environmental protection in China has currently reached a total of 1,598 items. There are 67 standards for environmental monitoring, 45 standards for environmental reference material, and 14 standards for cleaner production. There are 35 newly formulated emission standards and almost 250 standards are related to prevention and control of heavy metal pollution involving 20 heavy metals. There are 11 environmental quality standards involving heavy metals in the current environmental standard system, one for atmosphere [38], five for water [46–50], and three for soil [51–53]. Lead is the only heavy metal mentioned in the atmosphere environmental quality standard [38]. Eight heavy metals are referred to in the soil environmental quality standard, As, Cd, Cr, Cu, Hg, Ni, Pb and Zn. Eighteen kinds of heavy metals are referred to in the water environmental quality standard, As, Ba, Cd, Co, Cr, Cr^{6+}, Cu, Fe, Hg, Mn, Mo, Ni, Pb, Sb, Se, Tl, V and Zn. The pollutant emission standard in China consists of a comprehensive standard and an industry standard. Thirty standards refer to heavy metals in the pollutant emission standard, 17 standards for water pollutant emission [54–69], two standards for atmosphere emission [70, 71], four standards for both water and atmosphere [72–75], and seven standards for solid waste pollution control [76–82]. Seven heavy metal pollutants are referred to in the atmosphere emission standards, including Pb and its compounds, Hg and its compounds, Cd and its compounds, Ni and its compounds, As and its compounds, and chromic acid. In the water pollutants discharge standards, 17 kinds of heavy metals are mentioned., Ag, As, Ba, Cd, Co, Cr, Cr^{3+}, Cr^{6+}, Cu, Fe, Hg, Mn, Ni, Pb, Se, Zn and methyl mercury.

Development of current environmental quality standards involving different levels of atmosphere, water, and soil has been a complex task. Different industries have different heavy metal emission standards, but lack daily supervision and management for heavy metal pollution control and testing requirements. Related standards are given priority in determining whether the pollution discharged can meet the standards, while the cumulative effects of the heavy metals in the environment and human body are not considered. For example, a given company that is certified by the environmental protection department and whose main pollutant emissions have met the national standard may fail to meet emission standards for lead blood levels in children. The regional emission standard mostly requires

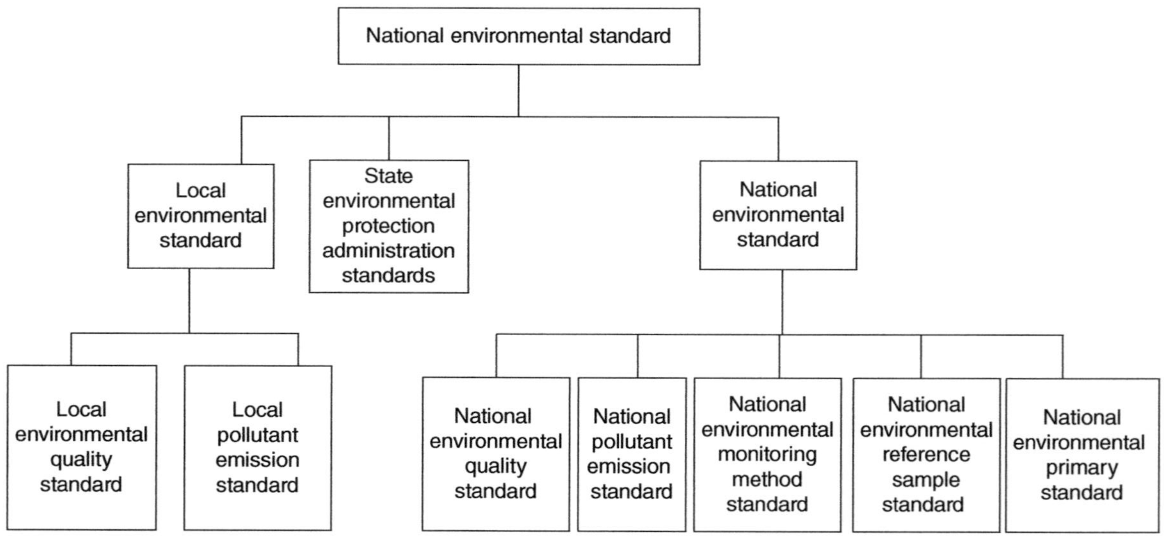

Figure 2 *Chinese environmental standard system*

measurement of a pollutant concentration, but rarely the total quantity, which focuses on environmental quality and human health. The problem is that the environmental standard system is not perfect and additional adjustments are required. At the same time, problems exist in the establishment of environmental standards. For example, China's industrial development may be challenged to adapt, since part of the national or industrial standard may be too strict for enterprises to reach and may thus gradually become a burden. Such a situation can lead to non-compliance.

In the process of legislative innovation and transformation, the key work in China is to perfect the environmental laws and regulation system. This is a work in progress that has three components. First, new laws and regulations must be formulated for prevention and control of heavy metal pollution, which includes the soil pollution prevention law, administrative laws and regulations or department regulations, and local laws and regulations. Second, existing laws and regulations must be modified as required. Third, environmental standards and health standards require improvement with regard to pollution prevention and control.

8.2.2.2 Policies for Heavy Metal Pollution Prevention and Control

Concerning prevention and control of heavy metal pollution, the central government of China follows a policy of "new for old, integrated control". Adhering to the approach of preventing pollution from the source, China takes measures such as enforcing market admittance, optimizing industrial structure, reducing pollution intensity, and controlling admittance of new pollutants. In addition, the central government pays strict attention to strengthening supervision of existing pollution sources, encouraging replacement of backward production capacity, and carrying out comprehensive treatment. At the same time, China works hard on digesting pollution inventory, paying off old scores, ensuring safety, and prevention of risks.

In operation, the government coordinates pollution prevention and industrial development, coordinates management of existing pollution sources, and attempts to solve pollution problems left over from the past. Essential to this effort is the highlighting of heavy metal pollution prevention and control of key areas, industries and enterprises. China promotes pollution prevention nationwide by division, classification and staging.

The 12th five-year plan lists 14 key provinces, 138 key areas, 5 key industries and 4,452 key enterprises for the control of heavy metal pollution. The 14 key provinces (areas) are Inner Mongolia, Jiangsu, Jiangxi, Henan, Hubei, Hunan, Guangdong, Guangxi, Sichuan, Yunnan, Shaanxi, Gansu and Qinghai. The five key industries involved are heavy non-ferrous metal mining, nonferrous metal smelting, lead battery manufacturing, leather manufacturing, and chemical raw materials and chemical products manufacturing. At the same time, the government gives full play to the guiding role of providing policy and institutional guarantees for heavy metal pollution prevention and control, strictly implementing the enterprise's main body responsibility, strengthening responsibility, and achieving stable discharge of the standard. In addition, the government encourages social participation in order to strengthen environmental information disclosure and public opinion input.

a. Eliminating the Drain on Production Capacity of High Energy Consumption and High Pollution The drain on production capacity should be eliminated by law wherever possible.

To reduce emissions, China should take legal measures to adjust the industrial structure, implement industrial policies, adjust revitalization plans for relevant industries, and develop and implement for key industries by region replacement of antiquated equipment and processes by state-of-the-art processes. The 12th five-year plan for the prevention and control of heavy metal pollution requests China's Ministry of Industry and Information Technology to expand the scope of eliminating inefficient production and promote the use of modern technological equipment to improve the effectiveness of these actions in minimizing heavy metal pollution.

Excess production capacity has been a problem for China's non-ferrous metals industry in recent years. The market price of non-ferrous metals was high, which encouraged investment in the industry and resulted in the flow of money into it increasing construction capacity. The situation of overcapacity gradually improved under the policy of closing down inefficient production facilities. The 12th five-year plan for energy conservation and reduction of emissions, published in 2012, clearly stipulated the comprehensive alternating current consumption of aluminum ingots, comprehensive energy consumption of copper smelting, and many other energy conservation and emission reduction targets. This plan promotes elimination of non-ferrous industries having high energy consumption and a record of high pollution. In July and September of 2012, the ministry announced closure of two batches of enterprises with inefficient production capacity. These batches involved more than 120 non-ferrous metals enterprises with 2.3 million tons of production capacity.

Currently, the investment priority for non-ferrous metal enterprises has been transferred to the western part of China. In northwest China, electrolytic aluminum production capacity has expanded quickly, and the non-ferrous metal industry still has difficulty in controlling total capacity. However, the percentage of production capacity of large-scale advanced enterprises has increased significantly. Non-ferrous metal industries that are small-scale and have low efficiency are gradually withdrawing from the market, changing the structure of this industry.

Published in October 2013, the excess capacity dissolving policy guidance of the State Council [83] stated that "we should make important progress in resolving the overcapacity problem of iron and steel, cement, electrolytic aluminum, flat glass and ship industries through 5 years of effort, and encourage these industries to achieve the goal of reasonable production scale, markedly improved development quality, establishment of long-term mechanisms." From 2010 to 2013, China has eliminated excess capacity of 85 million tons of iron and 57 million tons of steel. The goal of eliminating backward iron and steel production capacity was 27 million tons in 2013 (Table 4). It is expected that the production capacity of compression steel will be above 80 million tons to achieve a reasonable level from 2013 to the end of 2017.

In August 2014, the rare-earth industry first entered the list of industries of inefficient production capacity and excess capacity released by China's Ministry of Industry and Information Technology. In total, China had a backward production capacity of 103,710 tons of rare earths, including 62,210 tons in Inner Mongolia (accounting for more than 60%), 16,000 tons in Hunan Province (more than 15%), Jiangsu Province 12,000 tons (more than 12%), Sichuan Province 7,000 tons (more than 7%) and Gansu Province 6,500 tons (6%). Significant actions taken were to form six large rare earth groups for collecting and storage of rare earths and to subsidize the rare earth industry chain. China achieved the goal of protecting resources by eliminating backward production capacity, tightening supply, and enhancing industry concentration.

Table 4 *Elimination of Inefficient Production Facilities in China in 2012–2013 [84, 85]. Source: Adapted from [84–85]*

Units, ten thousand tons	2013	2012	increase or decrease	the change
iron industry	618	1078	–460	–42%
steel industry	884	937	–53	–5.6%
coke industry	2400	2493	–93	–3.73%
iron alloy	210	326	–116	–35.5%
calcium carbide	118	132	–14	10.6%
electrolytic aluminium	27	27	0	0%
copper smelting	86	75.8	10.2	13.4%
lead smelting	96	134	–38	–28.3%
zinc smelting	19	32.9	–13.9	–42.2%
lead battery(plate and assembly) 10^4 KVAh	2840	2971	–131	–4.41%

b. Improving Industry Admittance Threshold to Limit Projects with Heavy Metal Emissions China strictly enforces access rules and optimizes the industrial layout. Other government activities include formulating and improving the admittance threshold into industries and key areas to further improve energy savings, promote environmental protection, enhance security, monitor proper land use, and insure access to occupational safety and health standards. Strict environmental assessment, land and safety production license approval, implementation of heavy metals industry, and an admittance system of announcement are essential to the process.

The central government fully promotes strategic environmental assessment, limits environmental assessment of projects related to heavy metal emissions and takes measures such as "Regional limits for approval, industrial limits for approval." From 2008, the ministry of environmental protection has rejected 332 projects that involved high resource and energy consumption and duplicate constructions at low level or excess capacity. The total investment in these projects reached 1.1 trillion yuan. The government is developing and implementing environmental economic policies, and a preliminary environmental policy framework including green credit, insurance, trade, electricity, securities, taxation, and others has been established.

The ministry in December 2011 issued a *Non-Ferrous Metals Industry 'Twelfth Five-Year' Development Plan* stating that during the '*Twelfth Five-year*' period, industry access thresholds should be further improved to limit projects that discharge heavy metals. This plan intends to strengthen the cohesion of non-ferrous metal industry policy and policies on taxation, finance, trade, land, environmental protection, and safety production. In China, new construction, reconstruction and extension projects that will increase heavy metal emissions will be banned in nature reserves, drinking water source protection zones, and other areas in need of special protection, such as cities and their suburbs, residential areas, and other regions with high environmental conditions.

Copper smelting industry standard conditions, issued in April 2014, proposed that the smelting capacity of newly constructed or reform enterprises using copper concentrates and copper secondary resources should reach 100,000 tons/year or above. The production scale of enterprises that use existing copper secondary resources as raw materials should not be less than 50,000 tons/year. At present, China's copper smelting and processing enterprises are small but their number is large: the percentage of large copper enterprises is less than 6% while small businesses account for 80% of the total. As the industry admittance threshold and environmental requirements are further improved, backward production capacity will be eliminated faster to speed up integration of the copper smelting industry.

Lead and zinc industry standard conditions, issued in May 2015, claimed that the scale of new, small, single lead and zinc mines shall be not less than 100 thousand tons/year (300 tons/day), and the longevity of the mine should be 10 years. Ore mining production should be more than 300,000 tons/year (1,000 tons/day). For individual treatment of zinc or zinc oxide minerals, secondary resource projects and construction and renovation projects, the scale of fire-method processes must be 15 thousand tons of zinc metal/year or above. The scale of wet single-series must be 30 thousand tons of zinc metal/year or above. For projects that recycle precious metals through individual handle and smelt-off-scum processes, the scale of single-series spoil disposal should be greater than 50 thousand tons/year. The scale of single-series electrolytic production line of lead-bismuth alloy should be at least 20,000 tons/year. In addition, *Standard conditions for lead and zinc industry (2015)* standardizes the industry in terms of technology and equipment, energy consumption, resource consumption and integrated utilization, environmental protection, safety production, and occupational disease prevention.

The non-ferrous metals industry *Twelfth Five Year Plan development plan* also mentioned that new construction, reconstruction, and extension projects that increase emissions of heavy metal pollutants will be prohibited in areas like natural reserves and drinking water source protection zones, as well as areas with strict requirement for environmental protection, such as large and medium cities, their suburbs and residential areas.

c. Regional Comprehensive Improvement The Chinese government focuses on key areas and puts forward plans for comprehensive improvement of heavy metal polluted areas containing pollution-intensive industries, a legacy of pollution problems, and hidden risks. Focused planning for the prevention and control of pollution is made that highlights regional characteristics and strengthens measures like the adjustment of industrial structure, cleaner production, and treatment of pollutants. At the same time, the plan puts forward pollution prevention and control countermeasures and relevant supporting policies based on the principle of having a consistent pollution control policy and guidance in implementing this policy.

Since heavy metal control in the Xiangjiang River Basin was listed as the number-1 key project, the Hunan provincial government accelerated the treatment of heavy metal pollution. At present, three three-year action plans for protecting and treating the Xiangjiang River have been started. The Xiangjiang river basin of Hunan province, where urbanization is high and economic and social culture are most developed, has a high GDP accounting for about 76.42% of the entire province GDP value (2008). The task of preventing heavy metal pollution in the Xiangjiang River Basin was divided into five parts, Qingshuitang (Zhuzhou city), Zhubugang (Xiangtan city), Xikuangshan (Loudi city), Sanshiliuwan (Chenzhou city), and Shuikoushan (Hengyang city). According to the characteristics of the five zones and the principle of "Different districts with different policies", the local government explored a new way of 'one area, one policy' and determined how to set up different functional departments to guide the prevention and control work. Since the Qingshuitang area has many heavy-pollution factories, it was determined that the provincial development and reform commission is responsible for guiding the closure or relocation of polluting enterprises in this area. Shuikoushan and Zhubugang have high environmental risk, so it is determined that the provincial department of environmental protection is responsible for guiding environmental comprehensive improvement for the Shuikoushan area and the heavy pollution challenge for the Zhubugang area. The provincial water resources department is responsible for guiding remediation of the Sanshiliuwan district, which faces issues such as severe ecological destruction, drinking water safety problems, and severe river channel sedimentation. Last but not least, the Office of Land and Resources is responsible for the renovation and the solving of old, continuing problems in Xikuangshan. Integrating the advantages of each department gives full play to the role of local government and enterprise and to the exploration of a new mode of department linkage.

In recent years, 18 billion yuan have been spent to resolve heavy metal pollution problems in the Xiangjiang river basin. 893 enterprises with serious heavy metal pollution have been eliminated, and 255 pollution treatment projects were completed in 2014. Since 2014, due to these effective measures, heavy metal levels in the Xiangjiang river mainstream and in water sources in the river basin have met the required water quality standards. The environmental quality of this area is significantly improved.

8.3 Metal Sustainability in China

Recovery and recycling of metals has become a global issue due to dwindling natural resources and energy, the world's growing population, and the continued deterioration of the environment. The recovery rate of copper, aluminum and lead in developed countries has exceeded 50%, while in China the recovery rate is about 30%. Recycled metals account for a growing share of total metal consumption and have become a significant part of metal usage, which has an important influence on the world economy.

At present, recycling in the non-ferrous metal industry in China is in an important transition. With the world economy in a state of unabated turbulence and Chinese economy undergoing dramatic shifts (such as shifting economic growth, structure adjustment and policy transformation), the recycling industry confronts prominent contradictions and issues. Strengthening contact with the world's renewable non-ferrous metal industry and achieving sustainable development, through recycling have become the inevitable choice of China's renewable non-ferrous metal industry.

8.3.1 Metal Recycling in China

In the 21st century, with the rapid economic development, the lack of sufficient mineral resources and people's increasing need for metal resources have caused a gradual change from using raw resources to recycled resources and from underground to above-ground mining. Recycling of non-ferrous metals has gradually becoming a mainstream global industry. Recycling in the non-ferrous metal industry in China started relatively late and is in the primary stage of development. Although the industry has made great progress, the proportion of non-ferrous metals available from recycling is still far below the world average. In the meantime, since it is a large producer of non-ferrous metals and a large user of non-ferrous metal scrap, China has huge development potential and good investment prospects.

From 1949 to 2009, China produced a total of 43.69 million tons of recycled non-ferrous metals. Between 2000 and 2009 its production was 31.50 million tons, accounting for 80% of the total production of the 60 years from 1949 to 2009. In 2009, the recycled non-ferrous metal production was 6.33 million tons, a 19.4% increase year-on-year, which accounted for 24.3% in 10 kinds of non-ferrous metal production, 3.3% higher than that of 2008. The year 2012 was a big one for recycling scrap non-ferrous metals in China. Total recovery was up to 10.45 million tons, accounting for 28% of the total amount of non-ferrous metals produced that year. In 2012, recycled copper accounted for 26.5%, recycled lead for 13.5%, recycled aluminum for 46.2% and recycled zinc for 13.8% of the total [86]. Other metals such as nickel, magnesium, tin and other renewable metal resources had been recycled also, but the yield was small. Thus, it is seen that compared with the world average recycling level and industry recycling in developed countries, China's non-ferrous metal recycling industry still has a long way to go. At the same time, this industry in China also has vast room and potential for development.

At present, the Pearl River Delta, Yangtze River Delta, Bohai Bay Economic Rim and Chengdu Chongqing Economic Zone and other key industrial clusters of non-ferrous metal recycling have been formed. The world's largest aluminum recycling enterprise which produces 0.65 million tons a year, a copper recycling enterprise which produces 0.40 million tons a year, and a lead recycling enterprise which produces 0.65 million tons a year have emerged in those industrial clusters. Some primary non-ferrous metallurgy enterprises have also entered the field of non-ferrous metal recycling and constructed large-scale demonstration projects.

During the "12th five-year plan" period, major non-ferrous metal recycling goals of China are (1) to have production of recycled non-ferrous metals reach 11.1 million tons by 2015, including 3.8 million tons of copper, 5.8 million tons of aluminum and 1.5 million tons of lead; (2) to have the percentage of recycled copper reach 40% of the total copper produced; (3) to have recycled aluminum and recycled lead reach 30% or more, each, of the total amounts of these metals produced; and (3) to further optimize non-ferrous metal industrial production while maintaining environmental protection technology. Enhanced resource utilization and energy-saving emission reductions will be essential in achieving positive results. In order to achieve these goals, Government and industry have to be more involved in international exchanges and cooperation, and more active in promoting development of measures that will increase recycling in the domestic metals industry.

8.3.2 Metal Recycling from Wastewater, Solid Waste and Flue Gas

8.3.2.1 Metal Recycling from Wastewater

Different technical methods are used for treatment of water polluted by heavy metals. Which method is employed depends on the characteristics of the polluted source, such as concentration of the heavy metal pollutant, the speciation of the heavy metal, and the source of the wastewater [87]. For example, in the case of non-ferrous metal mining and smelting enterprises, high concentration sludge processes, electric flocculation, membrane technology, and ionic exchange resins were applied to treat wastewater, and for waste lead-acid batteries recycling enterprises, dry method and wet process of ore desulfurization-electrolytic deposition were applied to treat wastewater containing lead and mercury pollutants.

As metallurgical and chemical industries were developed and leaner ore exploited, increased contamination by arsenic was found. Ninety percent of arsenic pollution is a by-product of the smelting process. In China, smelting plants discharged more than 100,000 tons of arsenic into waste water every year. The arsenic stabilization methods could no longer meet China's environmental requirements. Now, arsenic oxides recycled in large and medium-sized smelting plants total only 10,000 tons annually, 500 tons of which is recycled by Guixi Smelter, Jiangxi Province. Other smelters rarely recovered arsenic. Large amounts of arsenic, which is discharged into wastewater during the use of arsenic and the treatment of arsenic waste (slag), could be effectively treated by a stable precipitation technique. Also, treatment of large-scale arsenic polluted water bodies (at low arsenic concentrations) using modified low-cost adsorption material or bioconversion technology can be effective [88].

Wastewater discharged from hot zinc plating plants contains high concentrations of zinc and iron, with the concentration of zinc reaching 10% to 15% and the concentration of iron 5% to 10%. Iron is in the divalent state in galvanized wastewater. By controlling the pH, iron can be oxidized and precipitated; followed by addition of base to precipitate zinc hydroxide. This method can be used to recover 90% or more of the zinc from waste water discharged from the hot dip galvanizing plant [89]. Photographic industry, electroplating industry, laboratory chemical analysis, and many other operations produce wastewater containing silver. The photographic industry and electroplating industry are two major industrial sources of wastewater containing silver. The main recycling methods used to recycle silver from wastewater are precipitation, reduction, electrolysis, ion exchange, adsorption, and various combined methods [90].

Traditional neutralization-sedimentation methods for heavy metal recovery from wastewater have disadvantages, such as high cost, difficult to control, low efficiency, and production of large amounts of slag. As more stringent environmental regulations and standards are promulgated and implemented, technology improvements are needed for heavy metal recovery from wastewater. These improvements include, (1) development and application of new, non-toxic chemicals (2) development and application of new economic, efficient, environmentally friendly technologies; (3) development and application of combined technologies; 4) increased use of automatic procedures, and (5) development and application of a two-way technology for improvement of wastewater reuse and heavy metal recycling.

8.3.2.2 Metal Recycling from Solid Waste

The main sources of solid waste containing heavy metals are tailings and waste rocks produced from metal mining, ore dressing operations, smelting and other activities, industrial sludge, some municipal sludge and solid waste, and fly ash containing heavy metals that are collected by the exhaust gas treatment process. Heavy metals can be recycled from solid waste by appropriate treatment methods chosen according to metal speciation and chemical properties.

At present, a wet process, which uses a low carbon polyhydric alcohol alkaline solution as a leaching agent, has good prospects for the treatment of smelting waste. During this process, copper, lead, bismuth and antimony can be leached from the smelting waste such as anode slime. After this process, an electro-deposition or neutralizing hydrolysis method is used to recover valuable metals from the alkoxide solution, and regeneration of the low carbon polyol is achieved [91, 92].

Zinc is easily recycled, and secondary zinc resources have become important raw materials for zinc production. Globally, 30% of the zinc supply comes from secondary zinc resources, and the annual output of recycled zinc is 290 million tons. Secondary zinc utilization in China started late. The annual output of secondary zinc was less than 100,000 tons, less than 5% of fine zinc production [93, 94].

Raw materials for zinc recycling mainly [95–97] are scum or bottom slag, and pan slag produced from hot dip galvanizing plants, zinc-containing dust produced from iron and steel factories, scrap zinc and zinc alloy parts, waste galvanized products, and used batteries. In general, zinc accounts for more than 80% of scum, and the generation rate of scum produced during the hot-dip galvanizing process is about 23%; zinc accounts for 97% of bottom slag, and the generation rate of bottom slag is about 10% from the zinc ingot used for zinc galvanizing. Secondary zinc production from Chinese enterprises is evolving. There are more than 200 enterprises in all of China, including family-owned small smelters. The production capacity of most of these enterprises is hundreds of tons to 1,000 tons, only two enterprises can produce over 10,000 tons.

For recycling of waste containing mercury, China mainly focuses on the recovery of mercury from energy-saving lamps. In recent years, a few fluorescent lamp recycling companies have been established in China to achieve the recovery of mercury from waste fluorescent lamps. By the end of 2008, four companies were engaged in waste disposal of mercury recovered from energy-saving lamps, the annual handling capacity of these companies are 1,500, 31,500, 1,000, and 484 tons. These companies obtained business certificates from the Ministry of Environmental Protection. However, the actual handling capacity is only 1/3 [98] of the treatment ability, because of low value, high recycling cost, and difficulties in centralizing and recovering the mercury product. With increasing consciousness of energy saving and increasing need for environmental protection, recovering mercury from lamp waste will become one of the most important ways of producing this metal.

The largest consumption of mercury in China is as a mercury catalyst. As a catalyst for the reaction of acetylene and hydrogen chloride to form vinyl chloride (VCM), mercuric chloride catalyst is massively used in the production of polyvinyl chloride (PVC), which use accounted for 60% of total mercury consumption. Therefore, there is great incentive to recycle mercury from waste mercuric chloride catalyst material. It is desirable in the PVC industry to increase efforts to recover spent mercuric chloride catalyst, promote the use of catalysts with minimal mercury content, devote more effort to research aimed at developing new green chemistry catalysts; strengthen the control and management of production processes, and reduce mercury loss and discharge to the environment. By the end of 2008, five companies engaged in waste disposal of spent catalysts containing mercuric chloride had obtained hazardous waste operating permits from the Ministry of Environmental Protection in China.

The lead recycling industry in China began in the 1950s [99] and has has made significant progress, especially in the past 10 years. The production capacity of this industry reached 5.59 million tons by the end of 2011, and the output was 1.35 million tons in 2011. There are significant disadvantages for many lead recycling enterprises, such as small scale, use of backward technology, high energy consumption, low lead recovery rate, and serious pollution. In China, recovery of more than 70% of recycled lead is centralized in Jiangsu, Shandong, Hebei, Henan, and Hubei. There are nearly 300 waste lead-acid battery recycling plants in China, the production capacity of which range from dozens of tons to several thousands of tons. Only 12 plants have production capacities of more than 10,000 tons. In the past 10 years, efforts have been made to improve secondary lead technology. However, there is still a large gap between the secondary lead smelting technology in China and that in the world's developed nations where advanced technologies are used. This gap is reflected in the co-existence in China of advanced and traditional technologies. The productivity of traditional technology accounted for 60% of the total secondary lead production. Promotion of sealed maintenance-free lead-acid batteries encourages the production of colloidal lead-acid batteries and mercury-free alkaline zinc manganese batteries. It is also beneficial to promote the replacement of nickel-cadmium batteries by lithium ion batteries and MH-Ni batteries, and to promote the replacement of zinc-manganese batteries by mercury-free zinc-manganese batteries. To achieve a closed-cycle in the process of producing and recycling of lead-acid batteries, the principle of "who produces must recover; who pollutes must control" needs to be insisted on. At the same time, lead smelting enterprises and lead-acid battery enterprises need to be encouraged to extend their knowledge and experience to recycling of the lead in waste lead-acid batteries. By 2015, the recycling rate of lead-acid batteries in China is expected to reach 90%.

At present, there are three main methods for reclamation of waste land, i.e., physical method, chemical method and phytoremediation. Phytoremediation is a hot issue in the international academic field. Compared with other technologies, it has advantages of low cost, no secondary pollution, more ecological and aesthetic value, and the ability to repair contaminated soil permanently. Since the 1950s, China has carried out census work on mine vegetation. In the 1950s, China mainly focused on finding indicator plants for metallic ore deposits and paid less attention on hyper-accumulators of metals, resulting in very few of these plant species being found and confirmed. In 1999, the hyperaccumulator *Pteris vittata* (accumulate arsenic [1]) was discovered. Since then, *Solanum nigrum* and lobar wellhead edge grass (accumulate arsenic[2]), *Sedum* (accumulate zinc[3]), Baoshan violet (accumulate cadmium[4]) and pokeweed (cumulative manganese[5]) have been discovered.

8.3.2.3 Metal Recycling from Flue Gas

In China, reduction in the emission of heavy metals is accomplished mainly through controlling the sources of particulates, since most heavy metals are discharged into the atmosphere in the form of particulates. Research shows that dust-separation facilities, desulfurization of flue gas and de-nitrification processes are able to remove heavy metals to a certain extent [105] thanks to the great achievements in the 11th five-year plan in desulphurization and removal of flue gas from chimneys. Although China has experienced a sharp increase in the amount of coal consumption in recent years, the increase of heavy metal emissions is far lower than the increase of coal consumption, particularly in the power sector. The desulphurization proportion of Chinese thermal power generating units was expected to increase from 14% in 2005 to 86% in 2010 with thermal power generating units being installed using high efficiency electrostatic dust precipitation facilities [106]. In 2011, China issued new thermal power plant air pollutant emission standards [107]. As part of these standards, China will implement stringent particulate matter emissions restrictions to further reduce the heavy metal emissions of the thermal power industry. In the near future, China will also constitute or revise emission standards for the concrete, iron and steel, and other industries, which is conducive to the control of heavy metal emissions in these industries.

Generally speaking, there will be economic value in recycling indium from flue gas when the indium content in raw materials is above 0.002%. By using a hydrometallurgical process to recycle valuable metals from the gold or copper smelting converter dust, the indium recycling rate could reach 60.88%, and the purity of the indium ingot is not less than 99.99% [108]. At the same time, the hydrometallurgical process avoided the secondary pollution from dust produced from the sulfate roasting-extraction process for recovering indium [109]. Through the combined process "reduction & volatilization-neutral leaching of zinc-leaching indium in high temperature and high acidity-extraction -reextraction-replacement-smelting-electrolytic refining to produce refined indium, refined bismuth and zinc", valuable metals like indium, zinc and bismuth in gas ash can be recovered, and the recovery rate of indium could reach 50–60% and the recovery rate of zinc could reach 70–75% [110].

Blast furnace dust discharged from the iron-making process mainly contains iron, coke and coal powder. As seen in Table 5 [111], as a by-product of the iron and steel industry, blast furnace dust contains a high content of zinc. One ton of steel produced means production of about 20 kg of blast furnace dust containing 10%–20% zinc. In 2013, the output of steel was about 0.78 billion tons in China, so about 15 million tons of blast furnace dust was produced, which contained 1.5–3 million tons of zinc that is equivalent to about 50 million tons of zinc ore, if mined. There are different methods of zinc recovery in different iron and steel plants. At present, blast furnace dust is treated mainly by physical, fire, wet, microwave, plasma, and combined methods [112].

During the process of recovering precious metals in a lead smelting plant, we often use a fire method to treat anode slime. In the roasting and flame smelting process, most of the arsenic is discharged into furnace gas and enriched in the dust in oxidized form. Due to the similar properties of antimony and arsenic, dust containing concentrated arsenic and antimony as well as other nonferrous metals like Ag, Pb, and Bi will be formed, in which the content of As, Sb, Pb, and Bi is 4.26%–26.61%, 38.42%–51.2%, 1.5%–10.32% and 1.03%–7.27% [113], respectively. For the treatment of high arsenic and antimony

Table 5 *Production and composition of blast furnace dust in China in iron and steel enterprises*

	Blast furnace dust production/ kg·ton steel^{-1}	TFe/%	Zn/%	MgO/%	CaO/%	SiO_2/%	C/%
Domestic	15–30	15–30	3–17	1	2–5	4–5	25–55

materials, the major problem is treating arsenic and recycling antimony at the same time. Treatment of arsenic- and antimony-bearing dust is mainly achieved by using the fire and wet methods. Adopting the sodium hypophosphite reduction technology to remove arsenic after acid leaching and then using a chlorination-hydrolyzation method to recover antimony is an effective way to separate As and Sb and to achieve the recycling of As in the low- toxic form and Sb in the form of antimony oxide.

8.3.2.4 Metal Recycling from E-waste

With the development of information technology and the acceleration of the upgrading of electronic products, life cycles of these products become shorter and shorter. E-waste is produced at an amazing growth rate of 4% annually [114] and has become the fastest growing garbage stream in the world [115]. In China, the total output of E-waste has increased from 230 million tons in 2010 to 560 million tons in 2015 [116]. E-waste has become a new and important environmental pollutant source associated with chemical industry, metallurgy, papermaking and the printing and dyeing industry [117]. As a developing country, China has become a dumping ground for E-waste from developed countries because of cheap labor and lagging environmental regulations. Of the E-waste exported to Asia from America, 90% enters dismantling centers in China [118]. Two typical dismantling centers are the Taizhou area in Zhejiang Province and Guiyu town in Guangdong Province [119].

In the late 1970s, Taizhou city [120] began to dismantle waste materials like scrap transformers, motors, circuit boards, and so forth. From the early 1990s, Taizhou city began to recycle foreign waste and became the largest collection and distribution center of imported E-waste and the largest dismantling and processing base of solid waste metal in China. After years of development, Taizhou has formed an industry chain of E-waste recycling which involves 50,000 employers and over 15 million practitioners with an annual production value of more than 20 billion yuan. The E-waste recycling industry has become an important part of the economy of Taizhou.

Guiyu town is the largest E-waste dismantling center in China and, probably, in the world. This city is one of the many locations globally where the environment is seriously polluted by the toxic components of end-of-life electronic products [121]. At the end of the 1980s and in the early 1990s, residents of Guiyu began to dismantle old hardware appliances with large profits, and gradually enlarged the scale of this emerging industry. About 100,000 tons of copper are recycled in Guiyu per year. Currently, there are 21 villages, more than 300 private enterprises, 5,500 household workshops and over 60,000 employers engaged in dismantling and recycling of electrical and electronic waste. One and one-half million tons of electrical and electronic waste and plastics are dismantled and processed, which provide more than 1.5 million tons of renewable metal, plastics, second-hand electronic components and associated products for the country every year. In 2003, the total output value of the waste electrical and electronic equipment from the plastic recycling, and dismantling, processing industry was 6 billion yuan in Guiyu, In 2006, the value reached nearly 12 billion yuan, accounting for more than 90% of the annual industrial output of the town. The waste electronic products dismantling and processing industry has become a pillar industry and the main source of income for farmers in the local economy [122].

8.4 Metal Sustainability in China: Future Prospects

Treatment of heavy metal pollution has been given great importance by the Chinese government, industrial enterprises and Chinese people, but the pollution status in China is still only vaguely understood by Chinese society and by the global population. Pollution sources and the level of pollution in those sources is incompletely understood. The effect of soil, water, and atmosphere pollution on the environment and the dynamic changes occurring are not clear. Basic data are lacking. Improved heavy metal pollution control in China requires attention to the following issues.

1. *Identify the pollution source and find specific ways to alleviate the problem.* If we do not first identify the heavy metal pollution source, we will fall into the vicious cycle of endless treatment of pollution year after year. Taking the governance of polluted farmland near the mine which is to shut down, for example, the pollution source may be agricultural irrigation water polluted by mine or groundwater that increases the water level during the rainy season. Pollution sources may be from land mines and bombs buried in the ground, which may explode at any time, threatening people's health. If the pollution source is not well controlled, money spent on soil remediation will be for naught. However, finding the source is not an easy thing, it requires long-term monitoring and a large number of basic data. Only when we find out the main sources of heavy metal pollution, can we determine the key directions to go and the proper methods of governance.
2. *Establish a continuous large-scale monitoring network and pay great attention to the dynamic changes of pollutants.* The pollution status varies with the environment. For example, the constituent of heavy metal in the Xiangjiang River varies with the seasons. If Xiangjiang water is used for irrigation in the dry season (the content of heavy metal increases in the dry season), it will lead to increased concentrations of heavy metals in soil. Some irrigation water will be mixed with sewage from factories, whose water quality is uncertain. Even if we monitor the water quality of irrigation water once a day, it is difficult to understand its water quality fluctuations. Therefore, China needs to establish a large-scale water, air and soil automatic monitoring network that can give real-time feedback on the concentrations of heavy metals and other pollutants and can take necessary action on control measures based on the pollutant concentration level.
3. *Formulate a more scientific and reasonable environmental standard system.* At present, China's environmental standard system is not perfect and some standards are detailed to specific processes, which hinders industry's technological innovation. Some standards are too stringent (more stringent than in Europe and the U.S.), often making them technically infeasible and not cost-effective. In such cases, compliance will put a heavy burden on the industrial organization. In some circumstances, there are no industry standards upon which to base compliance. When formulating environmental standards, it is essential to have a scientific basis grounded in experimental data for their establishment. As to heavy metal recycling and reuse industry, China needs to reduce the amount of waste produced and develop advanced recycling technologies, improve the industrial concentration and expand the industrial scale.
4. *Reduce the amount of waste generated and develop advanced metal recycling and recovery technologies.* Currently, huge amounts of waste rock, tailings, waste water, and

flue gas are produced in China making heavy metal recycling and recovery challenging tasks. Low recycling rates cause waste of resources and environmental pollution. Target metal recovery can be achieved by recycling heavy metals from huge amounts of waste. After recycling, the total amount of heavy metals in the waste is still high. Therefore, China needs to reduce the amount of waste by improving metal recovery and recycling rates. Impetus needs to be given to the development and application of advanced recycling technologies. In the industrial field, the full implementation of circular production methods and cleaner production is needed to promote the reduction of waste produced from the source.

5. *Improve industry concentration and do reasonable capacity planning* The non-ferrous metals recycling industry in China has a high degree of dispersion, and is running into the iceberg of overcapacity. Improvement of the industry concentration is conducive to opening up recycling channels to give full play to the role of market competition. Rational capacity planning, on the one hand, is conducive to the elimination of lagged output capacity and will reduce secondary environmental pollution caused by heavy metal recycling. On the other hand, such planning will give impetus to the development and the application of advanced recycling technologies, which is good for the healthy development of the industry in the long-term.

Heavy metal pollution control and metal recovery support and influence each other. Both are important components of metal sustainability. Successful accomplishment of both has a positive influence on environmental and economic sustainability and is closely related to people's health and quality of life. The serious environmental pollution that brings huge challenges for the Chinese people is the price that China has paid for its rapid economic development without sufficient concern for collateral consequences. As a developing country and without, as yet, sufficient economic development, China has not adequately shouldered the burden of heavy metal pollution control. Development of adequate pollution control in China will be a long and arduous process which needs the joint efforts of all Chinese people and the active participation of the global community. An encouraging sign is the active involvement of the Chinese government in laying the groundwork for pollution control by enactment and enforcement of legislative laws.

References

1. Cheng J H, 2010, Status of and Reflection on the High Consumption of Mineral Resources During Chinese Industrialization. *Journal of China University of Geosciences (Social Science Edition)*, 10(04), 45–48.
2. Tianfu Morning Paper, 2008, *450 People's Urinary Arsenic Exceeded for Arsenic Pollution in Hechi, Guangxi Province,* http://morning.scol.com.cn/new/html/tfzb/20081011/tfzb198638.html, accessed 20 August 2015.
3. Tecent, 2008, *Focus on the Liuyang cadmium pollution*, http://view.news.qq.com/a/20090803/000022.htm, accessed 20 August 2015.
4. Xinhua Network 2014, *Wastewater in Shunde (Guangxi province) Discharged into Water Source and Threatened Hundreds of Thousands of People's Drinking Water Safety,* http://news.xinhuanet.com/environment/2014-08/19/c_126890717.htm, accessed 20 August 2015.
5. Yunnan Network, 2008, *Yangzonghai River Arsenic Pollution Event*, http://special.yunnan.cn/feature/node_4520.htm, accessed 21 August 2015.

6. People's Network, 2011, *Qujing Cadmium Pollution Event*, http://env.people.com.cn/GB/211746/228644/, accessed 21 August 2015.
7. Soho Network, 2009, *Fengxiang (Shanxi Province) Pollution Event,* http://news.sohu.com/s2009/xueqian09/, accessed 21 August 2015.
8. Southern Weekly, 2011, *The Catastrophe of the Chromium Slag*, http://www.infzm.com/content/64051, accessed 21 August 2015.
9. Xinhua Network, 2009, *Related Enterprises Were Punished Severely for Arsenic Accident in Linyi* (Shandong province), http://www.sd.xinhuanet.com/news/2009-08/08/content_17339294.htm, accessed 20 August 2015.
10. Lang, W, 2008, Emergency Treatment and Considerations on Arsenic Pollution in Xinqiang River, *Hunan Hydro & Power*, 5, 51–53.
11. NetEase News, 2010, *Zijin Copper Acid Leaking Accident,* http://news.163.com/special/00014IT2/zjkyfsxlsg.html, accessed 21 August 2015.
12. China Youth News 2012, *Heavy Metal Pollution Happened Frequently*, http://zqb.cyol.com/html/2012-02/01/nw.D110000zgqnb_20120201_5-07.htm, accessed 21 August 2015.
13. Yao, Z.X, Xin, X.D, Wei, S, *et al.,* 2011, Research Progress of Heavy Metal Detection Method, *Analysis and Testing Technology and Instruments*, 25(1), 29–35.
14. Lin, C.H, Zhang, H.Y, Wang, Y, *et al.,* 2010, Analysing Flow Injection in Environmental Water Quality Detection of Heavy Metal Application, *Chemical Analysis and Meterage*, 4, 23–28.
15. Xu, X Q, Zhu, Y, Yang, T, *et al.,* 2007, Pollution Damage and Remediation Technology of Heavy Metals in Water Body, *Pollution Prevention Technology*, 20(4), 69–67.
16. The National Bureau of Statistics of the People's Republic of China, 2010, *The first national pollution census bulletin*, 2010/2
17. Liu, Y C, Gao, L. Li, Z G, *et al.,* 2010, Xiangjiang Heavy Metal Pollution Status, *Pollution Causes Analysis and Countermeasures Discussion, Environmental Protection Science*, 04, 26–29.
18. The Ministry of Environmental Protection of the People's Republic of China, 2013, *The National Environmental Quality Status in the First Half of 2013*, 2013/8.
19. Sun, G F, 2004, Arsenic Contamination and Arsenicosis in China, *Toxicology and Applied Pharmacology*, 198, 268–271.
20. Guo, Z H, Zhu, Y G, 2004, Typical Mining and Smelting Circumjacent Area Soil Heavy Metal Pollution and Effective Content, *Ecological Environment*, 13(4), 553–555.
21. Zhang D Y, Pan X L, Mu G J, 2010, Antimony pollution in water and soil and its ecotoxicological effects on plants and microorganisms, *Chinese Journal of Application and Environmental Biology*, 16(6), 891–894.
22. Zhang L, Xu S, Qiu G Y, *et al.,* 2008, Detection and analysis of the content of antimony in the water and oil around an Antimony Ore Factory in Weishan, *Journal of Dali University*, 7(12), 37–38.
23. The Ministry of Environmental Protection of the People's Republic of China (2014) National Soil Survey Bulletin 2014/4
24. Chen H M, 1996, *Heavy Metal Pollution in Soil-Plant System*, Science Press, Beijing
25. Wei C Y, Chen T B., 2001, Hyperaccumulators and phytoremediation of heavy metal contaminated soil a review of studies in China and abroad. *Acta EcologicaSinica*, 21(7), 1196–1203.
26. Duan Y C, 2014, Ministry of environmental protection will focus on arsenic pollution, *Oriental Morning Post*, http://www.zhongguowangshi.com/info.aspx?id=43433, accessed 20 August 2015.
27. Liu C P, Luo C L, Gao Y, *et al.* (2010) Arsenic contamination and potential health risk implications at an abandoned tungsten mine, southern China, *Environment Pollution*, 158, 820–826.
28. Wang W, 2005, *Heavy Metals in the Environment Polluted Chromium Plant*, Master's thesis, Tianjin Normal University.

29. Oriental Morning Post, 2013, Mercury pollution solutions in China, http://epaper.dfdaily.com/dfzb/html/2013-08/20/content_806698.htm, accessed 20 August 2015.
30. Bao Z D,Wang J X, Feng X B, Shang L H., 2011 Distribution of mercury speciation in polluted soils of Wanshan mercury mining area in Guizhou. *Journal of Ecology*, 30(5), 907–913.
31. Chen Z L, Qiu R L., 2002, Heavy metal contaminated soil remediation technology. *Environmental protection*, 29(6), 21–23
32. He M C, Ji H B, Zhao C Y, *et al.*, 2002, Preliminary study of heavy metal pollution in soil and plant near antimony mine area. *Journal of Beijing Normal University (Natural Science)*, 38(3), 417–420.
33. Tong F P, Xu Y P, Long Y Z, *et al.*, 2008, Study on physical and chemical character of forestry soil polluted by heavy metals in antimony mine of Len-Shuijiang City, *Chinese Agricultural Science Bulletin*, 24(12), 179–183.
34. Ning Z P, Xiao T P, Zhou L B, *et al.*, 2009, Preliminary study of pollution in soil and risk assessment of environmental quality for soil near antimony mine area. *Acta Mineralogica Sinica*, (S1), 402–404.
35. Song S Q, Liang L F, Zhou Y Z, *et al.*, 2003, The situation and remedial measures of the cropland polluted by heavy metals from mining along the Diaojiang River. *Bulletin of Mineralogy, Petrology and Geochemistry,* 22(2), 152–155.
36. Xiang M, Zhang G P, Li L, *et al.*, 2010, The characteristics of heavy metals in soil around the Hechi antimony -lead smelter, Guangxi, China. *Earth and Environment*, 38(4), 495–500.
37. Tian H Z, Cheng K, Wang Y, *et al.,* 2012, Temporal and spatial variation characteristics of atmospheric emissions of Cd, Cr and Pb from coal in China, *Atmospheric Environment*, 50(1), 157–163.
38. GB3095-2012.(2012) *Ambient Air Quality Standard*, The Ministry of Environmental Protection of the People's Republic of China.
39. Hu X, Zhang, Ding Z H, *et al.*, 2012, Bioaccessibility and health risk of arsenic and heavy metals(Cd, Co, Cr, Cu, Ni, Pb, Zn and Mn) in TSP and PM2.5 in Nanjing, China, *Atmospheric Environment*, 57(1), 146–152.
40. Hu Z J, Shi Y L, Niu H Y, *et al.*, 2002, Synthetic musk fragrances and heavy metals in snow samples of Beijing urban area, China. *Atmospheric Research*, 104, 302–305.
41. Tan J H, Duan J C, 2013, Heavy metals in aerosol in China, pollution, sources, and control strategies, *Journal of Graduate University of Chinese Academy of Sciences*, 30(2), 145–155.
42. Zhu B Q. Lead pollution quantitative assessment and prevention measures of the southeast coastal areas, in *The national lead pollution monitoring and control technology seminar of 2007*, April 24, 2007, Beijing, China.
43. Feng X B, Qiu G L, Fu X W *et al.*, 2009, Mercury pollution in the environment. *Progress in Chemistry*, 21(2/3), 436–457.
44. Wang M, Zheng B, Wang B *et al.*, 2006, Arsenic concentrations in Chinese coals. *Science of the Total Environment*, 357, 96–102.
45. Shraim A, Cui X, Li S, *et al.*, 2003. Arsenic speciation in the urine and hair of individuals exposed to airborne arsenic through coal-burning in Guizhou, *PR China, Toxicology Letters*, 137, 35–48.
46. GB 3838-2002. (2012) *Surface water quality standard,* The Ministry of Environmental Protection of the People's Republic of China, http://kjs.mep.gov.cn/hjbhbz/index.htm. (search the standard number GB 3838 at the web page)
47. GB 3097-1997, 1997, Sea water quality standard, The Ministry of Environmental Protection of the People's Republic of China, http://kjs.mep.gov.cn/hjbhbz/index.htm.
48. GB 5084-92, 1992, Irrigation water quality criteria, The Ministry of Environmental Protection of the People's Republic of China, http://kjs.mep.gov.cn/hjbhbz/index.htm.
49. GB 11607-89, 1989, Fishery water quality standards, The Ministry of Environmental Protection of the People's Republic of China, http://kjs.mep.gov.cn/hjbhbz/index.htm.

50. HJ 350-2007, 2007, The soil environment quality evaluation standards for land use in the exhibition (provisional), The Ministry of Environmental Protection of the People's Republic of China, http://kjs.mep.gov.cn/hjbhbz/index.htm.
51. HJ 332-2006, 2006, Environmental quality assessment standard of edible agricultural products, The Ministry of Environmental Protection of the People's Republic of China, http://kjs.mep.gov.cn/hjbhbz/index.htm.
52. GB 15618-1995, 1995, Soil environment quality standards, The Ministry of Environmental Protection of the People's Republic of China, http://kjs.mep.gov.cn/hjbhbz/index.htm.
53. GB 30486-2013, 2013, Leather and fur processing industrial water pollutant discharge standards, The Ministry of environmental protection of the People's Republic of China, http://kjs.mep.gov.cn/hjbhbz/index.htm.
54. GB 30484-2013, 2013, Battery industrial pollutants emission standards, The Ministry of Environmental Protection of the People's Republic of China, http://kjs.mep.gov.cn/hjbhbz/index.htm.
55. GB 4287-2012, 2012, Wastewater Pollutants Discharge Standard for Textile Printing Industry, The Ministry of Environmental Protection of the People's Republic of China, http://kjs.mep.gov.cn/hjbhbz/index.htm.
56. GB 28661-2012, 2012, Iron Ores Mining & Dressing industrial pollutants emission standards, The Ministry of Environmental Protection of the People's Republic of China, http://kjs.mep.gov.cn/hjbhbz/index.htm.
57. GB28666-2012, 2012, Ferroalloy industrial pollutants emission standards, The Ministry of environmental protection of the People's Republic of China, http://kjs.mep.gov.cn/hjbhbz/index.htm.
58. GB13456-2012, 2012, Discharge standard of water pollutants for iron and steel industry, The Ministry of Environmental Protection of the People's Republic of China, http://kjs.mep.gov.cn/hjbhbz/index.htm.
59. GB26451-2011, 2011, Emission Standards of Pollutants from Rare Earths Industry, The Ministry of Environmental Protection of the People's Republic of China, http://kjs.mep.gov.cn/hjbhbz/index.htm.
60. GB 26452-2011, 2011, Vanadium industrial pollutants emission standards, The Ministry of Environmental Protection of the People's Republic of China, http://kjs.mep.gov.cn/hjbhbz/index.htm.
61. GB 27632-2011, 2011, Emission standard of pollutants for rubber products industry, The Ministry of Environmental Protection of the People's Republic of China, http://kjs.mep.gov.cn/hjbhbz/index.htm.
62. GB 25463-2010, 2010, Ink industrial water pollutant discharge standards, The Ministry of Environmental Protection of the People's Republic of China, http://kjs.mep.gov.cn/hjbhbz/index.htm.
63. GB 25464-2010, 2010, Emission Standard of Pollutants for Ceramics Industry, The Ministry of Environmental Protection of the People's Republic of China, http://kjs.mep.gov.cn/hjbhbz/index.htm.
64. GB 25465-2010, 2010, Emission Standard of Pollutants for Aluminum Industry, The Ministry of Environmental Protection of the People's Republic of China, http://kjs.mep.gov.cn/hjbhbz/index.htm.
65. GB 25466-2010, 2010, Emission standard of pollutants for lead and zinc industry, The Ministry of Environmental Protection of the People's Republic of China, http://kjs.mep.gov.cn/hjbhbz/index.htm.
66. GB 25467-2010, 2010, Emission standard of pollutants for copper,nickel,cobalt industry, The Ministry of Environmental Protection of the Oeople's Republic of China, http://kjs.mep.gov.cn/hjbhbz/index.htm.

67. GB 25468-2010, 2010, Magnesium, titanium industrial pollutants emission standards, The Ministry of Environmental Protection of the People's Republic of China, http://kjs.mep.gov.cn/hjbhbz/index.htm.
68. GB 3544-2008, 2008, Discharge standard of water pollutants for pulp and paper industry, The Ministry of Environmental Protection of the People's Republic of China, http://kjs.mep.gov.cn/hjbhbz/index.htm.
69. GB 21900-2008, 2008, Electroplating pollutants emission standards, The Ministry of Environmental Protection of the People's Republic of China, http://kjs.mep.gov.cn/hjbhbz/index.htm.
70. GB 9078-1996, 1996, Air pollutants emission standards of industrial furnaces, The Ministry of Environmental Protection of the People's Republic of China, http://kjs.mep.gov.cn/hjbhbz/index.htm.
71. GB 4915-2013, 2013, Cement industry air pollutants emission standards, The Ministry of Environmental Protection of the People's Republic of China, http://kjs.mep.gov.cn/hjbhbz/index.htm.
72. GB 30484-2013, 2013, Battery industrial pollutants emission standards, The Ministry of Environmental Protection of the People's Republic of China, http://kjs.mep.gov.cn/hjbhbz/index.htm.
73. GB 26451-2011, 2011, Emission Standards of Pollutants from Rare Earths Industry, The Ministry of Environmental Protection of the People's Republic of China, http://kjs.mep.gov.cn/hjbhbz/index.htm.
74. GB 25467-2010, 2010, Copper, nickel, cobalt industrial pollutants emission standards, The Ministry of Environmental Protection of the People's Republic of China, http://kjs.mep.gov.cn/hjbhbz/index.htm.
75. GB 25466-2010, 2010, Lead and zinc industrial pollutants emission standards, The Ministry of Environmental Protection of the People's China, http://kjs.mep.gov.cn/hjbhbz/index.htm.
76. GB 30485-2013, 2013, Cement kiln collaborative treatment of solid waste pollution control standards, The Ministry of Environmental Protection of the People's Republic of China, http://kjs.mep.gov.cn/hjbhbz/index.htm.
77. GB 16889-2008, 2008, standard for pollution control on the landfill site of municipal solid waste, The Ministry of Environmental Protection of the People's Republic of China, http://kjs.mep.gov.cn/hjbhbz/index.htm.
78. GB 18484-2001, 2001, Hazardous waste incineration pollution control standards, The Ministry of Environmental Protection of the People's Republic of China, http://kjs.mep.gov.cn/hjbhbz/index.htm.
79. GB 18485-2014, 2014, Living garbage burning pollution control standards, The Ministry of Environmental Protection of the People's Republic of China, http://kjs.mep.gov.cn/hjbhbz/index.htm.
80. GB 18598-2001, 2001, Hazardous waste landfill pollution control standards, The Ministry of Environmental Protection of the People's Republic of China, http://kjs.mep.gov.cn/hjbhbz/index.htm.
81. GB 8172-87, 1987, Urban garbage agricultural control standards, The Ministry of Environmental Protection of the People's Republic of China, http://kjs.mep.gov.cn/hjbhbz/index.htm.
82. GB 4284-84, 1984, In the agricultural sludge pollution control standards, The Ministry of Environmental Protection of the People's Republic of China, http://kjs.mep.gov.cn/hjbhbz/index.htm.
83. Peoples Network, 2013, Excess capacity dissolving policy guidance of the State council, http://politics.people.com.cn/n/2013/1015/c1001-23210728.html, accessed 20 August 2015.
84. Ministry of Industry and Information Technology of The People's Republic of China (2014) Performance on closing down outdated capacities in 2013, http://www.miit.gov.cn/n11293472/n11293832/n11293907/n11368223/16320601.html, accessed 15 August 2015.

85. The Ministry of Industry and Information Technology of The People's Republic of China (2014) Performance on closing down outdated capacities in 2012, http://www.miit.gov.cn/n11293472/n11295023/n11310717/15753451.html,accessed 15 August 2015
86. Chinese government website (2013) Chinese renewable non-ferrous metal industry development report, http://www.gov.cn/xinwen/2014-04/28/content_2668225.htm, accessed 15 August 2015.
87. Dong R N, Li S P, Li G Q, 2011, In analysis on water pollution status of heavy metal and treatment technology, *China Environmental Science Society Academic Annual Conference Essays of 2011* (Vol. 2).
88. Bai J Z, 2009, Study on the recovery of arsenic from waste water containing arsenic in waste water, doctoral thesis, Northeastern University, Shenhang City, Liaoning, China.
89. Li J C., 2014, Heat recovery of Zn^{2+} in wastewater reuse technology. *Benxi Technology*, 01, 42–43.
90. Cao M Z, Zhang R, 2013, Research on the recovery of silver from waste liquid containing silver. *Henan Technology,* 13, 57–58.
91. Pan Z Q, Deng X H, Bing W D, *et al.*, 2011, Enrichment of gold and silver in the by-product of smelting. *Precious Metals,* 22(4), 17–22.
92. Peng G F, Zhang X F, 1990, Aluminum smelting by-product lime in gold and silver enrichment methods, glycerol, alkali leaching study on the removal of lead and antimony, *Nonferrous Metals (Metallurgy),* (2), 33–35.
93. Zhang J H, Lu W, 2007, Zinc renewable resources and recycling pathway and the present situation of resources of regenerated zinc, *Chinese Science,* 03, 86–93.
94. Kong M, Wang Y, 2010, China regenerated zinc industry, *Nonferrous Metals (Metallurgy)*, 05, 51–54.
95. Jiao P, Zhao B L, 2011, The recovery of zinc from zinc containing waste residue, *Gansu Metallurgy*, 06, 49–51.
96. Yan H J. (2010) Zinc scrap types and recovery technology research and application of. *Guangdong Chemical Industry*, 06, 283–284.
97. Liao Xiaohua, 2011, Research on the technology and application of the type and recovery of zinc containing waste, *Technology Information*, 24, 118.
98. Yin L P, Zhao J B, 2006, Status of heavy metal pollution of surface soil in Xi'an major functional areas of evaluation, Shanxi Normal University, *Natural Science*, 34(3), 109–112.
99. An S Y., 2012, The development status and recovery of lead acid battery use, *Heilongjiang, Science and Technology Information*, 13, 83.
100. Chen T B, Wei C Y, Huang Z C, *et al.*, 2002, The arsenic accumulation characteristic of Pteris vittata, *Chinese Science Bulletin*, 47(3), 207–210
101. Wei C Y, Chen T B, Huang Z C, *et al.*, 2002, A new arsenic accumulation plant, lobar wellhead edge grass, *Acta Ecologica Sinica*, 22(5), 777–778
102. Long X X, Yang X E, Ye Z Q, *et al.*, 2002, Zinc accumulation difference research of four kinds of Sedum, *Journal of Plants (English Version),* (2), 152–157
103. Wei L., 2004, Pb, Cd accumulation characteristic and mechanism of several kinds of violets, Master's thesis, Sun Yat-Sen University, Guangzhou, Guangdong province, China.
104. Xu X H, 2006, Mn accumulation mechanism research of superaccumulation plant pokeweed, Master's thesis, Zhejiang University, Hangzhou, Zhejiang province, China.
105. Tian He Z, Cheng K, Wang Y, *et al.*, 2012, Temporal and spatial variation characteristics of atmospheric emissions of Cd, Cr and Pb from coal in China. *Atmospheric Environment*, 50(1), 157–163.
106. Zhou S X, 2010, Play decisive battle, planning new development, actively explore China environmental protection in the new road. http://www.Mep.gov.cn/zhxx/hjyw/201001/t20100129_185134.html, accessed 18 August 2015

107. GB13223-2011, 2011, Fossil-fuel power plant atmospheric pollutant emission standard, The Ministry of Environmental Protection of the People's Republic of China, http://kjs.mep.gov.cn/hjbhbz/index.htm.
108. Xing X Z, Liu N, 2012, Research on the recovery process of indium in the copper smelting converter. *World Nonferrous Metals*, 06, 44–45.
109. Cao Y K, 2005, Recovery of indium from copper smelting, *Hunan Nonferrous Metals*, 21(1), 5–8.
110. Yaoping Zhu, 2009, The recovery of indium and zinc bismuth in blast furnace gas ash. *Nonferrous Metals (Smelting)*, 06, 14–16.
111. She X F, Xue Q G, Wang J S, *et al.*, 2010, Comparison of comprehensive utilization of zinc dust in iron and steel plant and related treatment technology, *Ironmaking*, 29(4), 56–62
112. Zeng D L, Liu S L, Zhang Q, Luo S H, Wang G H, Su M., 2014, From the blast furnace dust recovery zinc status of the process. *China Nonferrous Metallurgy*, 06, 57–60.
113. Hu X, 2013, Experimental study on the comprehensive recovery of arsenic and antimony from arsenic and antimony smoke, Doctoral thesis, University of Kunming, Science and Technology.
114. Fu J J, Wang Y W, Zhou L J, Zhang A Q, Jiang G B, 2011, Typical of China's electronic waste dismantling lasting toxic chemical contamination status. *Progress in Chemistry* (8), 1755–1768.
115. Sharpe M, 2005, Climbing the e-waste mountain, *Journal of Environmental Monitoring,* 7(10), 933–936.
116. Zhang H M, 2014, Gray industrial chain formation of electronic garbage siege China, http://www.ceh.com.cn/UCM/wwwroot/zgjjdb/cjpd/2014/01/297712.shtml, accessed 18 August 2015
117. Peng P A, Sheng G Y, Fu J M, 2009, Pollution problem of electronic waste, *Chemical Progress* (Z1), 550–557.
118. UNEP, 2005, E-waste, the Hidden Side: Equipment Manufacturing and Use, Early Warning on Emerging Environmental Threats, Nairobi, United Nations Environment Programme.
119. Zhang, W, 2013, *PCBs and Heavy Metal Pollution and Ecological Risk Assessment of a Waste Electronic Waste Dismantling Area in Taizhou*, Master's thesis, Zhejiang University of Technology, Hangzhou, Zhejiang Province, China.
120. Pan, H Y, Jiang, Y M, Yang, Y L, Lu, Q X, 2015, Progress of Electronic Waste Recycling and Heavy Metal Pollution in Taizhou Area, *Sichuan Environment*, 01, 127–132.
121. Yang, J X, 2014, The World's Largest "Quest Guiyu e-Waste Treatment Plant", *Renewable Resources*, 03, 33–36.
122. Mu, H S, Yang, J, Bai, X. Liu, D H, 2008, Based on Cluster Innovation of Circular Economy Development Model of Guiyu Dismantling of Electronic Waste Recycling Industry Development Experience and Revelation, *Studies in Dialectics of Nature,* 1, 75–80.

9

Mercury Mining in China and its Environmental and Health Impacts

Guangle Qiu, Ping Li, and Xinbin Feng

State Key Laboratory of Environmental Geochemistry, Institute of Geochemistry, Chinese Academy of Sciences, Guiyang 550081, China

9.1 Introduction

Mercury (Hg) is a non-essential and toxic element. The primary metal Hg has been extracted from cinnabar ores. Because of the well established toxicity and human health effects of Hg as well as environmental concerns, the demand for Hg has decreased sharply since the end of the last century. In October of 2013, the Minamata Convention on Mercury was signed in Kumamoto, Japan; this treaty aims to protect human health and the environment from anthropogenic emissions and releases of Hg and Hg compounds.

China is rich in cinnabar ore resources and ranks the third in total Hg resources in the world. Large quantities of Hg mines are distributed throughout the Nation, especially in the southwest and central parts of China, including Guizhou, Shaanxi, Chongqing, Hunan, and Yunnan provinces. Among those provinces, Guizhou accounts for approximately 38.3% of the total cinnabar ore reserve in China. Owing to abundant cinnabar ore resources, China has a long history of Hg mining and Hg usage that may date back more than 2000 years. Most large-scale industrial Hg mining activities ceased in 2004; the Xunyang Hg mine, Shaanxi, is the only active one for metallic Hg production in China.

The presence of methylmercury (MeHg) in the environment is of great concern, because it is readily bioavailable and can be bioaccumulated and biomagnified in food chains. It is widely accepted that fish and its products are the dominant pathway of human exposure to MeHg. Recently, rice rather than fish has been verified as the major pathway of MeHg

Metal Sustainability: Global Challenges, Consequences, and Prospects, First Edition. Edited by Reed M. Izatt.

exposure to the residents living in Hg mining regions in China. The aim of this chapter is to review state-of-the-art knowledge on the distribution of Hg in the environment of Hg mining areas in China and its related environmental and human health impacts.

9.2 Mercury Mines and Mining

9.2.1 Mercury Mines

Three global mercuriferous belts have been discovered, circum-Pacific, Mediterranean-central Asia, and Mid-Atlantic Ridge. These belts are mainly distributed along global plate tectonic boundaries [1]. Most world large-scale Hg mines are distributed in those three belts, such as Almadén Hg mine, Spain; Idrija Hg mine, Slovenia; Monte Amiata Hg mine, Italy; New Almadén Hg mine, United States; Palawan Hg mine, Philippines; and Wanshan Hg mine, China.

China is located in the circum-Pacific global mercuriferous belt and has abundant cinnabar ore resources. Cinnabar ore reserves in China rank third in the world, after Spain and Italy. Approximately 31 Hg mines with cinnabar ore reserves more than 500 tons are found in China (http://www.chinamining.com.cn/report/default.asp?V_DOC_ID=1140). Most Hg mines are distributed in southwestern China, especially in Guizhou province, and the current total Hg reserves in Guizhou account for 38.3% of the total cinnabar ore reserve in China. The Wanshan Hg deposit, the largest in Asia and the third largest in the world in Hg reserves, lies in the eastern part of Guizhou province, which is considered to be the "Mercury Capital" in China.

9.2.2 Mercury Production

The primary metal Hg has mainly been extracted from cinnabar ores. Hylander and Meili [2] report that approximately a million tons of metallic Hg was produced during the past 500 years, more than 75 % from Europe and North America. The Almadén Hg mines in Spain alone contributed one third of the total metallic Hg production in the world.

Statistics showed that about 61,000 tons of metallic Hg were produced during the long history of Hg mining in China. The Hg mine in Wanshan, Guizhou, has alone contributed two thirds of the total reported national production. Different developing stages had different production of Hg in China. It developed rather rapidly in the Tang Dynasty (7th to 10th century), with the highest rate of development in history [3]. Large-scale Hg mining activities occurred in the 20th century. In the early 1960s, for instance, primary Hg production increased more than 2,000 tons. Since then to late 1980s production reached about 1,000 tons per year on an average. Annual Hg production in China decreased in the late 1990s to less than 500 tons, but has increased in recent years to more than 1,000 tons.

Even though Hg mining activities decreased and the regulations on the international trade of Hg became stricter in recent years, domestic Hg production has increased steadily each year since 2002 to meet the growing internal demand in China. According to the *Yearbook of Non-ferrous Metals Industry in China* 2010 (http://www.chinabookshop.net/yearbook-nonferrous-metals-industry-china-2010-p-10730.html), total mercury production in 2009 was estimated to be approximately 1,400 tons, more than half of the global total production.

9.2.3 Mercury Usage

In ancient China, cinnabar was used for red pigments, drugs, and preservatives. In old cities, such as Beijing, Nanjing, and Xi'an, red pigments can be observed in old constructions and former palaces. Ancient people also used metallic Hg by an amalgamation technique to extract gold.

Currently, the main areas of Hg demand in China are identified as acetylene polyvinyl chloride (PVC) production and the manufacture of thermometers and sphygmomanometers, batteries, and lighting. Jian et al. [4] reported that the total metallic Hg supply was about 1,563 tons in 2007. The PVC industry accounted for approximately of 60.1%, followed by manufacture of thermometers, 14.9%, batteries, 13.1%, sphygmomanometers, 6.27%, lighting, 4.93%, and others, 0.768% (Figure 1).

During the last 20 years, significant reduction in Hg use has occurred in the battery area. Total Hg demand was about 600 tons in 1995 and then decreased to less than 150 tons in 2008. However, the metallic Hg supply remained at about 1,000 tons per year, while global Hg production is about 2,000 tons annually [5]. China, therefore, plays an important role in controlling and reducing the global supply and trade of Hg.

9.3 Mercury in the Environment

Most large-scale industrial Hg mining activities recently ceased in China. However, abandoned Hg mines continue to impact the environment through the release of Hg-enriched mine-waste tailings (calcines), Hg-rich mine drainage, and Hg vapor. Therefore, highly elevated Hg concentrations were widely observed in air, soils, waters, and biota in Hg mining areas. Environmental concerns related to Hg mines are contamination of soils, sediments, and waters by mine wastes; Hg vapor released during ore processing and from mine tailings and contaminated soils; and Hg mine drainage release into watersheds. Generally, the large quantities of mine-waste tailings and heavily contaminated soils were considered the major sources for both soluble and particulate Hg in the surroundings.

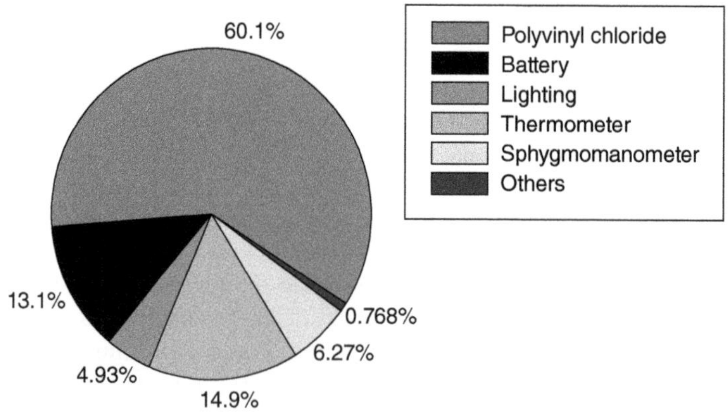

Figure 1 *Categories of Hg supply and usage in 2007 in China [4]. (See insert for color representation of the figure.)*

Most Hg mining is localized in remote regions where there is a lack of farmlands. The former farmlands occupied by Hg mining activities were reclaimed without any treatment after those Hg mines closed. The reclamation of contaminated farmlands increases the chance that Hg will enter into terrestrial food chains as well as the transformation of MeHg from inorganic forms under certain conditions. Hence, accumulation and biomagnifications of Hg, especially MeHg, in food chains poses a potential health risk to the residents in the region.

9.3.1 Air

9.3.1.1 Levels

The total gaseous mercury (TGM) concentrations in ambient air measured in Hg mining regions were characterized by significant variability and by extremely high values, ranging from 7.4 to 40,000 ng/m^3 in China (Table 1). The highest concentration of TGM, observed in Wuchuan Hg mining district, was related to artisanal Hg retorting activities [6]. The Wanshan Hg mining region also exhibited high TGM values of up to 2,100 ng/m^3 even after the Hg mines had ceased operation for more than five years.

TGM concentrations in ambient air in the Xunyang Hg mining region, the largest scale active Hg mine in China, were heavily impacted by anthropogenic activity. Elevated concentrations of TGM in ambient air were observed near the Hg retorting facility, reaching up to 410 ng/m^3 on average. Those high values decreased with distance from the retorting facility and reached a background value at 18 km in the downwind direction.

TGM concentrations in Hg mining regions were significantly elevated compared to the values reported for remote areas in China. Fu et al. [14–17] reported that TGM concentrations in ambient air at remote sites, such as Mt. Gongga, Mt. Changbai, Mt. Leigong, and Shangri-La, ranged from 1.6 to 4.0 ng/m^3.

Table 1 *Total gaseous mercury (TGM) concentrations in ambient air in Hg mining districts and remote regions in China*

Location	Time	Method	TGM (ng/m^3)	Reference
Wanshan, Guizhou	Nov 2002; Jul-Aug 2004	Tekran 2537A	18–1,100	[7]
	Feb 2003	Lumex RA-915+	30–1,950	
	Sep 2009	Lumex RA-915+	18–190	[8]
	May 2010	Lumex RA-915+	17–2,100	[9]
Yanwuping, Guizhou	Jul-Sep 2011	Lumex RA-915M	12–185	[10]
Wuchuan, Guizhou	Dec 2003; Dec 2004	Tekran 2537A	19.5–2,110	[11]
	Jun 2005	Lumex RA-915+	10–40,000	[6]
Lanmuchang, Guizhou	Dec 2002; May 2003	Tekran 2537A	24.5–210	[12]
Xunyang, Shaanxi	Sep 2009	Lumex RA-915+	7.4–410	[13]
Remote regions, China		Tekran 2538A	1.6–4.0	[14–17]

9.3.1.2 Emission Sources

Contributions from cinnabar ore retorting, widespread calcines, and heavily contaminated terrestrial and aquatic systems could represent major sources of Hg emissions to the atmosphere in Hg mining regions. Li et al. [18] found that the Hg emission factors during artisanal Hg retorting activities in Wuchuan ranged from 6.6% to 14.5% with an average of 9.8%, resulting in approximately 1.3–2.7 tons of Hg release per year into the atmosphere. These results showed that artisanal Hg retorting activity is an important atmospheric Hg source.

Wang et al. [7] conducted surveys on Hg emission flux from contaminated soils in Wanshan and found an average Hg emission flux of 162–27,827 ng/m^2/h and average Hg dry deposition flux of 0–9,434 ng/m^2/h, with average TGM concentrations of 17.8–1,100 ng/m^3 in ambient air. Even 10 years after the closure of Hg mining activities in the Lanmuchang Hg mining region in southwestern Guizhou province, Hg emission fluxes from contaminated soils reached up to 10,500 ng/m^2/h [12]. These results indicated that heavily contaminated soils are an important Hg emission source in abandoned Hg mining regions.

In addition, the mine-waste tailings generated by Hg mining and retorting activities have usually been emitted directly in the surroundings. Since the calcine is the ignited residue of cinnabar ores roasted at high temperatures, it contains a large quantity of secondary Hg minerals [19–23] that are readily released into the environment. In summer, calcines exhibited high emission of elemental Hg with a flux reaching up to 16,070 ng/m^2/h. This indicated that calcine piles are another important source of atmospheric Hg.

9.3.2 Mine-waste Tailings (Calcines)

The roasting process of cinnabar ores produced significant quantities of mine-waste tailings (calcines) that were directly piled nearby the adits or along the valleys and riverbanks. Calcines contain an abundance of soluble Hg compounds, such as elemental Hg, and Hg salts [24–26], resulting in release of large quantities of Hg into the surroundings via runoff, wind, and precipitation. Concentrations of total Hg in calcines differed widely in various Hg mining regions in China. Due to the inefficient and incomplete processing of cinnabar roasting, Hg concentrations in calcines ranged from 5.7 to 4,450 μg/g[27–29], comparable to the values reported by Gray et al. [20–23,30] and Bailey et al. [31].

These Hg-containing minerals generally have small particle sizes, i.e., micron or nanometer levels, and are easily attached to the surface of fine particles or dissolved in water in surface runoff. Thus, leachate of calcines generally contain elevated concentrations of Hg. For instance, the leachate of the calcines from Wanshan Hg mine can contain up to 4.46 μg/L of total Hg [32]. THg concentrations in calcines leachate of Almadén Hg mining district in Spain reached up to 120,000 μg/L, of which more than 70% was in dissolved Hg form [23].

Large amounts of soluble Hg compounds make calcines and its surroundings favorable sites for Hg methylation. Concentrations of MeHg measured in calcine samples collected from Wanshan Hg mining region ranged from 0.17 to 3.9 ng/g [27], comparable to the values found in Palawan Hg mining region but well below the 3,100 ng/g found in the Almadén Hg mining region, Spain [23][30]. Gray et al. [23] found that in calcines of the Almadenejos Hg mine, Spain the net Hg methylation rate is up to 11,100 ng/g/day, thousands of times higher than that of sediments, which were 1.7 ng/g/day. Generally, environmental conditions such as high temperatures, wet conditions, abundant organic carbon, and high sulfur concentrations favor Hg methylation.

9.3.3 Soil

9.3.3.1 Levels

Surface soils affected by retorting activities in Hg mining regions in China exhibit a wide range from background levels to hundreds of μg/g in THg concentration in agricultural soils. Most samples at sites adjacent to calcines contained levels greater than 10 μg/g in soil in Yanwuping Hg mining regions [10], suggesting large impacts from those ignited residues. A significant negative correlation ($r = -0.73$, $p < 0.05$) of THg concentration and distance from the mine-waste tailings was observed in the Wanshan Hg mining region [33]. Bao et al. [34] reported that even 15 km away from the calcines tailings, THg in paddy soils still reached 15 μg/g, probably resulting from irrigation water that was heavily contaminated with Hg.

Soils in Hg mining areas were also found to contain elevated MeHg. The concentrations of MeHg in soil in the Wanshan mining areas ranged from 1.27 to 22.5 ng/g [35], and that in Wuchuan varied from 0.69 to 20 ng/g [28]. Similar to THg data, most elevated MeHg concentrations were observed at sites proximal to the calcine piles, indicating that calcine piles are potential sources of MeHg. Therefore, more attention should be paid to the calcine piles to reduce production of MeHg, and its drainages and leachates should be properly treated.

Generally, soil MeHg concentrations slightly increased with THg concentration increase ($r = 0.43$, $p < 0.0001$, Figure 2a), which indicates that the soil in Hg mining area favors Hg methylation and the concentration of bioavailable Hg species increases with the soil THg concentration. In the environment of Hg mining areas, the increase of bioavailable Hg concentration in soil might result from atmospheric Hg deposition. However, the contribution of atmospheric Hg in the soil to THg was much lower than that from calcines. A significant negative correlation was observed between THg concentration in soil and the percentage of MeHg in THg in soil ($r = -0.84$, $p < 0.0001$, Figure 2b).

The mercury methylation process in the environment depends on a number of factors, such as temperature, organic matter, soil pH, soil microbes, and redox conditions. Studies found that the change of seasons can also affect the net rate of Hg methylation: the peak rate of Hg methylation often appears in summer.

9.3.3.2 Spatial Distribution

The contour maps of Hg concentrations in soils collected from the Lanmuchang Hg mining district, Guizhou province [11,36,37] provide overall information on Hg dispersion, accumulation, and transport trends (Figure 3). Most of the highest THg concentrations were localized at numerous point sites, such as mining adits, retorting sites, and mine-waste tailings generated by historical artisanal Hg mining activities. Highly elevated concentrations of THg in soil observed near Hg mine-waste tailings and Hg concentrations in soil rapidly decreased beyond 4–8 km from the tailings. The southeast direction of the Qingshui River upstream, which flows through the mining district, seems to be the dominant trend in Hg dispersion.

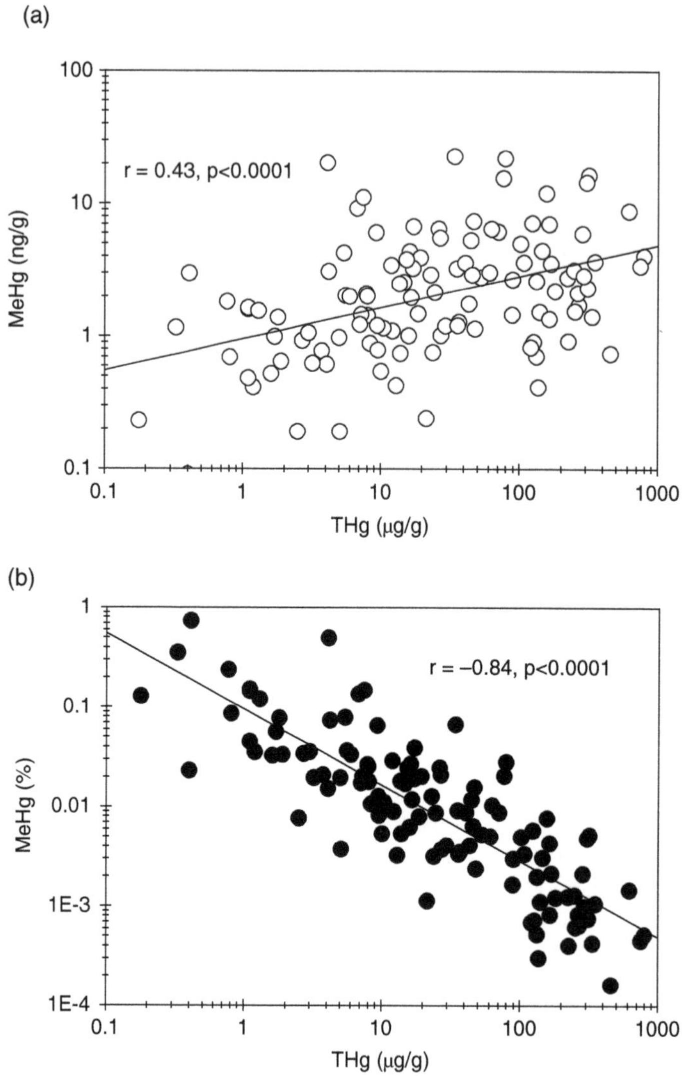

Figure 2 *Relations among THg, MeHg and its ratios in Hg mining areas in China*

In contrast to THg distribution in soil, the highest MeHg concentrations in soil were scattered along creeks and could be found upstream, midstream, and downstream of the tailings, indicating that methylation occurs readily in the soil along the river banks. Flooded paddy soil is an active net Hg methylation site. Typically, the soil near the streams on both banks floods readily and the anaerobic environment generated in the paddy soil is favorable to Hg methylation. At the same time, the moist environment is very conducive to microbial activity, resulting in high MeHg levels in soil along the banks of the river.

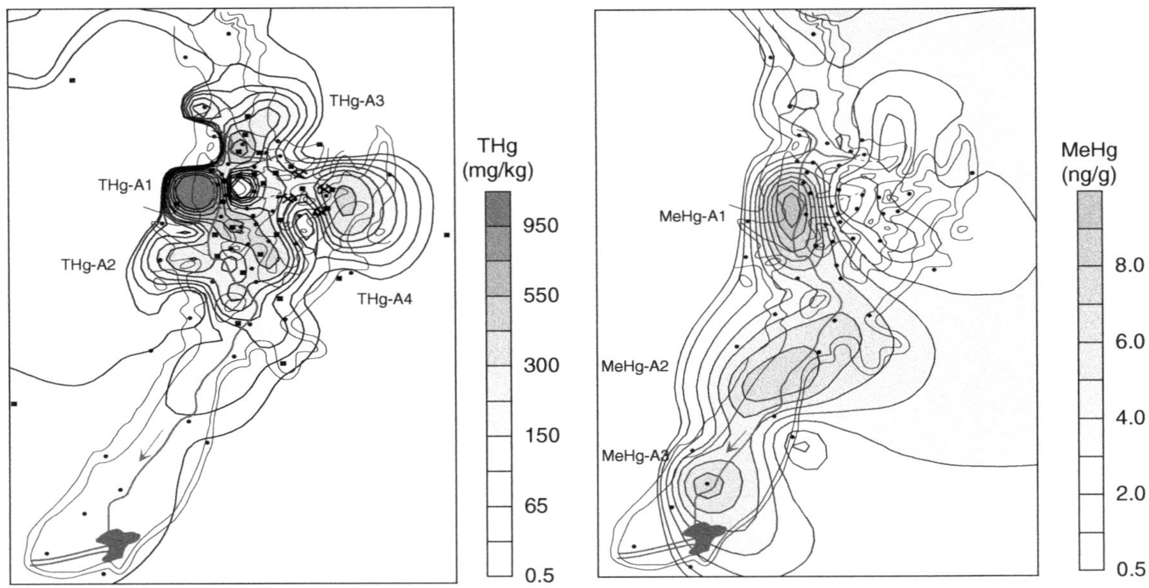

Figure 3 *Spatial distribution of THg (left) and MeHg (right) in surface soil with sampling locations in the Lanmuchang Hg mining district, China [37]. Reprinted from [37] with permission of Elsevier. (See insert for color representation of the figure.)*

9.3.3.3 Bioavailability

THg concentrations in soil reflect the degree of soil pollution, but do not indicate the potential environmental risk of Hg pollution [38]. Speciation information for soil Hg can indicate how its bio-toxic effects depend on its speciation in environmental media [39]. Using a modified sequence extraction procedure, we investigated the speciation of Hg in soil and sediment in Wanshan Hg mining areas. The total leachable Hg concentrations in sediment and soil reached up to 480 μg/g and 130 μg/g, respectively. The exchangeable Hg and strongly-bound Hg accounted for 10–30% and 20–40% of the total Hg [34,40]. Although soluble Hg and exchangeable Hg accounted for a small portion of THg, the absolute concentrations of these species in polluted soil were much higher than those found in soil at reference sites.

9.3.4 Water

Surface water affected by Hg mining contained elevated THg concentrations, ranging between 6.2 and 23,500 ng/L in China (Table 2). Peak concentrations of THg in surface water were generally found at sites directly affected by leachates from the calcines piles, verifying that calcine piles are important sources of Hg in surface waters. Horvat et al. [35] found that the THg concentration in leachate in Wanshan calcines was up to 10,580 ng/L, of which dissolved Hg (DHg) concentrations were in the range of 300–1,900 ng/L. In Xunyang Hg mining region, the leachate displayed extremely high THg levels of 23,500 ng/L [41]. Generally, however, the high THg concentrations show the characteristics of high particulate Hg levels, suggesting that most of the Hg transported downstream from the Hg mines occurred in suspended matter, probably as cinnabar particles.

Since the building of the Wukeng settling pond in the Wanshan Hg mining region, total Hg concentrations in the pond water were continuously monitored from 2000 to 2012; the resulting data are shown in Figure 4. Generally, THg concentrations showed a decreasing trend with values varying between >5.0 μg/L and 0.5 μg/L. In 2000, THg concentrations in

Table 2 *Mercury concentrations in surface water in Hg mining areas worldwide (ng/L)*

Location	THg*	DHg**	TMeHg***	References
Wanshan Hg mines, China	15–10,600	11–1,900	0.31–25	[32,35]
Tongren Hg mines, China	92–2,300		2.6–7.9	[42]
Wuchuan Hg mines, China	22–362	8.2–75.7	0.21–5.7	[28]
Lanmuchang Hg mines, China	25–7,020	<3–1,300		[37,43]
Xunyang Hg mine, China	6.2–23,500	0.21–3,500	0.022–3.7	[13,41]
Almadén Hg mines, Spain	7.6–20,300		0.41–30	[23,44]
Idrija Hg mines, Slovenia	2.8–322	0.03–134	0.01–0.6	[45,46]
Alaska Hg mines, USA	0.1–2,500	1.3–50	0.01–1.2	[20]
California Hg mines, USA	2–45,000	0.2–140	<0.003–47	[19,47–49]
Nevada Hg mines, USA	2.1–2,107	2–56	0.039–0.92	[50]
Palawan Quicksilver Mine, Philippines	120–31,000	8–30,000	<0.02–3.1	[30]

* Total mercury concentration
** Dissolved mercury concentration
*** Total methylmecury concentration

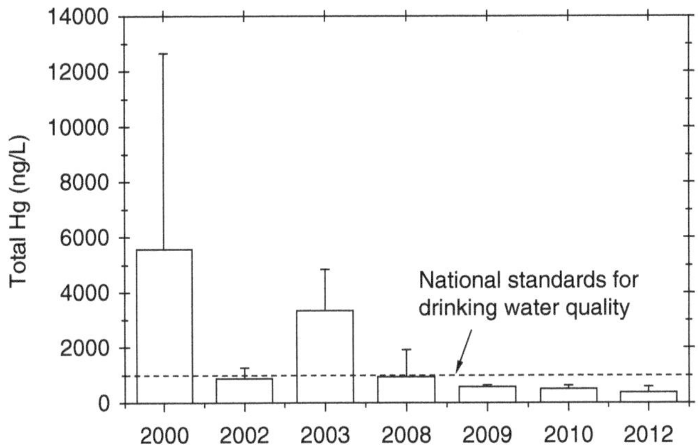

Figure 4 *Concentrations of THg in the water collected from Wanshan Hg mine Wukeng settling pond for the last 12 years [32,35,52]*

water exceeded 5.0 μg/L. The values decreased to <1.0 μg/L in 2002, but increased to 3.0 μg/L as a result of a rainstorm in 2003. In 2010, the values reached 0.5 μg/L, below the guideline value for National Standards for Drinking Water Quality set by the Ministry of Health, People's Republic of China [51].

Surface water in Hg mining regions also contained high concentrations of MeHg, reaching 25 ng/L [32]. The MeHg concentration in water showed a positive correlation with the dissolved Hg concentration in water, suggesting that MeHg in surface water mainly existed and migrated in the dissolved form, which is different from THg [53]. Affected by calcines leachates, the surface water environment favoured Hg methylation processes. The high MeHg in water adjacent to calcines would likely become a new pollution source for the downstream ecosystems.

9.3.5 Biota

9.3.5.1 Fish

Since fish are rare in streams of most Hg mining regions in China, few data on Hg concentrations in fish are available. Qiu et al. [32] reported that THg and MeHg concentrations in fish muscles collected from Wanshan Hg mining region were in the range of 0.061–0.68 μg/g and 0.024–0.098 μg/g, respectively. Similar results were observed in the Wuchuan Hg mining region with a range of 0.019 to 0.24 μg/g for THg and of 0.0061 to 0.11 μg/g for MeHg. The average THg concentrations in fish were 0.29 μg/g in Wanshan and 0.086 μg/g in Wuchuan, respectively, both of which values did not exceed the nation's guideline for acceptability of 0.50 μg/g.

Results for THg in fish from Wanshan were comparable to those of fish collected from Alaska Hg mines, but MeHg concentrations were much lower [23]. High average ratios of MeHg to THg (expressed as percent) were generally more than 90% in Alaska, while these ratios were as low as 28% and 41% on average in Wanshan and Wuchuan, respectively. Similar low MeHg-to-THg ratios were also reported for fish inhabiting areas impacted by Hg-contaminated sediments [54].

Obviously, factors such as species, age, sex, and feeding habits might impact MeHg to THg ratios. A weakly positive correlation ($r = 0.37$, $p < 0.5$) observed between MeHg and THg in fish from Wanshan and Wuchuan Hg mining areas might suggest that factors besides Hg concentration are influencing MeHg concentrations in these fish. Given the limitations of available data, however, reasons for the unusually low MeHg fraction in fish from Hg mining regions in China remain uncertain.

9.3.5.2 Rice

Rice is a staple food in China as well as in other regions of Asia. Rice samples collected from Hg mining regions in China exhibited a wide range of 9.0–1,120 ng/g for THg. Compared to the National Permitted Limit of 20 ng/g for THg in cereals in China, THg concentrations in rice samples collected from Hg mining regions were highly elevated. This demonstrated that rices cultivated in Hg mining areas were heavily contaminated with Hg because of the historic mining and retorting activities there.

Rice was also reported to accumulate higher levels of MeHg than other grains [55]. Compared to the edible parts of other crops, MeHg concentrations are observed to be much higher in rice, showing rice's high ability to accumulate MeHg.In the rice paddies in Hg mining areas, for instance, MeHg concentrations in grain can reach up to 170 ng/g, 1 to 2 orders of magnitude higher than MeHg in paddy soil. In rice, MeHg can account for more than 90% of THg, but the highest MeHg concentration in corn in the same area is only 1.3 ng/g.

MeHg bioaccumulation factors (BCFs) in over 97% of the rice samples tested were higher than 1 with an average value of 5.5–5.6, while inorganic Hg BCFs of rice, similar to other plant species, were generally much lower than 0.5. Data show that rice MeHg BCF values were 3 to 4 orders of magnitude higher than inorganic Hg [52]. Since MeHg is more easily translocated than inorganic Hg forms into above-ground plants, elevated MeHg concentrations in paddy soils are the major source of MeHg in rice [56]. Due to its high ability to accumulate in rice, MeHg in the terrestrial food chains has received considerable attention [52,57].

The phenomenal ability of MeHg to be accumulated in agricultural plants is rarely reported in the literature. Studies have shown that at the soil-plant root interface a layer of "iron plaque" exists that can effectively prevent soil inorganic Hg and MeHg ions from entering into the plant. In Hg mining areas, high MeHg concentrations can be detected in rice seeds and MeHg BCF values for rice seeds were usually higher than 1. These results showed that rice is a hyper-accumulator for MeHg.

9.3.5.3 Other Crops

In Hg mining areas, other crops such as corn, tobacco, and vegetables also suffered significantly from Hg pollution. THg concentrations in cabbages grown in the Lanmuchang Hg mining area, Guizhou, can reach as high as 18,000 ng/g, significantly above the Chinese National Food Standard of 10 ng/g for total Hg. Uptake of Hg in plants was primarily through leaves rather than through the root system. In Hg mining areas, TGM concentrations in ambient air are elevated due to Hg emissions from calcine piles and Hg-contaminated soils. THg concentrations in the leaf tissues of plants in the downwind direction from contaminated sites were much elevated.

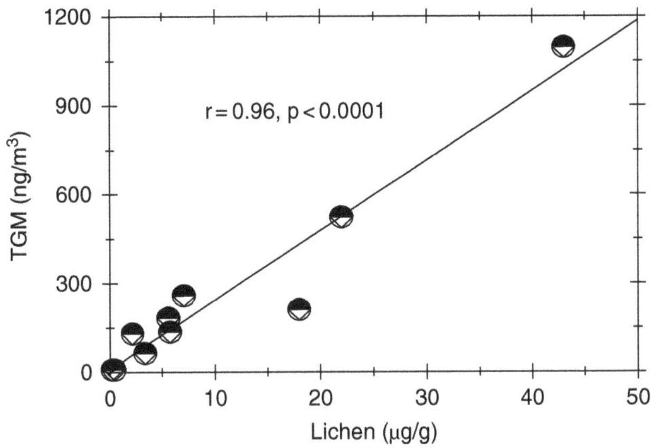

Figure 5 *Correlation between THg concentration in lichen and total gaseous Hg concentrations in ambient air. Reprinted from [37] with permission of Elsevier*

Lichen and moss are considered to be good biomarkers of air pollution. The Hg concentration of natural lichen grown on rock surfaces in the Wanshan Hg mining region reached up to 95,000 ng/g. Generally THg concentrations in lichens were significantly elevated at sites close to the Hg smelters and calcine tailings compared to sites far away from the contamination sources. Because nutrients of lichen are mainly from dry and wet deposition, the main source of Hg in lichen is also from atmospheric Hg deposition. THg concentrations in lichen were positively correlated with TGM concentrations in ambient air (Figure 5). The results also demonstrated that natural lichen plants on rock surfaces can serve as good atmospheric Hg pollution indicators.

9.4 Human Exposure and Health Risk Assessment

9.4.1 Human Exposure

As shown in Figure 6, in Wanshan Hg mine Li et al. [58] demonstrated a gradient of blood THg and MeHg concentrations at different sites with the distance from the pollution source in the two catchments. The highly elevated THg and MeHg levels in blood samples were observed in the upstream region (sites A, D, and E), which were seriously impacted by mine wastes. Blood THg and MeHg levels at site C and G basically represented the regional background level. In the total population, the overall mean of blood THg concentration was 12.2±15.0 µg/L (2.15–30.8, 95% confidence interval). The average of blood MeHg at different sites ranged from 2.20 to 9.36 µg/L. MeHg constituted 52.8±17.5% of THg in all blood samples on an average. About 72.6% (122/168) of blood THg concentrations exceeded 5.8 µg/L, which is the blood Hg level equivalent to the current RfD set by USEPA.

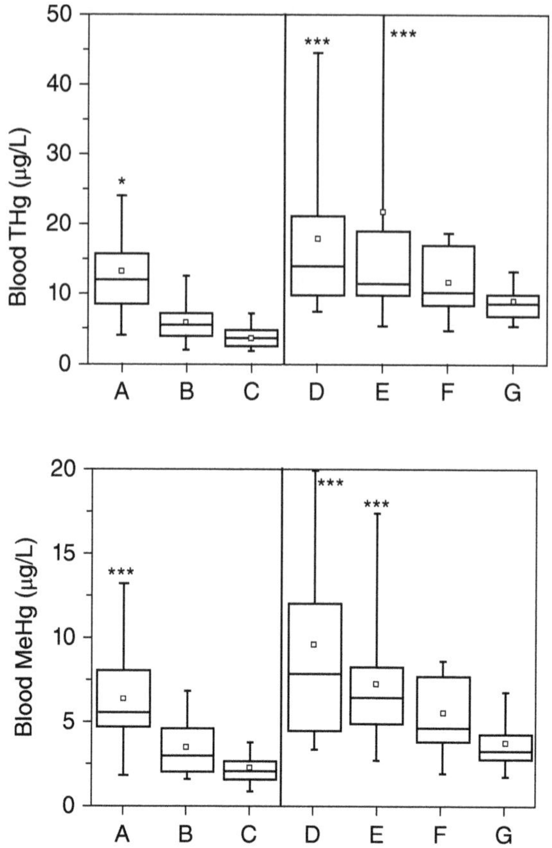

Figure 6 Comparison of blood Hg in the population from Wanshan Hg mine. Adapted from Li et al. (2015) with permission from American Chemical Society

9.4.1.1 Hair

Hair concentrations of THg and MeHg and percentages of THg as MeHg in the population from different Hg mines are summarized in Table 3. The average of hair THg for the control group was 0.78±0.28 μg/g [59], which is in the range considered "normal" by WHO (2 μg/g). Highly elevated THg concentrations were found in workers involved in artisanal mercury mining (AMM), with averages of 25.7±24.0 and 47.7±75.1 μg/g reported for workers in Wuchuan and Tongren Hg mines, respectively. Local residents in the Hg mine area also showed elevated hair THg concentrations, with average values of 5.04±7.57, 3.04±3.00, and 7.89±2.85 μg/g for residents in the Wanshan, Wuchuan, and Tongren Hg mines, respectively.

The average hair MeHg value for the control group was 0.65±0.25 μg/g [59]. In Tongren Hg mining areas, the average hair MeHg concentrations were 5.24±2.80 and 3.76±1.73 μg/g for AMM workers and local residents, respectively. Previous studies confirmed that rice can accumulate high levels of MeHg (170 ng/g) in Hg mining areas [55] and that rice, rather than fish, is the main route of human MeHg exposure in the Wanshan Hg mining area

Table 3 *Comparison of human hair Hg concentrations in different Hg mines*

Site	Subject	*n*	THg* (μg/g)	MeHg** (μg/g)	MeHg/THg (%)	Reference
Wanshan	Residents	95	5.04±7.57 (0.60–58.5)	1.85±1.16 (0.23–5.55)	55.3±22.5 (4.27–98.7)	[57]
Wuchuan	Workers	38	25.7±24.0 (2.03–93.1)	1.22±0.54 (0.52–2.53)	13.5±20.1 (1.10–96.8)	[60]
	Residents	100	3.04±3.00 (0.49–23.1)	1.25±0.74 (0.45–4.21)	53.7±27.8 (5.80–99.9)	
Tongren	Workers	18	47.7±75.1 (4.80–326)	5.24±2.80 (2.35–10.6)	28.5±23.3 (2.93–74.1)	[61]
	Residents	25	7.89±2.85 (3.52–13.1)	3.76±1.73 (1.87–9.23)	50.2±18.6 (23.6–99.3)	
Changshun	Residents	40	0.78±0.28 (0.32–1.72)	0.65±0.25 (0.26–1.38)	83.5±12.6 (52.7–99.9)	[59]

* Total mercury concentration
** Methylmercury concentration

[57], Guizhou Province [52] and the inland area of southern China [6]. The highly elevated hair MeHg levels indicated significant potential human health risk in Hg mining areas.

Considering the percentages of THg as MeHg in hair samples, the AMM workers showed the lowest ratios, with means of 13.5±20.1% and 28.5±23.3% for workers in Wuchuan and Tongren, respectively. Hair THg concentrations are considered a good biomarker of human MeHg exposure. For fish-eating populations, MeHg is the main form in the hair and can constitute from 80% to 98% of THg. Therefore, THg concentrations in hair samples are widely used as proxies for MeHg concentrations in human MeHg exposure. In this study, hair MeHg constituted only about 20% of THg on average for AMM workers, while the percentages of THg as MeHg in hair samples were about 50% for the local residents. As well, the ratios varied significantly for individuals.

Significant correlations were found between hair THg and UHg for AMM workers [59] and between inorganic mercury (IHg, which is defined as total Hg substrate MeHg) in hair and total gaseous mercury (TGM) concentrations in the ambient air [60] in Wuchuan Hg mine. The results from a simulated experiment indicated that the increases of hair THg concentrations were less than 1 μg/g under direct exposure of Hg vapor, which was negligible compared with hair THg concentrations in occupationally exposed workers [62]. The inhalation of Hg vapor constituted the major fraction of IHg exposure for the AMM workers [61], which resulted in highly elevated hair IHg concentrations. For the local residents, consumption of vegetables, rice, and meat accounted for >90% of the IHg exposure [63].

9.4.1.2 Blood

In the Chatian Hg mine, the mean Hg concentration in blood of the local residents was 6.09±3.26 μg/L (Table 4), significantly above those in the control area (3.67±0.82 μg/L [64]). The exposed groups in Wanshan Hg mine were characterized by high serum Hg concentrations (mean of 38.5 μg/L), again highly elevated compared with that of the control groups (mean of 0.91 μg/L) [65].

Table 4 Comparison of blood Hg (B-Hg) concentrations in the population from Hg mines

Site	Subject	*n*	B-Hg		Reference
			Mean±SD	Range	
Chatian	Local residents	54	6.09±3.26	1.29–15.07	[64]
	The control group	47	3.67±0.82	0.97–5.06	
Wanshan [a]	The Exposed group	37	38.5±61.5	1.85–210.3	[65]
	The control group	35	0.91±0.28	0.41–1.20	

[a] Serum Hg.

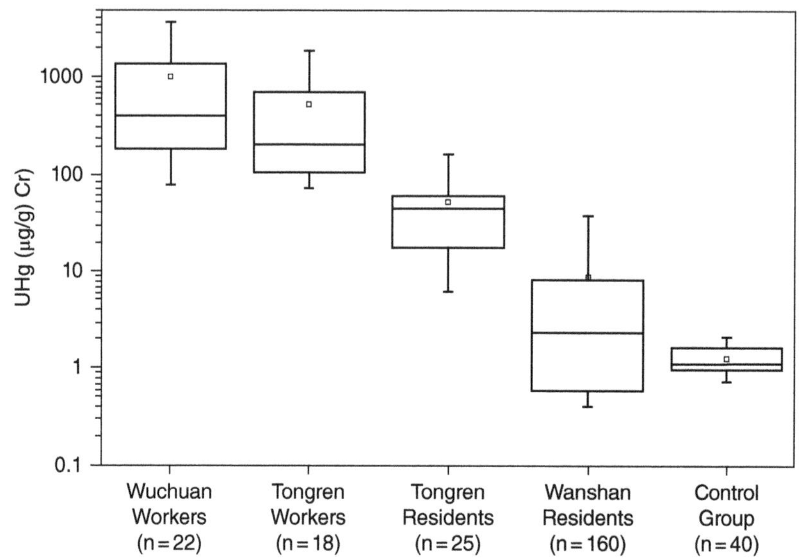

Figure 7 Comparison of UHg in the population from different Hg mines in China. Data from [59,61,63] with permission of Elsevier

9.4.1.3 Urine

Comparisons of total Hg concentrations in urine (UHg) in different populations from Hg mines are shown in Figure 7. The geometric mean of UHg for the control group was 1.30 μg/g creatinine (Cr) with a range of 0.68–2.32 μg/g Cr. However, the geometric mean for workers involved in artisanal mercury mining (AMM) in the Wuchuan Hg mine was 463 μg/g Cr, which was elevated by two orders of magnitude compared to that of the control group. The geometric means of UHg were 297 and 35.1 μg/g Cr for the AMM workers and the local residents from Tongren Hg mine, respectively. The geometric mean of UHg was 3.04 μg/g Cr for the local residents in Wanshan Hg mines. The maximum UHg concentration for occupational workers recommended by WHO [66] is 50 μg/g Cr and UHg levels for the general population should not exceed 5 μg/g Cr [67]. The highest UHg levels in workers involved in AMM from Wuchuan Hg mine reached 6,150 μg/g Cr, which is about 120 times higher than the occupational exposure limit level [66]. In conclusion, workers involved in AMM in Guizhou Hg mines are seriously exposed to IHg and the local residents are also exposed to IHg at a significant level.

Previous studies confirmed that AMM can emit a significant amount of Hg vapor to the ambient air [18], resulting in significant Hg pollution in the ambient air around AMM sites [6]. Workers involved in AMM in Hg mines were seriously exposed to Hg vapor through inhalation [59], resulting in highly elevated UHg. But for the local residents, dietary intake is the main pathway of IHg exposure [63], since the vegetables, rice, and meat samples are seriously polluted by IHg due to long-term Hg mining activities.

9.4.2 Health Risk Assessment

9.4.2.1 IHg Exposure

Occupational exposure to Hg vapor can have significant effects on the central nervous system. Exposure has resulted in erethism, irritability, excitability, excessive shyness, and insomnia as the principal features of a broad range of functional disturbances. With continuing exposure, a fine tremor develops, initially involving the hands and later spreading to the eyelids, lips, and tongue, causing violent muscular spasms in the most severe cases [68]. Gingivitis and typical dark lines on gums have been reported after high inhalation exposures [69]. Clinical symptoms include finger and eyelid tremor and gingivitis. Typical dark lines on gums were observed in six workers from the Wuchuan Hg mine [59] and three workers from the Tongren Hg mine [61]. These symptoms indicated that physical impairments had occurred and that workers were heavily exposed to Hg vapor during the process of cinnabar roasting.

Significant differences of urinary β2-microglobulin (β2-MG) concentrations were found between AMM workers in the Wuchuan Hg mine and the control group. β2-MG is a renal biomarker used to study human nephrotoxicity at an early stage. This biomarker is useful for assessing re-absorption function to indicate tubular injury. A significant correlation ($r = 0.85$, $p < 0.01$) was found between UHg and urinary β2-MG concentrations in the AMM workers, which confirmed that elevated urinary β2-MG concentrations in smelting workers does result from exposure to Hg vapor and supports the observation that Hg exposure results in serious adverse effects on the renal system in AMM workers.

For residents in the Wanshan Hg mine area, a significantly positive correlation is found between paired results for UHg concentrations and serum creatinine (SCr) ($p<0.05$). The SCr levels in the third (Q3) and fourth quartiles (Q4) of UHg were significantly higher than those from the first quartile (Q1) ($p < 0.05$ and $p < 0.01$, respectively). An increase of SCr levels was found in quartiles with the highest UHg concentrations, indicating that IHg exposure may result in renal impairment.

In conclusion, workers involved in AMM in Hg mines are extensively exposed to Hg vapor, which can cause neurologic damage and renal impairment. Local residents are also exposed to IHg in a certain level, which may have impact on renal function.

9.4.2.2 MeHg Exposure

The hair Hg threshold for onset of neurological symptoms in the human body is 50 μg/g [70], and the threshold for the onset of symptoms for MeHg is reported to be 10–14 μg/g in maternal hair [71]. The Joint Food and Agriculture Organization of the United Nations (FAO)/WHO Expert Committee on Food Additives [72] established a provisional tolerable weekly intake (PTWI) for MeHg of 1.6 μg Hg/kg bw/week (0.23 μg Hg/kg bw/day, equivalent to a hair Hg concentration of about 2.3 μg/g) using an uncertainty factor 6.4, where bw = body weight. The United States Environmental Protection Agency [68] set the limit at 0.1 μg Hg/kg bw/day (equivalent to a hair Hg concentration of 1.0 μg/g) as a reference dose (RfD) using an uncertainty factor of 10.

The geometrical mean of hair MeHg concentrations in the population from Hg mines was 1.38 μg/g (CI 95% 0.50–5.06 μg/g). About 24.6% of inhabitants in the Hg mines were exposed to MeHg with hair MeHg concentrations exceeding 2.3 μg/g and 62.0% were exposed to MeHg with hair MeHg concentrations exceeding 1.0 μg/g. In addition to Wanshan, there are 11 other Hg mining and smelting areas in Guizhou [53], with total population approximately 320,000 [73]. Approximately 20,467 residents in Guizhou (0.06% of the total population) are exposed to MeHg with resulting hair MeHg concentrations exceeding 2.3 μg/g, and approximately 198,400 residents (0.57% of the total population) are exposed to MeHg with resulting hair MeHg concentrations exceeding 1.0 μg/g.

Generally, inhabitants in the Hg mines are exposed to MeHg to a certain level. However, they are not under serious health risk. Nevertheless, some female participants may exceed the tolerable intake levels of MeHg for pregnant women established by USEPA.

9.5 Summary

China is the third largest Hg-producing country after Spain and Italy. Although most large-scale Hg mining and retorting activities in China recently ceased operation, abandoned Hg mines are continuing to impact local environments through historic mine-wastes present in nearby adits, valleys, and riverbanks as well as through drainages. A large quantity of Hg compounds continues to be released into the surroundings via runoff, wind power, and precipitation, resulting in extensive Hg contamination of the environment. Results for human exposures from both IHg and MeHg suggest that potential human health risks exist in Hg mining areas. In the future, proper management and remediation of Hg contaminated environments in Hg mining areas in China are urgently needed.

References

1. Gustin, M.S., Lindberg, S.E., Marsik, F., Casimir, A., Ebinghaus, R., Edwards, G., Hubble-Fitzgerald, C., Kemp, R., Kock, H., Leonard, T., London, J., Majewski, M., Montecinos, C., Owens, J., Pilote, M., Poissant, L., Rasmussen, P., Schaedlich, F., Schneeberger, D., Schroeder, W., Sommar, J., Turner, R., Vette, A., Wallschlaeger, D., Xiao, Z., Zhang, H. (1999). Nevada STORMS project: measurement of mercury emissions from naturally enriched surfaces. *J Geophys Res*, 104 (D17): 21831–21844.
2. Hylander, L.D. and Meili, M. (2003). 500 years of mercury production: global annual inventory by region until 2000 and associated emissions. *Science of the Total Environment,* 304: 13–27.
3. Liu, J. (2000). Introductory remarks on mercury pollution in China. *Journal of Environmental Sciences*, 12: 3–8.
4. Jian, X., Shen, Y., Yao, W., Wang, Y., Zhang, X. (2009). Status analysis and reduction countermeasures of China mercury supply and demand. *Research in Environmental Sciences*, 22(7): 787–792 (in Chinese with English abstract).

5. U.S. Geological Survey. (2013). Aluminum statistics [through 2012; last modified October 3, 2012], in Kelly, T.D., and Matos, G.R., comps., *Historical statistics for mineral and material commodities in the United States* (2013 version): U.S. Geological Survey Data Series 140, 4 p., accessed March 28, 2013, at http://minerals.usgs.gov/minerals/pubs/historical-statistics/.
6. Li, P., Feng, X., Qiu, G., Shang, L., Wang, S. (2012). Mercury pollution in Wuchuan mercury mining area, Guizhou, Southwestern China: The impacts from large scale and artisanal mercury mining. *Environmental International*, 42(SI): 59–66.
7. Wang, S., Feng, X., Qiu, G., Fu, X., Wei, Z. (2007a). Characteristics of mercury exchange flux between soil and air in the heavily air-polluted area, eastern Guizhou, China. Atmospheric Environment, 41(27): 5584–5594.
8. Zhang, C., Qiu, G., Anderson, C., Zhang, H., Meng, B., Liang, L., Feng, X. (2015). Effect of Atmospheric Mercury Deposition on Selenium Accumulation in Rice (*Oryza sativa* L.) at a Mercury Mining Region in Southwestern China. *Environmental Science & Technology*, 49(6), 3540–3547.
9. Dai, Z., Feng, X., Sommar, J., Li, P., Fu, X. (2012). Spatial distribution of mercury deposition fluxes in Wanshan Hg mining area, Guizhou province, China. *Atmospheric Chemistry and Physics*, 12: 6207–6218.
10. Qiu, G., Feng, X., Meng, B., Zhang, C., Gu, C., Du, B., Lin, Y. (2013). Environmental geochemistry of an abandoned mercury mine in Yanwuping, Guizhou Province, China. *Environmental Research*, 125:124–130.
11. Wang, S., Feng, X., Qiu, G., Shang, L., Li, P., Wei, Z. (2007b). Mercury concentrations and air/soil fluxes in Wuchuan mercury mining district, Guizhou province, China. *Atmospheric Environment*, 41(28): 5984–5993.
12. Wang, S., Feng, X., Qiu, G., Wei, Z., Xiao, T. (2005). Mercury emission to atmosphere from Lanmuchang Hg-Tl mining area, Southwestern Guizhou, China. *Atmospheric Environment*, 39(39): 7459–7473.
13. Qiu, G., Feng, X., Meng, B., Sommar, J., Gu, C. (2012). Environmental geochemistry of an active Hg mine in Xunyang, Shaanxi Province, China. *Applied Geochemistry*, 27(12): 2280–2288.
14. Fu, X., Feng, X., Zhu, W., Wang, S., Lu, J. (2008). Total gaseous mercury concentrations in ambient air in the eastern slope of Mt. Gongga, South-Eastern fringe of the Tibetan plateau, China. *Atmospheric Environment*, 42: 970–979.
15. Fu, X., Feng, X., Wang, S., Rothenberg, S., Shang, L., Li, Z., Qiu, G. (2009a). Temporal and spatial distributions of total gaseous mercury concentrations in ambient air in a mountainous area in southwestern China: Implications for industrial and domestic mercury emissions in remote areas in China. *Science of the Total Environment*, 407, 2306–2314.
16. Fu, X., Feng, X., Dong, Z., Yin, R., Wang, J., Yang, Z., Zhang, H. (2009b). Atmospheric total gaseous mercury (TGM) concentrations and wet and dry deposition of mercury at a high-altitude mountain peak in south China. *Atmospheric Chemistry and Physics Discussions*, 9: 23465–23504.
17. Fu, X., Feng, X., Sommar, J., Wang, S. (2012). A review of studies on atmospheric mercury in China. Science of the Total Environment. 421–422: 73–81.
18. Li, P., Feng, X., Qiu, G., Shang, L., Wang, S., Meng, B. (2009). Atmospheric mercury emission from artisanal mercury mining in Guizhou Province, Southwestern China. *Atmospheric Environment*, 43(14): 2247–2251.
19. Rytuba, J.J. (2000). Mercury mine drainage and processes that control its environmental impact. Science of the Total Environment, 260: 57–71.
20. Gray, J.E., Theodorakos, P.M., Bailey, E.A., Turner, R.R. (2000). Distribution, speciation, and transport of mercury in stream-sediment, stream-water, and fish collected near abandoned mercury mines in southwestern Alaska, USA. *Science of the Total Environment*. 260, 21–33.

21. Gray, J.E., Crock, J.G., Lasorsa, B.K. (2002a). Mercury methylation at mercury mines in the Humboldt River Basin, Nevada, USA. *Geochemistry: Exploration, Environment, Analysis*, 2, 143–149.
22 Gray, J.E., Crock, J.G., Fey, D.L. (2002b). Environmental geochemistry of abandoned mercury mines in West-Central Nevada, USA. *Applied Geochemistry*, 17, 1069–1079.
23. Gray, J.E., Hines, M.E., Higueras, P.L., Adatto, I., Lasorsa, B.K. (2004). Mercury speciation and microbial transformations in mine wastes, stream sediments, and surface waters at the almadén mining district, Spain. *Environmental Science & Technology,* 38: 4285–4292.
24. Biester, H., Gosar, M., Müller, G. (1999). Mercury speciation in tailings of the Idrija mercury mine. *Journal of Geochemical Exploration*, 65, 195–204.
25. Kim, C.S., Brown Jr, G.E., Rytuba, J.J. (2000). Characterization and speciation of mercury-bearing mine waste using x-ray adsorption spectroscopy. *Science of the Total Environment,* 261, 157–168.
26. Kim, C.S., Rytuba, J.J., Brown Jr, G.E. (2004). Geological and anthropogenic factors influencing mercury speciation in mine wastes: an EXAFS spectroscopy study. *Applied Geochemistry*, 19, 379–393.
27. Qiu, G., Feng, X., Wang, S., Shang, L. (2005). Mercury and methylmercury in riparian soil, sediments, mine-waste calcines, and moss from abandoned Hg mines in east Guizhou province, southwestern China. Applied Geochemistry, 20: 627–638.
28. Qiu, G., Feng, X., Wang, S., Shang, L. (2006a). Environmental contamination of mercury from Hg-mining areas in Wuchuan, northeastern Guizhou, China. *Environmental Pollution*, 142(3): 549–558.
29. Li, P., Feng, X., Qiu, G., Zhang, J., Meng, B., Wang, J. (2013). Mercury speciation and mobility in mine wastes from mercury mines in China. *Environmental Science and Pollution Research*, 20(12):8374–8381.
30. Gray, J.E., Greaves, I.A., Bustos, D.M., Krabbenhoft, D.P. (2003). Mercury and methylmercury contents in mine-waste calcine, water, and sediment collected from the Palawan Quicksilver Mine, Philippines. *Environmental Geology*, 43, 298–307.
31. Bailey, E.A., Gray, J.E., Theodorakos, P.M. (2002). Mercury in vegetation and soils at abandoned mercury mines in southwestern Alaska, USA. Geochemistry: Exploration, Environment, Analysis, 2, 275–285.
32. Qiu, G., Feng, X., Wang, S., Fu, X., Shang, L. (2009). Mercury distribution and speciation in water and fish from abandoned Hg mines in Wanshan, Guizhou province, China. *Science of the Total Environment*, 407(18): 5162–5168.
33. Dai, Z., Feng, X., Zhang, C., Wang, J., Jiang, T., Xiao, H., Li, Y., Wang, X., Qiu, G. (2013). Assessing anthropogenic sources of mercury in soil in Wanshan Hg mining area, Guizhou, China. *Environmental Science and Pollution Research*, 20(11): 7560–7569.
34. Bao, Z., Wang, J., Feng, X., Shang, L. (2011). Distribution of mercury speciation in polluted soils of Wanshan mercury mining area in Guizhou. *Chinese Journal of Ecology,* 30(5): 907–913 (in Chinese with English abstract).
35. Horvat, M., Nolde, N., Fajon, V., Jereb, V., Logar, M., Lojen, S., Jacimovic, R., Falnoga, I., Qu, L.Y., Faganeli, J., Drobne, D. (2003). Total mercury, methylmercury and selenium in mercury polluted areas in the province Guizhou, China. *Science of the Total Environment*, 304: 231–256.
36. Xiao, T.F., Guha, J., Boyle, D., Liu, C.Q., Chen, J.A. (2004). Environmental concerns related to high thallium levels in soils and thallium uptake by plants in southwest Guizhou, China. *Science of the Total Environment*, 318: 223–244.
37. Qiu, G., Feng, X., Wang, S., Xiao, T. (2006b). Mercury contaminations from historic mining to water, soil and vegetation in Lanmuchang, Guizhou, southwestern, China. *Science of the Total Environment*, 368(1): 56–68.

38. Millán, R., Gamarra, R., Schmid, T., Sierra, M.J., Quejido, A.J., Sánchez, D.M., Cardona, A.I., Fernández, M., Vera, R. (2006). Mercury content in vegetation and soils of the Almadén mining area (Spain). *Science of the Total Environment*, 368, 79–87.
39. Clarkson, T.W. (1998). Human toxicology of mercury. *Journal of Trace Elements in Experimental Medicine*, 11, 303–317
40. Lin, Y., Larssen, T., Vogt, R.D., Feng, X. (2010). Identification of fractions of mercury in water, soil and sediment from a typical Hg mining area in Wanshan, Guizhou province, China. *Applied Geochemistry.* 25, 60–68.
41. Zhang, L., Jin, Y., Lu, J., Zhang, C. (2009). Concentration, distribution and bioaccumulation of mercury in the Xunyang mercury mining area, Shaanxi Province, China. *Applied Geochemistry*, 24, 950–956.
42. Li, P., Feng, X., Shang, L., Qiu, G., Meng, B., Liang, P., Zhang, H. (2008a). Mercury pollution from artisanal mercury mining in Tongren, Guizhou, China. *Applied Geochemistry*, 23(8): 2055–2064.
43. Xiao, T.F., Boyle, D., Guha, J., Rouleau, A., Hong, Y.T., Zheng, B.S. (2003). Groundwater-related thallium transfer processes and their impacts on the ecosystem: southwest Guizhou Province, China. *Applied Geochemistry*, 18: 675–691.
44. Berzas, Nevado J.J., García Bermejo, L.F., Rodríguez Martín-Doimeadios, R.C. (2003). Distribution of mercury in the aquatic environment at Almadén, Spain. *Environmental Pollution*, 122: 261–271.
45. Horvat, M., Jereb, V., Fajon, V., Logar, M., Kotnik, J., Faganeli, J., Hines, M.E., Bonzongo, J.C. (2002). Mercury distribution in water, sediment and soil in the Idrijca and Soča river systems. *Geochemistry: Exploration, Environment, Analysis*, 2: 287–296.
46. Hines, M.E., Horvat, M., Faganeli, J., Bonzongo, J-C.J., Barkay, T., Major, E.B., Scott, K.J., Bailey, E.A., Warwick, J.J., Berry, L.W. (2000). Mercury biogeochemistry in the Idrija river, Slovenia, from above the mine into the Gulf of Trieste. *Environmental Research,* Sect; 83: 129–139.
47. Ganguli, P.M., Mason, R.P., Abu-Saba, K.E., Anderson, R.S., Russell, F.A. (2000). Mercury speciation in drainage from the New Idria mercury mine, California. *Environmental Science & Technology*, 34: 4773–4779.
48. Domagalski, J. (2001). Mercury and methylmercury in water and sediment of the Sacramento River Basin, California. *Applied Geochemistry*, 16:1667–1691.
49. Thomas, M.A., Conaway, C.H., Steding, D.J., Marvin-DiPasquale, M., Abu-Saba, K.E., Russell. F.A. (2002). Mercury contamination from historic mining in water and sediment, Guadalupe River and San Francisco Bay, California. *Geochemistry: Exploration, Environment, Analysis*, 2:1–7.
50. Bonzongo, J.C., Heim, K.J., Warwick, J.J., Lyons, W.B. (1996). Mercury levels in surface waters of the Carson river-Lahontan reservoir system, Nevada: Influence of historic mining activities. *Environmental Pollution*, 92: 193–201.
51. GB 5749-2006. (2006). *Standards for drinking water quality*. Ministry of Health People's Republic of China, issued 29th Dec., 2006.
52. Zhang, H., Feng, X., Larssen, T., Qiu, G., Vogt, R.D. (2010). In inland China, rice, rather than fish, is the major pathway for methylmercury exposure. *Environmental Health Perspectives*, 118(9): 1183–1188.
53. Feng, X. and Qiu, G. (2008). Mercury pollution in Guizhou, Southwestern China: an overview. *Science of the Total Environment*, 400(1-3): 227–237.
54. Lasorsa, B. and Allen-Gil, S. (1995). The methylmercury to total mercury ratio in selected marine, freshwater, and terrestrial organisms. *Water, Air, and Soil Pollution,* 80:905–913.
55. Qiu, G., Feng, X., Li, P., Wang, S., Li, G., Shang, L., Fu, X. (2008). Methylmercury accumulation in rice (Oryza sativa L.) grown at abandoned mercury mines in Guizhou, China. *Journal of Agricultural and Food Chemistry*, 56(7): 2465–2468.

56. Meng, B., Feng, X., Qiu, G., Liang, P., Li, P., Chen, C., Shang, L. (2011). The process of methylmercury accumulation in rice (Oryza sativa L.). *Environmental Science & Technology*, 45(7): 2711–2717.
57. Feng, X., Li, P., Qiu, G., Wang, S., Li, G., Shang, L., Meng, B., Jiang, H., Bai, W., Li, Z., Fu, X. (2008). Human exposure to methylmercury through rice intake in mercury mining areas, Guizhou province, China. *Environmental Science & Technology*, 42(1): 326–332.
58. Li, P., Feng, X., Chan, H.M., Zhang, X., Du, B. (2015) Human Body Burden and Dietary Methylmercury Intake: The Relationship in a Rice-Consuming Population. *Environmental Science & Technology*, 49: 9682–9689.
59. Li, P., Feng, X., Qiu, G., Li, Z., Fu, X., Sakamoto, M., Liu, X., Wang, D. (2008b). Mercury exposures and symptoms in smelting workers of artisanal mercury mines in Wuchuan, Guizhou, China. *Environmental Research*, 107: 108–114.
60. Li, P., Feng, X., Qiu, G., Shang L., Wang, S. (2008c). Mercury exposure in the population from Wuchuan mercury mining area, Guizhou, China. *Science of the Total Environment*, 395: 72–79.
61. Li, P., Feng, X., Shang, L., Qiu, G., Meng, B., Zhang, H., Guo, Y., Liang, P. (2011a). Human co-exposure to mercury vapor and methylmercury in artisanal mercury mining areas, Guizhou, China. *Ecotoxicology and Environmental Safety*, 74: 473–479.
62. Li, P., Feng, X., Qiu, G., Wan, Q. (2011b). Hair can be a good biomarker of occupational exposure to mercury vapor: Simulated experiments and field data analysis. *Science of the Total Environment*, 409: 4484–4488.
63. Li, P., Du, B., Chan, H.M., Feng, X. (2015a). Human inorganic mercury exposure, renal effects and possible path ways in Wanshan mercury mining area, China. *Environmental Research*, 140:198–204.
64. Li, Y., Zhang, B., Yang, L., Li, H. (2013). Blood mercury concentration among residents of a historic mercury mine and possible effects on renal function: a cross-sectional study in southwestern China. *Environmental Monitoring and Assessment*, 185, 3049–3055.
65. Chen, C.Y., Yu, H.W., Zhao, J.J., Li, B., Qu, L.Y., Liu, S.P., Zhang, P.Q., Chai, Z.F. (2006). The roles of serum selenium and selenoproteins on mercury toxicity in environmental and occupational exposure. *Environmental Health Perspectives*, 114(2), 297–301.
66. WHO. (1991). *Environmental Health Criteria 118-Inorganic Mercury*. World Health Organization, Geneva.
67. UNIDO. (2003). *Protocols for environmental and health assessment of mercury released by artisanal and small-scale gold miners* (ASM). Vienna: United Nations Industrial Development Organization.
68. WHO. (1990). *Environmental Health Criteria 101: Methylmercury*. World Health Organization, Geneva.
69. Barregard, L., Quelquejeu, G., Sallsten, G., Haguenoer, J.M., Nisse, C. (1996). Dose-dependent elimination kinetics for mercury in urine observations in subjects with brief but high exposure. *International Archives of Occupational and Environmental Health,* 68, 345–348.
70. U.S. EPA. (1997). *Mercury Study Report to Congress*, EPA 452/R-97-0003; U.S. Environmental Protection Agency: Washington, DC, 1997.
71. National Research Council. (2000). *Toxicological Effects of Methylmercury*; National Academy Press: Washington, DC, 2000.
72. JECFA. (2003). *Safety Evaluation of Certain Food Additives*. The Joint Food and Agriculture Organization of the United Nations (FAO)/WHO Expert Committee on Food Additives; Geneva, 2003.
73. Bureau of Guizhou Statistics. (2012). *Guizhou Statistical Yearbook 2011*. Guiyang, China: China Statistics Press.

10

Effects of Non-essential Metal Releases on the Environment and Human Health

Peter G.C. Campbell[1] *and Jürgen Gailer*[2]

[1]*Institut national de la Recherche scientifique, INRS-ETE, Centre Eau Terre Environnement, Québec, G1K 9A9, Canada*

[2]*University of Calgary, Department of Chemistry, Calgary, Alberta, T2N 1N4, Canada*

10.1 Introduction

This chapter focuses on the elements As, Cd, Pb and Hg, which will be also referred to as 'non-essential metals' throughout. First, the effects that human activities exert on the global biogeochemical cycles of these elements will be briefly summarized. Then, the effects of environmentally relevant non-essential metal species on the aquatic freshwater environment will be elaborated on using an ecological risk assessment approach. Finally, the chronic low-level exposure of human populations to environmentally abundant species of these non-essential metals will be discussed and possible disease links will be explored. Our overall goal is to present a perspective on realistic threats that anthropogenic emissions of these non-essential elements into the environment cause today and will continue to cause in the foreseeable future.

Metal Sustainability: Global Challenges, Consequences, and Prospects, First Edition. Edited by Reed M. Izatt.

10.2 Metal Biogeochemical Cycles

10.2.1 Natural and Anthropogenic Sources

The biogeochemical cycle of a metal is normally initiated by a natural process (e.g., wind erosion in desert environments; entrainment of sea spray in marine systems; weathering of solid geological phases; lava flows from volcanoes in terrestrial or marine seabed locations; release of gaseous metals in volcanic emissions) and is closely linked to the hydrological cycle. Such cycles are usually depicted as a series of compartments linked by arrows representing the movement of the metal from one compartment to another. To characterize such a cycle, one needs to identify the main compartments, to determine metal concentrations (and forms) in each compartment, and finally to estimate the flux of metal along each of the arrows [1]. Water acts both as a chemical reagent (solvent) and as a transport medium (river flow; groundwater movement).

The natural biogeochemical cycle of a metal can be perturbed by anthropogenic activities linked to the isolation of the metal (i.e., mining activities), its use and its eventual disposal. This perturbation results in increased metal fluxes between biogeochemical compartments and potentially greater environmental exposure to metals for living organisms, including humans. Nriagu [2] was among the first researchers to attempt to quantify the relative importance of natural and anthropogenic emissions; he focused on emissions to the atmosphere and identified Cd and Pb as non-essential metals[1] for which anthropogenic activities had greatly modified their natural biogeochemical cycle. The approach used by Nriagu has been refined by subsequent researchers [3, 4], who have expanded the analysis to include metals released not only to the atmosphere but also to the aquatic and soil environments. Based on these studies, human activities perturb the global biogeochemical cycles of at least 11 elements globally (including As, Cd, Hg and Pb), 15 elements regionally and 35 elements locally [5].

When looking for evidence of the perturbation a biogeochemical cycle by anthropogenic activities, early workers focused on the atmospheric compartment; background concentrations of metals in the atmosphere, either as fine particulates or in gaseous form, are normally very low, and thus it is relatively easy to pick up a signal from anthropogenic inputs in this compartment [6]. Alternatively, one can consider metal fluxes between compartments. For example, in their study of over 70 elements in the periodic table (including 67 metals), Klee and Graedel [7] estimated the anthropogenic mobilization fluxes of metals from mining, combustion of fossil fuels and human-initiated biomass burning. The sum of these fluxes was then compared to the natural mobilization of each metal by weathering of the continental crust and fluvial transport, by the entrainment of sea spray into the atmosphere and by uptake into net primary production. In the most recent iteration of this approach, the mobilization factor, defined as the ratio of anthropogenic mobilization to natural mobilization, varied from > 10,000 (for Au and Pt) to ~0.001 (for Na and Mg) [8].

In addition to the *mobilization* factor described above, a number of other criteria have been used to identify metal biogeochemical cycles that have been perturbed by anthropogenic activities. For example, one can determine metal concentrations in atmospheric

[1] The term "non-essential" refers here and elsewhere in this text to metals having no known biological function in mammals.

particulate matter collected at remote sites, normalize these concentrations with respect to the Al concentration in the particulate matter, and then calculate an *enrichment* factor for each metal by comparing the [M]/[Al] ratio in the atmosphere to that in the earth's crust [9]. It is also possible to examine *historical trends* in the atmospheric deposition of metals, using such archives as glaciers [10, 11], peat bogs [12], soils and lake sediments [13]. Historical trends in global biogeochemical cycles can also be inferred from the body burdens of non-essential elements in living organisms that were sampled in the past and carefully preserved (see section 10.5.1). Finally, one also has to consider the actual concentrations observed in the receiving environment and the inherent toxicity of each metal [14]. For example, the mobilization factor for a rare earth metal may be very high, but the actual concentrations in wet precipitation or surface waters may be far below known toxicity thresholds.

An early example of this approach can be found in the review by Galloway *et al.* [6], who identified metals that posed potential risks for humans (Hg, Pb) and aquatic life (Cd, Hg, Pb, Cu and Zn). In more recent rankings of problematic metals, Hg and Pb remain near the top of the list with mobilization factors of 500 and 20, respectively [8], despite the major efforts that have been deployed to control their emissions, and As is now also recognized as a major human health risk in many parts of the world (see section 10.5.1). In the present chapter we consider these three non-essential metals (As, Hg, Pb) and include Cd, another largely unessential element with a somewhat lower environmental and human health profile, but one that is of emerging concern [15].

10.2.2 Notions of Metal Speciation

To understand the environmental fate of the problematic metals identified above and their bioavailability, it is important to introduce the notion of "metal speciation" [16]. The speciation of a metal refers to its distribution among various chemical forms differing in isotopic composition, oxidation state, and/or complex or molecular structure. For example, Cd may exist in natural waters as the free hydrated cation, $Cd(H_2O)_6^{2+}$; normally the molecules of water in the hydration sphere around the cation are not shown and the free cation is written as M^{z+}, where 'M' refers to a generic cation and 'z+' = the positive charge (+1, +2 or +3). As indicated in Figure 1, the cation can interact with other electron-donating solutes in solutions (normally called 'ligands') to form metal complexes; potential ligands include both inorganic anions (e.g., Cl^-, CO_3^{2-}, SO_4^{-2}) and organic molecules (e.g., simple monomers such as cysteine or citrate, or complex natural molecules such as fulvic and humic acids). The distribution of the metal of interest among the various chemical forms shown in Figure 1 will depend upon the nature and concentration of the various ligands present in solution, the relative affinities of these ligands for the metal (expressed as the complexation constant, K_{ML}), the solution acidity (pH) and also the concentrations of competing cations (e.g., Ca and Mg). In aqueous solutions of known composition, and when the relevant complexation constants are known, chemical equilibrium models can be used to calculate the relative importance of the free metal ion and the various metal complexes [17]:

$$Cd^{2+} + Cl^{-1} \overset{K_{CdCl}}{\rightleftarrows} CdCl^{-1} \qquad (10.1)$$

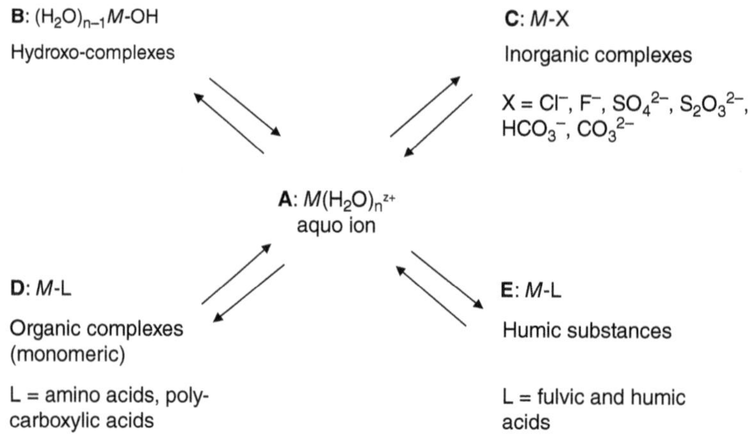

Figure 1 *Speciation of a typical cationic metal in a natural water*

The chemical bond formed between the cation and the ligand is a coordinate bond, and it is much weaker than a covalent bond. As a result, the different chemical forms shown in Figure 1 will normally be in chemical equilibrium. For example, if the chloride concentration increased, equation 1 would be driven to the right and the concentration of the free Cd^{2+} cation would tend to decrease. In other words, the speciation of a metal in solution is dynamic, not constant, and is influenced by such factors as the pH, salinity, and temperature as well as by the nature and concentration of the ligands present. It follows that when a metal is mobilized from the geosphere and enters an aqueous solution, whether *in vitro* or *in vivo*, its speciation will normally be determined by its new environment, not its source.

Organometallic compounds represent an exception to the previous statement. In addition to the metal complexes shown in Figure 1, a small number of metals can form covalent bonds with carbon (e.g., As, Hg, Pb). Such species, referred to as organometallic compounds, tend to be more stable than the metal-ligand complexes shown in Figure 1. As a consequence, when organometallic compounds enter an aquatic environment, their carbon-metal bond is not subject to the rapid equilibrium redistribution depicted in Figure 1. They are, however, subject to chemical and microbiological degradation, and to partitioning to the suspended solids present in the receiving water.

10.2.3 Environmental Fate of Metals

Having established that anthropogenic activities can perturb the normal biogeochemical cycle of metals, let us now consider briefly the environmental fate of these mobilized metals. Metals introduced into the atmosphere, for example from fossil fuel combustion or metal smelting, eventually return to the earth's surface via wet or dry deposition. Their rate of removal from the atmosphere depends on many factors, such as the height at which they entered the atmosphere (stack height), the physical form in which they entered the air mass (gaseous or particulate), their (photo)reactivity while in the atmosphere and the frequency of rain- or snowfall. Studies downwind from metal-sulfide smelting operations have shown that the metal footprint of the smelter (e.g., Cd, Cu, Zn) can often be detected more than 100 km downwind from the point of emission [18]. Metals that can enter the atmosphere in gaseous form, such as Hg and Pb, will be much more widely dispersed than those that remain in the aquatic and terrestrial compartments; mercury's status as a truly global contaminant is in large measure due to its volatility when in elemental form (Hg^0). Similarly, the addition of tetraethyl-lead to gasoline as an anti-knocking agent, beginning in the 1920s, contributed greatly to its dispersal via the atmosphere; Pb concentrations in atmospheric deposition decreased markedly in North America following its removal from gasoline in the early 1980s [19].

In contrast to metals that enter the atmosphere, those that enter the terrestrial environment (e.g., from atmospheric deposition, from weathering of the parent geological material, or from landfill operations) are much less mobile. Cationic metals have a strong affinity for the solid phases present in soil (e.g., clays, Fe and Mn oxyhydroxides, and detrital organic matter such as humus) and their movement downward towards the water table is normally very slow; in other words, these metals tend to remain close to the site of contamination. Cations present in the rooting zone of the soil can, however, be taken up by the terrestrial vegetation and returned to the Earth's surface environment, where they may be consumed by herbivores. The mobility of some metals is known to increase as the soil pH decreases (e.g., in response to acid precipitation) and in shallow soils the metals mobilized by this pH decrease can in fact attain the local surface waters (Al, Mn, Zn, Cd, but not Hg or Pb) [20]. Arsenic does not conform to this simple picture, however, as in its highest oxidation state (V) it is present as an oxyanion (arsenate, AsO_4^{-3}), which is more mobile in soils than are typical cations but does tend to bind to iron oxyhydroxides if present in the soil matrix; in its lower oxidation state (III), As is normally present as a neutral but polar species ($As(OH)_3$), which is more mobile in soils than arsenate.

Metals released to the aquatic environment tend to adsorb to suspended particles and as a result their normal fate is to settle to the bottom of the water body; the suspended particles may be derived from erosion of the watershed or from the growth and demise of aquatic (phyto)plankton. Normally, metals that reach the bottom sediments in a lake remain there and there is little or no remobilization into the overlying water column [21]. However, in hydrodynamically more active environments, such as a river or a dredged harbour, the metal-laden particles are subject to episodic resuspension events and may in this manner become more widely dispersed [22].

The observation that aquatic sediments constitute the major repository for metals mobilized from the terrestrial environment does *not* mean that they are no longer of environmental importance. On the contrary, from a microbiological perspective aquatic sediments constitute a key interface that controls the fluxes of nutrients (e.g., nitrogen and phosphorus) and contaminants (e.g., methyl-mercury; CH_3HgX, where X is normally Cl or OH) between the overlying water and the river or lake bottom. Similarly, aquatic sediments are recognized as a biologically important habitat that hosts a great diversity of freshwater organisms. The accumulation of anthropogenically mobilized metals at the sediment-water interface may be detrimental to the benthic organisms themselves, and in some cases these organisms may serve as a route for the trophic transfer of the metal to animals that feed on benthic prey. For example, Hg(II) is subject to methylation by microorganisms living at the oxic-anoxic interface in sediments. The resulting CH_3Hg-X, being lipophilic and highly bioavailable, tends to bioaccumulate in the invertebrates living at the sediment-water interface and when these Hg-laden invertebrates are consumed as prey, the Hg effectively "escapes" from the sediment and bioaccumulates in the food chain. Mercury can also return from the hydrosphere to the atmosphere via reduction (Hg(II) → Hg(0)), the reduced metallic form being volatile.

In the following section, we leave the abiotic external environment and consider how non-essential metals behave within living (aquatic) organisms.

10.3 Metal Environmental Toxicology

10.3.1 How Do Metals Interact with Aquatic Freshwater Organisms?

Environmental toxicology involves the study of the effects of toxic substances on biological organisms living in their native environment. It has developed as a distinct field of research since the 1960s, with an initial focus on organic contaminants (especially the now-banned organochlorine pesticides). This early work led to the widespread acceptance of the PBT framework for evaluating the potential environmental impact of a particular substance [23], where P = persistence, B = bioaccumulation, and T = inherent toxicity. However, this framework proved unworkable for metals [24, 25], notably with reference to the "persistence" and "bioaccumulation" criteria. Since metals do not "degrade" in the way that many organic contaminants do, metals would all be classified as infinitely persistent and the criterion would no longer be useful for ranking purposes. Similarly, the uptake of organic contaminants normally occurs by simple passive diffusion across a cell membrane. Given the lipidic and thus hydrophobic nature of this membrane, the octanol-water partitioning coefficient of an organic contaminant (K_{ow}) is often a good predictor of its tendency to bioaccumulate. In contrast, as discussed in the section on metal speciation, metals in the aquatic environment are generally present in hydrated and hydrophilic forms, which cannot cross biological membranes by simple diffusion, and their uptake normally occurs by facilitated transport involving membrane carriers or channels.[2] The K_{ow} values of these hydrophilic metallic species thus have no bearing on the relative facility with which they can cross biological membranes.

The failure of the PBT approach for metals reflects the fact that life on earth has evolved in their presence [26]. Many metals and metalloids are required for biological life (e.g., Cu, Co, Fe, Mn, Mo, Ni, Se, and Zn) and therefore their presence in the ambient environment is essential and their bioaccumulation is a natural process necessary for metabolism and growth. It follows that different environmental toxicology approaches are needed for organic and inorganic (metallic) contaminants. For metals, two types of models have proven to be useful for predicting metal uptake and toxicity: *equilibrium* models (e.g., the "Biotic Ligand Model" or BLM) and *kinetic* models (DYM-BAM). They are described in the following sections.

[2] Mercury is an important exception to this generalization. In its methylated forms, mercury is lipophilic and can indeed cross biological membranes by simple passive diffusion.

10.3.2 The Biotic Ligand Model (Chemical Equilibrium Approach)

The biotic ligand model (BLM) is designed to predict the response of aquatic organisms exposed to dissolved metals. The BLM focuses on the interaction of the metal with an external (epithelial) membrane. In the case of a unicellular alga or bacterium the membrane would the plasmalemma or cell membrane, whereas for a higher organism it might be a gill or olfactory membrane. In both cases, the approach consists of taking the equilibria described in Figure 1 and introducing a "biotic" ligand into the system. In other words, it is assumed that the epithelial membrane can be considered to be in equilibrium with the exposure medium, as shown below (where X is a membrane-bound metal transporter, M is a metal and L is a ligand in solution — Figure 2).

The BLM assumes that metals enter cells as cations, via facilitated transport, and that the plasma membrane is thus the primary site for metal interactions with living organisms. This interaction is assumed to occur via a ligand exchange reaction, yielding M-X-cell, and the biological response is predicted to vary as a function of the concentration of the M-X-cell surface complex. If the equilibrium condition is fulfilled, it can be shown that the proportion of X-cell sites bound to metal M can be calculated from the concentration of the free metal ion in solution, provided that possible competitive interactions between X-cell and the proton (H^+) or the water-hardness cations (Ca^{2+}, Mg^{2+}) are taken into account [27]. This result implies that the biological response of the exposed organism should be predictable on the basis of chemical equilibrium principles.

For laboratory exposures of aquatic organisms to cationic metals and their hydrophilic complexes, in chemically defined media and in experiments run at constant pH and constant water hardness, there is good evidence that indeed the free-metal ion concentration in the exposure medium is usually the best predictor of metal uptake and toxicity [28], although some exceptions have been documented [29–31]. However, there are far fewer results for experiments run under natural conditions or in the laboratory in the presence of natural dissolved organic matter (DOM). Results from the few quantitative experiments that have been carried out under these conditions are inconclusive, being more or less evenly divided between examples that conform to the predictions of the BLM and others that appear to be in contradiction.

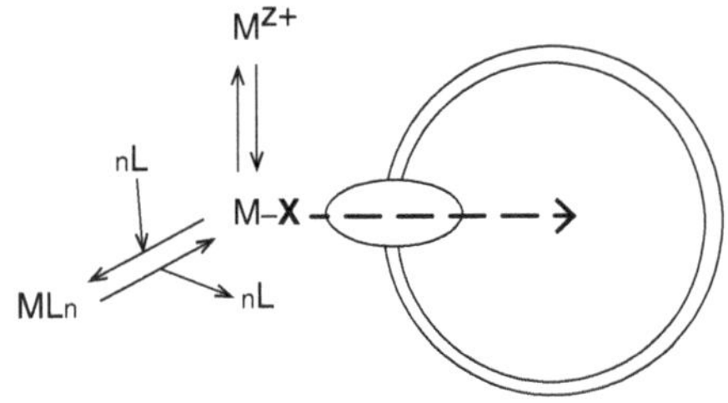

Figure 2 *Interaction of a typical cationic metal at a biological surface*

10.3.3 The Dynamic Multi-Pathway Bioaccumulation Model (Chemical Kinetics Approach)

The dynamic multi-pathway bioaccumulation model (DYM-BAM) is an alternative way to predict metal accumulation. Unlike the BLM, which is limited to aqueous metal exposures, the DYM-BAM model explicitly takes into account metal exposure from both water and ingested solids (food). In addition, it considers not only metal uptake via these two routes, but also metal elimination and growth dilution. At steady state, the bioaccumulated metal concentration can be predicted according to the following equation [32]:

$$\frac{d[\mathrm{M}]_{\mathrm{organism}}}{dt} = k_u\left[\mathrm{M}^{z+}\right] + \left(\mathrm{AE} \times \mathrm{IR} \times [\mathrm{M}]_{\mathrm{food}}\right) - k_e[\mathrm{M}]_{organism} - k_g[\mathrm{M}]_{\mathrm{organism}} \quad (10.2)$$

where k_u, k_e and k_g are rate constants for M-influx from water ($L^{-1}\cdot g^{-1}\cdot d^{-1}$), M-efflux ($d^{-1}$) and growth ($d^{-1}$), respectively, $[M^{z+}]$ is the free metal-ion concentration in water ($nmol\cdot L^{-1}$), $[M]_{organism}$ is the M concentration in the organism of interest ($nmol\cdot g^{-1}$ dry weight), $[M]_{food}$ is the metal concentration in the food consumed by the organism ($nmol\cdot g^{-1}$ dry weight), AE is the efficiency (%) with which the organism assimilates metal M from its food, and IR is the rate at which food is ingested by the organism (g prey$\cdot g^{-1}$ animal$\cdot d^{-1}$). As can be deduced from the equation, the model integrates information from the animal's habitat (metal concentrations in water and food) and physiological data (influx rates from water, influx rates from food, rate constants of loss, and growth rates) to yield estimates of the steady-state metal concentration in the target organism.

In a review of the use of biodynamic modeling to predict species-specific metal bioaccumulation in the field, Luoma and Rainbow [32] compared model predictions with field observations for six metals and one metalloid (Ag, Cd, Co, Cu, Cr, Zn; Se) and 14 animal species. The whole-organism metal concentrations varied over seven orders of magnitude, yet when all the data were log-transformed and pooled, most forecasts agreed with observations within a factor of two. On a log-log scale, the overall coefficient of determination (r^2) between predictions and observations was an impressive 0.98.

In principle, the dynamic modelling approach offers the possibility of estimating the internal dose of the metal and then relating this dose to a metal-induced effect. However, experimental data from field experiments suggest that if the internal dose is reached slowly, the organism has a better chance to survive than if the same dose is reached more rapidly. In other words, when the internalization flux of the metal is reasonably slow, the organism can "handle" the metal (e.g., by sequestering it in a non-bioavailable form or by accelerating its excretion rate), whereas if the influx rate is high, the organism is unable to handle all the incoming metal and it succumbs [33, 34]. Thus the application of the DYM-BAM approach to predict metal toxicity will have to include not only the calculated steady-state dose, but also the rate at which this dose is reached.

Before leaving this section of predictions of metal bioaccumulation and toxicity, it should be noted that the biological parameters in both the BLM and the DYM-BAM models are considered to be modelling "constants". These constants are normally determined in the laboratory on metal-naïve test organisms and are considered to be unaffected by antecedent exposure conditions. However, recent research suggests that the key epithelial properties of aquatic organisms that govern metal accumulation and toxicity are not constant, i.e., that exposed organisms can alter the characteristics of their epithelial membranes and influence both the affinity of the metal-binding sites and the rate at which the metal is internalized [35]. Clearly the application of these models to real-world cases of chronic exposure to metals remains a work in progress.

10.3.4 Metal Detoxification

Some metals exert their toxicity by binding to an external epithelial site (e.g., by blocking the activity of gill membrane transporters [36] or inhibiting an olfactory membrane sensor [37]), but in the vast majority of cases the metal must traverse a cell membrane and enter the organism in order to cause deleterious effects. Once this occurs, the metal encounters an intracellular environment in which the physicochemical conditions differ markedly from those in the external environment and its speciation will also change. Because of the high concentration of potential metal binding ligands, especially thiols, the free intracellular metal concentration is expected to be vanishingly small (e.g., $< 10^{-15}$ M Cu^{2+}, see Finney and O'Halloran [38]). Intracellular trafficking of essential metals is thought to involve metal "chaperones" that capture the incoming metal and then bind to specific receptor molecules and relinquish the metal in a ligand exchange reaction. If the entering metal is non-essential, it has the potential to displace essential metals from their normal binding sites and perturb cellular metabolism [39]. To avoid this outcome, the organism can impede the entry of the non-essential metal, accelerate its excretion, or sequester it within the cell. If the non-essential metal bioaccumulates, then this latter mechanism obviously assumes great importance.

One way of determining the extent to which a non-essential metal has been successfully detoxified is to carefully homogenize the tissue, without damaging the organelles, to separate the various subcellular fractions (often by differential centrifugation) and finally to determine the concentration of the metal in each fraction [40]. If the detoxification is effective, the non-essential metal should be found in the form of insoluble metal-rich granules or bound to metallothionein-like heat-stable peptides, and its concentration in cytosolic proteins or in organelles such as the mitochondria should be very low [41]. As we shall see in the following case study, metal detoxification in the field rarely seems to achieve this objective.

10.4 Case Study: Cadmium

To illustrate the environmental toxicology of metals in the freshwater environment as presented in the preceding section, we have chosen a case study in which Cd proved to be a major environmental stressor [42–44]; details of the study can be found in these three publications and in the papers cited therein. The chief biomonitor species was yellow perch (*Perca flavescens*), a non-migratory freshwater fish that is known to be acid- and metal-tolerant. The fish were collected from lakes in the vicinity of metal smelters that have been in operation at two locations, Rouyn-Noranda in northwestern Quebec (Cu, Zn) and Sudbury in northern Ontario (Cu, Ni), for more than 70 years. The smelter footprints also include Cd, which was present in the smelter feedstock at both locations and was emitted to the atmosphere for many years. The sampled lakes were located upwind and downwind from the smelters and thus constituted a marked metal-exposure gradient, both for the metals of commercial interest (Cu, Ni, Zn) and for Cd, as illustrated by the metal concentrations measured in the lake water and sediments.

10.4.1 Bioaccumulation (BLM vs. DYM-BAM)

In response to this abiotic gradient, Cd concentrations increased in the invertebrates that serve as prey for yellow perch early during their development, as illustrated by the positive relationship between aqueous Cd^{2+} and Cd bioaccumulated by the phantom midge (*Chaoborus* sp.). Croteau *et al.* [45] showed that this relationship could be modeled successfully with the BLM, provided that the pH gradient was also considered. For juvenile yellow perch, Cd concentrations also increased as a function of the aqueous Cd^{2+} concentration in the Rouyn-Noranda lakes (Figure 3) and the relationship improved slightly if competition with the ambient Ca^{2+} was included in the BLM [46]). However, the relationship was not as tight as for the midge and could not be transposed to the Sudbury lakes [47].

We also explored the potential of the dynamic modeling approach as a tool to predict Cd accumulation in juvenile yellow perch [48], using the results of field experiments designed to measure Cd uptake rates from water and food as well as Cd depuration rates and growth rates. The DYM-BAM model simulations for hepatic Cd agreed well with the observed time course of Cd accumulation in the liver, but for the gills and gut the predicted accumulations after 70 days were about three times higher than the observed values; this result

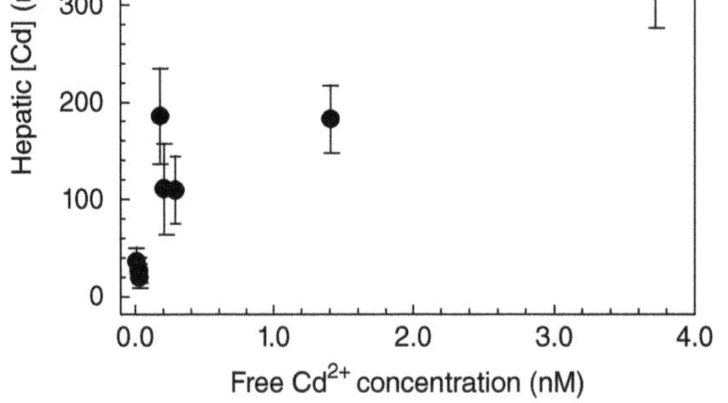

Figure 3 Relationship between the aqueous free Cd^{2+} concentration and the hepatic Cd concentration in juvenile yellow perch living in the Rouyn-Noranda lakes. Each point corresponds to fish collected from a different lake. Data replotted from Giguère et al. [46] with permission

suggests that these latter organs, which were in direct contact with the ambient water (gills) or with the ingested food (gut), can alter their ability to take up and (or) eliminate Cd. We conclude that both the BLM and DYM-BAM models show potential as predictors of Cd bioaccumulation in a field setting, but additional (physiological) factors need to be taken into account if these models are to be used with confidence to predict Cd bioaccumulation in indigenous biota in response to increased environmental exposure to Cd.

10.4.2 Subcellular Partitioning

To examine how yellow perch were handling their Cd burden along the metal contamination gradient, we focused on the liver as the target organ, this being the organ with the highest Cd concentrations, and subjected it to a subcellular partitioning protocol similar to that originally proposed by Wallace *et al.* [40]. Total hepatic Cd concentrations increased 14-fold along the geochemical gradient, and there was a virtually linear relationship between total hepatic Cd and the concentration of Cd bound to metallothionein-like peptides (Figure 4a). However, despite this clear evidence that yellow perch had mounted a detoxification response (i.e., fish from the more contaminated lakes had maintained a higher level of hepatic metallothionein), this response was not completely effective. Along the geochemical gradient, Cd concentrations also increased in the cytosolic heat-denaturable protein fraction, in the mitochondria and in the microsome + lysosome fraction (Figure 4b), suggesting that Cd detoxification in yellow perch liver was incomplete [49]. Given their essential metabolic roles, the presence of Cd in these fractions is indicative of chronic Cd toxicity in the yellow perch living in the more contaminated lakes.

Interestingly, there was no threshold exposure below which yellow perch succeeded in keeping the accumulated Cd from binding to "inappropriate" sites within their liver cells, a

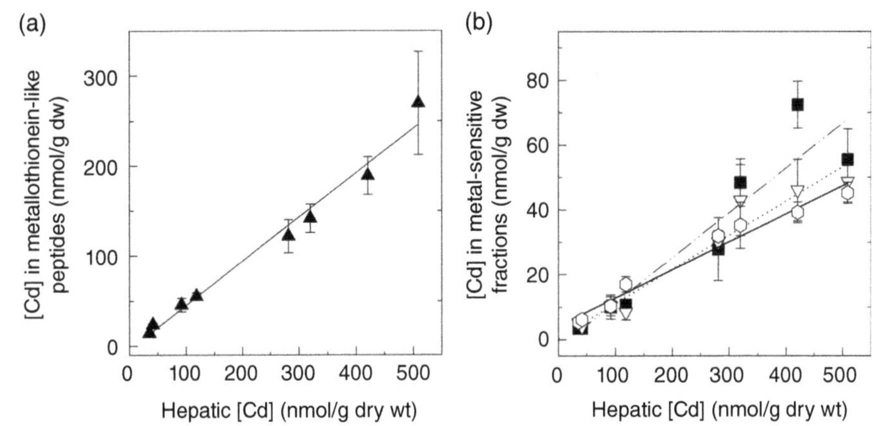

Figure 4 *(a) Cadmium detoxification response in yellow perch as indicated by the increase in metallothionein-bound Cd as a function of total hepatic Cd; (b) Cd binding to putative metal-sensitive sites in yellow perch liver increases despite the detoxification response. Symbols: solid triangles = metallothionein-like peptides; solid squares = mitochondria; inverted open triangles = lysosomes + microsomes; open hexagons = heat-denatured proteins. Data are replotted from Giguère et al. [49] with permission from Elsevier, copyright (2006)*

result in conflict with the classical threshold metal toxicology model for nonessential metals [41]. Note that yellow perch are not unique in this respect; Cd detoxification is also incomplete in other aquatic fauna living in smelter-affected lakes in the Rouyn-Noranda and Sudbury areas (*Pyganodon grandis*, a unionid bivalve, and *Chaoborus sp.*, the phantom midge) [50,51].

10.4.3 Evidence for Cd-Induced Effects in Aquatic Organisms

Consistent with the foregoing diagnosis of incomplete Cd detoxification, a variety of progressive responses (biochemical, physiological, morphological, demographic, etc.) were detected in yellow perch collected along the metal-concentration gradient and could be related to the internalized metal dose. For example, adult and juvenile perch collected from the more contaminated lakes exhibited endocrine impairment, as characterized by an attenuated cortisol stress response *in vivo* and a lower secretory capacity in response to adrenocorticotropic hormone *in vitro* [52–54]. Metabolic anomalies were detected in fish collected from the more contaminated lakes, leading to abnormal seasonal cycling of glycogen and triglycerides [55], and the aerobic capacity of yellow perch collected from contaminated lakes in the Sudbury area was shown to be reduced, as indicated by the activity of enzymes involved in aerobic metabolism as well as by swim performance and respiration rates [56].

As a caveat, it should be noted that in field studies such as those summarized here, exposure to a single metal (e.g., Cd) is rarely observed; for example, in the present case Cd, Cu and Zn tend to co-vary in the Rouyn-Noranda lakes whereas Cd, Cu and Ni co-vary in the Sudbury lakes. However, the internalized dose of Cd varies over a much wider range than do the internalized concentrations of the essential metals (Cu, Zn). For a more complete discussion of the evidence for metal-induced effects and for the central role played by Cd, the interested reader is invited to consult the following synthesis papers: Campbell *et al.* [42,44] and Rasmussen *et al.* [43].

To close this case study on a positive note, we should point out that although the evidence for metal-induced stress in yellow perch is persuasive, there was no evidence from our field study of an *upper* concentration threshold beyond which Cd detoxification completely ceased to function, i.e., steady-state metallothionein concentrations continued to increase in a dose-dependent manner along the exposure gradient. In other words, within the range of exposure conditions represented by the Rouyn-Noranda and Sudbury lakes, our biomonitor species appears to be metal-stressed but is nevertheless able to survive, mature and reproduce successfully.

10.5 Chronic Low-Level Exposure of Human Populations to Non-Essential Metals

Having focused in the previous sections on the behaviour of metals in the (aquatic) environment, we now consider the possible health consequences of the non-occupational exposure of humans to non-essential metals. In principle, there is a fundamental relationship between the health and wellbeing of every organism and the very ecosystem it inhabits (Figure 5). In the following sections we explore the mechanisms that constitute this link between the ecosystem and human health.

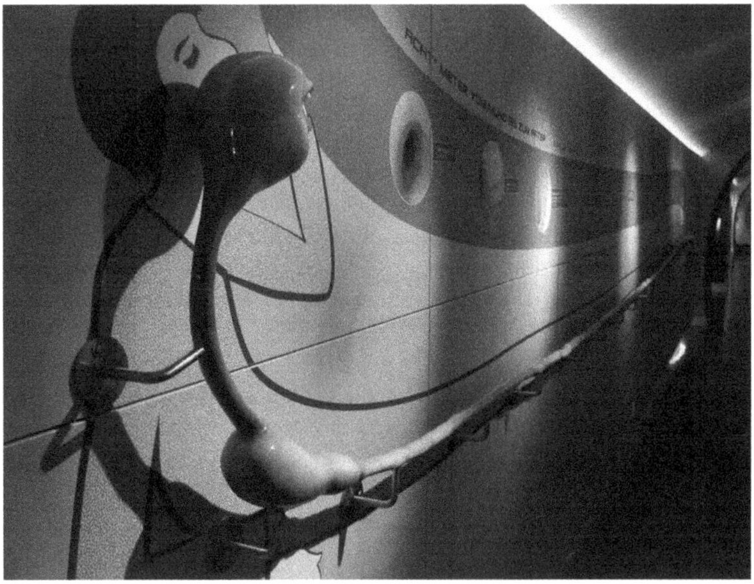

Figure 5 The gastrointestinal tract in a human adult is an astonishing 8 m long and represents the umbilical cord by which every human being is connected to its geochemical environment via the ingestion of food/drinking water and the non-essential metals therein throughout life. This picture was obtained from the exhibition "Reise in den menschlichen Körper" in the Haus der Natur, Salzburg, Austria (See insert for color representation of the figure.)

10.5.1 Historical Perspective

Man has been unwittingly exposed to non-essential metals since he first learned to use metals — in fact, since he first used fire [57]. Whereas Pb and Hg were known to ancient civilizations, As was isolated in ca. 1250 and Cd was discovered in 1817. Since these four elements are inherently toxic, it is not surprising that they are also among the oldest poisons known to mankind [58]. With the start of the industrial revolution in the 1760s, new manufacturing processes eventually increased the production of these elements. Hg, for example, was used to make mirrors, barometers and thermometers, Cd and Pb were transformed into the paint pigments CdS (yellow) and $PbCO_3$ (white), and As was used for the manufacturing of pesticides, wood preservatives and salvarsan — the first chemoptherapeutic agent — paving the way to modern anticancer chemotherapy [59]. Early on it was realized that the exposure of industrial workers to these elements and/or their compounds may cause neuropsychiatric disturbances and neurodevelopmental disorders [60]. The lesson that entirely different anthropogenic activities may also result in human exposure to these particular inorganic pollutants, however, has been learned the hard way over the past 100 years.

Between 1850 and 1930, exposure of humans to Pb from water pipes caused disease (eclampsia) and death of millions of people [61]. In the late 1950s, the inadvertent exposure of people near an industrial plant in Minamata, Japan to the neurotoxin $CH_3Hg\text{-}X$ [62] and the accidental ingestion of wheat that had been treated with a related mercurial as a seed dressing (to prevent spoilage) in Iraq [63] resulted in some of the most notorious pollution disasters of all time [64]. Another calamity that unfolded in the mid-1960s involved the utilization of Cd-laced effluent from a zinc mine near the Jintsu river, Japan to irrigate rice downstream in Toyama Bay; the ingestion of the latter caused severe chronic Cd poisoning — itai-itai disease (meaning ouch-ouch in Japanese) — which was associated with considerable pain [65]. Over the last 25 years in Bangladesh and West Bengal, India, the well-intentioned installation of ground-water wells to avoid exposure to bacteria-infested surface waters resulted in the unintended exposure of up to 100 million people to the carcinogen arsenite, As(III) — the biggest case of mass poisoning in history [66]. These incidents contributed to a revision of the existing maximum permissible concentrations of non-essential metals at the workplace as well as in air, soil, waste sites and drinking water [67, 68].

Today, human activities have such a profound impact on earth's ecosystems [3] that former U.N. Secretary General Kofi Annan was prompted to suggest that "The very basis for life is declining at an alarming rate". Owing to the increasing anthropogenic mobilization of As, Cd, Pb and Hg from the earth's crust into the biosphere (see Section 10.2.1), certain human populations are unwittingly exposed to higher daily doses of non-essential metals via air, food, drinking water and consumer products than ever before [69]. Given that geochemical exposure spans the lifetime of every human being, the low-level exposure to non-essential metals is of considerable public health concern [70], particularly with regard to children [71], who likely have a reduced capability to excrete certain non-essential metal species [72]. Direct experimental evidence in support of a 47-fold increase of the human exposure to Cd has been obtained by comparing its concentration in the kidney cortex of 61 specimens from 1897 to 1939 (year of death) with that of 207 autopsy tissue samples collected in 1980 [73].

Looking forward, the consumption of coal, which is laden with non-essential metals [5] that can severely affect human health [74], is projected to increase by 20% over the next 25 years [75]. In addition, non-essential-metal-laden electronic waste (e-waste) accumulates at a rate of 20–25 million t per year and is mostly disposed of in landfills [76], from which As and Pb can leach to groundwater [77]. The shredding of e-waste in recycling facilities can also release Cd and Pb, which may adversely affect human health [78]. Owing to these activities and ongoing world-wide mining, the environmental exposure of humans to non-essential metals will increase unless efficient abatement policies are implemented [79]. Combined with the exceptionally long residence times of non-essential metals in the Earth's surface environment [80], this trajectory will result in the insidious pollution of agricultural soils with these pollutants [81]; this is already evident in China, where 8.3% of the country's arable land is contaminated [82] and certain food crops exceed the maximum permissible concentrations of Cd and Pb [83]. Furthermore, one needs to consider that the low-level exposure of humans to harmful chemicals can result in epigenetic changes on the DNA that can promote diseases in future generations [84]. To minimize the exposure of a considerable fraction of the growing world population will require a stricter enforcement of existing environmental regulations [85], the implementation of government-mandated abatement strategies [67, 79] and — if necessary — the opposition of civil society to industries that are causing pollution [86]. To put things into perspective, the aforementioned itai-itai disease was attributed to daily Cd doses that were only approximately three times higher than those of the average Japanese citizen today [65]! Although human exposure to Cd is higher in Japan than in Europe and the United States [87], complacency is clearly not an option in addressing this important public health issue in our high-tech society [88]. Furthermore, metals cannot be degraded and may adversely affect human populations [89] — potentially for millennia — and at considerable economic costs [90–92]. The fact that ~5% of the U.S. population has urinary Cd concentrations close to the level that is associated with subtle kidney injury [93] has contributed to the fact that "Cadmium is likely to become an increasing factor in international trade negotiations as countries establish standards to control cadmium residues in food" [94]. These facts provide a clear incentive to develop policies aimed at reducing human exposure to non-essential metals [95].

10.5.2 Assessment of Human Exposure to Non-Essential Metals

Three approaches can be employed to assess human exposure to non-essential metals/metal species via air, drinking water, food and/or the use/application of consumer products (Figure 6). The first approach involves their quantification in air [96], food [97], beverages [98] and consumer products, such as children's toys [99], lip products [100], traditional medicines from India [101] and skin-lightening creams [102]. Considering that environmentally abundant species of Hg and As differ significantly in their bioavailability and/or toxicity [69], as many molecular forms as possible should be quantified. While this task is easily accomplished for drinking water [103], it is more difficult in case of air where non-essential metal species that are volatile [96, 104] and those that are adsorbed on particulate matter [96, 105] must be quantified. With regard to consumer products, some children's toys contained very high concentrations of Cd (37% [w/w] and Pb (65%)[99] and out of 32 lip products 24 contained Pb (average concentration 0.36 ppm) and 11 Cd (average concentration 1.16 ppm)[100]. Of the 549 investigated skin-lightening products, 33 had Pb concentrations >1000 ppm [102]. Some of the 42 investigated traditional Indian medicines contained up to 140 ppm As and 52,000 ppm Pb [101]. Since total element concentrations are insufficient to predict the bioavailability of non-essential metals (e.g. due to accidental ingestion of a toy), bioaccessibility studies based on the leaching of metals with simulated gastric/intestinal fluids can provide valuable insight [99, 101]. From an analytical perspective, food represents the most challenging matrix [98] because comparatively few methods are available that can determine all species of one non-essential metal [106], let alone those of all other non-essential metal species. Even if one were able to quantify all non-essential metal species, however, this information is useful only to estimate how much of each of these species will be absorbed into the bloodstream if fiber and/or complexing agents (e.g. phytic acid) are absent, as they can decrease gastrointestinal absorption [107].

The second approach involves the quantification of non-essential metals in blood, which is also referred to as 'biomonitoring' [93,108]. This approach is more convenient for assessing exposure because it will identify *all* pollutants that are bioavailable and are therefore of principal health concern. If blood samples from a population are analyzed regularly, valuable information about the population's exposure to environmental pollutants over time can be obtained [109]. The actual pollutant concentration in the blood of an individual, however, cannot indicate whether it will result in a particular disease [110], as only relative differences compared to other individuals in the same age group, sex and ethnicity can be determined [93]. Another conceptual shortcoming of this approach is that certain pollutants may be enriched in organs compared to the bloodstream, so as to

Principal methodologies to assess human exposure to non-essential metals

	Consumer products*			Natural + anthropogenic sources		
	skin	lungs	GI-tract	air/ $PM_{2.5}$	food	drinking water
$As^{III} + As^{V}$	+	+	+	+	+	+
Cd^{2+}	(+)#	+	+	+	+	+
Pb^{2+}	(+)#	–	+	+	–	+
Hg^{0}	–	–	–	+	–	–
Hg^{2+}	+	–	+	+	+	+
CH_3Hg^{+}	–	–	–	–	–	–

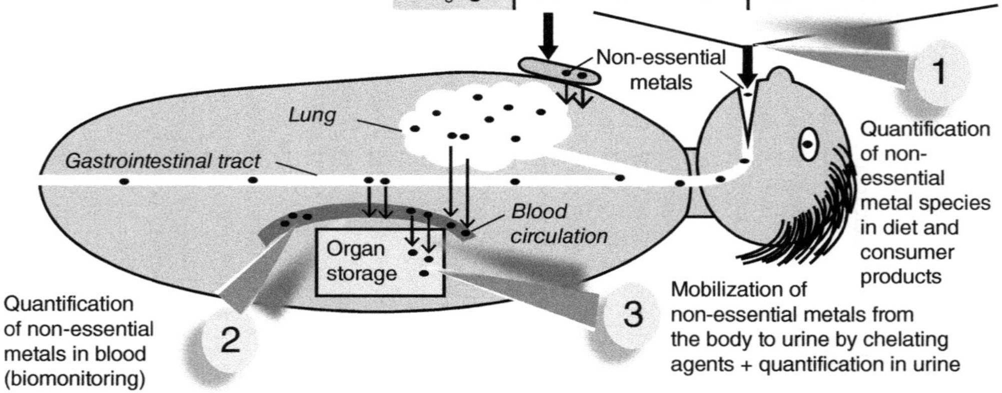

* Children's toys, skin & lip products, herbal medicines, broken light bulbs, necklaces

Extent of dermal absorption unknown

Figure 6 *Overview of possible approaches to establish human exposure to non-essential metals*

underestimate the exposure. Conversely, the slow clearance of detoxification products formed in the bloodstream (e.g. after a recent human exposure to Hg(II) [111]) may overestimate the exposure, although the detoxification product may be entirely non-toxic [112]. These shortcomings could be somewhat rectified if this approach were extended to organ concentrations, which may provide a better estimate of life-time exposure [73, 113]. The analysis of hair has also been employed to assess human exposure to non-essential metals and offers the considerable advantage of revealing changes in exposure over weeks to months [114].

The third approach involves the chelating-agent-driven mobilization of endogenous non-essential metals from bodily tissues to urine. Although this approach is frequently employed to treat patients suffering from excess metal exposure [115], it is discussed here solely as a tool to assess exposure. The oral administration of 300 mg of 2, 3-dimercapto-1-propanesulfonic acid (DMPS) to volunteers with and without dental amalgams followed by the analysis of urine for total Hg, for example, revealed a 19-fold increase in the non-amalgam group and a 24-fold increase in the amalgam group [116] and indicated the existence of a 'labile Hg pool' in both groups. DMPS has also been employed to assess the As body burden of environmentally exposed people [117]. Whereas the related chelating agent 2,3-dimercaptosuccinic acid (DMSA) is clinically used to mobilize Pb into urine [118], Cd was mobilized to urine only if DMSA was administered immediately after the injection of the metal. More recent studies, however, have demonstrated that DMSA appears to be effective to treat children less than 1 year of age [119]. Owing to the potential side-effects of the chelating agents and the lack of mobilization of certain non-essential metals (e.g. Cd), this approach is deemed impractical.

Based on these considerations, a comprehensive assessment of the exposure of humans to non-essential metals should involve the analysis of air/diet/consumer products *and* the collection of biomonitoring data. It is extremely difficult, however, to use the actual blood concentrations of a pollutant to quantitatively assess the health risk [120] and thus to link these concentrations to the etiology of particular diseases.

10.5.3 Bioavailability of Non-Essential Metal Species

Three pathways must be considered by which As(III), As(V), Cd, Pb, Hg(0), Hg(II) and CH_3Hg-X may invade human tissues: the lungs, the gastrointestinal (GI) tract (Figure 5), and the skin. The following discussion aims to assess these exposure routes in terms of their contribution to the total body burden of an element, but this ordering does not imply that a minor exposure route (e.g. inhalation, dermal exposure) is less health relevant.

10.5.3.1 Respiratory System

The uptake of non-essential metals (e.g. Cd) by the olfactory epithelium in the nose can result in a reduced sense of smell (hyposmia) or its complete loss (anosmia) and is particularly relevant with regard to their translocation to the brain [121]. In fact, active and passive inhalation of tobacco smoke is a major exposure source for Cd [94]. Hg-vapor, Hg(0), which is released from dental amalgams [122] and broken energy-saving compact fluorescent lamps [123], for example, easily crosses biological membranes and is retained almost entirely in the alveoli [124]. Within erythrocytes catalase mediates the oxidation of Hg(0) to Hg(II) [125]. This oxidation, however, is slow compared to the rapid translocation of dissolved Hg(0) across the blood-brain barrier, which explains the significant regression that has been observed between the number of dental amalgam surfaces (amalgam can contain up to 50% Hg) and the Hg-concentration in the brain [122].

The first studies into the distribution of non-essential metals in particulate matter of different size, reported in the early 1970s [105, 126], revealed that the highest concentrations of Pb, Cd and As were associated with the smallest particles (1.1–2.1 μm), which are respirable and therefore most dangerous [105]. In the inner city of New Orleans, the inhalation by children of Pb-contaminated dust liberated from soil was found to be as important as the exposure to dust from Pb-based paint [127]. Airborne particles from Huelva, Spain exceeded the maximum concentration of 6 ng As m^{-3} proposed by the European Union and revealed the preferential accumulation of the carcinogen As (70–80%) in $PM_{2.5}$ particles [128]. Although the total As body burden that is attributed to inhalation exposure is small compared to the dietary exposure, it can contribute up to 10–20% if As-laden coal is burned inside dwellings [129] and even more if people live in the vicinity of smelters, glass factories or certain chemical plants [130]. The exposure of rats to coal fly-ash particles (<5 μm) for 28 days (4 h daily, 5 days a week) revealed that Cd and Pb were translocated to the liver and the kidneys, where they may cause extrapulmonary health effects [131].

10.5.3.2 Gastrointestinal System

The GI tract (Figure 5) represents the most important route by which non-essential metals invade human tissues [5]. Studies with human volunteers have demonstrated that >80% of As(III), >90% of As(V), 100% of CH_3Hg-X and ~7% of Hg(II) are absorbed from the gastrointestinal tract [69], whereas the absorption of Cd ranged from 0.5-8% [132] and that of Pb was ~8% [133].

The global extent of naturally occurring non-essential metals in groundwater and the associated health effects have been the focus of a comprehensive recent review [134]. In general, Pb-exposure via drinking water is negligible compared to dietary exposure today, unless Pb water pipes contaminate the water supply [135] as unfolded in 2015 in Flint/Michigan. At present, As(III)-contaminated drinking water poses a problem of unimaginable proportions in West Bengal, India [66, 134, 136] and numerous other countries where drinking water exceeds the WHO guideline of 10 μg As/L [137]. Exposure to As from drinking water wells also represents an emerging problem near natural gas extraction sites [138]. In addition, effluents discharged from coal combustion residue settling ponds exceeded the U.S. EPA guidelines for drinking water [139]. The total Hg concentration in unpolluted fresh-water bodies is <5 ng/L, but may reach 3200 ng/L near mining sites [140].

Food is the primary source for the exposure to Pb, Cd and Hg [97] and also As if food items are dried over open coal-burning stoves fueled with As-laden coal [74, 129]. Rice has also been recently identified as a significant source of dietary As [141]: the consumption of 0.56 cups rice/day was found comparable to drinking 1 L of water containing 10 μg As/L [142]. Since Asian Americans consume more than 2 cups of rice a day, rice consumption must be considered in the design of As reduction strategies. Traditional medicines can also contain bioavailable Pb and As, which may exceed the standard for daily intake at the recommended oral dose [101]. In France, organic foodstuffs such as carrots and buckwheat contained As and Pb above the maximum levels proposed by the European Union [143]. The most important dietary source for Hg is the ingestion of fish, shellfish, marine mammals, crustaceans and poultry fed fish meal [144]. In Hg-polluted areas in China, the concentration of Hg in rice grains can reach up to 569 μg/kg, of which 145 μg/kg was CH_3Hg-X [145]. Although the Hg-concentration in high-fructose corn syrup is ≤ 0.57 ppm, its average consumption of 50 g per day suggests that one must take this dietary source into account in children and sensitive populations [146]. The predominant source for dietary exposure to Cd are cereals, vegetable, root vegetables, mollusks and crustaceans [94]. Fertilizers can contain Cd as a contaminant at concentrations up to 200 mg Cd/kg [147], and this may accumulate in the top soil and be absorbed by crops [83, 147]. The total dietary intake of non-essential metals in humans is estimated at 37–261 μg As/day [83,97] (this includes non-toxic arsenobetaine and <4 μg/day from drinking water [148]), 8–25 μg Cd/day [87], 8–59 μg Pb/day [97] and 4–19 μg Hg/day [97], although this value can reach 185 μg Hg/day in Inuit communities [149].

10.5.3.3 Skin

Although long-term exposure to As-containing pesticides can cause skin and lip cancer, their transdermal absorption has not been assessed [150]. Inorganic As compounds are primary skin irritants [151] that are worryingly often present in herbal medicines (in one case the daily dermal dose was 210 mg of As_2O_3!) [152]. With regard to Cd, seven of 17 jewelry items had Cd concentrations that did not comply with regulations in the United States and Canada [99]. The analysis of 32 lip products revealed that 24 contained Pb and 12 contained Cd, but their estimated daily intakes by this exposure route were well below the acceptable daily intake [100]. With regard to the dermatological exposure to Hg, one has to consider cosmetics [152] and/or the use of red cinnabar (HgS) in tattoos. Of 549 investigated skin-whitening creams, 6% contained Hg >1000 ppm and 45% of these contained Hg >10,000 ppm [102]. Overall, the dermal absorption of non-essential metals is insignificant for the general population but may be relevant if certain skin products are frequently applied.

10.5.4 Metabolism of Non-Essential Metals

With regard to the biotransformation of Hg(0), Hg(II), CH_3Hg-X, Cd, As(III) and As(V), one must consider reduction/oxidation reactions [relevant to Hg(0), As(III) and As(V)] [153], biomethylation reactions [relevant to As(III) and Hg(II) [154]], demethylation reactions [relevant to CH_3Hg-X [155]] and the conversion of Cd and Hg(II) to selenides in the bloodstream [156].

10.5.4.1 Blood Circulation

Upon entering the bloodstream, non-essential metal species may either bind to plasma transport proteins or undergo erythrocyte-driven biotransformations, such as the catalase-mediated oxidation of Hg(0) to Hg(II) [125]. Erythrocytes also mediate the reaction of free selenite, Se(IV) with Hg(II) [112] and As(III) [157] to form species with As-Se and Hg-Se bonds. Since the latter compounds are either non-toxic or rapidly excreted [69], the formation of species with As-Se and Hg-Se bonds will decrease the transport of highly toxic

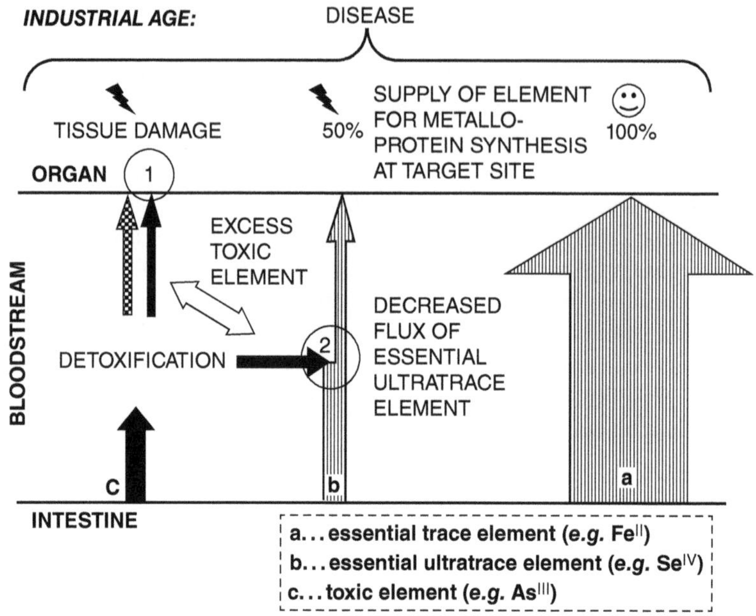

Figure 7 Conceptual depiction of two feasible pathways by which the chronic human ingestion of comparatively small daily doses of a non-essential metal (e.g., As(III)) may eventually result in adverse health effects. Whereas mechanism 1 has been extensively studied, much less is known about mechanism 2, which is associated with the perturbation of the metabolism of essential ultratrace elements (e.g., Se). Note that even though sufficient amounts of total Se may be ingested via the diet, the decreased amount of Se that is available for metalloprotein synthesis in organs may be insufficient to maintain health and eventually result in disease Adapted from [69].

non-essential metal species to target organs in the short term [69, 158]. The chronic ingestion of Hg(II), As(III), and Cd will therefore *cumulatively* affect the metabolism of the essential element selenium (Se), which may leave insufficient amounts of dietary Se available for the assembly of biochemically vital selenoproteins. Thus, the formation of species with As-Se and Hg-Se bonds in the bloodstream is a biochemically feasible mechanism by which chronic exposure to comparatively small daily doses of non-essential metals may be connected to the etiology of certain human diseases (Figure 7).

10.5.4.2 Organs

Although the metabolism of Hg(II), CH_3Hg-X, As(III) and As(V) has been extensively studied in mammals [69], several important aspects pertaining to their biochemistry in the bloodstream [75] and their organ-based metabolism remain unanswered [159, 160]. In this context, the advent of state-of-the-art synchrotron-based imaging techniques hold great potential to provide new insight into their subcellular localization [161] and their speciation in the brain [162], which may, in turn, provide important clues for elucidating their role in the etiology of neurodegenerative diseases [163].

10.5.5 Linking Non-Essential Metal Exposure to the Etiology of Human Diseases

Table 1 provides an overview of known adverse health effects of non-essential metal species, but the underlying biomolecular mechanisms are in many cases unknown [75]. Their elucidation, however, is regarded by many as one of the grand challenges in science, technology and public policy in the post-genomic era [70, 164]. The quest to link an exposure to infinitesimally small daily doses of a particular pollutant or a pollutant mixture with a particular disease requires one to uncover the biochemical mechanism(s) that provide the required leverage. Conceptually, these mechanisms may involve the direct effect of a particular non-essential metal species on a critical protein within a target organ [165] and/or may involve the non-essential-metal-induced perturbation of the metabolism of an essential trace element, such as Se [69] and possibly others [166]. Finding these elusive mechanisms represents a gargantuan task that is intricately tied to a better understanding of the metabolism of non-essential metals in mammals. Specifically, this will require a deeper insight into the transport and the biotransformation of non-essential metals in the bloodstream and the identification of the molecular species that are translocated to target organs [167], as well as the elucidation of the biochemical mechanisms by which the aforementioned molecular species adversely affect target organ cells at the genetic level [168, 169]. Finally, investigations into the application of nutritional amendments to increase the excretion of non-essential metals in affected people represents a neglected, but — in light of recent findings — rather promising avenue for further research [170].

Table 1 *Human exposure to specific non-essential metal species and known adverse health effects as well as potential disease links*

Non-essential metal species	Established adverse human health effect(s)	Reference	Potential link to human diseases	Reference
As(III)	human carcinogen	**[148,171]**	Alzheimer's disease	**[172]**
	cardiovascular disease	**[148]**		
	neurotoxicity	**[171]**		
	diabetes	**[148,171]**		
	hyperpigmentation	**[171]**		
Cd	human carcinogen	**[173]**	stroke/heart failure	**[174]**
	osteoporosis	**[175]**		
	renal toxicity	**[176]**		
	diabetes	**[176]**		
Hg(II)	autoimmunity	**[177]**	Parkinson's disease	**[178,179]**
	nephrotoxicity	**[180]**	Alzheimer's disease	**[181]**
	gastrointestinal toxicity	**[182]**		
CH_3Hg-X	neurotoxicity	**[183]**	Autism	**[184]**
Pb	neurotoxicity	**[183]**	Parkinson's disease	**[178,179,185]**

10. 5.6 Global Ecosystem Contamination by Arsenic, Cadmium, Lead and Mercury as an Underestimated Threat to Human and Ecosystem Health: A Summary

From a conceptual point of view, chemical matter continually 'flows' through organisms in the form of food and drinking water. Gradually increasing concentrations of non-essential metals will therefore at some point result in adverse health effects in ecosystems as well as in human populations.

Lead has had relatively little impact on the aquatic and terrestrial environments, largely because Pb concentrations in surface, soil and ground waters are kept very low by sorption and/or precipitation reactions. In contrast, and as discussed earlier, Pb has historically posed serious human health problems. Lead emissions to the environment are now largely controlled (banning of Pb-based paints and additives to gasoline; replacement of Pb plumbing) and Pb concentrations in precipitation, aquatic sediments and biota have declined markedly since the 1980s.

The situation for As is somewhat similar to that of Pb, in that impacts on human health appear to outweigh those on the fauna and flora. However, given its various oxidation states and its ability to form covalent arsenic-carbon bonds, the environmental chemistry of As is far more complex than that of Pb. The environmental effects of As are poorly documented, but it is now recognized as a major threat to human health, in large measure because of its presence in groundwater used as a source of drinking water. Unlike the situation with Pb, the threat posed by As to human health is current, not historical.

Mercury has been recognized as a global pollutant for over 40 years, in large measure because of its presence in fish and seafood and its effects on human consumers. Only quite recently has it been recognized that Hg can also have subtle but potentially important neurological effects on aquatic organisms themselves. In recognition of the pervasive presence of Hg and its neurotoxicity, the United Nations Environment Programme (UNEP) is currently spearheading a transition to a low-Hg world and this initiative resulted in a global treaty to reduce mercury emissions that was signed by 141 countries in October 2013. Reducing Hg emissions is certainly a worthy initiative, but given the unique biogeochemical properties of Hg (e.g., its propensity to change oxidation state and to be methylated) and its consequent ability to escape from aquatic sediments and re-enter the food web, it is destined to remain a major problem for the foreseeable future.

Turning finally to Cd, its threat to human health has been recognized for over 25 years, driven largely by early worries about Cd in fertilizer or biosolids finding its way into the edible portions of plants; more recently, the presence of Cd in seafood has led to the closure of major fisheries. In contrast, there has been much less attention paid to the effects of Cd on aquatic and terrestrial fauna and flora. As illustrated in the case study in section 10.4, this lack of interest in Cd ecotoxicology is misplaced [15]. We suggest that Cd should receive more attention as it likely plays an important role in the etiology of emerging diseases [174, 186] and as an environmental stressor.

When we think about future threats to the environment, we often picture cars and smokestacks, but in reality it is our need for clean air, food and drinking water that poses one of the biggest dangers to human health [187]. If we do not want to continue to 'fail as stewards', we must implement more stringent regulatory frameworks to minimize the unwitting chronic low-level exposure of human populations to non-essential metals in order to protect children, adults [188] and — via epigenetic effects to the DNA — future generations alike [189, 190].

References

1. Likens, G.E. and Borman, E.H. (1995) Biogeochemistry of a Forested Ecosystem, 2nd ed. Springer-Verlag, New York, NY, USA.
2. Nriagu, J.O. (1979) Global inventory of natural and anthropogenic emissions of trace metals to the atmosphere. *Nature*, **279**, 409–411.
3. Pacyna, J.M., Scholtz, M.T. and Li, Y.F.A. (1995) Global budget of trace metal sources. *Environmental Reviews*, **3**, 145–159.
4. Pacyna, E.G., Pacyna, J.M., Steenhuisen, F. and Wilson, S. (2006) Global anthropogenic mercury emission inventory for 2000. *Atmospheric Environment*, **40**, 4048–4063.
5. Pacyna, J.M. (1996) Monitoring and assessment of metal contaminants in the air, in *Toxicology of Metals* (ed L.W. Chang), CRC Press, Boca Raton, FL, USA, pp. 9–28.
6. Galloway, J.N., Thornton, J.D., Norton, S.A., Volchok, H.L. and McLean, R.A.N. (1982) Trace metals in atmospheric deposition—a review and assessment. *Atmospheric Environment*, **17**, 1677–1700.
7. Klee, R.J. and Graedel, T.E. (2004) Elemental cycles: a status report on human or natural dominance. *Annual Reviews of Environment and Resources*, **29**, 69–107.
8. United Nations Environment Program (UNEP) (2013) *Environmental Risks and Challenges of Anthropogenic Metals Flows and Cycles*. Nairobi, Kenya.
9. Duce, R.A., Hoffman, G.L. and Zoller, W.H. (1975) Atmospheric trace metals at remote northern and southern hemisphere sites: pollution or natural? *Science*, **187**, 59–61.
10. Lee, K., Do Hur, S., Hou, S., Burn-Nunes, L.J., Hong, S., Barbante, C., Boutron, C.F. and Rosman, K.J.R. (2011) Isotopic signatures for natural versus anthropogenic Pb in high-altitude Mt. Everest ice cores during the past 800 years. *Science of the Total Environment*, **412**, 194–202.
11. Schuster, P.F., Krabbenhoft, D.P., Naftz, D.L., Cecil, L.D., Olson, M.L., Dewild, J.F., Susong, D.D., Green, J.R. and Abbott, M.L. (2002) Atmospheric mercury deposition during the last 270 years: A glacial ice core record of natural and anthropogenic sources. *Environmental Science and Technology*, **36**, 2303–2310.
12. Shotyk, W., Norton, S.A. and Farmer, J.G. (1997) Summary of the workshop on peat bog archives of atmospheric metal deposition. *Water Air and Soil Pollution*, **100**, 213–219.
13. Thevenon, F., Graham, N.D., Chiaradia, M., Arpagaus, P., Wildi, W. and Poté, J. (2011) Local to regional scale industrial heavy metal pollution recorded in sediments of large freshwater lakes in central Europe (Lakes Geneva and Lucerne) over the last centuries. *Science of the Total Environment*, **412–413**, 239–247.
14. Wood, J.M. (1974) Biological cycles for toxic elements in the environment. *Science*, **183**, 1049–1052.
15. Campbell, P.G.C. (2006) Cadmium — a priority pollutant. *Environmental Chemistry*, **3**, 387–388.
16. Templeton, D.M., Ariese, F., Cornelis, R., Danielsson, L.G., Muntau, H., van Leeuwen, H.P. and Lobinski, R. (2000) Guidelines for terms related to chemical speciation and fractionation of elements: definitions, structural aspects, and methodological approaches. *Pure and Applied Chemistry*, **72**, 1453–1470.
17. Batley, G.E., Apte, S.C. and Stauber, J.L. (2004) Speciation and bioavailability of trace metals in water: progress since 1982. *Australian Journal of Chemistry*, **57**, 903–919.
18. Bonham-Carter, G.F., Henderson, P.J., Kliza, D.A. and Kettles, I.M. (2006) Comparison of metal distributions in snow, peat, lakes and humus around a Cu smelter in western Quebec, Canada. *Geochemistry - Exploration Environment Analysis*, **6**, 215–228.
19. Eisenreich, S.J., Metzer, N.A., Urban, N.R. and Robbins, J.A. (1986) Response of atmospheric lead to decreased use of lead in gasoline. *Environmental Science and Technology*, **20**, 171–174.

20. Campbell, P.G.C. and Tessier, A. (1987) Metal speciation in natural waters: influence of environmental acidification, in *Sources and Fates of Aquatic Pollutants ACS Advances in Chemistry Series,* 216 (eds R.A. Hites and S.J. Eisenreich), American Chemical Society, Washington, DC, USA, pp. 185–207.
21. Farley, K.J., Carbonaro, R.F., Fanelli, C.J., Costanzo, R., Rader, K.J. and Di Toro, D.M. (2011) Ticket-UWM: A coupled kinetic, equilibrium, and transport screening model for metals in lakes. *Environmental Toxicology and Chemistry*, **30**, 1278–1287.
22. Burton, G.A. (2010) Metal bioavailability and toxicity in sediments. *Critical Reviews in Environmental Science and Technology*, **40**, 852–907.
23. US EPA (1992) *Framework for Ecological Risk Assessment.* Washington, DC, USA: U.S. Environmental Protection Agency, Risk Assessment Forum.
24. Campbell, P.G.C., Chapman, P.M. and Hale, B.A. (2006) Risk assessment of metals in the environment. *Issues in Environmental Science and Technology*, **21**, 102–131.
25. Chapman, P.M. and Wang, F. (2000) Issues in ecological risk assessment of inorganic metals and metalloids. *Human and Ecological Risk Assessment*, **6**, 965–988.
26. Frausto da Silva, J.J.R. and Williams, R.J.P. (2001) *The biological chemistry of the elements: the inorganic chemistry of life*, Oxford University Press, Oxford, UK.
27. Meyer, J.S., Santore, R.C., Bobbitt, J.P., Debrey, L., Boese, C.J., Paquin, P.R., Allen, H.E., Bergman, H.L. and Di Toro, D.M. (1999) Binding of nickel and copper to fish gills predicts toxicity when water hardness varies, but free-ion activity does not. *Environmental Science and Technology*, **33**, 913–916.
28. Campbell, P.G.C. (1995) Interactions between trace metals and organisms: a critique of the free-ion activity model, in *Metal Speciation and Bioavailability in Aquatic Systems.* IUPAC Series on Analytical and Physical Chemistry of Environmental Systems (eds A. Tessier and D. Turner), J. Wiley & Sons, Chichester, UK, pp. 45–102.
29. Campbell, P.G.C. and Fortin, C. (2013) Biotic Ligand Model, in *Encyclopedia of Aquatic Ecotoxicology* (eds J.F. Férard and C. Blaise), Springer-Verlag, Heidelberg, pp. 237–245.
30. Hassler, C.S., Slaveykova, V.I. and Wilkinson, K.J. (2004) Some fundamental (and often overlooked) considerations underlying the free ion activity model and the biotic ligand models. *Environmental Toxicology and Chemistry*, **23**, 283–291.
31. Niyogi, S. and Wood, C.M. (2004) Biotic Ligand Model, a flexible tool for developing site-specific water quality guidelines for metals. *Environmental Science and Technology*, **38**, 6177–6192.
32. Luoma, S.N. and Rainbow, P.S. (2005) Why is metal bioaccumulation so variable? Biodynamics as a unifying principle. *Environmental Science and Technology*, **39**, 1921–1931.
33. Andrès, S., Baudrimont, M., Lapaquellerie, Y., Ribeyre, F., Maillet, N., Latouche, C. and Boudou, A. (1999) Field transplantation of the freshwater bivalve *Corbicula fluminea* along a polymetallic contamination gradient (River Lot, France): I. Geochemical characteristics of the sampling sites and cadmium and zinc bioaccumulation kinetics. *Environmental Toxicology and Chemistry*, **18**, 2462–2471.
34. Baudrimont, M., Andrès, S., Metivaud, J., Lapaquellerie, Y., Ribeyre, F., Maillet, N., Latouche, C. and Boudou, A. (1999) Field transplantation of the freshwater bivalve *Corbicula fluminea* along a polymetallic contamination gradient (river Lot, France): II. Metallothionein response to metal exposure. *Environmental Toxicology and Chemistry*, **18**, 2472–2477.
35. Niyogi, S., Couture, P., Pyle, G., McDonald, D.G. and Wood, C.M. (2004) Acute cadmium biotic ligand model characteristics of laboratory-reared and wild yellow perch (*Perca flavescens*) relative to rainbow trout (*Oncorhynchus mykiss*). *Canadian Journal of Fisheries and Aquatic Sciences*, **61**, 942–953.
36. Wood, C.M. (2001) Toxic responses of the gill, in *Target organ toxicity in marine and freshwater teleosts* (eds D. Schlenk and W.H. Benson), Taylor & Francis, London, pp. 1–89.

37. Green, W.W., Mirza, R.S., Wood, C.M. and Pyle, G.G. (2010) Copper binding dynamics and olfactory impairment in fathead minnows (*Pimephales promelas*). *Environmental Science and Technology*, **44**, 1431–1437.
38. Finney, L.A. and O'Halloran, T.V. (2003) Transition metal speciation in the cell: insights from the chemistry of metal ion receptors. *Science*, **300**, 931–936.
39. Mason, A.Z. and Jenkins, K.D. (1995) Metal detoxification in aquatic organisms, in *Metal Speciation and Bioavailability in Aquatic Systems*. IUPAC Series on Analytical and Physical Chemistry of Environmental Systems (eds A. Tessier and D. Turner), J. Wiley & Sons, Chichester, UK, pp. 479–608.
40. Wallace, W.G., Lee, B.G. and Luoma, S.N. (2003) Subcellular compartmentalization of Cd and Zn in two bivalves. I. Significance of metal-sensitive fractions (MSF) and biologically detoxified metal (BDM). *Marine Ecology Progress Series*, **249**, 183–197.
41. Campbell, P.G.C. and Hare, L. (2009) Metal detoxification in freshwater animals. Roles of metallothioneins, in *Metallothioneins and Related Chelators. Metal Ions in Life Sciences* (eds A. Sigel, H. Sigel and R.K.O. Sigel), Royal Society of Chemistry, Cambridge, UK, pp. 239–277.
42. Campbell, P.G.C., Hontela, A., Rasmussen, J.B., Giguère, A., Gravel, A., Kraemer, L., Kovesces, J., Lacroix, A., Levesque, H. and Sherwood, G. (2003) Differentiating between direct (physiological) and food-chain mediated (bioenergetic) effects on fish in metal-impacted lakes. *Human and Ecological Risk Assessment*, **9**, 847–866.
43. Rasmussen, J.B., Gunn, J.M., Iles, A., Gagnon, A., Lacroix, A., Campbell, P.G.C. and Hontela, A. (2008) Direct and indirect (foodweb-mediated) effects of metal exposure on the growth of yellow perch (*Perca flavescens*): implications for ecological risk assessment. *Human and Ecological Risk Assessment*, **14**, 317–350.
44. Campbell, P.G.C., Kraemer, L.D., Giguère, A., Hare, L. and Hontela, A. (2008) Subcellular distribution of cadmium and nickel in chronically exposed wild fish: inferences regarding metal detoxification strategies and implications for setting water quality guidelines for dissolved metals. *Human and Ecological Risk Assessment*, **14**, 290–316.
45. Croteau, M.N., Hare, L. and Tessier, A. (1998) Refining and testing a trace metal biomonitor (*Chaoborus*) in highly acidic lakes. *Environmental Science and Technology*, **32**, 1348–1353.
46. Giguère, A., Campbell, P.G.C., Hare, L., McDonald, D.G. and Rasmussen, J.B. (2004) Influence of lake chemistry and fish age on cadmium, copper, and zinc concentrations in various organs of indigenous yellow perch (*Perca flavescens*). *Canadian Journal of Fisheries and Aquatic Sciences*, **61**, 1702–1716.
47. Giguère, A., Campbell, P.G.C., Hare, L. and Cossu-Leguille, C. (2005) Metal bioaccumulation and oxidative stress in yellow perch (*Perca flavescens*) collected from eight lakes along a metal contamination gradient (Cd, Cu, Zn, Ni). *Canadian Journal of Fisheries and Aquatic Sciences*, **62**, 563–577.
48. Kraemer, L.D., Campbell, P.G.C. and Hare, L. (2008) Modeling Cd accumulation in indigenous yellow perch (*Perca flavescens*). *Canadian Journal of Fisheries and Aquatic Sciences*, **65**, 1623–1634.
49. Giguère, A., Campbell, P.G.C., Hare, L. and Couture, P. (2006) Sub-cellular partitioning of cadmium, copper, nickel and zinc in indigenous yellow perch (*Perca flavescens*) sampled along a polymetallic gradient. *Aquatic Toxicology*, **77**, 178–189.
50. Rosabal, M., Hare, L. and Campbell, P.G.C. (2012) Subcellular metal partitioning in larvae of the insect *Chaoborus* collected along an environmental metal exposure gradient (Cd, Cu, Ni and Zn). *Aquatic Toxicology*, **120–121**, 67–78.
51. Bonneris, E., Perceval, O., Masson, S., Hare, L. and Campbell, P.G.C. (2005) Sub-cellular partitioning of Cd, Cu and Zn in tissues of indigenous unionid bivalves living along a metal exposure gradient and links to metal-induced effects. *Environmental Pollution*, **135**, 195–208.

52. Laflamme, J.S., Couillard, Y., Campbell, P.G.C. and Hontela, A. (2000) Interrenal metallothionein and cortisol secretion in relation to Cd, Cu, and Zn exposure in yellow perch, *Perca flavescens*, from Abitibi lakes. *Canadian Journal of Fisheries and Aquatic Sciences*, **57**, 1692–1700.
53. Levesque, H., Dorval, J., Van Der Kraak, G.J., Campbell, P.G.C. and Hontela, A. (2003) Hormonal, morphological and physiological responses of yellow perch (*Perca flavescens*) to chronic environmental metal exposures. *Journal of Toxicology and Environmental Health*, **66**, 657–676.
54. Gravel, A., Campbell, P.G.C. and Hontela, A. (2005) Disruption of the hypothalamo-pituitary-interrenal axis in 1+ yellow perch (*Perca flavescens*) chronically exposed to metals in the environment. *Canadian Journal of Fisheries and Aquatic Sciences*, **62**, 982–990.
55. Levesque, H.M., Moon, T.W., Campbell, P.G.C. and Hontela, A. (2002) Seasonal variation in carbohydrate and lipid metabolism of yellow perch (*Perca flavescens*) chronically exposed to metals in the field. *Aquatic Toxicology*, **60**, 257–267.
56. Rajotte, J.W. and Couture, P. (2002) Effects of environmental metal contamination on the condition, swimming performance, and tissue metabolic capacities of wild yellow perch (*Perca flavescens*). *Canadian Journal of Fisheries and Aquatic Sciences*, **59**, 1296–1304.
57. Nriagu, J.O. (1996) A history of global metal pollution. *Science*, **272**, 223–224.
58. Emsley, J. (2005) *The elements of murder*, Oxford University Press, Oxford.
59. Gibaud, S. and Jaouen, G. (2010) Arsenic-based drugs: from Fowler's solution to modern anticancer chemotherapy, in *Medicinal Organometallic Chemistry. Topics in Organometallic Chemistry* (eds G. Jaouen and N. Metzler-Nolte), Springer Verlag Berlin-Heidelberg, pp. 1–20.
60. Lucchini, R.G., Riva, M.A., Sironi, V.A. and Porro, A. (2012) Torvis oculis: Occupational roots of behavioral neurotoxicology in the last two centuries and beyond. *Neurotoxicology*, **33**, 652–659.
61. Troesken, W. (2008) *The great lead water pipe disaster*, MIT Press, Cambridge, MA.
62. Kurland, L.T., Faro, S.N. and Siedler, H. (1960) Minamata disease. The outbreak of a neurologic disorder in Minamata, Japan, and its relationship to the ingestion of seafood contaminated by mercury compounds. *World Neurology*, **1**, 370–391.
63. Bakir, F., Damluji, S.F., Amin-Zaki, L., Murtadha, M., Khalidi, A., Al-Rawi, N.Y., Tikriti, S., Dhahir, H.I., Clarkson, T.W., Smith, J.C. and Doherty, R.A. (1973) Methylmercury poisoning in Iraq. *Science*, **181**, 230–241.
64. Normile, D. (2013) In Minamata, mercury still divides. *Science*, **341**, 1446–1447.
65. Inaba, T., Kobayashi, E., Suwazono, Y., Uetani, M., Oishi, M., Nakagawa, H. and Nogawa, K. (2005) Estimation of cumulative cadmium intake causing itai-itai disease. *Toxicology Letters*, **159**, 192–201.
66. Smith, A.H., Lingas, E.O. and Rahman, M. (2000) Contamination of drinking-water by arsenic in Bangladesh: a public health emergency. *Bulletin of the World Health Organization*, **78**, 1093–1103.
67. Mohapatra, S.P., Nikolova, I. and Mitchell, A. (2007) Managing mercury in the Great Lakes: an analytical review of abatement policies. *Journal of Environmental Management*, **83**, 80–92.
68. World Health Organization (2011) *Guidelines for drinking-water quality*. Geneva.
69. Gailer, J. (2007) Arsenic-selenium and mercury-selenium bonds in biology. *Coordination Chemistry Reviews*, **251**, 234–254.
70. Omenn, G.S. (2006) Grand challenges and great opportunities in science, technology, and public policy. *Science*, **314**, 1696–1704.
71. Balbus, J.M., Barouki, R., Birnbaum, L.S., Etzel, R.A., Gluckman, S.P.D., Grandjean, P., Hancock, C., Hanson, M.A., Heindel, J.J., Hoffman, K., Jensen, G.K., Keeling, A., Neira, M., Rabadán-Diehl, C., Ralston, J. and Tang, K.-C. (2013) Early-life prevention of non-communicable diseases. *Lancet*, **381**, 3–4.

72. Ballatori, N. and Clarkson, T.W. (1982) Developmental changes in the biliary excretion of methylmercury and glutathione. *Science*, **216**, 61–63.
73. Drasch, G.A. (1983) An increase of cadmium body burden for this century - an investigation on human tissues. *Science of the Total Environment*, **26**, 111–119.
74. Finkelman, R.B., Belkin, H.E. and Zheng, B. (1999) Health impacts of domestic coal use in China. *Proceedings of the National Academy of Sciences of the United States of America*, **96**, 3427–3431.
75. Gailer, J. (2012) Probing the bioinorganic chemistry of toxic metals in the mammalian bloodstream to advance human health. *Journal of Inorganic Biochemistry*, **108**, 128–132.
76. Robinson, B.H. (2009) E-waste: An assessment of global production and environmental impacts. *Science of the Total Environment*, **408**, 183–191.
77. Kiddee, P., Naidu, R., Wong, M.H., Hearn, L. and Muller, J.F. (2014) Field investigations of the quality of fresh and aged leachates from selected landfills receiving e-waste in an arid climate. *Waste Management*, **34**, 2292–2304.
78. Tsydenova, O. and Bengtsson, M. (2011) Chemical hazards associated with treatment of waste electrical and electronic equipment. *Waste Management*, **31**, 45–58.
79. CSIRO Land and Water. (2008) *Australia cadmium minimisation strategy* [updated 7 February 2008]. http:/www.cadmium-management.org.au/about.html.
80. Coulthard, T.J. and Macklin, M.G. (2003) Modeling long-term contamination in river systems from historical metal mining. *Geology*, **31**, 451–454.
81. Wuana, R.A. and Okieimen, F.E. (2011) Heavy metals in contaminated soils: a review of sources, chemistry, risks and best available strategies for remediation *International Scholarly Research Network, ISRN Ecology*, **2011**, Article ID 402647.
82. Liu, Y., Wen, C. and Liu, X. (2013) China's food security soiled by contamination. *Science*, **339**, 1382–1383.
83. Zhuang, P., McBride, M.B., Xia, H., Li, N. and Li, Z. (2009) Health risk from heavy metals via consumption of food crops in the vicinity of Dabaoshan mine, South China. *Science of the Total Environment*, **407**, 1551–1561.
84. Skinner, M.K. (2014) A new kind of inheritance. *Scientific American*, **August**, pp. 45–51.
85. Mortvedt, J.J. (1996) Heavy metal contaminants in inorganic and organic fertilizers. *Fertilizer Research*, **43**, 55–61.
86. Holden, W.N., Jacobson, R.D. and Moran, K. (2007) Civil society opposition to nonferrous metals mining in Montana. *Voluntas*, **18**, 266–292.
87. Jarup, L. and Åkesson, A. (2009) Current status of cadmium as an environmental health problem. *Toxicology and Applied Pharmacology*, **238**, 201–208.
88. Izatt, R.M., Izatt, S.R., Bruening, R.L., Izatt, N.E. and Moyer, B.A. (2014) Challenges to achievement of metal sustainability in our high-tech society. *Chemical Society Reviews*, **43**, 2451–2475.
89. Filippelli, G.M., Morrison, D. and Cicchella, D. (2012) Urban geochemistry and human health. *Elements*, **8**, 439–444.
90. Attina, T.M. and Trasande, L. (2013) Economic costs of childhood lead exposure in low- and middle-income countries. *Environmental Health Perspectives*, **121**, 1097–1102.
91. Trasande, L. and Liu, Y. (2011) Reducing the staggering costs of environmental disease in children, estimated at $76.6 billion in 2008. *Health Affairs*, **30**, 863–870.
92. Flanagan, S.V., Johnston, R.B. and Zheng, Y. (2012) Arsenic in tube well water in Bangladesh: health and economic impacts and implications for arsenic mitigation. *Bulletin of the World Health Organization*, **90**, 839–846.
93. Centers of Disease Control and Prevention (2009) Fourth National Report on Human Exposure to Environmental Chemicals. Atlanta, GA, USA (http://www.cdc.gov/exposurereport/pdf/FourthReport.pdf).

94. Mead, M.N. (2010) Cadmium confusion. Do consumers need protection? *Environmental Health Perspectives*, **118**, A528–A534.
95. Liu, J. and Diamond, J. (2005) China's environment in a globalizing world. *Nature*, **435**, 1179–1186.
96. Poissant, L., Pilote, M., Beauvais, C., Constant, P. and Zhang, H.H. (2005) A year of continuous measurements of three atmospheric mercury species (GEM, RGM and Hg_p) in southern Quebec, Canada. *Atmospheric Environment*, **39**, 1275–1278.
97. Borchers, A., Teuber, S.S., Keen, C.L. and Gershwin, M.E. (2010) Food safety. *Clinical Reviews in Allergy and Immunology*, **39**, 95–141.
98. Taylor, A., Day, M.P., Hill, S., Marshall, J., Patriarca, M. and White, M. (2014) Atomic spectrometry update: Review of advances in the analysis of clinical and biological materials, foods and beverages. *Journal of Analytical Atomic Spectrometry*, **29**, 386–426.
99. Guney, M. and Zagury, G.J. (2013) Contamination by ten harmful elements in toys and children's jewelry bought on the North American market. *Environmental Science and Technology*, **47**, 5921–5930.
100. Liu, S., Hammond, S.K. and Rojas-Cheatham, A. (2013) Concentrations and potential health risks of metals in lip products. *Environmental Health Perspectives*, **121**, 705–710.
101. Koch, I., Moriarty, M., House, K., Sui, J., Cullen, W.R., Saper, R.B. and Reimer, K.J. (2011) Bioaccessibility of lead and arsenic in traditional Indian medicines. *Science of the Total Environment*, **409**, 4545–4552.
102. Hamann, C.R., Boonchai, W., Wen, L., Sakanashi, E.M., Chu, C.-Y., Hamann, K., Hamann, C.P., Sinniah, K., Hamann, D. and Linda, L. (2013) Spectrometric analysis of mercury content in 549 skin-lightening products: is mercury toxicity a hidden global health hazard? *Journal of the American Academy of Dermatology*, **70**, 281–287.
103. Donard, O.F.X. and Martin, F.M. (1992) Hyphenated techniques applied to environmental speciation studies. *Trends in Analytical Chemistry*, **11**, 17–26.
104. Killelea, D.R. and Aldstadt, J.H. (2002) Identification of dimethylchloroarsine near a former herbicide factory by headspace solid-phase microextraction gas chromatography-mass spectrometry. *Chemosphere*, **48**, 1003–1008.
105. Natusch, D.F.S., Wallace, J.R. and Evans, C.A.J. (1974) Toxic trace elements: preferential concentration in respirable particles *Science*, **183**, 202–204.
106. Harris, H.H., Pickering, I.J. and George, G.N. (2003) The chemical form of mercury in fish. *Science*, **301**, 1203.
107. McLaughlin, M.J., Parker, D.R. and Clarke, J.M. (1999) Metals and micronutrients and food safety issues. *Field Crops Research*, **60**, 143–163.
108. Schulz, C., Conrad, A., Becker, K., Kolossa-Gehring, M., Seiwert, M. and Seifert, B. (2007) Twenty years of German environmental survey (GerES): human biomonitoring - temporal and spatial (West Germany/East Germany) differences in population exposure. *International Journal of Hygiene and Environmental Health*, **210**, 271–297.
109. Jones, R.L., Homa, D.M., Meyer, P.A., Brody, D.J., Caldwell, K.L., Prirkle, J.L. and Brown, M.J. (2009) Trends in blood levels and blood lead testing among U.S. children aged 1-5 years, 1988-2004. *Pediatrics*, **123**, e376–e385.
110. Stokstad, E. (2004) Pollution gets personal. *Science*, **304**, 1892–1894.
111. Eybl, V., Sýkora, J. and Mertl, F. (1969) Einfluß von Natriumselenit, Natriumtellurit und Natriumsulfit auf Retention und Verteilung von Quecksilber bei Mäusen. *Archives of Toxicology*, **25**, 296–305.
112. Gailer, J., George, G.N., Pickering, I.J., Madden, S., Prince, R.C., Yu, E.Y., Denton, M.B., Younis, H.S. and Aposhian, H.V. (2000) Structural basis of the antagonism between inorganic mercury and selenium in mammals. *Chemical Research in Toxicology*, **13**, 1135–1142.
113. Il'yasova, D. and Schwartz, G.G. (2005) Cadmium and renal cancer. *Toxicology and Applied Pharmacology*, **207**, 179–186.

114. Kempson, I.M. and Lombi, E. (2011) Hair analysis as a biomonitor for toxicology, disease and health status. *Chemical Society Reviews*, **40**, 3915–3940.
115. Blanusa, M., Varnai, V.M., Piasek, M. and Kostial, K. (2005) Chelators as antidotes of metal toxicity: Therapeutic and experimental aspects. *Current Medicinal Chemistry*, **12**, 2771–2794.
116. Aposhian, H.V., Bruce, D.C., Alter, W., Dart, R.C., Hurlbut, K.M. and Aposhian, M.M. (1992) Urinary mercury after administration of 2,3-dimercaptopropane-1-sulfonic acid: correlation with dental amalgam score. *FASEB Journal*, **6**, 2472–2476.
117. Aposhian, H.V., Arroyo, A., Cebrian, M.E., Razo, L.M.D., Hurlbut, K.M., Dart, R.C., Gonzalez-Ramirez, D., Kreppel, H., Speisky, H., Smith, A., Gonsebatt, M.E., Ostrosky-Wegman, P. and Aposhian, M.M. (1997) DMPS-arsenic challenge test. I: Increased excretion of monomethylarsonic acid in humans given dimercaptopropane sulfonate. *Journal of Pharmacology and Experimental Therapeutics*, **282**, 192–200.
118. Bradberry, S. and Vale, A. (2009) Dimercaptosuccinic acid (succimer; DMSA) in inorganic lead poisoning. *Clinical Toxicology*, **47**, 617–631.
119. Crinnion, W.J. and Tran, J.Q. (2010) Case report: heavy metal burden presenting as Bartter Syndrome. *Alternative Medicine Review*, **15**, 303–310.
120. Brown, S.L. (1985) Quantitative risk assessment of environmental hazards. *Annual Review of Public Health*, **6**, 247–267.
121. Sunderman, F.W. (2001) Nasal toxicity, carcinogenicity, and olfactory uptake of metals. *Annals of Clinical and Laboratory Science*, **31**, 3–24.
122. Nylander, M., Friberg, L. and Lind, B. (1987) Mercury concentrations in the human brain and kidneys in relation to exposure from dental amalgam fillings. *Swedish Dental Journal*, **11**, 179–187.
123. Lim, S.-R., Kang, D., Ogunseitan, O.A. and Schoenung, J.M. (2013) Potential environmental impacts from the metals in incandescent, compact fluorescent lamp (CFL), and light-emitting diode (LED) bulbs. *Environmental Science and Technology*, **47**, 1040–1047.
124. Hursh, J.B., Clarkson, T.W., Cherian, M.G., Vostal, J.J. and Mallie, R.V. (1976) Clearance of mercury (Hg-197, Hg-203) vapor inhaled by human subjects. *Archives of Environmental Health*, **31**, 302–309.
125. Magos, L., Halbach, S. and Clarkson, T.W. (1978) Role of catalase in the oxidation of mercury vapor. *Biochemical Pharmacology*, **27**, 1373–1377.
126. Lee, R.E.J., Goranson, S.S., Enrione, R.E. and Morgan, G.B. (1972) National air surveillance cascade impactor network. II. Size distribution measurements of trace metal components. *Environmental Science and Technology*, **12**, 1025–1030.
127. Mielke, H.W. and Reagan, P.L. (1998) Soil is an important pathway of human lead exposure. *Environmental Health Perspectives*, **106**, 217–229.
128. Sanchez-Rodas, D., Sanchez de la Campa, A., Oliveira, V. and de la Rosa, J. (2012) Health implications of the distribution of arsenic species in airborne particulate matter. *Journal of Inorganic Biochemistry*, **108**, 112–114.
129. Liu, J., Zheng, B., Aposhian, H.V., Zhou, Y., Chen, M.-L., Zhang, A. and Waalkes, M.P. (2002) Chronic arsenic poisoning from burning high-arsenic-containing coal in Guizhou, China. *Environmental Health Perspectives*, **110**, 119–122.
130. Marcus, W.L. and Rispin, A.S. (1988) Threshold carcinogenicity using arsenic as an example, *in Advances in Modern Environmental Toxicology* (eds C.R. Cothern, M.A. Mehlmans and W.L. Marcus), pp. 133–158.
131. Mani, U., Prasad, A.K., Kumar, V.S., Lal, K., Kanojia, R.K., Chaudhari, B.P. and Murthy, R.C. (2007) Effect of fly ash on biochemical and histomorphological changes in rat liver. *Ecotoxicology and Environmental Safety*, **68**, 126–133.
132. Thévenod, F. and Lee, W.-K. (2013) Toxicology of cadmium and its damage to mammalian organs, in *Cadmium: from Toxicity to Essentiality*. 11 (eds A. Sigel, H. Sigel and R.K.O. Sigel), Springer Science + Business Media Dordrecht, New York, pp. 415–490.

133. Rabinowitz, M.B., Kopple, J.D. and Wetherill, G.W. (1980) Effect of food intake and fasting on gastrointestinal lead absorption in humans. *American Journal of Clinical Nutrition*, **33**, 1784–1788.
134. Mitchell, E., Frisbie, S. and Sarkar, B. (2011) Exposure to multiple metals from groundwater — a global crisis: geology, climate change, health effects, testing, and mitigation *Metallomics*, **3**, 874–908.
135. Thomas, H.F., Elwood, P.C., Welsby, E. and Leger, A.S.S. (1979) Relationship of blood lead in women and children to domestic water lead. *Nature*, **282**, 712–713.
136. Rahman, M.M., Chowdhury, U.K., Mukherjee, S.C., Mondal, B.K., Paul, K., Lodh, D., Biswas, B.K., Chanda, C.R., Basu, G.K., Saha, K.C., Roy, S., Das, R., Palit, S.K., Quamruzzaman, Q. and Chakraborti, D. (2001) Chronic arsenic toxicity in Bangladesh and West Bengal, India - A review and commentary. *Clinical Toxicology*, **39**, 683–700.
137. Lièvremont, D., Bertin, P.N. and Lett, M.-C. (2009) Arsenic in contaminated waters: biogeochemical cycle, microbial metabolism and biotreatment processes. *Biochimie*, **91**, 1229–1237.
138. Fontenot, B.E., Hunt, L.R., Hildenbrand, Z.L., Carlton Jr., D.D., Oka, H., Walton, J.L., Hopkins, D., Osorio, A., Bjorndal, B., Hu, Q.H. and Schug, K.A. (2013) An evaluation of water quality in private drinking water wells near natural gas extraction sites in the Barnett Shale Formation. *Environmental Science and Technology*, **47**, 10032–10040.
139. Ruhl, L., Vengosh, A., Dwyer, G.S., Hsu-Kim, H., Schwartz, G., Romanski, A. and Smith, S.D. (2012) The impact of coal combustion residues effluent on water resources: a North Carolina example. *Environmental Science and Technology*, **46**, 12226–12233.
140. Lin, Y., Larssen, T., Vogt, R.D. and Feng, X. (2010) Identification of fractions of mercury in water, soil and sediment from a typical Hg mining area in Wanshan, Guizhou provine, China. *Applied Geochemistry*, **25**, 60–68.
141. Williams, P.N., Price, A.H., Raab, A., Hossain, S.A., Feldmann, J. and Meharg, A.A. (2005) Variation in arsenic speciation and concentration in paddy rice related to dietary exposure. *Environmental Science and Technology*, **39**, 5531–5540.
142. Gilbert-Diamond, D., Cottingham, K.L., Gruber, J.F., Punshon, T., Sayarath, V., Gandolfi, J.A., Baker, E.R., Jackson, B.P., Folt, C.L. and Karagas, M.R. (2011) Rice consumption contributes to arsenic exposure in US women. *Proceedings of the National Academy of Sciences of the United States of America*, **108**, 20656–20660.
143. Malmauret, L., Parent-Massin, D., Hardy, J.-L. and Verger, P. (2002) Contaminants in organic and conventional foodstuffs in France. *Food Additives and Contaminants*, **19**, 524–532.
144. Counter, S.A. and Buchanan, L.H. (2004) Mercury exposure in children: a review. *Toxicology and Applied Pharmacology*, **198**, 209–230.
145. Horvat, M., Nolde, N., Fajon, V., Jereb, V., Logar, M., Lojen, S., Jacimovic, R., Falnoga, I., Liya, Q., Faganeli, J. and Drobne, D. (2003) Total mercury, methylmercury and selenium in mercury polluted areas in the province Guizhou, China. *Science of the Total Environment*, **304**, 231–256.
146. Dufault, R., LeBlanc, B., Schnoll, R., Cornett, C., Schweitzer, L., Wallinga, D., Hightower, J., Patrick, L. and Lukiw, W.J. (2009) Mercury from chlor-alkali plants: measured concentrations in food product sugar. *Environmental Health*, **8**, 1–6.
147. Grant, C., Flaten, D., Tenuta, M., Malhi, S. and Akinremi, W. (2013) The effect of rate and Cd concentration of repeated phosphate fertilizer applications on seed Cd concentration varies with crop type and environment. *Plant and Soil*, **372**, 221–233.
148. Abernathy, C.O., Thomas, D.J. and Calderon, R.L. (2003) Health effects and risk assessment of arsenic. *Journal of Nutrition*, **133**, 1536S–1538S.
149. US EPA (1997) *Mercury study report to Congress. Volume VI: An assessment of exposure to mercury in the United States.* Contract No.: EPA-452/R-97-006.
150. Śpiewak, R. (2001) Pesticides as a cause of occupational skin diseases in farmers. *Annals of Agricultural and Environmental Medicine*, **8**, 1–5.
151. Lansdown, A.B.G. (1995) Physiological and toxicological changes in the skin resulting from the action and interaction of metal ions. *Critical Reviews in Toxicology*, **25**, 397–462.

152. Ernst, E. (2000) Adverse effects of herbal drugs in dermatology. *British Journal of Dermatology*, **143**, 923–929.
153. Carter, D.E. (1995) Oxidation-reduction reactions of metal ions. *Environmental Health Perspectives*, **103** (Suppl. 1), 17–19.
154. Challenger, F. (1945) Biological methylation. *Chemical Reviews*, **36**, 315–361.
155. Shapiro, A.M. and Chan, H.M. (2008) Characterization of demethylation of methylmercury in cultured astrocytes. *Chemosphere*, **74**, 112–118.
156. Sasakura, C. and Suzuki, K.T. (1998) Biological interaction between transition metals (Ag, Cd and Hg), selenide/sulfide and selenoprotein P. *Journal of Inorganic Biochemistry*, **71**, 159–162.
157. Manley, S.A., George, G.N., Pickering, I.J., Glass, R.S., Prenner, E.J., Yamdagni, R., Wu, Q. and Gailer, J. (2006) The seleno bis (*S*-glutathionyl) arsinium ion is assembled in erythrocyte lysate. *Chemical Research in Toxicology*, **19**, 601–607.
158. Gailer, J. (2009) Chronic toxicity of As^{III} in mammals: the role of $(GS)_2AsSe^-$. *Biochimie*, **91**, 1268–1272.
159. Hoffmeyer, R.E., Singh, S.P., Doonan, C.J., Ross, A.R.S., Hughes, R.J., Pickering, I.J. and George, G.N. (2006) Molecular mimicry in mercury toxicology. *Chemical Research in Toxicology*, **19**, 753–759.
160. Leslie, E.M. (2012) Arsenic-glutathione conjugate transport by the human multidrug resistance proteins (MRPs/ABCCs). *Journal of Inorganic Biochemistry*, **2012**, 141–149.
161. McRae, R., Bagchi, P., Sumalekshmy, S. and Fahrni, C.J. (2009) In situ imaging of metals in cells and tissues *Chemical Reviews*, **109**, 4780–4827.
162. Korbas, M., O'Donoghue, J.L., Watson, G.E., Pickering, I.J., Singh, S.P., Myers, G.J., Clarkson, T.W. and George, G.N. (2010) The chemical form of mercury in human brain following poisoning or environmental exposure. *ACS Chemical Neuroscience*, **1**, 810–818.
163. Charlet, L., Chapron, Y., Faller, P., Kirsch, R., Stone, A.T. and Baveye, P.C. (2012) Neurodegenerative diseases and exposure to the environmental metals Mn, Pb, and Hg. *Coordination Chemistry Reviews*, **256**, 2147–2163.
164. Lahner, B., Gong, J., Mahmoudian, M., Smith, E.L., Abid, K.B., Rogers, E.E., Guerinot, M.L., Harper, J.F., Ward, J.M., McIntyre, L., Schroeder, J.I. and Salt, D.E. (2003) Genomic scale profiling of nutrient and trace elements in *Arabidopsis thaliana*. *Nature Biotechnology*, **21**, 1215–1221.
165. Markovac, J. and Goldstein, G.W. (1988) Picomolar concentrations of lead stimulate brain protein kinase C. *Nature*, **334**, 71–73.
166. Ames, B.N. (2001) DNA damage from micronutrient deficiencies is likely to be a major cause of cancer. *Mutation Research*, **475**, 7–20.
167. Morris, T.T., Keir, J.L.A., Boshart, S.J., Lobanov, V.P., Ruhland, A.M.A., Bahl, N. and Gailer, J. (2014) Mobilization of Cd from human serum albumin by small molecular weight thiols. *Journal of Chromatography B*, **958**, 16–21.
168. Benton, M.A., Rager, J.E., Smeester, L. and Fry, R.C. (2011) Comparative genomic analyses identify common molecular pathways modulated upon exposure to low doses of arsenic and cadmium. *BMC Genomics*, **12**, 173.
169. Sanders, A.P., Smeester, L., Rojas, D., DeBussycher, T., Wu, M.C., Wright, F.A., Zhou, Y.-H., Laine, J.E., Rager, J.E., Swamy, G.K., Ashley-Koch, A., Miranda, M.L. and Fry, R.C. (2014) Cadmium exposure and the epigenome. *Epigenetics*, **9**, 212–221.
170. Kim, H., Lee, H.J., Hwang, J.-Y., Ha, E.-H., Park, H., Ha, M., Kim, J.H., Hong, Y.-C. and Chang, N. (2010) Blood cadmium concentrations of male cigarette smokers are inversely associated with fruit consumption. *Journal of Nutrition*, **140**, 1133–1138.
171. Mandal, B.K. and Suzuki, K.T. (2002) Arsenic round the world: a review. *Talanta*, **58**, 201–235.
172. Gong, G. and Bryant, S.E. (2010) The arsenic exposure hypothesis for Alzheimer's disease. *Alzheimer Disease & Associated Disorders*, **24**, 311–316.
173. Waalkes, M.P. (2003) Cadmium carcinogenesis. *Mutation Research*, **533**, 107–120.

174. Peters, J.L., Perlstein, T.S., Perry, M.J., McNeely, E. and Weuve, J. (2010) Cadmium exposure in association with history of stroke and heart failure. *Environmental Research*, **110**, 199–206.
175. Kazantzis, G. (2004) Cadmium, osteoporosis and calcium metabolism. *Biometals*, **17**, 493–498.
176. Satarug, S., Garrett, S.H., Sens, M.A. and Sens, D.A. (2010) Cadmium, environmental exposure, and health outcomes. *Environmental Health Perspectives*, **118**, 182–190.
177. Tchounwou, P.B., Ayensu, W.K., Ninashvili, N. and Sutton, D. (2003) Environmental exposure to mercury and its toxicopathologic implications for public health. *Environmental Toxicology*, **18**, 149–175.
178. Gorell, J.M., Johnson, C.C., Rybicki, B.A., Peterson, E.L., Kortsha, G.X., Brown, G.G. and Richardson, R.J. (1997) Occupational exposures to metals as risk factors for Parkinson's disease. *Neurology*, **48**, 650–658.
179. Seidler, A., Hellenbrandt, W., Robra, B.-P., Vieregge, P., Nischan, P., Joerg, J., Oertel, W.H., Ulm, G. and Schneider, E. (1996) Possible environmental, occupational, and other etiologic factors for Parkinson's disease: a case-control study in Germany. *Neurology*, **46**, 1275–1284.
180. Stachiotti, A., Borsani, E., Rodella, L., Rezzani, R. and Bianchi, R. (2003) Dose-dependent mercuric chloride tubular injury in rat kidney. *Ultrastructural Pathology*, **27**, 253–259.
181. Mutter, J., Curth, A., Naumann, J., Deth, R. and Walach, H. (2010) Does inorganic mercury play a role in Alzheimer's disease? A systematic review and an integrated molecular mechanism. *Journal of Alzheimer's Disease*, **22**, 357–374.
182. Clarkson, T.W. (1997) The toxicology of mercury. *Critical Reviews in Clinical Laboratory Sciences*, **34**, 369–403.
183. Chang, L.W. (1996) Toxico-neurology and neuropathology induced by metals in Toxicology of Metals (eds L.W. Chang, L. Magos and T. Suzuki), CRC Press, New York, pp. 511–535.
184. Bernard, S., Enayati, A., Roger, H., Binstock, T. and Redwood, L. (2002) The role of mercury in the pathogenesis of autism. *Molecular Psychiatry*, **7**, S42–S43.
185. Coon, S., Stark, A., Peterson, E., Gloi, A., Kortsha, G., Pounds, J., Chettle, D. and Gorell, J. (2006) Whole-body lifetime occupational lead exposure and risk of Parkinson's disease. *Environmental Health Perspectives*, **114**, 1872–1876.
186. Edwards, J.R. and Prozialeck, W.C. (2009) Cadmium, diabetes and chronic kidney disease. *Toxicology and Applied Pharmacology*, **238**, 289–293.
187. Foley, J. (2014) Feeding 9 billion. *National Geographic*, **May**, pp. 27–57.
188. Taylor, M.P., Davies, P.J., Kristensen, L.J. and Csavina, J.L. (2014) Licenced to pollute but not to poison: the ineffectiveness of regulatory authorities at protecting public health from atmospheric arsenic, lead and other contaminants resulting from mining and smelting operations. *Aeolian Research*, **14**, 35–52.
189. Wu, S., Powers, S., Zhu, W. and Hannun, Y.A. (2016) Substantial contribution of extrinsic risk factors to cancer development. *Nature*, **529**, 43–47.
190. Grandjean, P. (2014) Prenatal prevention. *Science*, **345**, 1462.

11

How Bacteria are Affected by Toxic Metal Release

Mathew L. Frankel, Sean C. Booth, and Raymond J. Turner

Department of Biological Sciences, University of Calgary, Calgary, Alberta, Canada

11.1 Introduction to Bacteria in the Environment

We live in a world teeming with life. Baas Becking's (1934) microbiological tenet, '*everything is everywhere*, but, *the environment selects*,' speaks to the idea that microbial life is ubiquitous in the world we live in and the particular conditions of a given setting determine what you can find. When Charles Darwin developed his theory of natural selection, diversity in macro-organisms (i.e. plants and animals) was attributed to speciation caused by geographic separation (O'Malley, 2008). For bacteria, however, geographic location has a minimal effect (O'Malley, 2008); instead, local physicochemical conditions (e.g. pH, salinity, temperature) determine bacterial biogeographic distribution (Cardozo, 2015; Chu *et al.*, 2010). In terrestrial environments, studies have shown that soil pH has the greatest effect on soil bacteria distributions from a range of ecosystems across the Arctic, North America, and South America (Fierer & Jackson, 2006; Lauber *et al.*, 2009; Chu *et al.*, 2010). Terrestrial, freshwater, and marine biomes are linked via the hydrological cycle among other modes of dispersal (i.e. wind, dust (Pearce *et al.*, 2009; Favet *et al.*, 2013)). A meta-analysis of 68 surface water studies' genetic data (Newton *et al.*, 2011) found that despite this interconnectivity of biomes, which results in an overlap of bacterial niches (freshwater, terrestrial, marine, and human biomes), local environmental chemistries (namely salinity) still determine which organisms will be present.

Along with surface soil and water, bacteria have also been found in subsurface environments (up to 1.6 km down), including hot (~80°C) subsurface oil reservoirs

Metal Sustainability: Global Challenges, Consequences, and Prospects, First Edition. Edited by Reed M. Izatt.

(L'Haridon *et al.*, 1995). Reservoir bacteria can be beneficial as well as harmful to oil production, depending on how they utilize available substrates and electron acceptors (Cai *et al.*, 2015). Reservoir bacteria have been the focus of biotechnological applications such as microbially enhanced oil recovery (MEOR), which utilizes the metabolic byproducts produced by microorganisms in the subsurface environment to enhance oil production (Le *et al.*, 2015; Sen, 2008). They can also be a detriment to oil production by causing oilfield souring (Gieg *et al.*, 2011) and microbially influenced corrosion of pipelines (Enning & Garrelfs, 2014).

Just as bacteria are found on the surface and depths of the Earth, they are similarly found on the skin and in the depths of the human body, described as the *human microbiome*. When most people consider interactions between bacteria and people, they often associate the relationship with disease. Buzzwords like 'germs' and their associated danger has led to household antibacterial products being marketed as offering health protection (Levy, 2001). This has led to the negative connotation associated with bacteria. In reality, typical healthy adults play host to thousands of species of bacteria with a diversity that varies considerably between individuals (Shreiner *et al.*, 2015). This human microbiome includes ~750,000 prokaryote genes or about 30 times the number of genes in the human genome (Li *et al.*, 2014). In the upper portion of the digestive tract (esophagus, stomach, and small intestine) bacteria are treated as harmful parasites or pathogens, while in the lower digestive tract bacteria are seen as allies in a commensal relationship (Brüssow, 2014). The relationship between humans and their bacteria is a reciprocal one in which the microbiome affects the individual, but the individual can also influence alterations to the microbiota via changes in diet or bowel function, or through the use of medications and antibiotics (Shreiner *et al.*, 2015). These human microcosm bacteria play a major role in human health by influencing the immune and digestive systems, protecting the body from pathogens, and contributing to basic metabolic functions (Shreiner *et al.*, 2015). Diseases in and outside of the gut have been associated with these dynamics, where dysbiosis of the human biome has been attributed to colorectal cancer (Kostic *et al.*, 2013), rheumatoid arthritis (Scher *et al.*, 2013), obesity (Le Chatelier *et al.*, 2013) and diabetes (Qin *et al.*, 2012).

Biodiversity is necessary for the maintenance of an ecosystem (Hector & Bagchi, 2007). Just as this is true for our own human microbiome, this commensalism is evident in agricultural systems where soil bacteria are responsible for acquiring nutrients for plants (van der Heijden *et al.*, 2008; Wagg *et al.*, 2014). Mycorrhizal fungi and nitrogen fixing bacteria are responsible for 5–20% (grassland and savannah) to 80% (temperate and boreal forests) of all nitrogen storage, and up to 75% of phosphorus acquired by plants annually (van der Heijden *et al.*, 2008). It has also been found that loss in soil biodiversity will result in impaired plant diversity, decomposition, nutrient retention, and nutrient cycling (Wagg *et al.*, 2014).

Just as Becking's tenet '*everything is everywhere*, but, *the environment selects*' is applied to microbiology, the same can be said about free metal ions. For metals, the activity of hydrogen ions (pH) as well as the activity of electrons (Eh, or redox condition) determine in what form you find metals in a given environment. The ubiquitous nature of metals and bacteria in the environment has inevitably led to their interacting and relying on each other to some degree. In this chapter we delve into these interactions, with a focus on how bacteria interact with metals in excess, which is very much a reality in some natural, but more often anthropogenic and industrial, environments.

11.2 Bacterial Interactions with Metals

11.2.1 Essential Metals

From a biological perspective, many metals are essential since they participate in biochemistries not attainable by organic molecules alone (Table 1 gives a brief summary). For the purposes of this discussion, we focus on alkaline earth metals, transition (d-block) metals, post-transition metals, and metalloids that are needed in trace amounts (generally μg–mg L^{-1} concentrations), and touch on bulk metals that are more prevalent on Earth and have therefore been incorporated in biological systems to a greater degree; see Figure 1.

Metals are commonly found as cofactors of proteins. The metal-loaded forms are referred to as *metalloproteins*; those that perform a catalytic function are called *metalloenzymes* (Lippard & Berg, 1994; da Silva & Williams, 2001; Waldron & Robinson, 2009; Reyes-Caballero *et al.*, 2011). The presence of metals in these metalloproteins and enzymes lead

Table 1 *Functions of essential metals in biological systems. General functions of the most common elements used throughout all domains of life. This is not an exhaustive list. Refer to da Silva and Williams (2001) for the uses of metals in proteins and biological systems*

Metal	General Role(s)	Specific Enzymatic Function(s)
Na	Charge carrier; osmotic balance	
K	Charge carrier; osmotic balance	
Mg	For chlorophyll in photosynthetic cyanobacteria; structure	Hydrolase; isomerase; DNA polymerase; ATPase; kinase
Ca	Structure; signalling; allosteric regulation; charge carrier	
V	Nitrogen fixation; free as V(III) or combined with low MW protein; bonded to N,O-ligands; Fe, V, S cluster; fixation of dinitrogen	Oxidase; peroxidase and catalase activity; cofactor of nitrogenase
Mo	Nitrogen fixation; oxygen atom transfer	Oxidase; nitrogenase; reductase
W	Redox enzymes	Dehydrogenase; ferredoxin-oxidoreductase
Mn	Photosynthesis; redox enzymes; counter ion to nucleotides	Oxidase; structure; catechol cleavage; superoxide dismutase
Fe	Dioxygen transport, metabolism, and storage; electron transfer; nitrogen fixation	Mono- and dioxygenasese; hydroxylase; superoxide dismutase; nitrogenase
Co	Alkyl group transfer; vitamin B12 in enzymes	Oxidase
Ni	Redox enzymes	Hydrogenase; dehydrogenase; hydrolase; urease; superoxide dismutase
Cu	Dioxygen transport; electron transfer; several periplasmic enzymes; various oxidation and reduction reactions, extracellularly, within the cytosol, and membrane	Oxidase; reductase; superoxide dismutase
Zn	Structure; catalysis; associated with ribosomes	Hydroxylase; DNA and RNA polymerase; DNA primase; superoxide dismutase
Se	Enzymes associated with hydrogen i.e. low potential metabolism	Reductases

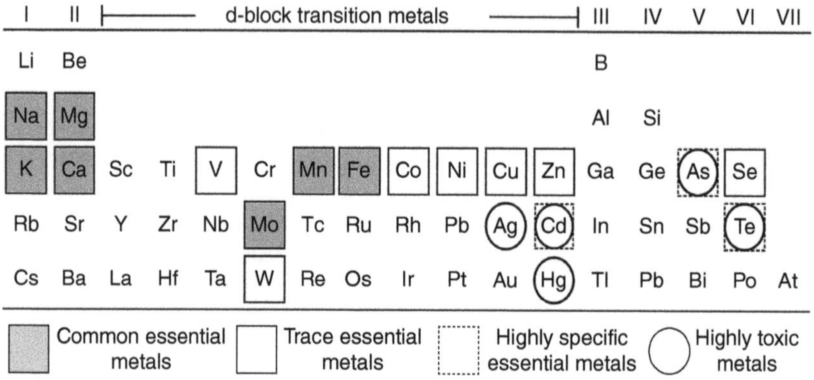

Figure 1 *Bacterial interactions with the metals of the periodic table. Bacterial interactions with metals are dependent on both the specific metal and its oxidative state. Some are needed at comparatively high concentrations (i.e. mg L^{-1}, grey boxes), others are needed in lower concentrations (i.e. μg–mg L^{-1}, white boxes); a few are used very rarely for specific purposes in particular bacteria (dashed outline boxes). Additionally, some metal elements are especially toxic to bacteria, even at very low concentrations (≤ μg L^{-1}, white circles)*

to their specialized properties, which have been exploited by nature to perform a range of roles in biological processes. This includes acting as essential cofactors in oxidation-reduction reactions, structural rearrangements of organic molecules, transfer of electrons, hydrolytic and acid base chemistry, and as structural centers that simply stabilize protein folding (Lippard & Berg, 1994; Reyes-Caballero *et al.*, 2011). Metal ions and their intrinsic properties play critical roles in the cell that cannot be filled by any other common atoms in nature; one-third of structurally characterized proteins have been estimated to require metals (Finney & O'Halloran, 2003; Waldron & Robinson, 2009), further highlighting their essential role for life.

11.2.2 Non-essential Metals

Nonessential metals (e.g. Hg) have no known physiological role and some are toxic at exceptionally low concentrations (≤ μg L^{-1}) (Ishaque *et al.*, 2006; Lemire *et al.*, 2013). All metals are toxic in excessive concentrations, including essential metals, which can be of consequence in environments where metals persist at elevated levels (e.g. due to anthropogenic or geological activity). Total prokaryotic cell numbers have been negatively correlated with metals such as Cd, Cu, Zn, and Pd in marine environments (Nogales *et al.*, 2011). In the environment metals can occur naturally through specific geological processes or as a result of anthropogenic activities.

Metals are omnipresent throughout the globe, but are naturally found at elevated concentrations in few environments. One such environment is in deep-sea hydrothermal vents (Jannasch & Mottl, 1985; Andreson *et al.*, 2015). In the absence of photosynthetic organisms, primary producing bacteria in this environment obtain energy by chemosynthesis, where CO_2 is assimilated with the use of reduced inorganic compounds as energy sources (Jannasch & Mottl, 1985). In the hot subsurface (~350°C), metals (i.e. Fe(II), Mn(II)) are leached from underlying basalts and deposited as minerals on the surface (Jannasch &

Mottl, 1985). These deposits are intermittently exposed to plumes of hydrothermal and bottom sea water (or mixes of the two) and covered by bacterial mats (Jannasch & Mottl, 1985). The presence of these bacteria has been attributed, to some degree, to their oxidation of available inorganics, especially Fe(III) and Mn(IV) (Jannasch & Mottl, 1985). These extreme environments have been shown to support diverse microbial life (Andreson *et al.*, 2015) in the presence of elevated metals, since they have persisted over geologic and evolutionary time. The adaptation to the high concentrations of metals in these environments has allowed these microbes to evolve unique metalloenzymes with diverse metabolic capabilities (Saito *et al.*, 2003).

Elevated presence of metals can be seen in areas where industrialization and urbanization are widespread (Wei & Yang, 2010). Studies of lake sediments linked the presence of elevated metal concentrations to the start of industrial activities as early as the late 1800s (Schneider *et al.*, 2014), and human activities have even been linked to the presence of trace metals in areas traditionally considered pristine, like the Arctic and Antarctica (Sokolova & Lannig, 2008).

A distinction must be made in the character of metal pollution into an environment, whether it is present from point pollution (i.e. discharge from industries like mining, refining, smelting, and electronic waste in landfills) or diffuse sources (i.e. harbor activity, atmospheric deposition, runoff from land-based activities), which can result in varying responses from the bacteria present in these environments. The ultimate effects on bacteria can depend on type (point versus diffuse), environment (soil versus water), and/or time scale (chronic versus acute) of the exposure. Computational approaches to understanding bacterial diversity in metal-laden soils have demonstrated that the biodiversity of bacterial communities can be reduced as much as 99.9% in metal-polluted soils compared to pristine soils (Gans *et al.*, 2005). Conversely, studies of marine environments have suggested that while the biodiversity of microbial communities is initially affected by metal pollution, over longer time scales of metal contamination (i.e. >80 years), the diversity of these communities can return to that of uncontaminated areas (Gillan *et al.*, 2005). This could be a reflection of microbial dispersion and the inherent diversity in different environments, as well as an indication that certain microbial phyla might be favored in heavy-metal-polluted sites (Nogales *et al.*, 2011) and therefore can proliferate and dominate the population over time.

11.3 Bacterial Response to Toxic Metals

11.3.1 What Are the Toxicity Levels of Metals to Bacteria?

The U.S. Environmental Protection Agency (EPA) (U.S.EPA Epa/542/R-04/015, 2004) has estimated that 294,000 sites need to be cleaned up in the U.S., costing a total of $209 billion, of which 1,244 have been prioritized on the National Priorities List (NPL or Superfund sites). Most sites have similar contaminants: solvents and other organics, petroleum products and metals. Many sites are former mines for metals, nonmetal minerals, and hydrocarbon extraction (coal, bitumen, natural gas) that continue to threaten human health and the environment as a result of the concentrated materials left behind in the soil and water. Of the twelve contaminants most frequently found in polluted sites, more than half are metals (primarily As, Pb, Zn, Ni, Cd) and have been reported to be the most prevalent contaminants of concern (77%) in NPL sites. Metals prioritized by U.S., Canadian, and European environmental agencies can be found on Table 2, with bacterial minimum inhibitory concentrations (MICs) for the known metal-tolerant bacterium, *Cupriavidus metallidurans* (previously *Ralstonia* sp., *Alcaligenes eutrophus* or *R. eutrophus*) (Monsieurs *et al.*, 2011), as well as the historically utilized model organism *Escherichia coli* (Nies, 1999). *C. metallidurans* is known to possess diverse metal resistance mechanisms that are reviewed in the following sections. *E. coli* data were also included to illustrate metal concentrations inhibitory to bacterial growth for the most well understood and characterized bacterium that is not considered particularly resistant to metals (Nies, 1999).

Bacterial metal-mediating interactions can be seen in one of the oldest existing groups of bacteria, cyanobacteria. Cyanobacteria have been found to be effective at accumulating metals, affecting metal speciation and therefore bioavailability and representing a biological sink for metals in aquatic environments (Baptista & Vasconcelos, 2006). Their ability to utilize metals is likely a product of this ancient organism's interactions with metals since the origins of life. The emergence of cyanobacteria has been attributed to changes in the Earth's ocean-atmospheric system, near the Archaean-Proterozoic boundary (~2.4 bya), from a reducing to oxidative environment via photosynthetic production of O_2 (Johnston *et al.*, 2009; Reinhard *et al.*, 2013). Prior to this, early Earth was characterized by extraterrestrial impacts and volcanic activity that affected the appearance, radiation, episodic extinction, and reemergence of biogenic habitats (Glikson, 2006), and probably increased available metals on early Earth. At this time, Earth's ancient oceans were sulfidic, and due to metal solubility in the presence of sulfide, some metals were more abundant (Fe, Ni, Co) and others were scarcer (Cu, Zn); some have suggested the evolution of metalloenzymes was a response to metals present (availability) and the functions they fulfilled (Saito *et al.*, 2003). As cyanobacteria oxygenated the early Earth, the changing redox state removed the available iron in the ancient oceans as the iron was oxidized and reacted with available sulfides (Saito *et al.*, 2003). Overall, the utilization of metals in enzymes is the result of their availability in the environment in which the bacteria evolved (Saito *et al.*, 2003).

11.3.2 Resistance Mechanisms of Bacteria to Metals

As bacteria have had to interact with metals throughout evolutionary history, genetic adaptations have accrued and have also been exchanged laterally between species. Bacterial resistance was first found on mobile genetic elements called plasmids (Nies, 2003). Genomic resistances have since been reported for the metals Cr(VI), Co(II), Ni(II), Cu(II), Zn(II), As(III,V), Ag(I), Cd(II), Sb(III), Te(IV), Hg(II), Tl(I) (as reviewed by Silver & Phung (2005)). These genetic determinants provide resistance towards toxic metals through diverse mechanisms. This diversity in strategies is likely reflective of the different physicochemical properties of metals, as no single strategy has been found to confer resistance to all toxic metals (Lemire *et al.*, 2013). Characterized resistance strategies include: altering a metal's chemical speciation state to less toxic or bioavailable forms, alteration of membrane protein activity to reduce metal uptake, active export of metals from cells (efflux), bypassing metabolic pathways that have been disrupted, intra- or extracellular sequestration of metals by proteins/polymers/chelating molecules, and repair of functional groups that have been damaged by redox reactions with metals (Bruins *et al.*, 2000; Lemire *et al.*, 2013) (summarized in Table 3). Among these mechanisms, the active export of the metal

Table 2 *Acceptable standards for metal contaminants in soil, water, and sediment and minimum inhibitory concentrations (MIC) for C. metallidurans and E. coli*

Metal	EPA water (μg L^{-1})	EU water (μg L^{-1})	CEQG water (μg L^{-1})	CEQG sediment (mg kg^{-1})	CEQG soil (mg kg^{-1} dry weight)	*C. metallidurans* MIC (mg L^{-1})	*E. coli* MIC (mg L^{-1})
Arsenic	10		5 – 12.5	5.9 – 41.6	12		
Lead	15	14		30.2 – 112	70 – 600	207.2	1036
Mercury	2	0.07	0.016 – 0.026	0.13 – 0.70	6.6 – 50	0.54	2
Cadmium	5	≤0.45 – 1.5	0.09 – 0.12	0.6 – 4.2	1.4 – 22	449.6	56.2
Cobalt					40 – 300	1473.3	58.9
Nickel		34			50	763	58.7
Chromium (VI)	100		1 – 1.5		0.4 – 1.4	20.8	10.4
Silver			0.1		20 – 40	0.05	2.2
Zinc			30	123 – 271	200 – 360	784.6	65.4
Copper	1300			18.7 – 197	63 – 91	190.6	63.5
Barium	2000				500 – 2000		

Presented here are the maximum allowable concentrations of metals in various environments from American, Canadian and European government regulatory bodies. MIC values from an 'average' (*E. coli*) and resistant (*C. metallidurans*) bacterium are presented for comparison. EPA water: U.S. Environmental Protection Agency (EPA) metals from the top 135 priority list of hazardous substances (EPA, 2013), and maximum contaminant levels for drinking water obtained from (California Department of Public Health, 2008); EU water: European Union (EU) environmental quality standards for maximum allowable concentration of each metal for inland and other surface waters from the top 45 pollutants, reported in (Official Journal of the European Union, 2013); CEQG water/sediment/soil: Canadian Environmental Quality Guidelines (CEQG) for the protection of aquatic life, environmental and human health (Canadian Council of Ministers of the Enviornment, 2014) from the top 100 toxic substance listed in the Canadian Environmental Protection Act (CEPA) (Minister of Justice, 1999); values are the maximum allowable concentration or range for the metal in the selected environment (water, soil, or sediment). *C. metallidurans* MIC: MIC of each metal to inhibit planktonic growth of the known metal-resistant bacterium *C. metallidurans* strain CH34, as reported in Monsieurs et al. (2011); *E. coli* MIC: MIC of each metal to inhibit planktonic growth of *E. coli* strain K38, which is not considered to be highly metal tolerant or sensitive, as reported by Nies (1999). Cells lacking values indicate that no data were reported.

Table 3 *Overview of microbial metal resistance mechanisms*

Category	Description	Metal	Source/Review(s)
Chemical modification	Intra- and/or extracellular enzymes can change redox state of metals, converting the toxic metal species to a less toxic or bioavailable form. These can form a crystal precipitate, become chelated to organic groups, or alter the metal to a redox neutral form.	Hg(II), Te(IV), As(V)	Silver & Phung (2005); Bruins *et al.* (2000)
Reduced uptake and efflux	Mobile plasmids and chromosomally-encoded protein families mediate metal transport in and out of the cell via membrane transporters that can bind metals with sensitivities to detect and regulate at as low as zeptomolar affinities. These can be driven by proton gradients, ATP hydrolysis, or chemiosmotic potentials. Active efflux systems are the largest reported category of metal resistance in microorganisms.	Mn(II), Zn(II), Ni(II), Fe(II), Ca(II), K(I), Cd(II), Ag(I), Co(II), Pb(II), Cr(VI), Cu(I,II), As(III)	Silver & Phung (2005); Bruins *et al.* (2000); Nies (2003); Nies (1999)
Metabolic bypass	Alternate metabolic pathways can be taken to meet biological functions via proteins whose catalytic cores do not bind with the metal toxicant. This can also be achieved by upregulating metabolites/enzymes of alternative pathways to avoid sensitive components, or bypassing the disrupted pathway altogether.	Al(III), Cd(II)	Lemire *et al.* (2010); Bruins *et al.* (2000)
Sequestration	Siderophores (high affinity Fe(III) chelating compounds) and extracellular polymers that have functional groups that bind with metal ions are upregulated and expressed on the cell surface. By chemical modification, these can precipitate metal ions on the cell surface, as well as through interactions with proteins or cell associated polysaccharides. Some bacteria also precipitate metals intracellularly, as metal-oxides, sulfides, and protein aggregates and/or as elemental crystals. These precipitates are found associated with the cytoplasmic membrane, and hence, these metals tend to bioaccumulate in the cytoplasm or periplasmic space.	Cu(II), Cd(II), Fe(II), Zn(II), Co(II), Ni(II), Cd(II), Cr(III), Cr(VI), Ag(I)	Aryal et al. (2015); Hullebusch *et al.* (2004); Bruins *et al.* (2000); Veglio' (1997)
Altering metal sensitivity of cellular targets and repair	Altering metal sensitivity can be achieved by increasing production of essential cellular components to stay ahead of metal arrest. Mechanisms of DNA repair provide some protection to genomic DNA and plasmids. Cellular glutathione has been suggested to provide protection from nonessential metals, as well as suppressing free radical accumulation.	Ag(I), Cu(I,II), Cd(II), Hg(II), Fe(II)	Harrison *et al.* (2009); Bruins *et al.* (2000)

ion out of the cell via efflux pumps appears to be the most common. Heavy metal resistance in prokaryotes has been found to be the result of multiple overlapping layers of redundancy in resistance systems and substrate specificities (Nies, 2003). When the efflux systems of 51 genome-sequenced bacteria were compared to that of the known metal-resistant bacterium *C. metallidurans* (Nies, 2003), it was found that some resistance systems are more general and work as a basic defense system with overall changes in cell physiology and biochemistry, while other less common defenses are more specialized to specific metals and biochemical processes. It is these specially adapted systems that lead to novel, greatly increased resistance that allow a bacterium to be considered 'metal-resistant.'

11.4 How Are Metals Toxic to Bacteria?

There is surprisingly little known about the specific mechanisms of toxicity for each metal and how they are bacteriostatic or bactericidal. Unfortunately, a lot of assumptions are made in the literature regarding the mechanisms of metal toxicity, yet only a few direct experimental cases have been shown. Here we present the most frequent mechanisms of toxicity.

11.4.1 Reactive Oxygen Species

One of the main mechanisms by which metals have been found to be detrimental to bacteria is by causing an increased production of reactive oxygen species (ROS). Whilst mainly produced by Fe(II), other metals have been shown to produce ROS, including Cu(II), As(III), Cr(VI), Al(III) and Te(IV) (Lemire *et al.*, 2010, 2013). Reactive by-products of aerobic respiration (i.e. hydrogen peroxide (H_2O_2) and superoxide ($O_2^{\bullet-}$)) react with cellular Fe in a process that involves the Fenton (1) and Haber-Weiss (2) reactions, thus producing highly reactive hydroxyl radicals ($OH^{\bullet}$) (Valko *et al.*, 2005):

$$\mathrm{Fe(II)} + \mathrm{H_2O_2} \rightarrow \mathrm{Fe(III)} + \mathrm{OH^-} + \mathrm{OH^{\bullet}} \qquad \text{(1) Fenton equation}$$

$$\mathrm{Fe(III)} + \mathrm{O_2^{\bullet-}} \rightarrow \mathrm{Fe(II)} + \mathrm{O_2}$$

$$\mathrm{O_2^{\bullet-}} + \mathrm{H_2O_2} \rightarrow \mathrm{O_2} + \mathrm{OH^-} \qquad \text{(2) Haber-Weiss (net) reaction}$$

In the net reaction, Fe(II) is regenerated, allowing further production of hydroxyl radicals. The instability of the unpaired electron results in $OH^{\bullet}$ quickly reacting with any available biochemical moiety, including proteins, lipids, and DNA, as well as producing more radicals, resulting in a cascading chain of oxidation (Valko *et al.*, 2005); see Figure 2.

11.4.1.1 Disruptive Reactions of ROS.

While Fe is an essential element, most cellular Fe is maintained in some bound form due to its ability to catalyze such damaging reactions, and its influx/efflux is carefully modulated to prevent excess amounts of free Fe being available in the cell (Andrews *et al.*, 2003). Since other metals can catalyze Fenton-like chemistry and ROS production as well as

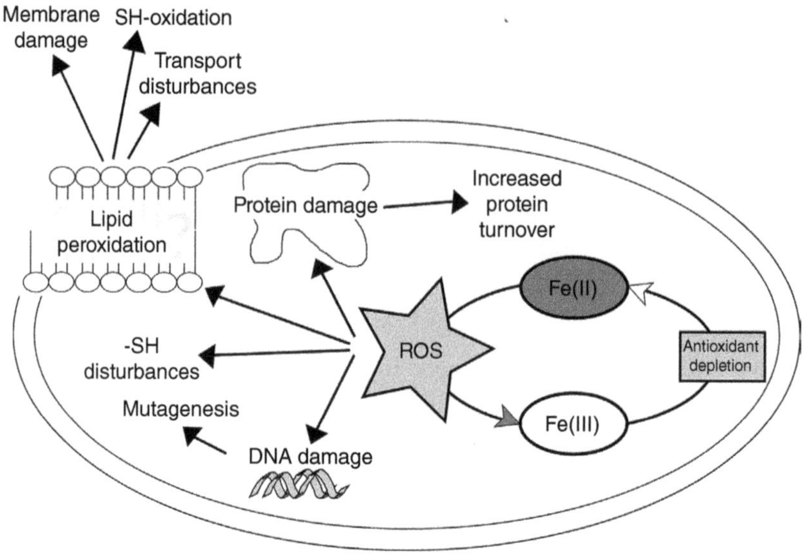

Figure 2 *Disruptive reactions of ROS*

mediate the release of bound Fe(II) (Strlič *et al.*, 2003; Valko *et al.*, 2005), thereby inducing Fenton reactions, it can be difficult to discern the specific modes of toxicity of individual metals. ROS protective mechanisms include producing antioxidants and enzymes that metabolize potentially damaging products of free radical reactions, which if overwhelmed can result in cellular injury (Cabiscol *et al.*, 2000).

11.4.2 Thiol Chemistry

Oxidative stress can be amplified by the depletion of compounds that normally maintain the reduced environment of the cytoplasm. The cellular redox climate is largely determined by the ratio of reduced (GSH) to oxidized (GSSG) glutathione (Filomeni *et al.*, 2002). A cellular environment favoring GSSG too heavily will drain electrons from the electron transport chain to maintain a reduced condition and, with that, destroy the proton motive force of the cell membrane (Harrison *et al.*, 2007), causing cell death. Some metal cations, as well as metal oxyanions, have electron-sharing affinities that favor thiols (S containing groups) (Lemire *et al.*, 2013). This can lead to the formation of protein disulfides (i.e. RSSR) and metal trisulfides (RS-M-SR) or higher-order thiol metal complexes (Harrison *et al.*, 2007), thus depleting cellular antioxidant reserves (particularly GSH); this has been attributed to the toxicity of metals such as Ag(I), Te(IV), Cd(II), Cu(II), Cr(VI), Co(II), Zn(II), and As(III) (Helbig *et al.*, 2008; Lemire *et al.*, 2013). GSH is also a cofactor for antioxidant enzymes (e.g. GSH peroxidase) (Filomeni *et al.*, 2002); it has also been demonstrated to provide residual metal resistance to *E. coli* cells lacking metal efflux systems (Helbig *et al.*, 2008), further underscoring the role of reduced cellular thiols in mitigating metal toxicity.

11.4.3 Replacement of Co-factor Metals in Metalloproteins

Metals can interact with particular prosthetic groups, exerting their toxic action in a semi-specific manner towards proteins that contain the specific cofactor. Exposed iron-sulfur clusters, particularly [4Fe-4S] clusters of dehydratases and respiratory enzymes, are susceptible to specific metals (Calderón *et al.*, 2009; Macomber & Imlay, 2009; Xu & Imlay, 2012). Tellurite (TeO_3^{2-}), Ag(I), Cd(II), Cu(I), Hg(II), and Zn(II) were all found to damage [Fe-S]-containing proteins. Interestingly, the metals listed here show clear chemical differences, TeO_3^{2-} being an oxyanion and Ag(I) and Cu(I) being monovalent, while the rest are divalent cations. Only TeO_3^{2-} damaged [Fe-S] clusters in an oxygendependent manner, indicating that its toxicity is actually indirect since the production of ROS (or intermediates) was likely disabling [Fe-S] clusters (Calderón *et al.*, 2009). In contrast, the inactivating action of the other metals was confirmed under anaerobic conditions. *In vitro* studies have shown that divalent thiophilic metals can compete with Fe(II) ions for the metal binding site in the cluster (Xu & Imlay, 2012); metal substitutions can also take place at non-catalytic metal-binding sites, inhibiting enzyme activity (Lemire *et al.*, 2013). It is less clear how Ag(I) and Cu(I) damage [Fe-S] clusters, but for all cases only dehydratases, such as isopropylmalate isomerase and fumarase, that have exposed [Fe-S] clusters were inactivated, whereas enzymes with buried clusters (e.g. glutamate synthase) were not (Macomber & Imlay, 2009). Studies suggest that this protein damage is limited to the [Fe-S] clusters, as opposed to the whole protein, as [Fe-S] cluster-repairing proteins have been demonstrated to restore the activity of metal-treated enzymes (Lemire *et al.*, 2013).

11.4.4 Mutagenic Effects

Genotoxicity screening of bacterial cells has historically been used to evaluate the mutagenic (and the associated carcinogenic) potential of metal interactions on animal and human cells (Green *et al.*, 1976; Asmuss *et al.*, 2000; Asakura *et al.*, 2009). Mutagenic susceptibility testing suggests that Cr(VI), Mn(II), Co(II), As(III), Mo(IV), Sb(III), and some organomercury complexes are all mutagenic to bacteria (Nishioka, 1975; Green *et al.*, 1976; Wong, 1988; Asakura *et al.*, 2009). Some metals demonstrate increased mutation frequency when combined with other mutagenic agents such as UV light, alkylating agents, and others (Asmuss *et al.*, 2000), yet their mechanism is poorly understood. Two modes of action seem to predominate mutagenic effects in bacteria: oxidative damage to DNA and disruption of DNA repair processes (Nishioka, 1975; Asakura *et al.*, 2009; Linley *et al.*, 2012). Damaging oxidation of DNA has been ascribed to Fe-mediated Fenton chemistry (Linley *et al.*, 2012), as well as intracellular reduction of Cr(VI) to the less toxic Cr(III), by which free radicals are formed (Ramírez-Díaz *et al.*, 2008). While the underlying mechanisms of repair inhibition are unclear, some suggest ionic competition by toxic metals may be of importance. So-called 'zinc finger' proteins are considered a target of this replacement mechanism (Asmuss *et al.*, 2000). Zinc fingers are a structural motif characterized by zinc ions stabilizing protein folding by coordinating four cysteines and/or histidines, known to bind DNA, RNA, and/or proteins (Asmuss *et al.*, 2000). While frequently found to be important in eukaryotic metal toxicity due to their use in DNA-binding proteins, this motif has also been identified in bacterial DNA repair enzymes (Hartwig, 2001). Formamidopyrimidine-DNA glycosylase (Fpg) is a bacterial nucleotide excision repair protein with an important role in the removal of oxidative DNA base modifications (Asmuss *et al.*, 2000; Hartwig, 2001). Fpg inhibition has been linked to both ionic replacement of zinc by thiophilic metals and oxidative damage to cysteine thiol groups (Asmuss *et al.*, 2000; Hartwig, 2001).

11.4.5 Other Mechanisms for Metal Toxicity

Bacterial membranes and associated proteins are also vulnerable to the toxic effects of metals. Given the charged nature of lipid head groups, many metals can interact through electrostatic interactions and affect lipid packing in the cell membrane. As mentioned above (11.4.2), a metal-induced oxidized environment in the cell can interfere with oxidative phosphorylation (Harrison *et al.*, 2007). Metals can cause proton leakage across the cell membrane, thereby interrupting membrane chemiosmotic potential (Li *et al.*, 2010; Lemire *et al.*, 2013) and resulting in cell death. Cell membrane damage has been associated with Cu(II), Cd(II), and Ag(I) toxicity in bacteria (Hong *et al.*, 2012; Lemire *et al.*, 2013) based on microscopy and observations of thiobarbituric acid reactive substances (TBARS). TBARS are byproducts of lipid peroxidation and have been shown to increase with metal exposure. However, these interpretations must be taken with caution, since the unsaturated lipids that make up bacterial membranes cannot propagate the radical chemistry needed for this mechanism to occur (Imlay, 2003). Thus, lipid peroxidation is expected to be possible only in the membranes of photosynthetic bacteria (Imlay, 2003) that contain the polyunsaturated lipids with which this mechanism has been observed to interact (Catalá, 2006; Lemire *et al.*, 2013). Since the TBARS assay and other analytical methods of studying lipid peroxidation in membranes measure indirect compounds from membrane damage (Catalá, 2006), the indirect nature of these techniques has raised questions on how metals actually damage cell membranes (Lemire *et al.*, 2013). Hong *et al.* (2012) in particular addressed the possible shortcomings of the TBARS assay and concluded that, despite the monounsaturated nature of bacterial membranes, ROS-mediated lipid peroxidation is indeed possible and occurs upon metal exposure.

Nutrient assimilation can also be the target of metal toxicity. In biological systems Fe(III) is required for bacterial growth, though when both are available, its chemical similarity to Ga(III) makes them indistinguishable to cells (Kaneko *et al.*, 2007). However, unlike Fe(III), Ga(III) cannot be reduced in the cell and thus cannot serve the biochemical role that comes with Fe reduction. Ga has been found to interfere with Fe uptake by repressing *pvdS*, an Fe-responsive transcription regulator responsible for the selective uptake of Fe (Kaneko *et al.*, 2007); this suggests that the dynamics of Ga and Fe are more complex than simply competition for uptake systems. It was also recently shown that Cd exerts its toxicity in *Streptococcus pneumonia* by reducing the uptake of Zn and Mn (Begg *et al.*, 2015). This presents itself as increased oxidative stress, as Mn is a regulator and cofactor of superoxide dismutase (SOD) (Geslin *et al.*, 2001; Waldron & Robinson, 2009). SODs are one of the major cellular defense mechanisms against oxidative stress (Geslin *et al.*, 2001). Cd causes decreased uptake of these other metals by mimicking their chemistry and binding to regulators, exerting toxic effects when Cd uptake occurs at the expense of Mn (Begg *et al.*, 2015).

11.5 Conclusions

We now are beginning to appreciate that Earth's biological/ecological systems require a diverse microflora (Hector & Bagchi, 2007). Yet we still consider microbes 'bad' and we tend to be in a constant battle against bacteria. Mankind's historical paranoia against microbes has led to the overuse of antibiotics. We are already seeing this with the new generation of metal-ion based antimicrobials. For example, silver nanoparticles (AgNP) are embedded in some commercial textiles (i.e. shirts, socks, and sleepwear) to kill odor-causing bacteria and fend off infections. While this provides an aesthetic olfactory benefit, AgNPs are not selective in the bacteria they affect, and by reducing the diversity of our skin microbiota we are effectively reducing the ability of our skin to be a first line of defense against pathogens. Additionally, this antimicrobial misuse is well documented to lead to bacterial resistance (Levy & Marshall, 2004). Similarly soil, plant, and marine environments have a sustained and beneficial microbial ecology. Anthropogenic release of metals into the environment through mining, E-waste, or other means reduces the inherent biodiversity in that environment (Gans *et al.*, 2005). This leads to increased ecosystem stress. Whether it is our human microbiome, a plant microbiome or complete marine microbiome, high levels of metal ion exposure, acute or chronic, can lead to serious imbalance with dire consequences.

References

Andreson, R.E., Sogin, M.L. & Baross, J.A. (2015), "Biogeography and ecology of the rare and abundant micribial lineages in deep-sea hydrothermal vents", *FEMS Mcrobialogy Ecology*, Vol. 91 No. 1, pp. 1–11.

Andrews, S.C., Robinson, A.K. & Rodríguez-Quiñones, F. (2003), "Bacterial iron homeostasis", *FEMS Microbiology Reviews*, Vol. 27 No. 2–3, pp. 215–237.

Aryal, M. & Liakopoulou-Kyriakides, M. (2015), "Bioremoval of heavy metals by bacterial biomass", *Environmental Monitoring and Assessment*, Vol. 187 No. 1, pp. 1–26.

Asakura, K., Satoh, H., Chiba, M., Okamoto, M., Serizawa, K., Nakano, M. & Omae, K. (2009), "Genotoxicity studies of heavy metals: Lead, bismuth, indium, silver and antimony", *Journal of Occupational Health*, Vol. 51 No. 6, pp. 498–512.

Asmuss, M., Mullenders, L.H., Eker, A. & Hartwig, A. (2000), "Differential effects of toxic metal compounds on the activities of Fpg and XPA, two zinc finger proteins involved in DNA repair", *Carcinogenesis*, Vol. 21 No. 11, pp. 2097–2104.

Baptista, M.S. & Vasconcelos, M.T. (2006), "Cyanobacteria metal interactions: Requirements, toxicity, and ecological implications", *Critical Reviews in Microbiology*, Vol. 32 No. 3, pp. 127–137.

Becking, B.L.G.M. (1934), *Geobiologie of inleiding tot de milieukunde*, Den Haag: W.P. Van Stockum & Zoon.

Begg, S.L., Eijkelkamp, B.A., Luo, Z., Couñago, R.M., Morey, J.R., Maher, M.J., Ong, C.Y., McEwan, A.G., Kobe, B., O'Mara, M.L., Ptaon, J.C, & McDevitt, C.A. (2015), "Dysregulation of transition metal ion homeostasis is the molecular basis for cadmium toxicity in Streptococcus pneumoniae", *Nature Communications*, Vol. 6, doi:10.1038/ncomms7418.

Bruins, M.R., Kapil, S. & Oehme, F.W. (2000), "Microbial resistance to metals in the environment", *Ecotoxicology and Environmental Safety*, Vol. 45 No. 3, pp. 198–207.

Brüssow, H. (2014), "Microbiota and the human nature: know thyself", *Environmental Microbiology*, Vol. 17 No. 1, doi:10.1111/1462-2920.12693.

Cabiscol, E., Tamarit, J. & Ros, J. (2000), "Oxidative stress in bacteria and protein damage by reactive oxygen species", *International Microbiology*, Vol. 3 No. 1, pp. 3–8.

Cai, M., Yu, C., Wang, R., Si, Y., Masakorala, K., Yuan, H., Yao, J. & Zhang, J. (2015), "Effects of oxygen injection on oil biodegradation and biodiversity of reservoir microorganisms in Dagang oil field, China", *International Biodeterioration & Biodegradation*, Elsevier Ltd, Vol. 98, pp. 59–65.

Calderón, I.L., Elías, A.O., Fuentes, E.L., Pradenas, G.A., Castro, M.E., Arenas, F.A., Pérez, J.M. & Vásquez, C.C. (2009), "Tellurite-mediated disabling of [4Fe-4S] clusters of Escherichia coli dehydratases", *Microbiology*, Vol. 155 No. 6, pp. 1840–1846.

California Department of Public Health. (2008), *Maximum contaminant levels and regulatory dates for drinking water: U.S. EPA vs California*, available at: http://www.cdph.ca.gov/certlic/drinkingwater/Documents/DWdocuments/EPAandCDPH-11-28-2008.pdf.

Canadian Council of Ministers of the Enviornment. (2014), *Canadian Environmental Quality Guidelines*, available at: http://st-ts.ccme.ca/en/index.html.

Cardozo, M. (2015), "Soil bacterial communities are shaped by temporal and environmental filtering: E=evidence from a long-term chronosequence", *Environmental Microbiology*, doi:10.1111/1462-2920.12762.

Catalá, A. (2006), "An overview of lipid peroxidation with emphasis in outer segments of photoreceptors and the chemiluminescence assay", *International Journal of Biochemistry and Cell Biology*, Vol. 38 No. 9, pp. 1482–1495.

Le Chatelier, E., Nielsen, T., Qin, J., Prifti, E., Hildebrand, F., Falony, G., Almeida, M., Arumugam, M., Batto, J-M., Kennedy, S., Leonard, P., Li, J., Burgdorf, K., Grarup, N., Jørgensen, T., Brandslund, I., Nielson, H.B., Juncker, A.S., Bertalan, M., Levenez, F., Pons, N., Rasmussen, S., Sunagwa, S., Tap, J., Tims, S., Zoetendal, E.G., Brunal, S., Clément, K., Doré, J., Kleerebezem, M., Kristiansen, K., Renault, P., Sicheritz-Ponten, T., de Vos, W.M., Zucker, J-D., Raes, J., Hansen, T., MetaHIT consortium, Bork, P., Wang, J., Ehrlich, S.D. & Pedersen, O. (2013), "Richness of human gut microbiome correlates with metabolic markers", *Nature*, Vol. 500 No. 7646, pp. 541–546.

Chu, H., Fierer, N., Lauber, C.L., Caporaso, J.G., Knight, R. & Grogan, P. (2010), "Soil bacterial diversity in the Arctic is not fundamentally different from that found in other biomes", *Environmental Microbiology*, Vol. 12 No. 11, pp. 2998–3006.

Enning, D. & Garrelfs, J. (2014), "Corrosion of iron by sulfate-reducing bacteria: New views of an old problem", *Applied and Environmental Microbiology*, Vol. 80 No. 4, pp. 1226–1236.

EPA. (2013), *Summary data for 2013 priority list of hazardous substances*, available at: http://www.atsdr.cdc.gov/SPL/resources/ATSDR_2013_SPL_Detailed_Data_Table.pdf.

Favet, J., Lapanje, A., Giongo, A., Kennedy, S., Aung, Y.-Y., Cattaneo, A., Davis-Richardson, A.G., Brown, C.T., Kort, R., Brumsack, H-J., Schnetger, B., Chappell, A., Kroijenga, J., Beck, A., Schwibbert, K., Mohamed, A.H., Kirchner, T., de Quadros, P.D., Triplett, E.W., Broughton, W.J., & Gorbushina, A.A. (2013), "Microbial hitchhikers on intercontinental dust: catching a lift in Chad", *The ISME Journal*, Vol. 7, pp. 850–867.

Fierer, N. & Jackson, R.B. (2006), "The diversity and biogeography of soil bacterial communities", *Proceedings of the National Academy of Sciences of the United States of America*, Vol. 103 No. 3, pp. 626–631.

Filomeni, G., Rotilio, G. & Ciriolo, M.R. (2002), "Cell signalling and the glutathione redox system", *Biochemical Pharmacology*, Vol. 64 No. 5-6, pp. 1057–1064.

Finney, L.A. & O'Halloran, T. V. (2003), "Transition metal speciation in the cell: Insights from the chemistry of metal ion receptors", *Science*, Vol. 300 No. 5621, pp. 931–936.

Fraústo da Silva, J.J.R. & Williams, R.J.P. (2001), *The biological chemistry of the elements: The inorganic chemistry of life*, Oxford University Press, New York, 2nd ed.

Gans, J., Wolinsky, M. & Dundar, J. (2005), "Computational improvements reveal great bacterial diversity and high metal toxicity in soil", *Science*, Vol. 309 No. 5739, pp. 1387–1390.

Geslin, C., Llanos, J., Prieur, D. & Jeanthon, C. (2001), "The manganese and iron superoxide dismutases protect Escherichia coli from heavy metal toxicity", *Research in Microbiology*, Vol. 152 No. 10, pp. 901–905.

Gieg, L.M., Jack, T.R. & Foght, J.M. (2011), "Biological souring and mitigation in oil reservoirs", *Applied Microbiology and Biotechnology*, Vol. 92 No. 2, pp. 263–282.

Gillan, D.C., Danis, B., Pernet, P., Joly, G. & Dubois, P. (2005), "Structure of sediment-associated microbial communities along a heavy-metal contamination gradient in the marine environment", *Applied and Environmental Microbiology*, Vol. 71 No. 2, pp. 679–690.

Glikson, A. (2006), "Extraterrestrial impact episodes and Archaean to early Proterozoic (3.8–2.4 Ga) habitats of life", *Comets at the Origin and Evolution of Life*, Springer, Berlin Heidelberg, Vol. 28, pp. 253–283.

Green, M.H., Muriel, W.J. & Bridges, B.A. (1976), "Use of a simplified fluctuation test to detect low levels of mutagens", *Mutation Research*, Vol. 38 No. 1, pp. 33–42.

Harrison, J.J., Ceri, H. & Turner, R.J. (2007), "Multimetal resistance and tolerance in microbial biofilms", *Nature Reviews Microbiology*, Vol. 5 No. 12, pp. 928–938.

Harrison, J.J., Tremaroli, V., Stan, M.A., Chan, C.S., Vacchi-Suzzi, C., Heyne, B.J., Parsek, M.R., Ceri, H., & Turner, R.J. (2009), "Chromosomal antioxidant genes have metal ion-specific roles as determinants of bacterial metal tolerance", *Environmental Microbiology*, Vol. 11 No. 10, pp. 2491–2509.

Hartwig, A. (2001), "Zinc finger proteins as potential targets for toxic metal ions: Differential effects on structure and function", *Antioxidants & Redox Signalling*, Vol. 3 No. 4, pp. 625–634.

Hector, A. & Bagchi, R. (2007), "Biodiversity and ecosystem multifunctionality", *Nature*, Vol. 448 No. 7150, pp. 188–190.

Van der Heijden, M.G.A., Bardgett, R.D. & Van Straalen, N.M. (2008), "The unseen majority: soil microbes as drivers of plant diversity and productivity in terrestrial ecosystems", *Ecology Letters*, Vol. 11 No. 3, pp. 296–310.

Helbig, K., Bleuel, C., Krauss, G.J. & Nies, D.H. (2008), "Glutathione and transition-metal homeostasis in Escherichia coli", *Journal of bacteriology*, Vol. 190 No. 15, pp. 5431–5438.

Hong, R., Kang, T.Y., Michels, C.A. & Gadura, N. (2012), "Membrane lipid peroxidation in copper alloy-mediated contact killing of Escherichia coli", *Applied and Environmental Microbiology*, Vol. 78, pp. 1776–1784.

Hullebusch, E.D. Van, Zandvoort, M.H. & Lens, P.N.L. (2004), "Metal immobilisation by biofilms: mechanisms and analytical tools", *Review in Environmental Science and Biotechnlogy*, Vol. 2 No. 1, pp. 9–33.

Imlay, J.A. (2003), "Pathways of oxidative damage", *Annual Review of Microbiology*, Vol. 57 No. 1, pp. 395–418.

Ishaque, A.B., Johnson, L., Gerald, T., Boucaud, D., Okoh, J. & Tchounwou, P.B. (2006), "Assessment of individual and combined toxicities of four non-essential metals (As, Cd, Hg and Pb) in the microtox assay", *International Journal of Environmental Research and Public Health*, Vol. 3 No. 1, pp. 118–120.

Jannasch, H.W. & Mottl, M.J. (1985), "Geomicrobiology of deep-sea hydrothermal vents", *Science*, Vol. 229 No. 4715, pp. 717–725.

Johnston, D.T., Wolfe-Simon, F., Pearson, A. & Knoll, A.H. (2009), "Anoxygenic photosynthesis modulated Proterozoic oxygen and sustained Earth's middle age", *Proceedings of the National Academy of Sciences*, Vol. 106 No. 40, pp. 16925–16929.

Kaneko, Y., Thoendel, M., Olakanmi, O., Britigan, B.E. & Singh, P.K. (2007), "The transition metal gallium disrupts Pseudomonas aeruginosa iron metabolism and has antimicrobial and antibiofilm activity", *Journal of Clinical Investigation*, Vol. 117 No. 4, pp. 877–888.

Kostic, A.D., Chun, E., Robertson, L., Glickman, J.N., Gallini, C.A., Michaud, M., Clancy, T.E., Chung, D.C., Lochhead, P., Hold, G.L., El-Omar, E.M., Brenner, D., Fuchs, C.S., Meyerson, M., &

Garrett, W.S. (2013), "Fusobacterium nucleatum potentiates Intestinal tumorigenesis and modulates the tumor-immune microenvironment", *Cell Host and Microbe*, Elsevier Inc., Vol. 14 No. 2, pp. 207–215.

L'Haridon, S., Reysenbacht, A.-L., Glénat, P., Prieur, D. & Jeanthon, C. (1995), "Hot subterranean biosphere in a continental oil reservoir", *Nature*, Vol. 377, pp. 223–224.

Lauber, C.L., Hamady, M., Knight, R. & Fierer, N. (2009), "Pyrosequencing-based assessment of soil pH as a predictor of soil bacterial community structure at the continental scale", *Applied and Environmental Microbiology*, Vol. 75 No. 15, pp. 5111–5120.

Le, J., Wu, X., Wang, R., Zhang, J., Bai, L. & Hou, Z. (2015), "Progress in pilot testing of microbial-enhanced oil recovery in the Daqing oil field of north China", *International Biodeterioration & Biodegradation*, Elsevier, Vol. 97, pp. 188–194.

Lemire, J., Mailloux, R., Auger, C., Whalen, D. & Appanna, V.D. (2010), "Pseudomonas fluorescens orchestrates a fine metabolic-balancing act to counter aluminium toxicity", *Environmental Microbiology*, Vol. 12 No. 6, pp. 1384–1390.

Lemire, J.A., Harrison, J.J. & Turner, R.J. (2013), "Antimicrobial activity of metals: mechanisms, molecular targets and applications", *Nature Reviews. Microbiology*, Vol. 11 No. 6, pp. 371–384.

Levy, S.B. (2001), "Antibacterial household products: cause for concern", *Emerging Infectious Diseases*, Vol. 7, pp. 512–515.

Levy, S.B. & Marshall, B. (2004), "Antibacterial resistance worldwide: causes, challenges and responses", *Nature Medicine*, Vol. 10 No. 12 Suppl, pp. S122–S129.

Li, J., Jia, H., Cai, X., Zhong, H., Feng, Q., Sunagawa, S., Arumugam, M., Kultima, J.R., Prifti, E., Nielsen, T., Juncker, A.S., Manichanh, C., Chen, B., Zhang, W., Levenez, F., Wang, J., Xu, X., Xiao, L., Liang, S., Zhang, D., Zhang, Z., Chen, W., Zhao, H., Al-Aama, J.Y., Edris, S., Yang, H., Wang, J., Hansen, T., Nielson, H.B., Brunak, S., Kristiansen, K., Guarner, F., Pederson, O., Dore, J., Ehrlich, S.D., MetaHIT Consortium, Bork, P. & Wang, J. (2014), "An integrated catalog of reference genes in the human gut microbiome", *Nature Biotechnology*, Vol. 32 No. 8, pp. 834–841.

Li, W.R., Xie, X.B., Shi, Q.S., Zeng, H.Y., Ou-Yang, Y.S. & Chen, Y. Ben. (2010), "Antibacterial activity and mechanism of silver nanoparticles on Escherichia coli", *Applied Microbiology and Biotechnology*, Vol. 85 No. 4, pp. 1115–1122.

Linley, E., Denyer, S.P., McDonnell, G., Simons, C. & Maillard, J.Y. (2012), "Use of hydrogen peroxide as a biocide: new consideration of its mechanisms of biocidal action", *Journal of Antimicrobial Chemotherapy*, Vol. 67 No. 7, pp. 1589–1596.

Lippard, S.J. & Berg, J.M. (1994), *Principles of Bioinorganic Chemistry*, University Science Books, Mill Valley, CA.

Macomber, L. & Imlay, J.A. (2009), "The iron-sulfur clusters of dehydratases are primary intracellular targets of copper toxicity", *Proceedings of the National Academy of Sciences*, Vol. 106 No. 20, pp. 8344–8349.

Minister of Justice. (1999), *Canadian Environmental Protection Act (CEPA)*, http://laws-lois.justice.gc.ca/eng/acts/c-15.31/, pp. 235–241.

Monsieurs, P., Moors, H., Van Houdt, R., Janssen, P.J., Janssen, A., Coninx, I., Mergeay, M. & Leys, N. (2011), "Heavy metal resistance in Cupriavidus metallidurans CH34 is governed by an intricate transcriptional network", *BioMetals*, Vol. 24 No. 6, pp. 1133–1151.

Newton, R.J., Jones, S.E., Eiler, A., McMahon, K.D. & Bertilsson, S. (2011), "A guide to the natural history of freshwater lake bacteria", *Microbiology and Molecular Biology Reviews*, Vol. 75 No. 1, pp. 14–49.

Nies, D.H. (1999), "Microbial heavy-metal resistance", *Applied Microbiology and Biotechnology*, Vol. 51 No. 6, pp. 730–750.

Nies, D.H. (2003), "Efflux-mediated heavy metal resistance in prokaryotes", *FEMS Microbiology Reviews*, Vol. 27 No. 2–3, pp. 313–339.

Nishioka, H. (1975), "Mutagenic activities of metal compounds in bacteria", *Mutation Research/Environmental Mutagenesis and Related Subjects*, Vol. 31 No. 3, pp. 185–189.

Nogales, B., Lanfranconi, M.P., Piña-Villalonga, J.M. & Bosch, R. (2011), "Anthropogenic perturbations in marine microbial communities", *FEMS Microbiology Reviews*, Vol. 35 No. 2, pp. 275–298.

O'Malley, M.A. (2008), "'Everything is everywhere: but the environment selects': ubiquitous distribution and ecological determinism in microbial biogeography", *Studies in History and Philosophy of Biological and Biomedical Sciences*, Vol. 39 No. 3, pp. 314–325.

Official Journal of the European Union. (2013), *Directive 2013/39/EU of the European parliament and of the council on 12 August 2013 amending Directives 2000/60/EU and 2008/105/EC as regarding priority substances in the field of water policy*, pp. 1–17.

Pearce, D.A., Bridge, P.D., Hughes, K.A., Sattler, B., Psenner, R. & Russell, N.J. (2009), "Microorganisms in the atmosphere over Antarctica", *FEMS Microbiology Ecology*, Vol. 69 No. 2, pp. 143–157.

Qin, J., Li, Y., Cai, Z., Li, S., Zhu, J., Zhang, F., Liang, S., Zhang, W., Guan, Y., Shen, D., Peng, Y., Zhang, D., Jie, Z., Wu, W., Qin, Y., Xue, W., Li, J., Han, L., Lu, D., Wu, P., Dai, Y., Sun, X., Li, S., Tang, A., Zhong, H., Li, X., Chen, W., Xu, R., Wang, M., Feng, Q., Gong, M., Yu, J., Zhang, Y., Zhang, M., Hansen, T., Sanchez, G., Raes, J., Falony, G., Okuda, S., Almeida, M., Le Chatelier, E., Renault, P., Pons, N., Batto, J-M., Zhang, Z., Chen, H., Yang, R., Zheng, W., Li, S., Yang, H., Wang, J., Ehrlich, S.D., Nielsen, R., Pederson, O., Kristiansen, K. & Wang, J. (2012), "A metagenome-wide association study of gut microbiota in type 2 diabetes", *Nature*, Vol. 490 No. 7418, pp. 55–60.

Ramírez-Díaz, M.I., Díaz-Pérez, C., Vargas, E., Riveros-Rosas, H., Campos-García, J. & Cervantes, C. (2008), "Mechanisms of bacterial resistance to chromium compounds", *BioMetals*, Vol. 21, pp. 321–332.

Reinhard, C.T., Planavsky, N.J., Robbins, L.J., Partin, C.A., Gill, B.C., Lalonde, S.V., Bekker, A., Konhauser, K.O. & Lyons, T.W. (2013), "Proterozoic ocean redox and biogeochemical stasis", *Proceedings of the National Academy of Sciences*, Vol. 110 No. 14, pp. 5357–5362.

Reyes-Caballero, H., Campanello, G.C. & Giedroc, D.P. (2011), "Metalloregulatory proteins: Metal selectivity and allosteric switching", *Biophysical Chemistry*, Elsevier B.V., Vol. 156 No. 2–3, pp. 103–114.

Saito, M.A., Sigman, D.M. & Morel, F.M.M. (2003), "The bioinorganic chemistry of the ancient ocean: The co-evolution of cyanobacterial metal requirements and biogeochemical cycles at the Archean-Proterozoic boundary?", *Inorganica Chimica Acta*, Vol. 356, pp. 308–318.

Scher, J.U., Sczesnak, A., Longman, R.S., Segata, N., Ubeda, C., Bielski, C., Rosteron, T., Cerundolo, V., Pamer, E.G., Abramson, S.B., Hutternhower, C. & Littman, D.R. (2013), "Expansion of interstinal Prevotella copri correlates with enhanced susceptability to arthritis", *eLIFE*, Vol. 2, doi:10.7554/eLife.01202.

Schneider, L., Maher, W., Potts, J., Batley, G., Taylor, A., Krikowa, F., Chariton, A., Zawadzki, A., Heijnis, H. & Gruber, B. (2014), "History of metal contamination in Lake Illawarra, NSW, Australia", *Chemosphere*, Vol. 119C, pp. 377–386.

Sen, R. (2008), "Biotechnology in petroleum recovery: the microbial EOR", *Progress in Energy and Combustion Science*, Vol. 34 No. 6, pp. 714–724.

Shreiner, A.B., Kao, J.Y. & Young, V.B. (2015), "The gut microbiome in health and disease", *Current Opinion in Gastroenterology*, Vol. 31 No. 1, pp. 69–75.

Silver, S. & Phung, L.T. (2005), "A bacterial view of the periodic table: genes and proteins for toxic inorganic ions", *Journal of Industrial Microbiology and Biotechnology*, Vol. 32 No. 11–12, pp. 587–605.

Sokolova, I.M. & Lannig, G. (2008), "Interactive effects of metal pollution and temperature on metabolism in aquatic ectotherms: Implications of global climate change", *Climate Research*, Vol. 37 No. 2, pp. 181–201.

Strlič, M., Kolar, J., Šelih, V.S., Kočar, D. & Pihlar, B. (2003), "A comparative study of several transition metals in fenton-like reaction systems at circum-neutral pH", *Acta Chimica Slovenica*, Vol. 50 No. 4, pp. 619–632.

U.S.EPA Epa/542/R-04/015. (2004), *Cleaning Up the Nation's Waste Sites: Markets and Technology Trends*, available at: http://clu-in.org/download/market/2004market.pdf.

Valko, M., Morris, H. & Cronin, M.T.D. (2005), "Metals, toxicity and oxidative stress", *Current Medicinal Chemistry*, Vol. 12 No. 10, pp. 1161–1208.

Veglio, F. & Beolchini, F. (1997), "Removal of metals by biosorption: a review", *Hydrometallurgy*, Vol. 44 No. 3, pp. 301–316.

Wagg, C., Bender, S.F., Widmer, F. & van der Heijden, M.G.A. (2014), "Soil biodiversity and soil community composition determine ecosystem multifunctionality", *Proceedings of the National Academy of Sciences of the United States of America*, Vol. 111 No. 14, pp. 5266–5270.

Waldron, K.J. & Robinson, N.J. (2009), "How do bacterial cells ensure that metalloproteins get the correct metal?", *Nature reviews. Microbiology*, Vol. 7, pp. 25–35.

Wei, B. & Yang, L. (2010), "A review of heavy metal contaminations in urban soils, urban road dusts and agricultural soils from China", *Microchemical Journal*, Elsevier B.V., Vol. 94 No. 2, pp. 99–107.

Wong, P.K. (1988), "Mutagenicity of heavy metals", *Bulletin of Environmental Contamination and Toxicology*, Vol. 40 No. 4, pp. 597–603.

Xu, F.F. & Imlay, J.A. (2012), "Silver(I), mercury(II), cadmium(II), and zinc(II) target exposed enzymic iron-sulfur clusters when they toxify Escherichia coli", *Applied and Environmental Microbiology*, Vol. 78 No. 10, pp. 3614–3621.

12

Application of Molecular Recognition Technology to Green Chemistry: Analytical Determinations of Metals in Metallurgical, Environmental, Waste, and Radiochemical Samples

Yoshiaki Furusho[1], Ismail M.M. Rahman[2], Hiroshi Hasegawa[3], and Neil E. Izatt[4]

[1]*GL Sciences Inc., Shinjuku, Tokyo, 163-1130, Japan*
[2]*Institute of Environmental Radioactivity, Fukushima University, Fukushima, 960-1296, Japan*
[3]*Institute of Science and Engineering, Kanazawa University, Kanazawa, 920-1192, Japan*
[4]*IBC Advanced Technologies, Inc., American Fork, Utah, 84003, U.S.A.*

12.1 Introduction

Rapid, simple, and accurate metal analysis capability is important in many industries, since processing and financial decisions are based on these analyses. Metals to be analyzed may occur in complex ores, end-of-life (EOL) products, environmental samples, or radioactive waste and may coexist with multiple interfering elements that must be removed prior to final analysis. Secondary materials such as low-grade waste, contaminated water, or EOL products containing metals can also exhibit complex matrices making it time- and labor-intensive to separate the desired metal for analysis [1]. Because the target metal to be determined is usually present in the sample at low, ~mg L^{-1} (ppm), concentrations, often in the presence of g L^{-1} concentrations of other potentially interfering ions, the metal

Metal Sustainability: Global Challenges, Consequences, and Prospects, First Edition. Edited by Reed M. Izatt.

must be selectively concentrated prior to analysis. Precious metal and other laboratories generally use conventional methods such as sedimentation, liquid-liquid extraction, or ion exchange (IX) extraction to separate metals prior to analysis with inductively coupled plasma mass spectrometry (ICP-MS) and inductively coupled plasma atomic emission spectrometry (ICP-AES), also referred to as inductively coupled plasma optical emission spectrometry (ICP-OES), procedures. These separation techniques are time-consuming and labor-intensive, requiring multiple extraction steps to obtain a pure sample, which increases laboratory costs and generates much waste. In addition, conventional methods become less effective in separating metals as metal concentrations decrease, especially if complex matrices are present. At mg L^{-1} or lower concentration levels, these methods become relatively ineffective.

Molecular recognition technology solid phase extraction (MRT-SPE) products, termed AnaLig®, provide a tested means for rapid, highly selective, and quantitative separations of metals from complex solution matrices, even when the metal is present at mg L^{-1} or lower concentration levels [1–11]. Following a washing step, the bound target metal is eluted using a small amount of eluent. The metal eluted is pure since it was selectively removed from the original solution by the AnaLig® product and subsequently concentrated in the elution process. Large and selective metal concentration factors can be achieved in this way. These concentrated metal solutions can be analyzed using ICP instrumentation or spectrophotometric procedures.

This chapter describes methods used for pre-treatment of experimental solutions for subsequent determination of target metal(s). Examples are given for determination of metals in environmental samples, high matrix plating solutions, recycled materials, and radionuclide samples. A wide variety of metals is included in the examples shown. Advantages of using MRT-SPE columns and MRT Empore Rad Disks in making separations preparatory to analysis are presented and discussed.

12.2 Technologies Used for Green Chemistry Trace Element Analysis

Green chemistry is an important component of the environmental movement and is also consequential to the sustainable growth of chemical and process industries [12]. In particular, use of green chemistry principles in chemical analysis is becoming more important as environmental awareness increases. The features of the elemental analyzer have experienced significant innovation during the past decade. One of the key features that have seen improvement in the elemental analysis is the direct introduction of the analyzer target sample. Direct introduction is important because the target sample is often difficult to collect, and the sample size is very small. In the past, various techniques have been used for elemental analysis, and the field is continually improving. It is desirable to introduce the analyte sample into the elemental analyzer in the liquid form. If, however, the target sample is in solid form, the sample must be prepared by (1) acid digestion of the solid followed by (2) solid-liquid extraction. It is preferable, from a green chemistry point of view, to keep the decomposition target sample as small as possible. Furthermore, it is desirable to use minute volumes of reagents for the acid decomposition step. On the other hand, if the target sample is a liquid, it is usual for the sample pre-treatment method to be either (1) liquid-liquid extraction (LLE) or (2) SPE.

LLE is widely used for dissolving trace amounts of elements into the liquid sample as a way to separate and concentrate these elements [13–15]. A green chemistry goal is to concentrate these elements from a liquid sample without using harmful organic solvents. It is desirable to avoid or minimize the use of such solvents as xylene or methyl isobutyl ketone (MIBK). SPE overcomes the disadvantages of LLE by avoiding the use of organic solvents. SPE is now widely used for sample preparation for many analytical instruments, including high performance liquid chromatography (HPLC) and gas chromatography (GC). Traditionally, SPE has been used as a sample pretreatment method for separating and concentrating organic compounds. However, SPE is increasingly used for sample pretreatment for elemental analysis. A major advantage of SPE in this application is avoidance of organic solvents, such as xylene and MIBK. The development of trace elemental analysis using SPE methods is extremely beneficial from the viewpoint of green chemistry.

Before describing case studies, we discuss and summarize the primary detection methods for elemental analysis. Specifically, we describe the most generic types of detection instruments for elemental analysis: atomic absorption spectrometry (AAS), ICP-OES, and ICP-MS.

Recently, highly specific resins with chelating functional groups have been developed for highly selective metal ion separations. These resins, when combined with SPE techniques, are useful for a number of applications, including removal of salts from environmental water samples, e.g., seawater and waste water. Examples of SPE resins of high metal selectivity are those developed by IBC Advanced Technologies, Inc. (IBC) based on MRT [4,5]. These AnaLig® resins are used for a wide range of SPE commercial analytical separations, including in 3M's Empore Rad Disk, to separate specific radioactive elements, such as Sr, Ra, and Tc, as discussed in Section 12.6.5. These commercialized AnaLig® products are available for a wide range of metal separations.

12.3 Elemental Analysis Instrumentation

Commonly used elemental analysis techniques, with a description of each, are given in Table 1. AAS is very versatile and can be used for analyzing environmental, food or biological samples. It is useful for detecting alkali and alkaline earth metals. AAS requires a lamp light source and, for each metal, the correct light source lamp must be used in order to get the right absorbance. Therefore, AAS is useful for measuring concentrations of single metals. With ICP-OES, it is possible to measure multiple metals simultaneously. ICP-OES relies on the light emission of each element for detection, and can measure the emission spectrum of each element in a sample at the same time. Compared with AAS, the dynamic range for the emissions spectrum in the ICP-OES method is wide, ranging from low to high concentrations. ICP-MS is a technique for detecting an element by monitoring the mass number of ionized elements in high energy reactions.

The AAS instrument has very high sensitivity in comparison with the emission spectral method, making possible accurate trace elemental analysis. AAS analysis can measure element concentrations at the mg L^{-1} level, whereas ICP-OES can analyze at the mg L^{-1} to µg L^{-1} (ppb) level, and the ICP-MS can analyze at the range of ng L^{-1} (ppt) to sub-mg L^{-1}.

Samples introduced into the ICP-OES instrument need to be in liquid form. If the experimental sample is a solid, it must be pretreated, e.g., by acid decomposition or acid

Table 1 *Elemental Analysis Techniques*

Detection Method	Application of Method
flame analysis atomic spectrometry (FAAS)	determination of alkali and alkaline earth metals
atomic absorption spectrometry (AAS)	general analytical method for elements
inductively coupled plasma optical emission spectrometry (ICP-OES)	general analytical method for metals and some non-metals
inductively coupled plasma mass spectrometry (ICP-MS)	highly sensitive detection method for elements, allowing detection of individual isotopes according to mass number
X-ray fluorescence spectrometry (XRF)	qualitative or quantitative determination of metals in a solid sample, without need for sample preparation, such as acid digestion
electrochemical analysis: anodic stripping voltammetry (ASV)	highly sensitive on-site detection technique using a portable instrument
activation analysis	analysis method using a radioactive isotope
ion chromatography	detection method for anions and metal ions

extraction. If a liquid sample contains any solids after the addition of acid, these must be removed by filtration before analysis.

A disadvantage of the ICP-OES analysis method is the interference caused when impurity salts at high concentrations are present in the sample. The challenge, in these cases, is to resolve the spectral interference of coexisting elements (e.g., Na, K, Ca, Ba). When the salt concentration in the sample is high, spray nebulizer efficiency decreases. A decrease in the spray efficiency greatly affects the detection sensitivity of the instrument for the target element. This problem can be solved by using an MRT-SPE resin to selectively remove impurity metals from the sample prior to analysis.

A disadvantage of the ICP-MS analysis method is that the procedure is affected by spectral interference of polyatomic ions generated by the acid and Ar in the sample within the ICP. For example, the measurement is adversely affected by the spectral interference caused by oxoanions in the solvent. These interferences are especially important in analyses where the target metal is present at low-concentrations. These disadvantages of the ICP-MS method can be alleviated using one of the following techniques: (1) use of a double-focusing mass spectrometer of high resolution in place of the quadrupole mass spectrometer, (2) adoption of a cool plasma generation of Ar ions that causes the generation of polyatomic ions to be suppressed, or (3) use of a collision reaction cell, which acts to reduce polyatomic spectrum interference problems. For example, if it is desired to detect ^{75}As, ^{78}Se, and ^{56}Fe by ICP-MS, the following ion interferences must be prevented: ($^{40}Ar^{35}Cl{=}^{75}As$, $^{38}Ar^{40}Ar{=}^{78}Se$, and $^{40}Ar^{16}O{=}^{56}Fe$). In this case, equipping the ICP-MS with a collision/reaction cell reduces the spectrum interference of $^{40}Ar^{35}Cl$, $^{38}Ar^{40}Ar$, and $^{40}Ar^{16}O$ in the cell. After this reduction, analysis of ^{75}As, ^{78}Se, and ^{56}Fe proceeds well.

ICP-MS instruments make possible highly sensitive metal detection following removal of interfering metals by use of separation resin techniques. Metal speciation can be an important consideration in many cases. For example, toxicities of certain elements, such as

As, Hg, Se, and Cr, depend on their speciation, which, in turn, depend on environmental conditions. The ability to identify and determine these elements in experimental samples is important, especially in environmental chemistry. For example, speciation analysis of arsenic compounds can be done using chromatography instrumentation, such as HPLC, and ion chromatography (IC). These techniques are referred to as the hyphenated analysis (LC/ICP-MS, IC/ICP-MS) processes and are widely used in analyzing environmental, food, and biological samples.

12.4 Arsenic Speciation in Food Analysis

In the food matrix, it is necessary to remove interfering elements prior to determination of the target element. When target elements are present in trace amounts, it is desirable to increase the sampling size and to selectively separate and concentrate the desired element prior to analysis. Liquid-liquid extraction has been commonly used for separation of the desired element from the matrix and for its concentration. However, in recent years, use of the MRT-SPE method has increased due to its simplicity and advantages in selective separation and concentration of the desired element.

It is important to know the speciation of the metal in food analysis. This is especially true in the case of As, which can exist in several forms of varying degrees of toxicity. The speciation and determination of As compounds is very important for food sample and biological monitoring because of the toxic effects of arsenite and arsenate species. Development of ICP-MS analysis methods for As following HPLC separation is a common procedure. Knowledge of As speciation has been useful in clarifying the metabolic mechanism in the ecosystem and pharmacokinetics in animals, including humans [16,17]. Arsenic compounds identified have included methylarsonic acid, dimethylarsonic acid, trimethylarsine oxide, arsenobetaine, arsenocholine, tetramethylarsonium ion, and arsenosuger (Figures 1, 2). Various techniques have been used to determine the speciation of arsenic compounds, most relying on use of anion and cation exchange HPLC for separations of the various As species followed by detection of these species using ICP-OES or ICP-MS instruments. These procedures require a long analytical time, including two injections into each of the anion and cation exchange columns used for separation of all As compounds. We have evaluated an improved method for As separations that includes rapid analysis time, improved As separations, and an inexpensive As analysis method using a hyphenated technique of "semi-micro reverse-phase" HPLC coupled with an ICP-MS system [18]. The semi-micro separation techniques reduce the bat effect, which appears as an ArCl peak near the arsenate peak. The reverse phase chromatography technique drastically reduced the analysis cycle time. This technique made possible the analysis of eight As compounds in about twelve minutes. This method reduces Ar consumption and is useful for the determination of As species in biological and environmental samples.

12.5 Metal Separation Resins and Their Application to Elemental Analyses

Concentrations of both toxic and essential elements have increased in the environment in past decades, mostly due to increased population growth and anthropogenic activities related to mining and industrial development aimed at the attainment of improved living standards in

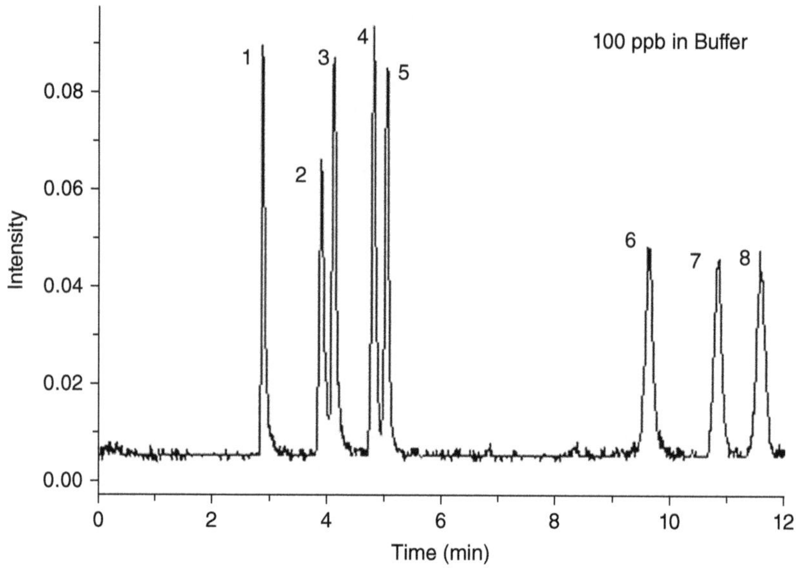

Figure 1 *Typical chromatogram of arsenic compounds in buffer solution. Analytical Column: Inetsil AS 3 mm 4.6 x 250 mm, Buffer solution: 10 mM sodium 1-butane sulfonate, 4 mM tetramethylammonium hydroxide, 4 mM malonic acid and 0.05% methanol in Milli-Q water with pH adjusted to 3.0 by HNO_3. (Analytical compounds: 1. Arsenate, 2. Arsenite, 3. Methylarsonic acid, 4. Dimethylarsinic acid, 5. Arsenobetaine, 6. Trimethylarsine oxide, 7. Tetramethylarsonium salt, 8. Arsenocholine.)*

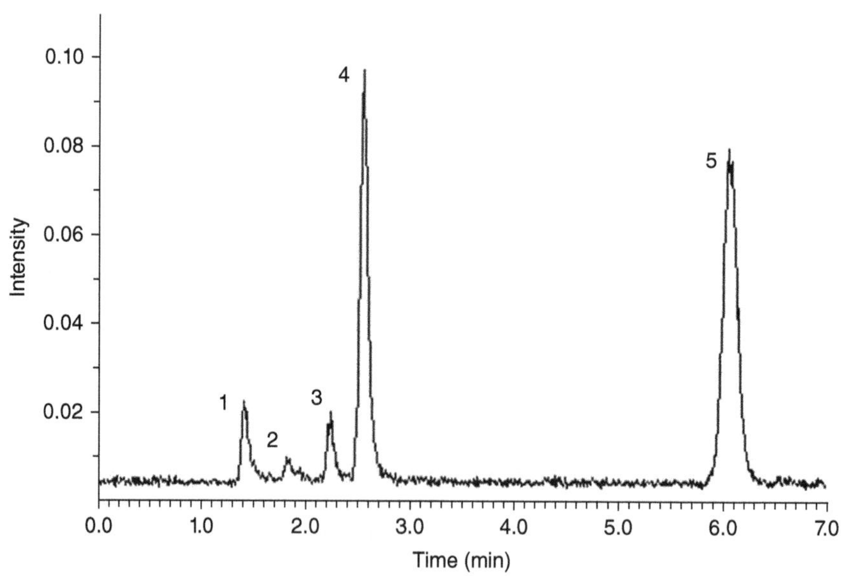

Figure 2 *Typical chromatogram of arsenic compounds in marine alga "Japanese food: Hijiki". Anaytical Column: Inertsil AS 3 mm 2.1 x 150 mm. (Analytical Compounds: 1. Arsenate, 2. Arsenite, 3. Dimethylarsinic acid, 4. Unknown peak, 5. Arseno Suger.)*

modern societies [19]. The growing extent of environmental metal pollution has been recognized and has resulted in a number of legislative measures, such as, the Restriction of Hazardous Substance (RoHS) directive [7], End of Life Vehicle (ELV) directive [8], and the European Union Council directive [9]. The RoHS and ELV directives have been imposed to specify limits for trace-element contamination in industrial products, while the EU directive has defined acceptable concentrations of contaminating elements in cultivable soils.

Sensitive analytical techniques such as flame AAS, electrothermal AAS, ICP-MS, and ICP-OES are available for precise analysis of trace elements in solution. An overall analytical process comprises a number of consecutive steps, including sampling, sample preparation, separation and quantification of the target element, and analysis of this element by an appropriate analytical technique. Among these steps, that involving sample preparation is the most important source of error due to one or all of the following: low concentrations of target species, heterogeneous distribution of analytes in the matrix, and complex nature of sample matrices. Therefore, a clean-up/separation step is usually necessary prior to the analytical determination of trace elements to avoid interferences from matrix ions. This sample preparation step should include, where possible, pre-concentration of the target species if this species is present at low concentration in the samples. The techniques commonly used for separation of elements from the matrix components are co-precipitation [10,11], solvent extraction or liquid-liquid extraction [20], cloud point extraction [21–24], membrane filtration [25–28], and solid-phase extraction (SPE) [29].

SPE uses resins containing a separating agent, such as is used in chromatography, in a disposable container as the stationary phase. SPE performs a "solid-liquid extraction" by contacting the liquid sample with the resin stationary phase. Use of this technique as a pre-treatment method for liquid chromatography began around 1970 [29]. Since then it has been used widely in many fields, including analysis of biological samples and as a pre-treatment method for HPLC analyses of pharmaceutical compounds in their various formulations. For many of these applications, the resins are composed of octadecyl groups chemically bonded to silica (ODS) or ion-exchange resins based on silica.

Interaction with the target substance is classified into the following three separation modes: (1) non-polar (reverse-phase mode), (2) polar (normal-phase mode), and (3) IX. From the relationship between the chemical properties of the target substance and the sample matrix, the optimal separation mode is selected from among these three modes.

For analysis of trace metals in samples containing complex matrices, several formats for SPE using IX or other resins have been developed. For samples containing complex matrices, with competing ions often at high concentrations, high selectivity for the target ion is required. Several classes of resin have been used to separate trace metals under high matrix conditions. In Table 2, specific separation modes and corresponding separation resins for several metal separations are shown. Chelating resins are modified IX resins, but with a unique bidentate functional group for separation of transition metal ions. MRT resins have high selectivity for metals [5,6].

12.5.1 Ion Exchange Resins

Ion exchange resins are widely used for separations of Cr(III) and Cr(VI) as well as As(III) and As(V). Cr(VI) is easily captured by anion exchange resins, while Cr(III) is captured by cation exchange resins. Ions containing hexavalent chromium, Cr(VI), are among the most

Table 2 *Commercial Products Containing Various Resin Types and Their Application to Metal Separations*

Separation Mode	Resin Type	Commercial Products[a]	Application
non-polar resins	C18 bonded silica (ODS) Styrene-DVB co-polymer DVB-methacrylate co-polymer alkylmethacrylate co-polymer	Empore Disk C18 InertSep PLS-2 InertSep RP-1 NOBIAS RP-SG1, RP-OD1	Organic metal compounds: organomercury, organotin; metal chelate compounds
ion exchange resins	anion exchange cation exchange	Empore Disk Anion SR Empore Disk Cation SR	speciation of Cr, As, Se; separation of metals in concentrated acid
chelating resins	iminodiacetate polyaminoacetate	Empore Disk chelate Altech maxiclean InertSep ME-1 NOBIAS chelate PA1, chelate-PB1	concentration of trace metals in aqueous solutions and biological fluids for XRF analysis
Highly selective MRT resin	Crown ether, cryptand, calixarene, and other organic ligands	TEVA resin, MetaSep, AnaLig®	metal purification from high matrix industrial solutions, radionuclide separations from environmental solutions

[a]Commercial products: ODS=Empore Disk C18, 3M product; DB=InertSepPLS-2, GL Sciences product; DVB-Methacrylate=InertSep RP-1, GL Sciences product.

important species to be determined in environmental samples, since these species are biological toxins. Identification and analysis of Cr(VI) in soil samples is becoming more important, as the Soil Contamination Countermeasures Law mandates the enforcement of Cr(VI) analysis in Japan. The official method of analysis has been mandated by the Ministry of the Environment of Japan (Ministry of the Environment Notifications No. 18 and 19). This analysis requires a dissolution test with water and an alkaline solution for indirect and direct risk assessment. Recognition of the importance of soil analysis is increasing in Japan, making compliance with laws and regulations more acceptable. For efficiency in analysis, it is desirable that rapid screening and simple field analysis methods be employed.

12.5.2 Chelating Resins

Chelating resins are widely used to separate bivalent transition metal ions from samples prior to ICP-MS or ICP-OES analysis. The Japanese Industrial Standards (JIS) regulation introduced the use of chelating resin SPE for sample preparation against eight metals (Fe, Ni, Cu, Zn, Co, Pb, Cd and U) in JIS K0102 [30]. The main benefit of a chelating resin is its ability to remove the salt contained in the sample.

12.5.3 Molecular Recognition Technology Resins

MRT processes involve use of pre-designed ligands, including macrocycles, immobilized on a solid-phase, such as silica gel, to separate and recover target metal ions from solutions [3,5,6,31,32]. These MRT AnaLig® resins are usually used in column mode and are capable of highly selective separation of target elements from feed solutions containing complex matrices. Following washing of the column to remove traces of feed solution, concentration of the bound elements is achieved by elution of the column with a small amount of eluent. The resulting eluate contains only the element of interest, the remainder of the feed solution having gone on to raffinate [31,32]. The separated element can then be analyzed by an appropriate instrument without interference from matrix ions.

12.6 Green Chemistry Analytical Applications of Metal Separation Resins

12.6.1 Analysis of Trace Levels of Rare Earth Elements in Rainwater in Suburban Tokyo, Japan [33]

Rare earth element (REE) concentrations were determined in 32 rainfall events collected in suburban Tokyo [33,34]. The REE were present in low amounts, i.e., sub ng L^{-1}, in the samples. ICP-MS cannot detect REE at these concentration levels. However, an SPE chelating resin ME-2 (GL Sciences) can be used to separate and concentrate REE prior to ICP-MS analysis. With this concentration procedure, levels of REE lower than ng L^{-1} in the original samples could be determined. The procedure involved use of a chelating resin SPE column MetSEP® (chelate resin disc was pre-packed, d =10 mm, l = 80 mm, v = 6 mL, GL Sciences Inc., Japan). Prior to the separation experiments with the rain samples, the retention characteristics of the ME-2 column as a function of pH were investigated using REE reagent standards [33].

At pH values of 3.7 to 4.0, recoveries of all REE, including Y, ranged from 99 to 101%. Using these results, the following separation procedure was adopted. A 15 mL portion of 2 M ammonium acetate solution was added to 300 mL of the rain sample. The pH of the sample was then adjusted to about 3.7 to 4.0. The rain sample was loaded onto the column with a flow rate of approximately 4 ml min^{-1}. The REEs were eluted by (1–2) mL of 2 M nitric acid and diluted to 10 mL by high-purity water. The concentration factor was 30-fold. All pre-concentration and separation procedures were performed under a class-10 clean bench in the clean room. The REE concentrations in the rain samples were determined by ICP-MS.

12.6.2 Analysis of Metal Pollutants in Aqueous Environmental Samples

Arsenic is a ubiquitous toxic trace element that has raised major toxicological and environmental concerns [3,35]. The speciation of As is dependent on the elemental oxidation state, ionic and molecular forms and concentration levels. Human exposure to high levels of As can have both cancerous and non-cancerous health effects. It is important to determine As concentrations in environmental samples. A combination of MRT SPE products containing immobilized As-selective ligands, i.e., AnaLig® TE-01, AnaLig® AN-01 Si, and AnaLig® As-01 PA, was used for selective separation of water-soluble As species, i.e., arsenite, arsenate, monomethylarsonic acid and dimethylarsinic acid [36]. A simple flow-based method was also developed for the selective separation of tri- and pentavalent As species from aqueous matrices using another MRT SPE product, i.e., AnaLig® An-02 [35]. Arsenic concentrations in experimental samples were determined by graphite furnace AAS. The selectivity characteristics of AnaLig® TE-01 towards As [36] and Fe [37] were further employed to formulate an approach for the treatment of spent iron-oxide coated sand from filters used in As removal [38].

Heavy metal contamination of water resources is causing serious ecological problems worldwide. MRT-SPE AnaLig® products are preferentially used for the separation of heavy metal ions from aqueous solutions prior to the determination of these metal ions, mainly due to the favorable characteristics of the products, i.e., simplicity, low cost, ease of recovery of the solid phase, and ability to attain a high concentration factor [5,6]. Application of MRT-SPE products to the selective separation of Cd [39] from a metal-affluent matrix and of Cr(VI) from wastewater [40] and conversion coatings [41] has been reported. Selective separation of the eco-toxic metal ions Co, Ni, Cu, Zn, and Cd from aqueous solutions was achieved using AnaLig TE-05 [42,43]. The AnaLig TE-01 MRT-SPE system was also used for nondestructive recovery of multiple elements, such as, As(V), Cd(II), Cr(III), Pb(II) and Se(IV), from the effluent containing excess aminopolycarboxylate chelant in solution [44,45].

12.6.3 Analysis of Trace Levels of Lead in High Matrix Plating Solutions

Lead is ubiquitous in nature, used extensively in industrial processes, and extremely poisonous, especially to children [46]. It is essential to have analytical procedures of high selectivity to separate the Pb from its matrix and to concentrate it for subsequent determination. One example is the need to separate trace levels of Pb from concentrated-matrix electroless nickel plating (ENP) waste solutions in order to meet increasingly stringent environmental regulations. Two MRT-SPE products, AnaLig® Pb-01 and AnaLig® Pb-02, were used for selective separation of trace levels of Pb from the waste discharge of ENP operations, followed by analysis with ICP-OES [46]. AnaLig® Pb-01 or AnaLig® Pb-02 packed in 3 mL polypropylene cartridges were used to treat synthetic metal-waste solutions designed to simulate typical metal mixtures in ENP bath waste (Figure 3). The sample solutions contained 100–1000 $\mu g\ L^{-1}$ of Pb in an HNO_3 matrix with pre-added Ni, Cu and other interfering elements (1000 $mg\ L^{-1}$). After loading the Pb from the feed solution, the MRT-SPE cartridges were washed with water and 0.1 M HNO_3, followed by elution with a small amount of 0.03 M EDTA to give a concentrated Pb solution. The matrix elements, i.e., Ni, Cu, were completely removed from the cartridge in the washing step and the bound Pb was then eluted. The concentration of Pb in the eluate was determined by ICP-OES. The detection limit of the proposed method was 2.6 $\mu g\ L^{-1}$.

'Real' samples from commercial ENP operations were used to assess the validity of this method, and almost quantitative Pb recovery was observed [46]. The excellent Pb selectivity of the SPE–MRT system indicates the potential of the proposed technique for trace-level Pb separation from the Pb-containing high matrix aqueous waste discharge.

The effect of co-existing ions on Pb separation and analysis was studied [46]. AnaLig® Pb-01 and AnaLig® Pb-02 products were used for the selective separation of Pb in the presence of other co-existing ions by treating sample solutions containing 20 elements, each at

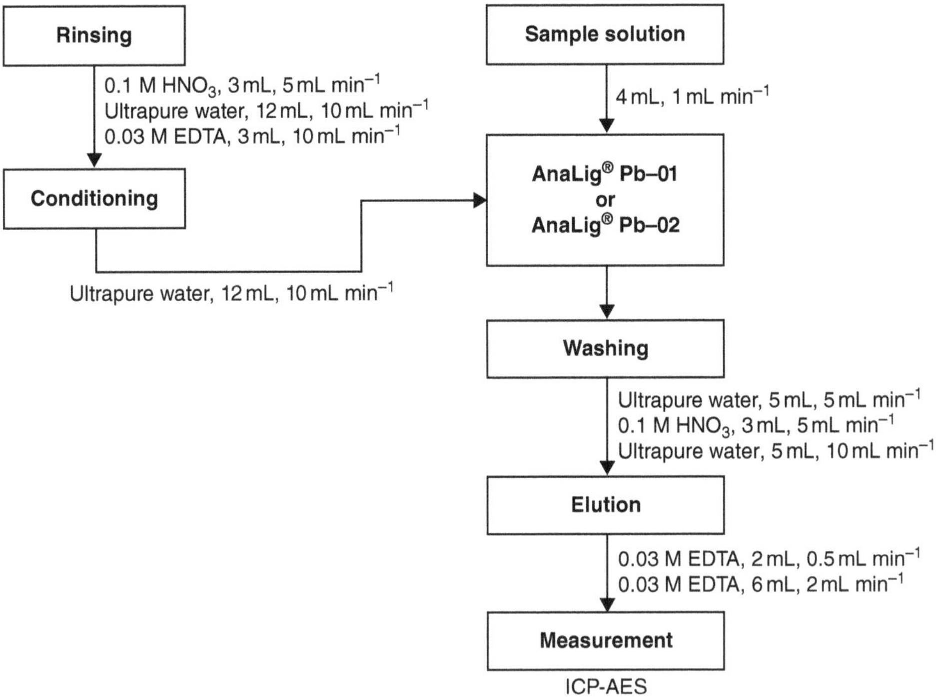

Figure 3 *Procedure for Pb separation using MRT-SPE products (AnaLig® Pb-01, AnaLig® Pb-02). Rahman, 2011 [46]. Reproduced with permission from Elsevier*

a concentration of 100 μg L^{-1} (Figure 4). The elements were added individually to the synthetic sample solutions, and the final solutions were allowed to equilibrate for 24 h before use. The study was carried out in a non-competitive environment by passing 4 mL of the sample at the optimized flow rate through the cartridge. The Pb was retained by the AnaLig® product while the remaining metals passed to raffinate. Following a washing step to remove traces of feed solution, the Pb was eluted using an appropriate eluent. As observed in Figure 4, the recovery rates (%) of Pb were 97.6 ± 3.2 with AnaLig® Pb-01 and 99.3 ± 4.1 with AnaLig® Pb-02, whereas the recovery rates of other elements were less than 37%. Therefore, both AnaLig® Pb-01 and AnaLig® Pb-02 have excellent selectivity and stronger affinity for Pb than for other elements in these acidic matrices. The recovery of Pb was fairly free from interference resulting from the co-existing ions.

Separation of Pb from synthetic high matrix metal waste solutions was studied [46]. Synthetic metal-waste samples (4 mL), mimicking typical metal mixtures in aqueous ENP bath discharge and called 'model' ENP solutions hereafter, were loaded onto SPE-MRT cartridges followed by elution with 0.03 M EDTA (8 mL). The recoveries (%) of Pb from the Cu-matrix 'model' ENP solution were found to be 87.1 ± 2.3 and 94.9 ± 1.8 after treatment with Pb-01 and Pb-02, respectively. Lead recoveries were 75.7 ± 3.2 % for AnaLig® Pb-01 and 93.8 ± 4.7% for AnaLig® Pb-02 from the Ni-matrix 'model' ENP solution.

The Pb separation efficiencies of the AnaLig® Pb-01 and AnaLig® Pb-02 MRT-SPE cartridges when using a matrix with a high level of background Ni were determined using

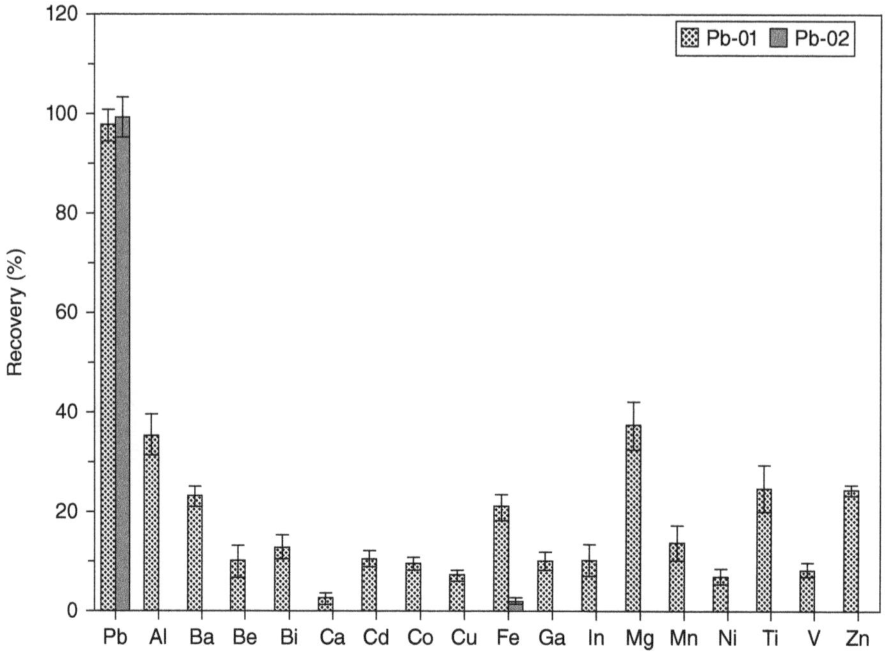

Figure 4 *Effect of co-existing ions on the performance of SPE-MRT cartridges. Sample solution: 100 μg L^{-1} (Pb and other elements), volume: 4 mL, loading flow rate: 1 mL min^{-1}, eluent: 0.03 M EDTA, and elution flow rate: 2 mL at 0.5 mL min^{-1} and 6 mL at 2 mL min^{-1}. Values reported are averages of three determinations in each case. Rahman, 2011 [46]. Reproduced with permission from Elsevier*

'real' ENP bath solutions. 'Real' ENP bath solutions usually contain about 1000 μg L^{-1} of Pb and several g L^{-1} of Ni and other components. Although the exact compositions of the ENP bath solutions of particular companies are proprietary, typical components include nickel sulfate (225 to 400 g L^{-1}), nickel chloride (30 to 60 g L^{-1}), and boric acid (30 to 45 g L^{-1}). The plating operation is typically conducted at temperatures of 44 to 66 °C and pH values of 2 to 4.5 [46]. Two different sample types were available for analysis: freshly prepared and old. After the separation treatment with the AnaLig® Pb-01 or AnaLig® Pb-02 MRT-SPE cartridges, the Pb recovery (%) was about 88.7 ± 1.8 and 104 ± 4.7, respectively, for freshly prepared ENP bath solution, while it was 87.6 ± 3.6 and 97.6 ± 4.1 for old ENP bath solution.

12.6.4 Analysis of Trace Levels of Precious Metals in Recycled Materials

Precious metals include platinum group metals (PGM), Au, and Ag. These metals are scarce and valuable, and are used in a broad range of applications including catalysts, fuel cells, jewelry, and electronics [47]. The demand for these metals in these and other applications is increasing. Recycling of precious metals as well as their extraction from base metal by-product refining is also becoming more important.

Coupled with increased usage of precious metals, there is a critical need for their accurate analysis. Analysis requires efficient and selective sample preparation prior to instrumental determination of the target metal(s) by ICP-MS or ICP-OES procedures. For example, PGM usually occur together in complex ores containing many elements, from which it is often difficult to separate the individual PGM. At the same time, secondary materials containing PGM can also exhibit complex matrices, from which it is difficult to separate individual PGM for analysis. If impurities remain with the separated PGM, subsequent analysis using ICP instruments is compromised. Therefore, it is necessary for individual precious metals to be separated from complex matrices containing base metal and other interferences prior to analysis.

Precious metal laboratories use methods such as sedimentation, liquid-liquid extraction, or IX extraction to separate individual target metals prior to ICP analysis. However, these techniques require many extraction steps, take much time and space, and can be labor-intensive, driving up laboratory costs [43]. These separation issues are successfully addressed by using MRT-SPE AnaLig® products, which are capable of achieving selective separation of precious metals at high purity levels from complex matrices. Following concentration during elution of the bound metal from the AnaLig® product, the precious metal(s) can be readily determined by ICP-OES, without impurity metal interference. A further benefit of the MRT-SPE procedure is that the sample preparation method can be automated, further simplifying the sample preparation procedure. Capital and operating expenses in the analytical laboratory are greatly lowered using the MRT-SPE process (Figure 5). MRT procedures are also effective in the separation and recovery of precious metals on a commercial scale [48–55].

One conventional method of sample preparation for precious metal analysis involves using an open column loaded with an IX resin. The IX resin is not highly selective for precious metals over each other or over accompanying base metals. This low selectivity results in interfering metal ions remaining in the solution, which complicates subsequent precious metal determination by the ICP-OES instrument. MRT-SPE changes the current paradigm by reducing or eliminating these interferents. This elimination is beneficial because time and labor are not required to remove interfering species downstream.

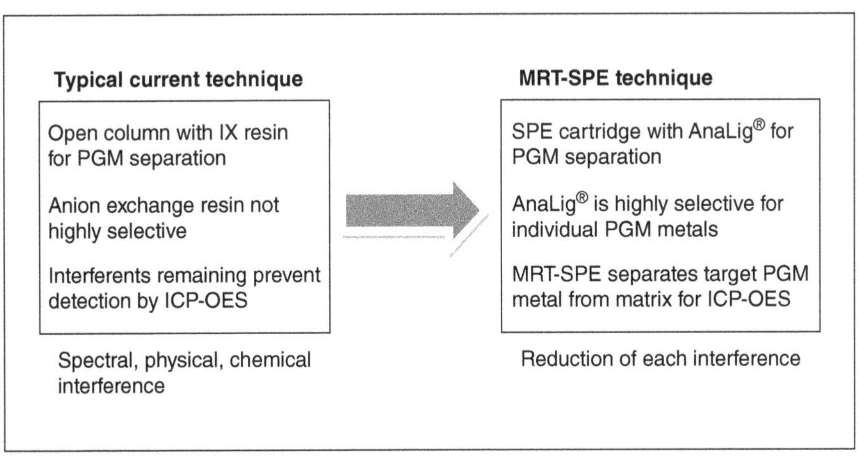

Figure 5 *MRT-SPE Method for Extraction of PGMs from High Matrix Solutions*

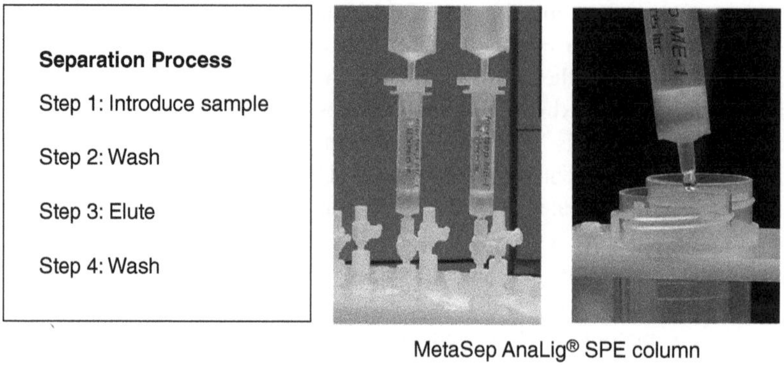

Figure 6 *Typical SPE procedure for precious metals. Furusho, 2008 [56]. Adapted with permission from IPMI. (See insert for color representation of the figure.)*

The MRT-SPE method uses a cartridge format, with highly selective AnaLig® products, as shown in Figure 6. The high selectivity of particular AnaLig® products for individual precious metals, even in the presence of high amounts of interfering base metals, allows preparation of a solution ready for analysis, free from interfering metals.

The MetaSep AnaLig® separation process consists of four steps: (1) introduction of sample, (2) washing of column, (3) elution of metal from column, and (4) washing column for repeat cycle beginning with step 1. In step 1, the sample feed solution is introduced into the column, which contains an AnaLig® product that is selective for the particular precious metal to be determined. After feeding the matrix sample through the column, a wash step removes remaining feed solution preparing the column for elution of the bound precious metal. In the elution step, the precious metal is stripped from the AnaLig® product with a small quantity of eluent, allowing collection of the precious metal in concentrated form, free of matrix interferents. Some eluents are fairly selective for elution of certain precious metals vs. others, e.g., thiourea for Pd, Au, and/or Pt. The pure eluate is then analyzed using the ICP-OES instrument. A final wash step prepares the column for the next loading. The column is regenerable, which allows it to be used repeatedly without decreased function.

The MetaSep AnaLig® process is depicted in Figure 7 [56]. The SPE cartridges containing the AnaLig® product are loaded into a manifold for easy sample preparation and automated handling. DigiTUBEs and drain racks are used in this laboratory setup to more easily collect the solutions used in the process workflow.

Capabilities, feed requirements, and elution reagents for several AnaLig® materials are provided in Table 3. Each AnaLig® product requires a separate workflow, starting with the feed solution required. All AnaLig® products perform separations well in acid solutions, but some of these products are effective in the neutral–slightly basic range as well. Each of the washes (pre-loading and pre-elution) requires an acid solution, with most being 6 M HCl. Thiourea and NH_4^+ are common elution options. Some products will allow elution in hot water (AnaLig® PM-12) or NaCl (AnaLig® PM-10).

While there are very few interferents because of the high selectivity of the AnaLig® products, Pb can interfere, when present in excess in a majority of cases, as seen in Table 3.

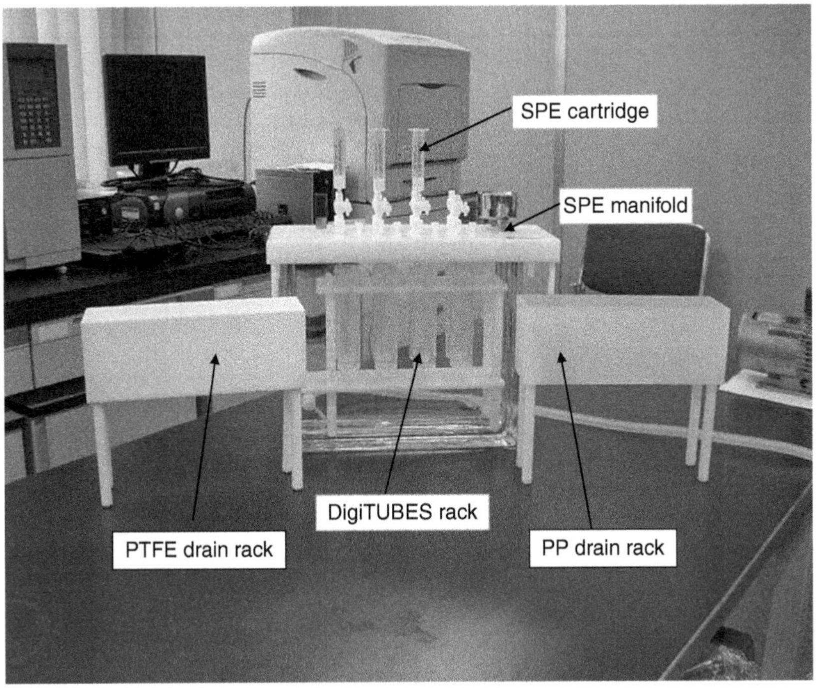

Figure 7 *Typical SPE vacuum manifold for precious metal separation. Furusho, 2008 [56]. Adapted with permission from IPMI. (See insert for color representation of the figure.)*

Cu, Fe and Pt (in excess) may interfere in selected cases. The general lack of interferents is a key advantage of the AnaLig® process.

The excellent results obtained are due to the high selectivity of the AnaLig® resin for the target metal leading to a single metal in the eluent, the high binding affinity of the AnaLig® resin for the target metal, and the simple, accurate, and rapid determination of the separated target metal by the ICP-OES. The results indicate that the method can be extended to wider metal concentration ranges and to lower metal concentrations in the $\mu g\ L^{-1}$ range. The latter is an important consideration, because traditional separation technologies become much less effective as the concentration of the metal to be separated decreases.

This analytical capability using the ICP-OES instrument should have wide applicability for precious metal analysis in the precious metals field. An important potential application is in developing nations, where there is a need for inexpensive and simple-to-operate analytical equipment capable of determining metal concentrations at sub-mg L^{-1} levels, especially where traditional ICP instruments are not available. An inexpensive and portable procedure for making such analyses is available [1,6]. This procedure combines the MRT-SPE process for separation with an inexpensive, portable spectrophotometric instrument, the Ultra Compact Elemental Analyzer, MH-6000, for determination of the separated metal. This process gives analytical results comparable to those obtained using ICP instrumentation.

Table 3 *Precious Metal Capabilities and Feed Requirements for Analig® Materials*

AnaLig®	Feed, pH Range	Precious Metals Recoverable	Other Feed Matrix Requirements	Elution Reagents	Competing Elements
PM-01	Pd, Au: 0-9 Pt, Rh, Ru, Ir: <1	Pd, Au, Pt, Rh, Ru, Ir	Pt: Cl^- Rh, Ru, Ir: >5 M Cl^-	thiourea: Au, Pd, Pt >5 M NH_4^+ or K^+ :Ir, Ru, Rh	Pb in excess in HCl
PM-02	Pd, Au: 0-9 Pt, Ru: <1	Pd, Au, Pt, Ru	Pt: Cl^- Rh, Ru, Ir: >5 M Cl^-	thiourea: Au, Pd, Pt, Ru	Pb in excess in HCl, Cu in large excess
PM-03	0-4	Pd, Au	ORP control	thiourea: Au, Pd	Hg in low Cl^-,Pt at low ORP
Pm-05	0-4	Pd, Au	ORP control	thiourea or NH_3 buffer: Au, Pd	Hg in low Cl^-,Pt at low ORP
PM-06	Pd, Au: 0-9 Pt, Rh, Ru, Ir: <1	Pd, Au, Pt, Rh, Ru, Ir	Pt: Cl^- Rh, Ru, Ir: >5 M Cl^-	thiourea: Au, Pd, Pt >5 M NH_4^+ or K^+ :Ir, Ru, Rh	Pb in excess in HCl
PM-07	Pd, Au: 0-9 Pt, Ru: <1	Pd, Au, Pt, Ru	Pt, Ru: Cl^-	thiourea: Au, Pd, Pt, Ru	Pb in excess in HCl, Cu in large excess
PM-08	Pd, Au: 0-9 Pt, Rh, Ru, Ir: <1	Pd, Au, Pt, Rh, Ru, Ir	Pt: Cl^- Rh, Ru, Ir: >5 M Cl^-	thiourea: Au, Pd, Pt >5 M NH_4^+ or K^+ :Ir, Ru, Rh	Pb in excess in HCl
PM-09	Pd, Au: 0-9 Pt, Ru: <1	Pd, Au, Pt, Ru	Pt, Ru: Cl^-	thiourea: Au, Pd, Pt, Ru	Pb in excess in HCl
PM-10	<1	Rh, Ru(III), Ir(III)	Rh, Ru, Ir: >5 M Cl^-	>5 M NH_4^+ or K^+ or Na^+: Rh, Ru, Ir	Pb in excess in HCl, Pt in large excess
PM-12	0-5	Au, Pt, Ru(IV), Ir(IV)	Cl^- :6 M H^+ or 0.01 M K^+ or 0.01 M NH_4^+ or 0.1 M Na^+	H_2O: Pt, Ru, Ir hot H_2O or thiourea: Au	Fe in high HCl concentration

12.6.5 Analysis of Radioactive Strontium and Other Radionuclides using MRT Rad Disks

The MRT-SPE separation procedure is effective in separations involving a range of radionuclides, including Cs, Sr, Ra, Tc, Cs, and Pu [4,57–71]. Combined with ICP instrumentation, MRT-SPE makes possible rapid, accurate analysis of these metals. This green chemistry procedure for rapidly separating and concentrating these species has significant advantages over conventional procedures. Disadvantages of traditional methods, i.e., IX, precipitation, co-precipitation, and extraction, for separation of Sr-90 from matrices prior to determination

Figure 8 *MRT-SPE column and MRT Empore RadDisk. SPE cartridge: AnaLig® Sr-01 500 mg/3 mL (IBC and GL Sciences); SPE Disk: Empore Sr Rad Disk 47 mm (3M). (See insert for color representation of the figure.)*

by ICP or AA are summarized by Ometacova *et al.* [57] as follows: time required for separation, use of concentrated acids, and generation of large amounts of liquid waste. These authors indicate that use of commercially available products (3M Empore Sr Rad Disk, AnaLig® Sr-01 resin) shortens the separation time and generates little waste. Similar results have been reported by Goken *et al.* [4], who indicate that the Sr-90 Rad Disk method for Sr separation and concentration requires six steps and approximately 20 minutes to complete. This compares to the published EPA Method 905 for radioactive strontium determination that lists over fifty steps are listed in the process and can take days to complete.

The need for rapid, accurate methods for radionuclide analysis is highlighted by the Fukushima Daiichi nuclear disaster in Japan in 2011. In situations such as this, analytical methods should be deployed quickly, results should be obtained rapidly, and minimal waste should be generated. Strontium isotope ratio analysis is also important for evaluation of Sr-80 uptake in foods. It is important that trace element ratio analysis of plant (vegetable) components match the analytical results for trace element ratios in components of cultivated soil.

For radionuclide analysis, two types of MRT-SPE formats are available: cartridge (column) and Empore Rad disk (Figure 8).

An MRT-SPE column using an appropriate AnaLig® product, is useful for separating radionuclides in small volume samples. In contrast, MRT-SPE disks, i.e., Empore Rad Disks, are useful for separating radionuclides in large volume samples. Although the working protocol with column or disk format MRT-SPE's are very much similar (Figure 9), the disk format has the advantage of high-speed separation for waste water analysis in the radionuclide plant, and acid digestion of the waste sample. For example, for Sr analysis, the Empore Sr Rad Disk using an AnaLig® Sr-01 resin selectively removes Sr from 1 mol L^{-1} –8 mol L^{-1} HNO_3 matrices, without removal of other interference metals present (which pass to raffinate). The performance of the two Sr-selective formats: AnaLig® Sr-01 and Empore Sr Rad Disk have been evaluated [57].

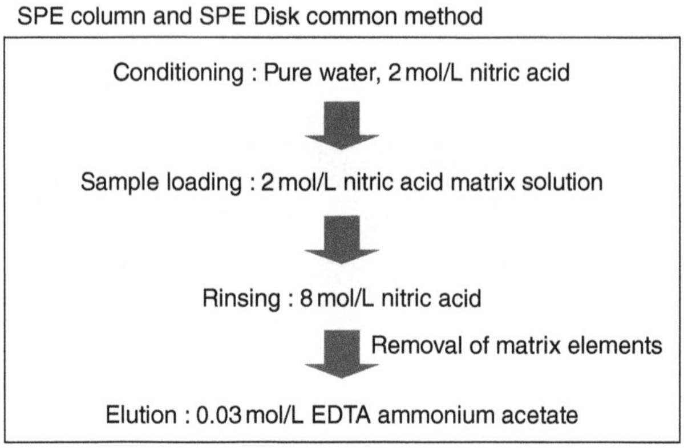

Figure 9 *Conditioning, sample loading, and elution steps for SPE columns and SPE disks*

A key feature of the Sr Rad Disk is its rapid separation of Sr from a large sample volume compared with the MRT-SPE column format (Figure 10).

MRT-SPE-systems have been used for the separation of uranium from nuclear power plant waste [60]. Various MRT-SPE-systems with selectivity for various radioactive elements such as, Sr, Tc, Cs and Ra have been reported [61–67] and evaluated for large-scale applications [68–71]. The demonstrated effectiveness of MRT-SPE systems in the selective separation of a wide variety of radionuclides from complex matrices make them ideal candidates for wide use in the nuclear energy field.

12.7 Conclusions

The need for rapid, reliable, green chemistry methods for element determinations in environmental, industrial, commercial, and radioactive chemical fields is increasing as the global use of metals in a diverse number of products increases. Selective separation of target metals from complex matrices prior to their analytical determination poses a challenge to conventional procedures, such as IX, metal extraction, and precipitation. These procedures are lengthy, use hazardous chemicals and solvents, generate much waste, and become increasingly less effective as target metal concentration levels decrease in the mg L^{-1} region. MRT-SPE systems are based on green chemistry principles and are simple; rapid; use mild chemicals, such as dilute acids or water, for wash and elution steps; and generate minimal waste. MRT-SPE systems are capable of concentrating metals at sub-mg L^{-1} levels in experimental samples many-fold for subsequent ICP or AA metal determination. Conventional separation systems are increasingly challenged as target metal concentrations decrease, especially if complex matrices are present. Going forward, use of MRT-SPE systems ensures greater efficiency in metal analyses and is consistent with rising global expectations of a cleaner environment.

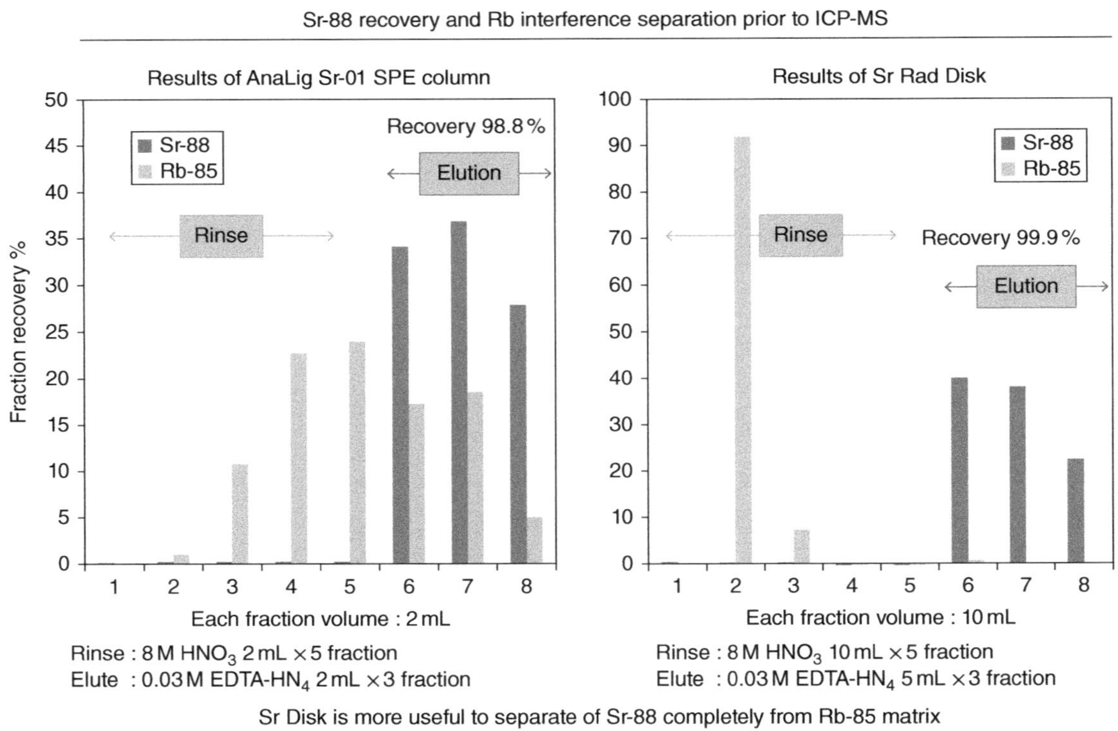

Figure 10 *Comparison of column and Rad disk for recovery of strontium-88 in the presence of rubidium-85 interferent. (See insert for color representation of the figure.)*

References

1. Rahman, I.M.M., Furusho, Y., Begum, Z.A., Sato, R., Okumura, H., Honda, H., Hasegawa, H., 2013, Determination of Lead in Solution by Solid Phase Extraction, Elution, and Spectrophotometric Detection using 4-(2-Pyridylazo)-Resorcinol, *Central European Journal of Chemistry*, 11, 672–678.
2. Hasegawa, H., Rahman, I.M.M., Umehara, Y., Sawai, H., Maki, T., Furusho, Y., Mizuatani, S., 2013, Selective Recovery of Indium from the Etching Waste Solution of the Flat-panel Display Fabrication Process, *Microchemical Journal*, 110, 133–139.
3. Rahman, I.M.M., Begum, Z.A., Hasegawa, H., 2013, Selective Separation of Elements from Complex Solution Matrix with Molecular Recognition plus Macrocycles Attached to a Solid Phase: a Review, *Microchemical Journal*, 110, 485–493.
4. Goken, G.L., Bruening, R.L., Krakowiak, K.E., Izatt, R.M., 1999, Metal- Ion Separations Using SuperLig or AnaLig Materials Encased in Empore Cartridges and Disks, *Metal-Ion Separation and Preconcentration; Progress and Opportunities*, Bond, A.H.; Dietz, M.L.; Rogers, R.D., Eds., ACS Symposium Series 716, American Chemical Society, Washington, D.C., Chapter 17, 251–259.
5. Izatt, R.M., Bradshaw, J.S., Bruening, R.L., Bruening, M. L., 1994, Solid Phase Extraction of Ions of Analytical Interest Using Molecular Recognition Technology, *American Laboratory*, 28C–28M.
6. Izatt, S.R., Izatt, N.E., Bruening, R.L., Furusho, Y., Miyabayashi, Y.T., Ono, M., Kitade, T., Kamimura K., Yamamoto, T., 2013, Analysis of Precious Metals by Solid Phase Extraction Using Highly Selective Molecular Recognition Resins, paper presented at the International Precious Metals Institute 37th Annual Conference, Phoenix, AZ, June 22–25.
7. *RoHS Regulations*, 2010, Government Guidance Notes, Department for Business, Innovation & Skills, Eco-design and Product Regulation Unit, Environmental & Technical Regulation Directorate, London, UK.
8. Anonymous, 2003, Council Directive 2000/53/EC on End of Life Vehicles, in *Environment 315 Directorate*, European Commission, Brussels, Belgium.
9. Council of European Communities, 1986, Council Directive 86/278/EEC on the Protection of the Environment, and in Particular of the Soil, When Sewage Sludge Is Used in Agriculture, *Official Journal of the European Communities*, No. L181, Brussels, Belgium.
10. Patnaik, P., 2004, *Dean's Analytical Chemistry Handbook*, Second Edition, McGraw-Hill, New York.
11. Townshend, A., Jackwerth, E., 1989, Precipitation of Major Constituents for Trace Preconcentration: Potential and Problems, *Pure and Applied Chemistry* 61, 1643–1656.
12. Anastas, P., Eghbali, N., 2010, Green Chemistry: Principles and Practice, *Chemical Society Reviews*, 39, 301–312.
13. Pena-Pereira, F., Lavilla, I., Bendicho, C., 2009, Miniaturized Preconcentration Methods Based on Liquid–Liquid Extraction and their Application in Inorganic Ultratrace Analysis and Speciation: a Review, *Spectrochimica Acta Part B: Atomic Spectroscopy* 64, 1–15.
14. Anthemidis, A.N., 2009, Recent Developments in Homogeneous and Dispersive Liquid–Liquid Extraction for Inorganic Elements Determination: a Review, *Talanta* 80, 413–421.
15. Nishihama, S., Hirai, T., Komasawa, I., 2001, Review of Advanced Liquid–Liquid Extraction Systems for the Separation of Metal Ions by a Combination of Conversion of the Metal Species with Chemical Reaction, *Industrial and Engineering Chemistry Research*, 40, 3085–3091.
16. Shiomi, K., Arsenic in Marine Organisms: Chemical Forms and Toxicological Aspects. In: *Arsenic in the Environment, Part II: Human Health and Ecosystem Effects*, J. O. Nriagu (Ed.), John Wiley, New York, 1994, pp. 261–282.

17. Ichikawa, S., Kaise, T., Hanaoka, K., Hamano, M., Maitani, T., 2004, Intake and Excretion of Arsenic Compounds in Edible Brown Algae in Mice, *Trace Nutrients Research*, 23, 128–133.
18. Yamada, M., Kamidate, Y., Furusho, F., Kuroiwa, T., Kaise, T., Fujiwara, K., 2000, A Rapid Method for the Detection of Arsenic Compounds in Biological Samples with Semimicro Column and an Inductively Coupled Plasma Mass Spectrometer, *Biomedical Research on Trace Elements*, 11, 445–446.
19. Izatt, R.M., Izatt, S.R., Bruening, R.L., Izatt, N.E., Moyer, B.A., 2014, Challenges to Achievement of Metal Sustainability in Our High-Tech Society, *Chemical Society Reviews*, 43, 2451–2475.
20. Müller, E., Berger, R., Blass, E., Sluyts, D., Pfennig, A., 2000, Liquid–liquid Extraction, *Ullmann's Encyclopedia of Industrial Chemistry*, Wiley-VCH Verlag GmbH & Co. KGaA, Weinheim.
21. Stalikas, C. D., 2002, Micelle-Mediated Extraction as a Tool for Separation and Preconcentration in Metal Analysis, *TrAC Trends in Analytical Chemistry*, 21, 343–355.
22. Bezerra, M.d.A., Arruda, M.A.Z., Ferreira, S.L.C., 2005, Cloud Point Extraction as a Procedure of Separation and Pre-Concentration for Metal Determination Using Spectroanalytical Techniques: a Review, *Applied Spectroscopy Reviews*. 40, 269–299.
23. Tani, H., Kamidate, T., Watanabe, H., 1997, Micelle-Mediated Extraction, *Journal of Chromatography A*, 780, 229–241.
24. Sanz-Medel, A., Fernandez de la Campa, M.d.R., Gonzalez, E.B., Fernandez-Sanchez, M.L., 1999, Organised Surfactant Assemblies in Analytical Atomic Spectrometry, *Spectrochimica Acta Part B*, 54, 251–287.
25. Chang, I.-S., Lee, C.-H., 1998, Membrane Filtration Characteristics in Membrane-Coupled Activated Sludge System — The Effect of Physiological States of Activated Sludge on Membrane Fouling, *Desalination*, 120, 221–233.
26. Eriksson, P., 1988, Nanofiltration Extends the Range of Membrane Filtration, *Environmental Progress*, 7, 58–62.
27. Ueda, T., Hata, K., 1999, Domestic Wastewater Treatment by a Submerged Membrane Bioreactor with Gravitational Filtration, *Water Research*, 33, 2888–2892.
28. Wu, J., Eiteman, M.A., Law, S.E., 1998, Evaluation of Membrane Filtration and Ozonation Processes for Treatment of Reactive-Dye Wastewater, *Journal of Environmental Engineering*, ASCE 124, 272–277.
29. Liska, I., 2000, Fifty Years of Solid-Phase Extraction in Water Analysis — Historical Development and Overview, *Journal of Chromatography A*, 885, 3–16.
30. JIS K0102: Testing methods for industrial wastewater (URL: http://www.jisc.go.jp/eng/jis-act/index.html), accessed on October 31, 2015.
31. Izatt, R.M., 1997, Review of Selective Ion Separations at BYU Using Liquid Membrane and Solid Phase Extraction Procedures, *Journal of Inclusion Phenomena and Macrocyclic Chemistry*, 29, 197–220.
32. Bradshaw, J.S., Izatt, R.M., 1997, Crown Ethers: The Search for Selective Ion Ligating Agents, *Accounts of Chemical Research*, 30, 338–345.
33. Iwashita, M., Saito, A., Arai, M., Furusho, Y., Shimamura, T., 2011, Determination of Rare Earth Elements in Rainwater Collected in Suburban Tokyo, *Geochemical Journal*, 45, 187–197.
34. Suzuki, Y., Hikida, S., Furuta, N., 2011, Cycling of Rare Earth Elements in the Atmosphere in Central Tokyo, *Journal of Environmental Monitoring*, 13, 3420–3428.
35. Rahman, I.M.M., Begum, Z.A., Furusho, Y., Mizutani, S., Maki, T., Hasegawa, H., 2013, Selective Separation of Tri- and Pentavalent Arsenic in Aqueous Matrix with a Macrocycle-Immobilized Solid-Phase Extraction System, *Water, Air, & Soil Pollution*, 224, 1526.
36. Rahman, I.M.M., Begum, Z.A., Nakano, M., Furusho, Y., Maki, T., Hasegawa, H., 2011, Selective Separation of Arsenic Species from Aqueous Solutions with Immobilized Macrocyclic Material Containing Solid Phase Extraction Columns, *Chemosphere*, 82, 549–556.

37. Hasegawa, H., Rahman, I.M.M., Kinoshita, S., Maki, T., Furusho, Y., 2011, Separation of Dissolved Iron from the Aqueous System with Excess Ligand, *Chemosphere*, 82, 1161–1167.
38. Rahman, I.M.M., Begum, Z.A., Sawai, H., Maki, T., Hasegawa, H., 2013, Decontamination of Spent Iron-Oxide Coated Sand from Filters Used in Arsenic Removal, *Chemosphere*, 92, 196–200.
39. Hasegawa, H., Rahman, I.M.M., Begum, Z.A., Umehara, Y., Maki, T., Furusho, Y., Mizutani, S., 2013, A Silica Gel-Bound Macrocycle System for the Selective Separation of Toxic Cadmium from Metal-Affluent Aqueous Matrix, *Central European Journal of Chemistry*, 11, 341–347.
40. Furusho, Y., Ono, M., Yamada, M., Kitade, T., Motomizu, S., 2009, Determination of Cr(VI) by Solid-Phase Extraction with Highly Selective Resin, *Bunseki Kagaku*, 58, 147–152.
41. Kamaya, M., Yanagi, S., Nagashima, K., Wakui, J., 2008, Simple and Sensitive Determination of Chromium (VI) in Conversion Coatings, *Journal of The Surface Finishing Society of Japan*, 59, 749–752.
42. Noguchi, O., Oshima, M., Motomizu, S., 2009, Enrichment with Air-Sandwiched Method of On-Line Collection/Concentration Using Chelating Resin and Simultaneous Determination of Trace Elements by Inductively Coupled Plasma Atomic Emission Spectrometry, *Talanta*, 78, 1235–1239.
43. Rahman, I.M.M., Furusho, Y., Begum, Z.A., Sabarudin, A., Motomizu, S., Maki, T., Hasegawa, H., 2011, Selective Separation of Some Ecotoxic Transition Metal Ions from Aqueous Solutions Using Immobilized Macrocyclic Material Containing Solid Phase Extraction System, *Central European Journal of Chemistry*, 9, 1019–1026.
44. Hasegawa, H., Rahman, I.M.M., Nakano, M., Begum, Z.A., Egawa, Y., Maki, T., Furusho, Y., Mizutani, S., 2011, Recovery of Toxic Metal Ions from Washing Effluent Containing Excess Aminopolycarboxylate Chelant in Solution, *Water Research*, 45, 4844–4854.
45. Hasegawa, H., Rahman, I.M.M., Kinoshita, S., Maki, T., Furusho, Y., 2010, Non-Destructive Separation of Metal Ions from Wastewater Containing Excess Aminopolycarboxylate Chelant in Solution with an Ion-Selective Immobilized Macrocyclic Material, *Chemosphere*, 79, 193–198.
46. Rahman, I. M.M., Furusho, Y., Begum, Z. A., Izatt, N., Bruening, R., Sabarudin, A., Hasegawa, H., 2011, Separation of Lead from High Matrix Electroless Nickel Plating Waste Solution Using an Ion-Selective Immobilized Macrocycle System, *Microchemical Journal*, 98, 103–108.
47. Wilburn, D.R., 2012, "Global Exploration and Production Capacity for Platinum-Group Metals from 1995 through 2015 (version 1.1)," U.S. Geological Survey Scientific Investigations Report 2012–5164, 26 pp., at http://pubs.usgs.gov/sir/2012/5164/, Accessed 20 November 2015.
48. Izatt, R.M., Izatt, S.R., Izatt, N.E., Krakowiak, K.E., Bruening, R.L., Navarro, L., 2015, Industrial Applications of Molecular Recognition Technology to Green Chemistry Separations of Platinum Group Metals and Selective Removal of Metal Impurities from Process Streams, *Green Chemistry*, 17, 2236–2245.
49. Izatt, S.R. Bruening, R.L., Izatt, N.E., 2012, Status of Metal Separation and Recovery in the Mining Industry, *Journal of Metals*, 64, 1279–1284.
50. Izatt, S.R., McKenzie, J.S., Bruening, R.L., Izatt, R.M., Izatt, N.E., Krakowiak, K.E., Selective Recovery of Platinum Group Metals and Rare Earth Metals from Complex Matrices Using a Green Chemistry – Molecular Recognition Technology Approach, In *Metal Sustainability: Global Challenges, Consequences, and Prospects*, Izatt, R.M. (Ed), Wiley, Oxford, Publication date 2016.
51. Izatt, S.R., Dale, J.B., Bruening, R.L., 2007, "Examples of Novel Commercial Precious Metal Separations and Recovery Using Molecular Recognition Technology (MRT)." *Symposium on Precious Metals Processing: Advances in Primary and Secondary Operations, Sponsored by SME, TMS, and IPMI,* Tucson, AZ, October 3–6.

52. Izatt, S.R., Dale, J.B., Bruening, R.L., 2007, "The Application of Molecular Recognition Technology (MRT) to Refining of Platinum and Ruthenium," *International Precious Metals Institute, 31st Annual Meeting,* Miami, FL, June 9–12.
53. Izatt, S.R., Mansur, D.M., 2006, "Environmentally Friendly Recovery of Precious Metals from Spent Catalysts," *International Precious Metals Institute Petroleum Seminar,* Houston, TX, November 13–14.
54. Black, W.H., Izatt, S.R., Dale, J.B., Bruening, R.L. 2006, "The Application of Molecular Recognition Technology (MRT) in the Palladium Refining Process at Impala and Other Selected Applications," *International Precious Metals Institute Conference,* Las Vegas, NV, June 10–13.
55. Ichiishi, S., Izatt, S. R., Bruening, R.L., Izatt, N.E., Bruening, M.L., Dale, J.B., 2000, "A Commercial MRT Process for the Recovery and Purification of Rhodium from a Refinery Feed stream Containing Platinum Group Metals (PGMs) and Base Metal Contaminants," *24th Annual International Precious Metals Institute Conference,* Williamsburg, VA, June 11–14.
56. Furusho, A., Izatt, S.R., Bruening, R.L., Izatt, N.E., Motomizu, S., 2008, Use of AnaLig® Molecular Recognition Technology (MRT) Products for Analysis of Platinum Group Metals, *International Precious Metals Institute, 32nd Annual Conference,* Phoenix, AZ, June 7–10.
57. Ometáková, J.O., Dulanská, S., Mátel, L., Remenec, B., 2011, A Comparison of Classical ^{90}Sr Separation Methods with Selective Separation using Molecular Recognition Technology Products AnaLig® SR-01 gel, 3M Empore™ Strontium Rad Disk and Extraction Chromatography Sr Resin®, *Journal of Radioanalytical and Nuclear Chemistry*, 290, 319–323.
58. Remenec, B., 2006, "The Selective Separation of ^{90}Sr and ^{99}Tc in Nuclear Waste Using Molecular Recognition Technology Products," *Czechoslovak Journal of Physics*, Vol. 56, Suppl. D, pp. D645–651.
59. Izatt, S.R., Bruening, R.L., Krakowiak, K.E., Izatt, R.M., 2003, "The Selective Separations of Anions and Cations in Nuclear Waste using Commercially Available Molecular Recognition Technology (MRT) Products," *WM'03 Conference,* Tucson, AZ, February 23–27.
60. Izatt, S.R., Bruening, R.L., Izatt, N.E., Dale, J.B., 2009, The Application of Molecular Recognition Technology (MRT) in the Nuclear Power Cycle: From Uranium Mining and Refining to Power Plant Waste Separation and Recovery, As Well As Element Analysis and Isotope Purification, *WM '09 Conference,* Phoenix, AZ, March 1–5.
61. Vajda, N., Kim, C.-K., 2010, Determination of Radiostrontium Isotopes: a Review of Analytical Methodology, *Applied Radiation and Isotopes*, 68, 2306–2326.
62. Kameo, Y., Katayama, A., Fujiwara, A., Haraga, T., Nakashima, M., 2007, Rapid Determination of ^{89}Sr and ^{90}Sr in Radioactive Waste Using Sr Extraction Disk and Beta-Ray Spectrometer, *Journal of Radioanalytical and Nuclear Chemistry*, 274, 71–78.
63. Kameo, Y., Katayama, A., Hoshi, A., Haraga, T., Nakashima, M., 2010, Simple Determination of ^{99}Tc in Radioactive Waste Using Tc Extraction Disk and Imaging Plate, *Applied Radiation and Isotopes*, 68, 139–143.
64. Condomines, M., Rihs, S., Lloret, E., Seidel, J.L., 2010, Determination of the Four Natural Ra Isotopes in Thermal Waters by Gamma-Ray Spectrometry, *Applied Radiation and Isotopes*, 68, 384–391.
65. Lee, D.D., Walker, J.F., Taylor, P.A., Hendrickson, D.W., 1997, Cesium-Removal Flow Studies Using Ion Exchange, *Environmental Progress*, 16, 251–262.
66. King, W.D., Spencer, W.A., Hamm, L.L., Aleman, S.E., Duffey, C.E., Pettis, M.A., 2005, Column Performance Testing of Superlig® 639 Resin with Simulated Hanford Waste Supernates: Identification of the Primary Sorbing Species and Detailed Characterization of their Desorption Profiles, *Separation Science and Technology*, 40, 383–394.

67. Fiskum, S.K., Blanchard, D.L., Arm, S.T., Peterson, R.A., 2005, Cesium Removal from Simulated and Actual Hanford Tank Waste Using Ion Exchange, *Separation Science and Technology*, 40, 51–67.
68. Peterson, R.A., Fiskum, S.K., Arm, S.T., Blanchard, D.L., 2006, Cesium Removal Demonstration Using Selected Actual Waste Samples from the Hanford Reservation Tank Farm, *Separation Science and Technology*, 41, 2361–2371.
69. Devol, T.A., Clements, J.P., Farawila, A., O'Hara, M.J., Egorov, O.B., Grate, J.W., 2009, Characterization and Application of SuperLig® 620 Solid Phase Extraction Resin for Automated Process Monitoring of ^{90}Sr, *Journal of Radioanalytical and Nuclear Chemistry*, 282, 623–628.
70. Fiskum, S.K., Blanchard, D.L., Steele, M.J., Wagner, J.J., 2006, Analysis of Spent SuperLig® 644 Resin Used for Cesium Removal from Hanford Tank Wastes, *Solvent Extraction and Ion Exchange*, 24, 65–79.
71. Adu-Wusu, K., Hassan, N.M., Nash, C.A., Marra, J.C., 2006, Sorption of Cesium from Aqueous Waste Solution on a SuperLig® 644 Resin, *Journal of Radioanalytical and Nuclear Chemistry*, 267, 381–388.

13

Ionic Liquids for Sustainable Production of Actinides and Lanthanides

Paula Berton[1,2], *Steven P. Kelley*[1,2], *and Robin D. Rogers*[1,2]

[1]*Department of Chemistry, The University of Alabama, Tuscaloosa, AL, 35487, USA*
[2]*Department of Chemistry, McGill University, 801 Sherbrooke St. West, Montreal, QC, H3A 0B8, Canada*

List of Abbreviations

General Abbreviations

IL	Ionic Liquid
LLE	Liquid-liquid extraction
PUREX	Plutonium uranium redox extraction
sc-CO_2	Supercritical CO_2
SLM	Supported liquid membrane
SPE	Solid phase extraction
TALSPEAK	Trivalent actinide lanthanide separation by phosphorous reagent extraction from aqueous komplexes

Ions and Molecules

[A336]Cl	Aliquat 336 ($[N_{x,x,x,x}][Cl]$ where x = 8 or 10)
$[BF_4]^-$	Tetrafluoroborate
$[C_3mpip]^+$	Propylmethylpiperidinium
$[C_4mpyrr]^+$	Buthylmethylpyrrolidinium
$[CA\text{-}100]^-$	*s*-Nonylphenyloxy acetate

Metal Sustainability: Global Challenges, Consequences, and Prospects, First Edition. Edited by Reed M. Izatt.

$[CA\text{-}12]^-$	*s*-Octylphenyloxy acetate
$[C_n mim]^+$	1-Alkyl-3-methylimidazolium
$[Cyanex272]^-$	Bis(2,4,4-trimethylpentyl)phosphinate
$[DEHEHP]^-$	Di(2-ethylhexyl)-2-ethylhexylphosphonate
$[DEHP]^-$	Di-(2-ethylhexyl)phosphate
$[DGA]^-$	Diglycolamide
$[N_{1,8,8,8}]^+$	Trioctylmethylammonium
$[NTf_2]^-$	Bis(trifluoromethanesulfonyl)amide
$[OTf]^-$	Trifluoromethanesulfonate
$[P507]^-$	Di-(2-ethylhexyl) orthophosphinate
$[P_{6,6,6,14}]^+$	Trihexyl(tetradecyl)phosphonium
$[PF_6]^-$	Hexafluorophosphate
$[TBP]^-$	Tributylphosphate
$[TS]^-$	Thiosalicylate
CMPO	*N*-octyl(phenyl)-*N,N*-diisobutylcarbonoylmethylphosphane oxide
DODGAA	*N,N*-dioctyldiglycolamic acid
HDBP	Dibutylphosphoric acid

13.1 Introduction

Lanthanides and actinides, the f-elements, have emerged as some of the most important metals in modern energy security. Lanthanides have unique optical and magnetic properties that make them critical elements in the manufacture of high-tech and clean-energy products, together with defense applications such as jet fighter engines, missile guidance systems, anti-missile defense, and satellite and communication systems (Humphries 2013). Actinides, specifically uranium and plutonium, are the only fuel source usable in nuclear power generation (Choppin *et al.* 1996), which presents a number of attractive features: It avoids air pollution and greenhouse gas emissions during production (limiting climate change), it is largely immune to the intermittency and unpredictability exhibited by wind and solar energy, and it uses fuels with very high energy density (OECD 2011). Energy and national security in the future will very likely depend on establishing sustainable supply chains for these elements.

For any f-element production to be sustainable, the economic, environmental, and socio-political aspects need to be considered. Currently, China dominates the lanthanide supply from primary mining (>90% of the global production in 2011; Humphries 2013). The global effect of reduction of Chinese exports of lanthanides in 2011 revealed the fragility of the lanthanide supply chain (Martina 2011). Viable uranium reserves (the natural resource which is ultimately the source of all presently used nuclear fuel) are estimated to be exhausted within 100 years at current rates of use (WNO 2011). The mining and processing of actinides and lanthanides is a chemical- and energy-intensive process, and the large amounts of caustic chemicals and generation of radioactive ore tailings are a major environmental concern (Izatt *et al.* 2014; Johnson 2014). Consequently, there is an increased demand for technologies to recycle lanthanides and actinides and thus reduce the mining of virgin ore (OECD 2011, IAEA 2015).

Strategic metals such as U, Pu, minor actinides (primarily Np and Am), and even valuable metals including lanthanides are also found in spent nuclear fuel (Vasudeva Rao *et al.*

2008; Busquim e Silva *et al.* 2010). Unused U and Pu are recovered from spent fuel, but other elements, including approximately 1 wt% of actinides Am, Cm, and Np, are left behind as high-level waste (HLW). Separations that recover these actinides are necessary to improve the sustainability of the fuel cycle, increase the amount of energy extracted from natural resources, and permanently dispose of minor actinides (Kolarik 2013). Current separations in fuel processing use volatile, combustible, and toxic solvents, and traces of solvent components are carried into effluent streams (Spietelun *et al.* 2014). More efficient separations are thus needed to overcome waste generation associated with lanthanide production and nuclear fuel processing.

Since the late 1990s, ionic liquids (ILs) have emerged as solutions for developing sustainable technologies, especially for separations (Huddleston *et al.* 1998). ILs are salts with low melting points, a category that has grown to include an extremely broad range of compounds. The numerous possibilities in IL structure design have led to the enormous number and diverse nature of suggested IL applications in different fields, as reflected in the recent reviews published in different areas including energy storage (MacFarlane *et al.* 2014), pharmaceuticals (Hough & Rogers 2007), energetic materials (Smiglak *et al.* 2007), catalysis (Hallet & Welton 2011), and separations (Joshi & Anderson 2012), among others (Wasserscheid & Stark 2013). IL properties can be "tuned" depending on the choice of ions. While the universal assumption that ILs are "greener" than organic solvents is no longer accepted, for certain applications their physical properties and chemical tunability can truly lead to advantages in sustainability. Consequently, ILs are becoming important options in separations not only due to their physical properties, but also for improving the sustainability of metal ion production through unprecedented extraction efficiencies or new strategies for solubilizing metallic complexes in reduced volumes of non-volatile liquids.

Here we discuss several of the directions in which ILs can be applied towards better and sustainable separations for f-element production. The most obvious and well-studied of these is the application of ILs to liquid-liquid extractions, which predominates in both the ore and nuclear fuel processing industries. However, it is important to recognize that the application of ILs is not restricted to replacing a component in an established system. ILs have properties which deviate from those of water and organic solvents, leading to interesting possibilities in ore dissolution and non-aqueous electrochemistry. Lastly, we discuss the application of ILs in the extraction of uranium from seawater to illustrate the beneficial impact on each other of tandem development of these two technologies.

13.2 f-Element Chemistry in Ionic Liquids

A discussion of aspects of inorganic chemistry in ILs is necessary to understand how these can be applied towards more sustainable f-element separations. Chemistry in ILs can differ from what is observed in organic solvents and higher-temperature molten salts, and ILs have proven to be useful across a wide range of applications in inorganic chemistry (Taubert 2010). The chemistry of f-elements itself is rather well developed, going beyond separations and into materials chemistry. f-Element chemistry in ILs has been reviewed multiple times recently (Binnemans 2007; Mudring & Tang 2010; Billard *et al.* 2011a), so it is discussed here only briefly.

Because ILs are composed of ions, pure ILs can often complex with and dissolve f-element salts without the need for any complexing agent. Some coordinating IL ions are shown in Figure 1. For ILs with non-coordinating cations, the anions typically coordinate to the metal cation to form an anionic complex, which has been observed both in crystal structures of metal complexes grown from ILs with halide (Nikitenko *et al.* 2005; Nikitenko *et al.* 2007a; Hines *et al.* 2008) and nitrate anions (Cocalia *et al.* 2010; Qu *et al.* 2014; He *et al.* 2015) and in spectroscopic studies on IL solutions (Gaillard *et al.* 2005; Gaillard *et al.* 2007; Gaillard *et al.* 2012). Anions such as bis(trifluoromethanesulfonyl)amide ($[NTf_2]^-$) that are considered to be non-coordinating in most solvents have been shown to coordinate Ln^{3+} ions when used in ILs (Mudring *et al.* 2005). IL cations will coordinate if they contain a basic functional group. In many cases, metal-selective extractants are grafted onto the core of a typical IL-forming cation, such as imidazolium, to make what are commonly known as task-specific ILs (TSILs) (Visser *et al.* 2001). Functional groups used in f-element separations such as carbamoylmethylphosphine oxide have been used to make TSILs for f-element selective separations (Mohapatra *et al.* 2013). Cations functionalized with acidic groups such as carboxylic acids (Nockemann *et al.* 2006a) or oximes (Barber *et al.* 2012) will exchange the proton for a metal cation to form zwitterionic complexes.

The complexation of neutral ligands to metal ions in ILs is of particular importance in separations and can differ from organic solvents in a number of ways. Unlike organic solvents, ILs often favor the formation of charged complexes, such as $[Sr(dibenzo\text{-}18\text{-}crown\text{-}6)(OH_2)_2]^{2+}$ (Jensen *et al.* 2003) and $[UO_2(NO_3)(CMPO)(OH_2)_x]^+$ (CMPO, n-octyl(phenyl)-*N,N*-diisobutylcarbonoylmethylphosphane oxide) (Visser *et al.* 2003). ILs tend to be much more hygroscopic than water-immiscible organic solvents, and water molecules in ILs are able to compete with weakly coordinating ions for metal binding (Maerzke *et al.* 2013). Also, despite their coordinating nature, ILs have proven useful for isolation of hard-to-obtain f-element complexes with neutral ligands, including Ln^{3+} complexes with tetrahydrofuran (Vasudevan *et al.* 2011) and uranyl ($[UO_2]^{2+}$) complexes with 1-methylimidazole (Gutowski *et al.* 2007).

f-Elements can also be incorporated into ILs as constituent ions. This is in spite of the fact that, due to their high coordination numbers, anionic f-element complexes tend to be highly charged. Dialkylimidazolium salts of partially hydrated Ln^{3+} thiocyanate complexes were the first examples of f-element containing ILs (Nockemann *et al.* 2006b). Since then, ILs based on lanthanide halides (Tang *et al.* 2008), nitrates (Singh *et al.* 2006, He *et al.* 2015), and more complex ligands such as dicyanonitrosomethanide (Chesman *et al.* 2012) have been developed. Anionic $[UO_2]^{2+}$ thiocyanate complexes have also been incorporated into ILs (Aoyagi *et al.* 2011). Most of these f-element ILs are investigated for materials applications, but they have interesting potential in separations due to the high loading of metal ions.

13.3 Applications of Ionic Liquids in f-Element Isolation

13.3.1 Liquid-Liquid Extractions

Liquid-liquid extraction (LLE) is a very important process for the separation and purification of lanthanides and actinides. The process is also of prime importance for the separation of fission products from reusable fissile material in spent nuclear fuel elements. The importance

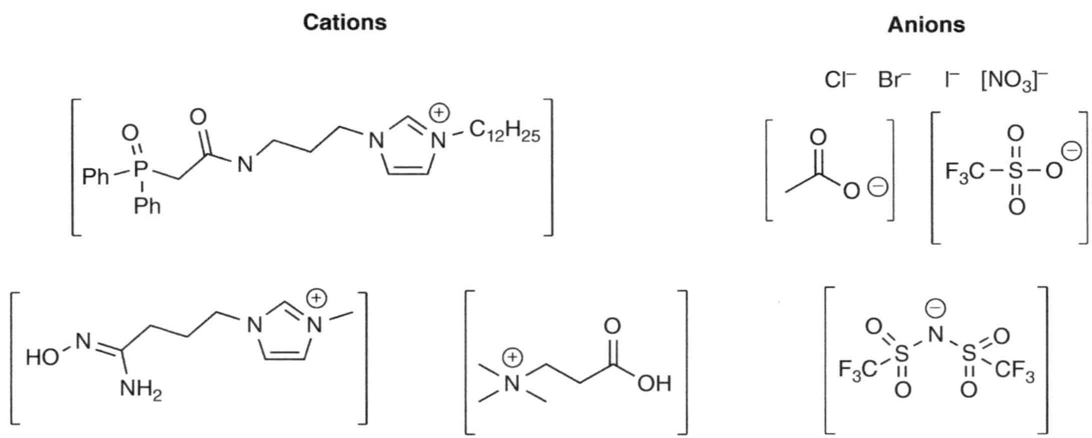

Figure 1 Examples of coordinating ions used to make ILs

of LLE in the nuclear processing industry led to the application of ILs as replacements for volatile organic solvents as non-aqueous phases in fuel-cycle relevant metal ion extractions (Dai *et al.* 1999, Rogers *et al.* 1999). Early into this research, ILs were generally found to have a strong effect on the extraction, particularly through promoting ion exchange mechanisms (Jensen *et al.* 2003) and increasing the solubility of hydrated ions (Visser *et al.* 2000). The extraction mechanism could also be controlled by choosing appropriate ions and extractants (Cocalia *et al.* 2005). Since then, much research has focused on using ILs not just to replace the non-aqueous phase but to significantly enhance the extraction process in ways not possible with organic solvents. The growth of this field is illustrated by the impressive number of reviews on the subject (Binnemans 2007; Billard *et al.* 2011a; Kubota *et al.* 2012; Sun *et al.* 2012a; Kolarik 2013; Takao *et al.* 2013). Here we discuss examples showing the wide spectrum of possibilities for increasing the sustainability of the LLE process through the use of ILs.

One way in which ILs can improve the sustainability of extraction processes is through achieving better extraction efficiencies, thereby reducing the volume of chemicals needed. Many extractants have been shown to extract f-elements from acidic feeds (HNO_3) more efficiently when dissolved in an IL than in a molecular solvent (Sun *et al.* 2012a). In many cases the improved efficiencies in ILs are correlated with ion exchange extraction mechanisms that are promoted by small, hydrophilic cations and suppressed by hydrophobic ones (Cocalia *et al.* 2005; Ansari *et al.* 2014; Rout & Binnemans 2015). The ion exchange mechanism does pose a disadvantage in that the IL itself is lost due to partitioning into the aqueous phase, but IL extraction systems can be designed so that the extraction mechanism may be similar to that of molecular solvents (Cocalia *et al.* 2005; Dietz & Stepinski 2008; Billard *et al.* 2011b). The ability to modify the side-chain functional groups of ILs, incorporating functional groups into the cation or anion of the IL, and thus design ILs for specific purposes, can also be exploited to increase sustainability through increasing extractant selectivity while avoiding ion-exchange. An interesting example was proposed by Vicente *et al.* (2012) in which phosphine oxide functionalized ILs reacted with 1,1,1,5,5,5-hexafluoroacetylacetonate and Eu^{3+} in methanol to form a tetrakis β-diketonate complex that crystalized out of the solution, thus extracting and isolating Eu^{3+} at the same time.

The addition of small amounts of IL to a conventional molecular solvent can also be used to improve extraction efficiency due to synergistic effects. Addition of ILs favors the extraction of ions and greatly increases metal distribution ratios. ILs dissolved in organic solvents can act as sources of hydrophobic anions which form neutral, extractable complexes with metal ions. A considerable synergistic effect for the extraction of lanthanides and actinides from nitric acid was observed when 1-alkyl-3-methylimidazolium bis(trifluoromethanesulfonyl)amide ($[C_n mim][NTf_2]$) ILs were dissolved within the organic phase with alkylphosphine oxides (Turanov *et al.* 2008; Pribylova 2011), and CMPO (Turanov *et al.* 2013) as complexing agents.

When using ILs to reduce the volume of chemicals consumed in an extraction process, another important trait is the ability to control the stability of IL extractants or receiving phases through careful selection of the ions. Many of the fluorinated anions used to make hydrophobic ILs can be hydrolytically unstable, including hexafluorophosphate ($[PF_6]^-$) and tetrafluoroborate ($[BF_4]^-$) (Swatloski *et al.* 2003; Freire *et al.* 2010). Such ions are unstable at high nitric acid concentrations (Zuo *et al.* 2008; Kolarik 2013), and the hydrolysis of $[PF_6]^-$ is accelerated by MCl_6^{2-} complexes (where M^{4+} is Th^{4+}, Np^{4+}, or Pu^{4+}), leading to the formation of soluble and insoluble species of M^{4+} (Nikitenko *et al.* 2007b). Hydrolytically stable ILs can be made just by using more stable anions such as $[NTf_2]^-$ and Cl^-.

When working with actinides, ILs should also be designed to enhance the stability of the extraction system against radiation (Mincher & Wishart 2014; Wishart 2014). ILs containing boron or chlorine are of particular interest owing to the thermal neutron scavenging ability of ^{10}B and ^{35}Cl (Lall-Ramnarine *et al.* 2009). Additionally, ILs based on biorenewable or biodegradable ions are available (Fukaya *et al.* 2007), and their use may avoid environmental concerns on their dissolution into the aqueous phase (Coleman & Gathergood 2010).

One of the most challenging and important separations tasks in fuel processing is the separation of minor actinides from lanthanides. This separation can be achieved using marginal differences between the affinities of the two groups towards certain reagents such as soft (N) donor ligands. Interest in applying ILs towards these separations has persisted since the first studies of f-element coordination chemistry with actinide-selective ligands such as dialkylphosphoric or dialkylphosphinic acids (Cocalia *et al.* 2005), but actual experimental data on actinide-lanthanide separations have only appeared more recently. The effects of a number of reagents used in actinide-lanthanide separations were first studied experimentally in the extraction of aqueous Eu^{3+}/Am^{3+} into $[C_8mim][NTf_2]$ (Rout *et al.* 2011). Exceptionally large separation factors (in the range of several thousands) were achieved using bis(5,6-dialkyl-1,2,4-triazinyl)pyridine dissolved in $[C_nmim][NTf_2]$ (where $n = 2, 3, 4, 6, 8$), even with 0.1 M HNO_3, through an ion-pair mechanism (Bhattacharyya *et al.* 2015). Eu^{3+}/Am^{3+} separations based on the well-studied Trivalent Actinide Lanthanide Separation by Phosphorous reagent Extraction from Aqueous Komplexes (TALSPEAK) process (Nilsson & Nash 2007) have also been performed in ILs by incorporating the actinide-selective ligand diethylenetriaminepentaacetic acid into a TSIL using $[NTf_2]^-$ ILs of choline (2-hydroxy-*N,N,N*-trimethylethanamonium) and betaine (*N,N,N*-trimethylglycine as diluents (Shkrob *et al.* 2014).

The work done with imidazolium ILs has been very important for elucidating IL-based extraction mechanisms, but it is unlikely that these particular ILs would be used in an industrial system. Though these are the most widespread ILs in use today, they are a small fraction of the possible ILs available for use in separations. Alternatively, non-fluorinated, hydrophobic ILs can be made using quaternary group 5A cations with long alkyl chains (abbreviated here using the naming convention $[R_{n1,n2,n3,n4}]^+$, where R is the group 5A atom and $n1$–$n4$ are the number of carbon atoms in the alkyl chains). These include trioctylmethylammonium chloride ($[N_{1,8,8,8}]Cl$), trioctylmethylammonium nitrate ($[N_{1,8,8,8}][NO_3]$), Aliquat 336 (A336, a technical grade mixture of $[N_{1,x,x,x}]Cl$, x = 8 or 10]) and trihexyl(tetradecyl) phosphonium chloride (Cyphos IL 101, or $[P_{6,6,6,14}]Cl$), which, due to hydrophobicity arising from the long alkyl chains on the cation, exhibit no cation exchange with the aqueous phase during extraction (Stockmann & Ding 2010; Bell & Ikeda 2011).

Several ILs derived from $[N_{1,8,8,8}]Cl$ have been synthesized and proposed as bifunctional ionic liquid extractants for the extraction of f-elements. Bifunctionality here refers to the strategy of incorporating separate functions in each ion: the cation is used to increase hydrophobicity while the anion is used for metal coordination, although the cation also participates to increase the hydrophobicity of the formed complex. Bifunctional ILs with extractant anions (diglycolamide ($[DGA]^-$), di-(2-ethylhexyl)phosphate ($[DEHP]^-$), di-(2-ethylhexyl) orthophosphinate ($[P507]^-$), s-octylphenyloxy acetate ($[CA\text{-}12]^-$), s-nonylphenyloxy acetate

([CA-100]⁻), thiosalicylate ([TS]⁻), or bis(2,4,4-trimethylpentyl)phosphinate ([Cyanex272]⁻)) have improved miscibility with traditional organic solvents compared to ILs such as $[N_{1,8,8,8}]$ $[NO_3]$ (Sun *et al.* 2010a; Rout *et al.* 2012). These ILs were evaluated for the extraction and separation of f-elements at pH values >2 using either organic solvents or the IL $[N_{1,8,8,8}][NO_3]$ as the receiving phase and showed remarkably higher extraction efficiencies than their acidic extractant precursors (Srncik *et al.* 2009; Sun *et al.* 2010b; Wang *et al.* 2011; Rout *et al.* 2012; Guo *et al.* 2014; Rout & Binnemans 2014). The selectivity of these ILs could be controlled by manipulating the anion. The ILs $[N_{1,8,8,8}][NO_3]$ (with the extractant di(2-ethylhexyl)-2-ethylhexylphosphonate, [DEHEHP]⁻, dissolved in the IL), $[N_{1,8,8,8}]$[CA-12], $[N_{1,8,8,8}]$[P507], and $[N_{1,8,8,8}]$[DEHP] were found to promote selective extraction of light lanthanides from heavy lanthanides (Sun & Waters 2014; Sun *et al.* 2012b; Sun *et al.* 2014, Zhu *et al.* 2015), while $[N_{1,8,8,8}]^+$ acid phthalate, and $[N_{1,8,8,8}]^+$ maltolate were used for selective extraction of U^{4+} and Th^{4+} (Biswas *et al.* 2014; Platzer *et al.* 2014).

Bifunctional ILs differ from their acidic extractant precursors in that both the anion and cation participate in the extraction and undergo multiple combinations to form supramolecular complexes. We have observed this type of interaction in the crystal structure of $[N_{4,4,4,4}]_2[LaCl_3$(*O,O′*-diethyldithiophosphate$)_2]$, which was crystallized by reacting $LaCl_3 \cdot 7H_2O$ with the IL $[N_{4,4,4,4}]$[*O,O′*-diethyldithiophosphate] (Kelley *et al.* 2015). In this structure, the quaternary ammonium counterions interact electrostatically with the complex anion to form a neutral structure reminiscent of reverse micelles. Incorporating acidic extractants into ILs as anions has also been proposed to increase the solubility and coordinative ability of the extractant group by disrupting hydrogen bonding which is present between acid extractant molecules (Figure 2, Sun *et al.* 2010b). The formation of supramolecular extractant complexes results in an increased viscosity of the organic phase, which can be considered an important drawback for their practical application. Modifiers including isooctanol or nitrate salts need to be added to the system, otherwise a stable third phase, composed mainly of the macromolecular extracted complexes, is formed and located between the aqueous and organic phases (Rout *et al.* 2012; Yang *et al.* 2012a).

LLE presents certain limitations, including relatively low achievable enrichment factors, extensive contact and settling times, and large volumes of reagent consumption/waste production (Pena-Pereira & Namieśnik 2014). Although most of the evaluated publications on LLE in ILs are based on the replacement of the traditional organic phase with an IL (hydrophobic and/or bifunctional ILs), we believe that this approach is just the first step to increasing the sustainability of the process. Due to their tunable properties, ILs can be incorporated in other systems besides LLE, thus motivating researchers to search for greener alternatives for extraction processes. For example, while the relatively high

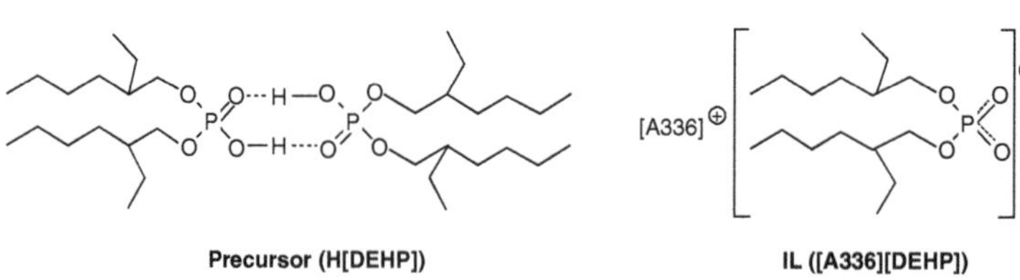

Figure 2 *Proposed dimeric structure of the precursor (left) and the monomeric IL based on [DEHP]- (right). Sun (2010b). Reproduced with permission from Elsevier*

viscosity of ILs is often cited as a disadvantage in LLE, the same high viscosity allows the use of IL-based supported liquid membranes (SLM) for the separation of Y^{3+}, Eu^{3+}, Dy^{3+}, and Nd^{3+} (Kubota *et al.* 2010; Baba *et al.* 2011). The SLM system is formed with an immiscible phase impregnated in the micropores of a polymer support, which separates two aqueous phases (the feed and receiving phases). The liquid membrane method requires only a small amount of the solvent as the membrane phase and allows the separation without agitation of viscous ILs. A carrier molecule selectively binds the target ions, simultaneously extracting and stripping them from one phase to the other. Quantitative transport of the lanthanide ions was achieved by using [C_8mim][NTf_2] supported on polyvinylidene fluoride film as the membrane phase and *N,N*-dioctyldiglycolamic acid (DODGAA) as the carrier. It was reported that DODGAA forms a neutral compound with lanthanides, where three DODGAA monomers complex one lanthanide ion (Yang *et al.* 2012b). The membrane was stable during the operation (more than 100 h) and allowed the separation of these ions from impurities such as Zn^{2+} and Fe^{2+} (Kubota *et al.* 2010; Baba *et al.* 2011).

Another strategy to increase sustainability is to increase the phase ratio, thereby minimizing the volume of the non-aqueous phase (the main source of chemical consumption and waste generation). Microextraction methodologies using ILs are widely developed for transition and main group metal ions, using different strategies to increase the extraction recoveries and reduce the extraction time (Martinis *et al.* 2014). However, only dispersive liquid-liquid microextraction techniques have been proposed for the extraction of uranium from aqueous solution using complexants dissolved in [C_4mim][PF_6] through a micelle formation mechanism. Here, an enrichment factor (calculated as the ratio between the analyte concentration in the IL phase and the initial concentration on the sample) of 448 was achieved at an aqueous to IL phase ratio of 250 (Kaykhaii & Ghasemi 2013). The development of further microextractions based on ILs already proposed for LLE is an open field that should be explored and incorporated in the discussion of sustainable f-element extraction processes.

Yet another strategy is to use ILs to modify the surface of solid materials as sorbents in solid-phase extraction (SPE), which simplifies the recovery and recycling of the non-aqueous phase. Solid phases prepared by the sol-gel sorption of phosphonium, imidazolium, and functionalized-imidazolium ILs were used for extraction of Y^{3+}, actinides, and lanthanides from acidic to neutral aqueous phases (Mokhodoeva *et al.* 2011; Gujar *et al.* 2015; Ruhela *et al.* 2015). As with LLE, special attention should be paid to the selection of ILs, since imidazolium ILs with hydrocarbon chain lengths less than hexadecyl have a higher solubility in the aqueous phase and are thus weakly retained in the solid support (Mokhodoeva *et al.* 2011). A step further to improve the analytical performance of SPE using ILs should be the development of solid phase microextraction techniques, and the use of polymeric ILs to modify the surfaces of the solid phases (Valkenberg 2002, Yuan *et al.* 2013) are fields widely explored for the extraction of other metals but missing in the f-elements field.

The stripping of the extracted metal ions from the non-aqueous phase is another sustainability challenge in LLE. Most of the systems previously discussed employ acidic solutions or organic complexants (such as ethylenediaminetetraacetic acid, EDTA) for this purpose. However, the properties of ILs allow alternatives for the stripping process. An uncommon advantage can be seen in the electrical conductivity of the ILs, which makes it possible to avoid the stripping step by electrochemically depositing separated components as solid products directly from the loaded organic phase. This possibility is discussed in the next section.

Supercritical CO_2 (sc-CO_2) could also serve as a less hazardous alternative to acid stripping, due to the poor solubility of ILs in sc-CO_2 (Blanchard *et al.* 1999). A two-step extraction process involving three phases (water, IL, and sc-CO_2) can be envisioned where metal ions are first concentrated in the IL by extraction from the aqueous phase and then transferred from the IL phase into the CO_2 phase. Preconcentration in an IL would have advantages such as greatly reducing the volume of radioactive materials to be treated by the high-pressure supercritical fluid process in nuclear fuel processing. Using this alternative, La^{3+}, Eu^{3+}, and $[UO_2]^{2+}$ were extracted from aqueous nitric acid to $[C_n mim][NTf_2]$ by using tributylphosphate ($[TBP]^-$) or CMPO as complexing agents, followed by sc-CO_2 stripping of the complex from the IL phase to the supercritical fluid phase. The solubility of these complexes in sc-CO_2 is very high (in the case of $[UO_2]^{2+}$ complexes, it is about 0.45 M), and the IL was recycled for repeated use after the back-extraction (Mekki *et al.* 2006; Wang *et al.* 2009; Fu *et al.* 2014).

Lastly, ILs may have a role in improving the sustainability of current industrial LLE technologies. For instance, ILs can be used to recover the different fractions of the components in traditional LLE used for spent fuel processing. Panja *et al.* used $[C_8mim]Cl$ to recover dibutylphosphoric acid (HDBP) and $[TBP]^-$ from degraded Plutonium Uranium Redox Extraction (PUREX) solvent (n-dodecane) through the formation of micelles and ion-pair formation. The IL could be reused for more than five cycles of operation and thus was considered to be more economically feasible (Panja *et al.* 2014).

13.3.2 Processing of Ore, Spent Fuel, and Recycling

13.3.2.1 Use of ILs for Dissolution of Metals and Metal Salts

Another burden facing sustainable extraction is processing the insoluble forms of the resources from which they are obtained. These include carbonate and phosphate minerals from which lanthanides and thorium are obtained (Cotton 2006), metal oxides such as pitchblende and spent nuclear fuel (Choppin *et al.* 1996), and metallic elements or alloys such as those in electronic waste. Currently, these materials are processed using hydrometallurgical techniques. This creates large volumes of acidic waste, and lanthanide mining has already created a significant legacy of acidic wastewater and gases (US EPA 2012). The strong acids used in hydrometallurgical processing also pose challenges in process design due to the harsh conditions required to digest metals and insoluble ores. This has had a particularly significant effect on nuclear waste processing, where relatively mature extraction technologies such as the TALSPEAK process still have not been applied on an industrial scale because of their incompatibility with the nitric acid used to dissolve the spent fuel (Nilsson & Nash 2007).

It would be almost erroneous to suggest ILs as an alternative to hydrometallurgical processing, as the ancient use of molten salts to process metal ore (i.e., by smelting) arguably makes them the original method. This area also includes some of the first research in modern ILs, such as the discovery of silicate ions in mixed melts of silica and metal oxides (Bockris *et al.* 1955). Nevertheless, hydrometallurgical processing is preferred where the energy cost and engineering difficulties associated with the high temperatures of typical molten salt processing become prohibitive. However, the relatively recent development of a wide range of ILs with melting points near or below room temperature has made possible molten salt processing at the same temperatures used in hydrometallurgical processing. The advantages of ILs here are reduction of solvent volume, which requires that the ILs be recyclable and have a high solubility limit for the metal oxide, and the use of electrochemical separations.

ILs designed to dissolve metals or metal oxides tend to have functional groups that facilitate reactive dissolution (Figure 3). Most metal oxides, including UO_2, UO_3, and Ln_2O_3, do not readily dissolve in non-coordinating ILs without the addition of acid (Bradley *et al.* 2002). Brønsted-acidic ILs which can dissolve metals and metal oxides effectively are made by incorporating protic ions such $[HSO_3]^-$ (Whitehead *et al.* 2004) or incorporating carboxylic acid functional groups into one of the ions (Nockemann *et al.* 2008).

Combinations of choline chloride with near-stoichiometric amounts of carboxylic acids give high ionic strength liquids which are also capable of dissolving metal oxides (Abbott *et al.* 2005). Such systems are known as deep eutectic solvents and share properties with ILs (such as low volatility), but their method of preparation, which often involves combining two commodity chemicals, overcomes some of the sustainability issues with ILs themselves revolving around purification and potential toxicity (Abbott *et al.* 2004). ILs with perhalide anions such as $[I_3]^-$ have also been used to reactively dissolve elements in their metallic form as a means for extraction from their ore (Rogers & Holbrey 2010).

One potentially interesting approach for dissolution of minerals in ILs that has not been well explored is the formation of oligomeric ions. It is known that the equivalence in basicity of the donor and acceptor groups results in especially strong hydrogen bonding between conjugate acid-base pairs, resulting in delocalization of the charge across both molecules (Gilli *et al.* 2009). Such complexes, termed oligomeric ions, have been used to prepare ILs of normally higher melting compounds (Bica *et al.* 2010) and increase the fluidity of low melting ILs (Johansson *et al.* 2008), as the increase in size-to-charge ratio of the oligomeric ion reduces the strength of interionic interactions. ILs incorporating anions such as $[H_2PO_4]^-$, $[HSO_4]^-$, $[HCO_3]^-$, or $[OH]^-$ could be useful for dissolving minerals containing phosphates, sulfates, carbonates, or oxides, respectively.

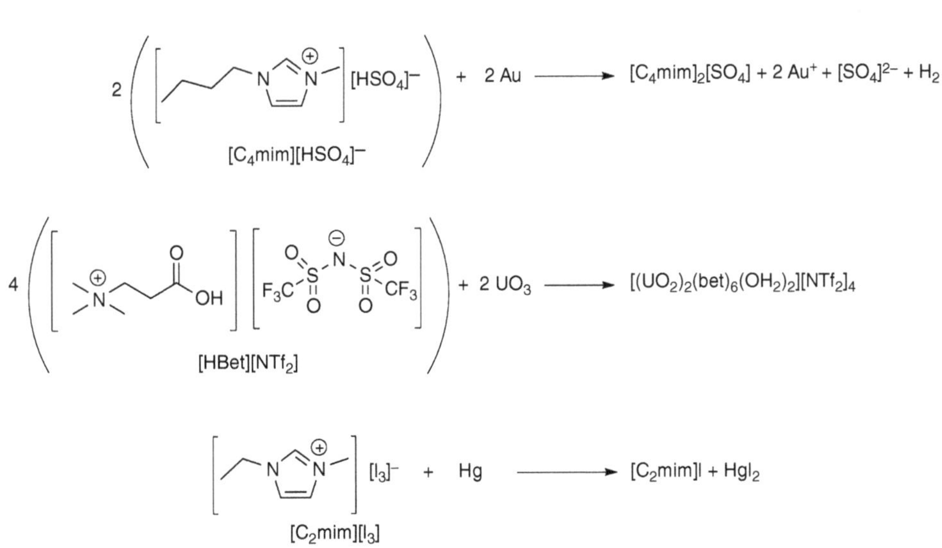

Figure 3 *Examples of reactive dissolution in functionalized ILs*

13.3.2.2 Strategies for Isolating f-Elements from Solid Resources

The potential for refining lanthanides using electrodeposition is one of the chief reasons for investigating non-aqueous processing. This would replace several chemical stages in the current extraction process with a single step, using electrons as the only reagent. Electrodeposition of highly electropositive metals from molten salts is already done on an industrial scale in aluminum refining. Investigation into the possible electrorefining of rare earths from ore was considered as far back as 1968 (Morrice *et al.* 1968), although high temperatures render the process impractical. The discovery of ILs with stable cations which allow the electrodeposition of such electropositive metals as lithium (Howlett *et al.* 2004) and aluminum (Liao *et al.* 1997) has led to the investigation of ILs for low temperature lanthanide electrodeposition. ILs used for this application typically have quaternary ammonium cations and hydrophobic anions (Figure 4). Significant differences in the reduction potentials of Ln^{3+} cations have been measured for different lanthanides in the same IL system (Bhatt *et al.* 2005), suggesting a possibility for electrochemical intra-group lanthanide separations as well. The differences between reported reduction potentials of f-elements in IL (Table 1) are also larger than the differences in lanthanide standard reduction potentials (Vanýsek, 2014). This may indicate an effect of the IL on the reduction potentials which could be tuned to enhance the separation.

There has already been considerable interest in using ILs to recover actinides electrochemically from spent nuclear fuel, with some of the first published results of f-element chemistry in ILs coming from this area (Anderson *et al.* 1991). Electrochemical reduction of uranium (Nikitenko *et al.*, 2005; Jagadeeswara Rao *et al.*, 2011) and thorium (Bhatt *et al.* 2006) has been published (Table 1). In addition to their wide electrochemical windows, ILs have properties that make them especially suitable for nuclear fuel reprocessing. As mentioned earlier, IL ions can be designed to be radiation resistant (Shkrob *et al.* 2012;

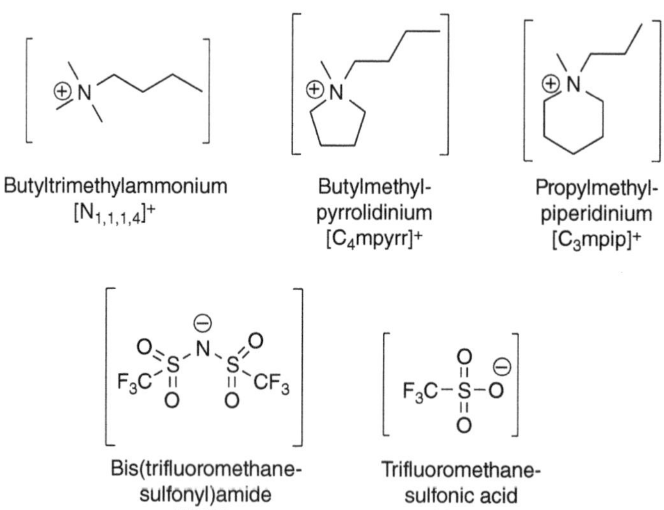

Figure 4 Structures of some ions used to make ILs for f-element electrodeposition

Table 1 Reported potentials for the reduction of f-element ions dissolved in ILs to their zero valent state

Ion	Potential (V)	IL	Reference
La^{3+}	–2.4	$[N_{1,1,1,4}][NTf_2]$	Bhatt *et al.* 2005
Sm^{3+}	–2.85	$[N_{1,1,1,4}][NTf_2]$	Bhatt *et al.* 2005
Eu^{3+}	–3.0	$[N_{1,1,1,4}][NTf_2]$	Bhatt *et al.* 2005
Gd^{3+}	–2.28	$[C_4mpyrr][OTf]$	Glukhov *et al.* 2010
Th^{4+}	–2.15	$[N_{1,1,1,4}][NTf_2]$	Bhatt *et al.* 2006
U^{4+}	–2.7	$[C_3mpip][NTf_2]$	Jagadeeswara Rao *et al.* 2011

Shkrob *et al.* 2013) and boron and chlorine-containing ions found in many ILs can reduce the risk of criticality (Harmon *et al.* 2001). The processes currently being proposed involve oxidative dissolution of the spent fuel rod in an IL, followed by electrochemical separation of U, Pu, and the minor actinides (Hatchett *et al.* 2014). Electrodeposition from ILs is also being investigated to recover uranium from other waste sources such as contaminated chemical traps from gaseous uranium processing (Ohashi *et al.* 2015).

Electronic waste also presents a source of metals, including lanthanides, that are usually in the form of alloys. Recovery of these resources is presently limited because there are no automatic processes for separating them from each other in such a heterogeneous matrix. This area has not received much attention from the IL community, but many of the ILs proposed for leaching gold and other metals from minerals may have promise in this area. ILs may be especially advantageous here because, despite often having a high affinity for even hard-to-dissolve metal salts and polar compounds, they are usually inert towards non-polar polymers and network solids such as silicon. ILs designed to solubilize only metals would be useful as non-volatile leachates for direct processing of post-consumer electronics.

13.3.3 Uranium from Seawater: A Case Study

Concerns for energy security have driven research efforts in several countries to recover U from seawater (Diallo *et al.* 2015). It is estimated that approximately 5×10^4 tons of U were produced worldwide in 2009, which can be compared with the total inventory of U in the oceans, estimated at 4.5×10^9 tons (Pohl 2011). This alternative source would avoid mining, transporting, crushing, and grinding ores, together with environmental costs associated with conventional mining, thus adding incentive from the point of view of sustainability to utilize brines wherever possible (Izatt *et al.* 2014). However, one of the major drawbacks of seawater as a resource not only for uranium, but also for many geochemically scarce metals, is the high metal dilution (the average U concentration is 3 $\mu g\ L^{-1}$, Pohl 2011), which presents a challenge for hydrometallurgical separations. Energy consumed in pumping water for processing can exceed the value of recoverable U, and thus passive techniques are desired (Kim *et al.* 2013). Currently, the preferred technology involves a synthetic fibrous polymer adsorbent functionalized with polyacrylamidoxime (Kim *et al.* 2013), which is the only material so far to have been demonstrated under actual marine deployment (Seko *et al.* 2003). However, polyethylene polymers are of major concern for anthropogenic marine pollution and are made from non-renewable petroleum.

ILs have been applied in this field not directly towards U extraction, but instead as a means to improve the sustainability of sorbent production by using chitin as a more sustainable alternative for preparing the sorbent backbone. This is based on another well-studied application of ILs in separations — their ability to extract biopolymers from biomass (Wang *et al.* 2012). The IL [C_2mim][OAc] was shown to extract high purity, high molecular weight chitin directly from shrimp shell waste in a one-pot process and pulled into fibers (Qin *et al.* 2010), or electrospun into a nanofibrous mesh (Barber *et al.* 2013). It was shown that the surfaces of these chitinous materials could then be covalently modified with amidoxime groups, resulting in greatly increased affinity of the material for aqueous $[UO_2]^{2+}$ ions (Barber *et al.* 2014). While this surface functionalization should, in principle, be applicable to any chitin source, it is only through the combination with IL-based liquid processing of chitin that high surface area materials can be made. These materials could have performance comparable to synthetic amidoxime resins, but have the sustainability advantages of renewability, biodegradability, and the need for smaller amounts of reagents during preparation.

13.4 Summary

In the field of lanthanide and actinide processing, LLE has been of predominant importance and sustainability concerns are abundant. This makes the area ripe for the application of ILs towards sustainable new technologies. ILs can be added to or used in place of organic solvents, where they can improve the efficiency of LLE, but their tunability allows them to be designed in roles beyond simple solvents, such as membranes, supported liquid phases, or micellar dispersions. Nor is the power of ILs as solvents is not limited to LLE. ILs can be potent solvents for inorganic materials, and the resulting non-aqueous solutions can be taken through processing methods (such as electrochemistry) that are not available in aqueous systems. In fact, in the case of uranium extraction from seawater, the role of the IL is not even restricted to the extraction of the element, but rather employed in sorbent preparation.

ILs have been dismissed at times for being too toxic, too expensive, too hard to use, etc. Such statements reflect the fact that ILs are not and cannot simply be used as substitutes for a step in an unsustainable process. Rather, the real advantage of ILs is that they allow (and in fact require) the development of wholly new approaches. Sustainability is more than just a series of technological improvements; it is a change in the way society views consumption and production in general. Similarly, sustainable technologies based on ILs are not dependent on finding the very best IL. They will require changing what is viewed as possible or necessary for the chemical processes themselves.

13.5 Acknowledgments

The authors would like to thank the U.S. Department of Energy Office of Science for supporting this work through the Office of Nuclear Energy Nuclear Energy University Programs (Project DE-NE0000672) and the Small Business Innovation Research (SBIR) grant (Award No. DE-SC0010152). This research was undertaken, in part, thanks to funding from the Canada Excellence Research Chairs Program.

References

Abbott, A.P., Boothby, D., Capper, G., Davies, D.L. & Rasheed R.K. (2004) Deep eutectic solvents formed between choline chloride and carboxylic acids: Versatile alternatives to ionic liquids. *Journal of the American Chemical Society*, **126** (29), 9142–9147.

Abbott, A.P., Capper, G., Davies, D.L., Rasheed, R.K. & Shikota, P. (2005) Selective extraction of metals from mixed oxide matrixes using choline-based ionic liquids. *Inorganic Chemistry*, **44** (19), 6497–6499.

Anderson, C.J., Deakin, M.R., Choppin, G.R., D'Olieslager, W., Heerman, L. & Pruett, D.J. (1991) Spectroscopy and electrochemistry of U(IV)/U(III) in basic aluminum chloride-1-ethyl-3-methylimidazolium chloride. *Inorganic Chemistry*, **30** (21), 4013–4016.

Ansari, S.A., Mohapatra, P.K., Sengupta, A., Nikishkin, N.I., Huskens, J. & Verboom, W. (2014) An insight into the complexation of pyrazine-functionalized calix[4]arenes with Am^{3+} and Eu^{3+} — solvent extraction and luminescence studies in room-temperature ionic liquids. *European Journal of Inorganic Chemistry*, **2014** (33), 5689–5697.

Aoyagi, N., Shimojo, K., Brooks, N.R., Nagaishi, R., Naganawa, H., Van Hecke, K., Van Meervelt, L., Binnemans, K. & Kimura, T. (2011) Thermochromic properties of low-melting ionic uranyl isothiocyanate complexes. *Chemical Communications*, **47** (15), 4490–4492.

Baba, Y., Kubota, F., Kamiya, N. & Goto, M. (2011) Selective recovery of dysprosium and neodymium ions by a supported liquid membrane based on ionic liquids. *Solvent Extraction Research and Development, Japan*, **18**, 193–198.

Barber, P.S., Griggs, C.S., Bonner, J.R. & Rogers, R.D. (2013) Electrospinning of chitin nanofibers directly from an ionic liquid extract of shrimp shells. *Green Chemistry*, **15** (3), 601–607.

Barber, P.S., Kelley, S.P. & Rogers, R.D. (2012) Highly selective extraction of the uranyl ion with hydrophobic amidoxime-functionalized ionic liquids via η^2-coordination. *RSC Advances*, **2** (22), 8526–8530.

Barber, P.S., Kelley, S.P., Griggs, C.S., Wallace, S. & Rogers, R.D. (2014) Surface modification of ionic liquid-spun chitin fibers for the extraction of uranium from seawater: Seeking the strength of chitin and the chemical functionality of chitosan. *Green Chemistry*, **16** (4), 1828–1836.

Bell, T.J. & Ikeda, Y. (2011) The application of novel hydrophobic ionic liquids to the extraction of uranium(VI) from nitric acid medium and a determination of the uranyl complexes formed. *Dalton Transactions*, **40** (39), 10125–10130.

Bhatt, A.I., Duffy, N.W., Collison, D., May, I. & Lewin, R.G. (2006) Cyclic voltammetry of Th(IV) in the room-temperature ionic liquid $[Me_3N_nBu][N(SO_2CF_3)_2]$. *Inorganic Chemistry*, **45** (4), 1677–1682.

Bhatt, A.I., May, I., Volkovich, V.A., Dollison, D., Helliwell, M., Polovov, I.B. & Lewin, R.G. (2005) Structural characterization of a lanthanum bistriflimide complex, $La(N(SO_2CF_3)_2)_3(H_2O)_3$, and an investigation of La, Sm, and Eu electrochemistry in a room-temperature ionic liquid. *Inorganic Chemistry*, **44** (14), 4934–4940.

Bhattacharyya, A., Ansari, S.A., Gadly, T., Ghosh, S.K., Mohapatra, M. & Mohapatra, P.K. (2015) A remarkable enhancement in Am^{3+}/Eu^{3+} selectivity by an ionic liquid based solvent containing bis-1,2,4-triazinyl pyridine derivatives: DFT validation of experimental results. *Dalton Transactions*, **44** (13), 6193–6201.

Bica, K., Rijksen, C., Niewenhuyzen, M. & Rogers, R.D. (2010) In search of pure liquid salt forms of aspirin: Ionic liquid approaches with acetylsalicylic acid and salicylic acid. *Physical Chemistry Chemical Physics*, **12** (8), 2011–2017.

Billard, I., Gaillard, C. & Hennig, C. (2007) Dissolution of UO_2, UO_3, and of some lanthanide oxides in $BumimTf_2N$: Effect of acid and water and formation of $UO_2(NO_3)_3^-$. *Dalton Transactions*, **2007** (37), 4214–4221.

Billard, I., Ouadi, A. & Gaillard, C. (2011a) Liquid-liquid extraction of actinides, lanthanides, and fission products by use of ionic liquids: From discovery to understanding. *Analytical and Bioanalytical Chemistry*, **400** (6), 1555–1566.

Billard, I., Ouadi, A., Jobin, E., Champion, J., Gaillard, C. & Georg, S. (2011b) Understanding the extraction mechanism in ionic liquids: $UO_2^{2+}/HNO_3/TBP/C_4$-mimTF$_2$N as a case study. *Solvent Extraction and Ion Exchange*, **29** (4), 577–601.

Binnemans, K. (2007) Lanthanides and actinides in ionic liquids. *Chemistry Reviews*, **107** (6), 2592–2614.

Biswas, S., Rupawate, V.H., Roy, S.B. & Sahu, M. (2014) Task-specific ionic liquid tetraalkylammonium hydrogen phthalate as an extractant for U(VI) extraction from aqueous media. *Journal of Radioanalytical and Nuclear Chemistry*, **300** (2), 853–858.

Blanchard, L.A., Hancu, D., Beckman, E.J. & Brennecke, J.F. (1999) Green processing using ionic liquids and CO_2. *Nature*, **399** (6731), 28-29.

Bockris, J.O'M., Mackenzie, J.D. & Kitchener, J.A. (1955) Viscous flow in silica and binary liquid silicates. *Transactions of the Faraday Society*, **51**, 1734–1748.

Bradley, A.E., Hatter, J.E., Nieuwenhuyzen, M., Pitner, W.R., Seddon, K.R. & Thied, R.C. (2002) Precipitation of a dioxouranium(VI) species from a room temperature ionic liquid medium. *Inorganic Chemistry*, **41** (7), 1692–1694.

Busquim e Silva, R., Kazimi, M.S. & Hejzlar, P. (2010) Nuclear fuel recycling: National and regional options for the US nuclear energy system. *Energy & Environmental Science*, **3** (8), 996–1010.

Chesman, A.S.R., Yang, M., Spiccia, N.D., Deacon, G.B., Batten, S.R. & Mudring, A.-V. (2012) Lanthanoid-based ionic liquids incorporating the dicyanonitrosomethanide anion. *Chemistry – A European Journal*, **18** (31), 9580–9589.

Choppin, G., Liljenzin, J.-O. & Rydberg, J. (1996) *Radiochemistry and Nuclear Chemistry*, Butterworth-Heinemann, Ltd., Oxford.

Cocalia, V., Smiglak, M., Kelley, S.P., Shamshina, J.L., Gurau, G. & Rogers, R.D. (2010) Crystallization of uranyl salts from dialkylimidazolium ionic liquids or their precursors. *European Journal of Inorganic Chemistry*, **2010** (18), 2760–2767.

Cocalia, V.A., Jensen, M.P., Holbrey, J.D., Spear, S.K., Stepinski, D.C. & Rogers R.D. (2005) Identical extraction behavior and coordination of trivalent or hexavalent f-element cations using ionic liquid and molecular solvents. *Dalton Transactions*, **2005** (11), 1966–1971.

Coleman, D. & Gathergood, N. (2010) Biodegradation studies of ionic liquids. *Chemical Society Reviews*, **39** (2), 600–637.

Cotton, S. (2006) *Lanthanide and Actinide Chemistry*, John Wiley & Sons, Ltd., Chichester.

Dai, S., Ju, Y.H. & Barnes, C.E. (1999) Solvent extraction of strontium nitrate by a crown ether using room-temperature ionic liquids. *Journal of the Chemical Society: Dalton Transactions*, **1999** (8), 1201–1202.

Diallo, M.S., Kotte, M.R. & Cho, M. (2015) Mining critical metals and elements from seawater: Opportunities and challenges. *Environmental Science & Technology*, **49** (16), 9390–9399.

Dietz, M.L. & Stepinski, D.C. (2008) Anion concentration-dependent partitioning mechanism in the extraction of uranium into room-temperature ionic liquids. *Talanta*, **75** (2), 598–603.

Fu, J., Chen, Q. & Shen, X. (2014) Stripping of uranium from an ionic liquid medium by TOPO-modified supercritical carbon dioxide. *Science China Chemistry*, **58** (3), 545–550.

Fukaya, Y., Iizuka, Y., Sekikawa, K. & Ohno, H. (2007) Bio ionic liquids: Room temperature ionic liquids composed wholly of biomaterials. *Green Chemistry*, **9** (11), 1155–1157.

Gaillard, C., Billard, I., Chaumont, A., Mekki, S., Ouadi, A., Denecke, M.A., Moutiers, G. & Wipff, G. (2005) Europium(III) and its halides in anhydrous room-temperature imidazolium-based ionic liquids: A combined TRES, EXAFS, and molecular dynamics study. *Inorganic Chemistry*, **44** (23), 8355–8367.

Gaillard, C., Chaumont, A., Billard, I., Hennig, C., Ouadi, A. & Wipff, G. (2007) Uranyl coordination in ionic liquids: The competition between ionic liquid anions, uranyl counterions, and Cl^- anions investigated by extended X-ray absorption fine structure and UV-visible spectroscopies and molecular dynamics simulations. *Inorganic Chemistry*, **46** (12), 4815–4826.

Gaillard, C., Klimchuk, O., Ouadi, A., Billard, I. & Hennig, C. (2012) Evidence for the formation of $UO_2(NO_3)_4^{2-}$ in an ionic liquid by EXAFS. *Dalton Transactions*, **41** (18), 5476–5479.

Gilli, P., Pretto, L., Bertolasi, V. & Gilli, G. (2009) Predicting hydrogen-bond strengths from acid-base molecular properties. The pK_a slide rule: Toward the solution of a long-lasting problem. *Accounts of Chemical Research*, **42** (1), 33–44.

Glukhov, L.M., Greish, A.A. & Kustov, L.M. (2010) Electrodeposition of rare earth metals Y, Gd, Yb in ionic liquids. *Russian Journal of Physical Chemistry A*, **84** (1), 104–108.

Gujar, R.B., Ansari, S.A. & Mohapatra, P.K. (2015) Spectacular enhancements in actinide ion uptake using novel extraction chromatography resins containing TODGA and ionic liquid. *Separation and Purification Technology*, **141**, 229–334.

Guo, L., Chen, J., Shen, L., Zhang, J., Zhang, D. & Deng, Y. (2014) Highly selective extraction and separation of rare earths(III) using bifunctional ionic liquid extractant. *ACS Sustainable Chemistry & Engineering*, **2** (8), 1968–1975.

Gutowski, K.E., Cocalia, V.A., Griffin, S.T., Bridges, N.J., Dixon, D.A. & Rogers, R.D. (2007) Interactions of 1-methylimidazole with $UO_2(CH_3COO)_2$ and $UO_2(NO_3)_2$: Structural, spectroscopic, and theoretical evidence for imidazole binding to the uranyl ion. *Journal of the American Chemical Society*, **129** (3), 526–536.

Hallett, J.P. & Welton, T. (2011) Room-temperature ionic liquids: Solvents for synthesis and catalysis. 2. *Chemical Reviews*, **111** (5), 3508–3576.

Harmon, C.D., Smith, W.H. & Costa, D.A. (2001) Criticality calculations for plutonium metal at room temperature in ionic liquid solutions. *Radiation Physics and Chemistry*, **60** (3), 157–159.

Hatchett, D., Czerwinski K. & Droessler, J. (2014). Room temperature electrodeposition of actinides from ionic solutions. PCT Int. Appl. WO 2014124428 A1 August 14, 2014.

He, L., Ji, S.-P., Tang, N., Zhao, Y. & Tao, G.H. (2015) Synthesis, structure, and near-infrared photoluminescence of hexanitratoneodymate ionic liquids. *Dalton Transactions*, **44** (5), 2325–2332.

Hines, C.C., Cordes, D.B., Griffin, S.T., Watts, S.I., Cocalia, V.A. & Rogers, R.D. (2008) Flexible coordination environments of lanthanide complexes grown from chloride-based ionic liquids. *New Journal of Chemistry*, **32** (5), 872–877.

Hough, W.L. & Rogers, R.D. (2007) Ionic liquids then and now: from solvents to materials to active pharmaceutical ingredients. *Bulletin of the Chemical Society of Japan*, **80** (12), 2262–2269.

Howlett, P.C., MacFarlane, D.R. & Hollenkamp, A.F. (2004) High lithium metal cycling efficiency in a room-temperature ionic liquid. *Electrochemical and Solid State Letters*, **7** (5), A97–A101.

Huddleston, J.G., Willauer, H.D., Swatloski, R.P., Visser, A.E. & Rogers, R.D. (1998) Room temperature ionic liquids as novel media for 'clean' liquid-liquid extraction. *Chemical Communications*, **1998** (16), 1765–1766.

Humphries, M. (2013) Rare Earth Elements: The Global Supply Chain. Congressional Research Service, USA. Available at: http://fas.org/sgp/crs/natsec/R41347.pdf (Accessed May 05, 2015).

International Atomic Energy Agency. Nuclear Fuel Cycle & Waste Technology https://www.iaea.org/OurWork/ST/NE/NEFW/Technical-Areas/NFC/home.html. (Accessed May 24, 2015).

Izatt, R.M., Izatt, S.R., Bruening, R.L., Izatt, N.E. & Moyer, B.A. (2014) Challenges to achievement of metal sustainability in our high-tech society. *Chemical Society Reviews*, **43** (8), 2451–2475.

Jagadeeswara Rao, Ch., Venkatesan, K.A., Nagarajan, K., Srinivasan, T.G. & Vasudeva Rao, P.R. (2011) Electrodeposition of metallic uranium at near ambient conditions from room temperature ionic liquid. *Journal of Nuclear Materials*, **408** (1), 25–29.

Jensen, M.P., Neuefeind, J., Beitz, J.V., Skanthakumar, S. & Soderholm, S. (2003) Mechanisms of metal ion transfer into room-temperature ionic liquids: The role of anion exchange. *Journal of the American Chemical Society*, **125** (50), 15466–15473.

Johansson, K.M., Izgorodina, E.I., Forsyth, M., MacFarlane, D.R. & Seddon K.R. (2008) Protic ionic liquids based on the dimeric and oligomeric anions $[(AcO)_xH_{x-1}]^-$. *Physical Chemistry Chemical Physics*, **10** (20), 2972–2978.

Johnson, J. Decades for decommissioning. *Chemical & Engineering News*, June 2 2014, pp. 22–23.

Joshi, M.D. & Anderson, J.L. (2012) Recent advances of ionic liquids in separation science and mass spectrometry. *RSC Advances*, **2** (13), 5470–5484.

Kaykhaii, M. & Ghasemi, E. (2013) Room temperature ionic liquid-based dispersive liquid–liquid microextraction of uranium in water samples before spectrophotometric determination. *Analytical Methods*, **5** (19), 5260–5266.

Kelley, S.P., Nuss, J.S. & Rogers, R.D. (2015) Using crystal structures of ionic compounds to explore complexation and extraction of rare earth elements in ionic liquids, in *Application of Ionic Liquids on Rare Earth Green Separation and Utilization* (ed. J. Chen), Springer, Hiedelberg, pp 21–42.

Kim, J., Tsouris, C., Mayes, R.T., Oyola, Y, Saito, T., Janke, C.J., Dai, S., Schneider, E. & Sachde, D. (2013) Recovery of uranium from seawater: A review of current status and future research needs. *Separations Science and Technology*, **48** (3), 367–387.

Kolarik, Z. (2013) Ionic liquids: How far do they extend the potential of solvent extraction of f-elements? *Solvent Extraction and Ion Exchange*, **31** (1), 24–60.

Kubota, F., Baba, Y. & Goto, M. (2012) Application of ionic liquids for the separation of rare earth metals. *Solvent Extraction Research and Development, Japan*, **19**, 17–28.

Kubota, F., Shimobori, Y., Koyanagi, Y., Shimojo, K., Kamiya, N. & Goto, M. (2010) Uphill transport of rare-earth metals through a highly stable supported liquid membrane based on an ionic liquid. *Analytical Sciences*, **26** (3), 289–290.

Lall-Ramnarine, S.I., Castano, A., Subramaniam, G., Thomas, M.F. & Wishart, J.F. (2009) Synthesis, characterization and radiolytic properties of bis(oxalato)borate containing ionic liquids. *Radiation Physics and Chemistry*, **78** (12), 1120–1125.

Liao, Q.; Pitner, W.R., Stewart, G., Hussey, C.L. & Stafford, G.R. (1997) Electrodeposition of aluminum from the aluminum chloride-1-methyl-3-ethylimidazolium chloride room temperature molten salt + benzene. *Journal of the Electrochemical Society*, **144** (3), 936–943.

MacFarlane, D.R., Tachikawa, N., Forsyth, M., Pringle, J.M., Howlett, P.C., Elliott, G.D., Davis Jr, J.H., Watanabe, M., Simon, P. & Angell, C.A. (2014) Energy applications of ionic liquids. *Energy & Environmental Science*, **7** (1), 232–250.

Maerzke, K.A., Goff, G.S., Runde, W.H., Schneider, W.F. & Maginn, E.J. (2013) Structure and dynamics of uranyl(VI) and plutonyl(VI) cations in ionic liquid/water mixtures via molecular dynamics simulations. *Journal of Physical Chemistry B*, **117** (37), 10852–10868.

Martina, M. (2011) China Halts Rare Earth Production at Three Mines (Sep 06), Reuters [Online], http://www.reuters.com/article/2011/09/06/china-rare-earth-idUSL3E7K620P20110906 (Accessed May 08, 2015).

Martinis, E.M., Berton, P. & Wuilloud, R.G. (2014) Ionic liquid-based microextraction techniques for trace-element analysis. *TrAC Trends in Analytical Chemistry*, **60**, 54–70.

Mekki, S., Wai, C.M., Billard, I., Moutiers, G., Burt, J., Yoon, B., Wang, J.S., Gaillard, C., Ouadi, A. & Hesemann P. (2006) Extraction of lanthanides from aqueous solution by using room-temperature ionic liquid and supercritical carbon dioxide in conjunction. *Chemistry — A European Journal*, **12** (6), 1760–1766.

Mincher, B.J. & Wishart, J.F. (2014) The radiation chemistry of ionic liquids: a review. *Solvent Extraction and Ion Exchange*, **32** (6), 563–583.

Mohapatra, P.K., Kandwal, P., Iqbal, M., Huskens, J., Murali, M.S. & Verboom, W. (2013) A novel CMPO-functionalized task specific ionic liquid: Synthesis extraction and spectroscopic investigation of actinide and lanthanide complexes. *Dalton Transactions*, **42** (13), 4343–4347.

Mokhodoeva, O.B., Myasoedova, G.V. & Zakharchenko, E.A. (2011) Solid-phase extractants for radionuclide preconcentration and separation. New possibilities. *Radiochemistry*, **53** (1), 35–43.

Morrice, E., Shedd, E.S. & Henrie, T.A. (1968) Direct electrolysis of rare earth oxide to metals and alloys in fluoride melts. Technical Report. Report Number CONF-670502–. Bureau of Mines, Reno Metallurgy Research Center, Reno, NV, USA.

Mudring, A.-V. & Tang, S. (2010) Ionic liquids for lanthanide and actinide chemistry. *European Journal of Inorganic Chemistry* **2010** (18), 2569–2581.

Mudring, A.-V., Babai, A., Arenz, S. & Giernoth, R. (2005) The "noncoordinating" anion Tf_2N^- coordinates to Yb^{2+}: A structurally characterized Tf_2N^- complex from the ionic liquid [mppyr][Tf_2N]. *Angewante Chemie (International Edition)*, **44** (34), 5485–5488.

Nikitenko, S.I., Hennig, C., Grigoriev, M.S., Le Naour, C., Cannes, C., Trubert, D., Bossé, E., Berthon, C. & Moisy, P. (2007a) Structural and spectroscopic studies of the complex $[BuMeIm]_2[UCl_6]$ in the solid state and in hydrophobic room temperature ionic liquid [BuMeIm][Tf_2N]. *Polyhedron*, **26** (13), 3136–3142.

Nikitenko, S.I., Berthon, C. & Moisy, P. (2007b) Instability of actinide(IV) hexachloro complexes in room-temperature ionic liquid $[BuMeIm]PF_6$ due to hydrolysis of the hexafluorophosphate anion. *Comptes Rendus Chimie*, **10** (10-11), 1122–1127.

Nikitenko, S.I., Cannes, C., Le Naour, C., Moisy, P. & Trubert, D. (2005) Spectroscopic and electrochemical studies of U(IV)-hexachloro complexes in hydrophobic room-temperature ionic liquids [BuMeIm][Tf_2N] and [$MeBu_3N$][Tf_2N]. *Inorganic Chemistry*, **44** (25), 9497–9505.

Nilsson, M. & Nash, K.L. (2007) A review of the development and operational characteristics of the TALSPEAK process. *Solvent Extraction & Ion Exchange*, **25** (6), 665–701.

Nockemann, P., Thijs, B., Parac-Vogt, T.N., Van Hecke, K., Van Meervelt, L., Tinant, B., Hartenbach, I., Schleid, T., Thi Ngan, V., Tho Nguyen, M. & Binnemans, K. (2008) Carboxyl-functionalized task-specific ionic liquids for solubilizing metal oxides. *Inorganic Chemistry*, **47** (21), 9987–9999.

Nockemann, P., Thijs, B., Pittois, S., Thoen, J., Glorieux, C., Van Hecke, K., Van Meervelt, L., Kirchner, B. & Binnemans, K. (2006a) Task-specific ionic liquids for solubilizing metal oxides. *Journal of Physical Chemistry B*, **110** (42), 20978–20992.

Nockemann, P., Thijs, B., Postelmans, N., Van Hecke, K., Van Meervelt, L. & Binnemans, K. (2006b) Anionic rare-earth thiocyanate complexes as building blocks for low-melting metal-containing ionic liquids. *Journal of the American Chemical Society*, **128** (42), 13658–13659.

Ohashi, Y., Harada, M., Asanuma, N. & Ikeda, Y. (2015) Feasibility studies on electrochemical recovery of uranium from solid wastes contaminated with uranium using 1-butyl-3-methylimidazolium chloride as an electrolyte. *Journal of Nuclear Materials*, **464**, 119–127.

Organization for Economic Co-operation and Development (OECD), Nuclear Energy Agency, (2011) Nuclear Development. Trends towards Sustainability in the Nuclear Fuel Cycle. Executive Summary, https://www.oecd-nea.org/ndd/reports/2011/trends-nuclear-fuel-cycle-ex.pdf. (Accessed May 26, 2015).

Panja, S., Misra, S.K., Tripathi, S.C., Bindu, M. & Gandhi, P.M. (2014) Fractionation of pure solvent components from degraded PUREX solvent using room temperature ionic liquids. *Separation and Purification Technology*, **122**, 67–72.

Pena-Pereira, F. & Namieśnik, J. (2014) Ionic liquids and deep eutectic mixtures: Sustainable solvents for extraction processes. *ChemSusChem*, **7** (7), 1784–1800.

Platzer, S., Sap, O., Leyma, R., Wallner, G., Jirsa, F., Kandioller, W., Krachler, R. & Keppler, B.K. (2014) Extraction of natural radionuclides from aqueous solutions by novel maltolate-based task-specific ionic liquids. *Journal of Radioanalytical and Nuclear Chemistry*, **303** (3), 2483–2488.

Pohl, W.L. (2011) *Economic Geology: Principles and Practice*. John Wiley & Sons, Chichester, UK.

Pribylova, G.A. (2011) Influence of ionic liquids on actinides extraction by diphenyl(dibutyl)carbamoylmethylphosphine oxide in different solvents from nitric acid solution. *Journal of Radioanalytical and Nuclear Chemistry*, **288** (3), 693–697.

Qin, Y., Lu, X., Sun, N. & Rogers, R.D. (2010) Dissolution or extraction of crustacean shells using ionic liquids to obtain high molecular weight purified chitin and direct production of chitin films and fibers. *Green Chemistry*, **12** (6), 968–971.

Qu, F., Zhu, Q.-Q. & Liu, C.-L. (2014) Crystallization in ionic liquids: Synthesis properties, and polymorphs of uranyl salts. *Crystal Growth & Design*, **14** (12), 6421–6432.

Rogers, R.D. & Holbrey, J. (2010) Ionic solvents of perhalide type for metals for extraction of radioactive metals from mineral ores. PCT Int. Appl. (2010), WO 2010116167 A1 October 14, 2010.

Rogers, R.D., Visser, A.E., Swatloski, R.P. & Hartman, D.H. (1999) Metal ion separations in room temperature ionic liquids: potential replacements for volatile organic diluents, in *Metal Separation Technologies Beyond 2000: Integrating Novel Chemistry with Processing, Proceedings of a Symposium, Kahaku, Hawaii, June 13-18, 1999*. (Liddell, K.C., ed.) TMS Publications, Warrendale, pp. 139–147.

Rout, A. & Binnemans, K. (2014) Solvent extraction of neodymium(III) by functionalized ionic liquid trioctylmethylammonium dioctyl diglycolamate in fluorine-free ionic liquid diluent. *Industrial & Engineering Chemistry Research*, **53** (15), 6500–6508.

Rout, A. & Binnemans, K. (2015) Influence of the ionic liquid cation on the solvent extraction of trivalent rare-earth ions by mixtures of Cyanex 923 and ionic liquids. *Dalton Transactions*, **44** (3), 1379–1387.

Rout, A., Karmakar, S., Venkatesan, K.A., Srinivasan, T.G. & Vasudeva Rao, P.R. (2011) Room temperature ionic liquid diluent for the mutual separation of europium(III) from americium(III). *Separation and Purification Technology*, **81** (2), 109–115.

Rout, A., Venkatesan, K.A., Srinivasan, T.G. & Vasudeva Rao, P.R. (2012) Ionic liquid extractants in molecular diluents: Extraction behavior of europium (III) in quaternary ammonium-based ionic liquids. *Separation and Purification Technology*, **95**, 26–31.

Ruhela, R., Iyer, N., Yadav, M., Singh, A.K., Hubli, R.C. & Chakracartty, J.K. (2015) *N,N,N*-Trimethyl benzyl ammonium bis-(2-ethylhexyl)-phosphonate grafted polymer — a solid supported ionic liquid for the separation of uranium from aqueous processing streams. *Green Chemistry*, **17** (2), 827–830.

Seko, N., Katakai, A., Hasegawa, S., Tamada, M., Kasai, N., Takeda, H., Sugo, T. & Saito, K. (2003) Aquaculture of uranium in seawater by a fabric-adsorbent submerged system. *Nuclear Technology*, **144** (2), 274–278.

Shkrob, I.A., Marin, T.W. & Jensen, M.P. (2014) Ionic liquid based separations of trivalent lanthanide and actinide ions. *Industrial & Engineering Chemistry Research*, **53** (9), 3641–3653.

Shkrob, I.A., Marin, T.W. & Wishart, J.F. (2013) Ionic liquids based on polynitrile anions: Hydrophobicity, low proton affinity, and high radiolytic resistance combined. *Journal of Physical Chemistry B*, **117** (23), 7084–7094.

Shkrob, I.A., Marin, T.W., Chemerisov, S.D., Hatcher, J. & Wishart, J.F. (2012) Toward radiation-resistant ionic liquids: Radiation stability of sulfonyl imide anions. *Journal of Physical Chemistry B*, **116** (30), 9043–9055.

Singh, R.P., Verma, R.D., Meshri, D.T. & Shreeve, J.M. (2006) Energetic nitrogen-rich salts and ionic liquids. *Angewandte Chemie (International Edition)*, **45** (22), 3584–3601.

Smiglak, M., Metlen, A. & Rogers, R.D. (2007) The second evolution of ionic liquids: From solvents and separations to advanced materials — energetic examples from the ionic liquid cookbook. *Accounts of Chemical Research*, **40** (11), 1182–1192.

Spietelun, A., Marcinkowski, Ł., de la Guardia, M. & Namieśnik, J. (2014) Green aspects, developments and perspectives of liquid phase microextraction techniques. *Talanta*, **119**, 34–45.

Srncik, M., Kogelnig, D., Stojanovic, A., Körner, W., Krachler, R. & Wallner, G. (2009) Uranium extraction from aqueous solutions by ionic liquids. *Applied Radiation and Isotopes*, **67** (12), 2146–2149.

Stockmann, T.J. & Ding, Z. (2010) Hydrophobicity of room temperature ionic liquids assessed by the Galvani potential difference established at micro liquid/liquid interfaces. *Journal of Electroanalytical Chemistry*, **649** (1–2), 23–31.

Sun, X. & Waters, K.E. (2014) Development of industrial extractants into functional ionic liquids for environmentally-friendly rare earth separation. *ACS Sustainable Chemistry & Engineering*, **2** (7), 1910–1917.

Sun, X., Do-Thanh, C.-L., Luo, H. & Dai, S. (2014) The optimization of an ionic liquid-based TALSPEAK-like process for rare earth ions separation. *Chemical Engineering Journal*, **239**, 392–398.

Sun, X., Ji, Y., Hu, F., He, B., Chen, J. & Li, D. (2010b) The inner synergistic effect of bifunctional ionic liquid extractant for solvent extraction. *Talanta*, **81** (4–5), 1877–1883.

Sun, X., Ji, Y., Liu, Y., Chen, J. & Li, D. (2010a) An engineering-purpose preparation strategy for ammonium-type ionic liquid with high purity. *AIChE Journal*, **56** (4), 989–996.

Sun, X., Luo, H. & Dai, S. (2012a) Ionic liquids-based extraction: A promising strategy for the advanced nuclear fuel cycle. *Chemical Reviews*, **112** (4), 2100–2128.

Sun, X., Luo, H. & Dai, S. (2012b) Solvent extraction of rare-earth ions based on functionalized ionic liquids. *Talanta*, **90**, 132–137.

Swatloski, R.P., Holbrey, J.D. & Rogers, R.D. (2003) Ionic liquids are not always green: Hydrolysis of 1-butyl-3-methylimidazolium hexafluorophosphate. *Green Chemistry*, **5** (4), 361–363.

Takao, K., Bell, T.J. & Ikeda, Y. (2013) Actinide chemistry in ionic liquids. *Inorganic Chemistry*, **52** (7), 3459–3472.

Tang, S., Babai, A. & Mudring, A.-V. (2008) Europium-based ionic liquids as luminescent soft materials. *Angewandte Chemie (International Edition)*, **47** (40), 7631–7634.

Taubert, A. (2010) Heavy elements in ionic liquids, in *Ionic Liquids* (ed. B. Kirchner), Topics in Current Chemistry, vol. **290**, Springer Berlin Heidelberg, pp. 127–159.

Turanov, A.N., Karandashev, V.K. & Baulin, V.E. (2008) Extraction of rare-earth elements from nitric acid solutions with bidentate neutral organophosphorus compounds in the presence of l-butyl-3-methylimidazolium hexafluorophosphate. *Radiochemistry*, **50** (3), 266–273.

Turanov, A.N., Karandashev, V.K. & Yarkevich, A.N. (2013) Extraction of REEs(III), U(VI), and Th(IV) from nitric acid solutions with carbamoylmethylphosphine oxides in the presence of an ionic liquid. *Radiochemistry*, **55** (4), 382–387.

United States Environmental Protection Agency (2012) *Rare Earth Elements: A Review of Production, Processing, Recycling, and Associated Environmental Issues*. EPA 600/4-12/572.

Valkenberg, M.H., deCastro, C. & Hoelderich, W.F. (2002) Immobilization of ionic liquids on solid supports. *Green Chemistry*, **4** (2), 88–93.

Vanýsek, P. (2014) Electrochemical series, in *CRC Handbook of Chemistry and Physics*, 95th edn (ed. W.M. Haynes), CRC Press, LLC, Boca Raton, pp. 5-81–5-89.

Vasudeva Rao, P.R., Venkatesan, K.A. & Srinivasan, T.G. (2008) Studies on applications of room temperature ionic liquids. *Progress in Nuclear Energy*, **50** (2–6), 449–455.

Vasudevan, K.V., Smith, N.A., Scott, B.L., McKigney, E.A., Blair, M.W., Gordon, J.C. & Muenchausen, R.E. (2011) An ionic-liquid mediated route to cerium(III) bromide solvates. *Inorganic Chemistry*, **50** (10), 4627–4631.

Vicente, J.A., Mlonka, A., Gunaratne, H.Q.N., Swadźba-Kwaśny, M. & Nockemann, P. (2012) Phosphine oxide functionalized imidazolium ionic liquids as tunable ligands for lanthanide complexation. *Chemical Communications*, **48** (49), 6115–6117.

Visser, A.E., Swatloski, R.P., Reichert, W.M., Mayton, R., Sheff, S., Wierzbicki, A., Davis, J.H., Jr. & Rogers, R.D. (2001) Task specific ionic liquids for the extraction of metal ions from aqueous solutions. *Chemical Communications*, **2001** (1), 135–136.

Visser, A.E., Swatloski, R.P., Reichert, W.M., Griffin, S.T. & Rogers, R.D. (2000) Traditional extractants in nontraditional solvents: Groups 1 and 2 extraction by crown ethers in room temperature ionic liquids. *Industrial & Engineering Chemistry Research*, **39** (10), 3596–3604.

Visser, A.E., Jensen, M.P., Laszak, I., Nash, K.L., Choppin, G.R. & Rogers, R.D. (2003) Uranyl coordination environment in hydrophobic ionic liquids: An in situ investigation. *Inorganic Chemistry*, **42** (7), 2197–2199.

Wang, H., Gurau, G. & Rogers, R.D. (2012) Ionic liquid processing of cellulose. *Chemical Society Reviews*, **41** (4), 1519–1537.

Wang, J.S., Sheaff, C.N., Yoon, B., Addleman, R.S. & Wai, C.M. (2009) Extraction of uranium from aqueous solutions by using ionic liquid and supercritical carbon dioxide in conjunction. *Chemistry — A European Journal*, **15** (17), 4458–4463.

Wang, W., Yang, H., Cui, H., Zhang, D., Liu, Y. & Chen, J. (2011) Application of bifunctional ionic liquid extractants [A336][CA-12] and [A336][CA-100] to the lanthanum extraction and separation from rare earths in the chloride medium. *Industrial & Engineering Chemistry Research*, **50** (12), 7534–7541.

Wasserscheid, P. & Stark, A. (2013) *Handbook of Green Chemistry, Green Solvents, Ionic Liquids*, vol. **6**, Wiley-VCH Verlag GmbH & Co. KGaA, Weinheim, Germany.

Whitehead, J.A., Lawrance, G.A. & McCluskey, A. (2004) "Green" leaching: Recyclable and selective leaching of gold-bearing ore in an ionic liquid. *Green Chemistry*, **6** (7), 313–315.

Wishart, J.F. (2014) Ionic Liquid Radiation Chemistry, in *Ionic Liquids Further UnCOILed: Critical Expert Overviews* (eds N.V. Plechkova and K.R. Seddon), John Wiley & Sons, Inc., Hoboken, NJ, pp. 259–274.

World Nuclear Organization (WNO). *Supply of Uranium*. http://www.world-nuclear.org/info/inf75.htm (Accessed May 24, 2015).

Yang, F., Baba, Y., Kubota, F., Kamiya, N. & Goto, M. (2012b) Extraction and separation of rare earth metal ions with DODGAA in ionic liquids. *Solvent Extraction Research and Development, Japan*, **19**, 69–76.

Yang, H., Wang, W., Cui, H., Zhang, D., Liu, Y. & Chen, J. (2012a) Recovery of rare earth elements from simulated fluorescent powder using bifunctional ionic liquid extractants (Bif-ILEs). *Journal of Chemical Technology and Biotechnology*, **87** (2), 198–205.

Yuan, J., Mecerreyes, D. & Antonietti, M. (2013) Poly (ionic liquid)s: An update. *Progress in Polymer Science*, **38** (7), 1009–1036.

Zhu, M., Zhao, J., Li, Y., Mehio, N., Qi, Y., Liu, H. & Dai, S. (2015) An ionic liquid-based synergistic extraction strategy for rare earths. *Green Chemistry*, **17** (5), 2981–2993.

Zuo, Y., Liu, Y., Chen, J. & Li, D.Q. (2008) The separation of cerium(IV) from nitric acid solutions containing thorium(IV) and lanthanides(III) using pure [C_8mim]PF_6] as extracting phase. *Industrial & Engineering Chemistry Research*, **47** (7), 2349–2355.

14

Selective Recovery of Platinum Group Metals and Rare Earth Metals from Complex Matrices Using a Green Chemistry/Molecular Recognition Technology Approach

Steven R. Izatt[1], *James S. McKenzie*[2], *Ronald L. Bruening*[1], *Reed M. Izatt*[1,3], *Neil E. Izatt*[1], *and Krzysztof E. Krakowiak*[1]

[1]*IBC Advanced Technologies, Inc., American Fork, UT, 84003, USA*
[2]*Ucore Rare Metals, Inc., Bedford, Nova Scotia, B4A 0H3, Canada*
[3]*Department of Chemistry and Biochemistry, Brigham Young University, Provo, UT, 84602, U.S.A.*

14.1 Introduction

Platinum group metals (PGM, singular or plural) and rare earth elements (REE, singular or plural) share a common important involvement as essential components of high technology devices in our modern global economy [1,2]. Rapidly increasing demand for PGM and REE in electronic products, clean energy applications, biomedical equipment, electric automobiles, automotive catalytic converters and many other applications has accelerated their production from native ores during the past half-century [3,4], with important environmental and resource depletion consequences [4]. Of key importance, individual metals in both sets are necessary for optimal product or system performance and usually cannot be replaced without loss of function, making it imperative that their sustainability be considered by involved stakeholders.

Metal Sustainability: Global Challenges, Consequences, and Prospects, First Edition. Edited by Reed M. Izatt.

Despite the critical need for continuing supplies of these metals, in both cases recycling rates are low, <1% for REE, with only economically valuable PGM being recycled at 50% or higher rates in cases where closed cycles exist [3,5]. As a result, large amounts of these metals are continually being produced, used once, and discarded, usually in landfill, with the end-of-life product containing them. This open cycle requires that additional ore be mined and processed to meet increasing product demand, with resulting resource depletion and negative environmental consequences [3]. The discarded metals are, for the most part, not recovered and persist in the environment with uncontrolled effects, since metals are indestructible. The amounts of metals involved are huge, as evidenced by estimates of global electronic waste generation, for example, which range from 25 to 50 million tons per year with a growth rate of about 4–5% [6,7].

The mining industry is the largest producer of solid wastes globally [8]. It has been estimated [3] that >80% of global mine production of PGM since 1900 has occurred since 1980, and the production rate of PGM is increasing. The relatively low grade of PGM ores, approximately a few grams per ton of ore, means that >99% of the ore becomes solid waste with environmental and health consequences. As available ore grades decline, new sources must be found. These sources often involve extending mining to greater depths with attendant increases in water usage, labor costs, and energy input, largely through increased use of electricity [1,8]. In many cases, electricity is derived through coal combustion, with consequent carbon dioxide generation and emission of dangerous particulates and gases, such as Hg, unless care is taken.

The growth of the rare earth industry since 1960 resulted, by the turn of the century, in China being the sole global source of these valuable metals, both mining and production [4]. By 2010, about 120,000 metric tons of REE were being produced annually, whereas in 1960 production was approximately 5,000 metric tons annually [4]. Production of REE in China is associated with large negative environmental and health problems [9]. A major source of the environmental problems associated with individual REE production in China is the use of low-selectivity solvent extraction (SX) separation systems, which require many stages for separation of these chemically similar metals and the generation of large quantities of waste.

Anastas and Eghbali [10] have defined green chemistry as "the design of chemical products and processes to reduce or eliminate use and generation of hazardous substances." In their analysis of chemical systems, these authors note that "it is better to prevent waste than to treat or clean up waste after it is formed." They also cite work indicating that organic solvents account for most industrial waste and that it is preferable to use systems that do not require such solvents. Finally, they suggest that "the design of chemical reactions or systems that do not require intensive energy use is highly desirable." The introduction of processing and separation systems based on green chemistry principles in commercial processes could have great benefits in reducing waste generation. These benefits with regard to the mining industry have been discussed [11].

Major challenges in the attainment of green chemistry operations lie in the separations area [4]. Separations, as used in industrial chemistry processes, often involve large volumes of organic solvents that are discarded after use, frequently with significant environmental consequences, as in Chinese REE operations [4,9]. In addition, conventional separation systems of low metal selectivity such as SX, ion exchange (IX), and precipitation are less effective in recovering metals as concentration levels decrease [4,12]. Consequently, metal loss can be significant in separation processes using these procedures. Poor metal recoveries during processing or in recycling processes require that additional ore must be mined and processed to meet increased demand for metals, especially for use in high-technology products. Mining ore requires large expenditures of energy and water, with resultant environmental concerns [3,13]. Thus, meeting the demand for metals by recycling and by use of green chemistry processes at each level of the metal life cycle to conserve metals has tremendous benefits to industry and society. Looking forward, it is inevitable that green chemistry methods of increased selectivity will displace traditional energy-consuming and low-selectivity waste-generating procedures in ore beneficiation and recycling processes. Driving this transformation are the lower capital and operating expenses resulting from clean, effective processing and separation processes, especially when negative externality factors associated with many present mining operations are taken into account. These factors are presently borne by society, as seen, for example, in Chinese REE operations [9].

In this chapter, Molecular Recognition Technology (MRT) is presented as a highly selective green chemistry process for selective separations involving PGM and REE at several stages in their life cycles. The basis of MRT is given together with its strengths in metal separations. Examples are presented of the use of MRT in separations and recoveries involving both sets of metals. An analysis of the capex and opex costs associated with use of MRT and SX in separation and recovery of REE is presented to emphasize the benefits to be derived from use of green chemistry processes.

14.2 Molecular Recognition Technology

During the past two decades, a breakthrough in achieving extremely high selectivity in metal separations has been made through introduction of molecular recognition principles in designing separation systems that employ pre-designed metal-selective ligands [12,14]. These systems involve incorporation by chemical binding of ligands on solid supports such as silica gel or polymer substrates, making their use in a solid phase extraction mode possible. The supported ligand product is termed SuperLig®. No organic solvents are used in MRT systems. MRT has numerous advantages over other separation technologies. These advantages primarily arise from large ligand-target metal binding energies (log K), rapid reaction kinetics, high degree of selectivity by the ligand for the target metal even in complex matrices such as highly acidic or basic solution and/or in solutions containing high concentrations of competing ions, and the ability to carry out selective metal separations and to recover pure metals over wide metal concentration levels ranging from g/L to mg/L.

SuperLig® products consist of small (0.5 mm) particles packed into fixed-bed columns. These columns are present in skid-mounted modular form, and are fully automated for continuous operation. MRT processes operate on a system-cycle basis. A complete system cycle consists of the following sequence: (1) loading phase — target ion is loaded from feed solution onto an appropriate metal-selective SuperLig® product charged into the column(s); (2) pre-elution wash phase — any remaining feed solution is washed from the column; (3) elution phase — target ion is eluted with a small amount of eluent, forming an eluate solution containing the concentrated metal product; and (4) post-elution wash phase — any remaining eluent is washed out of the column. The cycle begins again with Step (1). Small elution volumes allow the production of eluate solutions highly concentrated in the desired metal.

14.3 Strengths of Molecular Recognition Technology in Metal Separations

MRT provides an approach to chemical separations that is consistent with green chemistry principles [10]. A description of the MRT process, including its ability to perform green chemistry separations in an environmentally friendly fashion and several of its commercial applications, has been given [11,12,14]. The importance of selectivity in making green chemistry separations and the need for improvement in separation technologies to achieve greater global metal sustainability have been presented [4,11]. The MRT approach forms a basis for improving metal recovery, recycling, and end-product fabrication processes during metal life cycles with significant operating advantages and environmental gains over traditional procedures. The important strengths of MRT processes discussed below lead to improved separations behavior.

14.3.1 Significant Improvement in Process Conditions

High selectivity of SuperLig® products for specific metal-containing species results in large column loading capacities and no retention on the column of unwanted metals, thus eliminating further expensive and time-consuming separations required by conventional separation technologies. The MRT process is simple in design, operates at room temperature (T) (~25 °C) and pressure (P) (~0.1 MPa), and is effective over a wide metal concentration range from g/L to mg/L or lower. Thus, recoveries on-site for streams with wide metal concentration ranges can be carried out under widely varying process conditions. A range of solid support systems provides choices for the best SuperLig® product(s) to operate effectively depending on plant requirements. A significant advantage of MRT systems is the ability to handle large volumes of solution at high flow rates on a continuous and automated basis, switching between multiple columns as required, making rapid, large scale separations possible. MRT systems can be incorporated into flow sheets on a real time basis. These systems can be designed and constructed for any size, concentration, and volume-throughput requirement.

14.3.2 Short Process Time

Time is a major concern in metals refining, and MRT processes offer significant time advantages. First, resin loading times are rapid. Second, combined with rapid elution times, the turnaround time for consecutive runs is fast compared to that of other separation technologies. Third, incorporation of these advantages into existing flow sheets can markedly shorten processing times and increase productivity, providing significant returns to the metal owner.

14.3.3 High Selectivity for Target Species

The distinguishing feature of SuperLig® products is their pre-determined high selectivity for target metal species [11,12]. This selectivity makes possible high loading on the column of the target metal by direct recovery of that metal in high yield from matrices containing competing metals that otherwise would contaminate separated products. Matrices encountered in metallurgical operations typically include high acid or base concentrations, metals with chemical properties similar to those of the metal to be separated, and high concentrations of common metal salts. Negative effects of these matrices on the ability to make desired selective separations increase as concentrations of metals to be separated decrease toward mg/L levels. Conventional hydrometallurgical methods do not generally have sufficiently high metal selectivity under these challenging conditions, resulting in gross inefficiency in the form of poor recovery, low concentration factors, and impure products [4,11]. Performance can be improved with multiple stages or even use of more than one separation process, but costs mount due to greater needs for space, chemicals, energy, equipment, and labor. Worker safety and health are compromised. The amount of secondary waste production also increases resulting in higher costs for disposal or recycling. Elution of the selectively separated metal from the column in the MRT process provides a pure and concentrated target metal product without need for further purification. Simple production of end-products is easily attained. Examples below illustrate the ability of MRT to accomplish difficult PGM separations and to recover and separate individual REE from each other.

14.3.4 Availability of SuperLig® Products for a Wide Range of Species

Commercially proven SuperLig® products are available for a wide range of metals, nonmetals, and anions, giving great versatility in the design of selective separations and metal recovery. Most metals in the periodic table are included and anions include halides; oxyanions such as nitrate, sulfate and perrhenate; and many complex ions such as the chloroanions of Rh(III) and Ru(III). This versatility makes possible use of MRT to separate and recover metals with very similar chemical properties, such as PGM and REE.

14.3.5 Significant Operating Advantages

Significant operating advantages result from the use of MRT. First, the MRT plant is compact, easy to install and easy to operate. SuperLig® products can be used over multiple cycles and do not introduce contaminants into the separation process. Commercial MRT systems have smaller floor space requirements than conventional separation systems. Due to their high selectivity, relatively small quantities of SuperLig® products are required, resulting in much smaller installations than pyrometallurgical and traditional hydrometallurgical operations. Hydrometallurgical plants using MRT can accomplish the same separations as smelters without the procedural, environmental, high energy consumption, and carbon footprint problems associated with smelters [12,15]. Second, the operator usually has a choice of low-cost, elution reagents (e.g., common acids, NaCl, water) that are compatible with the operation. These reagents normally work effectively at ambient temperatures and pressures, although increased temperatures may be required in some cases. MRT operates in a closed system of columns, thus minimizing worker exposure to process chemicals. Third, efficient SuperLig® elution and wash procedures are used that are rapid and amenable to automated online multi-cycle operation. Fourth, high selectivity, high loading capacity, and rapid loading and release kinetics of the SuperLig® products result in attractive capital and operating costs compared to those required for other separation procedures. Fifth, metal inventory time is shorter than for traditional hydrometallurgical operations based on SX, IX, and precipitation. Sixth, chemicals such as organic solvents, which are expensive and can have negative environmental, safety, and health effects, are not used. MRT is often superior to IX [12], particularly when considering the elution and wash chemicals; rapid metal binding and release; speed of extracting and recovering the metal; metal purity; metal losses; and metal inventory time. This results in a very attractive overall separations process.

14.3.6 Environmentally and Ecologically Friendly Processes

SuperLig® products selectively and rapidly separate, concentrate, and recover target and accompanying metals in pure form with minimal carbon footprint. Resulting raffinate streams, with negligible target metal concentration levels, can be efficiently treated within a plant's flow sheet without need for further processing to recover the metal of value. Non-target metals can be retained for value or disposed of in an environmentally friendly manner.

14.3.7 Cost Effectiveness

The economy of MRT has been proven in a number of commercial applications involving both base and precious metals. One perceived advantage of less selective technologies, such as IX and SX, over MRT is the low price of the resins or reagents involved. However, this price is not indicative of the true costs incurred by these technologies. As compared to MRT, the negative externality costs associated with less selective technologies are paid for in more complex capital equipment systems; larger system footprints; limited or no resin reusability; extensive use of sub-optimal and/or hazardous and flammable chemicals, such as solvents, that introduce substantial risk into the process and require complicated operational and environmental protocols; higher energy costs; lack of system flexibility to target specific, commercially important metals early in the flowsheet; higher number of separation stages; increased volumes of eluates, washes, and wastes; larger and more complex waste treatment systems; slower metal binding and release; incomplete and slow extraction and recovery of the target metal; complex pre- and post-treatment regimens; lower metal purities; lower metal recoveries; higher metal losses; and longer retention of valuable metals in the process, resulting in high metal inventories and working capital costs. Low selectivity means that traces of impurity metals follow target metals, and this translates into the creation of multiple side streams that then need to be processed, resulting in higher costs, greater environmental liabilities and increased worker exposure. All the above factors must be taken into account to determine the true cost effectiveness of a separations technology.

14.4 Applications of Molecular Recognition Technology to Separations Involving Platinum Group Metals

MRT SuperLig® products are used worldwide for refining PGM from both primary and secondary sources [12,16]. Selective separations and recoveries are now described for (1) Ir from Ir/Rh feed solutions containing base metals; (2) Rh from feed solutions containing base metals, Rh, Pt, Pd, Ru, Ir, Au, and Ag; (3) individual, sequential, selective separations of Pd, Pt, and Rh from feed solutions containing these metals and base metals; and (4) individual, sequential, selective separations of Pd, Pt, Ir, and Rh from feed solutions containing these metals and base metals [16]. The separations described in Figures 1–4 are all commercially proven. These flow sheets are selected to illustrate features of the MRT

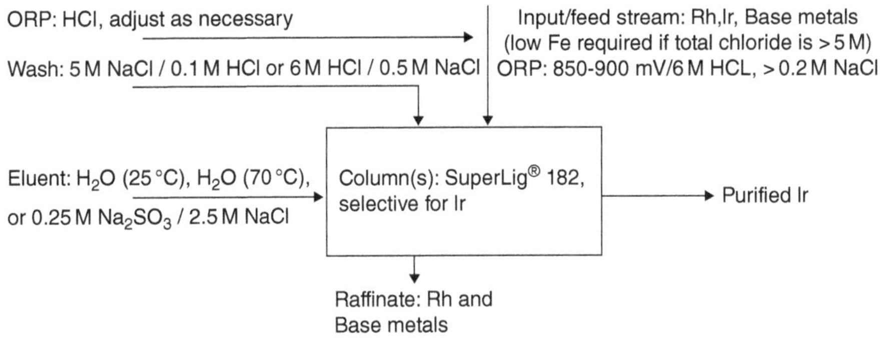

Figure 1 *Flow sheet for selective separation of iridium from rhodium and base metals. ORP = oxidation-reduction potential. Izatt, 2014 [16]. Adapted with permission from International Precious Metal Institute (IPMI).*

separations process that make it a green chemistry procedure. Table 1 summarizes green chemistry principles that apply to the selective PGM separations illustrated in Figures 1–4.

In Figure 1, selective separation of Ir from Rh and base metals is shown. SuperLig® 182 has high selectivity for Ir over Rh and the base metals present and, thus, Ir is separated in high purity in one pass through the column. Use of appropriate SuperLig® products allows further separation of Rh from the base metals. Three different eluents are shown, H_2O (25 °C), H_2O (70 °C), and 0.25 M Na_2SO_3/2.5 M NaCl solution. Which eluent is chosen depends on the process conditions, such as the Ir speciation present. Similar options are available for wash solutions.

In Figure 2, the selective separation of Rh from Pd, Pt, Ir, Ru, Au, Ag, and base metals is shown. This flow sheet shows the separation of small amounts of Pt that follow Rh and are eluted with Rh from the SuperLig® 190 column. The Pt polishing is achieved using a SuperLig® 133 column to polish the Pt, leaving pure Rh. Washes and eluents, consisting of H_2O, HCl and NaCl (or KCl), are compatible with PGM refinery flow sheets.

In Figure 3, individual, sequential, selective separations of Pd, Pt, and Rh from a feed solution containing these PGM and base metals are presented. Wash and elution solutions are compatible with overall PGM plant operations. ORP values are adjusted as needed. Selective separations of the PGM are made in sequential steps using SuperLig® products designed for each individual PGM. Each separation step results in an eluate solution containing the pure, concentrated target PGM product and a raffinate solution containing the remaining PGM and impurity metals. Since the separation at each step is selective and complete, the remaining metals pass on to the raffinate without loss.

In Figure 4, individual, sequential, selective separations of Pd, Pt, Ir, and Rh in a base metal matrix are presented. These results illustrate the separation of PGM from each other using SuperLig® products having affinity for each individual metal.

The flow sheets in Figures 1–4 are representative of those developed at IBC Advanced Technologies, Inc. (IBC) for PGM separations from each other and from a wide range of metal impurities. A key aspect of the separation process is the use of SuperLig® products designed to bind selectively with specific PGM species in the presence of competing metals. The important role of high selectivity in separations was discussed in Section 14.3.3. SuperLig® product design must take into account the speciation of the PGM as a function of the anion present and

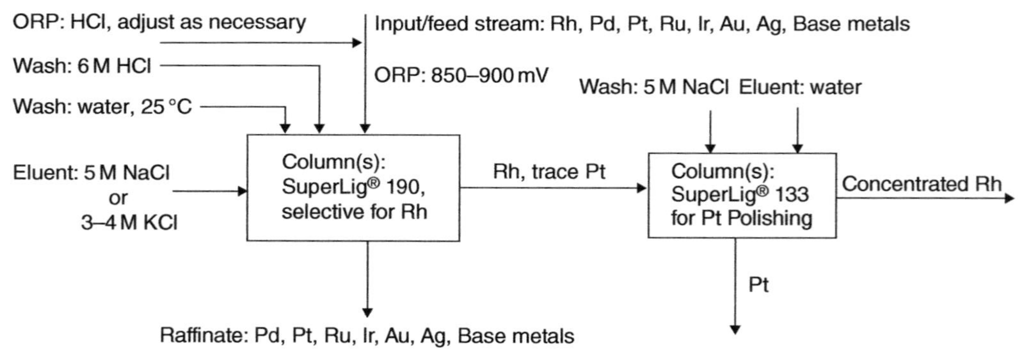

Figure 2 *Flow sheet for selective separation of Rh from Pd, Pt, Ru, Ir, Ag, Au, and base metals. ORP = oxidation-reduction potential. Izatt, 2014 [16]. Adapted with permission from IPMI.*

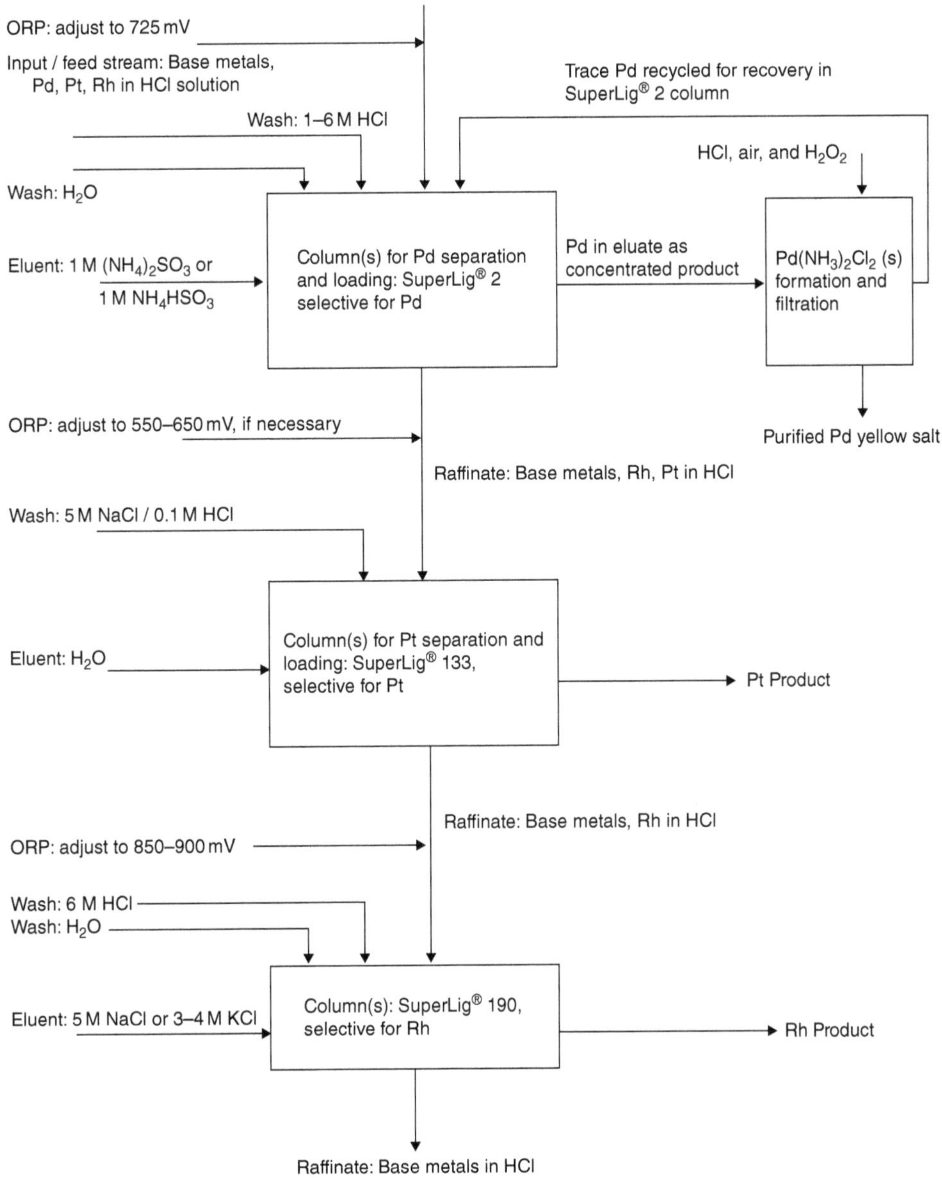

Figure 3 *Individual, sequential, selective separations of Pd, Pt, and Rh from a feed solution containing these metals and base metals in an HCl solution. ORP = oxidation-reduction potential. Izatt, 2014 [16]. Adapted with permission from IPMI.*

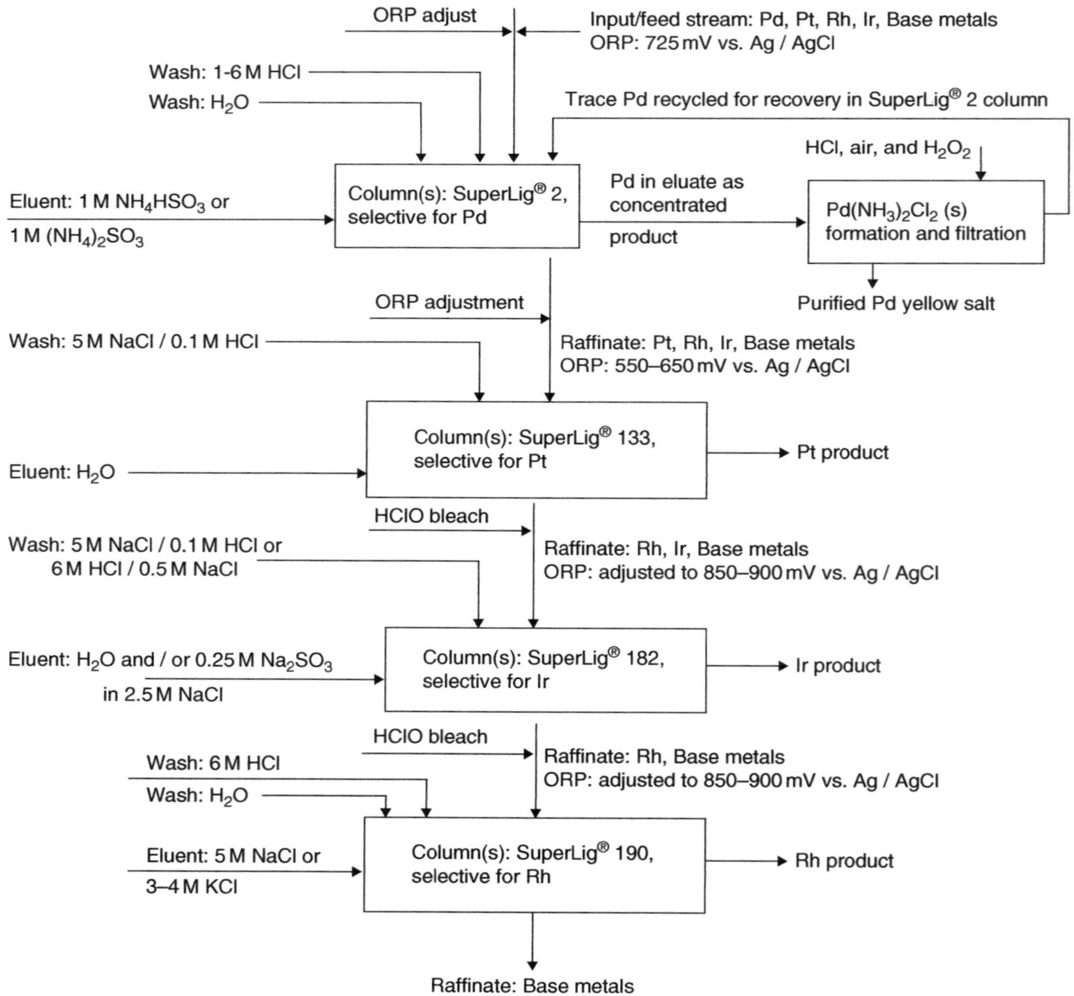

Figure 4 *Individual, sequential, selective separations of Pd, Pt, Ir, and Rh from a base metal matrix. ORP = oxidation-reduction potential. Izatt, 2014 [16]. Adapted with permission from IPMI.*

Table 1 Green Chemistry Principles Applied to MRT Platinum Group Metal Separations

(i)	No organic solvents are used in the separation process that operates at room T (~25 °C) and P (~0.1 MPa).
(ii)	No contaminants are added to the process stream by the SuperLig®.
(iii)	Wash and eluent solutions are as simple as possible while being compatible with the overall PGM refining plant operations. Washes and eluents used include H_2O, HCl, NaCl, KCl, Na_2SO_3, $(NH_4)_2SO_3$, and NH_4HSO_3.
(iv)	Target metals are recovered in pure concentrated form from eluate solutions following elution from the column with a small amount of eluent.
(v)	Eluates can be easily precipitated to final products using common reagents such as, for example, HCl, air, and H_2O_2 to form Pd yellow salt. Precipitated pure metals/metal compounds are collected by filtration. There is no addition of contaminating or hazardous reagents.
(vi)	Impurity metals are selectively separated either for value or safe disposal. Silver, Au, and base metals are examples mentioned in the flow charts. This procedure results in minimal waste.
(vii)	Selective separations of individual PGM are found in the various steps in the flow charts. This selectivity is of critical importance since it eliminates the need for multiple stages and use of hazardous/contaminating chemicals, as described earlier in the text.
(viii)	The PGM flow charts demonstrate that MRT is capable of making a variety of individual and/or group separations involving these elements even with high concentrations of impurity base metals, Ag, and Au present. These impurity metals are recovered in pure form for value or safe disposal (item 4). The MRT process comes close to zero hazardous waste generation. This achievement represents green chemistry applied to chemical separations.

the pH of the solution as well as the relative binding strengths of the several possible host-guest interactions. As seen in Figures 1–4, the selective separations are single pass resulting in efficient operations. The high selectivity shown in Figures 1–4 results in a marked reduction in the time required to make the indicated separations. Time is an important factor in chemical separations, as discussed in Section 14.3.2. A summary of major advantages of MRT compared to conventional technologies for PGM processing is given in Table 2.

14.5 Applications of Molecular Recognition Technology to Separations Involving Rare Earth Elements

The entire suite of sixteen individual REE, Pm excluded, has been separated at the bench level at high purity of >99% using a proprietary MRT procedure [17,18]. A pregnant leach solution (PLS) derived from Bokan-Dotson Ridge, Alaska, feedstock was the feed material for the separations. The PLS was prepared by Hazen Research Inc., of Golden, Colorado from beneficiated ore using a metallurgical process described in [17]. The Bokan deposit has high levels of certain of the 'heavy' REE such as Dy, Tb, and Y.

The entire set of sixteen REE was initially separated from the non-REE gangue metals present. This initial separation, accomplished at the >99% level, is an important first step in the process. Having the entire suite of REE separated from accompanying gangue material

Table 2 *Summary of Major Advantages of MRT in PGM Refining Versus Conventional Processes and Technologies*

- Single-pass separations for the target PGM of 99 + %
- Product purity of 99.95% - 99.99%
- Minimization of platinosis and other health and safety risks due to the system being self-contained, thereby greatly reducing worker exposure
- Rapid recycling of the target PGM — reduction of total processing time
- Increase in PGM recovery (reduction in processing losses)
- Lower PGM refining cost by eliminating and/or reducing the use of process chemicals and the number of process steps
- Lower maintenance costs
- Reduction in the processing pipeline, thus reducing metal financing costs and releasing metal earlier for sale
- Enhancement of effluent control measures
- Treatment of both dilute and concentrated feed streams
- Option of producing a salt product that may be sold or reduced to market-grade metal
- Rapid and semi continuous process
- Treatment of any solution volume
- High flow rates for both loading and elution
- Recycling of solutions
- Competitive capital costs with simple, compact equipment requiring minimal space
- Reduction of precious metal security risk as the system is self-contained, thus minimizing exposure
- Multi-cycle use of SuperLig® products

simplifies subsequent separation of individual REE since impurity metals are not present downstream. Furthermore, the >99% pure concentrated REE solution contains essentially all the REE present in the original PLS. This exceptionally high REE recovery rate is a key economic differentiator between MRT and other less selective processes. It maximizes the value obtained from the PLS and ensures that minimal amounts of REE are discarded to tailings.

Separation of the sixteen REE into individual elements at >99% purities was achieved in a three- step process as shown in Figure 5. In Step 1, Sc and Ce are removed. In Step 2, light REE (Y, La, Pr, Nd) are separated from the heavy REE (Sm through Lu). In Step 3, individual members in the light and heavy groups are separated from each other. Individual REE were collected as carbonates. Elution of the bound REE, as groups or individuals, from the SuperLig® columns by small amounts of eluent (acid) produces concentrated solutions of the pure metal(s) for easy and economical salt production.

Table 3 gives the purities of each individual, separated REE as a carbonate salt, or in solution prior to salt production. The purity of each individual REE in solution was determined at IBC by ICP analysis. Separate analyses of selected carbonate salts were made by an independent analytical laboratory. In all cases, purities exceeding 99% were obtained.

A pilot plant has been constructed [19–23] that enables the separation system to be scaled up to commercial production of individual REE. The commercial MRT system has several advantages over conventional processes for REE separations. First, the MRT system is based on green chemistry principles throughout. Minimal waste is generated. No solvents are used. Few chemicals are used and those used in washing and elution of

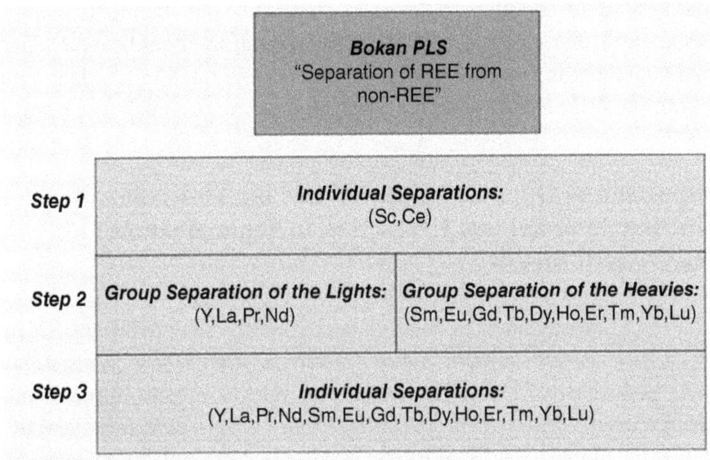

Figure 5 *Scheme for separation of sixteen rare earth elements from Bokan-Dotson Ridge pregnant leach solution and from each other [17]*

Table 3 *Purity levels of sixteen separated rare earth element salts [17,18]*

Rare Earth Element	Purity of Carbonate Salt, %
Lanthanum (La)	99.9
Cerium	99.5[1]
Praseodymium	99.9
Neodymium	99.9
Samarium	99.2[1]
Europium	99.9
Gadolinium	99.2[1]
Terbium	99.1
Dysprosium	99.9
Holmium	99.5
Erbium	99.9
Thulium	99.6
Ytterbium	99.0
Lutetium	99.2[1]
Yttrium	99.4
Scandium	99.1[1]

[1] Purity reported for solution prior to salt production

columns are as benign as possible. Energy requirements are minimal with separations performed at room T (~25°C) and P (~0.1 MPa). Second, the high selectivity of the SuperLig® products makes possible separation and recovery of REE from the PLS at >99% levels and subsequent separation of individual REE at >99% levels. Results of these high recovery levels are achievement of maximum economic value, minimal loss of REE to the tailings, many fewer stages required to achieve given REE purities, and reduced use of auxiliary substances. Third, the entire MRT operation is physically and operationally simple, reducing capital costs. Fourth, the MRT operation requires fewer personnel, equipment, and chemicals, thereby reducing operating costs. Fifth, the green chemistry MRT operation should readily meet stringent environmental standards in the U.S. Sixth, Dy can be separated in a targeted manner from the REE without the need to first separate low-value rare earths such as Ce and La.

14.6 Comparison of Opex and Capex Costs for Molecular Recognition Technology and Solvent Extraction in Separation and Recovery of Rare Earth Metals

Items contributing to capex and opex costs associated with use of MRT and SX in separation and recovery of REE are presented in Table 4. Comparison of these items shows that costs are much lower in the case of MRT. The lower costs for MRT result from this process being simpler in design, more efficient in operation, more conservative in its use of chemicals, and more environmentally friendly in its use of simple inorganic chemicals, such as dilute acids. No solvents are used in the MRT process. High capex and opex costs result when SX processes are used in REE beneficiation and separation processes.

Elimination of solvents in chemical production processes is one of the most active areas of green chemistry research because of the difficulties they present [10]. Solvents often account for the vast majority of mass wasted in syntheses and processes. Many conventional solvents are toxic, flammable, and/or corrosive, requiring large capex and opex expenditures to ensure that they can be used safely and for their disposition after use. Their volatility and solubility properties have contributed to extensive waste generation resulting in air, water and land pollution, as seen in the REE mining and production facilities in China [9]. Use of solvents in ore beneficiation has increased the risk of workers' exposure, and has led not only to serious accidents, but to fires that destroy facilities and shut down production for extended periods of time. Recovery and reuse of solvents, when possible, is often associated with energy-intensive distillation and sometimes cross contamination. When environmental regulations are non-existent, ignored, or not enforced, spent solvents are often discarded into the commons, with serious environmental and health consequences. These negative externalities endanger the wider population as well as the workers associated with

Table 4 *Comparison of MRT and SX Capex and Opex Costs*

ITEM	MRT	SOLVENT EXTRACTION
	CAPEX	
Floor space	Small	Large
Equipment cost	Small	Large
	OPEX	
Processing costs	Small	Large
Labor	Small	Large
Safety provisions	Small	Large
Spent chemical and waste water discard	Small	Large
REE processing time	Rapid – low metal inventories/ working capital	Slow – high metal inventories/ working capital
REE recovery rates	High	Low

the processing of REE. Chinese REE mining and production facilities provide an example of the effects of pollution from use of solvents [9]. Such problems, and the resulting economic inefficiencies, were in part a cause of the closure of the Mt. Pass REE facility in California in 2002 and again in 2015.

A major reason for the problems associated with the use of SX in REE processing is the low selectivity and slow reaction kinetics of SX for target REE in separating these metals either from gangue metals or from each other. These deficiencies result in low recovery rates, many separation stages, extended time periods for effective separations, and the inability to separate individual REE in a targeted manner, resulting in the need to first separate low-value rare earths such as Ce and La. The results are increased capex and opex costs, including low recovery rates and extended processing times, that result in high working capital requirements due to 'in-process' REE inventories. MRT, on the other hand, has high selectivity, high recovery rates, and rapid processing times for REE separations both from gangue and from each other, as shown in Figure 5 and Table 4.

The high selectivity of the MRT process and its use of benign chemicals result in markedly lowered capex and opex costs. The number of stages involved in each separation is much lower for MRT than SX because of MRT's higher selectivity. The result, as seen in Table 4, is reduced space requirements as well as lower equipment, processing, labor and waste treatment costs. High REE recovery rates coupled with rapid processing times maximize the financial rate of return and minimize working capital needs due to low metal inventories. MRT is a green chemistry process of the type needed in 21st century REE separations where increasingly stringent environmental and health standards must be achieved in concert with intensifying demands for economic efficiency as REE applications continue to expand.

14.7 Conclusions

MRT has been shown to be an effective green chemistry procedure for the efficient, economic, and environmentally friendly separation from complex matrices of PGM and REE into their individual components. Major benefits of MRT include: (1) high selectivity and high capacity of the SuperLig® products for the target metal resulting in simple and rapid separations, (2) no loss of the target metal during the loading phase, (3) very high feed flow rates during individual metal loading, (4) production of concentrated eluate solutions containing the individual target metal allowing the metal to be readily recovered in pure form, (5) recovery of the target metal for reuse, avoiding wasteful discharge into the environment and reduction in the need to mine replacement virgin ore, (7) smaller facilities as a result of the high selectivity and simplicity of the process compared to those required when using other methods, (8) no use of organic solvents, and (9) use of simple eluent and wash solutions compatible with plant operations.

References

1. Wilburn, D.R., 2012, "Global Exploration and Production Capacity for Platinum-Group Metals from 1995 through 2015 (version 1.1)," U.S. Geological Survey Scientific Investigations Report 2012–5164, 26 pp., at http://pubs.usgs.gov/sir/2012/5164/, accessed 16 November 2016.
2. USDOE, Critical Materials Strategy; DOE/PI-0009; December 2011. http://energy.gov/sites/prod/files/DOE_CMS2011_FINAL_Full.pdf, accessed 16 November 2016.

3. Hagelüken, C., 2014, Recycling of (Critical) Metals: In *Critical Metals Handbook*, Gunn, G., Ed., Wiley, Oxford, pp 41–69.
4. Izatt, R.M., Izatt, S.R., Bruening, R.L., Izatt, N.E., Moyer, B.A., 2014, Challenges to Achievement of Metal Sustainability in Our High-Tech Society, *Chemical Society Reviews*, 43, 2451–2475.
5. Reck, B.K., Graedel, T.E., 2012, Challenges in Metal Recycling, *Science*, 337, 690–695.
6. Osibanjo, O., Nnorom, C.I., Sdie, G.U., Ogundiran, M.B., Oketola-Adeyi, B., 2016, Global Management of Electronic Wastes: Challenges Facing Developing and Economy-in-Transition Countries, this volume.
7. Williams, I.D., 2016, Global Metal Reuse, and Formal and Informal Recycling from Electronic and Other High-Tech Wastes, this volume.
8. Mudd, G.M., 2012, Sustainability Reporting and the Platinum Group Metals: A Global Mining Industry Leader, *Platinum Metals Reviews*, 56, 2–19.
9. Yang. X.J., Lin, A., Li, X.-L., Wu, Y., Zhou, W., Chen, C.Q. 2013, China's Ion-adsorption Rare Earth Resources, Mining Consequences and Preservation, *Environmental Development*, 8, 131–136.
10. Anastas, P., Eghbali, N., 2010, Green Chemistry: Principles and Practice, *Chemical Society Reviews*, 39, 301–312.
11. Izatt, R.M., Izatt, S.R., Izatt, N.E., Krakowiak, K.E., Bruening, R.L., Navarro, L., 2015, Industrial Applications of Molecular Recognition Technology to Green Chemistry Separations of Platinum Group Metals and Selective Removal of Metal Impurities from Process Streams, *Green Chemistry*, 17, 2236–2245.
12. Izatt, S.R., Bruening, R.L., Izatt, N.E., 2012, Status of Metal Separation and Recovery in the Mining Industry, *Journal of Metals*, 64, 1279–1284.
13. Norgate, T.E., 2010, Deteriorating Ore Resources, In *Linkages of Sustainability*, Graedel, T.E., van der Voet, E., Eds., The MIT Press, Cambridge, MA, pp. 131–148.
14. Izatt, N.E., Bruening, R.L., Krakowiak, K.E., Izatt, S.R., 2000, Contributions of Professor Reed M. Izatt to Molecular Recognition Technology: From Laboratory to Commercial Application, *Industrial & Engineering Chemistry Research*, 39, 3405–3411.
15. Izatt, S.R., Mansur, D.M., 2006, Environmentally Friendly Recovery of Precious Metals from Spent Catalysts, International Precious Metals Institute Petroleum Seminar, Houston, TX, November 13–14.
16. Izatt, S.R., Bruening, R.L., Izatt, N.E., 2014, Green Chemistry Approach to Platinum Group Metals Refining, International Precious Metals Institute 38th Annual Conference, Orlando, Florida, June 7–10, 2014.
17. Press release, March 2, 2015, Ucore Successfully Separates Entire Suite of Individual Rare Earth Elements at High Purity, http://ucore.com/press-releases Accessed June 15, 2016.
18. Press release, April 28, 2015, Ucore Updates on the Separation of Individual Rare Earth Elements, http://ucore.com/press-releases Accessed June 15, 2016.
19. Press release, July 8, 2015, Ucore Commissions Design and Construction of SuperLig®-One Pilot Plant, http://ucore.com/press-releases Accessed June 15, 2016.
20. Press release, March 7, 2016, Ucore Completes Construction of SuperLig®-One Pilot Plant, http://ucore.com/press-releases Accessed June 15, 2016.
21. Press release, April 5, 2016, Ucore Commissions SuperLig®-One Pilot Plant and Accepts Initial Feedstock, http://ucore.com/press-releases Accessed June 15, 2016.
22. Press release, June 7, 2016, Ucore Separates HREE and LREE Classes at 99%+ Purity via SuperLig®-One Pilot Plant, http://ucore.com/press-releases Accessed June 15, 2016.
23. Press release July 5, 2016, Ucore Separates Dy and Ho Sub-Groups at 99%+ Purity via SuperLig®-One Pilot Plant, http://ucore.com/press-releases Accessed July 5, 2016.

15

Refining and Recycling Technologies for Precious Metals

Tetsuya Ueda, Satoshi Ichiishi, Akihiko Okuda, and Koichi Matsutani

TANAKA KIKINZOKU KOGYO K.K., Kanagawa, Japan

15.1 Introduction

"Precious metals" is the collective term for a total of eight elements that include gold (Au) and silver (Ag) in addition to the six platinum group metals, platinum (Pt), palladium (Pd), rhodium (Rd), iridium (Ir), ruthenium (Rh) and osmium (Os). Excluding osmium, these precious metals are used for both jewelry and industrial products. Gold and silver have long been known, but the chemical and physical similarities among the six platinum group metals have made separation from each other difficult. It is only in recent times that all elements in the platinum group have been discovered. As a result of their extremely low yields and high prices, the term "precious metals" has therefore been applied to these eight elements. For gold in particular, only about 160,000 tons have ever been mined in the 6000 years of human history, which equates to three Olympic-sized swimming pools.

The characteristics of precious metals, which have good chemical resistance, do not corrode, have good electrical conductivity, low toxicity, good thermal resistance and catalytic properties, have resulted in their use in many fields that demand reliability, such as the automotive and electronics industries. Technological innovations have enabled reduced use of precious metals and use of substitute materials instead, but the value of precious metals is not expected to change in the future because of expanded use of precious metals by new technologies.

New supply of platinum group metals from mines is limited to production from a few countries only, such as South Africa, so there is a potential risk to the supply source. These metals are in use in many different industries, with usage volumes forecast to increase even more in

Metal Sustainability: Global Challenges, Consequences, and Prospects, First Edition. Edited by Reed M. Izatt.

the future where reliability and high functionality are required. Under conditions such as these, it is impossible to rely on output from mines alone, so it is important to recycle the precious metals already widely distributed in the marketplace and to reuse them again as resources — the mission, we believe, of everyone handling these precious metals.

15.2 Precious Metals Supply and Demand

15.2.1 Supply

15.2.1.1 Platinum

Table 1 shows platinum supplies. One particular area of note in regard to platinum supply from mining over recent years is the impact of a large decrease in production at mines in South Africa, which accounts for approximately 75% of total world output. Issues such as unstable power supplies, labor strikes, safety-related shutdowns and closures of small-scale mines and reprocessing plants are behind decreased production in the country. Despite some progress toward resolving these issues, supply has continued on its downward trend since 2012. Platinum production in Russia is experiencing a modest reduction with the deteriorating quality of platinum group metals obtained from ores mined in the nickel mines. In Zimbabwe, on the other hand, production is increasing in line with plans of the South African parent companies operating the mines there, based on the comparatively cheaper production costs.

Table 1 *Platinum Supply and Demand*

Year		Unit:ton				
		2009	2010	2011	2012	2013
Supply	South Africa	144.2	144.2	151.2	127.2	128.1
	Russia	24.4	25.7	26.0	24.9	24.3
	North America	8.1	6.2	10.9	9.6	9.8
	Zimbabwe	7.2	8.7	10.6	10.6	12.4
	Others	3.6	3.4	3.1	3.4	3.9
	Total Supply	**187.4**	**188.2**	**201.7**	**175.7**	**178.5**
Gross Demand by Application	Autocatalyst	68.0	95.6	99.1	99.2	97.2
	Chemical	9.0	13.7	14.6	14.0	16.8
	Electrical	5.9	7.2	7.2	5.1	6.4
	Glass	0.3	12.0	16.0	5.0	7.3
	Investment	20.5	20.4	14.3	14.2	23.8
	Jewelry	87.4	75.3	77.0	86.5	85.2
	Medical & Biomedical	7.8	7.2	7.2	7.3	7.3
	Petroleum	6.5	5.3	6.5	6.4	4.8
	Others	5.9	9.3	10.0	12.1	13.1
	Total Gross Demand	**211.3**	**245.9**	**251.8**	**249.8**	**261.9**
Recycling	Autocatalyst	(25.8)	(33.7)	(38.6)	(35.1)	(39.7)
	Electrical	(0.3)	(0.3)	(0.3)	(0.6)	(0.8)
	Jewelry	(17.6)	(22.9)	(25.2)	(27.7)	(24.1)
	Total Recycling	**(43.7)**	**(56.9)**	**(64.1)**	**(63.5)**	**(64.5)**
	Total Net Demand	**167.6**	**189.0**	**187.7**	**186.3**	**197.4**

Source: Platinum 2013,Johnson Matthey PLC

Platinum supply from recycling of autocatalysts, electronics products and other sources is increasing yearly. However, since recyclers adjust scrap output depending on the price of platinum group metals, there is a tendency for the volume of recovered metals to be low when prices are low. In addition, while end-of-life autocatalysts make up approximately 60% of total recycled platinum, the actual amounts recovered currently are decreasing because of low demand in the new vehicle market brought about by economic instability in the European region, where use of platinum for autocatalysts is high.

15.2.1.2 Palladium

Table 2 shows palladium supplies. An important factor in regard to palladium supply is the significant reduction in volumes sold from Russian stockpiles since 2012. Supply has decreased from the level of 20–30 tons per year until 2011 to the current level of a couple of tons per year. The actual size of Russia's stockpiles is a state secret, so no detailed information is available, but since the country had already been selling down for many years, it is assumed the cycle is mostly coming to an end. Like platinum, palladium supply from mines in Russia has also been decreasing in recent years because of the deteriorating quality of platinum group metals obtained from ores. Again like platinum, the South African palladium supply is also continuing to decline because of strikes and other issues causing reduced output. Even if production volumes returned to normal in South Africa it would still not balance the reduced sell-down of the Russian stockpile, so it is difficult to expect total palladium supply from mining to return to its 2011 level.

Table 2 *Palladium Supply and Demand*

Year		Unit:ton				
		2009	2010	2011	2012	2013
Supply	South Africa	73.7	82.1	79.6	72.2	73.1
	Russia Primary	83.2	84.6	84.1	81.8	80.9
	Russia Stock Sales	29.9	31.1	24.1	8.1	3.1
	North America	23.5	18.4	28.0	27.8	28.9
	Zimbabwe	5.6	6.8	8.2	8.2	9.6
	Others	5.0	5.8	4.8	5.0	4.4
	Total Supply	**220.8**	**228.8**	**228.9**	**203.1**	**200.0**
Gross Demand by Application	Autocatalyst	126.0	173.6	191.4	208.5	216.8
	Chemical	10.1	11.5	13.7	16.5	16.5
	Dental	19.8	18.5	16.8	16.5	15.9
	Electrical	42.6	43.9	42.8	37.0	32.8
	Investment	19.4	34.1	(17.6)	14.6	2.3
	Jewelry	24.1	18.5	15.7	13.9	12.1
	Others	2.2	2.8	3.4	3.1	3.1
	Total Gross Demand	**244.2**	**302.8**	**266.2**	**310.1**	**299.5**
Recycling	Autocatalyst	(30.0)	(40.7)	(52.7)	(51.9)	(57.9)
	Electrical	(12.3)	(13.7)	(14.9)	(13.4)	(13.1)
	Jewelry	(2.2)	(3.1)	(6.5)	(5.9)	(5.6)
	Total Recycling	**(44.5)**	**(57.5)**	**(74.2)**	**(71.2)**	**(76.5)**
	Total Net Demand	**199.7**	**245.3**	**192.1**	**238.9**	**223.0**

Source: Platinum 2013, Johnson Matthey PLC

In regard to supply from recycling, the amounts of palladium recovered from end-of-life autocatalysts have increased significantly since 2011 and further increases are predicted going forward. However, like platinum, recovered volumes are impacted by prices and economic conditions. Recovery of palladium from post-consumer electronics products is expected to increase as a result of the WEEE (Waste Electrical and Electronic Equipment) Directive of the European Parliament and Council, but the trend is toward a decrease in palladium usage as electronics components become smaller, usage amounts decline and substitute metals are used, so total recovery volumes are likely to be impacted.

15.2.1.3 Gold

Gold supply from mining is experiencing record high levels through increased production output around the world, with significantly high increases in China and Canada in particular. Supply is predicted to increase from regions currently in full-scale development of mines, but development itself may be postponed due to the price of gold.

From its peak in 2009, supply from recycled scrap is continuing to trend downward, having dropped 25% between 2009 and 2013 because of a 15% drop in the US dollar-based annual average price of gold. China is the only region experiencing growth in supply. Going forward, the scrap markets in China and India are likely to have a large impact on overall supply.

15.2.2 Demand

15.2.2.1 Platinum

Table 1 shows platinum demand. Approximately 70% of demand for platinum comes from use in autocatalysts and jewelry. More details about demand for autocatalysts are provided in section 15.3, but overall demand remains fairly constant. Although Europe is expected to use more catalysts in diesel and compact vehicles than other regions, uncertain economic conditions are causing a reduction in vehicle sales, which is impacting demand for platinum. On the other hand, rebounding vehicle production in Japan and a significant increase in diesel vehicle production in India are balancing out the reductions in the European region. Growth in retailer networks in China and India are driving an increase in demand for platinum jewelry. Platinum is used for the magnetic layers of hard disks in the electronics products industry as well, but the trend is toward decreased usage because of a boom in the use of smartphones and other devices that employ solid-state storage instead of hard-disk recording media. Demand is predicted to grow in China and Asia in the future as well.

Table 2 shows palladium demand. Approximately 70% of demand for palladium comes from use in autocatalysts. More details about demand for autocatalysts are provided in section 15.3, but demand is likely to continue increasing because of increasing use as a catalyst in gasoline vehicles in China and the current high ratio of palladium usage in diesel vehicles when compared to platinum usage. Demand for use in electronic products centers mainly on its application in multilayer ceramic capacitors, which is increasing, but the trend is toward a decrease in palladium usage as usage amounts decline and substitute metals are used. On the other hand, palladium is also being used as a replacement for gold because it is a cheaper alternative for plating electronic products. Demand for use as catalysts in the chemical industry centers mainly on its application as a catalyst in the production of high-purity terephthalic acid. Like platinum, palladium demand is likely to be impacted by a slowdown in construction of new plants in China.

15.2.2.2 Gold

Demand for gold is experiencing year-on-year growth due to a strong demand for jewelry and small-scale investments because of a drop in the price of gold. Demand for jewelry in 2013 was at a multi-year high after a year-on-year increase of eighteen percent. Investment demand from the Asian markets, China and India in particular, is predicted to have an impact on gold demand. Jewelry production in China has overtaken India as the largest in the world. On the other hand, production in India has decreased slightly due to government regulations on gold supply and an increase in the rupee-based price of gold. With continuing substitution with copper as bonding wire, usage of gold in the electronics industry is trending downward. Likewise, substitution of other materials for gold is continuing in the dental and medical industries as well.

15.2.3 Outlook for Supply and Demand

South African production continues to have a significant impact on the supply of platinum group metals, and the price of those metals affects the supply from recycling. However, the systems for recovery from end-of-life autocatalysts are well established and operating, so recycling is likely to remain a stable source of supply going forward.

In light of tougher emission regulations for autocatalysts, which are the main application for platinum group metals, and its relative price against platinum, the ratio of palladium usage is expected to remain high.

Since supply from recycling is not balancing out the reduced production in South Africa and the reduced sell-down of stockpiles in Russia, the situation is changing from oversupply to undersupply. If this continues, it will have considerable impact on the price of platinum group metals going forward. Price increases and a balancing of supply and demand are likely to result in a return to past levels of supply from South Africa, which is currently in a slump but overall is certain to remain.

15.3 Autocatalysts (Pt, Pd, Rh)

15.3.1 Demand for Autocatalysts by Region

Since the first exhaust gas regulations for compact gasoline vehicles were established in the U.S. in 1970, regulations have gradually become stricter and have expanded globally. In 1983, three-way catalysts using platinum, palladium and rhodium were developed as a way of improving the safety of carbon monoxide, hydrocarbons and nitrogen oxide.

Figures 1–3 show the various amounts of platinum, palladium and rhodium used by region in the production of catalysts for purifying vehicle exhaust gases. The use of platinum in North America and Japan has remained fairly constant over the past thirty years, while the use of platinum in Europe and other regions with a high market share for compact diesel vehicles has increased rapidly due to enactment of the Euro 3 exhaust gas regulations in 2000. Nevertheless, the economic climate in Europe these days has resulted in fewer new

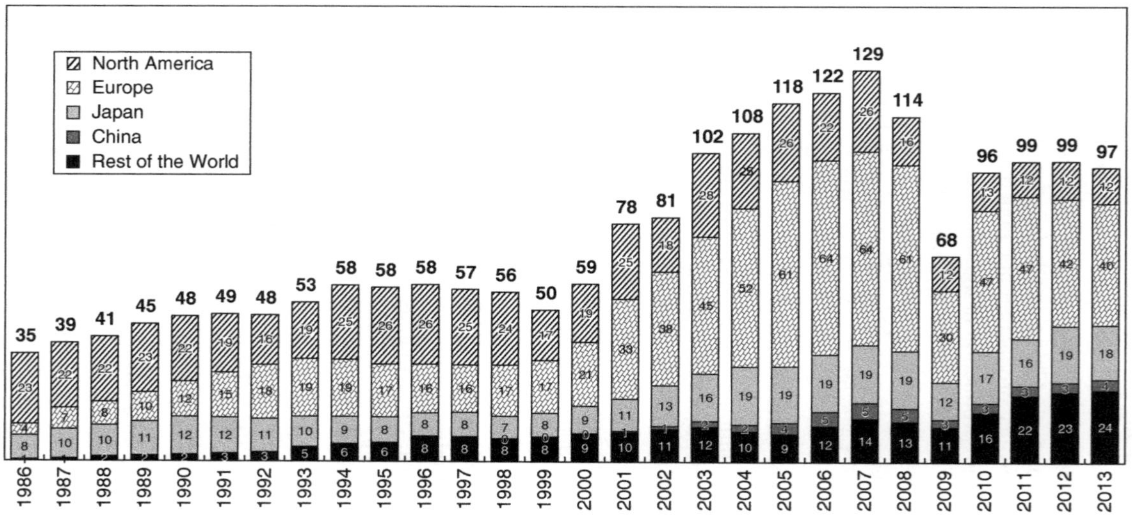

Figure 1 *Platinum Use by Region*

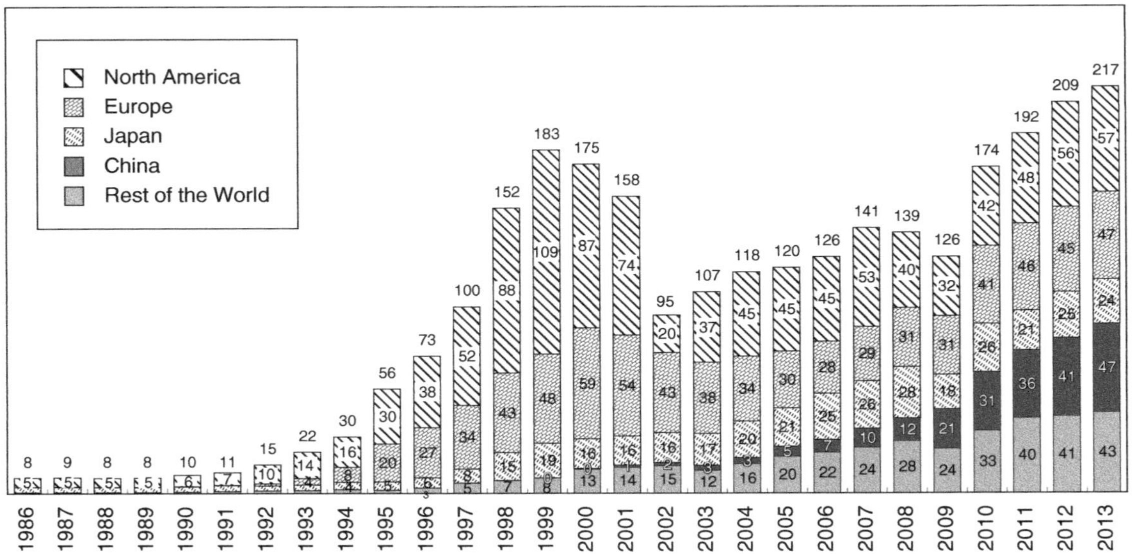

Figure 2 Palladium Use by Region

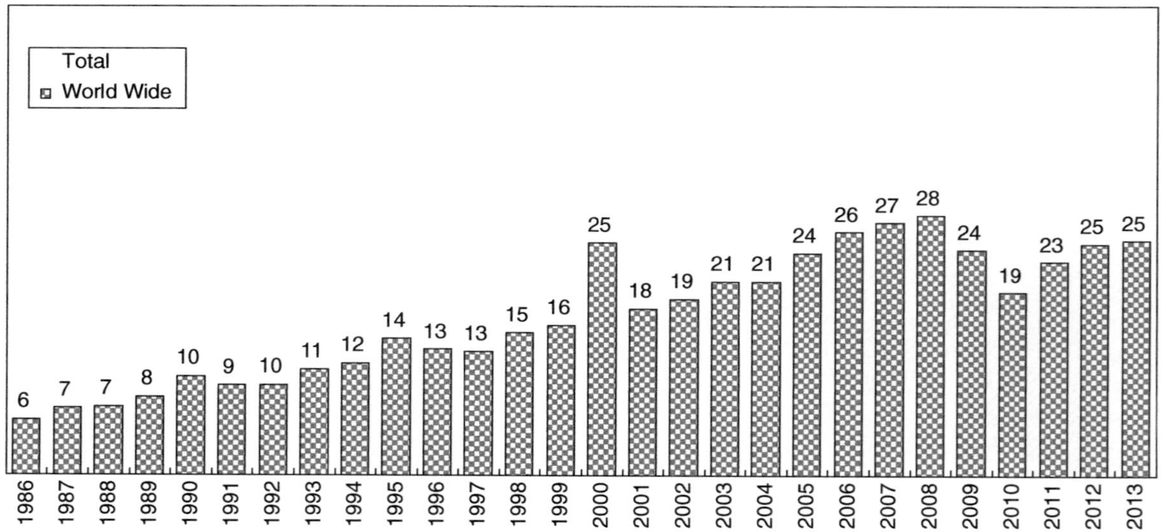

Figure 3 *Rhodium Use Worldwide*

vehicle sales, which has the effect of also reducing the use of platinum. However, Europe still uses more platinum than other regions. Catalyst manufacturers are using less platinum than previously or replacing platinum with palladium, thanks to innovations in catalyst technologies and improvements in the quality of diesel fuels. With further strengthening of NOx regulations and enforcement of the Euro 4 regulations, platinum usage as diesel vehicle catalysts is likely to increase going forward. The use of palladium as a substitute metal in North America was actively promoted in the latter half of the 1990s because of its low price compared to platinum, but that situation has now changed drastically due to steep rises in the price of palladium. With growth of the compact gasoline vehicle market in recent years, there has been a rapid increase in demand for palladium and a return to record high levels. Catalyst manufacturers are focused on improving and adopting palladium as a cheaper alternative to platinum, and in fact palladium is now the principal material used in autocatalysts. As more vehicles are manufactured, we will see increased use of rhodium as well. Nevertheless, new supply from mining is only one tenth that of platinum and palladium at approximately 20 tons per year, so in light of the associated value and price risk, manufacturers are avoiding active development of the material as a new autocatalyst. However, rhodium catalysts will potentially be used more often in the future as regulations become stricter.

Future demand from growth of the vehicle markets of China and India is predicted to have an impact on the amounts of platinum, palladium and rhodium used. Exhaust gas regulations in emerging nations are currently fairly relaxed, but much stricter controls are expected in the future with the enforcement of the China 4 standard in China and the Euro 5 standard in India, and this is the reason for the predicted growth in demand.

15.3.2 Recycling System for Autocatalysts

As explained earlier, the current trend is toward annually increasing use of platinum group metals for autocatalysts. Looking at the balance between supply and demand for platinum group metals, however, supply from mining alone is insufficient to meet that demand. Since the amount of material recovered from end-of-life autocatalysts compares well with supply from mining, recycling is an important source of supply.

Recycling systems for autocatalysts like that shown in Figure 4 are being built all over the world. With these systems in operation, manufacturers can achieve a stable source of supply. Vehicles being sold at present will usually be replaced or scrapped after a certain period (approximately ten years). Over recent years, the number of these vehicles being exported to a second life in emerging nations and other countries has increased, so the full service life of vehicles is likely to be fifteen or twenty years. In such a situation, the actual place they are finally scrapped could be anywhere in the world, so vehicle buyers, dismantlers and resource buyers are also located all around the world. After the vehicle buyers and dismantlers separate out the resource materials, such as iron, aluminum and catalysts, they sell them on to resource buyers. These people are fully aware of the value of precious metals in the autocatalysts, so they adjust their sales according to price. In this way, end-of-life autocatalysts are collected and separated, the platinum group metals are recovered, refined and reformed into bullion (or powder), and they are then reborn as platinum group metal compounds for use as the raw materials for new catalysts. The system is complete when the catalysts are rebuilt, installed into new vehicles and returned to the market.

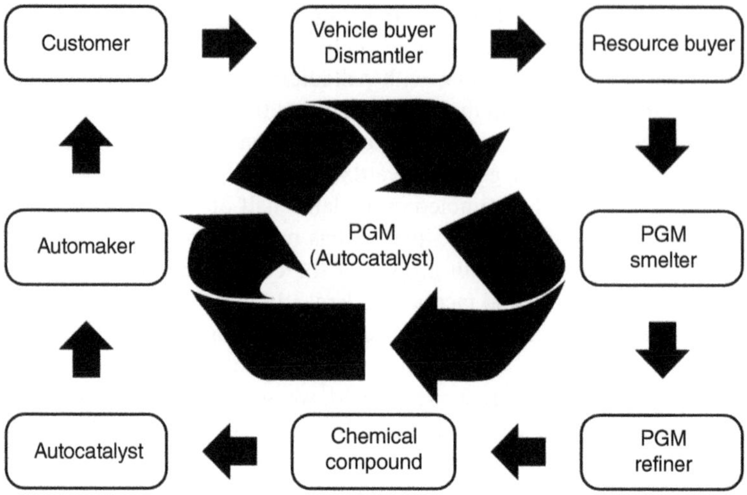

Figure 4 *Recycling System for Autocatalysts*

Recycling systems have always existed for vehicles because of the large amounts of resource materials they contain. This is probably the reason that precious metal catalysts are handled so easily by precious metals recycling systems after use in vehicles. Without these recycling systems in place, it is likely that the large amount of platinum group metals used in products would result in depletion of platinum group metal stocks and a rapid rise in platinum group metal prices.

15.3.3 Extraction and Refining Technologies for End-of-Life Autocatalysts

Autocatalysts are located inside the catalytic converters attached near the exhaust ports of engines. During recycling, the converters are dismantled and the materials are separated for trading as catalyst supports. The total concentrations of platinum group metals in end-of-life autocatalysts depend on the type of vehicle and when it was manufactured, but in general the amount is approximately 0.3 weight percent. Three of the available techniques for separating and recovering platinum group metals from end-of-life catalysts are (1) smelting the supported platinum group metals only, (2) smelting the supports only and then recovering the platinum group metals from the molten residue, and (3) smelting the supports and the supported platinum group metals and then separating out the platinum group metals for recovery. From the perspective of productivity and economy, technique (3) is the most used because of the specific amounts and platinum group metal concentration ranges in end-of-life autocatalysts. A dry process is generally used to separate and concentrate the platinum group metals, after which a wet process is used to separate and refine each platinum group metal element. In the dry process, collector metals such as iron and copper are used to separate and concentrate the platinum group metals from end-of-life autocatalysts, and after adding flux compositions to lower the melting point for ceramic supports and improve the flow characteristics, the mixtures are smelted. These collector metals and flux compositions vary according to each company's processes. Gas, electric heaters, high frequency induction heating and plasma are the heating methods generally employed.

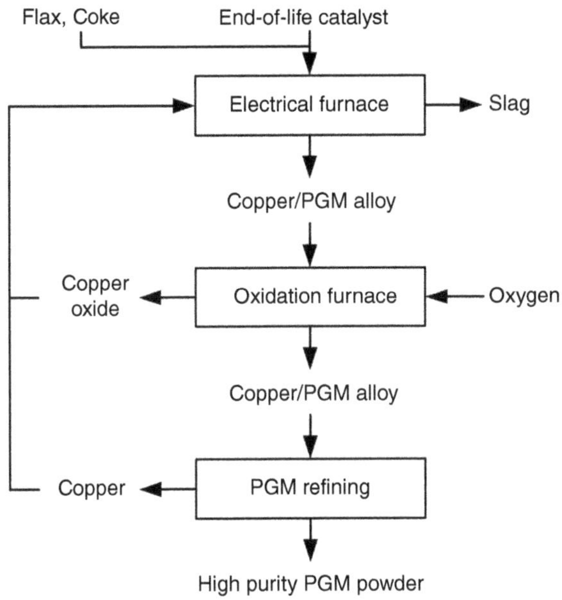

Figure 5 *Separation and Recovery of PGMs Using ROSE Process*

An example of this dry process, shown in Figure 5, is the ROSE process used by Nippon PGM for separation and recovery of platinum group metals [1]. Using this process, the end-of-life catalysts are placed in an electrical furnace together with a flux composition, coke and copper oxide. The mixture is melted down and the copper collector metal then extracts the platinum group metal composition from the ceramic supports. The platinum group metals absorbed into the copper then pass through the next oxidation process to be separated out as copper oxide and copper/platinum group metal alloys with high concentrations of platinum group metals. TANAKA KIKINZOKU KOGYO then uses a wet process to separate and refine each platinum group metal element from these copper alloys. The copper oxide is placed back in the electric furnace for reuse. Some characteristics of this technique are a high recovery rate for platinum group metals, low cost recovery, and a shorter time for recovery compared to using the wet process alone.

15.3.4 Outlook for Recycling

Figures 1 – 3 make it possible to predict future recycled volumes from end-of-life autocatalysts. Estimating fifteen to twenty years before vehicles are scrapped for catalyst recovery (the period between release on the market and final scrapping after reuse as used cars), it is likely that the catalysts currently being recycled come from vehicles manufactured between 1994 and 1999. The amount of platinum used during the applicable recycling period will steadily increase and peak in ten years' time, at which time it will likely have increased to approximately 80 tons per year. Palladium recycling volumes are also likely to peak in a few years' time at approximately 90 tons per year, after which they will drop slightly and then rapidly increase. Considering that the current trend is for increased use of palladium as a catalyst, it is possible that recycling volumes will increase to the level of 150 tons per year. Rhodium recycling volumes are likely to follow the same path as platinum and will reach 15 tons per year. The quality of palladium being recovered is higher over recent years, reflecting past changes in palladium use. It is predicted that recycling volumes from end-of-life autocatalysts will double from their current volumes, so it is possible that recycled supply will reach approximately 30% of total platinum group metal supply (total of mining and recycling volumes). Predictions such as this are likely to also impact mine development and management going forward. This and increasing processing capabilities among platinum group metal recovery and refining businesses will become important issues going forward.

15.4 Electronic Components

15.4.1 Demand for Electronic Components

Table 3 shows changing demand for precious metals in the electronic materials industry. Total demand for platinum for this sector in 2012 dropped by approximately 30% year-on-year, with a significant portion of that drop coming from reduced demand for target materials for hard disks. The reason is that demand for personal computers and DVD players that use hard disks, which in turn use platinum and ruthenium, is being overtaken by demand for tablet computers and smartphones that use solid-state storage technology, which does not use precious metals. However, with an increasing need for greater storage capacity to handle larger file sizes and greater use of digital content, it is expected that

Table 3 *Precious metal demand in the electronics industry*

Year		Unit: ton				
		2008	2009	2010	2011	2012
Platinum	Europe	0.6	0.6	0.5	0.6	0.5
	Japan	1.1	0.9	0.9	0.8	0.6
	North America	0.9	0.8	0.8	0.8	0.6
	China	0.9	0.6	0.9	0.9	0.8
	Rest of the World	3.6	3.0	4.0	4.0	2.6
	Total	**7.1**	**5.9**	**7.1**	**7.1**	**5.1**
Palladium	Europe	5.9	6.1	6.1	5.9	5.8
	Japan	10.0	8.4	9.2	9.3	9.6
	North America	5.3	5.3	5.0	4.5	4.4
	China	7.9	10.4	11.2	8.4	5.8
	Rest of the World	13.5	12.4	12.4	14.6	11.8
	Total	**42.6**	**42.6**	**43.9**	**42.7**	**37.4**
Rhodium	World Wide	0.1	0.1	0.1	0.2	0.2
	Total	**0.1**	**0.1**	**0.1**	**0.2**	**0.2**
Gold	World Wide	311.0	275.0	326.0	320.0	285.0
	Total	**311.0**	**275.0**	**326.0**	**320.0**	**285.0**

Source: Platinum 2013, Johnson Matthey PLC

more hard disks will be needed to meet that demand and therefore more platinum will be used. Total demand for palladium in 2012 dropped by approximately 13% year-on-year. In this sector, palladium is mainly used for multilayer ceramic capacitors, and while production of these capacitors is increasing along with increasing complexity of electronic devices such as smartphones, tablet devices and automotive electronic components, technological innovations are actually causing a decrease in the use of palladium by enabling reduced usage as electrode material and continuing replacement with nickel and copper. In this way, use of palladium for consumer applications is trending downward, although there are expectations for higher demand in military and aerospace applications because of their greater focus on reliability. Demand for plating applications for connectors, substrates and other electronic components should continue to show strong growth going forward because of palladium's price competitiveness and use as a substitute for gold. Total demand for gold in 2012 dropped by approximately 10% year-on-year but it still accounts for a huge volume of the precious metal market. Reduced sales and production volumes in many electronic component markets are also having an impact. The trend for high gold prices is accelerating the reduced use and replacement by palladium mentioned above.

15.4.2 Recycling System for Electronic Components [2]

Figure 6, a flowchart for the recycling of cell phones, shows recovered material (a) coming via the manufacturing processes used to manufacture each unit and component of a cell phone. This relates to scrap from punching lead frames and reject products. Recovered material (a) is comparatively well separated and has a clear product source, so it is easy to recycle. On the other hand, defective products, broken products, old model and post-consumer cell phones (b) are normally sent to collectors and dismantlers who disassemble and separate the printed circuit boards, displays, antennas, batteries and other components. The separated material (c) is then taken by various specialist recyclers, or recyclers (B), to be collected as rough recovered precious metals. In turn, the rough recovered precious metals (d) pass on to recycler/refiners (A) who refine the metals into recycled precious metal bullion, which is supplied to the market again as new materials and electronic components. Sometimes, instead of first separating cell phones into their separate components, they are crushed and separated into metals and plastics using machinery such as crushers, magnetic separators and air tables. The metals are then separated into group, such as copper or aluminum, and are collected by specialist businesses. In both of these precious metal recovery paths, effective use of scarce resources is a more important factor than environmental conservation or waste reduction. In fact, recovered precious metals are treated more like valuable resources than waste products, so a business category of precious metal recyclers has developed to balance the commercial viability of the transaction for both the businesses disposing of materials and the businesses receiving materials. However, even when precious metals are present, if the cost of recovery does not match the man-hours required to separate the metals, and recovered volumes are low, they are treated as waste anyway.

A fundamental element of all recycling, not just recycling of precious metals, is material separation. It is possible to recover a lot of valuable resources if materials have been separated, but mixed materials often only end up as waste.

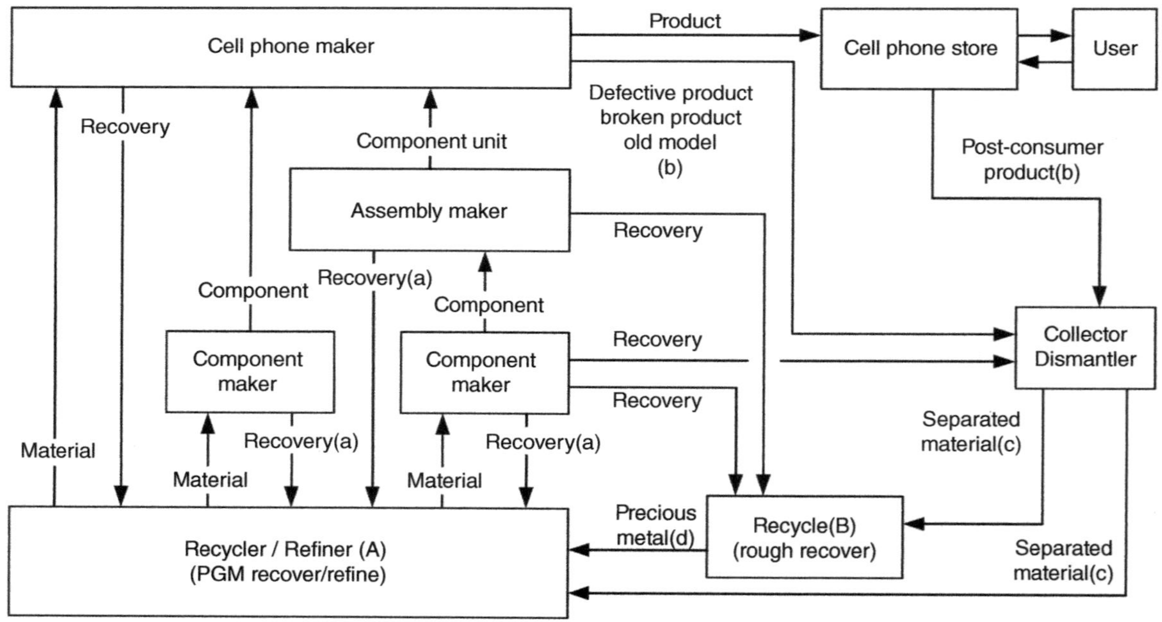

Figure 6 *Precious metal recycling flowchart for cell phones*

15.4.3 Extraction and Refining Technologies for Electronic Waste [3]

Precious metals recovered from electronic components can be generally divided into solid materials, such as post-consumer products and production scrap from manufacturing processes, and liquid materials, such as the solutions generated from the plating process when manufacturing components. As explained previously, separation of materials from recovered solids is an important factor in their treatment as a valuable resource. Materials recovered from the product manufacturing process are comparatively easy to separate, but many more man-hours are needed to separate materials from finished products.

One example of solid material recovery is components like PCBs (printed circuit boards) that use gold, silver, palladium and other materials. With recent high-density mounting technologies, it has become difficult to separate PCBs into their separate componentry so the PCBs are traded directly for recovery. Techniques to recover precious metals from multilayered PCBs and resin-molded PCBs include crushing the substrates and then subjecting them to repeated calcination (to remove organic compounds) and wet process, or smelting them in smelting furnaces operated by mining companies and others to collect the precious metals along with other valuable metals. In the wet recovery process, the pre-processing technologies of crushing and calcination have a large impact on the effectiveness of the dissolution process. Solutions used in the wet process can be mineral acids such as hydrochloric acid, nitric acid and sulfuric acid, or alkali cyanide solution with oxidizing agents. With this technique, the choice of solution enables selective dissolution of precious metals only, leaving alone any other metal components. It is particularly useful for plated parts, which often have precious metal plating on the surface only, so the precious metal can be dissolved leaving the base material there. Using alkali cyanide solution with oxidizing agents enables selective dissolution of precious metals, which can greatly reduce the post-processing burden of separating and refining precious metals from other metals (considered impurities in the collection of precious metals). This also enables the un-dissolved base materials to be directly recycled as valuable metal resources. However, cyanide compounds are extremely poisonous and must be controlled and handled with care.

Recovery from liquids refers to the plating solutions used when plating PCBs and resin with precious metals. Typical plating solutions include gold cyanide salt, gold sulfite salt and palladium salt. An increasing trend in recent years is the use of environment-friendly cyanide-free substances (gold sulfite salt) and substitute materials (palladium salt) instead of expensive gold. All manufacturer plating lines have different configurations, scales and parameters, but all follow the basic configuration shown in Figure 7 (plating tank, drag-out tank and rinse tank). When transferring workpieces to the next process after plating in the

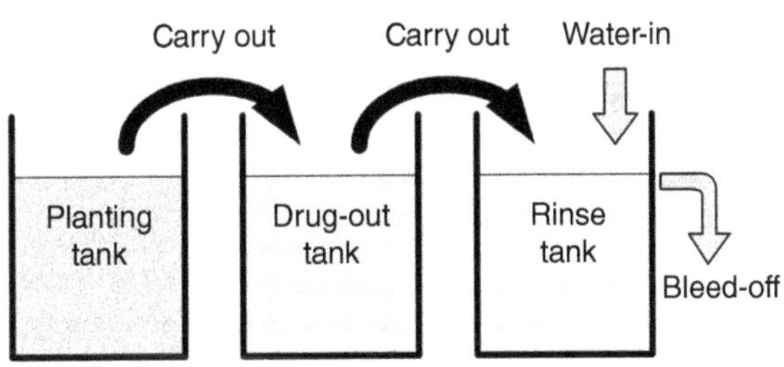

Figure 7 *Plating line configuration*

Table 4 Plating solutions and recovery techniques

Type of Plating solution	Recovery technique
gold cyanide salt solution	electroextraction, ion exchange
silver cyanide salt solution	electroextraction, ion exchange
gold sulfite salt solution	electroextraction, ion exchange
palladium salt, palladium alloy solution	electroextraction, ion exchange, activated carbon adsorption
electroless palladium solution	activated carbon adsorption, add flocculant
gold etching solution (iodine)	chemical reduction, ion exchange

plating tank, drag-out tanks and rinse tanks are used to prevent any plating solution remaining on the workpieces and plating racks from adversely affecting the next process. On plating lines like this, recoverable solutions are the plating solution, the drag-out solution and the rinse solution. Table 4 shows recovery techniques for each type of plating solution. The optimum recovery technique must be selected depending on the solution type, solution volume, precious metal concentration and impurity concentration. From the perspective of efficiency, electroextraction and wet reduction methods are used for solutions with comparatively high concentrations of precious metals (100 mg/dm^3 or higher). Rather than using one technique only, though, various combinations of methods are also used for recovery, such as combining electroextraction with ion exchange and activated carbon adsorption. Liquid for recovery is generated when the solution in the plating tank is replaced (initial makeup) because of a high concentration of impurities in the plating tank solution, degradation of the plating solution, or other reasons. The precious metal concentration in this recovery liquid is comparatively high, so the work of used plating solution recovery is often subcontracted to precious metal recovery businesses. Solutions in drag-out tanks and rinse tanks must also be periodically replaced as concentrations of precious metals increase with increased use. However, the precious metal concentration in these tanks is considerably lower than in the plating tank so the platers must do their own recovery locally from an economic perspective (whether there are any cost benefits to subcontracting recovery to external businesses, considering transport costs against precious metals recovered). As a result, they install their own electroextraction and ion exchange resin equipment and subcontract the collected precious metals only to the recyclers.

15.4.4 Outlook for Recycling

Recovery of precious metals used as electronic materials is easy when it comes in the form of production scrap and used target material generated by electronic components manufacturers, but more difficult when it comes in the form of post-consumer products, and the necessary recycling systems are often not in place. It is therefore important for the promotion of recycling that countries, regions, manufacturers and other groups join forces to build systems like the WEEE Directive in Europe and the Home Appliance Recycling Law in Japan. It is important for the operation of recycling systems that consumer electronics manufacturers, component manufacturers and product assemblers consider recyclability from the design stage, and that product dismantlers be properly educated. Improved and innovative recycling techniques are also necessary for the increasingly difficult recovery of precious metals from electronic components that have undergone miniaturization, multi-layering, reduced precious metal usage and other changes.

Demand for palladium and gold as electronic materials has been decreasing over recent years, but considering that they still account for ten percent of total demand, it is essential that recycling systems be built. With approximately thirty percent of the demand for palladium as electronic materials over recent years being met through recycling of electronic materials, there will certainly be a need for further establishment of recycling systems around the world going forward.

15.5 Catalysts for Fuel Cell Application

15.5.1 Platinum, Platinum/Cobalt Alloy/Carbon and Platinum Ruthenium Alloy/Carbon Catalysts for Polymer Electrolyte Membrane Fuel Cells

15.5.1.1 Fuel Cells

Fuel cells are devices that produce electricity from the energy generated in a chemical reaction between hydrogen and oxygen, which yields water (Figure 8). This is the opposite reaction to electrolysis, where electrical energy is used to produce hydrogen and oxygen from water.

Fuel cells have a long history: the principles were invented in 1801 by Sir Humphry Davy in England and the first successful experiment to generate electricity was carried out

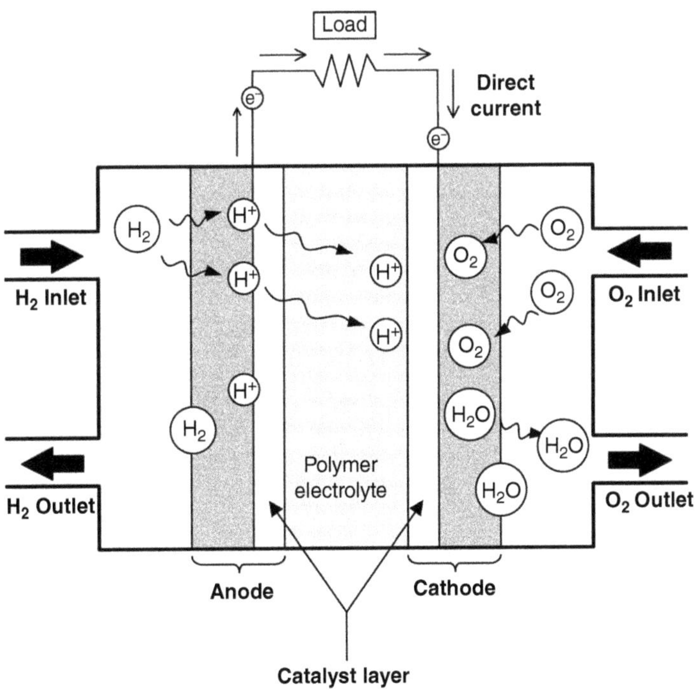

Figure 8 *Schematic diagram and operating principle of a solid oxide fuel cell (single cell)*

Table 5 *Types and characteristics of fuel cells*

Fuel cell type	Low temperature type		High temperature type	
	Polymer electrolyte membrane fuel cell	Phosphoric acid fuel cell (PAFC)	Molten carbonate fuel cell (MCFC)	Solid oxide fuel cell (SOFC)
Raw fuel	Methanol Natural gas	Methanol Natural gas	Methanol Natural gas Naphtha Coal gasification gas	Methanol Natural gas Naphtha Coal gasification gas
Usable fuel	H_2	H_2	H_2, CO	H_2, CO
Operating temperature	Room temp. to 363 K	433–483 K	873–973 K	1173–1273 K
Electrolyte	Hydrogen ion-exchange membrane	Highly concentrated phosphoric acid	Lithium-potassium carbonate	Zirconia-based ceramic
Charge carrier	Hydrogen ion	Hydrogen ion	Carbonate ion	Oxide ion
Waste heat utilization	Hot water	Hot water and steam	Steam turbine Gas turbine	Steam turbine Gas turbine
Electricity generation efficiency	30–40%	30–40%	Up to 60%	Up to 65%
Characteristics	Low temperature operation High output density	Low temperature operation	High electricity generation efficiency Internal reforming	High electricity generation efficiency Internal reforming

in 1839 by Sir William Robert Grove in England. Although known about for many years, these principles have only recently been developed into an actual product. In Japan, home fuel cells using reforming gas began being sold under the ENE-FARM label in 2009, while fuel cell vehicles first went on sale at the end of 2014.

There are currently four fuel cell types that are well known (Table 5). Of these, the two that use platinum group metals are the polymer electrolyte membrane fuel cell and the phosphoric acid fuel cell (used as the electricity-generating electrode catalyst in polymer electrolyte membrane fuel cells). Because these two fuel cell types operate at comparatively low temperatures of no more than 200 °C, they use platinum-based catalysts to achieve the required activity.

15.5.1.2 Highly Active Platinum and Platinum Alloy Catalysts for Cathodes (Air Poles)

The cathodes (air poles) and anodes (fuel poles) of both polymer electrolyte membrane fuel cells (PEFCs) and phosphoric acid fuel cells (PAFCs) use platinum and platinum alloys as the catalyst. To make the most effective use of the expensive rare metal platinum, it is

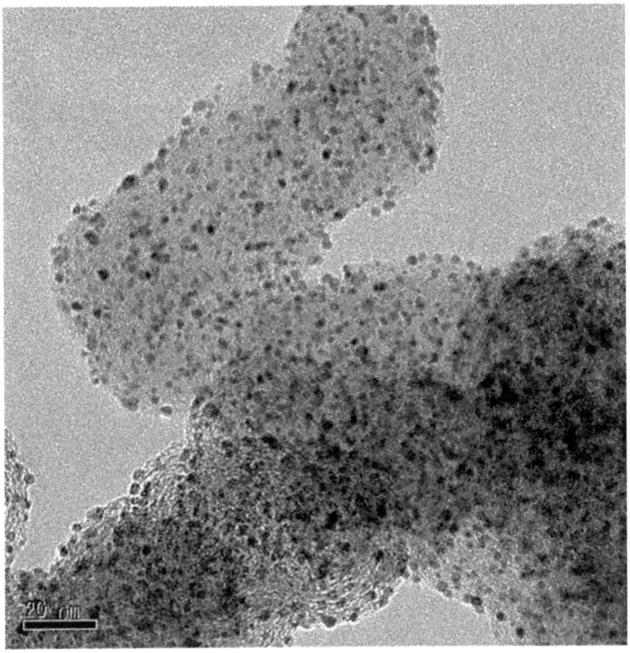

Figure 9 TEM image of highly dispersed platinum particles on carbon. (Source: TANAKA KIKINZOKU KOGYO K.K. internal report)

applied to carbon supports as fine particles of approximately 2–5 nm (Figure 9). Platinum particles are highly active and have a very large specific surface area, but reaction speed is very slow with PEFCs because they operate at low temperatures (up to 90 °C), so they require high platinum loading on the electrodes to achieve the required output. The platinum loading on automotive fuel cell electrodes is currently about 0.5–1.0 g/kW (50–100 g per vehicle), while that on stationary fuel cell electrodes is about 5–8 g/kW (3.75–6 g per unit), so further reductions in these figures are considered necessary to achieve further reductions in the cost of fuel cells. Platinum loading targets for 2020, as promoted by the New Energy and Industrial Technology Development Organization (NEDO), are 0.1 g/kW for FCVs and 1.0 g/kW for stationary cells. The first thing that is essential to meet objectives like this is a dramatic improvement in the activity of cathode platinum catalysts.

The materials that currently show the most catalytic activity are platinum/transition metal alloy catalysts. Compared to platinum catalysts, alloy catalysts of platinum and mainly cobalt, iron and nickel show very high levels of catalytic activity (Figure 10).

15.5.1.3 Highly Durable Platinum Catalysts and Platinum Alloy Catalysts for Cathodes (Air Poles)

The platinum loading on electrodes is also affected by the effective utilization rate and durability of the platinum. The platinum effective utilization rate is considered to be approximately 10%, so a high platinum loading is required to guarantee the required performance. In comparison with many different metals, platinum catalysts are stable in a fuel cell environment, but load fluctuations during operation cause them to dissolve and surface area

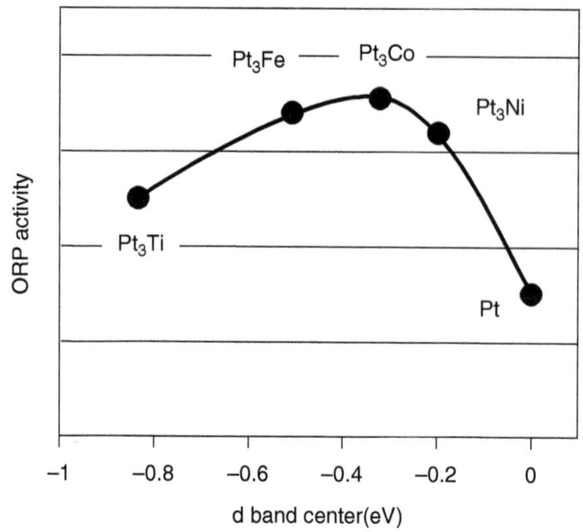

Figure 10 *Oxygen reduction reaction activity of platinum/transition metal alloy catalysts [4]. Stamenkovic, 2006 [4]. Reproduced with permission from Elsevier*

to be reduced. Because of this phenomenon, and in particular to guarantee performance at the expected end-of-life (after deterioration), the platinum loading must be increased. In addition to this, the support material also oxidizes and deteriorates in the fuel cell environment. Here as well, the resulting drop in gas diffusion and platinum surface area currently means an even greater platinum loading is required to meet the required lifetime.

TANAKA KIKINZOKU KOGYO provides platinum catalysts that have undergone high temperature heat-treatment (stabilization) to create highly dissolving-resistant products (Figures 11 and 12), as well as platinum catalysts and platinum alloy catalysts that use high-durability carbon with enhanced graphitization. Products such as these are not used just for cathodes but also as anode catalysts for FCVs with high start-stop frequencies.

The above high-performance catalysts and high durability catalysts contain one important contradiction. High-performance catalysts generally tend to have poor durability, and highly durable catalysts generally tend to have poor performance. Catalyst development going forward should concentrate on resolving this contradiction as much as possible.

15.5.1.4 Platinum/Ruthenium Alloy Catalysts

In home fuel cells, hydrogen is produced from a reforming reaction of hydrocarbon-based fuels such as city gas and propane gas, and that hydrogen is then supplied to the anode to generate electricity. During the reforming reaction, small amounts of carbon monoxide and carbon dioxide are formed in addition to the hydrogen. It is known that carbon monoxide is readily adsorbed by platinum, which causes a reduction in cell performance called carbon monoxide poisoning, so platinum/ruthenium alloy catalysts are used for the anodes for home fuel cells to reduce this effect.

TANAKA KIKINZOKU KOGYO provides platinum/ruthenium catalysts that have a good reputation for having enhanced carbon monoxide poisoning resistance. Figure 13 shows the voltage drop effect of carbon monoxide concentrations in hydrogen. A high

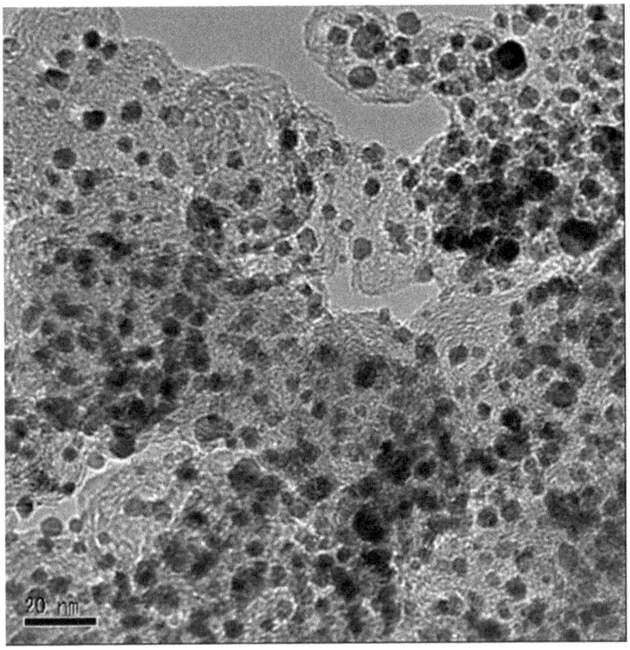

Figure 11 *TEM image of highly dissolving-resistant platinum catalyst. (Source: TANAKA KIKINZOKU KOGYO K.K. internal report)*

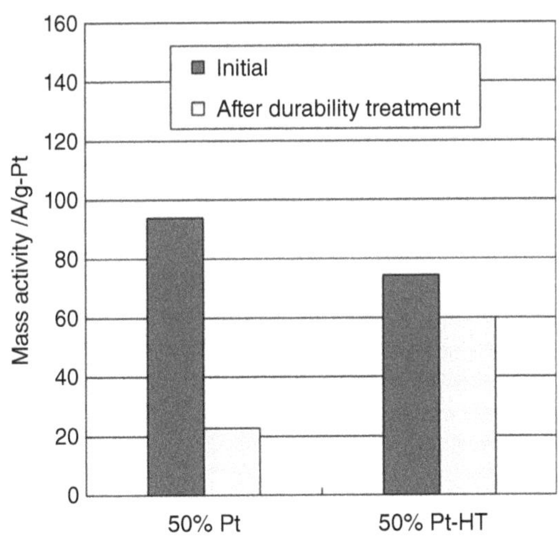

Figure 12 *Activity comparison before and after durability testing [5]: 50% Pt: Before stabilization treatment; 50% Pt-HT: After stabilization treatment. Matsutani K, 2010 [5]. http://www.technology.matthey.com/article/54/4/223-232/ CC0-4.0 public domain http://creativecommons.org/licenses/by-nc-nd/4.0/*

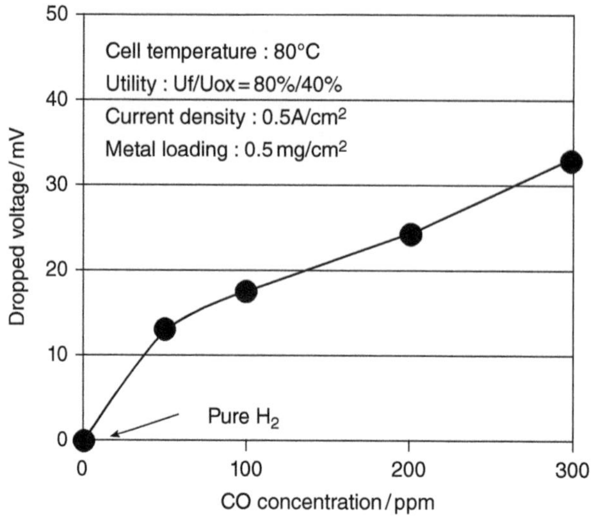

Figure 13 *Fuel cell voltage drop against carbon monoxide concentration. (Source: TANAKA KIKINZOKU KOGYO K.K. internal report)*

alloying degree is required to increase carbon monoxide poisoning resistance, with heat treatment at high temperatures being particularly effective. In general, high-temperature heat treatment causes the catalyst particles to aggregate and become enlarged, but their products are able to remain as highly dispersed fine particles (Figure 14).

15.5.2 Outlook for Recycling

The price of fuel cell systems is important for sustained adoption and use of the systems. From this perspective, recycling of the precious metals used for the electrode catalysts is an essential factor in the pricing structure.

Potential recyclability for fuel cells would include the production scrap generated when producing electrodes and electrode catalysts, and the end-of-life fuel cell systems themselves. Collection and recovery of production scrap is comparatively easy because the sources are limited to catalyst and fuel cell system manufacturers, whereas collection and recovery of end-of-life systems is more difficult because they could have been sold and used anywhere in the world. Wherever collection and recovery are not possible, valuable resources are destroyed and production costs down the line are affected. For reasons such as this, the most important factor when considering widespread take-up of fuel cell catalysts, as well as products using rare metals, is the building and operating of systems for collection, dismantling and separation to ensure stable supply of rare metals, which is similar to recycling systems in place for autocatalysts; this requires the cooperation of countries and industry groups around the world.

It is possible that platinum group metals recovery from production scrap and collected fuel cell catalysts (electrode catalysts) can be achieved using combinations of existing technologies. In this case, how fluororesins can be pre-processed and separated from precious metals, and metals containing precious metals, is important. One example of a

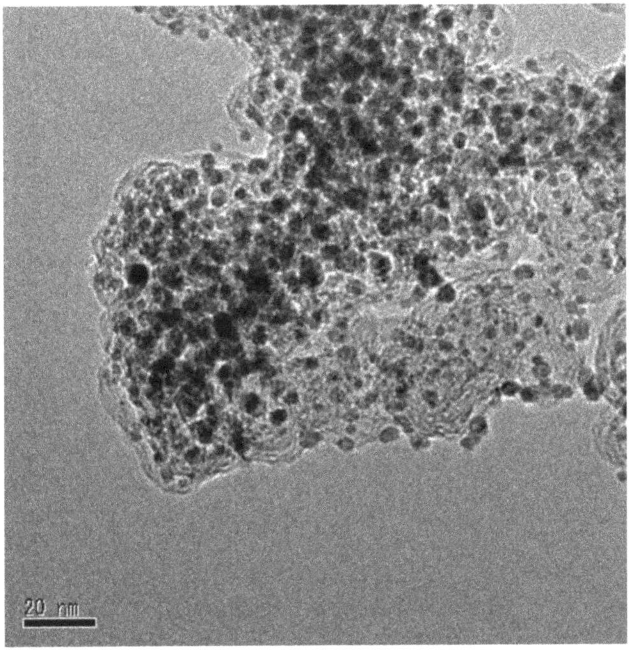

Figure 14 TEM image of platinum/ruthenium alloy catalyst. (Source: TANAKA KIKINZOKU KOGYO K.K. internal report)

process that can achieve effective separation is burning of the electrode catalysts, but from an environmental perspective, treatment of the generated exhaust gases and wastewater must also be considered. After using nitric acid, hydrochloric acid or other chemicals to create solutions of the separated platinum group metals, they can then be further separated and refined through various existing processes to be finally returned as raw materials (bullion, powder and chemical compounds) for new products.

15.6 Extraction and Refining Technologies for Precious Metals

15.6.1 Extraction Technologies

Recovery and refining techniques for recovered products containing precious metals can be generally divided into a wet process and a dry process. The technique selected depends on the form the recovered products come in, but a combination of both processes is often employed when building large-scale recovery and refining processes. In the example of recovery and refining of end-of-life autocatalysts, the total concentration of platinum group metals in the post-market end-of-life catalysts is approximately 0.3 weight percent, so from an economic and efficiency perspective, the platinum group metals are first extracted and concentrated by collector metals such as iron and copper using the dry process. The concentrated platinum group metals are then subjected to the wet process where individual platinum group metals are separated and refined. In this way, an optimal process must be built from a combination of various techniques depending on the form of the products to be recovered. A typical wet process used for recovery and refining of precious metals is described below.

15.6.1.1 Dissolving Precious Metals [6]

The optimal technique used for dissolution depends on such factors as concentrations and types of precious metals or non-precious metal impurities existing in the recovered products. With the wet process, the material is often initially reduced to an aqueous solution before being refined. In the case of recovered solids, materials undergo a leaching process using acids and alkalis. This requires the use of an economically effective method to maximize the dissolution of the precious metals. In industrial practice, low-cost mineral acids such as hydrochloric acid, nitric acid and sulfuric acid, or cyanide solution with oxidizing agents are used for dissolving the materials. When recovered products are plated with precious metals, another method enables the surface-plated precious metals to be selectively dissolved, leaving any other metal components as solids. If precious metals only can be selectively dissolved in this way, the post-processing burden of refining can be greatly reduced.

Nitric acid and aqua regia are often used to dissolve precious metals. Aqua regia is a mixed acid solution of nitric acid and hydrochloric acid that forms nitrosyl chloride (NOCl), a compound with a strong oxidizing power, to dissolve away precious metals. Alternatively, nitric acid can be avoided by using a hydrochloric acid solution with chlorine gas as the oxidizing agent. Platinum group metals are often difficult to dissolve with acids, however, so a technique can be used to form soluble salts by heating the metals together with sodium chloride (NaCl) in a chlorine gas flow. In this way, combinations of oxidizing agents and complexing agents can be used to dissolve precious metals.

15.6.1.2 Chemistry Behind Precious Metal Aqueous Solutions [6]

Platinum group metals, part of the precious metals group, all have similar chemical properties, so they are difficult to separate from each other. Since each platinum group metal forms different types of ion at the same time, in addition to the similar chemical properties they share in aqueous solution, it is important to understand and control their dissolved state. The concentration of ligands and oxidation-reduction potential of the aqueous solution in particular have a significant impact on the formation of ion types. Table 6 shows precious metal oxidation states and typical ion types for the regularly handled chloride ions. In many cases, precious metals are dissolved in aqua regia and hydrochloric acid or chloride. Understanding the dissolved state of the precious metals at this time enables optimization of methods and conditions applied to the recovery and refining process. For example, among platinum complexes, platinum (IV) chloro complex has an octahedral structure while platinum (II) has a planar structure. The differences between these structures is most evident during the ligand exchange process of extractive reactions, with the ligand exchange reaction of octahedral structures being slower than that of planar structures. Because precious metals normally dissolve in concentrated acid, they often dissolve as anion complexes. On the other hand, when the concentration of chloride ions is low, as in rinse fluids and such, they dissolve as cations. Understanding these precious-metal dissolved states therefore offers an effective guide when selecting ion exchange resins, ion exchange membranes, or reagents used for solvent extraction.

Table 6 *Typical valences of precious metals and chloride ion types*

Precious metal	Valence	Ion type	Important compound other than chloro complexes
Au	Au(I)		$Au(CN)_2^-$
	Au(III)	$[AuCl_4]^-$	
Ag	Ag(I)	$[AgCl_2]^-$	
Pt	Pt(II)	$[PtCl_4]^{2-}$	
	Pt(IV)	$[PtCl_6]^{2-}$	
Pd	Pd(II)	$[PdCl_4]^{2-}$	
Rh	Rh(III)	$[RhCl_6]^{3-}$	
Ir	Ir(III)	$[IrCl_6]^{3-}$	
	Ir(IV)	$[IrCl_6]^{2-}$	
Ru	Ru(III)	$[RuCl_6]^{3-}$	
	Ru(IV)		RuO_2
	Ru(VII)		RuO_4
Os	Os(IV)	$[OsCl_6]^{2-}$	
	Os(VII)		OsO_4

15.6.1.3 Ion Exchange Resin and Activated Carbon

Precious metals are recovered from dilute solution using an adsorption method with ion exchange resin and activated carbon. In many cases, the solution is a chloride solution or cyanide solution, with the precious metals being dissolved as anions. For this reason, anion exchange resins with strong-base functional groups are used to recover the precious metals. Large volumes of eluents are required because strong adsorption of these precious metals make them difficult to elute, so the resins are combusted to recover the precious metals. Some resins with functional groups and chelate groups containing sulfur and nitrogen are used because they can also selectively adsorb and elute precious metals. Chelate resins are more expensive than the ion exchange resins used for water treatment, so they are normally used repeatedly for adsorption and elution. Activated carbon is often used downstream of the ion exchange resin process for adsorption of precious metals from dilute solution. Activated carbon also adsorbs organic compounds together with precious metals, so it is also used for wastewater treatment. Adsorbed gold is combusted for recovery because often it cannot be eluted since it reduces to metal particles in the activated carbon pores.

15.6.2 Refining Technologies

The core method of separating and refining recovered precious metals is the wet process. It is uncommon for a single separation and refining operation to recover the precious metals completely, so refining usually consists of a combination of several processes. The wet process for separation and refining of precious metals is a composite process combining several methods that include the classic precipitation separation process, ion exchange resin, solvent extraction, electroextraction and oxidation-distillation.

15.6.2.1 Precipitation Crystallization

Precipitation separation is a precious metal recovery and refining method that has been used for many years. Despite being an old method, it is still used a lot in industrial circles today. Using the precious metal precipitation separation method, the precious metals are first leached

Table 7 *Precipitation separation method used during refining*

Precious metal	Precipitate	Method
Au	Au powder	Reduction with hydrazine and SO_2
Ag	AgCl	Precipitate formation with HCl and NaCl
Pt	$(NH_4)_2PtCl_6$	Precipitate formation with addition of saturated NH_4Cl solution
Pd	$PdCl_2(NH_3)_2$	Precipitate formation with addition of ammonia and HCl
Rh	$(NH_4)_3Rh(NO_2)_6$	Precipitate formation with addition of NH_4NO_2
Ir	$(NH_4)_2IrCl_6$	Precipitate formation with addition of saturated NH_4Cl solution

and dissolved from recovered valuable resource material using an acid or alkali solution, after which the target precious metals are refined through precipitation as low-solubility salts. Table 7 shows one example of this precipitation separation method of refining precious metals. Since the target purity cannot be achieved through a single refining process, the dissolution and precipitation crystallization process is repeated. However, precipitation crystallization separation uses many manual batch processing operations, so the trend is toward replacing the main separation and refining unit process with solvent extraction and ion exchange.

15.6.2.2 Solvent Extraction

Platinum group metals, part of the precious metals group, all have similar chemical properties so with traditional precipitation crystallization, the process had to be repeated in a wide variety of operations. With manual batch processing often being required, it posed the problems of inefficient productivity and inferior working environments. The treatment process is quite long as well, taking several months to process the materials for recovery and refining. As a result, these traditional separation and refining methods have been replaced in industrial circles by precious metal solvent extraction as a method that enables automation and continuous operation [7].

Solvent extraction of precious metals began to be used commercially in the 1970s, with a number of typical processes developed mainly by mining companies for separation and refining. Using di-butyl carbitol (DBC) and methyl isobutyl ketone (MIBK) as the extraction agent for gold, di-octyl sulfide (DOS) and di-hexyl sulfide (DHS) for palladium, and tri-n-butyl phosphate (TBP) for platinum and iridium, separation and refining is continuing on a commercial scale. Other processes investigated over recent years incorporate solvent-extraction methods, to thoroughly remove base metals, into the precious metal refining process. When smelting precious metals from their ores, there is little base metal contamination because anode slime is the starting material, so this incorporation is related to the wide range of base metal contamination involved in the recovery of precious metals from scrap. By including a base-metal removal process, the burden can be reduced and contamination by impurities can be prevented in the downstream precious metal refining process.

Each company employs a variety of processes for recovery and refining of precious metals. Existing equipment and technologies are naturally utilized when developing new processes, but for separation and refining of platinum group metals in particular, solvent extraction is the main process employed.

15.6.2.3 Molecular Recognition Technology (MRT)

IBC Advanced Technologies, Inc. (IBC) in the U.S. has developed and commercialized a process known as Molecular Recognition Technology (MRT) for selective separations of individual precious metals from mixtures of base and precious metals. The process uses SuperLig® products, which consist of metal-selective ligands attached chemically via a tether to a solid support, such as silica gel. Tailoring the ligands for different precious metals allows highly selective separations of individual metals to be made. The separations are normally carried out using columns packed with the SuperLig® material. In operation, feed solution is flowed through a column and the target precious metal is retained by the metal-selective SuperLig® material, with the remaining feed solution exiting the column to raffinate. Following washing of the column to remove traces of the feed solution, the target metal is eluted by a small amount of eluent. The eluate solution produced is concentrated in the target metal, which can be removed in high purity (99.99% or higher) by precipitation or reduction to the metal. The MRT process is characterized by its high target metal selectivity, rapid kinetics, and minimal generation of waste. These MRT systems are being used globally for commercial precious metal separations and recovery [8, 9]. Additional details are given in Chapter 14 of the present book.

15.6.2.4 Electrolytic Refining

Electrolytic refining is still often used as the refining method for gold and silver. It is used to further purify unrefined silver and gold (approximately 99% pure) that has been concentrated and recovered from valuable resource materials. Cast plates of each metal are used as anodes in the electrolytic refining process. One process widely used in electrolytic refining of silver is the Moebius process. In this process, anodes are surrounded by anode bags and pure silver plates or stainless steel plates are used for the cathodes, with electrolysis taking place at normal temperatures in a weakly acid silver nitrate electrolytic solution. During electrolysis, the silver is deposited on the cathode plates in the form of dendrite particles, which are scraped off as silver. Silver powder produced as dendrite particles during electrolysis is scrubbed and then cast as ingots for supply as silver bullion. At this time, the refined silver is 99.99% pure or higher.

In addition to the wet process of chemical selective reduction, Wohlwill electrolysis is used for refining gold. In the electrolytic gold refining process, unrefined gold is used for the anodes and pure gold plates are used for the cathodes, with electrolysis taking place in a weakly acid chloroauric acid electrolytic solution. To prevent passivation of the anodes during electrolysis, alternating and direct currents are used together during this electrolytic refining process. This method enables refining to a purity of 99.99% or higher, with the recovered electrolytic gold being cast as ingots for supply as gold bullion.

15.7 Conclusions

Precious metals are used in a large number of industries and have become an indispensable part of our modern world. However, new supplies of precious metals from mines are extremely unstable, so technologies for recycling and reusing those precious metals are important factors in avoiding that risk. Recycling systems are already in place in many countries for resource recovery from autocatalysts, which is the main application for platinum group metals. Recycling is being positioned as a stable source of supply, but with forecasts of future market growth in emerging nations and expansion of new uses for precious metals, such as fuel cell vehicles, it is important that similar recycling systems are built and operated. This requires the cooperation of countries and industry groups around the world. It is also predicted that technological hurdles to recycling (recovery and refining) of precious metals will become greater and greater as the amounts of precious metals used in individual products decrease. When building recycling systems, collection, dismantling and separation of end-of-life products containing precious metals are important processes that greatly impact system stability and the cost of recycling the precious metals. Developers of products that use precious metals therefore should consider this recycling process from the design stage.

With global announcements about research into new product dismantling and separation technologies, and precious metal recovery and refining technologies, new systems designed to link in with existing technologies are expected to be introduced going forward. At this time, engineers must remain focused on the goal of effective utilization not just of precious metals but of all our limited resources.

References

1. Suzuki S., Ogino M. and Matsumoto T. (2007) Recovery of Platinum Group Metals at Nippon PGM Co., Ltd., *Journal of the Mining and Materials Processing Institute of Japan*, **123**, 734–736.
2. Okuda A. (2000) Recycling Technology of Precious Metals Used for Electronic Parts, *Journal of The Japan Institute of Electronics Packaging*, **3**, 533–539.
3. Okuda A. and Ueda T. (2014) Processing of Precious Metals at Tanaka Kikinzoku Kogyo K.K., The Resources Processing Society of Japan, Paper presented at *Symposium Recycle sekkei-to-bunriseisei gijutsu 29th*, September 11, Tokyo.
4. Stamenkovic V., Mun B.S., Mayrhofer K.J.J., Ross P.N., Markovic N.M., Rossmeisl, J. Greeley J. and Nørskov J.K. (2006) Changing the Activity of Electrocatalysts for Oxygen Reduction by Tuning the Surface Electronic Structure, *Angewandte Chemie International Edition*, **45**, 2897–2901.
5. Matsutani K., Tada T. and Hayakawa K. (2010) Effect of Particle Size of Platinum and Platinum-Cobalt Catalysts on Stability Against Load Cycling, *Platinum Metals Review*, **54**, 223–232.
6. Shibata J. and Okuda A. (2002) Recycling Technology of Precious Metals, *Journal of the Mining and Materials Processing Institute of Japan*, **118**, 1–8.
7. Nishimura S. (1998) Present Situation of Solvent Extraction Technology, *Journal of the Resources Processing Society of Japan*, **45**, 37–45.
8. Izatt S.R., Bruening R.L. and Izatt N.E. (2012) Status of Metal Separation and Recovery in the Mining Industry, *Journal of Metals*, **64**, 1279–1284.
9. Izatt R.M., Izatt S.R., Izatt N.E., Krakowiak K.E., Bruening R.L. and Navarro L. (2015), Industrial Applications of Molecular Recognition Technology to Separations of Platinum Group Metals and Selective Removal of Metal Impurities from Process Streams, *Green Chemistry*, **17**, 2236–2245.

Chapter 1

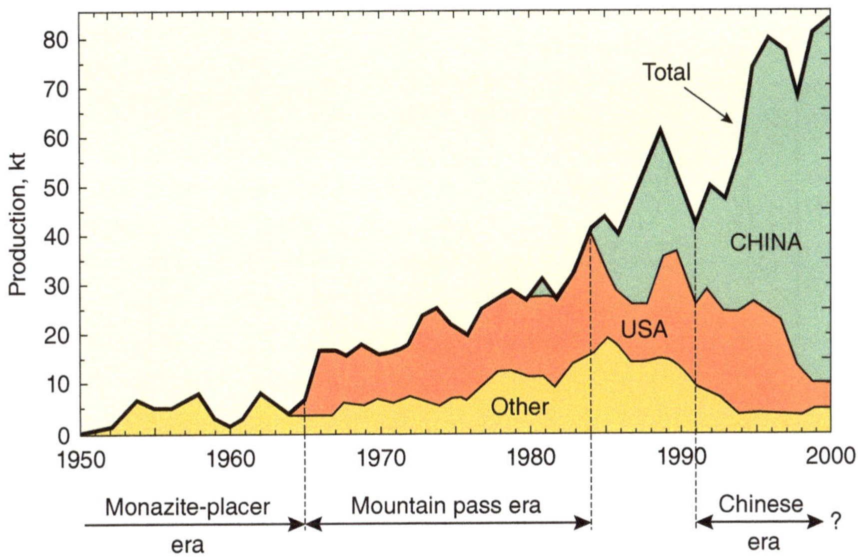

Figure 1 *Global rare earth oxide production trends, 1950–2000. The Mountain Pass deposit is in California, U.S.A. Reproduced from Wikipedia: https://en.wikipedia.org/wiki/Rare_earth_element Accessed from website March 10, 2016. kt = kilotonnes*

Metal Sustainability: Global challenges, Consequences, and Prospects, First Edition. Edited by Reed M. Izatt.

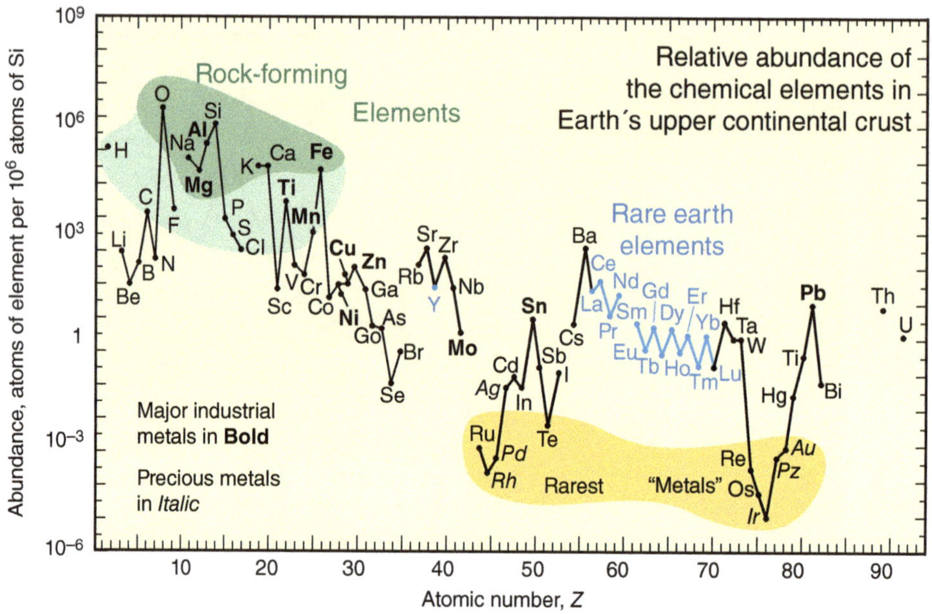

Figure 2 *Abundance (expressed as atoms of element per* 10^6 *atoms of Si) of the chemical elements in Earth's upper continental crust as a function of atomic number. Many of the elements are classified into (partially overlapping) categories: (1) rock-forming elements (major elements in green field and minor elements in light green field); (2) rare earth elements (lanthanides, La–Lu, and Y; labeled in blue); (3) major industrial metals (global production >~* 3×10^7 *kg/year; labeled in bold); (4) precious metals (italic); and (5) the nine rarest "metals" — the six platinum group elements plus Au, Re, and Te (a metalloid). Source: U.S. Geological Survey. http://pubs.usgs.gov/fs/2002/fs087-02/*

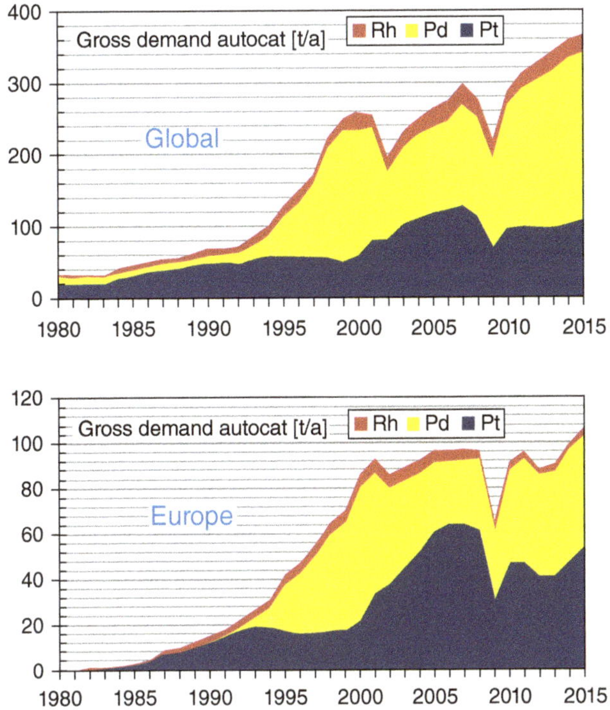

Figure 3 *Gross global and European demand for platinum, palladium, and rhodium for automotive catalysts from 1980 to 2015*

Chapter 3

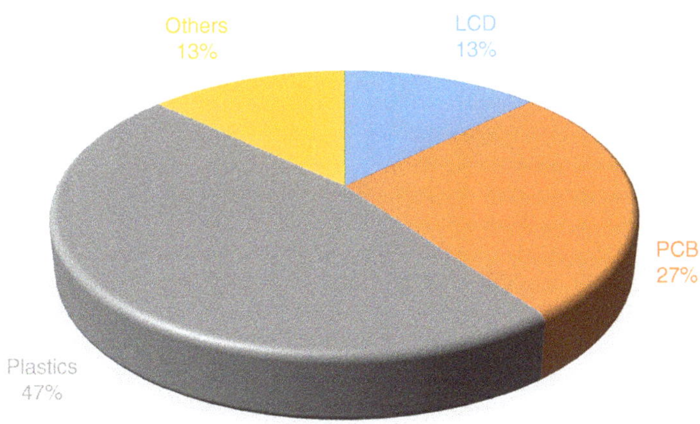

Figure 3 *Material composition of waste mobile phone*

Figure 7 *E-waste flows to developing countries*

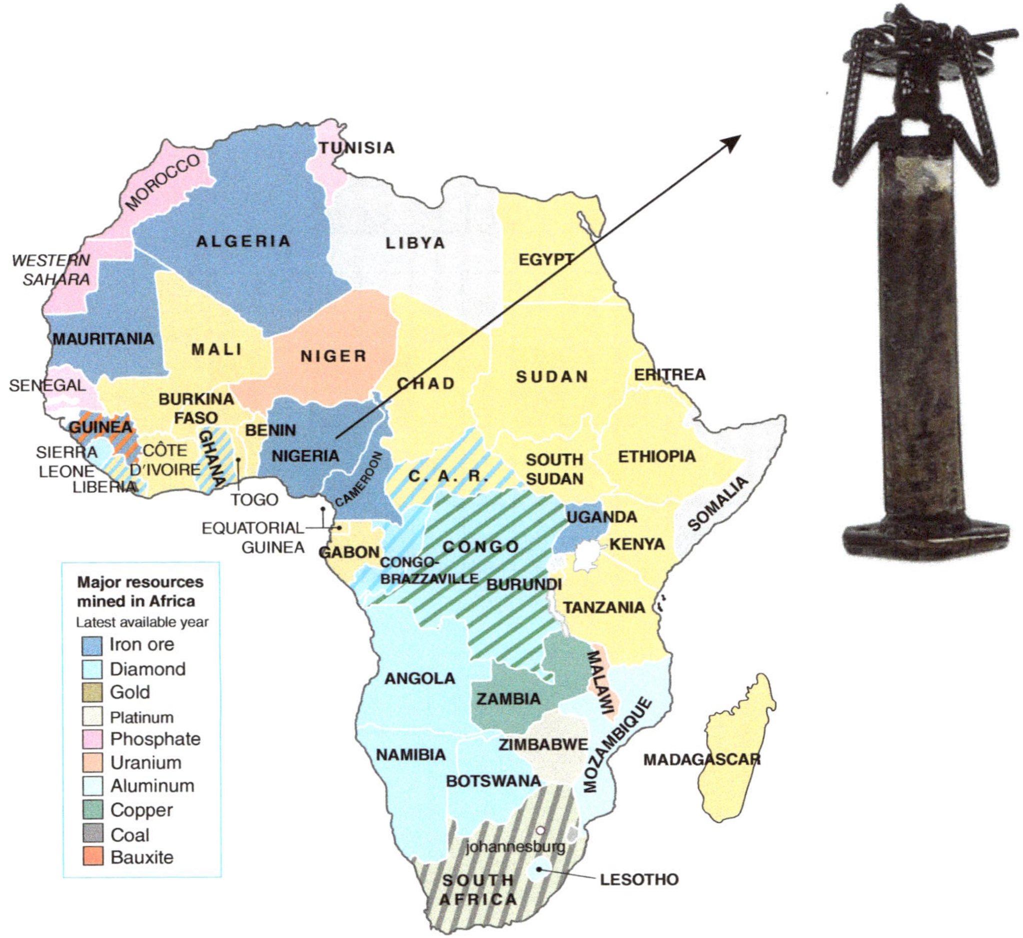

Chapter 4

Figure 2 *The African continent showing the major resources; on the right is an art piece made from scrap metal. (Source: Raw Material Group & Ernst Y (2009)*

Table (UNEP, 2011b).

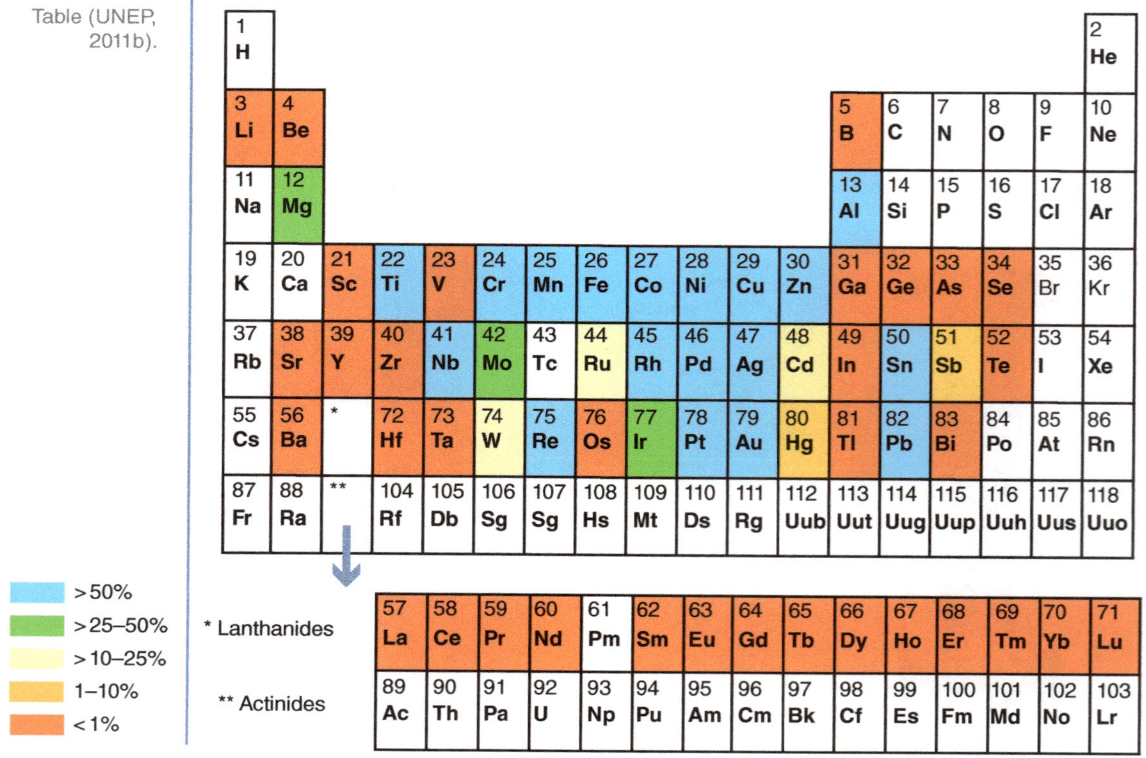

Figure 4 *Recycling rates of various metals at global level (Source: http://www.unep.org/resourcepanel/Portals/24102/PDFs/Metal_Recycling_Full_Report.pdf)*

Chapter 5

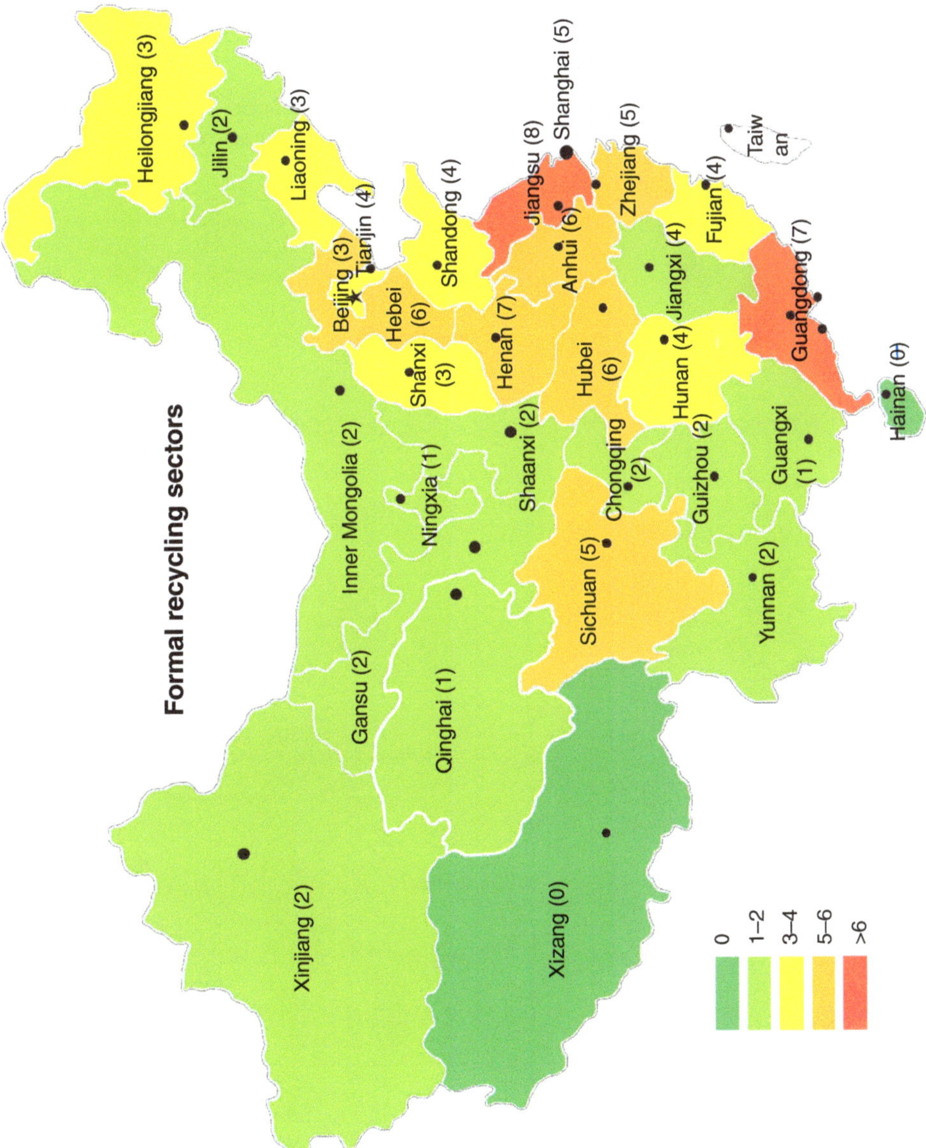

Figure 3 *Formal recycling enterprises in China as of December 2013*

Chapter 9

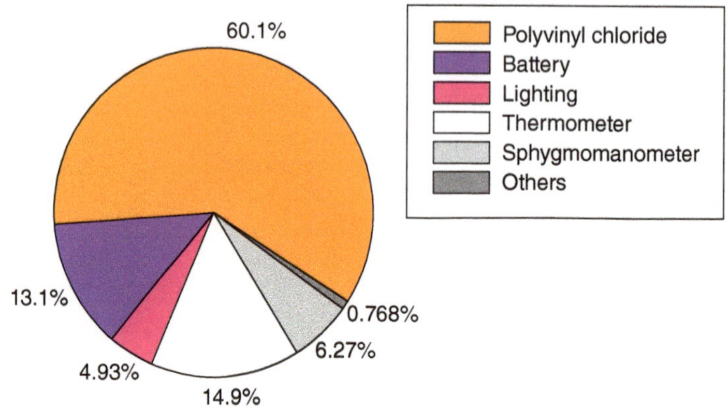

Figure 1 *Categories of Hg supply and usage in 2007 in China [4]*

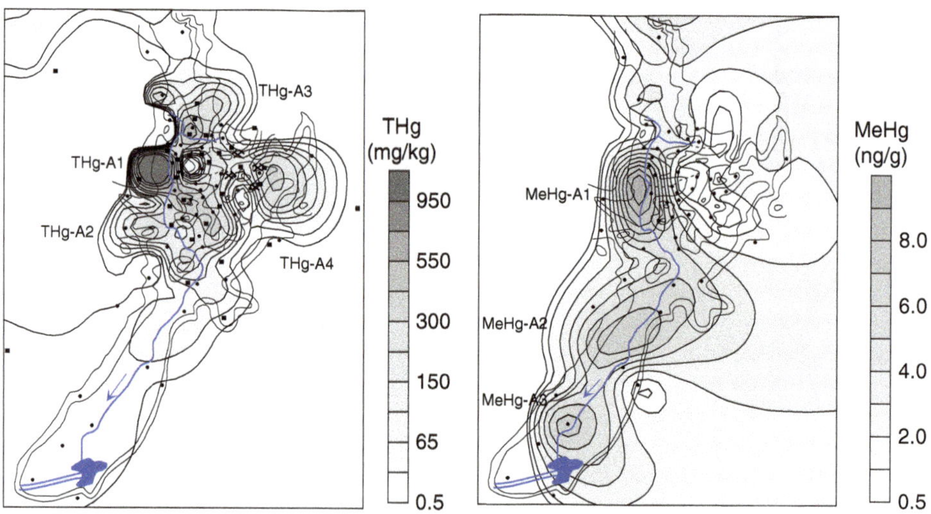

Figure 3 *Spatial distribution of THg (left) and MeHg (right) in surface soil with sampling locations in the Lanmuchang Hg mining district, China [37]. Reprinted from [37] with permission of Elsevier*

Chapter 10

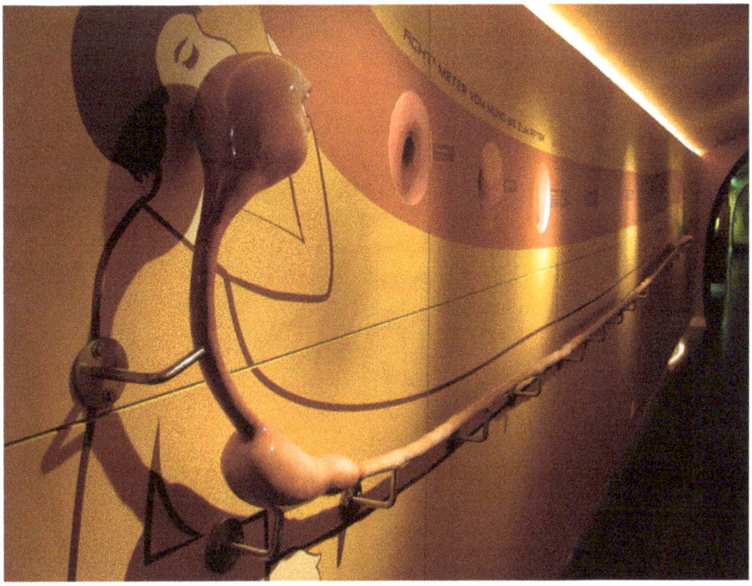

Figure 5 *The gastrointestinal tract in a human adult is an astonishing 8 m long and represents the umbilical cord by which every human being is connected to its geochemical environment via the ingestion of food/drinking water and the non-essential metals therein throughout life. This picture was obtained from the exhibition "Reise in den menschlichen Körper" in the Haus der Natur, Salzburg, Austria*

Chapter 12

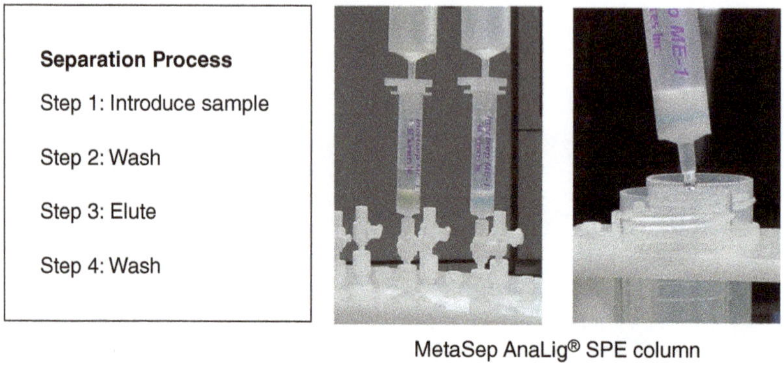

Figure 6 *Typical SPE procedure for precious metals. Furusho, 2008 [56]. Adapted with permission from IPMI*

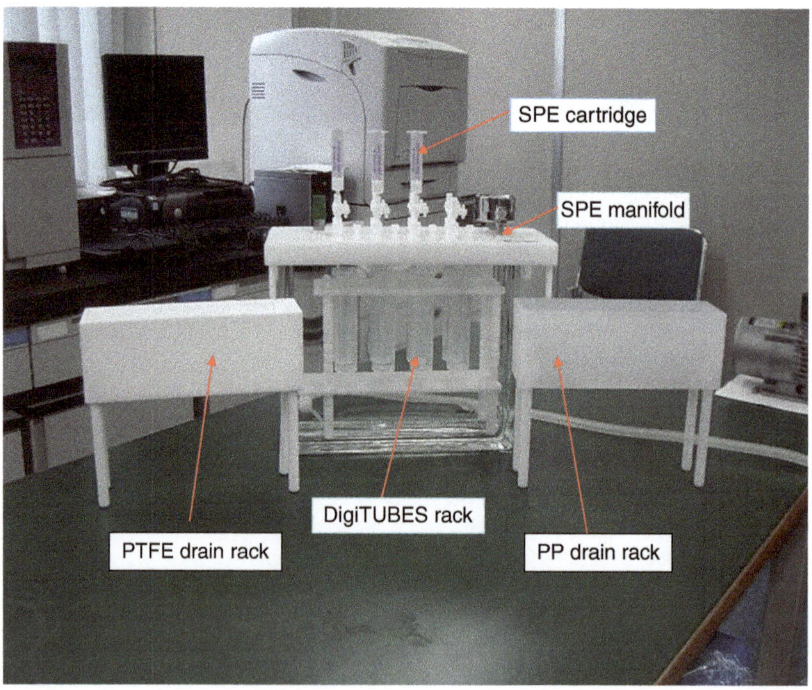

Figure 7 *Typical SPE vacuum manifold for precious metal separation. Furusho, 2008 [56]. Adapted with permission from IPMI*

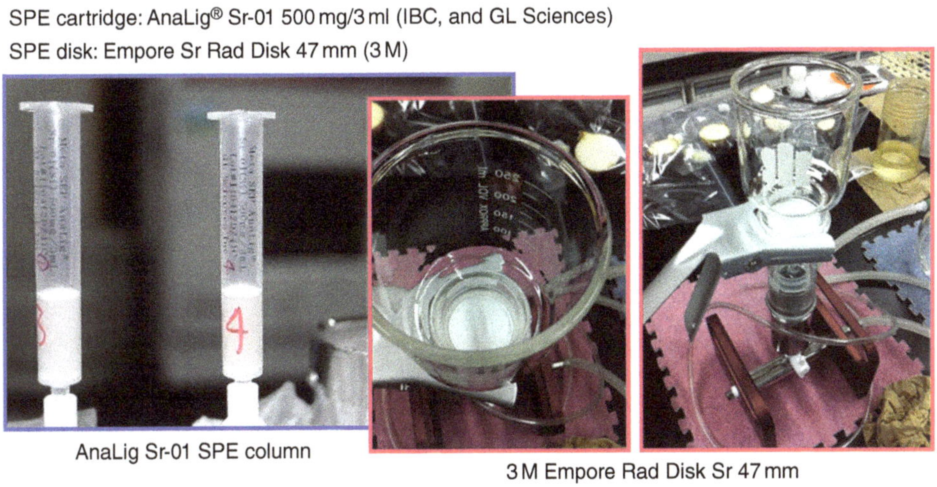

Figure 8 *MRT-SPE column and MRT Empore RadDisk. SPE cartridge: AnaLig® Sr-01 500 mg/3 mL (IBC and GL Sciences); SPE Disk: Empore Sr Rad Disk 47 mm (3 M)*

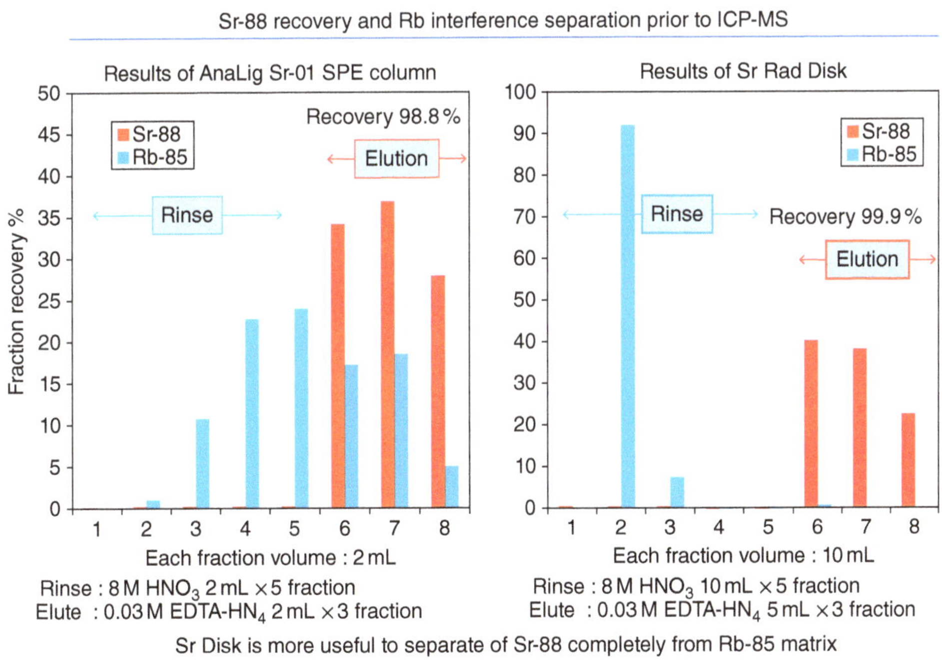

Figure 10 *Comparison of column and Rad disk for recovery of strontium-88 in the presence of rubidium-85 interferent*

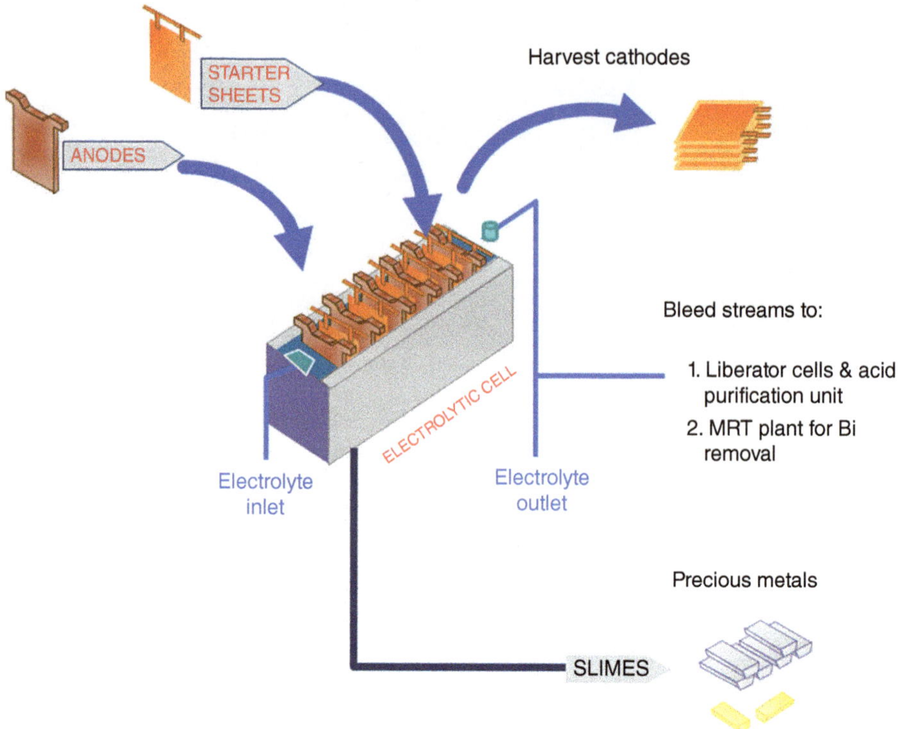

Chapter 17

Figure 2 *Schematic Diagram of the Copper Electrorefining Process*

Chapter 18

Figure 1 *Tracked and tired lightweight drill in operation. Reproduced with permission from CRC Press / Balkema*

Figure 2 *Closeup of rubber-tired, self- propelled combination diamond, reverse circulation and auger drill (Scout). Reproduced with permission from CRC Press / Balkema*

Figure 6 *Cactus salvage nursery*

Figure 7 *TSF before reclamation*

Figure 8 *Benches and side slopes before reclamation*

Figure 10 *Revegetation success: side slopes of TSF in year 3*

Figure 11 Avalon Solar Project

Chapter 19

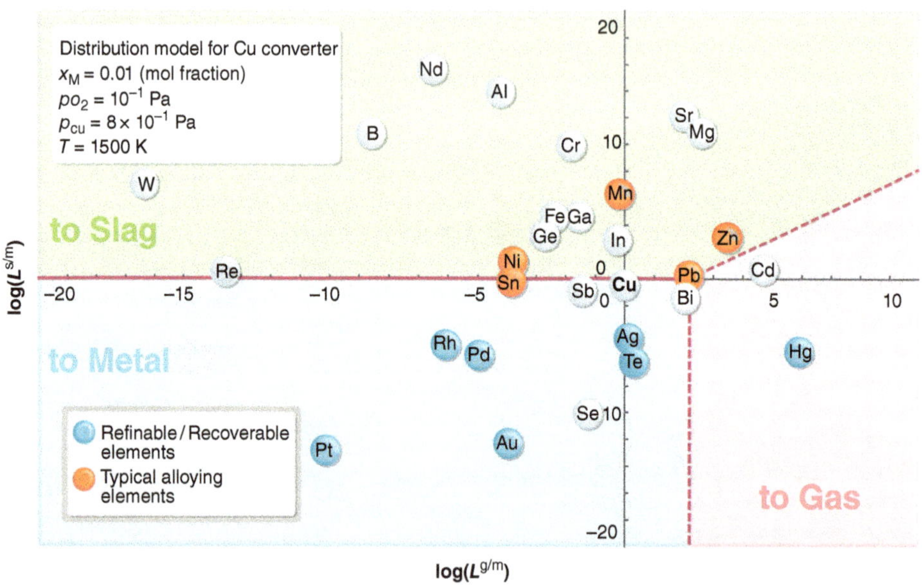

Figure 1 Distribution of elements in the metal, slag, and gas phases during copper smelting under the simulated atmosphere of the copper converter [5]. Reprinted with Permission from [5]. Copyright {2011} American Chemical Society

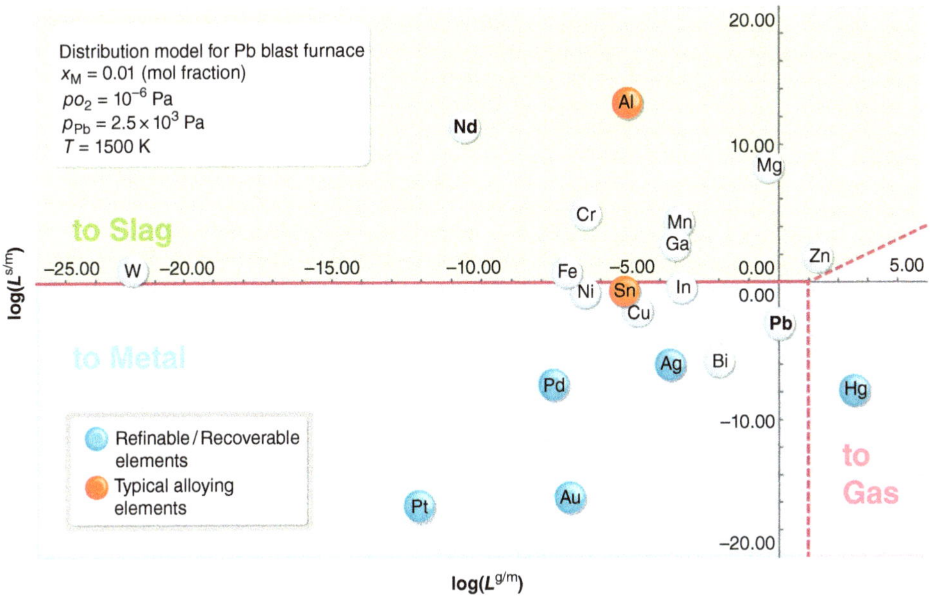

Figure 3 Distribution of elements in the metal, slag, and gas phases during lead smelting under the simulated atmosphere of the lead blast furnace [4]

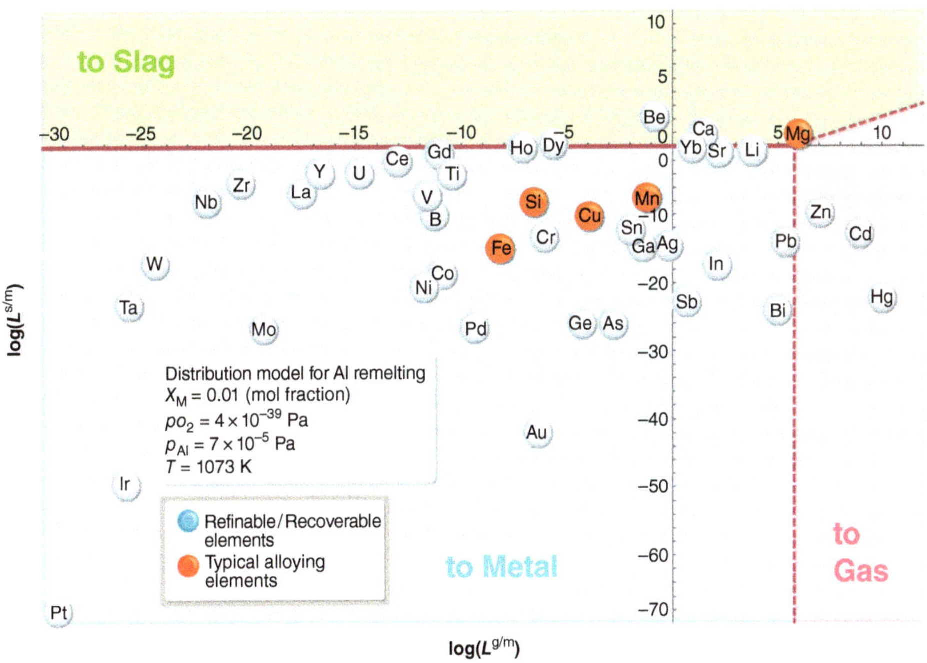

Figure 5 Distribution of elements in the metal, slag, and gas phases during the aluminum remelting process under a simulated atmosphere [6]. Reprinted with Permission from [6]. Copyright {2010} American Chemical Society

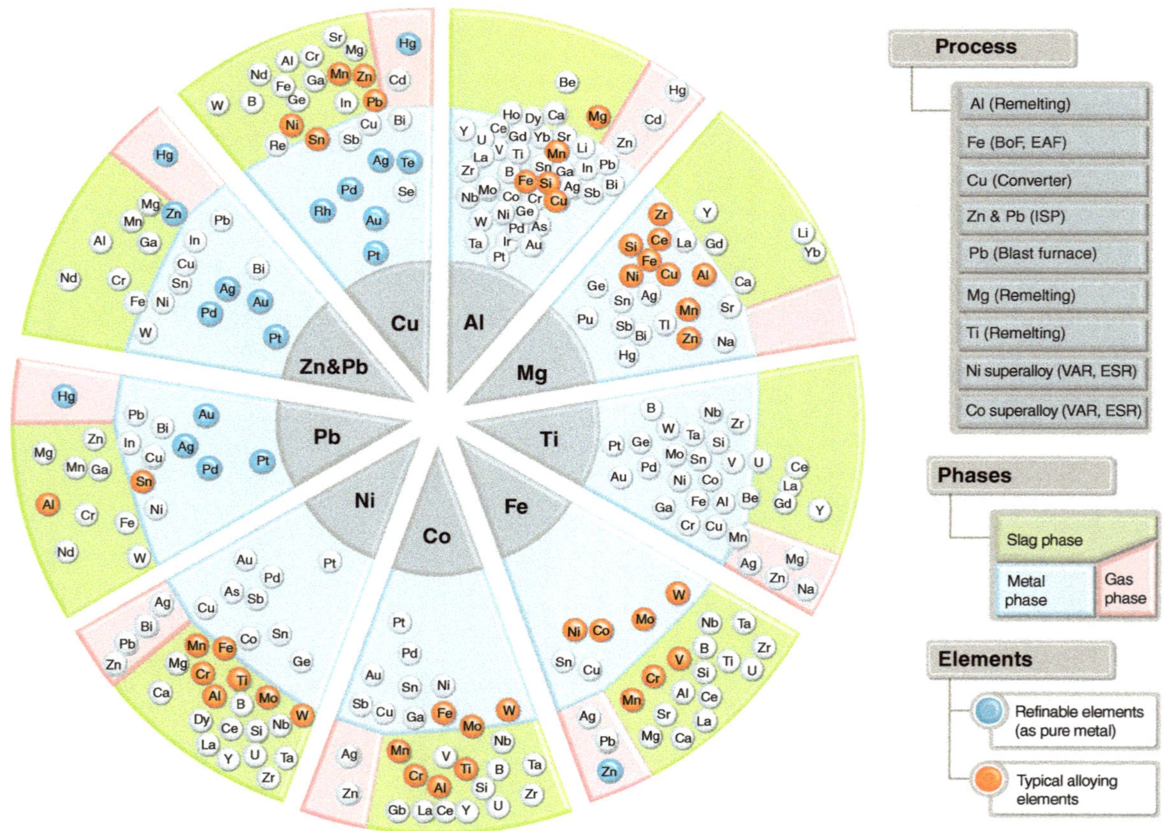

Figure 6 *Element radar chart for the recycling of Fe, Cu, Zn, Pb, Al, Ti, Mg and superalloys into the metal phase [4,6,7,28]. BOF, basic oxygen furnace; EAF, electric arc furnace; VAR, Vacuum arc remelting; ESR, Electro-slag remelting*

16

The Precious Metals Industry: Global Challenges, Responses, and Prospects

Michael B. Mooiman[1], Kathryn C. Sole[2], and Nicholas Dinham[3]

[1]*Franklin Pierce University, Manchester, New Hampshire, USA*
[2]*Consulting hydrometallurgist, Johannesburg, South Africa*
[3]*Platinum-group metals consultant, Johannesburg, South Africa*

16.1 Introduction: The Precious Metals Industry

The precious metals industry differs considerably from those of most other metals and minerals, being driven mainly by the high value and special properties of these metals. Precious metals (comprising gold, silver, platinum, palladium, rhodium, and iridium) occur rarely in nature and only in a few geographic regions. Even in high-value deposits, precious metals are typically found in very low concentrations (levels of parts per million, ppm) in the host minerals [1]. Precious metals have unique and highly valued properties, such as corrosion resistance, high conductivity, and catalytic capabilities [2], as well as aesthetic appeal. Gold and silver have been regarded as stores of value since antiquity; more recently, platinum and palladium are increasingly considered for investment purposes.

The precious metals industry is complex and today is confronted by many challenges. This chapter discusses issues of increased demand, coupled with decreasing grades, increasing prices, increasing production costs, and deleterious byproducts; geopolitics, public perception, and regulations surrounding the primary industry; labor unrest; artisanal and illegal mining; sustainability and water and energy usage. Various mitigating approaches are addressed, including an increased focus on recycling, the concepts of thrifting and substitution, and the emergence of new technologies. In conclusion, a positive prognosis for the future of this industry is offered.

Metal Sustainability: Global Challenges, Consequences, and Prospects, First Edition. Edited by Reed M. Izatt.

16.1.1 Structure of the Industry

As in many other metals-related industries, the key aspects of the precious metals business comprise mining, extraction, refining, product fabrication, and recycling (Figure 1). The business structure of the gold and silver industry is, however, quite different from that of the platinum-group metals (PGM, comprising Pt, Pd, Rh, Ir, Ru, Os). PGM refineries are more complex and expensive, and often use proprietary technology; as a consequence, they are usually intimately associated with mining companies and extractive operations. These include the refineries associated with PGM mining in South Africa (Anglo American Platinum, Impala Platinum, and Lonmin), the Krastsvetmet operations in Russia that treat PGM concentrates from Norilsk Nickel, and Vale's Acton refinery in the U.K. that treats concentrates from Vale's Canadian operations and from some smaller PGM producers. Some integrated base-metal smelters, such as Umicore's Hoboken facility (Belgium) and Aurubis (Germany), also operate their own refineries [3]. The inherent ability of base-metal processing routes to recover and concentrate PGMs means that such operations are often best suited to recover precious metals from low-grade materials. Stand-alone PGM refineries are usually associated with large, integrated precious metals companies, such as those of BASF, Heraeus, and Johnson Matthey. The companies operating in this arena tend to be very prominent on the product side of the business: their refining operations exist mainly to treat their own scrap generated during their product-fabrication operations, although some do undertake toll refining for other customers, such as the processing of concentrates from smaller PGM mining companies.

Because of their high value, PGMs are extensively recycled, so recycling forms a very important part of the industry. High-grade feedstocks, such as industrial scrap or old jewelry, are easily processed and often comprise the core of recycling businesses for stand-alone PGM refineries. Lower-grade materials are often recycled through base-metal smelters, but certain PGM mining companies, including Impala (South Africa) and Stillwater Mining Company (USA), have developed the ability to process a certain volume of secondary materials, particularly spent autocatalysts, along with their primary mine-based feedstock.

In contrast, gold and silver refining operations are seldom associated with mining and extraction: very few miners operate their own refineries. The main reason is that gold and silver refining is much simpler than PGM refining and, as a result, there are many independent gold and silver refiners that bid aggressively to mine doré (an impure gold–silver alloy produced by mine sites, typically containing about 65% Au and 30% Ag). Refining fees are low and advance payments are a routine component of the competitive nature of this business (clients typically receive 90% of their value within one or two days and the difference within about a week). Compared with the ease of sending doré to an independent refinery, it seldom makes sense for a mining company to operate its own refinery, with the attendant capital and operating costs and the cash-flow implications inherent in inventory holdup. A few gold operations have, at times, operated their own refineries, but these inevitably close down when the operating challenges, better pricing from refiners, and inventory holdup issues become apparent. Harmony Gold Minc, a South African primary producer, is a case in point. Following philosophical disagreements with the Rand Refinery, which was the only licensed seller of gold bullion in the country at the time, Harmony commissioned an on-site 24 t/a (300 kg/d) Au

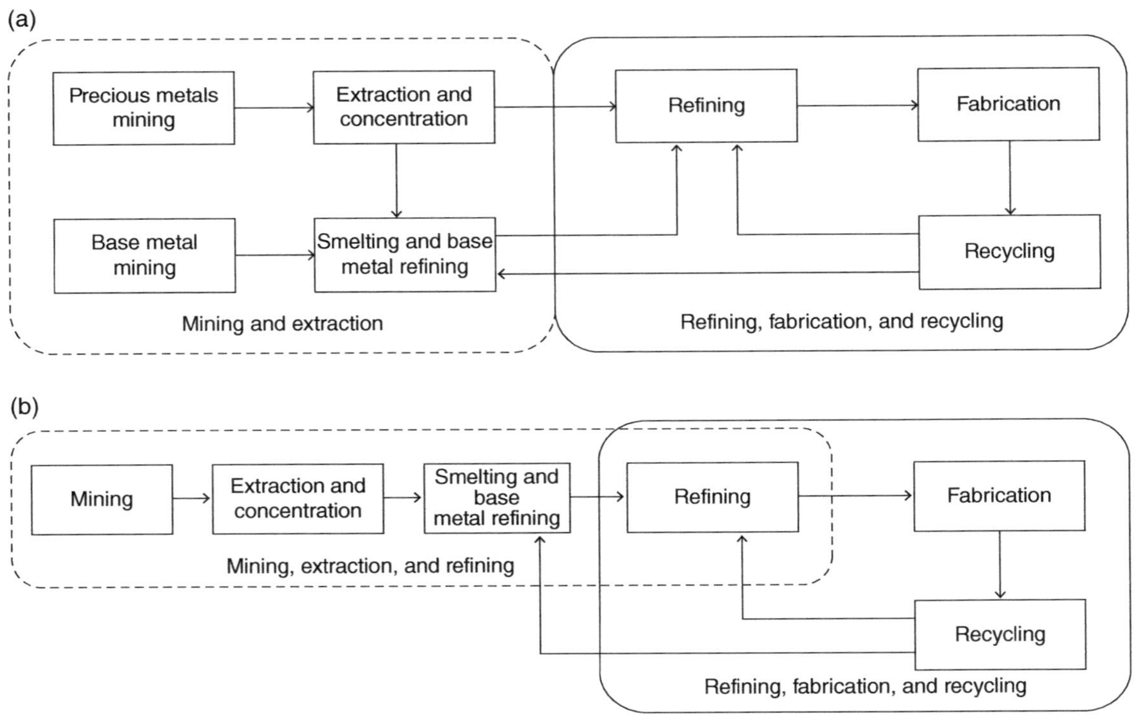

Figure 1 *Schematic structure of the precious metals industry. (a) Gold and silver refining, fabrication, and recycling occur largely separately from their mining and extraction activities, while (b) PGM refineries are typically more integrated with the upstream operations and often include recycling operations*

operation in 1997, which produced 12.5 kg good delivery bars of >99.99% purity [4]. However, security and logistical difficulties, compounded by the large gold inventory inherent to the solvent-extraction technology employed and lack of technical expertise, saw the mine return to sending impure doré for refining to the Rand Refinery within a few years.

Silver refining operations, usually associated with gold refining operations, operate on a similar principle, but some larger silver operations are associated with base-metals extraction and often operate their own refineries. Examples include the lead and zinc operations of Teck (Canada), Kazzinc (Kazakhstan), and Met-Mex Peñoles (Mexico).

16.1.2 Precious Metal Demand and Prices

Table 1 summarizes key information for the demand and pricing of precious metals over the period from 2004 to 2014. These are illustrative snapshots only, and cannot show the ebb and flow from the effect of changing investor interest or, in the case of gold, of central bank sales and purchases, which can vary significantly from one year to the next. For example, central banks sold off 480 t of gold in 2004 but purchased 408 t in 2013. Investor demand, through coins, bars, or physical metal exchange-traded funds (ETFs), can swing wildly as well: investor interest from 2004 to 2010 increased threefold but decreased by almost 50% in 2013 [9]. Some interesting conclusions regarding demand over the decade can, nevertheless, be drawn. Gold demand increased significantly, largely driven by investor and speculator interest, which resulted in a large price increase. Silver, riding on the coat-tails of gold, also experienced a large investor-driven price increase, but total demand over this period remained relatively flat, due mainly to the growth in digital imaging at the expense of silver-based photography. Platinum and palladium demand has grown steadily as a consequence of their increased use in autocatalysts and, more recently, physical metal interest through ETFs, both serving to drive up the price.

Figure 2 apportions the 2014 usages for gold, silver, platinum, and palladium. Investments form a significant proportion of the demand for gold and silver (and increasingly so for platinum), while the main uses of platinum and palladium are in autocatalysts and other industrial applications. These trends are discussed in further detail in section 16.2.1. Table 2 shows the changes in industrial demands for the same period, as well as the sharp increase in the extent of recycling, particularly for the PGMs (discussed in Section 16.3.1).

Table 1 *Variation in average annual demand and price of the precious metals [5–9]*

Metal	Annual demand (t)			Average annual price (USD)		
	2004	2014	ACGR* (%)	2004	2014	ACGR (%)
Au	2,040	3,924	6	410	1,266	11
Ag	28,526	33,188	1	7	19	10
Pt	203	258	3	845	1,386	5
Pd	205	332	4	229	803	12

**ACGR: annual compounded growth rate*

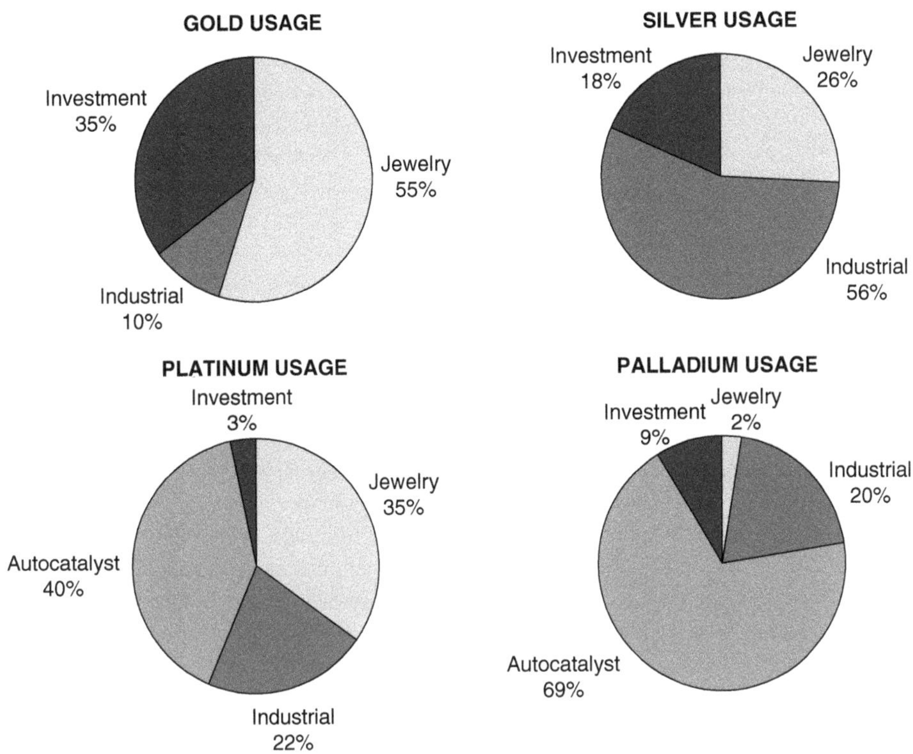

Figure 2 *Main categories of demand and use for gold, silver, platinum, and palladium in 2014 [5–9]*

Table 2 *Variation in annual non-investment uses and recycling demands for the precious metals [5–8]*

Metal	Industrial and jewelry use (% of demand)		Supply from recycling* (% of demand)	
	2004	2014	2004	2014
Au	100	65	27	29
Ag	94	82	22	16
Pt	~100	97	10	25
Pd	~100	91	7	26

* From end-of-life products, including industrial, jewelry, electronics, and catalyst.

16.2 Current and Emerging Challenges

16.2.1 Increased Demand

Demand for precious metals continues to grow. As shown in Table 1, gold demand has increased by 6% per year over the last decade and that for PGMs by some 3%. Silver growth has been lower, at about 1%.

Silver and especially gold are in large demand for investment, particularly for physical investment and speculative interest in coins and bars and in ETFs [8]. The growth in speculative interest and demand is now split between the derivative markets and ETFs holding physical metals. In 2014, investment/speculator demand comprised 35% and 18% of the markets for gold and silver, respectively (Figure 2). The supply of gold and silver far exceeds industrial and jewelry requirements and it is this investment/speculator demand that keeps prices high. In fact, demand for non-investment applications has actually decreased significantly as a consequence of the high prices. A benefit of strong investor/speculator demand for physical metal is that it drives up prices and creates an inventory stockpile, placing it into storage vaults and private homes around the world. This creates a useful buffer to mitigate future supply interruptions. It is estimated that 36% of all the gold ever mined is stored in vaults, with about 49% tied up in jewelry and 12% in industrial products [10].

Platinum and palladium — at least for the present — are compelled by traditional supply-and-demand considerations: prices go up when industrial demand increases, when supply is interrupted, or by geopolitical issues (such as concerns about sanctions on supply from Russia during 2014/2015). It is notable that, during the five-month strike in the South African PGM mining industry during early 2014, suppliers had sufficient inventory to keep the market supplied and price movements were relatively muted. Industrial demand, particularly that for autocatalysts (currently accounting for more than 95% of rhodium, 60% of palladium, and 40% of platinum annual demands [6]) and uncertainty in supply largely drives the prices of PGMs. Interest in physical platinum has grown in recent years and there is now similar emerging interest in palladium, largely due to the establishment of EFTs backed by physical holdings of these metals. More recently, a group of South African platinum miners has formed the World Platinum Investment Council to promote investment interest in platinum [11,12].

16.2.2 Increasing and Volatile Prices

The past decade has seen general increases in commodity prices. As shown in Table 1, precious metal prices have increased significantly — and greatly above the average global inflation rate of 4.6% during this period [14]. Price volatility has also been high over this period. Table 3 compares price volatility, as measured by the percentage standard deviation of average prices, of selected precious and base metals. This is a long period over which to measure volatility, but it does provide a useful point of comparison. The precious metals have generally shown larger price increases over this period when compared with the base metals. Gold and silver have the highest volatilities (on top of some of the biggest price increases over the period), while the PGM have shown much lower volatility.

Figure 3 illustrates the changes in the annual price volatility of the precious metals over the past decade. Volatility over a 12-month period generally fluctuates between 5 and 10%, interspersed with some extreme events. Although the platinum data show the lowest long-term volatility (Table 3), there were shorter periods of extremely high volatility, such as from 2008 to 2009.

Increased demand and higher prices promote the development of new and existing mining operations; however, the associated price volatility, in conjunction with the long lead-times required for mine development, create cash-flow forecasting and mine-financing challenges: since the price downturn in 2014 and continuing into 2015, marginal operations are no longer profitable and there has been a pull-back in mining activities.

Table 3 *Comparison of average prices and price volatility of selected precious and base metals between 2004 and 2014. (Price data taken from [8, 13])*

Precious metal	Price (USD/oz)		ACGR (%)	Price volatility*	Base metal	Price (USD/t)		ACGR (%)	Price volatility
	2004	2014				2004	2014		
Au	410	1,266	11	43	Al	1,719	1,847	1	19
Ag	6.7	19	10	51	Cu	2,863	6,863	8	30
Pt	845	1,386	5	24	Ni	13,821	16,893	2	39
Pd	229	803	12	51	Zn	1,048	2,160	7	35

* Price volatility = standard deviation/average price over period.

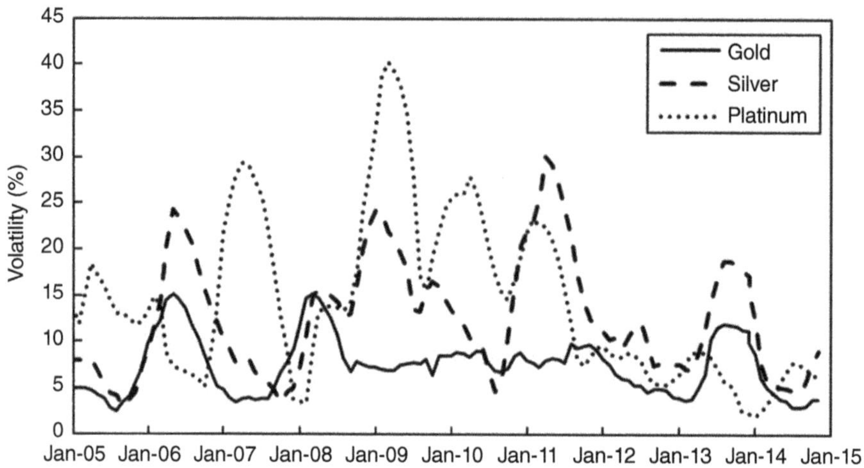

Figure 3 *Long-term trends in price volatility of gold, silver, and platinum from 2004 to 2014 (measured by the ratio of the moving 12-month standard deviation to the preceding 12-month average)*

16.2.3 Decreasing Grades and Increasingly Complex Mineralogy

As with most mineral-based resources, grades of precious metal ores are declining. High-grade resources are now very limited and the exploitable resources are often complex, multi-metal ores that require extensive processing and complicated metallurgical flowsheets. As a consequence, operating costs have increased and profit margins have declined.

In South Africa, until 1995 the world's No. 1 gold mining country, production has dropped from 80% of the world's supply in 1970 to only 6% in 2013 [15]. In North America, gold mining output is currently at 78% of 2002 levels [16]. Figure 4 shows the global decline in grades of primary gold ores. Some of this decline could be attributable to the larger scale of lower-grade operations and advances in technology, such as heap leaching and the carbon- and resin-in-pulp processes, which enable increasingly lower gold grades to be economically treated. However, there is concern that this situation is today reaching the stage of diminishing returns [17].

The 4E (Au, Pt, Pd, Rh) ore grades of Anglo American Platinum, which produces 40% of the world's primary PGM [19], have declined by almost 50% in the last 15 years [20, 21]. The rich Merensky reef was mined first, but due to limited mining access and reef exhaustion, its shafts are now being largely utilized to access the UG2 reef. In addition to significantly lower grades, UG2 also contains a significant portion of refractory chromite, which complicates processing [22]. The company's move towards processing a much higher proportion of ore from the open-pit Mogalakwena mine that has long life and higher margins (because of a low strip ratio) is an attempt to lower mining costs, increase safety, and maintain production outputs under a constrained operating environment [23]. Similar trends for declining PGM ore grades are reported for other major South African, North American, and Russian processing operations [20].

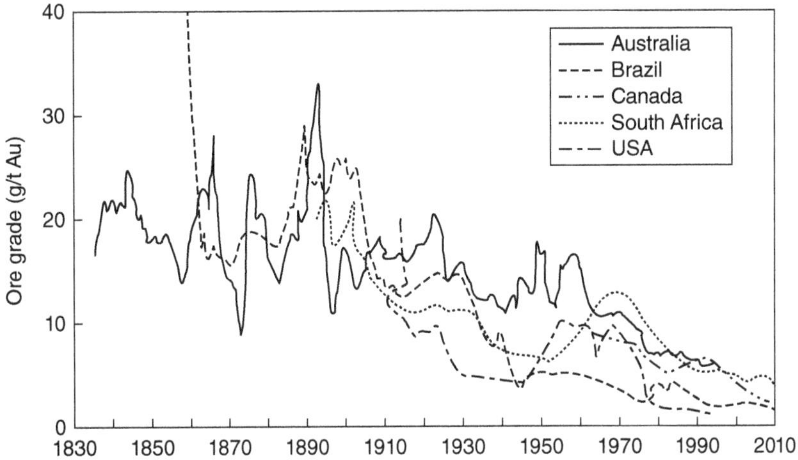

Figure 4 *Global trend of declining ore grades of gold mines from 1830 to 2010 (adapted from [18] with permission of Prior, T.)*

16.2.4 Increasing Production Costs

Lower ore grades, mining at greater depths, and the need to treat more complex mineralogies inevitably incur higher mining costs. AngloGold Ashanti's Mponeng mine in South Africa — the deepest mine in the world at 3.95 km and comprising almost 400 km of tunnels — has a rock face temperature of 66 °C (requiring 4000 t/d of ice for cooling). Miners commute more than one hour from surface to rock face, using the world's tallest elevator (five times the height of the tallest skyscraper) and must remove 3.5 t of rock to obtain a single ounce of gold [24].

Energy and chemical costs have increased and, in many developing countries, wage demands have escalated far above the rates of local inflation [25]. Mining companies, moreover, are dealing with increasing scope of responsibilities, including expectations from their host country governments for greater reporting, increased contributions to local communities, and higher royalties and taxes (see Section 16.2.6). Newmont Mining, the second-largest gold miner in the world, saw its average gold mining costs rise by 67% per ounce between 2002 and 2007 [16]. A recent report [26] indicates that the all-in costs of gold mining (including cash costs, depreciation, amortization, ongoing capital costs, and other indirect costs) are of the order of \$1200 to \$1300/oz Au: this puts extraordinary pressure on mining companies that produce a commodity with a price in a similar range. Cost pressures are also felt by primary silver producers but, because 71% of silver is produced as a byproduct of gold and base-metal mining [27], a host of other considerations, such as base-metal prices and extraction costs, also come into play.

With the recent fall in platinum price, rapidly increasing input costs (including electricity), reduced productivity, labor cost increases, increasing mine depth, more complex, lower-grade ores, and increasing capital intensity, PGM miners find themselves under similar pressures [28]. The average cost of producing an ounce of platinum increased by 18% annually from 2009 to 2014 in South Africa [29, 30]. A break-even analysis based on cash costs and maintenance capital expenditure reveals that, at an average price of USD 1,555/oz

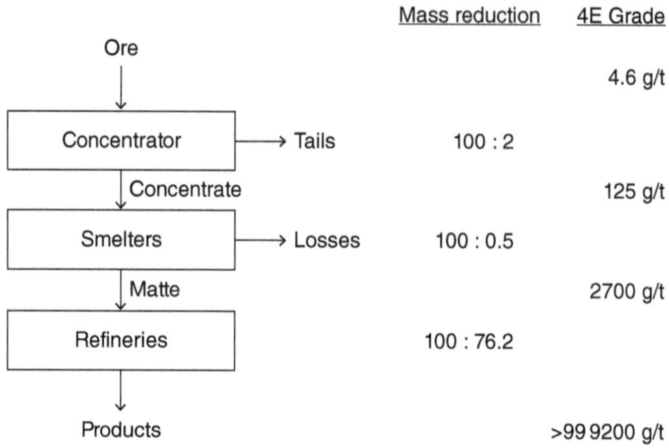

Figure 5 *High-level schematic of PGM refining from ore, showing the mass reduction at each main step of the process and the respective 4E (Au, Pt, Pd, Rh) grade*

Pt in 2012, 45% of platinum mines were in a loss-making or break-even position [30]. It is noteworthy that, in December 2015, the platinum price dropped as low as USD 850/oz.

On average, 1 oz of high-purity platinum requires the processing of 7 to 12 t ore, as shown by the schematic of PGM refining in Figure 5. Depending on the flowsheet configuration and ore mineralogy, PGM holdup times in the extraction and refining circuits can range from weeks to months. The costs associated with this long pipeline inventory are enormous — and very susceptible to the price volatility illustrated in Figure 3. Flowsheet developments and technological innovations are constantly being sought to reduce processing time, improve first-pass separations, allow the more valuable PGMs to be recovered earlier in the process, and reduce the overall inventory.

16.2.5 Deleterious Byproducts

The extraction of precious metals from ores and their recycling from industrial and end-of-life scrap is frequently associated with the presence of deleterious elements, including mercury, lead, cadmium, selenium, tellurium, bismuth, beryllium, and arsenic [31]. In many cases, insufficient attention is paid to the deportment of deleterious elements in the recovery and refining processes and, as a result, these can contaminate operations and expose operators to toxic substances.

Deleterious elements cause three types of problems: they (i) interfere with the sampling and assaying process, making evaluation of the precious metals content difficult; (ii) complicate the extraction and subsequent processing of the precious metals by reducing yields, introducing contaminants and the need for complex processing chemistries, and requiring expensive pollution control measures; and (iii) create health and safety concerns for workers, the surrounding communities, and environmental contamination.

As an example, mercury is a byproduct of gold mining in Nevada, USA. The sulfide ores contain 0.1 to 100 ppm of mercury, which is released during roasting and needs to be recovered. These mines are the largest source of new mercury in the USA. Estimates from 2006 indicate that approximately 110 t was produced by mining companies [32], much of which was sold and then exported. In 2013, a mercury export ban came into effect in the U.S.; all byproduct mercury must now be stored at a Department of Energy-approved facility and a storage fee is levied on the producer [33].

Mercury is also often present in recycled precious metals, such as dental amalgams and switches and sensors in electronic scrap. Correct treatment of such scrap is labor-intensive because each printed circuit board needs to be individually inspected and the mercury-containing components removed by hand. Mercury presents a notable processing and occupational health challenge for many precious metal recyclers and refiners, because they are often unaware of its presence in their feedstocks. Its presence only manifests during processing of the material, which can lead to inadvertent releases, exposure of personnel, and high costs to purge process equipment and operations of contamination [34].

Arsenic is also associated with gold in sulfidic ores. Arsenic can be released during the roasting of arsenopyrite [35] and, in the aftermath of mining, weathering of pyritic materials causes acid mine drainage that solubilizes arsenic and can lead to contamination of natural and drinking water supplies. Selenium is another element that is notorious for creating problems in gold and PGM refining circuits because it complicates separations and poses challenging wastewater treatment choices [36].

The trouble with many of these deleterious elements is that many refiners, recyclers, and some mining companies are unaware of their presence or are not able to deal with them. Integrated recycling operations, such as Umicore, with the technology to handle, recover, and refine a suite of minor and deleterious elements [2], are particularly well positioned to treat such complex feedstocks in a safe and environmentally effective manner.

Not only are the deleterious metals problematic; paradoxically, osmium, one of the so-called Other Precious Metals (OPMs), has few modern uses [37]. Osmium is used by itself or as an alloy for fingerprint detection and in pacemakers, fountain-pen nibs, and jewelry. A primary use for OsO_4 is for tissue fixation in electron microscopy. The South African and Russian PGM ores contain some 0.04 to 0.1 g/t Os, which needs to be removed during the production of high-purity PGMs. However, long-term storage of byproduct osmium is problematic because most of its compounds are unstable to varying degrees, releasing acutely toxic and strongly oxidizing OsO_4 gas. Consider, for example, the largest global primary platinum producer, Anglo American Platinum: its annual production of 2.4 Moz/a Pt generates approximately 750 kg/a Os that needs to be stored indefinitely.

16.2.6 Geopolitics, Public Perception, and Regulations

16.2.6.1 Government–Mining Company Interactions

Figure 6 schematically illustrates the evolution in the relationship between mining companies and governments over the last century, moving from the old system of minimal government interference in mining activities to the overt (sometimes excessively so) government involvement seen in many jurisdictions today. This trend has been accompanied by a move from minimal, voluntary corporate social responsibility (CSR) actions to mandated or required sharing of value across the economic, environmental, and social sectors in which the mining company operates [38].

The end game in this progression is resource nationalism, in which governments claim full control over their mineral assets. This can take the form of outright nationalization of mining companies (as has often been seen in the fossil fuel industry), but, in the minerals

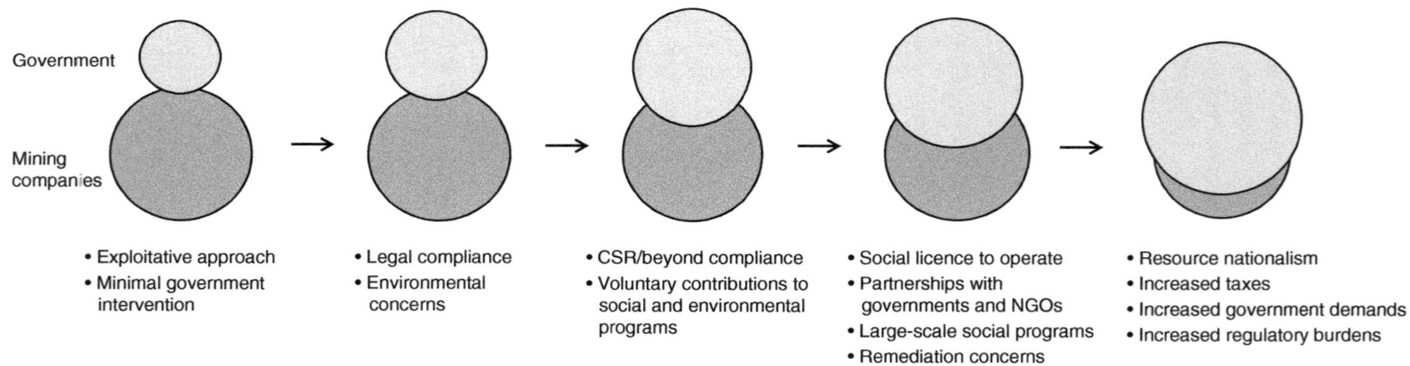

Figure 6 *Schematic showing the evolution of the extent of interaction between governments and mining companies over the past century*

industry, has more recently manifested in demands for higher royalty payments (recently implemented in Zambia [39] and Indonesia [40]), super-profit (windfall) taxes (Australia [41] and Mongolia [42]), more in-country processing (Democratic Republic of Congo [43] and Zimbabwe [44]), or higher degrees of local ownership. Examples are the indigenization of mining companies in Zimbabwe, where local ownership is now required to be 51% [45], and the South African Mining Charter, which requires a minimum of 26% ownership to be held by historically disadvantaged people and at least 10% representation by women in all mining activities [46]. Resource nationalism and its consequences now feature on risk assessments carried out by mining companies [47]. Recent complicating experiences in the precious metals industry include those of DRD Gold's Vatulkoula gold mine in Fiji, which was seized by a military government in 2007, the expropriation by the Uzbekistan government of a Newmont gold mine in compensation for supposed taxes owed, and its annulment of Oxus Gold's license to develop the Khandiza polymetallic deposit [48].

In some cases, greater and carefully considered government involvement has merit: in Botswana, for example, joint public–private ownership and prudent mining taxes have enabled profitable mining operations — with significant social development and upliftment in the country. Chile, through the state-owned mining company Codelco, has also been largely successful in nationalization outcomes. However, in many developing countries, with high turnovers of politicians and ongoing uncertainty regarding mining policies, mining companies face significant complications as they consider investing large amounts of capital into developing mines and extraction operations only to face a landscape of shifting regulations and governmental expectations.

16.2.6.2 Safety in Mining and Processing

The safety of mining and associated processing operations features heavily in public attitudes toward precious metals operations. Mining accidents receive enormous press attention, which has assisted in driving the enactment of significantly more mining safety, labor, and environmental laws in many countries. Safety has shown big improvements over the years and it is anticipated that increased automation will further improve worker health and safety. In China, today the world's top gold producer, however, safety standards at many sites remain far below international norms [49]. Primary PGM operations in South Africa and Russia still record fatalities each year, associated with the dangers of narrow-reef mining without using automated and robotic equipment. Platinum refining is associated with the health risk of platinum salt sensitivity, also known as platinosis, a debilitating condition that cannot be cured; relief is found only when the patient is permanently moved out of the processing environment.

16.2.6.3 Environmental Impacts

Precious metals operations also have the potential to create major impacts on local communities, both from the chemicals used and the byproducts. In 2000, 151 kg of mercury was spilt along a 60 km stretch of roadway in the villages of Choropampa, San Juan, and Magdalena in Peru from a truck transporting the byproduct metal from Newmont's Yanachocha gold mine [50, 51]. Villagers gathered, played with, and stored the mercury, unaware of its dangers. The spill ultimately contaminated the villages and its inhabitants, resulting in many claims of mercury exposure and associated illnesses. After a protracted series of lawsuits, taking almost ten years, Newmont ended up settling with most of the claimants.

Earlier that year, another precious metals-related catastrophe of a different sort took place in Romania. A dam storing spent cyanide leach solutions from a tailings reprocessing plant, operated by the Australian company Esmeralda Exploration, failed, releasing 100 000 m^3 of cyanide solution across farmlands and into the Somes River. The contaminated water eventually made its way to the Tisza River and then into the Danube, killing wildlife and fish and interrupting drinking water supplies across Europe [52, 53]. In 1994, a gold-mining tailings dam failed in South Africa, releasing some 2.5 million tonnes of liquefied tailings onto the village of Merriespruit, killing seventeen people and causing extensive damage to property [54]. The scale of the disaster and the findings of the subsequent inquest provided impetus to the mining industry and all those involved with the design and operation of tailings dams take stock of their management and disposal methods to ensure the safe storage of tailings.

16.2.6.4 Fungibility of Precious Metals

Due to the fungibility of precious metals, they have often been used as a means of transferring value along the criminal chain and to launder illegal gains from crime. Moreover, gold mined in conflict zones, often by forced labor, can be used to fund local insurgent groups. To circumvent this, precious metals companies are now required by law to know, understand, and take responsibility for their entire supply chain: they must determine the source of incoming metals and, to the best of their ability, ensure that they are not inadvertently supporting an armed group or criminal money-laundering scheme. Refiners and recyclers are particularly vulnerable in this regard because they accept materials from collectors and mining operations in various forms and from locations around the world and transform the contained precious metal into a fungible product.

Even though many participants and trade groups in the industry were concerned about these matters and had taken action, responsibility for the precious metals supply chain is now codified and enforced through various regulations. In the United States, the Patriot Act of 2001 [55] recognizes that precious metals are a conduit for money laundering and requires associated refiners to establish anti-money-laundering programs. Compliance with this Act involves the establishment of a compliance office, "Know Your Customer" due diligence, reporting of suspicious activities, training, and independent auditing. The Dodd-Frank Act (2010) [56] covers minerals originating from conflict zones in Africa, and specifically addresses the provenance of gold, tantalum, tungsten, and tin. The Act covers requirements for company policies, supply-chain risk assessment, due diligence on supply chains, independent audits, and reporting components. As important and well-intentioned as these initiatives and regulations are, concerns remain that they could drive supply chains into countries and jurisdictions where regulations are not as rigorous: the net outcome may be a further loss in transparency.

16.2.7 Labor Relations

Mining is still very labor-intensive in many developing countries, and the industry is heavily dependent on a steady stream of migrant labor drawn from far-afield non-mining regions. The migrant labor pool resides near the mine in formal dormitory-style housing or in local communities. Both situations create challenges for the social fabric of the home and host communities. Mine-site communities and close-quarter living often fuel labor tension and unrest.

Prolonged and severe industrial action in South Africa in 2013/4 had a profound effect on PGM production. Following the deaths of 44 miners and police in a stand-off with an emerging union demanding doubling of wages, the platinum industry — which supplies

90% (130 t/a Pt) of the world's primary production — and their almost 70,000 employees and an estimated 700,000 dependents suffered a strike lasting almost five months [57]. The strike later spilled over into gold mines. This was the longest and most expensive strike in South African history, estimated to have cost some 1.2 Moz in lost PGM production. Workers ultimately lost some USD 1.7B in earnings, the companies lost USD 2.4B in revenue, and the immediate and long-term damage to the economy has been enormous [58, 59]. Producers were able to draw on refined metals and in-process inventories to keep the market supplied during this time, but about two-thirds of South Africa's platinum mines were losing money or barely making a profit before the strike, because of rising costs in labor and other inputs eating into diminishing margins due to flat PGM prices during the previous five years. Subsequent developments are permanently changing the face of industrial relations in that country and provide a harbinger of trends in other emerging markets: inexpensive labor will no longer be the norm and a far greater degree of engagement and compromise will be required from all stakeholders to ensure fair and equitable labor practices as well as the continued profitability and economic viability of this industry. At first glance, this appears to be a good move in the right direction, but it should be recognized that it presents a societal conundrum: as wage requirements increase, technology advances and increased automation become increasingly economically viable, so there is associated reduction in the number of mineworkers, which, in turn, reduces opportunities for the growing pool of poorly educated workers who need to be employed.

Labor discontent is not limited to South Africa. Industrial relations issues have recently been reported in Zambia [60], Ghana [61], Indonesia [62], and South America [63], amongst others. The causes are generally similar: a combination of difficult and dangerous work, challenging living situations, social unrest triggered by the use of migrant labor, and rising wage expectations.

16.2.8 Artisanal and Illegal Mining

Small-scale informal mining, known as artisanal mining, is growing in importance worldwide. It is particularly prevalent in gold mining, where it is estimated that 25% of the world's gold supply comes from this sector [64]. This is less an issue in the PGM sector because the complicated recovery and separation processes do not readily lend themselves to small-scale mining. The challenges posed by the informal mining sector are numerous, including the following:

- Use of forced and child labor;
- Basis of support for armed groups;
- Clashes with formal mining sector;
- Dangerous conditions in informal mining operations;
- Environmental impact of informal mining;
- Extensive use of mercury for gold extraction.

Artisanal mining happens throughout the world and its interface with mining companies can be problematic. Artisanal miners work their own claims but often utilize shuttered mine operations and shafts that are no longer in operation. Artisanal miners occasionally intersect with formal mining in areas where the latter have established mineral claims and conflict can occur. For example, the North Mara mine in Tanzania, previously owned by Barrick Gold and now by Acacia Mining, has been plagued by invasions of artisanal miners into the open-pit workings, resulting in vandalism of equipment and tailings ponds, as well as armed response by security personnel and injury and deaths of informal miners. Fences, walls, relocation of the local population, and additional security personnel had limited success in limiting illegal access: decisions were recently made to proceed with underground mining to both minimize local community impact and limit access to illegal miners [65]. In South African gold mines, it is estimated that the number of illegal miners is approximately equal to that of legal miners [66]. Illegal mining of gold is estimated to have cost the country some USD 0.6B in 2014, while the value of platinum theft (from smelters, concentrators, and downstream processes) was some USD 0.5B [67]. The so-called *zama zamas* ('we are trying'), mainly migrant men but also comprising women and children, are highly organized, heavily armed, and operate in syndicates and gangs, often sabotaging working operations and causing danger to legal miners and their security. They are, themselves, subject to frequent dangers of working in old shafts, including those of toxic gases and rock falls, and more than 100 deaths per annum of illegal miners are estimated.

The use of mercury in the informal gold-mining sector has received a great deal of attention in the past decade. Mercury amalgamation technology is the easiest and quickest way to recover gold from concentrates produced by milling and gravity separation of ores. The problem is that the mercury is often just distilled off on a hot stove or with a blow torch, which exposes miners, their families, and their local environment to mercury pollution and its attendant health hazards. Various organizations have published information, aimed at the informal sector, on improved extraction practices that reduce mercury use and provide details of best practice for the small-scale distillation and recovery of mercury [68].

Even given these myriad challenges, the informal mining sector provides livelihoods for millions of miners and their families around the world. It is increasingly accepted by governments and formal companies that this sector has a legitimate role in the industry [69]. To deal with some of the difficulties associated with artisanal mining, several associations and non-governmental organizations (NGOs), such as the Artisanal Gold Council [70], have been established to reduce exploitation, introduce better and safer mining and extraction practices, and indeed to certify the gold from small mining operations that have implemented best practices [71].

16.2.9 Sustainability and Sustainable Development

The concepts of sustainability and sustainable development have become increasingly important in the mining industry. Mining companies are often active participants in sustainability reporting and transparency: the juxtaposition of the apparently contradictory terms *mining* and *sustainability*, although uncomfortable for some, is now well established [72].

A precise definition of sustainability is, however, hard to come by: it is, in essence, a holistic systems view of how business, society, and the environment interact with and influence one another. A frequently cited definition is that of the UN's Bruntlandt Commission (formerly the World Commission on Environment and Development) [73]: *"Sustainable development is development that meets the needs of the present without compromising the ability of future generations to meet their own needs."*

Determining, applying, and measuring sustainability in a business environment is complex: the concept of the *triple bottom line* (TBL) [74] is frequently invoked. This approach recommends that businesses focus not only on economic performance (i.e., profits), but also on a broader view of business success and also consider their environmental performance and responsibilities to society in general. In other words, a sustainable business needs to be concerned about the "planet, people, and profits" — the triple bottom line.

Today, many businesses have become sufficiently concerned about the environment, energy, natural resources, and the image they project to society that they have committed themselves, in one form or another, to sustainable business practices. These often fall under the umbrella of CSR activities, in which companies go beyond compliance and report as openly on their progress in social and environmental matters as they do on their financial performance. In fact, the terms *sustainability* and *corporate responsibility* are often used interchangeably. The irony is that many companies engaged in the active exploitation of natural resources, such as miners, have been very forward-thinking in this regard and can be viewed as having excellent CSR credentials.

There is, furthermore, an increasing recognition that mining companies need to do more than just dig holes, move rock, and recover the valuable metals: there is an imperative for a 'social license to operate', in addition to the legal license [75]. This involves investment not only in direct mining activities but also in benefits for the surrounding community, by initiatives such as:

- Minimizing the environmental impacts of their activities;
- Reducing energy and water consumption through efficiencies and new technologies;
- Creating infrastructure, such as roads, railroads, harbors, water-treatment facilities, electricity generation, sanitation, medical clinics, and schools;
- Establishing remediation plans for rehabilitation of the environment once mining ceases;
- Understanding of and active involvement in both the positive and negative impacts of current and future activities on local communities, and planning for the future of the community and workforce at the end of mine life.

Grappling with the philosophy and practicalities of the paradox of 'sustainable mining' requires much thoughtful consideration, initiative, and action [76]. Detailed frameworks have come out of the Mining, Minerals and Sustainable Development (MMSD) initiatives convened by the International Council on Mining and Metals (ICMM), a consortium of major companies and global mining and commodity associations [38]. The World Bank has also provided guidelines [77, 78]. Mining companies today have to deal with an overwhelming list of well-intentioned international initiatives and campaigns, such as those shown in Table 4. These, coupled with increasing stringent labor and environmental laws, represent growing philanthropic and legislative creep — with the associated responsibilities, reporting, and costs. Considerable progress has been made, understanding has improved, and sustainable development considerations are now deeply embedded in many mining companies. The biggest challenge, however, lies in implementing best practices on the ground, in different regions and cultures, and in creating knowledge of and support for these practices at the government level [38]. Mining companies often experience an evolution in their interactions with surrounding communities [93]: mining development frequently starts in sparsely populated locations, so the initial focus is on community and infrastructure building; with time, job creation in the area increases the local population, the surrounding communities grow, and sustainability issues move to those of much larger-scale and far more complicated urban support and development.

Table 4 *Major international initiatives and campaigns impacting precious metals and sustainability. Adapted from [38]*

Initiative	Description
Global Reporting Initiative (GRI)	Promotes sustainability development through the use of sustainability reporting. Performance indicators and guidelines are provided for organizations to measure and report on their economic, environmental, social, and governmental impacts [79].
Extractive Industries Transparency Initiative	A global standard promoting transparent and accountable management of natural resources. It requires full disclosure of taxes and payments made to governments by oil, gas, and mining companies [80].
Voluntary Principles on Security and Human Rights	Provides guidelines for extractive companies to maintain the safety and security of their operations while respecting human rights [81].
IFC Sustainability Framework	Guidelines established by the World Bank's International Finance Corporation [77] for the transparent consideration of sustainability initiatives, such as environment and social practices in development projects. Aligns with The Equator Principles [78].
Sustainable Development Framework	Binds ICMM member companies to the implementation and reporting on ten binding sustainable development principles. GRI reporting standards are used and independent auditing is required [82].
UN Guiding Principles on Business and Human Rights	Outlines state and corporations' responsibilities to protect human rights and access to remedies in the case of abuses [83].
OECD Guidelines for Multinational Enterprises	Principles and standards for responsible business practices for multinational corporations. Business ethics are applied to a range of issues, including consumer protections, human rights, taxation, bribery, environment, and competition [84].
Akwé: Kon Guidelines	Voluntary guidelines for developments impacting sacred sites and lands used by indigenous and local communities [85].
Framework for Responsible Mining	Set of recommendations addressing social, environmental, and human rights issues in the mining industry [86].
Natural Resources Charter	A set of 12 economic principles that can be used by governments and societies to manage utilization of natural resources to ensure sustained economic development [87].
Responsible Jewellery Council	A code of practices for diamond, gold, and PGM companies addressing human and labor rights, mining practices, environmental impact, disclosure, chain of custody, and other topics in the jewelry supply chain [88].
Jewelers Vigilance Committee	Long-established (since 1917) trade organization promoting ethical practices in the jewelry industry [89].
No Dirty Gold	An international campaign established to educate consumers, retailers, and fabricators about the impacts of irresponsible mining. Looking to improve gold mining practices [90].
Mineral Policy Institute	Australian NGO working with mining-affected communities to minimize impacts, improve mining practices, and ensure equitable distribution of benefits [91].
Mining Watch Canada	NGO focused on Canadian mining companies looking to reform mining law, improve environmental planning, protect water resources, and ensure corporate accountability [92].

16.2.10 Water and Energy Use

Incorporating concepts of long-term sustainability and environmental preservation, water and energy usage is becoming increasingly important. Due to the low ore grades of precious metals, their embodied energy per kilogram (the total of all direct and indirect energy inputs involved in all steps of mining, extraction, and refining) is orders of magnitude higher than that of the base metals [94]. Estimates indicate that the energy requirement of approximately one barrel of oil, 4.5 gigajoules (GJ), is required to recover one ounce of gold [95]. The energy consumption involved in extracting PGMs is larger, with estimates of 6.9 GJ/oz PGM [21]. The difference results from the more complicated mining and processing of the PGMs, which is only partially compensated for by their generally higher grades.

Deep-shaft mining, such as predominantly used for platinum and gold, relies primarily on electricity for its energy requirements. Electricity is the major energy source for the transport of personnel, material and ore, production machines, and mineral processing. Electricity is also critical for health- and safety-related applications, such as the pumping of water, ventilation, and refrigeration. In the gold sector, the refrigeration plant and the ventilation fans system consume around 27% of the total power used by these deep-level operations [96]. Electricity comprises some 12% of operating costs of South African platinum mines [96]; electricity is also the cost component that has registered the largest inflation from 2007 to 2012 in the platinum industry (with an average annual cost increase above 25%), ahead of diesel and reinforcing steel.

The reliability of energy supply, total energy consumption, and the associated costs are viewed as important business risks to be managed [21]. Continuity of energy supply is of profound concern for many mining operations, especially those in developing countries, remote locations, or far from an electrical grid. In such operations, electrical energy requirements for the entire operation are provided by diesel generators. Even mining operations traditionally connected to the grid install backup generator capacity to assist during supply interruptions.

Renewable energy today provides competitive and sustainable alternatives to onsite generation of electricity by diesel generators. Many opportunities exist, including installation of photovoltaic units, wind turbines, and biomass-based generation, but the challenge with renewable energy sources is one of providing steady base-load power. The solution is usually a hybrid of fossil fuel- and renewable source-generated electricity [97]. Implementation of renewable energy sources involves consideration of many factors, include the fluctuating costs of fossil fuels, the steadily decreasing cost of renewable technologies, and the life of the operation. It is sometimes difficult to justify renewable energy in a cheap fossil fuel environment but there are other benefits, such as lower carbon footprint and reduced logistical issues associated with the transport of fossil fuels, that can make implementation attractive. Renewable-friendly government regulations, such as net metering [98] or the introduction of carbon taxes, further encourage the implementation of renewable energy sources and incentivize reductions in the overall power consumption.

A recent report by the World Bank [99] suggests a different approach — that mining companies should not focus on self-sufficiency in terms of their energy needs; instead, it is recommended that companies operating in developing countries engage with electrical utilities and become anchor customers for new power-generation plants. In this way, the utilities (or even independent power producers) gain large, important customers that provide base-load earnings and then become better positioned to provide electricity to a growing market of residential and commercial customers. The mining companies benefit from being able to negotiate long-term power purchase agreements and being less dependent on on-site generation of power.

Energy costs comprise some 15 to 20% of total operating costs of precious metals fabrication and refining operations. Energy efficiency drives can result in significant savings, especially in high-cost, low-margin parts of the business. A key benefit of metal recycling is that it not only permits the recovery of high-value materials, but also results in considerable energy savings in terms of the embodied energy content of the recovered metals.

Water use is similarly under scrutiny throughout the mining industry. Mining uses water primarily for mineral processing, dust suppression, slurry transport, and employees' needs, and it is usually obtained from groundwater, streams, rivers and lakes, or through commercial water service suppliers [100]. Especially with underground mining, water may also need to be extracted and pumped away from a mine site, which can reduce the levels of ground water, deplete surface water, or cause pollution to local rivers. Water consumption per kg of product is many orders of magnitude higher for precious metals than for other metals, estimated at approximately 21.5 m^3/oz of gold recovered and 24.9 m^3/oz for PGMs [95, 101]. Water supply, consumption, and reuse are consequently of critical importance in the development and operation of both mines and processing facilities, and many precious metals companies are actively involved in massive grey-water projects.

16.2.11 Technology Cycles

Because of their unique properties, precious metals often find use in new technologies and applications. As such applications grow and the technology matures (for example, in the use of precious metals in compact disks and digital video recorders), the products become commoditized and competition grows. The production focus then shifts to reducing costs, which is when the precious metal content comes under the spotlight. Cost-reduction efforts manifest in two ways: (i) a reduction in the amount of precious metals used, usually by thrifting or substitution (see Section 16.3.2), or (ii) if thrifting or substitution are not options, then by removing variability in the fabrication process to reduce manufacturing waste. This often leads to tighter materials specifications and higher purity metals: the increasing use of five nines (99.999%) and even higher purities of gold and silver in electronic applications exemplifies this trend [102].

16.3 Responding to the Challenges: Mitigating Approaches and New Developments

As evident from the discussion in Section 16.2, the challenges facing the precious metals industries are many and complex. Although pressure on the mining segment of the business is particularly intense, these pressures are passed down the value chain, creating responsibilities, compliance requirements, and costs at all of the various stages. Although this industry is mature, it remains very creative, and operators have responded to the challenges in innovative ways. The key responses have been in recycling of precious metals, reduction in use through thrifting, and technical improvements along the entire chain.

16.3.1 Recycling of Precious Metals

The high monetary value of precious metals has always made recycling an important part of the business; this has become even more important with the price escalations of the past decade. There are two key components of precious metals recycling: the first, and largest, is the recycling of high-grade material (typically with precious metals contents of 10% or greater); the second is the recovery of precious metals (typically less than 1% of content) from lower-grade materials.

16.3.1.1 Recycling of High-Grade Materials

High-value scrap includes industrial byproducts from fabrication operations and old jewelry. It is relatively straightforward to recover precious metals from these feedstocks using a combination of hydro- and pyrometallurgical processes.

Precious metals refining comprises three key operations, viz., evaluation, separation, and conversion (Figure 7). The first task — and that of most importance to the customer — is evaluation, which involves the accurate determination of the precious metal content in the incoming material by sampling and assaying. The feedstock is homogenized, normally by melting, followed by removing a sample that is representative of the whole melt. This sample is then analyzed for precious metal content using a variety of instrumental and fire-assay techniques. Although the industry places much emphasis on accurate analytical techniques, effective sampling of the feedstock is just as important: even the most sophisticated analysis cannot compensate for poor sampling technique and a non-representative sample.

After evaluation, the customer is paid and the refiner is now responsible for separation of the gold, silver, and PGM from the feedstock. These separations are complex, involving an in-depth knowledge of chemistry and metallurgy, as well as the use of hazardous chemicals. The final value-adding step is to convert the refined precious metals into a saleable product. This often takes the form of precious metal bars weighing 1 kg, 100 oz, or 400 oz

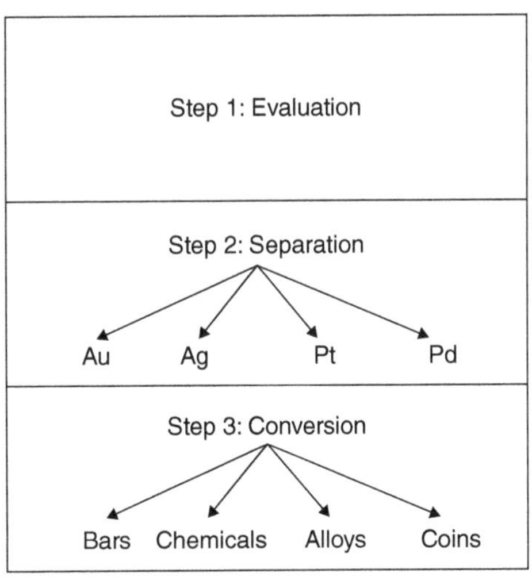

Figure 7 *The three main activities of a precious metals refiner*

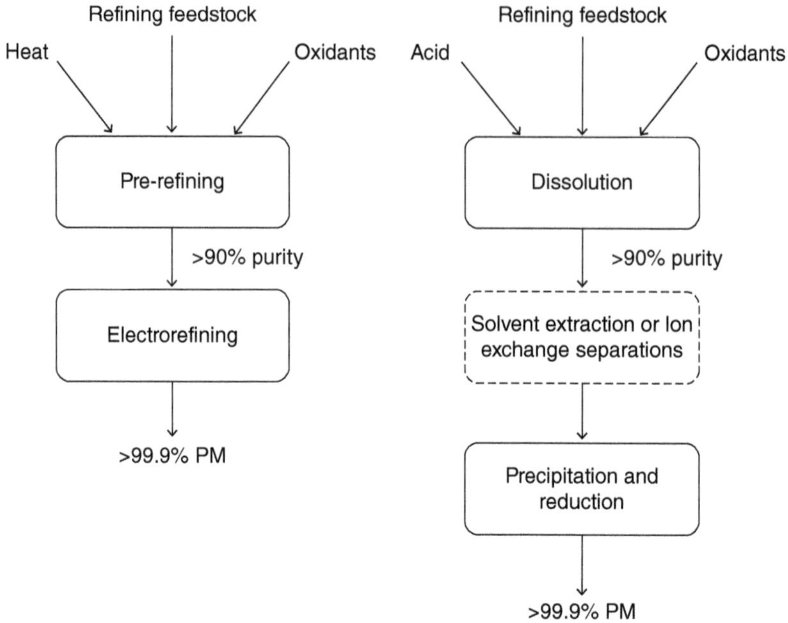

Figure 8 *Two basic processing schemes for the refining of gold, silver, and PGM*

for gold, 100 oz or 1000 oz for silver, or ingots for the PGM. Some refiners produce investment bars and coins, precious metal chemicals, alloys, grain, or fabricated parts.

Two basic separation schemes are used (Figure 8). The first is a combination of pyrometallurgical pre-refining followed by electrorefining, as used in the refining of gold using the pyrometallurgical Miller process and the Wohlwill electrorefining process or in the refining of silver in a convertor followed by electrorefining [103]. The second scheme is largely hydrometallurgical, involving the dissolution of feedstock into acid solutions, followed by a sequence of chemical separations, and, finally, by the precipitation and reduction of precious metals to their metallic form. Reductive precipitation processes are highly selective for gold so the intermediate separation steps can be avoided (although processes such as solvent extraction can improve the first-pass yield of refined gold [104]). PGM processing more typically employs ion exchange, solvent extraction, or precipitation (or a combination of all three) for separation of the closely chemically related PGMs from each other and from the base-metal contaminants. Some excellent reviews of the different PGM refining approaches have been published [105–108].

16.3.1.2 Recycling of Low-Grade Materials

The precious metals content of low-grade materials is typically below 10%, and often below 1%. Evaluation of low-grade materials usually involves incineration to burn off any combustible material and then crushing, screening, and blending to produce a homogeneous powder that can be sampled and analyzed. The separations and costs involved are daunting for a stand-alone precious-metals refining operation, so these materials are most commonly processed through an integrated base-metals smelter. The bulk of this material is refractory in nature and can contain considerable amounts of base metals, refractories, organics, and deleterious elements. Smelting provides conditions for the precious metals to collect in a base-metal matrix while the refractory material separates into the slag and the deleterious elements typically report to the bag-house dusts. The base metals, usually copper, are then electrorefined and the precious metals are concentrated into the insoluble anode slimes. After slimes treatment and consolidation, the upgraded material is treated in a precious-metals refinery.

The importance of access to smelting capability in recycling low-grade materials cannot be overstated. Integrated base-metal smelting operations perform a critical service in the recycling of precious metals. Their ability to do so has coined the term *urban mining*, because they are able to recover and recycle the small amounts of precious metals found in consumer electronic scrap, waste electronic equipment, autocatalysts, and other products.

Some primary refiners blend a certain volume of spent autocatalyst with their regular mine-based feedstock and concentrates, however challenges associated with collection networks, transportation, long lead times, and processing difficulties have raised questions regarding the profitability of treating these recycle streams.

16.3.1.3 Trends and Efficiencies in Precious Metals Recycling

Figure 9 shows the extent of recycling of the precious metals. Pertinent points include the following:

- Typically about one-third of the annual gold supply arises from recycled material, but this varies depending on the gold price and economic conditions: higher prices and economic downturns encourage recycling. During the 2009 Global Financial Crisis and periods of rising gold prices, over 40% of global supply came from recycled sources.
- A recent World Gold Council report [109] highlights that about 90% of recycled gold comes from high-grade material, such as jewelry.
- Silver recycling has remained steady for long periods, at about 22% of supply.
- The extent of platinum and palladium recycling has steadily climbed during this period, driven by increased use in autocatalysts (with increasing new vehicle sales, particularly in emerging markets), improved collection networks, and jewelry demand.
- The bulk of PGM recycling (60% of Pt and 70% of Pd) originates from low-grade autocatalyst material [6].

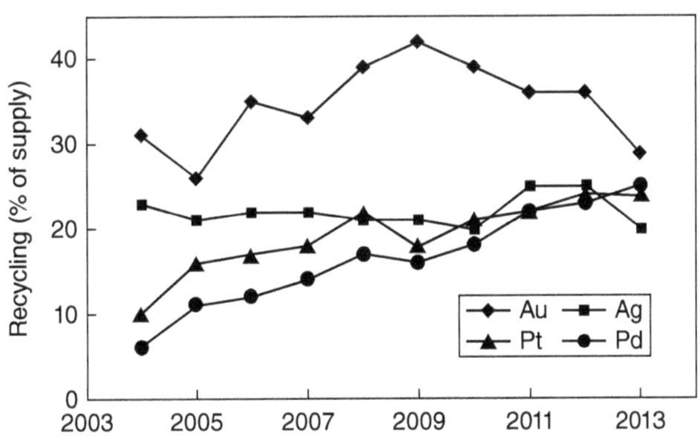

Figure 9 *Precious metal recycling as a proportion of supply*

Recycling rates are generally high for items such as jewelry and coins. In industrial applications, such as catalyst recycling, the precious-metal content may be low but if it occurs in a discrete, separate form, this makes recovery easier. Relatively high rates of PGM recycling occur in vehicle applications, where the valuable metal is confined to a single component, such as an autocatalyst, which can easily be separated from the bulk of the vehicle. The intrinsic value might be small and concentrations low, but the discrete nature of the component allows effective separation and collection. When metals such as gold and silver are dispersed between many different components in devices, such as switches, electronics, and connectors, separations are more complicated, costs are higher, and recycling rates are correspondingly low. Much of the recovery of precious metals for recycling is dependent on the effectiveness of collection networks, which is aided by high precious metal prices and government recycling mandates.

16.3.2 Thrifting and Substitution

In the developed world, and increasingly in developing countries, precious metals are a vital and necessary part of many electronic devices and automobiles. An increasingly wired society has made communication devices indispensable and seen growth of electronic and electrical components in automobiles and home appliances. There are two opposing forces: increasing use of precious metals in such devices and increasing numbers of devices, offset by the drive towards miniaturization to reduce cost, mass, energy consumption, and increase processing speed. Because cost is a key aspect, manufacturers are continually targeting reduced use of precious metals. The data for cellular phones, shown in Figure 10, clearly illustrate the decreasing trend of both unit mass and gold content.

Cost reductions by miniaturization are one approach to minimizing precious metals consumption; another is substitution. For example, palladium has been replaced in multi-layer ceramic capacitors (MLCC) by nickel- or silver-palladium alloys that contains as little as 2% Pd [111]. Substitution, driven at that time by the high price of palladium, was so successful that the demand for palladium in electronic applications halved within one

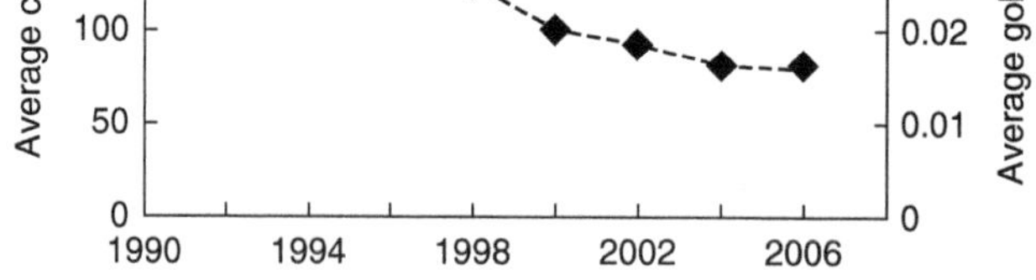

Figure 10 Average cellular phone handset mass and gold content between 1992 and 2006. (Adapted from [110] with permission of Geyer, R. and Blass, V.D.)

year. Substitution can be by a less-expensive precious metal, such as palladium replaced by silver as mentioned above or the replacement by palladium of more expensive platinum in autocatalyst applications, or it can be by another material completely, such as the replacement of precious metals in dental applications by ceramic composites [112].

Similar substitution initiatives are underway in the solar-panel industry, where lowering costs is a major driving force. Conductive silver pastes are an important component of silicon-based solar panels: high silver prices have catalyzed much research aimed at reducing silver content by replacement with tin, nickel, and copper [113].

Thrifting, however, presents difficulties for recycling. It creates lower-grade wastes and, as discussed in Section 16.3.1.3, this leads to the associated recovery challenges. As noted, high precious metal prices, government recycling mandates, and effective collection networks promote the recycling of precious metals from low-grade materials.

16.3.3 Mining and Recovery from Lower-Grade Materials

As demonstrated by the decade-long boom and bust of the US shale oil [114] and gold mining industries, an exploitation spiral occurs with many non-renewable resources. The basic sequence involves the following steps (Figure 11):

1. Demand for the resource increases due to an improving economy, new applications, or a growing population.
2. Demand can grow so much that the supply from traditional reserves cannot keep up and readily available supply decreases.
3. Prices then rise in response.
4. Reserves then increase, because lower-grade ores can now be economically exploited at higher commodity prices, and more lower-grade resources are developed.
5. With the exploitation of lower-grade resources, costs are higher, so technology is employed to improve mining and increase extraction.
6. With successful application of technology and some key advances, the costs of exploiting lower-grade resources decrease.

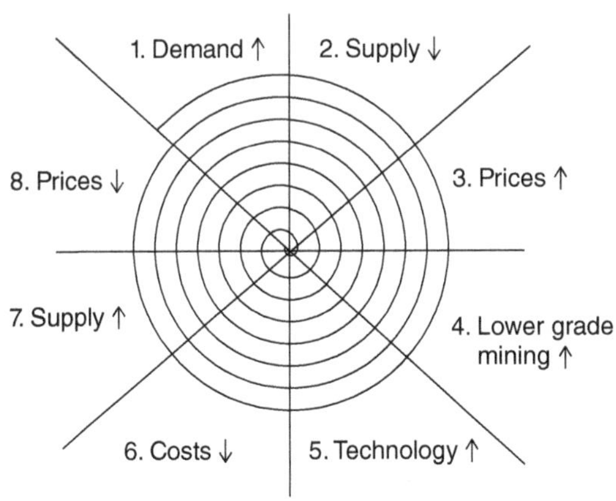

Figure 11 *Non-renewable resources spiral*

7. As lower-cost technologies are employed, more lower-grade resources are exploited and supply again increases beyond demand.
8. When supply exceeds demand, the commodity price decreases. In response, lower-grade operations are again driven to the side lines.

The non-renewable resource is then poised for another trip around the spiral. With each pass, lower- and lower-grade resources are exploited (including the reworking of tailings from older operations) and there is a general shift down the grade curve.

This exploitation spiral works, for the most part, with resources that are driven largely by supply and demand, such as PGMs and silver. With a commodity such as gold, where anomalies such as speculator and central bank hoarding/dishoarding can severely distort the supply-and-demand interactions, there can be leapfrogging over various steps of the spiral. This exploitation spiral incorporates the opportunity-cost paradigm of Tilton and Lagos [115, 116], who noted that mineral availability over the long term will most likely be determined by a balance of increasing costs, created by grade decreases, and decreasing costs, driven by the implementation of new technologies.

16.3.4 Improved Mining, Recovery, and Separation Technologies

Many technological advances have been implemented to deal with the challenges associated with declining grades and rising mining and extraction costs. Most of these focus on improving safety, lowering costs and electricity and water usage, and minimizing environmental impact.

Mining has moved away from being an industry built on the strong backs of a large labor force working at the rock face to one based on energy- and technology-intensive machinery and equipment. A few developments expected to become increasingly important in the future are highlighted:

- Application of data analysis via three-dimensional (3D) modeling and simulations to map ore bodies and create mining plans [117];
- Application of mechanization and automation in mining and ore movement, including mine designs that use automated equipment to reduce energy dependency and improve kWh/t ratios, such as embodied in the goals of AngloGold Ashanti's Technology Innovation Consortium [97]. Companies such as Impala Platinum are investigating new shaft designs with high-level specifications for energy efficiency and power management [96]. Mechanised mining has the potential to reduce the costs of refrigeration, cooling, and electricity.
- Autonomous mining equipment, such as drilling rigs and trucks transporting ore from the bottom of an open pit to an unloading point, which can be operated and directed from a control room that may be located several thousand kilometers from the mine itself;
- Advances in ore analysis, including applications and automation of techniques such as inductively coupled plasma (ICP)-based spectroscopy and X-ray analysis. These techniques are applied in portable hand-held devices that can be used on site to identify "pathfinder elements" that indicate the presence of precious metals [118] and in the real-time monitoring of ore and process streams. Automated image analysis and various methods of spectroscopic analysis are being used for ore-sorting in the PGM industry [119, 120], for the measurement of ore grind from milling circuits [121], and in optimization of

flotation circuits [122]. Recent advances in high-resolution X-ray tomography allows 3D characterization of particles, the mineral phases, and fracture, assisting with liberation analysis and thereby improving recoveries of valuable minerals [123, 124].

AngloGold Ashanti recently announced the development of a new reef-boring technology that extracts the gold-bearing ore from the reef and replaces it with cement and chemicals that stabilise the mining structure. This removes the need to blast rock, reducing the need to ventilate, which has a knock-on effect on electricity demand. Furthermore, only the reef itself is targeted, reducing electricity consumption by removing the energy costs associated with transporting tonnes of rock and material that would have required removal in the development of a conventional shaft [125].

The 1980s was a decade of innovation for the extraction of gold and silver. Heap leaching of low-grade ores with cyanide was widely applied and activated carbon-based technologies, such as carbon-in-leach, carbon-in-pulp, and carbon-in-column, and their resin equivalents were introduced. Innovations in bioleaching and the electrowinning of gold and silver from cyanide solutions led to the development of many low-grade deposits, improved recoveries, and the reworking of many old tailings dumps. Since this flurry of developments, improvements have continued but they have been more incremental in nature.

A similar period of innovation in the PGM industry occurred during the 1970s. Important advances in arc furnace technology and controls [126, 127] and the use of solvent extraction and ion exchange as PGM-separation technologies were implemented at many refineries [3]. Energy-saving initiatives have continued to dominate, with improvements in arc furnace technologies [128], high-intensity smelter technology for treating of high-chromium ores, and PGM recovery processes involving collection into iron-rich alloys instead of the traditional matte smelting approaches. Junior platinum producers have pioneered new production processes with substantial energy-saving potential. Pallinghurst is implementing the Kell process, a patented smelting technology innovation that can handle the chromium content of the PGM concentrate and removes the need to smelt [129]. The technology uses only 14% of the electricity used in conventional smelting methods and also cuts out the milling stage, further reducing power requirements [130]. Braemore Platinum, a wholly-owned subsidiary of Jubilee Platinum, implemented the ConRoast process [128] in 2012, which reportedly increased overall production by 45% [131]. This process, developed by Mintek, involves removing sulfur from metal sulfide concentrate prior to smelting in a direct current arc furnace (making it a more environmentally friendly process) and enables furnaces to accept any proportion of chromite, resulting in more efficient and cost-effective platinum production.

Perhaps the most significant recent advance in PGM refining has been the use of metal-selective ligands attached to a solid support, such as silica gel. The tailoring of the ligands for different precious metals allows highly selective separations from solutions containing a suite of base and precious metals [132]. Molecular Recognition Technology (MRT) has been implemented at numerous operations but, significantly, these MRT materials are now responsible for all the palladium separations at Impala Platinum's Germiston refinery, South Africa, as well as in applications at Tanaka in Japan and at several PGM refineries in China [133].

Solid functionalized materials offer several advantages over their liquid solvent-extraction counterparts, the most important being fire safety and worker exposure.

Development efforts have yielded some interesting materials. Functionalized silica materials have proved to be useful for scavenging precious metals from low-grade solutions [134, 135]; similarly, functionalized fibrous materials have been used in precious metal scavenging applications [136].

16.4 Concluding Remarks: A Long-Term View of the Precious Metals Industry

The challenges facing the global precious metals industry indeed present a long and growing list. However, in reflecting on the industry over the past decades, one cannot help but be impressed at the progress that has been made. Precious metals are now harvested from depths greater than 3.9 km underground, the ores are lower grade and their mineralogy more complex, yet recoveries have improved. Automation is being implemented in all phases of the industry — from the rock face, to the refinery, to the analytical laboratory. Safety has improved considerably. Mining companies have been at the forefront of the sustainability movement and many are actively engaged in their communities and going far beyond simple regulatory and environmental compliance.

Recycling has always been an important part of the industry and is growing in importance. As smaller amounts of precious metals are used in larger numbers of consumer devices, recycling of low-grade materials is taking on increasing significance. Base-metal smelting operations, which permit the collection and concentration of precious metals from low-grade sources, have become a very necessary and increasingly relevant part of the recycling chain.

Awareness and appreciation of the need to prevent the illegal use of precious metals are now prevalent in most participants in the value chain, who now actively work to understand the sources and uses of their feedstocks and products and increasingly collaborate to ensure that their counterparties do the same.

The industry will continue to apply technology to create safer, less polluting, and lower-cost operations. Although we can count on the steady implementation of technology and innovative approaches, volatile precious metal prices complicate matters, causing waves of expansion and troughs of consolidation for the industry. Such business cycles are inherent in the extractive industries but are perhaps more acutely felt for precious metals, where a significant portion of the metal inventories are held for investment purposes, often for long periods of time. Price volatility can cause revenues in precious metals operations to be highly variable: astute companies can compensate for fluctuating prices, to a degree, by implementing financial or technical changes, e.g., lowering operating costs, controlling the ore grades mined and processed, and adjusting their investment strategies.

The bigger challenges for the industry lie in the areas where technical or financial responses are less effective. Most notably, these include increasing governmental demands, labor unrest, and artisanal mining. Increased regulation, greater taxation, and government intervention to the point of nationalism are of great concern, and make the industry very cautious of starting up and funding operations in countries with uncertain or excessively onerous regulatory and policy environments.

Regulatory certainty is essential for a healthy, productive, and involved industry. Precious metals operations — whether mining, processing, fabrication, or recycling — create jobs, build communities, and permit modern lifestyles. They can accelerate economic development, provide government tax revenue, offer a base for other added-value industries, and stimulate community growth and upliftment. The direct economic contribution of the industry is considerable; the indirect contributions are likely many magnitudes larger. Continued growth, development, and contributions from the precious metals industry are anticipated as the 21st century unfolds further.

References

1. Vermaak, C.F. (1995) *The Platinum-Group Metals: A Global Perspective*, Mintek, Randburg, South Africa.
2. Umicore AG (2012) *Precious Materals Handbook*, 1st edition (ed. U. Sehrt), Umicore, Hanau-Wolfgang, Germany.
3. Crundwell, F.K., Moats, M.S., Ramachandran, V., Robinson, T.G. and Davenport, W.G. (2011) *Extractive Metallurgy of Nickel, Cobalt and the Platinum-Group Metals*, Elsevier, Oxford.
4. Sole, K.C., Feather, A., Watt, J., Bryson, L.J. and Sorensen, P.F. (1998) Commercialisation of the Minataur™ Process: Commissioning of Harmony Gold Refinery, in *EPD Congress 1998* (ed. B. Mishra), The Minerals, Metals and Materials Society, Warrendale, PA, pp. 175–186.
5. Johnson Matthey (2005) *Platinum 2005*, http://www.platinum.matthey.com/documents/market-review/2005/full-review/english.pdf (accessed 11 May 2015).
6. Johnson Matthey (2015) *Market Research*, http://www.platinum.matthey.com/documents/new-item/pgm%20market%20reports/pgm%20market%20report%20may%202015.pdf (accessed 9 December 2015).
7. The Silver Institute (2015) *World Silver Supply and Demand*, https://www.silverinstitute.org/site/supply-demand (accessed 9 December 2015).
8. Kitco.com (2015) *Historical Charts*, http://www.kitco.com/charts/ (accessed 9 December 2015).
9. World Gold Council (2015) *Gold Demand Trends, Full Year* 2014, http://www/gold.org/download/file/3691/GDT Q4 2014.pdf (accessed 9 December 2015).
10. Thomson Reuters (2012) *Gold Survey 2012*, Thomson Reuters GFMS Ltd, London, 124 pp.
11. World Platinum Investment Council (2015), http://www.platinuminvestment.com/(accessed 18 May 2015).
12. Cowley, M. (2014) Platinum Producers Plan to Polish up Precious Metal. *Wall Street Journal*, 18 November, http://www.wsj.com/articles/platinum-producers-plan-to-polish-up-precious-metal-1416312375? (accessed 10 May 2015).
13. Index Mundi (2015) *Commodities*, http://www.indexmundi.com/commodities/(accessed 28 December 2014).
14. World Bank (2015) *Inflation, GDP* Deflator (Annual %), http://data.worldbank.org/indicator/NY.GDP.DEFL. KD.ZG/countries? display=graph (accessed 10 May 2015).
15. Williams, L. (2013) South African Gold Output Continues its Decline — Where Will it End? *Mineweb,* (9 August), http://www.mineweb.com/archive/south-african-gold-output-continues-its-decline-where-will-it-end/(accessed 11 May 2015).
16. Bullion Vault (2015) *Gold Mining*, https://www.bullionvault.com/guide/gold/Gold-mining (accessed 10 May 2015).
17. McKeith, T. (2009) Exploration — a Gold Industry Perspective. Paper presented at the *World Mining Investment Congress*, London, https://www.goldfields.co.za/pdf/presentations/2009/03062009_world_ mining_conference.pdf (accessed 11 May 2015).
18. Giurco, D., Prior, T., Mudd, G.M., Mason, L. and Behrisch, J. (2010) *Peak Minerals in Australia: a Review of Changing Impacts and Benefits*, Institute for Sustainable Futures, University of Technology, Sydney and Department of Civil Engineering, Monash University, Sydney.

19. Anglo American Platinum (2015) *Products*, http://www.angloamerican.com/products/platinum (accessed 10 May 2015).
20. Mudd, G.M. (2012) Key Trends in the Resource Sustainability of Platinum Group Elements. *Ore Geology Reviews*, **46**, 106–117.
21. Mudd, G.M. (2012) Sustainability Report and the Platinum Group Metals: a Global Mining Industry Leader? *Platinum Metals Review*, **56** (1), 2–19.
22. Cramer, L.A., Basson, J. and Nelson, L.R. (2004) The Impact of Platinum Production from UG2 Ore on Ferrochrome Production in South Africa. *Journal of the South African Institute of Mining and Metallurgy*, **104** (9), 517–527, http://www.saimm.co.za/Journal/v104n09p517.pdf (accessed 10 May 2015).
23. Seccombe, A. (2013) Anglo American to Increase Output from Sishen, Mogalakwena. *Business Day,* (12 December), http://www.bdlive.co.za/business/mining/2013/12/12/anglo-american-to-increase-output-from-sishen-mogalakwena (accessed 10 May 2015).
24. Zhang, S. (2013) Terrifying Facts About the World's Deepest Gold Mine, http://gizmodo.com/terrifying-facts-about-the-worlds-deepest-gold-mine-1484301368 (accessed 10 May 2015).
25. International Labour Organization (2013) *Global Wage Report 2012/13: Wages and Equitable Growth*, International Labour Office, Geneva, http://www.ilo.org/wcmsp5/groups/public/@dgreports/@dcomm/@publ/documents/publication/wcms_194843.pdf (accessed 28 April 2015).
26. Thomson Reuters (2014) *Gold Survey 2013, Update* 2, http://share.thomsonreuters.com/assets/forms/Gold-Survey-2013-Update2.pdf (accessed 11 May 2015).
27. Thomson Reuters (2014) *World Silver Survey*, https://www.silverinstitute.org/wp-content/uploads/2011/06/WSS2104Summary.pdf>. [11 May 2015].
28. PWC (2014) *SA Mine*, 6th ed., http://www.pwc.co.za/en_ZA/za/assets/pdf/sa-mine-2014-6th-edition-21-nov.pdf (accessed 10 May 2015).
29. Baxter, R. (2014). *Platinum state of the Nation*, http://www.platinumwagenegotiations.co.za/assets/downloads/fact-and-figures/platinum-state-of-the-nation.pdf (accessed 25 May 2015).
30. Chamber of Mines of South Africa (2014) *Platinum Wage Talks. Changing Fundamentals….*, http://www.platinumwagenegotiations.co.za/assets/downloads/fact-and-figures/fact-sheet-changing-fundamentals.pdf (accessed 25 May 2015).
31. Mooiman, M.B. (2012) Deleterious Elements in Precious Metals Refining, in *Proceedings Recycling Metals from Industrial Waste: A Short Course and Workshop with Emphasis on Plant Practice*, Colorado School of Mines, Golden, CO.
32. Miller, G. (2007) *Byproduct Mercury Production in Modern Precious Metals Mines in Nevada*, http://www.epa.gov/mercury/archive/stocks/ByproductMercuryProductioninModernPreciousMetalsMinesinNevada.pdf (accessed 10 May 2015).
33. US Environmental Protection Agency (2014) *Mercury. Questions and Answers About the Mercury Export Ban Act of 2008*, http://www.epa.gov/mercury/exportban-ques. htm#q15 (accessed 10 May 2015).
34. Mooiman, M.B. and Apprahamian, V. (2012) Mercury Deportment in Precious Metals Refining Operations, in *Proceedings of the 36th International Precious Metals Conference*, International Precious Metals Institute, Pensacola, FL.
35. Marsden, J.O. and House, C.I. (2006) *Chemistry of Gold Extraction*, Society for Mining, Metallurgy & Exploration, Littleton, CO, pp. 205–22353.
36. Mooiman, M.B. and Cettou, P. (2013) Selenium Deportment in Precious Metals Refining Operations, in *Proceedings of 37th International Precious Metals Conference*, International Precious Metals Institute, Pensacola, FL.
37. Shott, I. (2003) Osmium. *Chemical & Engineering News,* http://pubs.acs.org/cen/80th/osmium.html (accessed 28 December 2014).

38. Buxton, A. (2012) MMSD+10. *Reflecting on a Decade of Mining and Sustainable Development, IIED Sustainable Markets.* Discussion paper, International Institute for Environment and Development, London, http://pubs.iied.org/16041IIE D (accessed 28 December 2014).
39. Hill, M. (2014) Zambian Mine Royalty Rise to Cost 12,000 Jobs, Chamber Says. *Bloomberg Business,* (19 December), http://www.bloomberg.com/news/articles/2014-12-19/zambia-mining-royalty-increase-to-cost-12-000-jobs-chamber-says (accessed 20 April 2015).
40. Deloitte (2014) *Taxation and Investment in Indonesia 2014. Reach, Relevance and Reliability*, https://www2.deloitte.com/content/dam/Deloitte/global/Documents/Tax/dttl-tax-indonesiaguide-2014.pdf (accessed 10 May 2015).
41. The Daily Bludge (2009) *Facts and Myths About the Mining Tax*, http://www.mining-tax.com.au/wp-contentuploads/2009/12/MiningTaxFacts.pdf (accessed 10 May 2015).
42. Shapiro, R. (2009) *Economic Modernization In Mongolia: the Impact of Tax and Regulatory Policies on the Mining Sector*, http://www.sonecon.com/docs/studies/Report_Mongolian_Mining-Shapiro-Jan2009.pdf (accessed 11 May 2015).
43. Vella, H. (2014) *Mining the DRC – Doing Business in the World's Second Least Developed Country*, http://www.mining-technology.com/features/featuremining-the-drc-doing-business-in-the-worlds-second-least-developed-country-4380854/(accessed 10 May 2015).
44. Baloyi, P. (2014) Zim Wants Minerals Processed Locally. *Mining Weekly*, http://www.miningweekly.com/article/zim-wants-minerals-processed-locally-2014-09-05 (accessed 11 May 2015).
45. Anon (2012) Resource Nationalism in Africa. African Governments Are Seeking Higher Rents and Bigger Ownership Stakes from Foreign Miners. *The Economist,* (11 February), http://www.economist.com/node/21547285 (accessed 11 May 2015).
46. Republic of South Africa (2010) *Amendment of the Broad-Based Socio-Economic Empowerment Charter for the South African Mining and Minerals Industry*, http://www.dmr.gov.za/publications/summary/108-mining-charter-downloads/128-amendedofbbseecharter.html (accessed 27 June 2013).
47. Ernst & Young (2014) *Business Risks Facing Mining And Metals 2014-2015*, http://http://www.ey.com/Publication/vwLUAssets/EY-Business-risks-facing-mining-and-metals-2014%E2%80%932015/$FILE/EY-Business-risks-facing-mining-and-metals-2014%E2%80%932015.pdf (accessed 28 December 2014).
48. Mills, R. (2009) *Country risk*, http://www.kitco.com/ind/Mills/mar112009.html (accessed 11 May 2015).
49. Ibrahim, O. (2012*) China's Appalling Mining Death Rate — Dealing with 'Disorderly' Management'*, http://www.mining-technology.com/features/featurechina-mine-death-rate-coal-safety/(accessed 10 May 2015).
50. Compliance Advisor Ombudsman (2000) *Investigation into the Mercury Spill of June 2, 2000 in the Vicinity of San Juan, Choropampa and Magdalena, Peru*, http://www.cao-ombudsman.org/cases/document-links/documents/June2000MercurySpill.pdf (accessed 11 May 2015).
51. Newmont Mining (2013) *Yanacocha Fact Sheet, Choropampa*, http://newmont.com/files/doc downloads/south america/yanacochc/Choroplama-Fact-Sheet-01-25-2013.pdf (accessed 5 April 2015).
52. Mergel, M. (2011) Baie Mare cyanide spill, *Toxipedia,* http://www.toxipedia.org/display/toxipedia/Baia+Mare+Cyanide+Spill (accessed 11 May 2015).
53. United Nations Environmental Programme (2000) *The Cyanide Spill at Baia Mare, Romania. Before, During and After*, http://archive.rec.org/REC/Publications/CyanideSpill/ENGCyanide.pdf (accessed 10 May 2015).
54. Van Niekerk, H.J. and Viljoen, M.J. (2005) Causes and consequences of the Merriespruit and other tailings-dam failures. *Land Degradation & Development*, **16** (2), 201–212.

55. Financial Crimes Enforcement Network (2015) *FAQ – Interim Final Rule. Anti-Money Laundering Programs for Dealer in Precious Metals, Stones or Jewels*, http://www.fincen.gov/statutes_regs/guidance/html/faq060305.html (accessed 10 May 2015).
56. Jewelers Vigilance Council (2013) *The Essential Guide to Conflict Minerals and the Dodd Frank Act*, http://www.jvclegal.org/JVC_Conflict_Materials.pdf (accessed 10 May 2015).
57. Seccombe, A. (2014) Special report: Platinum strike – An Industry Teetering on the Precipice. *Business Day,* (5 June), http://www.bdlive.co.za/business/trade/2014/06/05/special-report-platinum-strike--an-industry-teetering-on-the-precipice (accessed 10 May 2015).
58. Platinum Wage Negotiations (2014) *Financial Impact of Strike*, http://www.platinum wagenegotiations.co.za/(accessed 10 May 2015).
59. Singh, S. (2014) Effects of the Mining Strikes on the South African Economy. *Mining Review,* http://www.miningreview.com/effects-of-the-mining-strikes-on-the-south-african-economy/ (accessed 22 October 2014).
60. Bariyo, N. (2011) Zambian Miners Riot over Labor Dispute. *Wall Street Journal,* (19 January), http://www.wsj.com/articles/SB10001424052748703951704576091953050313980 (accessed 11 May 2015).
61. Reuters (2013) Gold Fields' Ghana Mines Hit by Wildcat Strike. *MineWeb*, (3 April), http://www.mineweb.com/archive/gold-fields-ghana-mines-hit-by-wildcat-strike/(accessed 11 May 2015).
62. Montlake, S. (2011) Freeport Struggles to Restart Indonesian Mine Amid Strike Unrest. *Forbes,* (31 October), http://www.forbes.com/sites/simonmontlake/2011/10/31/freeport-struggles-to-restart-indonesian-mine-amid-strike-unrest/(accessed 11 May 2015).
63. Reuters (2011) Strike Deal Elusive as Chile Escondida Talks in Limbo. *Reuters,* (3 August), http://af.reuters.com/article/idAFN1E77209520110803 (accessed 11 May 2015).
64. International Council on Mining and Metals (2012) Trends in the Mining and Metals Industry, in *Mining's Contribution to Sustainable Development,* https://www.icmm.com/document/4441 (accessed 10 May 2015).
65. Menon, R. (2014) UPDATE 1-Former African Barrick Gold to Go Underground at North Mara. *Reuters.* (Nov 27), http://www.reuters.com/article/2014/11/27/african-barrick-strategy-idUSL3N0TH2IE20141127 (accessed 11 May 2015).
66. Motala, M. (2014) *The Deafening Silence on the Lives and Deaths of the Zama Zama Miners.* South African Civil Society Information Service, http://sacsis.org.za/site/article/1935 (accessed 9 April 2015).
67. Chamber of Mines of South Africa (2015) *Illegal Mining Fact Sheet*, http://www.chamber ofmines.org.za/media-room/mining-factsheets (accessed 9 April 2015).
68. United Nations Environmental Programme (2015) *Reducing Mercury in Artisanal and Small-Scale Gold Mining (ASGM)*, http://www.unep.org/chemicalsandwaste/Mercury/GlobalMercury Partnership/ArtisanalandSmallScaleGoldMining/tabid/3526/language/en-US/Default.aspx (accessed 11 May 2015).
69. Rupprecht, S. (2015) Needs analysis for capacity building of artisanal miners in Central Africa, in *Proceedings Copper Cobalt Africa*, Southern African Institute of Mining and Metallurgy, Johannesburg, 45–52.
70. Artisan Gold Council (2015), http://www.artisanalgold.org/home (accessed 11 May 2015).
71. Alliance for Responsible Mining (2015) *Fairmined Gold Standard*, http://www.communitymining.org/en/section-fairmined-certification/arm-fairmined-gold (accessed 11 May 2015).
72. van Zyl, D. and Gagne, D. (2010) Mining, in *Berkshire Encyclopedia of Sustainability*, Volume 2, Berkshire Publishing Group, MA, pp. 332–336.
73. World Commission on Environment and Development (1987) *Our Common Future*. Report of the World Commission on Environment and Development. Annex to General Assembly Document A/42/427.

74. Elkington, J. (1998) *Cannibals with Forks: the Triple Bottom Line of 21st Century Business*, New Society Publishers, CT.
75. Fraser Institute (2012) *What is the Social License to Operate (SLO)* http://www.miningfacts.org/Communities/What-is-the-social-licence-to-operate/(accessed 18 May 2015
76. Kellogg Innovation Network (undated) *Catalyst Mining Company of the Future. Reinventing Mining: Creating Sustainable Value. Introducing the Development Partner Framework,* http://www.kinglobal.org.catalyst.php (accessed 19 August 2014).
77. International Finance Corporation (2015) *IFC Sustainability Framework*, http://www.ifc.org/wps/wcm/connect/topics_ext_content/ifc_external_corporate_site/ifc+sustainability/our+approach/risk+management/sustainability+framework (accessed 10 May 2015).
78. The Equator Principles Association (2011) *Equator Principles*, http://www.equator-principles.com/(accessed 10 May 2015).
79. Global Reporting Initiatives (2015), https://www.globalreporting.org/Pages/default.aspx (accessed 11 May 2015).
80. Extractive Industries Transparency Initiative (2015) *The EITI Standard*, https://eiti.org/ (accessed 10 May 2015.
81. Voluntary Principles on Security and Human Rights (2015), http://www.voluntaryprinciples.org/what-are-the-voluntary-principles/(accessed 11 May 2015).
82. International Council on Mining and Metals (2015) *Sustainable Development Framework*, https://www.icmm.com/our-work/sustainable-development-framework (accessed 10 May 2015).
83. Business and Human Right Resource Centre (2015) *UN Guiding Principles*, http://business-humanrights.org/en/un-guiding-principles (accessed 11 May 2015).
84. Organisation for Economic Co-operation and Development (2011) Declaration on *International Investment and Multinational Enterprises*, http://www.oecd.org/daf/inv/investment-policy/oecddeclarationoninternationalinvestmentandmultinationalenterprises.htm (accessed 11 May 2015).
85. Convention on Biological Diversity (2015) *Akwé: Kon Guidelines*, https://www.cbd.int/traditional/guidelines.shtml (accessed 10 May 2015).
86. Framework for Responsible Mining (2005) *A Guide to Evolving Standards*, http://www.frameworkforresponsiblemining.org/(accessed 10 May 2015).
87. Natural Resource Charter (2015) The Twelve Precepts, http://www.naturalresourcecharter.org/precepts (accessed 11 May 2015).
88. Responsible Jewellery Council (2015), http://www.responsiblejewellery.com/(accessed 10 May 2015).
89. Jewelers Vigilance Committee (2015), http://www.jvclegal.org/index.php?categoryid=10> (accessed 10 May 2015).
90. Earthworks (2015) *No Dirty Gold*, http://nodirtygold.earthworksaction.org/(accessed 18 May 2015).
91. Mineral Policy Institute (2015), http://www.mpi.org.au/(accessed 18 May 2015).
92. Mining Watch Canada (2012) *Key Issues*, http://www.miningwatch.ca/key-issues (accessed 11 May 2015).
93. Franks, D. (2009) Avoiding Mine-Community Conflict: from Dialogue to Shared Futures, in *Enviromine, First International Seminar on Environmental Issues in the Mining Industry*, (eds. J. Wiertz and C.J. Moran), BN Americas, Santiago.
94. Stilwell, L.C. (2008) Platinum, Electricity and the Economy, in *Third International Platinum Conference: Platinum in Transformation*, Southern African Institute of Mining and Metallurgy, Johannesburg, pp. 307–312.
95. Mudd, G.M. (2007) Resource Consumption Intensity and Sustainability of Gold Mining, in *Second International Conference on Sustainability, Engineering and Science*, New Zealand, http://www.thesustainabilitysociety.org.nz/conference/2007/papers/MUDD-Gold-Mining-v-Sustainability.pdf (accessed 11 May 2015).

96. Trade and industrial Policy Strategies (TIPS) (2014) The Impact of Electricity Price Increases on the Competitiveness of Selected Mining Sector and Smelting Value Chains in South Africa. Policy Paper prepared for the Economic Development Department and the Department of Trade and Industry, http://www.tips.org.za/files/tips_for_gggi_policy_paper_march_2015.pdf (accessed 25 May 2015).
97. AngloGold Ashanti Technology Innovation Consortium (2015), https://anglogoldashanti technology.com/about_us/?abtId=1 (accessed 25 May 2015).
98. Solar Energy Industries Association (2015) *Issues and Policies. Net-metering* http://www.seia.org/policy/distributed-solar/net-metering (accessed 11 May 2015).
99. Banerjee, S.G., Romo, Z., McMahon, G., Toledano, P., Robinson, P. and Pérez Arroyo, I. (2015) *The Power of the Mine: A Transformative Opportunity for Sub-Saharan Africa*, World Bank, Washington, DC, https://openknowledge.worldbank.org/handle/10986/21402 License: CC BY 3.0 IGO (accessed 5 May 2015).
100. Vella, H. (2013) *Managing Water Consumption in Mining*, http://www.mining-technology.com/features/feature-managing-water-consumption-mining-global-shortage/(accessed 25 May 2015).
101. Rankin, W.J. (2011) *Minerals, Metals and Sustainability*, CSIRO Publishing, pp. 191–196.
102. Klimkowich, D.G. (2015) Private Communication, President Advanced Materials Group, Materion Corporation (13 January).
103. Nadkarni, R.M., Kinneberg, D.K., Mooiman, M.B. and Cettou, P. (1990) Precious Metals Processing, in *Proceedings of the Elliott Symposium on Chemical Process Metallurgy*, AIME, pp. 93–128.
104. Feather, A.M., Sole, K.C. and Bryson, L.J. (1997) Gold Refining by Solvent Extraction–The Minataur™ Process. *Journal of the South African Institute of Mining and Metallurgy*, **97**, 169–173.
105. Al-Bazi, S.J. and Chow, A. (1984) Platinum Metals: Solution Chemistry and Separation Methods (Ion Exchange and Solvent Extraction). *Talanta*, **31**, 815–836.
106. Charlesworth, P. (1981) Separating the Platinum Group Metals by Liquid–Liquid Extraction. *Platinum Metals Review*, **25**, 106–112.
107. Harris, G.B. (1993) A Review of Precious Metals Refining, in *Precious Metals* 1993, International Precious Metals Institute, Allentown, PA, pp. 351–374.
108. Mooiman, M.B. (1993) The Solvent Extraction of Precious Metals: a Review, in *Precious Metals 93*, International Precious Metals Institute, Allentown, PA, pp. 411–434.
109. Hewitt, A., Keel, T., Tauber, M. and Le-Fiedler, T. (2015) The Ups and Downs of Gold Recycling. *World Gold Council,* http://www.gold.org/supply-and-demand/ups-and-downs-gold-recycling (accessed 10 May 2015).
110. Geyer, R. and Blass, V.D. (2010) The Economics of Cellphone Reuse and Recycling. *International Journal of Advanced Manufacturing Technologies*, 47, 515–525, http://escholarship.org/uc/item/8z18c5h6#page-1 (accessed 11 May 2015).
111. Cross, J. (2004) Precious Metals in Electronics: the Winners and Losers, in *Proceedings LBMA Precious Metals Conference 2004,* http://www.lbma.org.uk/assets/3c_b_cross_lbmaconf2004.pdf (accessed 11 May 2015).
112. Shenoy, A. (2008) Is it the End of the Road for Dental Amalgam? A Critical Review. *Journal of Conservative Dentistry*, 11 (3), 99–107, http://www.ncbi.nlm.nih.giv/pmc/articles/PMC2813106/(accessed 10 May 2015).
113. Ali-Oettinger, S. (2013) Silver Lining for Cost Savings. *PV Magazine,* http://www.pv-magazine.com/archive/articles/beitrag/silver-lining-for-cost-savings-100010408/572#axzz3YK3afeXi (Accessed 10 May 2015).
114. Olson, B. And Murtaugh, D. (2015) The Shale Oil Industry Has Already Gone Bust – At Least For Now. *Bloomberg,* (4 May), http://www.bloomberg.com/news/articles/20150503/the-shale-boom-has-already-gone-bust-at-least-for-now (accessed 12 April 2015).

115. Tilton, J.E. (2010) Is Mineral Depletion a Threat to Sustainable Mining? *Society of Economic Geologists Newsletter*, **82** (7), https://www.segweb.org/pdf/views/2010/07/SEG-Newsletter-Views-John-Tilton.pdf (accessed 10 May 2015).
116. Tilton, J.E. and Lagos, G. (2007) Assessing the Long-Run Availability of Copper. *Resources Policy*, **32**, 19–23.
117. Whittle 3D and Whittle 4D software (2015), http://www.miningsoftware.org/portfolio-items/whittle-software/(accessed 18 May 2015).
118. Somarin, A. (2014) *XRF Helps Meet the Platinum Group Metals Mining Challenge*, http://acceleratingscience.com/mining/xrf-helps-meet-the-platinum-group-metals-mining-challenge/(accessed 11 May 2015).
119. Bergmann, C. (2009) Developments in Ore Sorting Technologies, in *Proceedings Mintek 75 Conference,* http://www.mintek.co.za/Mintek75/Proceedings/M02-Bergmann.Pdf (Accessed 11 May 2015).
120. Walker, S. (2014) Ore Sorters and Analyzers: the Technology Develops. *Engineering & Mining Journal.* (April), http://www.e-mj.com/features/3897-ore-sorters-and-analyzers-the-technology-develops.html (accessed 11 May 2015).
121. Steyn, C.W., Keet, K. and Breytenbach, W. (2012) *Optimization and Control of a Primary SAG Mill Using Real-Time Grind Measurement*, http://www.bluecubesystems.com/images/stories/pdf/Com12PrimarySAGMillGrindControlFinal.pdf (accessed (11 May 2015).
122. Marais, C. and Aldrich, C.A. (2010) The Estimation Of Platinum Flotation Grade from Froth Image Features y Using Artificial Neural Networks, in *Fourth International Platinum Conference, Platinum in Transition 'Boom or Bust'*, Southern African Institute of Mining and Metallurgy, pp. 143–148, http://www.saimm.co.za/Conferences/Pt2010/143-148_Marais.pdf (accessed 11 May 2015).
123. Godel, B. (2013) High-resolution X-ray Computed Tomography and Its Application to Ore Deposits: from Data Acquisition to Quantitative Three-Dimensional Measurements With Case Studies From Ni-Cu-PGE Deposits. *Economic Geology*, **108** (8), 2005–2019.
124. Godel, B., Barnes, S.J., Barnes, S.-J. and Maier, W.D. (2010) Platinum Ore in Three Dimensions: Insights from High-Resolution X-Ray Computed Tomography. *Geology*, **38** (12), 1127–1130, http://www.researchgate.net/publication/257117772_Platinum_ore_in_three_dimensions_Insights_from_high-resolution_X-ray_computed_tomography (accessed 11 May 2015).
125. Creamer, M. (2013), Exciting New AngloGold Ashanti Produces First Gold. *Mining Weekly*, http://www.miningweekly.com/article/exciting-new-anglogold-ashanti-technology-produces-first-gold-2013-11-06 (accessed 25 May 2015).
126. Jones, R.T. (1999) Platinum Smelting in South Africa. *South African Journal of Science*, **95**, 525–534, http://www.mintek.co.za/Pyromet/Files/Platinum.pdf (accessed 11 May 2015).
127. Nell, J. (2004) Melting of Platinum Group Metal Concentrates in South Africa. *Journal of the South African Institute of Mining and Metallurgy*, **104** (7), 423–428, http://www.saimm.co.za/Journal/v104n07p423.pdf (accessed 10 May 2015).
128. Phillips, R.E., Jones, R.T. and Chennells, P. (2008) Commercialization of the ConRoast process, in *Third International Platinum Conference: Platinum in Transformation*, Southern African Institute of Mining and Metallurgy, Johannesburg, pp. 141–147.
129. Creamer, M. (2014) Pallinghurst Bringing in New Tech to Boost Platinum Business. *Mining Weekly,* http://www.miningweekly.com/article/pallinghurst-bringing-in-new-tech-to-boost-platinum-business-2014-04-10 (accessed 25 May 2015).
130. Liddell, K., Newton, T., Adams, M. and Muller, B. (2010) Energy Consumption for Kell Hydrometallurgical Refining Versus Conventional Pyrometallurgical Smelting and Refining of PGM Concentrates, in *Platinum in Transition "Boom or Bust"*, Southern African Institute of Mining and Metallurgy, Johannesburg, pp. 181–186.

131. Jubilee Platinum (2014), http://www.jubileeplatinum.com/investors-and-media/announcements/2014/20-jun-2014.php (accessed 25 May 2015).
132. Izatt, R.M., Izatt, S.R., Izatt, N.E., Krakowiak, K.E., Bruening, R.L. and Navarro, L. (2015) Industrial Applications Of Molecular Recognition Technology to Separations of Platinum Group Metals and Selective Removal of Metal Impurities from Process Streams. *Green Chemistry*, **17**, 2236–2245.
133. Izatt, S.R. (2016) Precious Metal Recovery from Complex Matrices using a Green Chemistry/Molecular Recognition Technology Approach, Wiley.
134. Henry, M. (2013) The Right Chemistry. *InBusiness*, (September), p. 23.
135. Mascini, G. (2013) New Technologies for PGM Recoveries from Refining Streams, in *Proceedings ALTA 2013*, ALTA Metallurgical Services, Melbourne.
136. Johnson Matthey (2014) *Scavenging Technologies. Smopex*, http://www.scavengingtechnologies.com/page-view.php?pagename=Smopex (accessed 11 May 2015).

17

Metal Sustainability from a Manufacturing Perspective: Initiatives at ASARCO LLC Amarillo Copper Refinery

Luis G. Navarro[1], Tracy Morris[1], Weldon Read[1], and Krishna Parameswaran[2]

[1]ASARCO LLC, Amarillo, Texas, 79109, USA
[2]tfgMM Strategic Consulting, Scottsdale, Arizona, 85262, USA

17.1 Introduction

The increasing importance given to competitiveness in mining and manufacturing has led companies to place much greater emphasis on cost/benefit analyses and process optimization in an effort to improve production efficiencies and lower unit cost while meeting increasingly stringent environmental regulations. In addition, companies have realized the importance of obtaining and maintaining the social license to operate. In this regard, the primary copper industry is no exception, especially because price for its primary product refined copper, is determined on markets by world supply and demand.

The implementation of sustainable development principles or sustainability considerations is one of the current priorities for the primary copper industry due to increasing demand for copper products driven mainly by the economic development in China and other developing countries such as Brazil and India and the increasing emphasis on the quality requirements for the raw material inputs used to manufacture refined copper products. In July, 2008 the European Commission presented the "Sustainable Consumption and Production and Sustainable Industrial Policy (SCP/SIP) Action Plan" [1], which included

Metal Sustainability: Global Challenges, Consequences, and Prospects, First Edition. Edited by Reed M. Izatt.

proposals (1) targeting improvements in the environmental performance of products and (2) increasing the demand for more sustainable goods and production technologies, while encouraging the industry to innovate and consumers to make informed choices with the ultimate objective of improving the competitiveness of the industrial economy.

Companies are responding to this and other earlier initiatives by integrating sustainability considerations in their operations and making the necessary investments in environmental emissions controls as regulations become increasingly more strict while at the same time preserving investments already made and protecting employment. Chapter 18 in this book discusses the concept of sustainable development from the perspective of a a mining company, ASARCO LLC (Asarco). That chapter illustrates how hard rock mining, from exploration and mine development to beneficiation, reclamation, closure and post-closure, can be sustainable. The current chapter discusses the sustainability initiatives that have been and are being implemented at Asarco's Amarillo Copper Refinery (ACR) and provides a thorough discussion of these initiatives at ACR, mainly in the areas of electrolyte purification and metals recovery.

17.2 General Overview of Sustainability from the Copper Industry Perspective

Copper and its alloys are very durable, reliable in products, often outlive aluminum, plastic and other substitute materials, and can be repeatedly recycled to specification metal with no significant loss of performance. Recycling, therefore, is an important attribute of sustainability, as discussed in Chapter 18. As a part of a voluntary risk assessment initiated in 2000, completed in 2008, and published in 2009, the European Copper Institute, working together with the European Commission and EU Member State experts, concluded that: [2]

- The use of copper in products is in general safe for the environment and public health.
- Copper is not a carcinogenic, mutagenic, reprotoxic or PBT (persistent, bio-accumulative, toxic) substance.
- Environmental risks are possible at industrial sites where there is insufficient on-site water treatment, or where the effluent goes into a water body with low dilution.
- Occupational health risks are possible at some industrial sites, specifically for workers involved in the production of copper chemicals and powders, but they can be mitigated with proper engineering controls and use of personal protection equipment.

In recent decades, the copper industry has made significant progress in improving both its environmental performance and energy efficiency, making it possible through resource conservation measures to recover valuable constituents in copper ores, such as gold, silver, platinum, palladium, selenium, nickel, tellurium, etc., cost effectively.

Even though the copper industry has been improving its performance, recent even more stringent environmental regulations are requiring companies to make multi-million dollar investments to reduce discharges to water, air emissions and releases to the land. It has become increasingly more difficult to obtain permit modifications for existing operations and permits for new operations. Therefore, it is becoming even more important to incorporate sustainable manufacturing practices. Doing so will provide the proper balance for the three pillars of sustainability — economic, environmental and social considerations — allowing for a healthy environment, a profitable enterprise and a thriving community with satisfied stakeholders.

The great challenge, however, is to integrate environmental sustainability while pursuing strong economic growth. In the copper industry this concept receives a great deal of attention since the final refined copper product specification calls for 99.99+% purity from available raw materials that are dictated by mine geology, i.e. ores containing as little as 0.5% copper, with minor constituents that might include precious metals like gold, silver, platinum; valuable recoverable constituents such as lead, zinc, rhenium, molybdenum, bismuth, nickel, selenium and tellurium, and constituents of environmental concern such as arsenic and cadmium. Sustainable consumption practices and production should lead to economic opportunities targeting present needs, while ensuring that future generations will be able to meet their own needs, the very essence of sustainable development.

17.3 A Brief History of ASARCO LLC

Asarco was organized in 1899 as **A**merican **S**melting **A**nd **R**efining **Co**mpany. Originally a consolidation of a number of lead-silver smelting companies, it evolved over the years into an integrated producer of primary non-ferrous metals, including copper, lead and zinc. Grupo Mexico, S.A. de C.V acquired Asarco in 1999. Today, Asarco is a fully integrated producer of refined copper and associated co-products. Asarco's domestic mines produce approximately 350–400 million pounds of copper in concentrates annually from the Mission, Silver Bell and Ray open-pit copper mines in Arizona.

Asarco has solvent extraction/electrowinning plants at the Ray and Silver Bell mines that produce about 375 million pounds of refined copper annually and an INCO copper flash smelter in Hayden, Arizona that produces almost half a billion pounds of anodes annually and feeds ACR, where rod wire, cake and cathodes are produced.

17.3.1 Asarco's Footprint in Amarillo, Texas

The history of the High Plains of Texas began with a quest for metal. Centuries before the Amarillo region became cattle and oil country, legends about gold lured Spanish adventurers to what is often called the Golden Spread. Now, hundreds of years after the conquistadors abandoned their search for metallic riches, another bright metal, copper, is helping shape the area's history and its future.

Asarco has been associated with Amarillo and Texas High Plains for more than ninety years, and the Amarillo Copper Refinery has been a vital economic force for Amarillo for the last 40 years and a major supplier of copper and silver of international repute.

Timeline of Asarco's presence in Amarillo, Texas:

1922 Zinc smelter construction begins
1924 Zinc smelter commences operation
1974 Copper refinery construction at a cost of $210 million
1975 First shipment of copper made; zinc smelter closed
1979 Patented Reatrol process (Reagent Control) increased production efficiency, enabling plant eventually to exceed design capacity

1993 Electrolyte purification facility installed, improving the quality of refined copper production
1994 Plant receives ISO-9002 certification for copper cathode
1995 Plant receives ISO-9002 certification for copper rod and cake; stormwater recycling and installation of Rod Zero stand
1996 Plant receives ISO-9002 certification for precious metals parting, Se/Te, nickel sulfate and anode casting
1997 Rod mill upgrade
1998 Plant receives ISO-9002 certification for anode/blister sampling, precious metals sampling and anode mold making
2000 Increased rod and cake facilities capacities
2001 Electrolyte purification upgrade
2003 Plant receives ISO 9000: 2000 certification for copper cathode, rod, cake, precious metals parting, Se/Te, nickel sulfate and anode casting
2006 Enviroalloy™ production and rod shaft furnace shell replacement
2008 Plan receives ISO 9001: 2008 certification
2012 Precious metal and selenium production suspended; installation of pallet handling robot
2015 Installation of robotic starter sheet stripping system

17.4 How Refined Copper Is Produced

Asarco operates three open-pit copper mines in Arizona: the Ray, Silver Bell and Mission Complex mines. Because ore grades are around 0.4–0.7% Cu, large quantities of earthen materials have to be moved. For example, the Mission Mine has an average grade of about 0.67% Cu, which means that one ton of ore has to be mined to produce 13 pounds of copper. Also, for every ton of ore mined, about three more tons of waste rock must be removed.

The first step in the mining process is to drill blast holes in a pattern determined generally by the hardness of the rock and the location of the ore body. The goal of blasting is to break the rock into small enough pieces for the large shovels to load into the haul trucks. After being drilled to the proper depth, the holes are loaded with a mixture of ammonium nitrate and diesel fuel called ANFO. Blasting technology has steadily improved and is becoming much more effective in terms of breaking the rock and getting to the ore. This is because blasting patterns are designed to direct explosive energy into fracturing the rock underground, thereby minimizing the creation of airborne dust.

Huge electric shovels load the blasted waste rock and ore into haul trucks ranging in capacity from 240 to 400 tons. This helps reduce the amount of dust generated by the haul trucks operating on mine haul roads. Asarco takes many proactive measures to control dust: mine roads are wetted by spraying them from water trucks, ranging from 10,000 to 30,000-gallon capacity, that are almost as big as haul trucks. Periodically dust-suppressant chemicals are also applied. This helps reduce the amount of dust generated by the haul trucks and other vehicular traffic on the mine roads.

The haul trucks transport the material to dump pockets that feed primary, secondary and tertiary crushers that reduce the ore size to 1/2" or less. After the crusher, the ore is ground to the consistency of fine talcum powder by large rotating mills called semi-autogenous grinding (SAG) or rod mills, which grind the ore with the help of steel balls or rods, followed by further size reduction in ball mills.

Only after intensive grinding are the copper minerals liberated from the rest of the rock, ready to be separated by froth flotation. The pulverized ore slurry is mixed with milk of lime (to raise the pH) and small amounts of special flotation reagents: a frother to create air bubbles and a collector chemical that causes the copper minerals to be water repellent, making it stick to those bubbles. Air is blown into the flotation tank and the mixture is vigorously agitated, as in a high-speed blender. Rising air bubbles carry the copper minerals up and over the edge of the flotation tank. The bubbles break soon after they flow over the edge. The collected copper minerals are then ground even finer and concentrated in additional flotation cells to produce a copper concentrate. The material that does not float is called tailings and drops to the bottom of the flotation cells. It is dewatered in thickeners and managed in engineered tailings storage facilities. About 80 percent of the water used in the milling process is reclaimed and reused. The rest remains with the tailings, keeping it wet and preventing wind-blown dust. Dust-suppressant chemicals are also used to control fugitive dusts from tailings storage facilities.

The dried copper concentrate of about 28% copper is shipped to the smelter. It represents less than one percent of the material removed from the mine.

At the Asarco copper smelter in Hayden, Arizona, copper concentrates from the Mission, Ray and Hayden concentrators, along with silica-bearing flux and technically pure oxygen, is blown into the INCO oxygen flash furnace through four burners. The furnace is initially heated with natural gas. When the copper concentrate is blown into the hot furnace it immediately ignites and melts. The silica flux binds with the oxidized iron and other impurities to form a slag and separates from the liquid copper matte containing copper, iron and sulfur. The matte and slag fall to the bottom of the furnace where they separate into two layers, the lighter slag at the top and the heavier matte at the bottom. The sulfur in the concentrate combines with oxygen to form sulfur dioxide gas; this combustion releases a tremendous amount of heat that sustains the high temperature in the furnace, making it auto thermal.

Sulfur dioxide gas from the flash furnace is captured and combined with water in an adjacent acid plant to produce sulfuric acid, which is sold as a by-product to chemical manufacturers in the United States. The Hayden smelter produces 2,500 tons per day of high-purity sulfuric acid.

The matte is tapped from a hole near the bottom of the furnace and poured into a huge ladle. An overhead crane picks up the ladle and pours the molten matte into another type of furnace called a converter.

The slag is a dense, glassy material of mostly iron and silica. It is tapped from a hole in the upper part of the furnace wall and carried outside to the slag dump to cool. The slag cools quickly and tends to break into pieces by itself. Slag containing higher amounts of copper is sent back through the Hayden concentrator to recover copper as a copper slag concentrate.

In the converter, air is blown into the matte to oxidize away the iron and most of the remaining sulfur. This oxidation releases more heat, which keeps the materials molten. The particulate matter and sulfur dioxide-bearing gases are collected and cleaned to keep them out of the air. After blowing air through the converter for a few hours, the entire furnace rotates on its side to pour the molten metal, now called blister copper, into another large ladle that carries it to the anode furnace. The blister copper is more than 98 percent pure but contains too much sulfur. Air

Figure 1 GV Asarco's Amarillo Copper Refinery (ACR) Tankhouse in Amarillo, Texas

is blown into the molten blister copper in the anode furnace to burn off the remaining sulfur and the excess oxygen is removed by blowing a mixture of natural gas and steam that produces a reducing gas. The anode furnace is rotated on its side to pour the molten copper, called anode copper, into an anode-casting machine to produce anodes of 99.4% purity and weighing between 750 and 850 pounds each.

Anodes are two-inch thick slabs of copper about three feet wide and three-and-a-half feet tall, including the two handles called lugs molded into the top. They are loaded into special railcars or bundled onto flatbed trucks and shipped to ACR in Texas. Figure 1 shows an aerial view of ACR plant site showing the Tankhouse (long building in the foreground) where copper refining takes place.

17.5 Introduction to Physical Chemistry of Copper Electrorefining

Electrorefining is much more common than electrowinning in primary copper production, requiring less energy than electrowinning. The process uses soluble copper anodes and the copper from the anode dissolves in the electrolyte and is plated on the cathode. A fairly constant copper concentration is maintained during the process, while in electrowinning, insoluble anodes are used and the desired metal is depleted from the solution. Fundamentally,

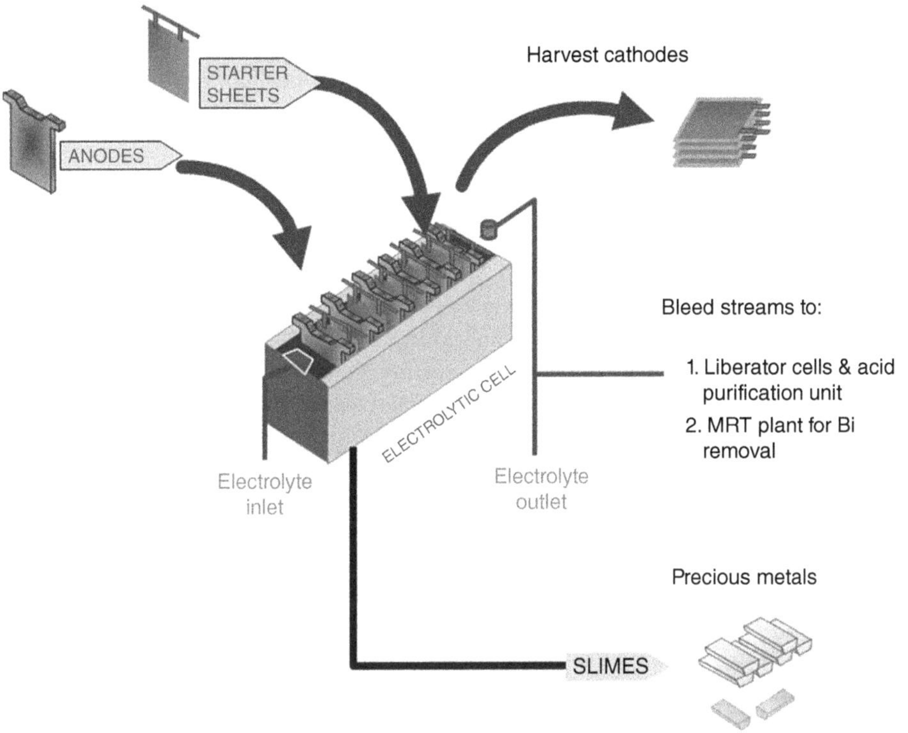

Figure 2 Schematic Diagram of the Copper Electrorefining Process (See insert for color representation of the figure.)

electrolytic refining of copper consists of an electrolytic cell filled with a copper sulfate, sulfuric acid electrolyte and pairs of copper anodes and cathodes. When a current is applied, copper in the anode dissolves and is plated onto the cathode. The electrode reactions can be represented as follows; at the copper anode:

$$Cu \rightarrow Cu^{2+} + 2e^{-} \tag{1}$$

and at the copper cathode:

$$Cu^{2+} + 2e^{-} \rightarrow Cu \tag{2}$$

The objective of Cu electrorefining is not only to eliminate unwanted impurities to produce a 99.99⁺% pure Cu cathode, but also to separate valuable constituents so they can be recovered as co-products. Ultimately, impurities in the anode, i.e., Ag, Au, Pt, Pd, Te, that are more noble than copper, do not dissolve and end up at the bottom of the cell; they are referred to as slimes (also “anode slimes” or “anodic slimes”). The remaining impurities dissolve and gradually increase in concentration in the electrolyte; if they are not removed periodically they can reach a concentration that will reduce cathode purity [3,4]. Figure 2 shows the electrorefining process at ACR where starter sheets are used as cathodes.

Even though the feed for ACR is mainly produced at the Hayden smelter in Arizona, there is considerable variability in the type and concentration of different metal species present in the anodes. This is due to the diversity of concentrates and smelting by-products used as raw materials in the smelter feed mix and the limited capability to remove impurities during the copper smelting, converting and anode refining process.

17.6 Electrolyte Purification

A major challenge in primary copper production is the requirement to produce high- purity refined copper from feed stock containing a diverse range of metal impurities that can adversely impact the refined copper product quality. Pyrometallurgical technologies alone cannot separate these metal impurities to the extent required. Specifications for high-purity copper are increasing as the need for this metal in new high-tech devices increases. Failure of the copper product to meet specifications results in severe economic loss due to the facilities, labor and time required to reprocess the copper, often using the same inadequate impurity-metal-removal processes. Metal impurities present in copper ores include Pb, As, Bi, and Sb. During electrorefining, their concentrations gradually increase in the electrolyte, necessitating the use of some means to maintain their concentrations at acceptable levels. The most critical impurities are Bi, As, and Sb, which cause grain-boundary cracks in wiredrawing [5]. Traditionally, a variety of methods including bleed streams, controlled As addition, solvent extraction, and ion exchange have been employed to maintain acceptable levels of these metal impurities in the electrolyte [5].

Environmental concerns associated with use of these methods are significant. New chemical reagents added to the system can increase the number of impurities present. Additional process steps are needed to separate the target metal impurity, often posing potential health concerns for workers that must be mitigated. Following separation, the additional chemical residue must be disposed of, usually in landfill or by deep well injection, potentially adding further environmental concerns. Altogether, the maintenance of acceptable concentrations of impurity metal ions in the copper electrolyte presents significant challenges.

Different technologies used to control impurities from copper refinery electrolyte at the ACR, including conventional methods (bleed streams and controlled arsenic additions). In 2009, ACR adopted molecular recognition technology (MRT) as its main process to control bismuth in the tankhouse electrolyte. In 2011, ACR installed an acid purification unit (APU) to help recover Ni and Cu and recycle arsenic and sulfuric acid back to the tankhouse. These processes are discussed in the following sections.

17.6.1 Conventional Methods for Electrolyte Purification

Bismuth is currently the main impurity that needs to be controlled at ACR, due to the high and fluctuating input of bismuth in the copper anodes coming from the Hayden smelter and because bismuth is very close to Cu in the electromotive series. This makes it possible for bismuth to plate on the copper cathode, adversely affecting the quality of the copper cathode. The constant fluctuations of Bi and other metal impurities in the anodes coming from the smelter make control of impurity levels in the electrolyte very difficult but necessary in order to meet ASTM standards or other customer specifications. Over the period

January 2010 through 2015, the amount of Bi processed in Cu anodes at ACR ranged from ~1360 to ~5900 kg per month, with most values being ~2700 to ~3600 kg per month.

Several methods for electrolyte purification have been studied. The method chosen by some other companies in North and South America is to adjust the mole fraction ratio of As/(Sb+Bi) (mol/mol) during the smelting process in order to convert Sb and Bi into the insoluble species antimonous and bismuth arsenate ($SbAsO_4$ and $BiAsO_4$) during the electro-refining process [4]. These precipitates report at the bottom of the tankhouse cell mixed with the anode slimes. Figure 3 shows the behavior of bismuth in the electrolyte at different ratios of As/(Sb+Bi) in the anode; we see that at a ratio above 1, Bi will stay below desired concentration levels in the electrolyte. This method has the potential to cause safety issues if not controlled carefully including the use of respirators inside the tank house, a practice that is undesirable due to the extra heat stress this can cause in a hot region like Amarillo, TX. Disposal or processing/treatment of the residues can also attract scrutiny from environmental agencies and raise public concern, since As and many of its compounds are potent poisons [6].

The conventional practice used at ACR is to bleed certain volume of copper electrolyte from the tankhouse and treat it in liberator cells using insoluble lead anodes and copper starter sheets as cathodes in a two-step process. In the first step of this process, the copper is plated, making a product that can be used either inside the refinery to produce final product or as a mix with anode scrap to produce new anodes. In the second step, some of the As, Sb and Bi impurities are co-deposited with copper and the rest is precipitated as a sludge; both these streams are shipped back to the Hayden smelter. A disadvantage of this method is that the smelter returns these impurities in the form of new anodes to ACR with attendant labor, time and shipping costs associated with the reprocessing of these residues.

ACR has adopted molecular recognition technology (MRT) as its main process to control bismuth in the tankhouse electrolyte. Recently ACR has installed an acid purification unit

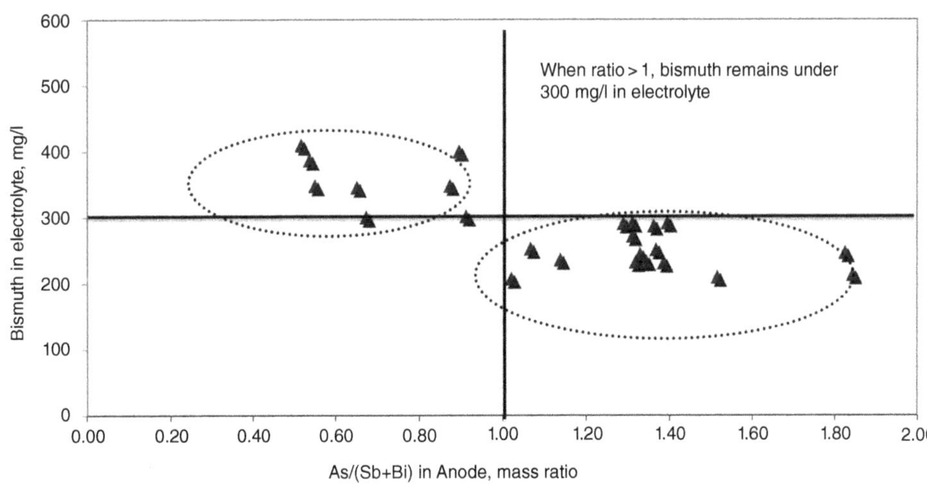

Figure 3 *Bismuth Behavior in Electrolyte vs the As/(Sb+Bi) Mass Ratio*

(APU) to help recover Ni and Cu and recycle arsenic and sulfuric acid back to the tankhouse. These processes are described in the following sections.

17.6.2 Molecular Recognition Technology (MRT)

In MRT processing, specially designed ligands such as organic chelating agents or macrocycles, chemically bonded by a tether to solid supports such as silica gel or polymers, are used to achieve highly selective metal separations. The assemblage of ligand, tether, and solid support is termed SuperLig®. MRT processes [7, 8] use "lock and key" or "host guest" chemistry as a basis for their high selectivities. Such MRT products are packed into fixed-bed columns that, in commercial operation, are built in skid-mounted modular form and are fully automated for continuous operation. Feed solution is passed through the column and the target metal is removed from the solution in a highly selective solid-phase extraction (SPE) process.

MRT products can be designed to bind selectively with ions based on multiple parameters such as size, coordination chemistry, and geometry. In contrast, separation methods based on conventional technologies generally recognize differences between ions based only on a single parameter (e.g., charge, solubility, size). MRT products can bind ions selectively even when the ions are present in highly acidic or basic solution and/or in solutions containing high concentrations of competing ions. High species selectivity, high binding factors, and rapid reaction kinetics result in efficient separations. Since small volumes of eluate are used, highly concentrated eluent solutions are produced. The MRT process operates on a system cycle basis consisting of a loading phase; pre-elution wash phase; elution phase, in which the target metal is eluted from the column with a small amount of eluent, forming a solution concentrated in the product; and a post-elution wash phase. The product is recovered and the cycle begins again with the loading phase. Cycle time depends on feed concentration, flow rate, and column size [9].

Notable features of MRT are that new reagents are not added during the procedure, the target metal can be recovered in pure form following separation for resale or environmentally safe disposal, separations are made on-line at or near ambient temperatures, impurity concentration levels from g/L to < mg/L can be treated, and a wide variety of impurity metals can be processed [10].

17.6.2.1 Use of MRT for Bismuth Removal at ACR

In order to achieve the desired copper cathode quality, ACR in 2009 developed a new electrolyte purification plant (EPP) based on the MRT approach by modifying an ion exchange process that had been implemented in 1993 (but was no longer in operation) mainly to remove Sb. The EPP is not located inside the tankhouse but adjacent to it, eliminating any potential contamination of the tankhouse electrolyte. Initially, one column of the new Bi removal system packed with SuperLig® 83 was used. A year later, a single column was found to be incapable of handling the amount of Bi coming into the refinery; the capacity was then increased to 3,100 kg per month by adding a second column.

The MRT process for bismuth removal has the following steps (Figure 4): (1) electrolyte feed from the tankhouse is loaded on the columns and Bi is selectively removed by the SuperLig® 83 material, (2) the loaded column is washed with two bed volumes (BV) of hot water to remove residual electrolyte, (3) 3.1 BV of eluent solution (9 M H_2SO_4) is fed to

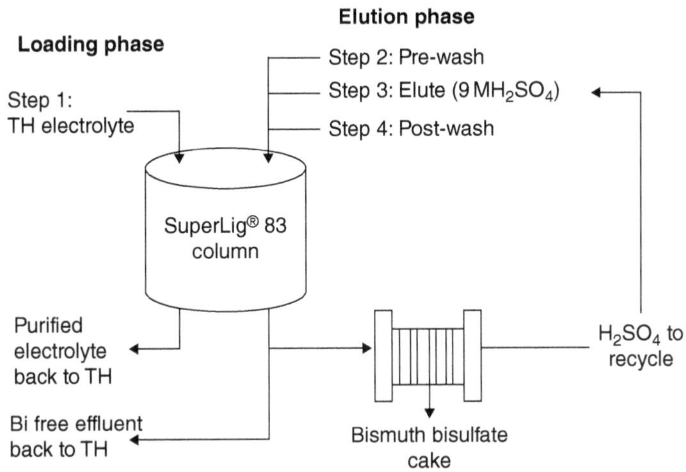

Figure 4 *Schematic Flow Diagram of the EEP for Bismuth Removal using MRT [11] (TH = Tankhouse)*

the column removing Bi as a pure concentrated product, and (4) the column is washed with 2BV of hot water to ready it for step (1). In step (2), a hot water wash is used. Hot water does not remove all the Cu, which may be desired in some Bi-removal operations. If removal of all Cu is required, a dilute H_2SO_4 wash may be substituted for the hot water wash in step (2).

In the actual operation part of the eluent, ~2.2 BV, is sent to the Bi precipitation tank and the remainder is recycled back into the eluent tank. After precipitation is triggered by dropping the temperature below 42°C, the product is sent to a filter press to recover bismuth sulfate. On a periodic basis, a 6 M HCl wash is performed to remove Pb and Sb that have built up on the SuperLig® 83 product. Pb and Sb cannot be removed effectively with 9 M H_2SO_4.

The performance of SuperLig83® is tracked by performing loading and elution profiles through the life of each of the different batches used in either of the two columns with capacity ~475 kg (±25 kg) and average BV 973 liters. The maximum loading capacity obtained in the process was 0.18 mmole Bi/kg of SuperLig® 83 [11]. Figure 5 shows that the loading capacity of SuperLig® 83 is reached after ~30 BV when the Bi concentration is >550 mg/L and ~45 BV when it is <450 mg/L.

The elution profile allows tracking the release of Bi from SuperLig 83® during the numerous stripping cycles. This profile is created by obtaining 50 mL samples drawn 24 times throughout the numerous stages of the stripping cycle and analyzing them analyzed for Bi and H_2SO_4 concentrations. A typical elution profile is shown in Figure 6. The Bi elution is sharp and rapid, and was accomplished with ~2.2 BV of 9M H_2SO_4 at 63°C. Strip 1 is the hot wash performed to push the electrolyte out of the column; no acid or Bi is found there. Strips 2-4 are related to the eluent, 9M H_2SO_4, passing through the column. Strips 2 and 3 go to the precipitation tank and strip 4 recirculates back to the eluent tank to be reused.

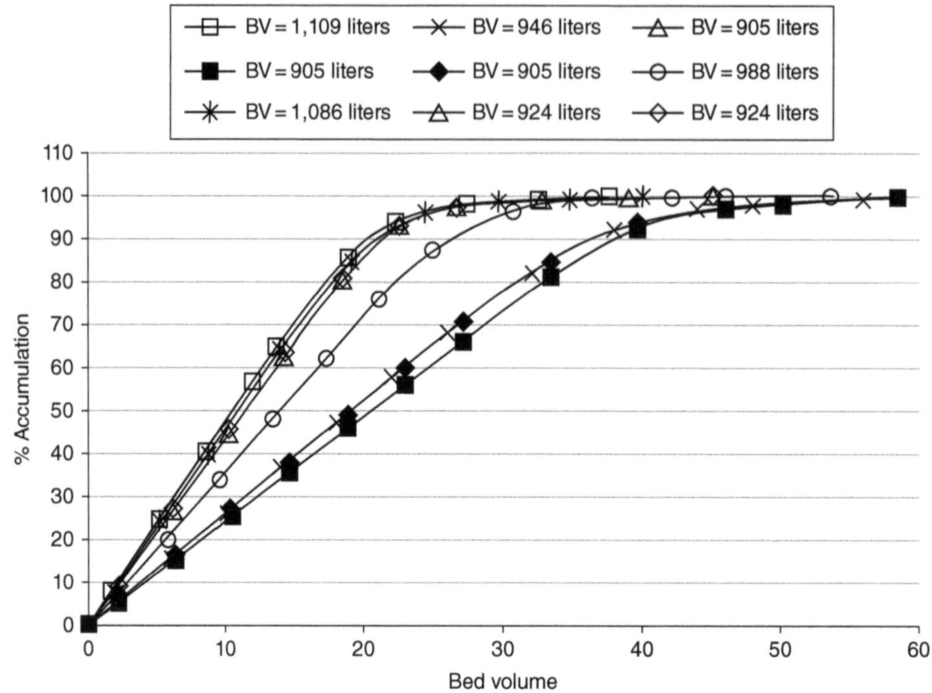

Figure 5 *Bi Loading Capacity (% accumulation) vs BV; Concentrations of Bismuth are <450 mg/L in Lower Three Curves and >550 mg/L in Remaining Curves [11]*

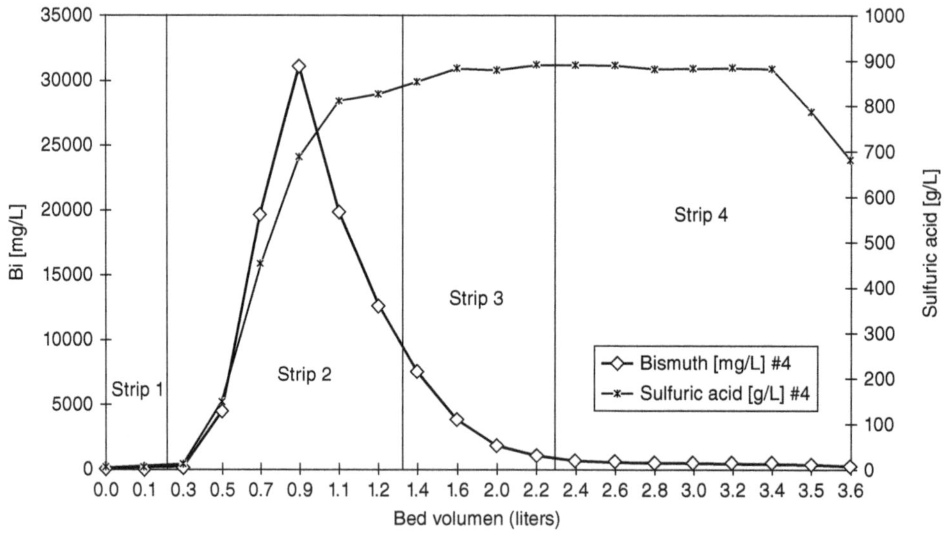

Figure 6 *Typical Bismuth Elution Profile using 9M H_2SO_4 [11]*

The MRT process has proven to be an important means for control of Bi impurity. Main benefits of the process include: (1) The SuperLig® 83 product has high selectivity for Bi and has a high capacity. (2) There is no pickup or loss of Cu during the loading phase. (3) Due to rapid reaction kinetics, very high feed flow rates are possible for Bi loading. (4) Following elution, the bismuth product in the eluate solution is very concentrated and readily precipitates to form a high-purity bismuth bisulfate precipitate that can be collected by filtration. (5) The bismuth bisulfate has a significant market value that makes a sizeable contribution to reduction of operating costs. (6) Recovery of Bi for sale avoids its wasteful discharge into the environment. (7) The number of BV used to remove Bi from SuperLig® 83 is low making it possible to have a smaller purification plant than would be the case when using other methods for Bi removal.

17.7 Recovery of Metals by Precipitation from Acidic Streams

Heavy metals content in aqueous solutions can be effectively reduced to low concentration levels by precipitating the metals into an insoluble form. Metals are typically precipitated as:

1. Hydroxides: Most common precipitation method, typically follows the reaction

$$M^{+n} + nOH^{-} \Leftrightarrow M(OH)_n \tag{3}$$

   Caustic materials are added to raise the pH forming insoluble precipitates; sodium hydroxide and calcium hydroxide are the most commonly used alkalis.
2. Sulfides or sometime sulfates: Metal sulfides are typically very insoluble. Therefore metals can be precipitated by adding sulfide ions (S^{-2}). It has the advantage that metal sulfides have much lower solubility than the corresponding metals hydroxides, so that by this method lower residual metal concentrations can be achieved. Sulfide precipitation is always conducted under alkaline conditions to promote sulfide ion formation.
3. Carbonates: Sodium carbonate is added to the acidic solution to raise the pH and allow the presence of carbonate ions (CO_3^{-2}) promoting the formation and co-precipitation of insoluble carbonates. High pH values also promote the precipitation of the metals as hydroxides, but values above 10 promote the formation of metal hydroxyl complexes that can increase metal solubility and reduce the precipitation effectiveness.
4. Cementation: Ions are reduced to zero valence state at a solid metallic interface by contacting with another metal causing it to precipitate in a heated environment. The metal with the more positive (oxidation) potential passes into solution and displaces a metal with a less positive potential, provided the solutions are dilute and the metal ion is uncomplexed. This process is often used to refine leach solutions.

In the following section we discuss the process to precipitate a nickel carbonate product when the de-acidified electrolyte is neutralized sequentially with sodium hydroxide and sodium carbonate. Later we discuss the process to recover tellurium from leach liquor by cementation using elemental copper.

17.7.1 Nickel Carbonate Recovery

The use of liberators cells is a common practice at copper refineries to control metal impurities in the electrolyte and to recover the remaining copper coming from different acidic streams (anodic Cu slimes leach liquor, Cu rod rinse). During this electrowinning process a de-copperized electrolyte is produced and if the concentration of some metals (mainly tellurium) is high enough that is not suitable for recycling back to the tankhouse, it must be discarded as a waste. Some refineries can send it to the heap leach if a mine site is close enough and the cost for transportation is low enough, but doing this means that valuable metals (nickel, tellurium, and selenium) are never recovered.

In order to recover these metals together with sulfuric acid, ACR installed an APU; after the electrowinning process, the de-copperized electrolyte solution is sent to this APU. In the APU sulfuric acid and arsenic are absorbed into the resin and then desorbed using water that is then returned to the tankhouse to be reused as acid makeup and to increase arsenic concentration in the electrolyte. A by-product stream is generated that is high in nickel and other valuable metals that can be further processed and the metals recovered. This process eliminates the production of black acid and the use of evaporators with their associated high energy consumption; in addition, more than 50% of the arsenic is returned with the sulfuric acid, making it feasible to modify the operating conditions on the final liberation stage to increase the concentration of Cu and As left in solution and minimize the As leaving the plant as a sludge to be recycled back to the smelter. Table 1 shows the average separation efficiencies achieved at ACR since the installation of the APU through 2015.

There are two steps in the APU process, the upstroke and the downstroke. During the upstroke, filtered de-copperized electrolyte is pumped into the bottom of the resin bed. Acid is adsorbed by the resin particles and the remaining de-acidified metallic salt solution, called byproduct stream, is collected from the top of the bed and sent to a holding tank. Next, during the downstroke, water is pumped into the top of the bed, desorbing the purified acid plus the arsenic from the resin so that a purified acid product is collected from the bottom of the bed and sent to a holding tank for further use (Figure 7).

17.7.1.1 Nickel Carbonate Precipitation

The method selected for the process was a carbonate precipitation preceded by a neutralization using sodium hydroxide (NaOH). Sodium carbonate (Na_sCO_3) was chosen for the metal precipitation since carbonate precipitates settle and can be dried more

Table 1 *Separation Efficiencies of APU at Amarillo Copper Refinery*

Streams	H_2SO_4 (g/L)	Ni (g/L)	Cu (g/L)	As (g/L)	Fe (g/L)	Bi (mg/L)
APU feed: de-copperized electrolyte	200.64	6.08	3.96	2.28	0.81	205.99
APU product	151.04	1.98	1.37	1.46	0.26	101.94
APU by-product	49.6	4.10	2.59	0.82	0.55	104.05
Recovery (%)	75.28	32.52	34.66	34.66	32.35	49.49

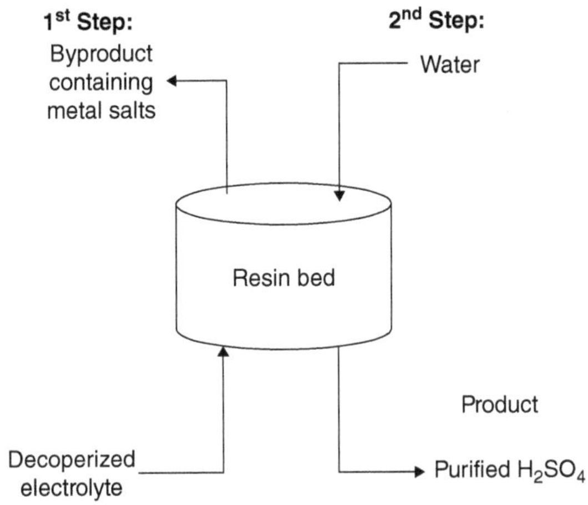

Figure 7 *Schematic Flow Diagram of the Acid Purification Unit Operating Cycle [12]*

easily than the corresponding hydroxide precipitates. For the matrix of metals species present in the de-copperized electrolyte, the following equations can be used to describe the precipitation reactions:

$$Na_2CO_3 + H_2SO_4 \rightarrow Na_2SO_4 + H_2O + CO_2 \tag{4a}$$

$$2NaOH + H_2SO_4 \rightarrow Na_2SO_4 + 2H_2O \tag{4b}$$

$$Na_2CO_3 + CuSO_4 \rightarrow Na_2SO_4 + CuCO_3 \tag{5}$$

$$3Na_2CO_3 + 2H_3AsO_4 \rightarrow 3H_2O + 2Na_3AsO_4 + 3CO_2 \tag{6}$$

$$3Na_2CO_3 + Bi_2\left(SO_4\right)_3 \rightarrow 3Na_2SO_4 + Bi_2O_3 + 3CO_2 \tag{7}$$

$$Na_2CO_3 + FeSO_4 \rightarrow Na_2SO_4 + FeCO_3 \tag{8}$$

$$Na_2CO_3 + NiSO_4 \rightarrow Na_2SO_4 + NiCO_3 \tag{9}$$

Following these reactions and using the chemical analysis in Table 1, it was determined that by using sodium carbonate around 80% of the total consumption will happen during the acid neutralization showed in equation (4a), and only 13% will be used to precipitate nickel carbonate described in equation (9), forming another reason to use a liquid reagent (NaOH) instead of a powder that is more difficult to handle.

Other possible reactions can occur when sodium hydroxide is added:

$$2NaOH + CuSO_4 \rightarrow Na_2SO_4 + Cu(OH)_2 \tag{10}$$

$$3NaOH + H_3(AsO_4) \rightarrow Na_3(AsO_4) + 3H_2O \tag{11}$$

$$6NaOH + Bi_2(SO_4)_3 \rightarrow 3Na_2SO_4 + 2Bi(OH)_3 \tag{12}$$

$$2NaOH + FeSO_4 \rightarrow Na_2SO_4 + Fe(OH)_2 \tag{13}$$

$$2NaOH + NiSO_4 \rightarrow Na_2SO_4 + Ni(OH)_2 \tag{14}$$

To prevent these reactions from occurring, sodium hydroxide addition should stop before all the acid is neutralized around pH 1.7.

The metals coming out of the APU process are in a stable, dissolved aqueous form and are unable to form solids. The goal of the carbonate precipitation is to adjust the pH of the solution so that the metals can form insoluble precipitates. Once the metals precipitate they can be separated by filtration. The process occurs in two purification steps: in the first purification tank sodium hydroxide is added until neutralization (pH around 2), then sodium carbonate is added until pH of 6.5–7 is reached (see Figures 8 and 9). The first filtration of the APU byproduct should occur in this pH range, thus removing several metals from the solution such as iron, copper, and arsenic. Once filtration of the solution at this pH has occurred, the solution is sent to a second purification tank where more sodium carbonate is added to the solution until a final pH of 9 is reached. This is where nickel has formed an insoluble precipitate, also known as nickel carbonate, that can then be sold. This process is

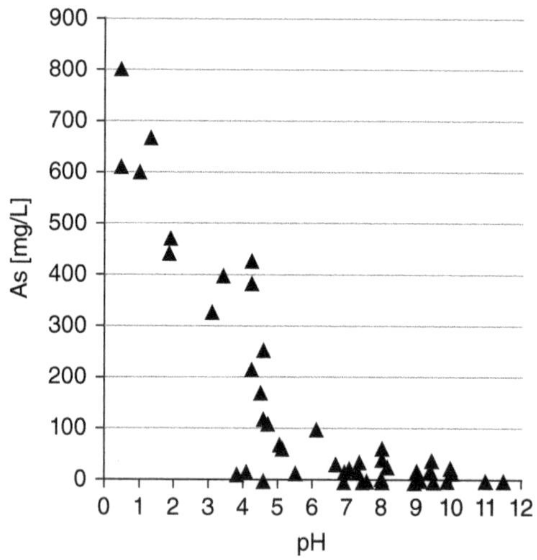

Figure 8 *Precipitation Curves for Arsenic Using Sodium Carbonate [12]*

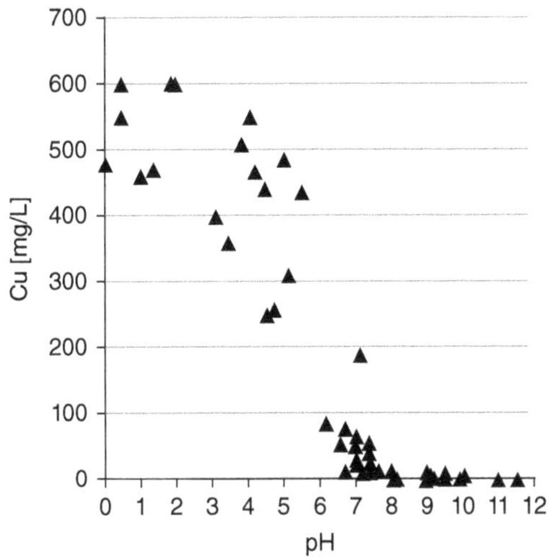

Figure 9 *Precipitation Curves for Copper Using Sodium Carbonate [12]*

exemplified in Figure 10 when the precipitation curves for the main elements (Ni, Cu, Fe, As, Bi) and the pH for both separations are shown.

In the pH range 6.5–7, most of the Cu, Fe, and As had formed an insoluble precipitate that can be filtered and sent back to Hayden for smelting operations.X-ray results showed that it can contain more than 30% As and 20% Cu, on dry basis. X-ray analysis showed that a near-pure nickel carbonate with less than 1% metal impurities can be obtained, Figure 11 shows a photograph of the nickel carbonate product.

The process can be modeled using the reactions in Eqs. 4a–9 with high accuracy, making it possible to predict the reagent consumption within 95% accuracy when only sodium carbonate is added [11] and within 80% when a combination of NaOH and Na_sCO_3 is used.

17.7.2 Tellurium Recovery

In August 2008, Asarco completed the installation of a fully automated high-pressure slimes leaching and de-tellurization processes in the precious metals department. Before its installation, the ACR used any of the three atmospheric pressure leach tanks available either for leaching slimes or to de-tellurize leach liquor before sending it back to the tankhouse. This process had no automatic control, making it difficult to achieve constant-quality, low-cost and time-efficient slimes processing. The modified process comprised one active autoclave that enhanced and optimized copper extraction; reduced cycle time, lowered reagent consumption, eliminated tying up precious metals in the process and lessened transportation costs by reducing furnace slag production. Together with the autoclave, two tellurium reactors were also installed to reduce tellurium recirculation within the department by making possible the production of copper telluride, which can be processed inside the plant for tellurium and copper recovery. During the leaching of anodic copper slimes, tellurium dissolves into the leach liquor together with copper and must be removed before the solution is sent to liberator cells for copper recovery.

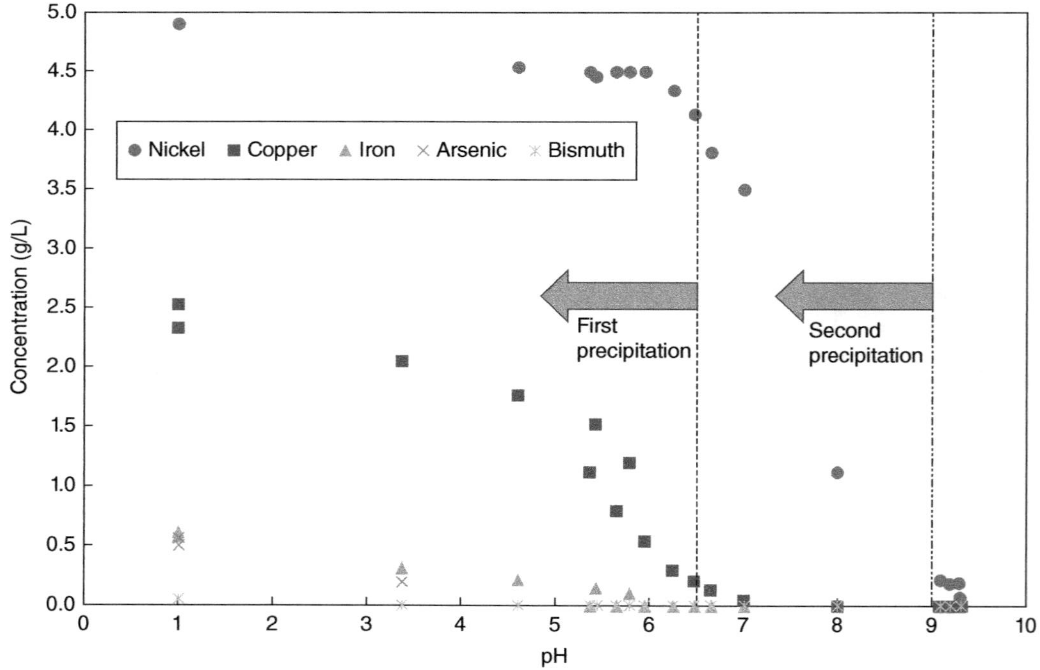

Figure 10 *Precipitation Curves Using Sodium Carbonate after Acid Neutralization with Sodium Hydroxide*

Figure 11 Nickel Carbonate Precipitate

17.7.2.1 Atmospheric Oxidizing Slimes Leaching Process

The raw slimes were pumped to the leaching tanks to start treatment. The first step, tellurium removal, consisted of the addition of tellurium-bearing solution coming from a previous batch, water, steam and the agitation of the mixture in the tank. Sometimes copper chips had to be added in order to increase the formation of copper telluride (Cu_2Te), which will precipitate into the slimes. This operation tried to deplete the tellurium concentration of the solution using raw slimes. After the tellurium concentration in the solution had been lowered to less than 10 mg/L, the treatment would be stopped and the remaining solids allowed to settle out. The now tellurium-poor solution would be decanted from the solids and sent to the liberator cells department for copper removal.

The raw slimes that settled in the bottom of the leach tank were again treated. The next process is a batch operation that seeks to reduce the copper content in these slimes. Water, sulfuric acid, steam, oxygen and agitation must be added to dissolve the copper into the solution by forming copper sulfate. After de-copperization is achieved, the aqueous solution, a mix of sulfates (mainly copper sulfate, tellurium sulfate), must be decanted and sent it to a holding tank for further tellurium removal. The remaining solid/liquid solution was fed into a washing filter press that returns the leach liquor to the same holding tank with the previous tellurium-bearing solution and dries the slimes, as illustrated in Figure 12. The de-copperized pressed slimes were processed through several pyrometallurgical and hydrometallurgical steps to produce pure Ag and Au ingots.

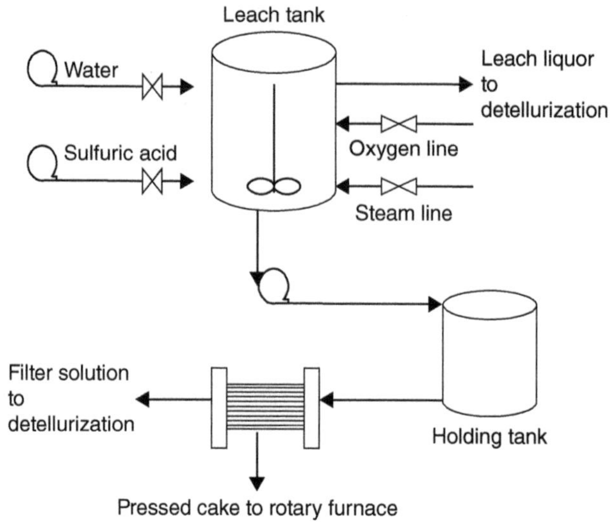

Figure 12 *Schematic Drawing of the Conventional Atmospheric Oxidizing Leaching Process [13]. Reproduced from [13] with permission of John Wiley & Sons*

17.7.2.2 Pressurized Leaching Process of Anodic Copper Slimes

As a solution to the low copper extraction from the raw slimes and to end the high recirculation of precious metals via copper and sharp slag, ACR modified the atmospheric leaching process by installing an autoclave that can reach higher temperatures and pressures and achieve a higher copper recovery. The modifications involved installation of an autoclave with a total capacity of 19 m^3, installation of a stainless steel flash tank with a total capacity of 38 m^3, redesign of an atmospheric leach tanks for use as a holding tank, installation of a mist eliminator to remove acid mist from process emissions, making possible the venting of this mist into the atmosphere, installation of a polishing filter to remove remaining particles before the de-tellurization process takes place, and the altering of two of the three remaining atmospheric tanks in order to use them as tellurium removal reactors.

The pressure leaching process is done as follows: slimes are pumped from the settling tanks to a holding tank. Once the tank is filled to a certain level and a homogeneous mix is achieved, a sample is taken so as to analyze the sulfuric acid and copper concentration of the aqueous solution and the copper content of the solids. Those parameters are the analytical inputs needed to be indicated into the computer program in order to determine the quantity of slurry, sulfuric acid and water to be fed automatically into the autoclave. Once the desired quantities of slimes slurry, water and sulfuric acid are achieved inside the autoclave, steam is added to increase the temperature to 65°C. Oxygen is injected until the exothermic reaction reaches a temperature plateau, indicating most of the readily leachable copper has been dissolved. The system will bleed gas to maintain a pressure less than 6.89 bars until the de-copperization is achieved. After this, the slurry mix is sent to a flash tank for rapid cooling by taking advantage of depressurization and the cooling effect of evaporation to prevent damage to the filter press. The filtrate solution, high in copper and tellurium, is then pumped to a holding tank for tellurium removal. The pressed slimes are processed in a rotary furnace to produce a high silver/selenium matte used to feed the

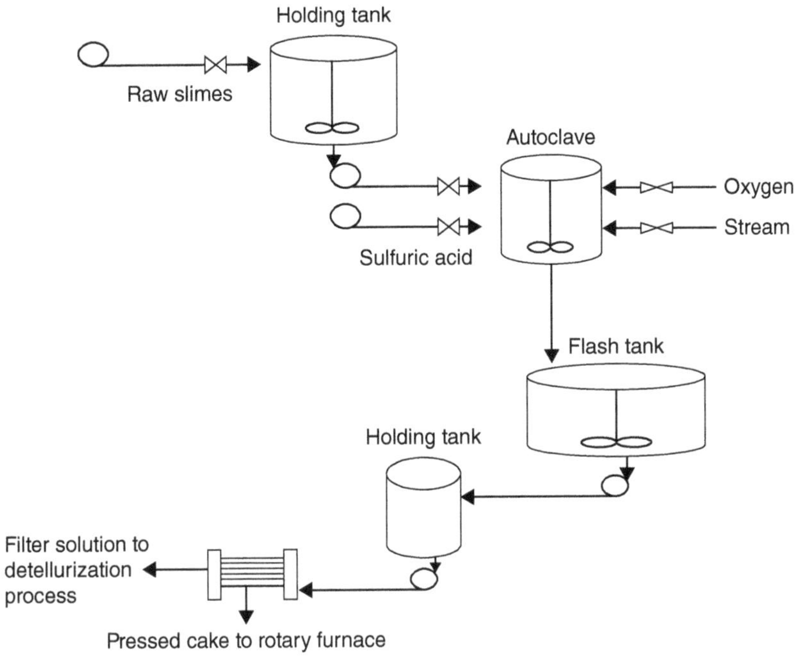

Figure 13 *Schematic Drawing of the Pressure Slimes Leaching Process [13]. Reproduced from [13] with permission of John Wiley & Sons*

bottom blown oxygen cupel to produce silver-rich doré metal. This doré is sent to Moebius cells for further refining in which silver crystals and gold sand are produced; the Au sand is refined using Wohlwill cells. Figure 13 illustrates the process described before.

17.7.2.3 Detellurization Process

During the slimes leaching process, tellurium is dissolved together with the copper, and it must be removed before the solution can be sent to the liberator cells for copper recovery. In the original process, the tellurium dissolved in solution was precipitated using raw slimes. This caused significant recirculation because the tellurium was being dissolved back into solution after the slimes were leached and only some tellurium was moving forward into the pressed slimes after decopperization.

Two ways were proposed to achieve tellurium removal. The first was to mix the solution with copper chips to facilitate the cementation and formation of copper telluride; the second was to use silver crystals instead of copper inside of the tellurium reactors, forming Ag_2Te that can be charged directly to a bottom-blown oxygen cupel (BBOC) in order to remove the tellurium by forming a soda slag compound and recover the silver as doré ingots.

The use of copper chips was selected because the use of silver crystals was a concern due to security and economic issues. The process occurs as follows: leach liquor from the slimes filter press is pumped into a tellurium reactor and copper chips are dropped into the top of the tank. Sulfuric acid is added to a desired concentration of 150 g/L. The solution is agitated and steam heated to a minimum temperature of 90°C (190° F). The de-tellurizing reactor is provided with variable-speed agitation that is designed to be at full speed (84 rpm) during the formation of copper telluride. When the solution reaches the target tellurium concentration of <10 mg/L, the agitation is slowed (20 rpm) to permit any remaining large copper particles to settle out while maintaining the finer copper telluride particles in suspension. The slurry is then pumped through a second filter press. The filtrate is sent to the liberator cells plant for copper removal and the remaining copper telluride cake is sent to the selenium tellurium (SeTe) department for further tellurium recovery. Figure 14 illustrates the possible processes to remove Te from the leach liquor solution at ACR.

During the de-tellurization process, tellurium is precipitated as copper telluride (Cu_2Te) via the following reaction [14]:

$$H_2TeO_3 + 4Cu + 2H_2SO_4 \rightarrow Cu_2Te + 2CuSO_4 + 3H_2O \quad (15)$$

Depending on the temperature at which the de-copperization takes place, the tellurium may be present in solution as either tetravalent or hexavalent tellurium. Hexavalent tellurium will increase the copper requirement according to the reaction.

$$H_2TeO_4 + 5Cu + 3H_2SO_4 \rightarrow Cu_2Te + 3CuSO_4 + 4H_2O \quad (16)$$

The detellurization operation should be done in closed vessels to avoid the ingress of air during the reaction. Air oxidation of the copper will readily occur and substantially increase copper consumption.

The 19 m^3 baffled tanks are made with 316L stainless steel, as are all the wetted parts. A Chemineer Model 4HTN-20 is used for the agitator drive for the 1.425 m diameter SC-3 single impeller. The tanks are lined with acid-resistant brick to act as a barrier against abrasion as well as corrosion. The brick also acts as an insulator.

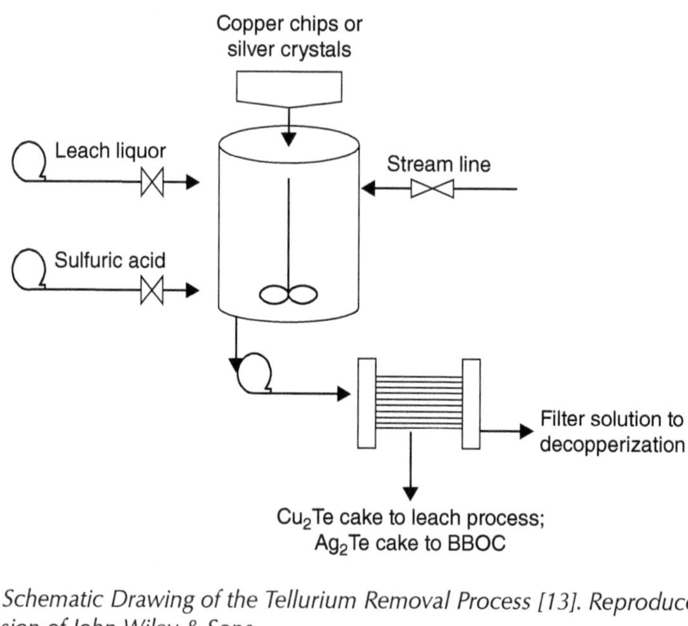

Figure 14 Schematic Drawing of the Tellurium Removal Process [13]. Reproduced from [13] with permission of John Wiley & Sons

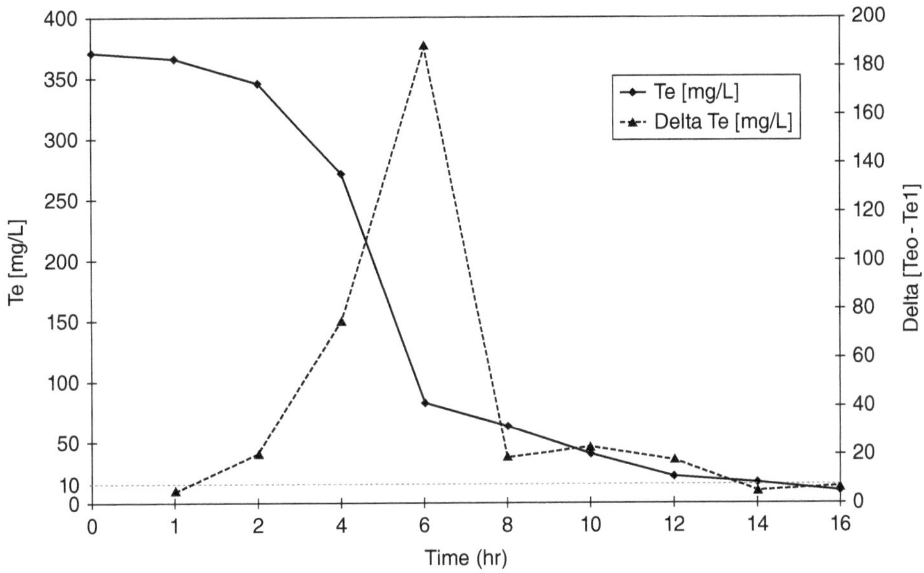

Figure 15 *Tellurium Concentration vs Time when Excess Copper is added at the Beginning of the Batch [13]. Reproduced from [13] with permission of John Wiley & Sons*

Figure 15 shows the behavior of the tellurium when an excess of copper is added at the beginning of the batch, which leads to a faster turnaround compared to when copper is added every two hours, as seen in Figure 16.

Reaction time is a function of the tellurium and copper coming into contact. With less copper being circulated at any one time, the chances of contact are lessened. Reaching the 10 mg/L level of tellurium in the treated solution can prove difficult at times; tests using other reagents to bring the tellurium content down more quickly at the tail end of the process are currently underway.

17.8 Other Sustainable Development Efforts at ACR

Asarco has implemented several corporate programs aimed at employee awareness and motivation, e.g. Vitality®, "better place to work", ASPIRE award and safety award.

Besides ACR's emphasis on green chemistry initiatives, it has implemented other programs aimed at making its operations more sustainable. One such program with the objective of waste reduction is the replacement of wood pallets with plastic pallets, implemented in 2013. This was a successful waste-reduction project since 75% of the wood pallets had to be replaced after one use while the plastic pallets can be reused 3–7 times. Moreover, the damaged plastic pallets can be returned to the manufacture for recycle, resulting in zero waste generation. Another benefit was that since the shape and dimensions of plastic pallets are more consistent, it was possible to automate the process using a robotic arm to supply pallets to the shipping dock.

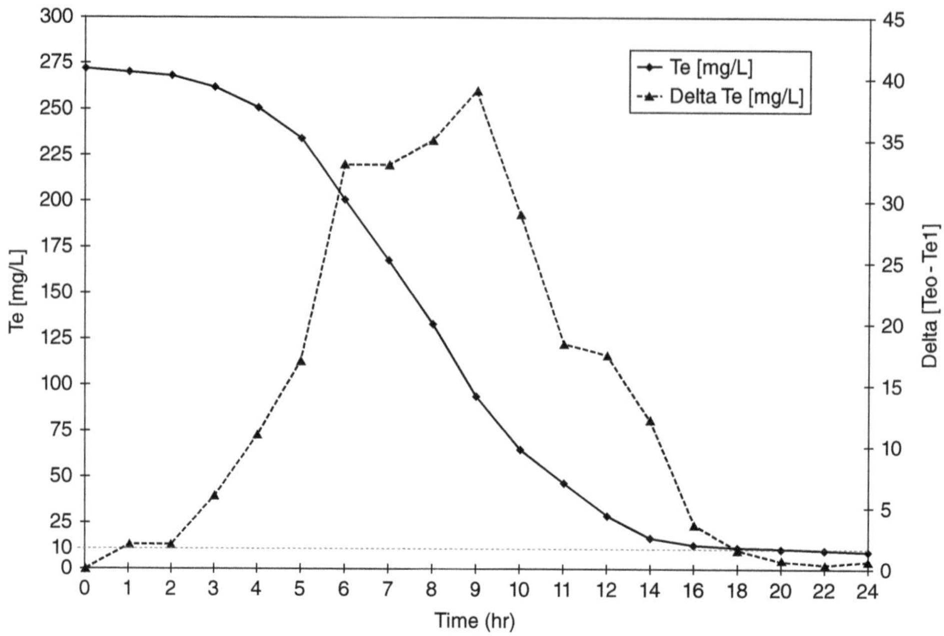

Figure 16 *Tellurium Concentration vs Time when Copper is added Every Two Hours [13]. Reproduced from [13] with permission of John Wiley & Sons*

Another advantage is that plastic material used for the pallets is not typically recyclable. For most plastics to be recycled, they have to be heated to a molten state and then injected into a mold that is then quickly cooled to solidify the plastic. When plastic is heated to a molten state, two things can happen. First, the plastic can release toxic gases in the workplace, requiring engineering controls or the use of personal protective equipment. Second, every time plastic is heated up it loses some of its properties, reducing its useful life. However, the pallet supplier is able to make the plastic bond together by compressing it at high pressure, with no additional heat other than that generated by friction in the process. Since the pallets are not heated up as in traditional plastic processes, they can be used multiple times without losing their properties.

Another waste-reduction project is the reuse of "rod rinse" solution. "Rod rinse" solution is produced when sulfuric acid at low concentration is used to remove oxides from the copper surface during rod production. After several cycles this solutions is discarded as waste. In 2011, several options for using/reusing this solution were studied. The option selected was to return controlled amounts to the tankhouse or to utilize it, instead of using fresh water, for washing cathodes.

ACR has also implemented new techniques in engineering projects, e.g., 3D modeling together with simulation software for tension analysis. 3D modeling makes the design more comprehensible and understandable, providing a vivid idea or concept about the project to be performed. Also, extensive use of computer-aided design (CAD) software is commonplace in modeling parts, equipment such as tanks, or complete systems, as shown in Figure 17.

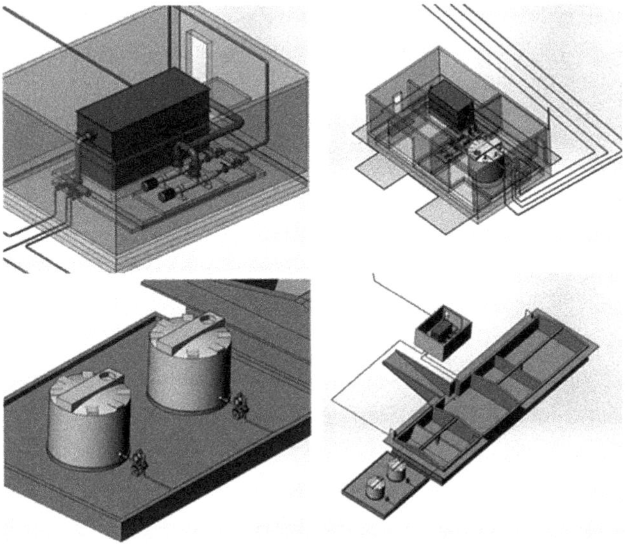

Figure 17 3D Models of Tanks, Storage and Containment Areas of an Oil Separator System

17.8.1 Implementation of Quality Management System

Over the last decade the primary copper industry has embraced the development and implementation of total quality management (TQM) programs and the utility of certification of quality management systems (QMS). The Amarillo Copper Refinery has obtained ISO 9001:2008 certification for most of its products, with the last recertification audit occurring in August, 2015. In addition to environmental aspects, Asarco and ACR had developed an EMS based on ISO 14001:1998 standard, but has not recertified to the more recent standards. The ACR programs, especially the management systems, provide a framework that can be expanded to an integrated environmental management system (EMS), that can include economic and social aspects as well as environmental aspects. This will allow developing a facility and company-wide sustainable development strategy and metrics to evaluate the success of its strategy.

17.9 Conclusions

Using risk assessment, the European Commission has determined that the use of copper products is, in general, safe for the environment and human health. Current trends show the importance of recycling for future world copper production and consumption. An imperative for companies is to address public awareness of the challenges facing society in its decision-making while striving for a sustainable future.

This chapter illustrates ACR's varied approaches to ensure sustainability as a viable copper producer, with initiatives related to the recovery of metal co-products (Bi, Ni, Te) and sulfuric acid using the cleanest and most economical processes currently available, while ensuring that cathode quality is maintained.

Electrolyte purification is an ongoing process that changes based on the impurities coming from the ore, but it is necessary to maintain quality of product. The MRT process has proven to be an important strategy to control Bi in the tankhouse electrolytes. Research will continue in trying to minimize reaction time for the current de-tellurization process by the addition of other reagents (e.g. thiourea) at the tail end of the process.

References

1. European Commission, (2008), Communication from the Commission to the European Parliament, the Council, the European Economic and Social Committee and the Committee of the Regions on the Sustainable Consumption and Production and Sustainable Industrial Policy Action Plan {SEC(2008) 2110}, {SEC(2008)2111}.
2. European Copper Institute (2009), *The Copper Voluntary Risk Assessment* [Online], Available at http://www.copperalliance.eu/docs/default-source/resources/copper_leaflet_vra_april_2009 (for_web).pdf [4/01/2009].
3. Braun, T.B., Rawling, J. R., Richards, K. J. (1976), "Factors Affecting the Quality of Electrorefined Cathode Copper," in *International Symposium on Copper Extraction & Refining*, Las Vegas, Nevada.
4. Hoffmann, J. E. (2004), "The Electrolyte Purification of Copper Refinery Electrolyte," *Journal of the Minerals, Metals & Materials Society*, 7, 30–33.
5. Wang, S. (2004), "Impurity Control and Removal in Copper Tankhouse Operations," *Journal of Metals*, 56, 2004, 34–37.
6. Pan, J., Chon, H.S., Cave, M.R., Gates, C.J., Plant, J.A. (2012), "Toxic Trace Elements," in *Pollutants, Human Health and the Environment: a Risk Based Approach*, J. A. Plant, N. Voulvoulis and K. V. Ragnarsdottir, Eds., Wiley-Blackwell, Oxford, UK, 87–114.
7. Izatt, N. E, Bruening, R. L., Krakowiak K. E., Izatt, S. R., (2000), "Contributions of Professor Reed M. Izatt to Molecular Recognition Technology: from Laboratory to Commercial Application," *Industrial & Engineering Chemistry Research*, 39, 3405–3411.
8. Izatt, S. R., Bruening, R. L., Izatt, N.E. (2012), "Status of Metal Separation and Recovery in the Mining Industry," *Journal of Metals*, Vol. 64, 1279–1284.
9. Izatt, R. M., Izatt, S. R., Izatt, N. E., Krakowiak, K. E., Bruening, R. L., Navarro, L. (2015), "Industrial Applications of Molecular Recognition Technology to Separations of Platinum Group Metals and Selective Removal of Metal Impurities from Process Streams," *Green Chemistry*, 17, 2236–2245.
10. Izatt, R. M., Izatt, S. R., Bruening, R. L., Izatt, N. E., Moyer, B. A. (2014), "Challenges to Achievement of Metal Sustainability in Our High-Tech Society," *Chemical Society Reviews*, 43, 2451–2475.
11. Navarro, L., Morris, T., Read, W., Izatt, N. E., Bruening, R. L., Izatt, S. R, (2013), "Control of bismuth in Tankhouse electrolytes at the ASARCO refinery," Proceedings of *Copper 2013, Electrowinning/Electrorefining*, Vol 5, 475–484.
12. Bidwell, J., Navarro, L., Read, W., Morris, T., (2013), "Treatment of Decopperized Electrolyte by Carbonate Precipitation," Proceedings of *Copper 2013, Electrowinning/Electrorefining*, 5, 497–507.
13. Navarro, L., Morris, T. and Read, W. (2012), "Detellurization Process of Copper Anodic Slimes Leach Liquor by Cementation of Tellurium Using Elemental Copper," in *T.T. Chen Honorary Symposium on Hydrometallurgy, Electrometallurgy and Materials Characterization*, 141–149.
14. Hoffmann, J. E. (1989), "Recovering Selenium and Tellurium from Copper Refinery Slimes," *Journal of the Minerals, Metals & Materials Society*, 41, 33–38.

15. Amer, A. M. (2002), "Processing of Copper Anode-Slimes for Extraction of Metal Values," *Physicochemical Problems of Mineral Processing*, 36, 123–134.
16. Kagawa, T., Masuda, N., Fuwa, A., 2007, "A View on Sustainable Development of Non-Ferrous Metals Production Technology," *Proceedings of the Sixth International Copper-Cobre Conference, Vol VI: Sustainable Development, HS&E and Recycling*, p 27.
17. Shibasaki, T., Abe, K., Takeuchi, H. (1992), "Recovery of Tellurium from Decopperizing Leach Solution of Copper Refinery Slimes by a Fixed Bed Reactor," *Hydrometallurgy*, 29, 399–412.
18. Zheng, Y, Xiao, F, Wang, Y, Xu, W., Jian, H, Ma, Y, (2008), "Industrial Experiment of Copper Electrolyte Purification by Copper Arsenite", *Journal of Central South University of Technology*, 15, 204–208.

18

Sustainability Initiatives at ASARCO LLC: A Mining Company Perspective

Dr. Krishna Parameswaran

tfgMM Strategic Consulting, Scottsdale, AZ, 85262

18.1 Introduction

The metal mining industry, which comprises mineral exploration, mine development, mining (extraction), mineral processing (also referred to as beneficiation, milling or concentration) and metallurgical processing (smelting and refining), is capital-, water- and energy-intensive and has the potential to cause significant environmental impact if the potential environmental impacts are not properly managed. The classic Brundtland Commission definition of sustainable development (SD) is "development that meets the needs of the present without compromising the ability of future generations to meet their own needs" [1].

ASARCO LLC (Asarco) was organized in 1899 as the **A**merican **S**melting **a**nd **R**efining **Co**mpany, a consolidation of lead-silver smelting companies, and evolved over the years to an integrated primary non-ferrous metals producer of copper, lead, zinc and associated co-products. Grupo México, S.A. de C.V acquired Asarco in 1999. Asarco today is a vertically integrated producer of primary refined copper, as copper cathode, rod and cake and associated co-products such as gold, silver, molybdenum selenium and tellurium. Asarco operates three open-pit mines (Ray, Mission and Silver Bell); four mills (Ray, Hayden and Mission-North and South); two leach, solvent extraction electrowinning plants (Ray and Silver Bell) and an INCO flash smelter (Hayden) in Arizona and a refinery in Amarillo, Texas.

Over the years, Asarco has had robust exploration programs and has been in the forefront of technological development: development and implementation of particulate air pollution control equipment such as, baghouses and electrostatic precipitators, and sulfur dioxide

Metal Sustainability: Global Challenges, Consequences, and Prospects, First Edition. Edited by Reed M. Izatt.

control technologies- sulfuric acid and dimethylamine (DMA) sulfur dioxide plants; process technology development such as development of the Asarco shaft furnace that is used all over the world to melt refined copper, and the Reatrol reagent control technology for increasing current efficiencies in copper refineries. In addition, Asarco has been involved in researching innovative reclamation methods to promote revegetation in the arid southwest U.S.A. and is looking for novel post-mining use of disturbed lands that it owns.

Resource conservation in the copper smelting step is accomplished by the re-smelting of copper-rich flue dusts or filter cake as part of the smelter charge or, if they are enriched with respect to lead or bismuth, they can be shipped to off-site facilities for metal recovery. Acidic scrubber solutions from Asarco's Hayden smelter are reused for leaching copper ores, for their acid and water content. Copper-bearing secondary materials from electrorefining of copper can be resmelted at the smelter. These include: "sharp" slag, precious-metals-bearing furnace bricks, copper telluride leach residue, foul cathodes, and copper residues. Sustainability initiatives at Asarco's Amarillo Copper Refinery are discussed by Navarro et al. in Chapter 17 of this book.

This chapter illustrates, using examples from Asarco's past and current operations, how hard rock mining (from exploration and mine development, through mineral processing operations, reclamation, closure, post-closure and post-mining) can be sustainable.

18.2 What is Sustainable Mining?

There is no simple answer to the question: what is sustainable mining? The concept of SD or sustainability is much more difficult to understand and explain when it is applied to a non-renewable resource. In this section various attributes of SD, as applied to mining, are discussed in order to come up with a working definition of sustainable mining practices. Two characteristics distinguish mining from other non-extractive industries:

> Perhaps, the two most significant elements that set mining apart from other industries are the finite nature of a mine's life and the high risk, and occasionally high reward in exploration and mining. Every ore body has a finite life and eventually the operation must close. In some areas, self-sustaining new industries will grow up around a mine, and will be able to support the community, as the mine itself declines and closes. For many mines, however, the remote site will leave the community totally isolated. Our corporate ethics are about caring for the community as the mine develops and as it finally closes. Again, this is where industry must set its own agenda. There will be no painless solution, but forward planning can make an enormous difference. [2]

Parameswaran [3] discussed the challenges and opportunities for the world mining industry at the advent of the 21st century, including the role of mining in SD. He noted that, at first glance, mining and SD appear to be incompatible. A commentator in an international financial journal a long time ago asserted that industrial development was fully compatible with sustainability "except, of course, for mining." Others have considered the concept of sustainability as applied to mining an oxymoron. The main reason for such views is that we are dealing with a finite, non-renewable resource and certain historic practices that are no longer acceptable have had serious negative impacts on the environment.

Mining can contribute to sustainable development of a particular community in a host of ways: by creating financial capital that is in turn usefully employed in a number of diverse enterprises; building human capital by increasing the levels of education and technical skills of local employees and others who provide goods and services to the operations; and generating social capital when a share of the wealth generated funds finds its way back into the community for the development of socially useful infrastructure and institutions [4].

Moreover, mining is important for the industrial and economic development of nations and, because everything in modern commerce is made from something that is grown or mined, it is difficult to conceive of development that is not to some extent based on mining and mineral development. Stigson provides a practical definition of what it takes for a company to be sustainable:

> Sustainable development is built on three pillars: economic growth, ecological balance and social progress. It is a continually evolving process in which society gradually moves to a position where the three elements of sustainable development act in a sustainable manner. For a company to become sustainable, it must be able to strike the right balance between these three pillars. This balance is referred to as "the triple bottom-line," and increasingly it is about strategic positioning and staying ahead of the competition.[5]

Attributes of mining that contribute to the sustainability of mining [6] [7] include the following:

- As demand for metals grows so do proven and probable resources. This is due, in part, to the discovery of new resources through continued exploration, even as mining proceeds, and, in part, to technological advances that allow the economic processing of lower grade ores and even wastes. Examples of technological advances in primary non-ferrous metals production include: froth flotation for the concentration of non-ferrous metal sulfide ores; Solvent Extraction-Electrowinning (SX-EW) for leaching of copper ores; and cyanide heap leaching for gold ores. Each of these advances has added considerably to the inventory of economically mineable reserves. Exploration by mining companies concurrent with production and research, development and adoption of new technologies furthers the sustainability of mining.
- Another attribute contributing to sustainability of metal mining is the recyclability of metals. Many metals once produced are capable of being recycled, smelted time and again to their original elemental form and refined to demanding specifications. Furthermore, recycling can conserve energy since recycling processes are much less energy intensive than the primary metals production processes. Recycling also contributes to the conservation of natural resources by providing an above-ground virtual mine.

Since a mine will at some point in time run out of ore and cease operations, incorporation of sustainability considerations becomes even more important because:

> Sustainability encompasses many more values than the continuing availability of the resource being developed. Indeed, it is the very fact that mineral development will end someday that makes integration of those other sustainability considerations into the mining process highly appropriate [8].

Mining is sustainable when it is conducted in a manner that balances economic, environmental and social considerations, i.e., by paying attention to the "triple bottom-line." It should be noted that these are not three separate bottom lines but a single integrated one. Another analogy that has been used is that of a three-legged stool, which needs to be balanced for it to serve its purpose. Sustainable mining practices are those that promote this balance. These practices begin with exploration, continue with mine development, mineral and metallurgical facility design, through mine operations and until reclamation and mineral and metallurgical facility closure is completed and post-mining land uses are implemented. The emphasis throughout a mining project, which could last several decades, is one of communication with the local community and all other stakeholders, employing technologies for efficient production and minimizing environmental impacts throughout the production chain.

18.3 Exploration

Over the years, Asarco has had robust exploration programs, some of which have involved exploration in sensitive environments. In this section we discuss the planning and permitting aspects of underground mine development for the Troy mine; exploration activities for the Rock Creek project, both in northwestern Montana in the U.S.; and the Camp Caiman gold exploration project in the rain forest in French Guiana, South America, respectively.

18.3.1 Montana, USA

The development of the Troy mine and exploration for the Rock Creek deposit represent contrasting case studies of developing mining operations in environmentally sensitive areas of northwestern Montana. The Troy project was proposed as an underground room and pillar copper-silver mine in an environmentally scenic area with no previous mining. However, the area had been logged and therefore impacted by human activity. The Rock Creek deposit, on the other hand, was largely adjacent to and beneath the Cabinet Mountains Wilderness Area and within the habitat for the grizzly bear, a threatened species under the Endangered Species Act (ESA). Despite the geological similarity of the orebodies and the fact that they would be mined and milled by nearly identical methods, the planning and permitting process and timelines were significantly different due to increased regulatory requirements and changes in public perceptions, largely media-promoted, about mining. These case studies in planning and permitting are discussed by Balla [9] in the proceedings of an international conference, *Minerals, Metals and the Environment*.

Both the Troy mine and the Rock Creek projects were acquired by Revett Mining Company, Inc. (Revett) from Asarco in 1999. In March 2015, Hecla Mining Company (Hecla) announced a merger agreement pursuant to which Hecla would acquire Revett. The transaction closed on June 15, 2015. As part of the acquisition, Hecla plans on permanently closing and reclaiming the Troy mine.

18.3.1.1 Troy Mine

Mt. Vernon, which contains the Troy orebody, has an elevation of 5,800 feet and is characterized by steep slopes (45–70%) with a dense coniferous tree cover. The climate is cold snow forest type with average annual precipitation over 100 inches per year and average snowfall in the mine area around 700 inches. In 1974, Asarco announced the acquisition of the Troy deposit and the commencement of environmental baseline studies and feasibility studies for developing the Troy deposit.

In 1976, Asarco submitted an application for an operating permit to develop the Troy mine. The mine was located on federally owned National Forest Land and private land acquired by Asarco. The proposal involved mine development, crushing and milling facilities, a seven-mile access road and a 450-acre tailings impoundment area. The proposed mine development consisted of driving a 15' by 30' service and conveyor adit to reach the lowest elevation of the orebody. A primary crusher would be installed at that location. The gently dipping orebody, on average 60 feet thick, would be mined by a room-and-pillar method, the ore would be hauled underground to the primary crusher, where after crushing it would be conveyed to the mill for further crushing and grinding and bulk floated to produce a concentrate containing 40% copper and 100 ounces per ton of silver for shipment to a smelter. The mine life was estimated at 20 years and would employ 300 people. Production over the mine life was estimated at 48 million short tons of ore, with a yield of 355,000 tons of copper and 74 million troy ounces of silver.

Permitting the mine development involved a number of federal and state laws and regulations, principally the National Environmental Policy Act (NEPA) and the Montana Environmental Policy Act. Three federal agencies — U.S. Forest Service, U.S. Bureau of Mines and U.S. Bureau of Land Management — and four state agencies — Department of State Lands, Department of Highways, Department of Health and Environmental Sciences and Department of Natural Resources and Conservation — were involved in the preparation of a joint environmental impact statement (EIS), covering nine federal laws:

- The Antiquities Act
- The Multiple Use Mining Act
- The National Historic Preservation Act
- The National Environmental Policy Act
- The Clean Air Act
- The Federal Water Pollution Control Act
- The Solid Waste Disposal Act
- The Endangered Species Act
- The Federal Land Policy and Management Act

Two other state agencies — Department of Game and Fish and Department of Community Affairs — participated in a consultative role.

In November 1976, the Montana Department of State Lands and the Kootenai National Forest (KNF) began the joint analysis of the environmental impact of the mine development proposal and published a draft EIS in June 1978. In this process alternatives to the mine proposal were considered, including: (1) alternative mining, milling and tailings disposal methods; (2) alternative location for mine-related facilities, including an underground milling operation; (3) alternative energy sources and (4) varying the mine life, extending the period of operations beyond the proposed life of the mine.

The EIS was finalized in October 1978 and eight major permits were issued in November 1978. It took almost six years from the commencement of the initial baseline studies until the issuance of the final permits, at a total cost of about $1 million. The project cost was around $88 million. The mine began full production of 8,500 tons per day since 1982, with an average grade of 0.74% copper and 1.54 troy ounce per ton silver, and operated until early 1993. Revett resumed production in late 2004, producing an additional 8.4 million ounces of silver and 69.5 million lbs. of copper. The company suspended operations in December 2012 due to unstable ground conditions and later placed the facility under care and maintenance.

The development of the Troy Mine was one of the first major mining operations successfully started since the environmental movement began in 1960s. It was a learning experience for Asarco as well as the federal, state and local regulatory agencies.

18.3.1.2 Rock Creek

Asarco acquired the Rock Creek project at the same time it acquired the Troy deposit. The property is 16 miles to the southeast of the Troy orebody and adjacent to and beneath the Cabinet Mountain Wilderness Area, a congressionally designated wilderness area under the Wilderness Act of 1964. The Cabinet Mountains are a range of mountains in northwestern Montana that vary in elevation from 2,880 feet to 8,738 feet in elevation. The Cabinet Mountains Wilderness Area consists of the higher portions of the mountain range, and contains the majority of the Rock Creek copper-silver deposit. The Cabinet Mountains are home to a wide assortment of wildlife, from deer, elk and moose to black bears and grizzly bears, designated a threatened species under the ESA.

From 1964 until December 31, 1983, mineral exploration and development was permitted in the wilderness areas. After that date, wilderness areas were closed to mineral exploration and development, subject to valid existing rights. In 1985, the U.S. Forest Service (USFS or Forest Service) concluded that Asarco had established valid existing rights to the Rock Creek deposit. Mineral patents were issued in 1989 affirming ownership of the mineral rights.

In 1979, a helicopter-supported drilling program was initiated inside the wilderness area, with encouraging results. At the same time, the environmental community began legal action to stop the drilling, asserting that exploration program would adversely affect the grizzly bear. Federal courts decided that the exploration program would not adversely impact the grizzly bear.

Exploration in a sensitive environment required institution of best management practices [10]. The exploration program was conducted mostly in alpine conditions, at or above the tree line, where trees do not grow. Due to the rugged terrain, it was not feasible to build roads at the higher elevations to access locations for diamond drill sites. The location of the drill sites was predicated upon the necessity to make a discovery of valuable minerals on each mining claim, which are 600 feet by 1,500 feet in dimensions, or approximately 20 acres in size. In order to minimize the number of drill sites, multiple drill holes were drilled from a common corner of adjacent mining claims. In order to ensure that the drill holes did not significantly deviate from their planned intersection with the mineral deposit, which on average is 1,000 feet beneath the surface, all of the drill holes were surveyed down the drill hole every 50 feet. A total of 124 diamond drill holes, totaling over 125,000 feet, were drilled, with over 100 drill holes utilizing helicopter support.

It was decided that, to be in compliance with The Wilderness Act of 1964, no permanent evidence of the mineral exploration program would remain behind once the exploration program was concluded. Accordingly, in order to minimize the impact of the drill rigs upon the vegetation and the exposed rocks, Astro Turf, a synthetic turf, was laid down first on the ground. Wooden pallets were then placed upon the Astro Turf. The diamond drill rigs, water containers, and all other necessary equipment were placed upon the pallets. All this equipment was flown to the drill sites by helicopter. In addition, all personnel working on the exploration program (drillers, geologists, geophysicists, surveyors) were transported in and out of the wilderness by helicopter. Time constraints were imposed by various federal agencies upon the exploration program so that the exploration program would not interfere with the habits of the endangered species, such as the grizzly bear.

Upon completion of the drilling program and removal of the equipment by helicopter, the drill holes were covered over with nearby rocks, thus covering and blending in with the immediate ecology of the area.

From 1980 to 1983, the helicopter-supported drilling continued inside of the wilderness and outlined the Rock Creek ore body. The Rock Creek orebody was 12,000 feet long, 600 feet wide and on average 27 feet thick. Ore reserves were estimated at 143.8 million tons, with an average grade of 0.68% copper and 1.65 troy ounces per ton of silver, with approximately 90% of the orebody within the wilderness.

In 1984, Asarco submitted a mine plan of operations to the USFS and the Montana Department of State Lands. The mine plan for the Rock Creek project was very similar to that for the Troy mine. Two parallel 25-feet-by-20-feet adits would be driven underground 9000 feet at a 13% grade from outside of the wilderness to reach the lowest elevation of the Rock Creek orebody, which is inside the wilderness. A room for a primary crusher would be excavated and a primary crusher installed. Room and pillar mining would commence at a rate of 10,000 tons per day or 3.6 million tons per year. Annual metal production would be 21,600 tons of copper and 5.3 million ounces of silver. Mill and all surface facilities would be built outside of the wilderness. The total surface disturbance outside of the wilderness would consist of mine and mill facilities, tailings and power corridor and the tailings pond. Total disturbed acreage would be about 640 acres. The disturbance within the wilderness would be minimal and would consist of an air intake ventilation adit driven from underground to the surface. The opening would have a gate to prevent access of wildlife or unauthorized personnel.

Asarco proposed an evaluation adit of the Rock Creek orebody in July 1992. The Montana Department of State Lands (DSL) and Kootenai National Forest (KNF) reviewed the plan and Asarco made revisions in October 1992 and February 1993 in response to agency review and comments on the Plan. The Evaluation Adit Plan was analyzed in an Environmental Impact Statement (EIS) by DSL's successor, the Montana Department of Environmental Quality (MDEQ). The Evaluation Adit Plan was approved as part of Alternative V in the Record of Decision (ROD) by the USFS and MDEQ in 2001.

Since 2001, the approved Evaluation Adit Plan has been modified by Revett to address final design changes, provide an alternate wastewater disposal method, lessen environmental impacts, and satisfy stipulations in the ROD. The permitting process culminated with the USFS and MDEQ issuing a revised ROD approving of development of the Rock Creek Project in June 2003.

A companion document to the FEIS and RoD is the Biological Opinion, which is completed by the U.S. Fish and Wildlife Service (USFWS) after thorough review of all potential impacts to wildlife. The Biological Opinion (BO) addresses potential impacts and appropriate mitigation measures for wildlife species listed as either threatened or endangered under the Endangered Species Act (ESA). The BO, issued in 2003 and revised several times with the last amendment in 2007, resulted in a comprehensive mitigation plan for grizzly bears and bull trout.

Litigation brought by environmental groups challenged both the USFS ROD and the Biological Opinion. Eventually the lawsuits were merged into one lawsuit due to overlapping issues and the Federal District Court issued an Order and Opinion in May 2010. The Order vacated the FEIS and ROD and remanded them back to the USFS for further consideration due to issues related to the NEPA process, with the Court recommending completion of a Supplemental EIS (SEIS) to correct the NEPA issues. A Draft Supplemental Environmental Impact Statement (DSEIS) was released for public comment on February 19, 2016 for a 45-day comment period, which was extended to April 19, 2016.

18.3.2 Camp Caiman Gold Exploration Project, French Guiana, South America

In this section sustainable practices and mitigation measures employed by Asarco at the Camp Caiman gold exploration project to minimize environmental impacts are discussed. These measures included the use of a rubber-tired, self-propelled combination diamond, reverse circulation and auger drill that could drive on forest litter and aggressive drainage control measures. An Environmental Advisory Committee provided the company advice on the environmental impacts of mineral exploration in a tropical rainforest environment [11–13].

The Camp Caiman gold project is located on the south side of the Kaw Mountains, 50 kilometers southeast of Cayenne, French Guiana in South America and about 500 kilometers north of the mouth of the Amazon River. The climate of the region is typical of the Amazon rainforest with an average annual rainfall of 4 meters (157 inches) and an average maximum daily temperature of 32°C. The project site was accessible by paved road to within 3 km of the exploration camp and then by forest road, requiring a four-wheel-drive vehicle.

In 1995, Asarco acquired two exploration permits covering 50 square kilometers over a gold anomaly from the local French administration following a public tender. The Camp Caiman area had been originally identified by the Bureau de Recherches Géologiques et Minières (BRGM) as an area where gold values in soil were very high. In addition, there were several abandoned campsites and small ditches that indicated prior small-scale alluvial mining. Three additional permits covering 27 km^2 were acquired in 1997 and an application for an additional permit was also submitted.

The five exploration permits were for areas adjacent to the Kaw Biotope zone. The project area covers wetland, which is a principal habitat for the Black Caiman alligator, and adjacent hills, where caves high on the north side of the Kaw Mountains are home to an elusive bird called the Cock-of-the-Rock. An extension of this zone and its classification as a natural reserve, reaches the southern edge of the five permits. Further, a regional park was planned for the area.

The Biotope zone is roughly coincident with the Kaw Wetland System, which is a Ramsar site, i.e., designated for inclusion in the Ramsar List of Wetlands of International Importance. These wetlands are identified for specific protective measures in the Convention on Wetlands signed in 1971 in Ramsar, Iran by 90 nations. Currently, the Convention has 169 contracting parties and lists more than 2241 wetland sites for special protection covering a surface area of 215,240,652 hectares.

Graybeal describes Asarco's initial exploration work, how the company modified its environmental practices and the formation and functioning of an independent environmental oversight group [11]. An exploration camp had been built and a moderate drilling program was underway by 1996. Some of the findings based on Asarco's initial exploration work were as follows:

- The use of bulldozers in land clearing and building of access roads resulted in removal of forest litter and soil layers, which were generally less than six inches thick. The rainwater runoff from the camp caused silting of the adjacent stream.

- In order to preserve the forest canopy, drill roads were located with care to avoid felling larger trees, which meant that the roads were shaded from the sun and, therefore, constantly muddy. This led to several reroutings of access roads from the camp to the drill sites. The rainwater runoff caused silting of small creeks at road crossings, although this was not as serious because the forest floor litter acted as an effective filter.
- The mitigation measures employed by Asarco's exploration staff included the filling and recontouring of exploration trenches after mapping. The campsite was also recontoured to divert some of the runoff to the adjacent forest floor in order to reduce the silting of nearby streams. Care was also taken in locating operations to ensure that streams draining from the camp and drill sites flowed away from the Kaw Biotope. Concrete pads were constructed for the containment of small oil spills from the servicing of machinery and equipment. A wastewater treatment plant was built and non-biodegradable trash was hauled to Cayenne.

Despite these mitigation measures, it was a constant challenge to minimize the environmental impacts of the project while controlling project expenditures, especially when the project viability remained uncertain, and at the same time being able to demonstrate that the project was advancing rapidly. In addition, by this time there was heightened public awareness on loss of tropical rainforests, loss of biodiversity and an increasing emphasis on sustainable development. Asarco was very much aware of these issues as well as the often negative perception of the mining industry presented by Non-Governmental Organizations (NGOs). Recognizing that the company lacked expertise to deal with environmental issues unique to rainforest environments, external advice was sought. An Environmental Advisory Committee (EAC or Committee) of five members was formed to advise Asarco on the environmental and social impacts associated with mineral exploration, mine planning, mining operations, mine closure and post-closure reclamation and monitoring. The intent was to make the Committee part of the planning process, not simply a review group after plans had been formulated.

The Committee members were biologists, botanists and specialists with expertise in tropical rainforest environments and social issues raised by mineral development. The Committee members operated as an independent group and would receive no compensation from Asarco for their work, other than reimbursement for travel expenses. Their objective was to have an open dialog that, while recognizing the potential environmental impacts of an open pit gold mine in the French Guiana rainforest, sought improvements in exploration practice and subsequently a more environmentally friendly mine, if exploration activities were successful.

According to Graybeal [12], the consultative process with the Committee appears to have been a success. As a result of advice from the Committee, the following actions were taken:

- The pending exploration permit application, within the proposed extension of the Biotope zone was withdrawn.
- The use of a new rubber-tired, self- propelled combination diamond, reverse circulation and auger drill that could drive on the forest litter was accelerated (see Figures 1 and 2).
- Clearing was accomplished by cutting trees instead of bulldozing; elevated wooden bridges were built (see Figure 3) and the size of drill sites were reduced in order to reduce silting in the streams.
- An intensive program for the revegetation of the campsite, bridge construction and resdesign of the principal access road was undertaken at significant cost.
- A full-time French environmental manager was hired in February 1998.

Figure 1 Tracked and tired lightweight drill in operation. Reproduced with permission from CRC Press / Balkema. (See insert for color representation of the figure.)

The Committee recommended that two studies be conducted, one for establishing the baseline for the natural environment and the other for the social environment, including public safety and health. They took an active role, advising on study design, coordination of studies and management framework. Their advice was instrumental in the final selection among several alternatives for the plant site for a pre-feasibility study.

Graybeal [12] also noted that the existence of the Committee improved Asarco's access to various NGOs who initially viewed the company and the project with great suspicion. Asarco has commented informally that the EAC approach was one of its most successful environmental initiatives.

Asarco sold the gold deposit to Ariane Gold Corporation, a Canadian company, which continued and expanded the Environmental Advisory Committee [14]. In 2003, Arianne Gold Corporation was merged into Cambior Inc., which was acquired by IAMGOLD Corporation (IAMGOLD) in November 2006.

In January 2008, IAMGOLD indicated that the French government had not approved the final permits necessary to start construction in anticipation of a moratorium on exploration and mining activities. Since 2009, the company has been in litigation challenging the improper denial of the company's mining permit and sought compensation for damages. These appeals and the compensation claims have been dismissed twice and in May 2012, the company filed an appeal to the Administrative Court of Appeals in Bordeaux and increased its compensation claims to 763 million Euros http://mininglifeonline.net/news_view_3020.html

Figure 2 *Closeup of rubber-tired, self- propelled combination diamond, reverse circulation and auger drill (Scout). Reproduced with permission from CRC Press / Balkema. (See insert for color representation of the figure.)*

Figure 3 *Elevated bridge crossing. Reproduced with permission from CRC Press / Balkema*

18.4 Innovative Reclamation Methods

This section examines the utilization of municipal sewage sludge (biosolids) and cattle in promoting revegetation in the reclamation of copper mine tailings. Over many decades, Asarco and other copper mining companies have demonstrated several innovative techniques for reclamation of copper mine tailings in the arid southwestern U.S. Bengson and Parameswaran have described the literature on the subject and discussed the efforts in utilizing biosolids and livestock to enhance the reclamation of copper mine tailings [15–17].

18.4.1 Use of Biosolids

The use of biosolids for land reclamation has been researched extensively for well over two decades. Sopper noted that as of 1993, there were around 72 sites in the United States that had been reclaimed using biosolids [18]. He also noted that many of the environmental concerns such as leaching of nitrates and heavy metal contamination were unfounded. Most of the research had been conducted at sites in the coal mining region of Pennsylvania and only limited research involved sites in the western U.S. [19]. More recently there has been growing recognition and interest in utilization of biosolids to restore metal-affected ecosystems [20]. There are, however, fewer references in the literature on the use of biosolids for the reclamation of metal mine wastes, e.g., taconite iron ore tailings in Minnesota [21], copper tailings in British Columbia, Canada [22] and copper tailings in Utah, [23].

The main objective of the use of biosolids for successful reclamation is to incorporate enough organic matter and nutrients into the tailings to produce a "growth medium" that can sustain plant growth without the need of natural "topsoil." It requires establishment of self-sustainable microbial and plant communities [24, 25]. Soil micro-organisms are essential to maintain a healthy ecosystem. Microbial communities promote plant growth through mineralization of inorganic nutrients [26], development of soil structure [27], nutrient cycling [28] and reduction in metal toxicity [29, 30]. Biosolids offer a viable and economical source of organic matter. For the municipalities, this use of the biosolids addresses an ever-increasing problem of finding a cost-effective alternative to the disposal of biosolids in landfills.

In 1996, the Arizona Department of Environmental Quality (ADEQ) adopted rules allowing the use of biosolids for reclamation. Rules promulgated by ADEQ allow the one-time use of biosolids for reclamation of up to 150 dry tons per acre.

Class A biosolids are biosolids that have been treated to be essentially pathogen free. Treatment processes include composting, heat drying, heat treatment, thermophilic aerobic digestion, beta or gamma ray irradiation and pasteurization. Since 1998, Asarco has been participating in a research project with the Pima County Regional Wastewater Reclamation Department and University of Arizona for the application of Class A biosolids on Tailing Storage Facility (TSF) No. 5 at Asarco's Mission Mine. The Class A biosolids were produced using solar beds at the Green Valley Wastewater Treatment Plant in Arizona. The mine is located in Sahuarita, Arizona (24 km.) south of Tucson, at an elevation of 900 meters within the arid Sonoran Desert environment, where average summer temperatures are 35^0 C and rainfall is minimal (less than 30 cm. annually).

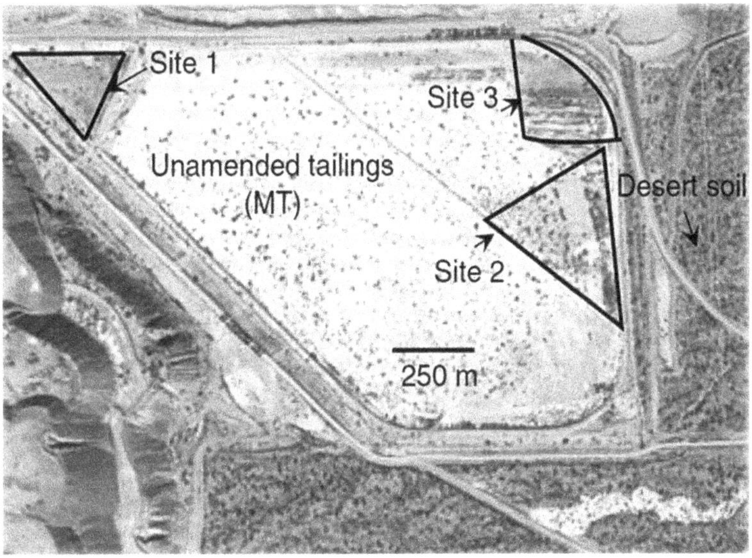

Figure 4 21 November 2009 Quickbird image of the study site showing the three biosolids plots, unamended tailings and undisturbed desert soil east of the tailings. Reproduced from [31] with permission of John Wiley & Sons

The first test with Class A biosolids was conducted in December 1998 in an isolated area of the tailing dam and continued through 2000. Earlier research and testing of mixing organics into tailings indicated that as little as 15–25% organic matter mixed into the surface (10–12-inches) could sustain vegetation. There was a period of inactivity until 2006, but in April 2006 approximately 340 tons of Class A biosolids were delivered and mixed into the surface of the tailings.

Figure 4 shows the locations of the three biosolids plots, unamended tailings and undisturbed desert soil east of the tailings. Biosolids containing 90% solids were added at Sites 1, 2 and 3 in December 1998, December 2000 and April 2006.

Site 1 is a 4.5-hectare tailing plot where Class A biosolids was surface applied at a rate of 248 tons per hectare in December 1998. Site 2 is a 12-hectare plot located close to Site 1 where biosolids were applied at a rate of 371 tons per hectare in December 2000. Another difference was that at Site 1 the biosolids were tilled into the soil whereas at Site 2 they were not tilled in but remained in a surface layer about 12–15 cms. thick. Site 3 is to the north of Site 2 and is about the size of Site 1; biosolids were applied at this site at a rate of 270 tons per hectare in April 2006.

Pepper et al. [31] have studied the effects of amendment of copper mine tailings at Asarco's Mission with Class A biosolids in relation to the bacterial populations within the amended tailings. Site 1 was sampled for microbial numbers — heterotropic plate counts and activity and microbial diversity 10 and 11 years respectively after biosolids application; Site 2 was sampled 8 and 9 years respectively after application and Site 3 was sampled for microbial diversity only 3 weeks and 3.5 years respectively after application.

The purpose of the study was to monitor the microbiological ecological structure and diversity as a function of time following biosolids land application onto copper mine tailings. The study premise was that a sustainable soil bacterial population could be established following amendment of copper mine tailings with biosolids and significant revegetation would follow. In particular, the long-term effects of amending copper mine tailings with biosolids on (1) bacterial numbers, (2) common microbial transformation or activity (nitrification, sulfur oxidation and dehydrogenase activity and (3) microbial diversity post-biosolid amendment was studied. The assays for bacterial numbers transformation were all carried out on Sites 1 and 2, whereas the diversity analysis were conducted for Sites 1, 2 and 3.

The study showed that use of Class A biosolids was effective in restoring disturbed soils. Long-term monitoring at one site for a decade and measurement of soil microbial characteristics were used to determine the success of the reclamation process. The data suggests that increased bacterial numbers, enhanced microbial activity and shifts in dominant bacterial population all occurred post-biosolid application. The amendment was able to establish a typical soil bacterial population capable of functioning as normal soil, which is a key factor for long-term ecosystem stability. All the measured changes were sustained over time, indicating that amendment of mine tailings with biosolids is an effective way of reclaiming copper mine tailings.

Pepper et al. [32] also studied whether the application of biosolids to the copper tailings at Mission would be effective in enhancing revegetation, in particular plant colonization of the tailings. The study also evaluated the sustainability of Class A biosolids amendment of copper mine tailings with respect to soil quality. Since Class A biosolids were used, biosolids associated pathogens and metals were not a concern. The data reported in this study allowed comparison of the influence of land application of biosolids ten years (Site 1) and eight years (Site 2) after the biosolids were amended into the tailings.

The results show that although the method of applications of relatively high amounts of biosolids at the two sites were different, both amended sites resulted in high eventual land cover. However, the method of application did appear to influence the amount of residual nitrates in the amended tailings. At Site 1, tilling the biosolids into the tailings aerated and loosened the soil (amended tailings), promoting denitrification and loss of soil moisture. Denitrification is expected to be low in the aerated plow layer due to lack of anaerobic sites. Consequently, nitrate accounted for the total soil nitrogen 10 years after biosolids application. By contrast, biosolids at Site 2 retained high soil moisture levels and denitrification was more rapid than at Site 1, resulting in a loss of nitrate. At Site 1 nitrate leached to a depth of 1.5 meters below the soil surface. Groundwater contamination in arid areas would be a concern if they leached through the soil into the vadose zone. In arid regions, this would happen only if excess irrigation was applied. At these sites plots were not irrigated. By comparison at Site 2, nitrate compositions were low and mainly confined to the top 0.3 meters. Based on this observation, surface application of biosolids, without incorporation, appears to the best management practice for minimizing nitrate leaching and maximizing plant growth.

Overall percentage revegetation at each site followed a similar progression, with Site 1 being two years ahead of Site 1. Both sites responded similarly to the annual amount of rainfall in a given year. At both sites similar ranges of percentage revegetation were found beyond the initial two-year period of establishment. Ranges were approximately 30–78%, with higher values found in the same years at both sites (2003, 2005 and 2008). By the end of the study, total vegetation covers at both sites were similar.

18.4.2 Use of Cattle

The use of cattle/livestock as a tool for reclamation of copper mine tailings stems from the concept of using livestock as a management tool for enhancing ecosystems. The original idea came from a wildlife biologist working in South Africa. He observed the impacts of large herds of ungulates on the vegetative communities. He noticed that as these large herds migrated across the arid plains they would severely disturb the soil surface, breaking up the heavy soil crust and opening it up for water infiltration and aeration, and trample dead litter and other organics into the soil. Also, the animals would graze old decadent plants, which then opened up the root crown to stimulate vigorous regrowth. His observations evolved into a natural resource management philosophy that became known as "Holistic Resource Management" or HRM. The concept of using livestock for tailings reclamation is a natural extension of this "holistic" theme.

The standard methods for reclamation of copper mine tailings rely almost exclusively on the use of mechanical and other high-tech methods. These include the use of heavy equipment for surface manipulation of slopes and top-soiling; as well as irrigation, fertilizers, and other soil amendments to try to induce plant growth on the tailings reclamation site. A far simpler, and perhaps more ecologically sound, approach is to use livestock. Livestock can trample organics into the tailings to build a soil-like medium that will support plant growth. Livestock also help stabilize the steep slopes with their hoof action trampling up and down the slopes. As plant communities are developed a self-sustaining ecosystem is established.

After some initial success at revegetating copper mine tailings by incorporating manure 6 inches into the surface at Asarco's Silver Bell mine in the late 1970s, Asarco put a small livestock feedlot on the top of the tailings to spread manure and incorporate organics into the tailings. A few years earlier, Ken Ludeke, another researcher, was having success on steep slopes at the Pima mine copper tailings by spraying a slurry of manure and sewage sludge and incorporating the organics by rolling a sheepsfoot roller up and down the slope. Then, in 1989, Noel Gillespie, working with a group of consultants from AZ Ranch Management at the Miami Mine, started to successfully use livestock on copper tailings to stimulate vegetative growth and to stabilize the steep slopes. Since these early beginnings many other mines have started using livestock for copper tailings reclamation.

Asarco started using "ASARCOws" (also called "flosbies" — four-legged organic soil builders by other companies) in 1994 (Figure 5). The livestock are concentrated on relatively small areas for a very short duration and fed hay. An abundance of organics is incorporated into the tailings by the hoof action of the animals. As the organics build-up in the sterile tailings, a soil-like medium is produced and enhances the reclamation of the tailings site.

The monitoring of the health of livestock did not indicate any problems. Analysis of blood samples from livestock before and after placement at the reclamation site show no problems with heavy metals. Rather, the blood levels of copper, molybdenum and zinc indicate deficiencies of these essential minerals and the diet of the cattle had to be supplemented with mineral blocks. Many healthy calves have been conceived and born at reclamation sites. Steers that have been taken to sale from reclamation sites have shown an average weight gain of 0.5 lbs./day.

Livestock have been successfully used at Asarco's Mission Mine and the Ray Complex Hayden tailings impoundment as well as the Pinto Valley Mine, Miami Mine, Sierrita Mine, Morenci Mine and Mineral Park Mine in Arizona; as well as several other mines in the western United States.

Figure 5 *Reclamation of copper tailings with livestock (ASARCOws). Reproduced with permission from CRC Press / Balkema*

18.5 Reclamation of San Xavier Tailings Storage Facilities and Waste Rock Deposition Areas

A portion of Asarco's Mission Complex is located within the boundaries of the San Xavier District (District) of the Tohono O'odham Nation. The San Xavier District is situated in the extreme southern area of Arizona within the Sonoran Desert ecosystem. Asarco and the Nation entered into a lease agreement in 1959 to allow use of the Nation's property for copper mining and related activities.

In 2008, Asarco, the Tohono O'odham Nation, the San Xavier District, the San Xavier Allottees Association and the United States of America finalized a comprehensive Mining and Reclamation Plan (MARP) and established a $33 million fund for reclamation construction and long-term maintenance activities. Reclamation activities commenced in 2009 and were completed in 2012. The MARP addressed mining and reclamation activities on disturbed District lands associated with the reclamation of tailings storage facilities (TSF) Nos. 1 through 3 that had reached their storage capacity by the time reclamation began in early 2009, the Mission North waste rock dumps and inactive open pit mines. The open-pit mine areas will be fenced and secured for safety but still accessible for future mining. These areas will be reclaimed once mining in active areas is completed.

The reclamation project scope involved capping with alluvium, rock armoring, construction of storm water controls and the revegetation of nearly 2200 acres of tailings storage facilities and waste rock deposition areas. The reclamation plan incorporated sustainable desert vegetation goals while minimizing long-term maintenance and respecting the significant cultural and religious values associated with the land. Balancing these factors into a comprehensive reclamation plan in a very arid region was challenging.

It was recognized that the severely altered ecosystems such as disturbed mine sites cannot be restored to their former climax vegetative ecosystems due to substantial difference in soils and topography, and therefore will be in a state of disequilibrium for some time. Reclamation was therefore geared to converting the land to an advanced seral stage of ecosystem development that is safe, stable, sustainable, functional and suitable for achieving the tribe's goal of creating a Sonoran native desert native plant community. This was accomplished by preserving existing ecosystems that constitute or are suitable for evolution towards the self-sustaining ecosystem, erosion controls, restricting pedestrian access to slopes and tops and establishment of financial mechanisms to allow long-term maintenance.

Asarco and the Nation formed a six-member panel (Panel or MARP Panel) to oversee and administer the implementation of the reclamation design and construction phases of the project. Although the Panel included a diverse team of technical personnel, it also relied on independent consultants and governmental organizations for input and guidance. The U.S. Bureau of Indian Affairs, the U.S. Bureau of Land Management and the University of Arizona were instrumental in developing a very specialized re-vegetation matrix of reclamation seeds, trees and cactus. A portion of the grass seed mix is being grown locally by tribal industry to generate a grass variety truly adapted to the reclamation area.

The surface of tailings dams were capped with a minimum of one foot of alluvium and seeded with a mix of native plant species. Steep slopes and benches graded to more stable 3:1 angle and covered with one foot of alluvium to reduce stormwater infiltration and one foot of rock to minimize soil erosion.

The waste rock deposition areas also comprised overburden and alluvial soil stockpiles. Even though the rock deposition areas are much less susceptible to erosion than the tailings storage facilities, the reclamation goal for these areas also includes soil capping for the establishment of vegetation on the flat areas. Surface of waste rock deposition areas were regraded to manage stormwater, capped with a minimum one foot of alluvium. Native varieties of cacti, grasses, shrubs and wildflowers adapted to the desert was used for revegetation. Figure 6 shows the cactus salvage nursery. Photographs of the top of the TSF and benches and slide slopes prior to reclamation are shown in Figure 7 and Figure 8 respectively. Figure 9 shows the success of revegetation on the tops of TSF and Figure 10 shows successful revegetation on the tailings slopes in the third year after revegetation respectively.

The success of the MARP project has generated much interest from regulatory agencies and the mining industry throughout the U.S. and in many foreign countries.

18.6 Fostering Renewable Energy Projects on Disturbed Lands

This section discusses Asarco's experience in hosting a solar farm on retired agricultural land that had been acquired for water rights purposes and was being used for grazing. Asarco is of the view that a solar project can be viable use of un-utilized buffer lands and post-mining use of disturbed mined lands. Asarco has been interested in locating renewable energy projects on disturbed lands that it owns.

Asarco, like other mining companies, has substantial land holdings, much of which is not utilized for operations. These lands include lands disturbed by mining for such purposes as tailings storage facilities and waste rock dumps; buffer lands, some of which might be leased for agriculture or grazing, and retired agricultural lands acquired for obtaining water rights. Moreover, mines represent large electricity loads and have the necessary critical infrastructure including transmission lines, roads and utilities. They also have access to heavy earthmoving equipment. The proximity to distribution lines makes for lower costs of connection to the electricity grid. The cost of developing, constructing and maintaining the solar facility is borne by the developer. Leasing the land is a source of revenue to the mine for land that is otherwise essentially idle. The revenue may defray the cost of maintenance of cap on reclaimed lands.

The Pima Mine Road Solar Generating Facility, also known as the Avalon Solar Project, is a utility-scale 35-megawatt (MW) single-axis tracking solar power photovoltaic (PV) facility. The facility is hosted by Asarco on approximately 500 acres of disturbed land in Pima County, Arizona near the Mission Complex.

The Avalon Solar project was born out of an Environmental Protection Agency (EPA) program established with the objective of siting renewable energy projects on disturbed lands, including mine sites. A unique collaboration among Asarco, Tucson Electric Power (TEP) and Clenera, LLC (Clēnera) made the project a reality. Clēnera was responsible for constructing the facility, which provided 300 construction jobs. Panasonic Eco Solutions is overseeing operations and maintenance of the project and Coronal Group is the operating manager for the project. Electricity generated is sold to Tucson Electric Power (TEP) under a 20-year power purchase agreement (PPA) and will help TEP meet Arizona's renewable energy mandate requiring that 15 percent of its power production come from renewable sources by 2025. The Pima Mine Road Solar Generating Facility should provide approximately $4 million in property-tax revenue to Pima County over the life of the project.

Figure 6 *Cactus salvage nursery. (See insert for color representation of the figure.)*

Figure 7 *TSF before reclamation. (See insert for color representation of the figure.)*

Figure 8 *Benches and side slopes before reclamation. (See insert for color representation of the figure.)*

Figure 9 *Revegetation success: top of TSF in year 3.*

Figure 10 *Revegetation success: side slopes of TSF in year 3. (See insert for color representation of the figure.)*

The Asarco site conforms to TEP's land use policies and is ideally suited to host a solar generating station. The site is located outside of the Pima County Land Conservation System and its flat topography and past agricultural use provides suitable terrain and soil conditions for constructing a solar facility. In addition, the site is located in proximity to an existing TEP substation, which increases the efficiency and minimizes the additional transmission infrastructure required to deliver power generated to the electricity grid.

Clēnera has also installed a small photovoltaic solar plant at the nearby Asarco Mineral Discovery Center (AMDC) using the same type of solar panels as the Avalon Solar Project. An exhibit explaining the benefits of solar photovoltaic energy generation and the Pima Mine Road Solar Generating Facility was placed at Asarco's Mineral Discovery Center. This adds yet another dimension to the Center's ongoing efforts to provide education of the copper mining industry.

Asarco signed the ground lease in May 2013. The groundbreaking ceremony for the project took place on May 14, 2013 and the facility commenced operation on December 23, 2014. Figure 11 shows a photograph of the Avalon Solar Project. The ribbon-cutting ceremony took place on February 27, 2015. The project will power 5,720 homes, or over 50% of the homes in Sahaurita. It will avoid discharging 57, 660 tons of carbon dioxide into the atmosphere, which is equivalent to taking 12,140 cars off the road. Phase 2 of the project, which increased the generating capacity to 56 MW, began commercial operation in February 2016.

The U.S. Environmental Protection Agency (EPA) conducted a preliminary evaluation of the potential for renewable energy development on Asarco reclaimed tailings areas in the San Xavier District on Tohono O'dham Nation lands [33]. The study found that "the ASARCO Mission Mine tailings area offers unique and potentially innovative opportunity to accommodate and potentially expand a utility-scale solar energy project at a single location. The tailings area has existing transmission capacity, roads, industrial zoning and other critical infrastructure in place for a utility-scale PV project."

Asarco's long-term vision is to evaluate the feasibility of locating PV solar facilities on mine disturbed lands such as reclaimed TSFs and waste rock dumps. A few tailings storage facilities on lands owned by Asarco will be closed in the next several years at the Mission Complex and Hayden Operations. Asarco intends to evaluate siting of a renewable energy project as a potential post-mining land use, building on studies already done by EPA on reclaimed tailings dams on tribal lands at Mission.

18.7 Utilization of Mining Wastes

An alternate approach to mining and metallurgical waste management is waste utilization, especially large volume wastes such as copper mine tailings and smelter slag (SG). One such project, conducted in collaboration with the University of Arizona Civil Engineering Department, studied alkali activation of copper mine tailings (MT) and low-calcium INCO flash furnace copper smelter slag (SG), two major types of wastes from the primary copper industry. Copper mine tailings from Asarco's Mission mine and flash furnace smelter slag were used in this study. The objective of the study was to treat copper MT and SG to produce a geopolymer that can substitute for ordinary Portland cement (OPC).

Alkali activation of blended copper MT and low-calcium SG was studied by Ahmari et al. through mechanical strength tests as well as microscopic and spectroscopic

Figure 11 *Avalon Solar Project. (See insert for color representation of the figure.)*

characterization [34]. The effect of SG content, curing temperature and curing time on the unconfined compressive strength (UCS) of the MT/SG-based geopolymer was investigated systematically. Scanning electron microscopy/energy-dispersive X-ray spectroscopy (SEM/EDX), X-ray diffraction (XRD), and Fourier transform infrared (FTIR) spectroscopy were also used to investigate the microstructure and elemental and phase composition of the MT/SG-based geopolymer.

The results showed that the addition of SG significantly improves the UCS and microstructure of the geopolymer. The improvement is mainly attributed to the high solubility of silica in the SG and the fine particle size of the SG. The optimum curing temperature (i.e., the curing temperature at the highest UCS) decreases with increasing SG content. For 0, 50 and 100% SG the optimum curing temperatures are 90°C, between 75 and 90°C and 90°C. This is attributed to the higher reactivity of SG compared to MT. In addition, the MT/SG-based geopolymer sets fast and gains a major portion of its ultimate strength within only seven days.

Based on the results, it was concluded that the MT/SG-based geopolymer is promising as a sustainable construction material for civil engineering applications. The paper also identifies the following areas for future research: (1) Study of physical properties such as water absorption and durability in aggressive environments such as freeze-thaw resistance of MT/SG-based geopolymers; (2) Environmental characterization of MT/SG-based geopolymers; a study by Ahmari and Zhang [35] indicates that MT-based geopolymers are environmentally acceptable; (3) Pilot production of MT/SG-based geopolymer products such as bricks to evaluate production process, costs and safety issues; (4) Government policy and public education related to waste recycling and sustainable development, since the MT/SG-based geopolymer products are produced from mining wastes and (5) standardization of MT/SG-based geopolymer products.

18.8 Conclusions

This chapter illustrates, using examples from Asarco's past and current operations, how hard rock mining, from exploration to reclamation and post-mining, can be sustainable. Looking to the future, Asarco will continue to pursue opportunities for sustainable development in its business.

Asarco is continuing its research initiative on the utilization of biosolids in mine tailings reclamation in collaboration with the University of Arizona and Pima County Wastewater Department. In particular, research will focus on Class A biosolids produced by the MagnaGro process.

Asarco will continue to evaluate the technical and economic feasibility of locating a renewable energy project on a reclaimed tailings storage facility. The evaluation will build on EPA studies to determine whether reclaimed tailings facilities are suitable for a solar PV project. If found suitable, the company will commence dialog with solar developers and utilities to see if a Power Purchase Agreement can be negotiated and financing can be obtained. If some of the projects that are being contemplated come to fruition, Southern Tucson, Arizona has the potential to become the solar capital of the United States.

Other research areas we consider worthwhile focusing on include: the recovery of cobalt, molybdenum, rhenium and rare earth metals from pregnant leach solutions; bioleaching of molybdenum from copper ores; use of copolymers/biopolymers for fugitive dust control and pursuing green chemistry initiatives, such as described by Navarro et al. in Chapter 17 of this book.

References

1. World Commission on Environment and Development, 1987, *Our Common Future*, Oxford University Press, Oxford, U.K.
2. West, M., 1998, Report on lecture by Rev. Michael West, *Mining Journal*, May 8, 1998.
3. Muth, R.J and Parameswaran, K, 1998, The 21st Century: Challenges and Opportunities for the Mining Industry, in *Proceedings of International Symposium on Environmental Management of Mining and Mineral Industries (EMOMAMI-'98)*, Natural Resources Development Foundation and Institute of Advance Technology & Environmental Studies, Bhubaneshwar, India, pp. 3–7, 13.
4. Ackerman, R., 1998, Is Mining Compatible with Sustainable Development? A World Bank Perspective," *International Council on Metals and the Environment (ICME) Newsletter*, 6, No. 2.
5. Stigson, B., 1998, "Sustainable Business: Performing Against the Triple Bottom Line," World Business Council for Sustainable Development (WBCSD), *International Council on Metals and the Environment (ICME) Newsletter*, 6, No. 3.
6. Rajaram, R. and Parameswaran, K., 2005, What Is Sustainable Mining? in *Sustainable Mining Practices: A Global Perspective*, Rajaram, V., Dutta. S. and Parameswaran, K. (Eds.), A.A. Balkema Publishers, Taylor & Francis Group, London, UK, 1–11.
7. Parameswaran, K., 2008, Sustainability in Mining, in *Proceedings International Symposium on Environment and Waste Management of Mining and Mineral Based Industries* (EWMMI-2008), Bhubaneswar, India, February 2008, 62–86.
8. Pring, G., 1998, Sustainable Development: Historical Perspectives and Challenges for the 21st Century, presented at United Nations Development Programme (UNDP) and United Nations Revolving Fund for Natural Resource Exploitation (UNRFNRE) Workshop for Sustainable Development of Natural Resources Towards the 21st Century, New York, October 15–16, 1998.
9. Balla, J.C., 1992, Planning and Permitting Exploration and Underground Mines In Environmentally Sensitive Areas, Troy and Rock Creek Projects, Northwestern Montana, USA, in *Minerals, Metals and the Environment*, published for the Institution of Mining and Metallurgy by Elsevier Science publishers Ltd., Essex, U.K., 535–546.
10. Balla, J.C., 2015, private communication.
11. Graybeal, F.T., 2001, Evolution of Environmental Practice During Exploration at the Camp Caiman Gold Project in French Guiana, in *Footprints in the Jungle Natural resource Industries, Infrastructure and Biodiversity Conservation*, Bowles, I.A. and Prickett, G.T. (Eds.),Oxford University Press, Inc., New York, 222–232.
12. Graybeal, F.T, 2005, Camp Caiman Project, in *Sustainable Mining Practices – A Global Perspective*, Rajaram, V., Dutta, S. and Parameswaran, K. (Eds.), A.A. Balkema Publishers, Leiden, Netherlands, 265–267.
13. Parameswaran, K., 2008, Sustainability in Mining, in *Proceedings International Symposium on Environment and Waste Management of Mining and Mineral Based Industries, (EWMMI-2008)*, Bhubaneswar, India, February 2008, 62–86.
14. Viens, F., Gaillou, J. and Jean-Elie, J., Exploration and Mining in the Rain Forest of French Guiana, Prospectors and Developers Association Annual Convention, Convention Program, 2003, 14.
15. Bengson, S.A., 1998, Innovative Reclamation Techniques, Internal Communication ASARCO LLC.
16. Bengson, S.A., 2005, Innovative Reclamation Practices in the Arid Southwestern United States, in *Sustainable Mining Practices-A Global Perspective*, Rajaram, V., Dutta, S. and Parameswaran, K. (Eds.), A.A. Balkema Publishers, Leiden, Netherlands, 268–271.
17. Parameswaran, K. and Bengson, S., 2009, Reclamation of Mine Tailings in the Arid Southwest USA, *Proceedings of International Symposium on Waste, Energy & Environmental Management of Mining and Mineral Based Industries* (WEEMMI-2009).
18. Sopper, W.E., *Municipal Sludge Use in Land Reclamation*, Lewis Publishers, Ann Arbor, 1993.
19. Thomson, T.L., and Rogers, M., 1999, *Reclamation of Acidic Copper Mine Tailings Using Municipal Biosolids*, Research report to Arizona Department of Environmental Quality.

20. Brown, S.L., Henry, C.I., Chaney, R., Compton, H. and DeVolder, P., 2003, Using Municipal Biosolids in Combination with Other Residuals to Restore Metacontaminated Mining Areas, *Plant Soil*, 248, 203–215.
21. Norland, M.R., Veith, D.L. and Dewar, S.W., Vegetation Response to Organic Soil Amendments on Coarse Taconite Tailing, in *Achieving Land Use Potential Through Reclamation*, Proceedings of the 9th Annual Meeting of the American Society for Surface Mining & Reclamation, Duluth, MN, June 14–18, 1992.
22. Wilson, S., Peddie, C., Salahub, D., and Murray, M., 1994, Environmental Effects of High Rate Biosolids Application for the Reclamation of Copper Tailings, *Proceedings of the 1994 Water Environmental Federation (WEF) Biosolids Specialty Conference*, Washington, D.C., June 19–22,1994, 11–25-11–36.
23. McNearney, R.L., 1997, Revegetation of a Mine Tailings Impoundment Using Municipal Biosolids in a Semi-arid Environment, Utah, U.S.A., *Mineral Resource Engineering*, Imperial College Press, U.K., 6, No. 3, 155–172.
24. Pond, A.P., White, S.A., Milcerak, M. and Thompson, L., 2005, Accelerated Weathering of Biosolids Copper Mine Tailings, *Journal of Environmental Quality*, 34: 1293–1301.
25. Rosario, K., Iverson, S.L., Henderson, D.A., Chatrand, S., McKeon, C., Glenn, E. P. and Maier, E.M., 2007, Bacterial Community Changes During Plant Establishment at the San Pedro River Mine Tailings Site, *Journal of Environmental Quality*, 36: 1249–1259.
26. Xu, J.G. and Johnson, R.L., 1995, Root Growth, Microbial Activity and Phosphorus Activity in Oil-Contaminated Remediated and Uncontaminated Soils Planted to Barley and Field Pea, *Plant Soil*, 173, 3–10.
27. Maier, R.M., Pepper, I.I. and Gerba, C.P., 2008, *Environmental Microbiology*, Academic Press, San Diego, 2008, 61–89.
28. Burd G.I., Dixon, D.G. and Glick, B.R., 2000, Plant Growth Bacteria that Decrease Heavy Metal Toxicity in Plants, *Canadian Journal of Microbiology*, 48, 237–245.
29. Salt, D.E., Benhamou, N., Leszcsyniecka, M., Raskin, I and Chet, I., 1999, A Possible Role for Rhizobacteria in Water Treatment by Plant Roots, *International Journal of Phytoremediation*, 1, 67–69.
30. Sandrin, T.R., Chech, A.M., and Maier, R.M., 2000, A Rahmnolipid Biosurfactant Reduces Cadmium Toxicity in Naphthalene Biodegradation, *Applied Environmental Microbiology*, 4585–4588.
31. Pepper, I.L., Zerzghi, H.G., Bengson, S.A., Iker, B.C., Banerjee, M.J., and Brooks, J.P., 2012, Bacterial Populations Within Copper Mine Tailings: Long Term Effects of Amendment with Class A Biosolids, *Journal of Applied Microbiology*, 113, 569–577.
32. Pepper, I.L., Zerzghi, H.G., Bengson, S.A. and Glenn E. P., 2013, Revegetation of Copper Mine Tailings Through Land Application of Biosolids: Long-Term Monitoring, *Arid Land Research and Management*, 27:3, 245–256, DOI:
33. U.S. EPA, 2011, Renewable Energy Development Opportunities, ASARCO Mission Mine Tailings Area, San Xavier District, Tohono O'odham Nation.
34. Ahmari, S., Parameswaran, K. and Zhang, L., 2014, Alkali Activation of Copper Mine Tailings and Low-Calcium Flash-Furnace Copper Smelter Slag, *Journal of Materials in Civil Engineering*, 10.1061/(ASCE) MT.1943-5533.0001159, 04014193.
35. Ahmari, S., Zhang, L., 2013, Durability and Leaching Behavior of Mine Tailings-Based Geopolymer Bricks, *Construction Building Materials*, 44, 743–750.

19

Recycling and Dissipation of Metals: Distribution of Elements in the Metal, Slag, and Gas Phases During Metallurgical Processing

Kenichi Nakajima[1], Osamu Takeda[2], Takahiro Miki[2], Kazuyo Matsubae[2], and Tetsuya Nagasaka[2]

[1]*Center for Material Cycles and Waste Management, National Institute for Environmental Studies, Tsukuba, Japan*

[2]*Graduate School of Engineering, Tohoku University, Sendai, Japan*

19.1 Introduction: Background, Motivation, and Objectives

This chapter presents the results of a chemical thermodynamic analysis of the distribution of elements in the smelting and remelting processes of metallic materials. These results can provide answers to the following questions: Which impurity elements can be removed in the pyrometallurgical process? To what extent can the impurity levels be controlled? Which elements can be recovered in the metallurgical process?

In their extensive studies, Verhoef et al. [1] and Reuter and Verhoef [2] suggested the importance of understanding relationships between metals during natural resource processing and introduced the concept of the "metal wheel", and Schaik and Reuter [3] pointed out limitations of recycling based on thermodynamic considerations. However, these representative studies do not provide information on the quantitative limitations of impurity removal or the recoverability of elements in recycling end-of-life (EoL) metal products.

Metal Sustainability: Global Challenges, Consequences, and Prospects, First Edition. Edited by Reed M. Izatt.

Here we present quantitative criteria for evaluating the extent to which impurities can be removed during metal recycling by considering relevant thermodynamic parameters.

One of the best tools for predicting and understanding the behavior of impurities in recycling EoL metal products is metallurgical physical chemistry. This well-established academic discipline has been used to study the production of primary metals or inorganic materials from natural minerals through pretreatment, smelting, and refining with hydro- and pyrometallurgical technologies. Using these technologies, it should be possible to predict whether impurities can be removed, or elements recovered, from remelted metal scrap.

We have previously discussed extensively the thermodynamic behavior of impurity elements in the metal, slag, and gas phases in the steel-making process (for both basic oxygen furnaces (BOF) and electric arc furnaces (EAF)) [4,5]; in copper converters [4,5]; in lead blast furnaces [4]; in the Imperial smelting process for zinc and lead [4]; in the aluminum remelting process [6,7]; in the magnesium remelting process [8]; and in the titanium remelting process [9]. In these metallurgical processes, impurities in the main metal product are mostly controlled by oxidation–reduction reactions and evaporation, except in some special cases. Taking this technological understanding into account, we quantitatively evaluated the distribution ratio of impurity elements in the slag/metal or gas/metal phases in this study. The most conventional operating conditions for each metal process were chosen as process parameters to simplify the discussion. To represent our results graphically, an "element radar chart" [6] plotting the distribution tendencies of the metal elements processed by the methods mentioned above is proposed.

In a previous report [4], the distribution ratio of impurity elements among the metal, slag, and gas phases that would occur in a smelting furnace under typical operating conditions was calculated as a first approximation. However, as described in this chapter, the removal of impurities from remelted metal by evaporation or oxidation is simultaneously affected by several thermodynamic parameters: the Gibbs free energy change of the oxidation reaction, the total pressure, the activity coefficients of the impurity in the metal and its oxide in slag, the oxygen partial pressure, and the temperature. Activity coefficients are also dependent on the composition of the metal and slag. We also theoretically evaluated the effects of process parameters (temperature, oxygen partial pressure, and slag composition) on the distribution ratio of elements in the steel-making process and a copper converter and discussed thermodynamic criteria for the impurity removal limit or for controlling impurities in these two processes [5].

19.2 Method: Chemical Thermodynamic Analysis of the Distribution of Elements in the Smelting Process

The most conventional pyrometallurgical processes, volatilization and oxidization, were selected. In each case, the distribution ratio of a given element M is thermodynamically obtained with the following equations:

$$\mathrm{M(l)=M(g)} \qquad \text{(Eq. 1)}$$

$$L^{g/m} = \frac{p_{\mathrm{M}} / p^{\circ}}{p_{\mathrm{Solv}} / p^{\circ}} = \frac{a_{\mathrm{M}} p_{\mathrm{M}}^{0}}{p_{\mathrm{Solv}}} = \frac{p_{\mathrm{M}}^{0} \gamma_{\mathrm{M}} x_{\mathrm{M}}}{p_{\mathrm{Solv}}} \qquad \text{(Eq. 2)}$$

$$M(l)+\frac{n}{2}O_2(g)=MO_n(\text{slag}) \quad \text{(Eq. 3)}$$

$$K_3=\frac{a_{MO_n}}{a_M\left(p_{O_2}/p^\circ\right)^{n/2}}=\frac{x_{MO_n}\gamma_{MO_n}}{\gamma_M x_M\left(p_{O_2}/p^\circ\right)^{n/2}} \quad \text{(Eq. 4)}$$

$$L^{s/m}=\frac{x_{MO_n}}{x_M}=\frac{K_3\gamma_M\left(p_{O_2}/p^\circ\right)^{n/2}}{\gamma_{MO_n}} \quad \text{(Eq. 5)}$$

where $p_M{}^\circ$, p_M, and p_{Solv} are the partial pressures of the pure element M (Pa), of M dissolved in the main metal product (solvent metal) (Pa), and of the solvent metal vapor (Pa) (p° is a conversion factor (101,325 Pa/atm)); a_M and γ_M are the activity and the activity coefficient of M in the pure liquid standard state; x_M is the mole fraction of M in the solvent metal; a_{MO_n} and γ_{MO_n} are the activity and the activity coefficient of the oxidation product MO_n in slag in the pure solid standard state; x_{MO_n} is the mole fraction of MO_n in slag; and p_{O_2} is the oxygen partial pressure (Pa). The symbols l, g, and slag in parentheses in Eqs. 1 and 3 denote the states of the species: M(l) is liquid solvent metal, M(g) and O_2(g) are gas-phase metal and gas-phase O_2, and MO_n(slag) is slag-phase MO_n. K_3 is the equilibrium constant of reaction (Eq. 3) and is obtained from the Gibbs free energy change of the reaction $\Delta G_3{}^\circ$ by Eq. 6:

$$K_3=\exp\left(-\frac{\Delta G_3^\circ}{RT}\right) \quad \text{(Eq. 6)}$$

where R and T denote the gas constant and temperature (K), respectively. In the oxidation reaction (Eq. 3), a monocationic oxide species is assumed as an oxidation product throughout this paper. For example, in the case of oxidation of aluminum in the solvent metal, the product is $AlO_{1.5}$ rather than the general stoichiometric form Al_2O_3. The dissociation reaction of the general stoichiometric oxide to the monocationic oxide is represented in Eq. 7:

$$M_xO_y(\text{slag})=x MO_n\ (n=y/x) \quad \text{(Eq. 7)}$$

The free energy change of this reaction is reasonably assumed to be zero. Therefore, $\Delta G_3{}^\circ$ and K_3 can be evaluated from general thermodynamic references.

The distribution ratios $L^{g/m}$ and $L^{s/m}$ quantitatively show the extent to which the vapor pressure of M is larger than that of solvent and the extent to which the concentration of oxidation product MO_n in slag is larger than that of M in the solvent metal at a given condition. Large distribution ratios favor movement of M from the solvent metal to the gas or slag phase.

Parameters from published thermodynamic data were used to calculate $L^{g/m}$ and $L^{s/m}$ in Eqs. 2 and 5 (see refs. 4–9 for details).

In the figures, the initial impurity content in the solvent metal was set at 1 mol % (x_M = 0.01), where γ_M is constant due to the dilute concentration. The activity coefficient of MO_n was assumed to be unity as a base case. The boundary between metal and gas was assumed to be the condition that satisfies both $\log(p_M/p_{Solv}) = 1$ and $\log(p_M) = -3$ [10]. Hence, the boundary between slag and gas was assumed to have a slope of –1 in the figures.

19.3 Element Distribution Tendencies in Recycling Metals

19.3.1 Copper Smelting

Figure 1 shows the distribution of 29 elements under the typical operating conditions of a copper converter, that is, 1500 K and $p_{O_2} = 1 \times 10^{-1}$ Pa [11]. Published activity coefficients γ_M were used to calculate $L^{s/m}$ and $L^{g/m}$ [12–18].

Tungsten, boron, aluminum, chromium, gallium, iron, germanium, manganese, indium, magnesium, zinc, neodymium, and strontium can be transferred to slag from molten copper by oxidation, and mercury can be removed by evaporation (Figure 1). Nickel, rhenium, tin, cadmium, and lead are partially transferred to the slag phase. In contrast, precious metals (silver, gold, platinum, palladium, and rhodium), bismuth, selenium, tellurium, and antimony remain in the molten copper phase. Most of the elements distributed within the metal phase can be removed or recovered by further processing, such as an electrorefining process (electrolysis) or a hydrometallurgical process, and some of the elements distributed within the gas phase can also be recovered by additional processing. In Figure 1, the

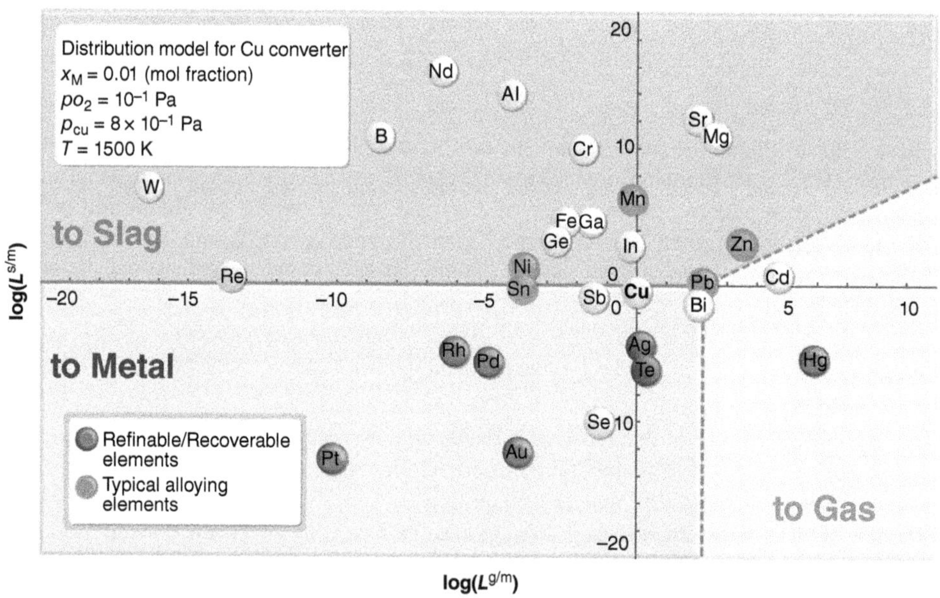

Figure 1 Distribution of elements in the metal, slag, and gas phases during copper smelting under the simulated atmosphere of the copper converter [5]. Reprinted with Permission from [5]. Copyright {2011} American Chemical Society. (See insert for color representation of the figure.)

recoverable elements are shown as white circles, and the gray circles denote typical alloying elements for industrial copper products. Most of the alloying elements can be removed or their concentrations can be managed by controlling the processing conditions.

The primary metallurgical processing of copper is generally followed by electrorefining and extraction processes (Figure 2), which efficiently remove and recover elements from crude metals. Figure 2 shows the flow associated with the copper production process. Copper, sulfur, and oxygen primary materials; minor elements; and gangue minerals are not included in the figure. Figure 2 shows that some valuable elements (bismuth, gold, silver, and platinum group metals (PGMs)) in copper from EoL electronic products (e-waste) can be extracted into an anode slime. In fact, metals nobler than copper, including PGMs, remain in copper after processing in a copper converter (Figure 1). These noble metals can be recovered from the anode slime by a hydrometallurgical (solvent extraction) process. Thus, contamination is not a major problem during copper smelting when scrap containing these noble metals is processed by an ordinary smelter. Even if PGMs are mixed in copper scrap from e-waste, they can be ultimately recovered via the anode slime (Figures 1 and 2).

19.3.2 Lead and Zinc Smelting

Figure 3 shows the distribution of 20 elements at the bottom of a lead blast furnace at 1500 K under $p_{O_2} = 10^{-6}$ Pa. The activity coefficient of MO_n was assumed to be unity, and the boundary between the metal and gas phases was drawn similarly to that for the copper converter. The activity coefficients of metals in molten lead were obtained from published reports [12,19–22]. The activity coefficients of iron, chromium, manganese, platinum, and tungsten at infinite dilution were assumed to be unity.

Iron, gallium, manganese, chromium, aluminum, magnesium, neodymium, and zinc can be transferred to slag from molten lead by oxidation. Zinc in ore is partially reduced and vaporized inside the blast furnace, but most of the zinc is reoxidized and transferred to slag. In some smelters, zinc oxide in the slag recovered from the blast furnace is reduced again to zinc vapor for secondary recovery in a slag-fuming process. Precious metals, copper, tin, nickel, indium, and bismuth remain in the molten lead.

In contrast with the lead blast furnace, in the Imperial smelting process (ISP) zinc as well as lead is recovered. Because molten lead accumulates at the bottom, the distribution of elements for the ISP is similar to that for the lead blast furnace. Zinc is distributed in the slag phase, but zinc vapor can be removed from the blast furnace in a real process because the activity of zinc is markedly lowered when zinc vapor dissolves in the large volume of molten lead in the splash condenser. Zinc is transferred to outside the blast furnace by the large difference in activity. The slag-fuming process has been used to recover zinc from molten zinciferous slags generated from a lead-smelting furnace. This reductive process involves adding or injecting a reductant source such as pulverized or lump coal or natural gas.

19.3.3 Aluminum Remelting

Although aluminum is recycled in large quantities, contamination by alloying elements is a problem in the recycling of EoL aluminum products. Aluminum has a high ionization tendency and thermodynamic reactivity (the standard electrode potential of aluminum in an aqueous solution at 25 °C is −1.676 V vs. the normal hydrogen electrode) [18]. Accordingly,

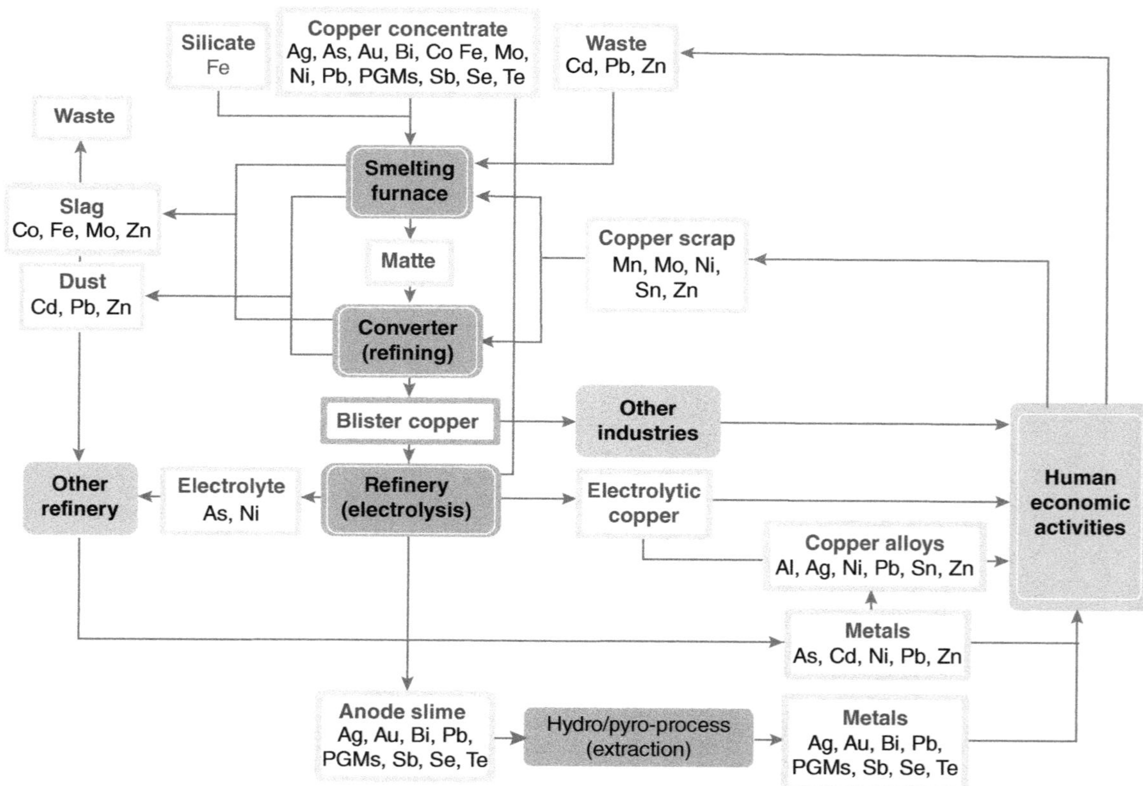

Figure 2 *Flow of elements associated with the copper production process [5]. Reprinted with Permission from [5]. Copyright {2011} American Chemical Society*

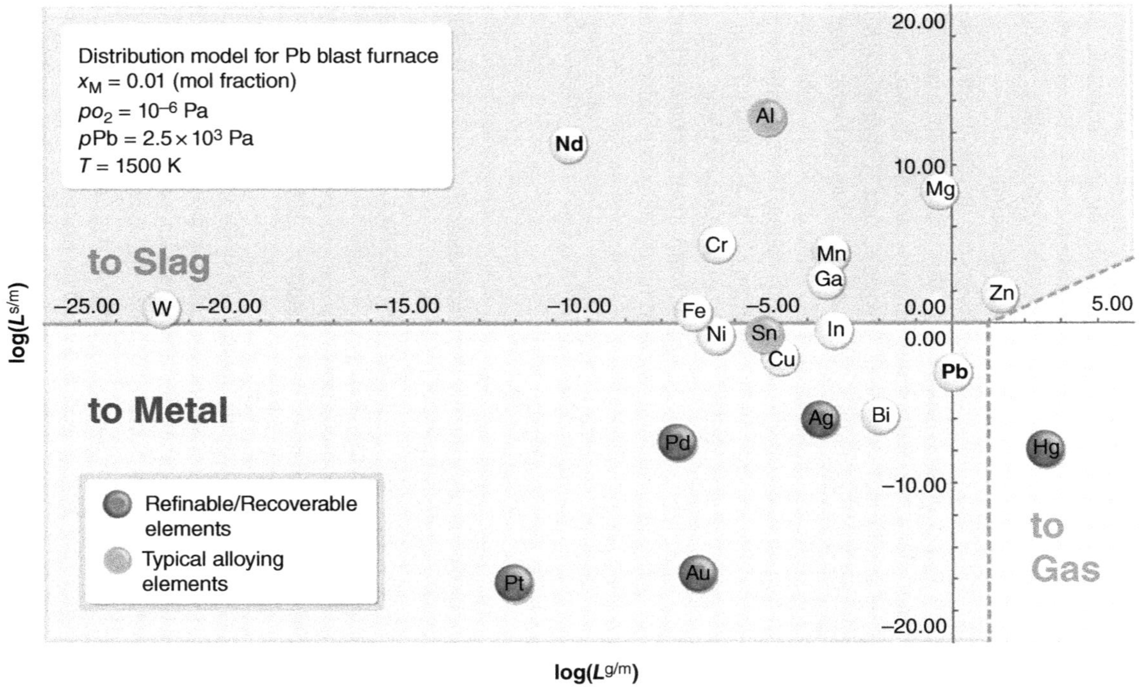

Figure 3 *Distribution of elements in the metal, slag, and gas phases during lead smelting under the simulated atmosphere of the lead blast furnace [4] (See insert for color representation of the figure.)*

when the oxidation method, which is the typical refining technique, is employed for the removal of impurities from aluminum scrap, aluminum tends to be preferentially oxidized and lost to dross or slag. In the recycling of aluminum beverage cans (can-to-can recycling), the scrap quality is carefully controlled to meet industrial standards. For example, both the end tab (JIS H4000, 5182) and the body (JIS H4000, 3004) of used aluminum cans must be exclusively recycled into aluminum beverage can bodies. These standards have been imposed owing to the severe limitations associated with impurity removal. Because of these limitations, most used aluminum undergoes cascade recycling. Because cast and die cast aluminum products can tolerate high impurity concentrations, these products effectively act as a final sink for relatively low-grade aluminum scrap.

Figure 4 shows a schematic diagram of the current system for aluminum production and recycling, including the target process analyzed in this study. Aluminum (the primary material), oxygen, minor elements, and gangue minerals are not shown. "Electrolysis" includes the Hall–Héroult and trinal electrolytic processes, and "Fabrication, Machinery" includes casting and die casting. In remelting, primary aluminum and scrap aluminum are used as the principal materials for the production of aluminum alloys, which are used for products such as plates and rods and for die casting and casting. During remelting, the alloy composition is adjusted to a target value by the addition of Cu, Mn, and other pure metals. The aluminum alloy scrap contains various kinds of impurities along with the alloy components. Therefore, the control of impurities and scrap composition is important.

The activity coefficient for element M in an Al–M binary alloy is obtained from Eq. 8:

$$RT\ln\gamma_{\mathrm{M}} = {}^{0}\Omega_{\mathrm{Al-M}}x_{\mathrm{Al}}^{2} + {}^{1}\Omega_{\mathrm{Al-M}}x_{\mathrm{Al}}^{2}\left(4x_{\mathrm{Al}}-3\right) + {}^{2}\Omega_{\mathrm{Al-M}}x_{\mathrm{Al}}^{2}\left(2x_{\mathrm{Al}}-1\right)\left(6x_{\mathrm{Al}}-5\right) + {}^{3}\Omega_{\mathrm{Al-M}}x_{\mathrm{Al}}^{2}\left(2x_{\mathrm{Al}}-1\right)^{2}\left(8x_{\mathrm{Al}}-7\right) \quad \text{(Eq. 8)}$$

For the 45 elements in molten aluminum, the thermodynamically assessed Redlich–Kister parameters $\Omega_{\mathrm{Al-M}}$ [23] were taken from the references in the Supporting Information. The standard Gibbs free energies for oxide formation and the vapor pressures of pure elements were taken from thermodynamic tables [24,25].

Figure 5 shows the distribution of elements in the various phases during the aluminum remelting process. The red circles denote typical additive elements. The figure indicates that Mg, Ca, and Be can be removed by oxidation (transferred to dross of slag) and that Zn, Cd, and Hg can be removed by evaporation. The removal of the other 39 elements, including Cu, Si, Fe, and Mn, is difficult; they remain in the metal phase. These results support the observation that most impurities contained in aluminum alloy scrap are not removed to any appreciable extent under actual operational conditions. Moreover, our previous results [6] indicate that the extent to which the process parameters (oxygen partial pressure, temperature, and flux composition) can be changed in aluminum production is quite limited compared to that for iron and copper production, owing to the relatively low melting point and strong oxygen affinity of aluminum.

However, in some cases, the removal of impurities may result in the formation of a solid aluminate instead of a pure oxide. For instance, $MgAl_2O_4$ may form instead of MgO during the removal of Mg from an aluminum alloy. In such a case, the activity of MgO will decrease from 1 to 0.015 at 1073 K [26], and Mg removal will be thermodynamically enhanced. In this regard, it is important to note that, in the decision tree model and the

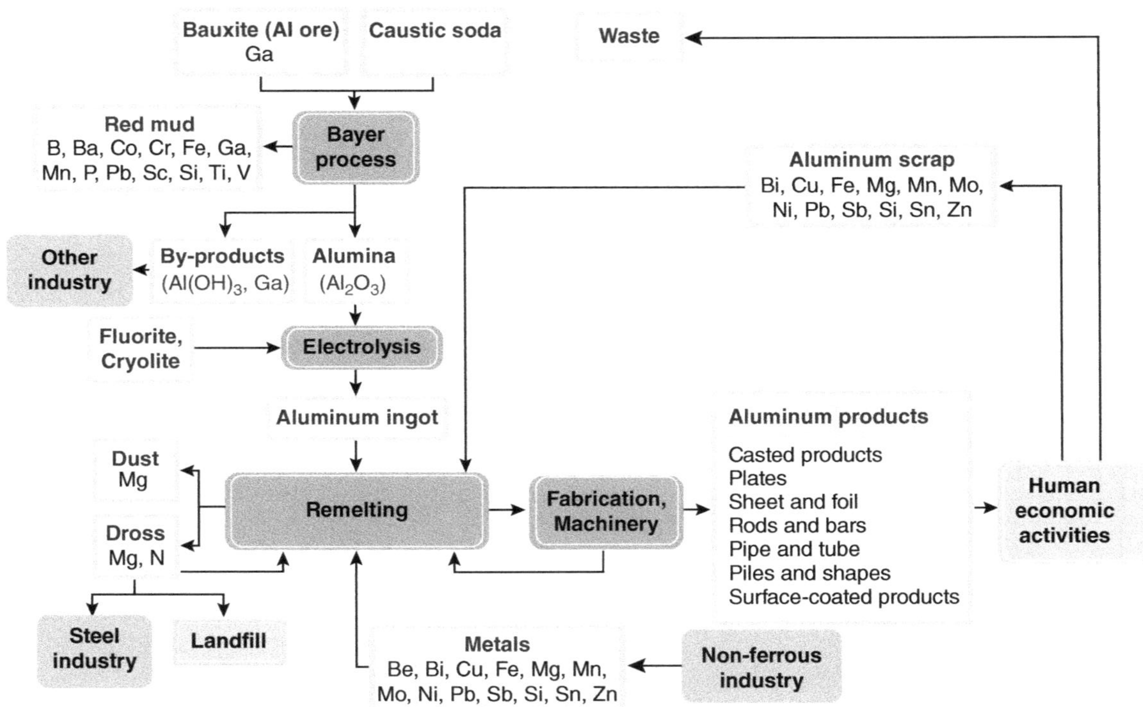

Figure 4 *Flow of elements associated with aluminum refining and remelting [3]. Reproduced from [3] with permission of Elsevier*

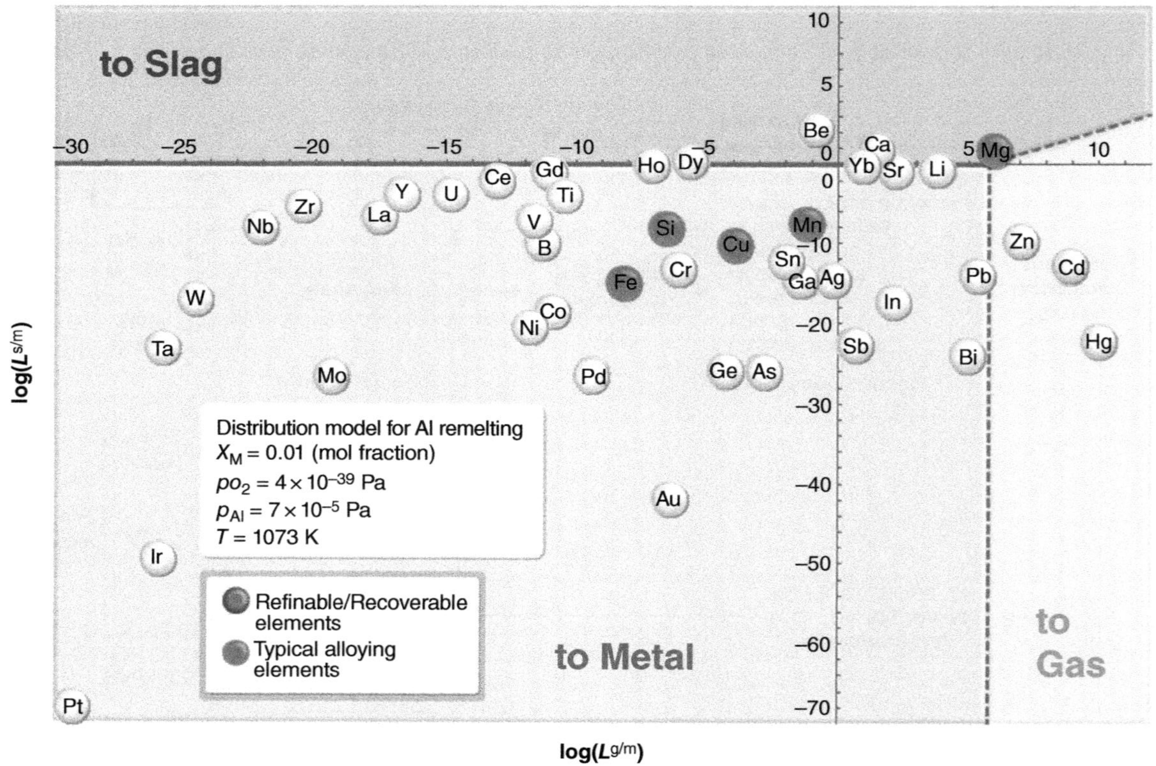

Figure 5 *Distribution of elements in the metal, slag, and gas phases during the aluminum remelting process under a simulated atmosphere [6]. Reprinted with Permission from [6]. Copyright {2010} American Chemical Society. (See insert for color representation of the figure.)*

THEMA model/matrix proposed by Castro et al. [27] for the evaluation of material combinations involved in the recycling of aluminum scrap, magnesium and zinc are classified as metals that should be separated from aluminum scrap before the scrap is subjected to a remelting process. Our analysis, however, demonstrates that, so far as the thermodynamics are concerned, magnesium and zinc are exceptional elements that can be separated from aluminum scrap in the remelting process. Our analysis can thus provide thermodynamic criteria for the highly innovative decision-tree model of Castro et al.

19.4 Metallurgical Knowledge for Recycling: Element Radar Chart for Metallurgical Processing

Metal recycling is expected to promote efficient use of resources. However, owing to the likely uncontrolled mixing of alloying elements, open-loop metal recycling systems currently used to recover metals from EoL electronic products and cars could result in material, quality, and dilution losses. Material loss includes separation and/or refining losses, the latter including shredding residues and slag, which are ultimately deposited in landfills. Quality losses occur when the quality of secondary materials no longer matches that of the original materials due to inadequate compositional control. Dilution losses occur when additional primary materials are required to lower the concentrations of contaminants to attain a quality target. Processing mixed metals with different inherent properties produces secondary metals whose properties are no longer equal to those of the original ones.

Precious and rare (minor) metals from e-waste are becoming increasingly important as secondary resources, and a recycling process has been established, mainly in the copper metallurgical industry. In addition, e-waste management is gaining interest from an environmental point of view at a global level.

Thermodynamics can assist in product design by determining which metals can be metallurgically removed from complex input or scrap. The element radar chart, which indicates how metals are separated in different carrier metal processes, graphically represents the distribution tendencies of elements in metallurgical processes and is useful for comparing the relative ease of removal of impurities among the base metals (Figure 6). In the element radar chart, radian direction indicates element distribution tendency for oxidation, and circular arc direction indicates that for volatilization (Figures 1, 3, and 5).

The disadvantage of the steel-making process and the remelting processes for aluminum, titanium, and magnesium is that these processes cannot effectively remove impurities. This thermodynamic feature strongly indicates that loss of or contamination by these uncontrolled elements could occur in conventional recycling systems for steel, aluminum, and magnesium. In contrast, the smelting of copper and lead includes an effective impurity removal process (electrolytic smelting) as a post-process that acts as an excellent refiner for a number of impurities. The removal or control of impurities is essential in this type of system, but control of elements in the slag phase, such as neodymium and tungsten, is difficult.

To effectively use these elements as important alloying elements, it is important to develop a presorting system for scrap aimed at alloy-to-alloy recycling and accurate scrap specifications that take material composition into account. Therefore, new metallurgical solutions should be developed to ensure a sustainable supply of high-quality metal products.

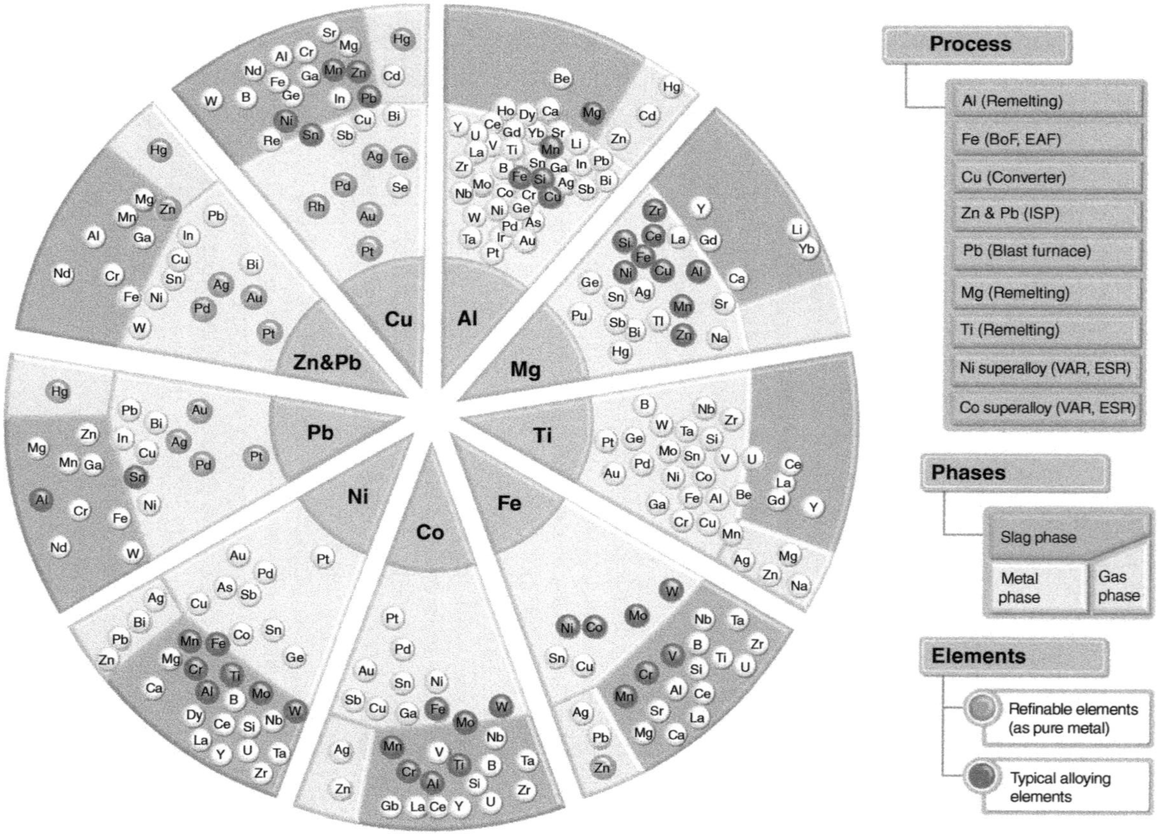

Figure 6 *Element radar chart for the recycling of Fe, Cu, Zn, Pb, Al, Ti, Mg and superalloys into the metal phase [4,6,7,28]. BOF, basic oxygen furnace; EAF, electric arc furnace; VAR, Vacuum arc remelting; ESR, Electro-slag remelting (See insert for color representation of the figure.)*

References

(1) Verhoef, E.V., Dijkema, G.P.J., Reuter, M.A. (2004) Process Knowledge, System Dynamics, and Metal Ecology. *Journal of Industrial Ecology* 8: 23–43.

(2) Reuter, M.A., Verhoef, E.V. (2004) A Dynamic Model for Assessment of the Replacement of Lead Solders. *Journal of ELECTRONIC MATERIALS* 33: 1567–1580.

(3) Schaik, A.V., Reuter, M.A. (2010) Dynamic Modeling of E-waste Recycling System Performance Based on Product Design. *Minerals Engineering* 23: 192–210.

(4) Nakajima, K., Takeda, O., Miki, T., Nagasaka, T. (2009) Evaluation Method of Metal Resources Recyclability Based on Thermodynamic Analysis. *Materials Transactions* 50: 453–460.

(5) Nakajima, K., Takeda, O., Miki, T., Matsubae, K., Nagasaka, T. (2011) Thermodynamic Analysis for the Controllability of Elements in the Recycling Process of Metals. *Environmental Science & Technology* 45: 4929–4936.

(6) Nakajima, K., Takeda, O., Miki, T., Matsubae, K., Nakamura, S., Nagasaka, T. (2010) Thermodynamic Analysis of Contamination by Alloying Elements in Aluminum Recycling. Environmental Science & Technology 44: 5594–5600.

(7) Hiraki, T., Miki, T., Nakajima, K., Matsubae, K., Nakamura, S., Nagasaka, T. (2014) Thermodynamic Analysis for the Refining Ability of Salt Flux for Aluminum Recycling. *Materials* 7: 5543–5553.

(8) Hiraki, T., Takeda, O., Nakajima, K., Matsubae, K., Nakamura, S., Nagasaka, T. (2011) Thermodynamic Criteria for the Removal of Impurities from End-of-Life Magnesium Alloys by Evaporation and Flux Treatment. *Science and Technology of Advanced Materials* 12: 035003 (10 pp).

(9) Lu, X., Hiraki, T., Nakajima, K., Takeda, O., Matsubae, K., Zhu, H.-M., Nakamura, S., Nagasaka, T. (2012) Thermodynamic Analysis of Separation of Alloying Elements in Recycling of End-of-Life Titanium Products. *Separation and Purification Technology* 89: 135–141.

(10) Azakami, T., Awakura, Y. (1999) Extractive Metallurgy, Vol. 87. The Japan Institute of Metals, Sendai, Japan.

(11) Kurokawa, H., Kemori, N. (2003) Reaction Mechanism in Copper-Making Stage of PS Copper Converter. *Shigen-to-Sozai (now Journal of MMIJ)* 119: 55–60.

(12) Hultgren, R. (1971) Selected Thermodynamic Values and Phase Diagrams for Copper and Some of Its Binary Alloys. International Copper Research Association, New York.

(13) The Japan Society for the Promotion of Science. (1988) The 19th Committee on Steelmaking: Steelmaking Data Sourcebook. Gordon and Breach, New York.

(14) Zhang, X., Han, Q., Chen, D. (1991) Dissolution Equilibrium of Magnesium Vapor in Liquid Iron. *Metallurgical and Materials Transactions B* 22B: 918–921.

(15) Su, X., Tang, N., Toguri, J.M. (2001) Thermodynamic Evaluation of the Fe–Zn System. *Journal of Alloys and Compounds* 325: 129–136.

(16) Vijayakumar, M., Sriramamurthy, A.M., Nagender Naidu, S.V. (1988) Calculated Phase Diagrams of Cu–W, Ag–W and Au–W Binary Systems. CALPHAD 12: 177–184.

(17) Risold, D., Hallstedt, B., Gauckler, L.J., Lukas, H.L., Fries, S.G. (1996) Thermodynamic Optimization of the Ca–Cu and Sr–Cu Systems. CALPHAD 20: 151–160.

(18) Bard, A.J., Faulkner, L.R. (2001) Electrochemical Methods — Fundamentals and Applications, 2nd ed. John Wiley & Sons, Hoboken, NJ, p. 808.

(19) Kawai, Y., Shiraishi, Y., Eds. (1988) Handbook of Physico-chemical Properties at High Temperatures. Iron and Steel Institute of Japan, Tokyo, Japan.

(20) Ueno, S., Waseda, Y. (1988) *Nihon Kinzoku Gakkaishi* 52: 204–210.

(21) Pomianek, T. (1986) Thermodynamic Properties of Lead in Ni-Pb Solutions. *Zeitschrift für Metallkunde* 77: 388–392.

(22) Zydzik, M., Smids, M., Ende, M.A.V., Jung, I.H. (2014) CALPHAD 46: 1–17.
(23) Hillert, M. (2008) Phase Equilibria, Phase Diagrams and Phase Transformations: Their Thermodynamic Basis, 2nd ed. Cambridge University Press, Cambridge, UK.
(24) Chase, M.W., Jr., Davies, C.A., Downey, J.R., Jr., Frurip, D.J., McDonald, R.A., Syverud, A.N. (1985) JANAF Thermochemical Tables, 3rd ed. American Chemical Society and American Institute of Physics (for National Bureau of Standards), Washington, DC.
(25) Barin, I. (1989) Thermochemical Data of Pure Substances. VCH Verlagsgesellschaft mbH, Weinheim, Germany.
(26) Fujii, K., Nagasaka, T., Hino, M. (2000) Free Energies of Formation of MgO, $MgO{\bullet}Al_2O_3$ and Activities of Constituents in Spinel Solid Solution. *ISIJ International* 40: 1059–1066.
(27) Castro, M.B.G., Remmerswaal, J.A.M., Reuter, M.A., Boin, U.J.M. (2004) A Thermodynamic Approach to the Compatibility of Materials Combinations for Recycling. *Resources, Conservation and Recycling* 43: 1–19.
(28) Lu, X., Matsubae, K., Nakajima, K., Nakamura, S., Nagasaka, T. (2016) Thermodynamic Considerations of Contamination by Alloying Elements of Remelted End-of-Life Nickel- and Cobalt-Based Superalloys, *Metallurgical and Materials Transactions B* 47: 1073–5615.

20

Economic Perspectives on Sustainability, Mineral Development, and Metal Life Cycles

Roderick G. Eggert

Colorado School of Mines and Critical Materials Institute

20.1 Introduction

Since the 1980s, *sustainability* has dominated public discussion of minerals and metals. Although it has many related but different meanings, sustainability — as it applies to mineral resources and metals — embodies the idea that current production and use of metals may not be consistent with sustaining (indefinitely) the natural environment, the social fabric of communities and societies, and human wellbeing.

A *metal life cycle*, in the context of sustainability, is a way of visualizing and organizing data and analysis of the stocks and flows of materials over their entire life cycles, as well as the environmental and social impacts of material production and use at each stage of a life cycle. A life cycle begins with characterization of a mineral resource, mineral development and mining, proceeds through manufacturing and use, and concludes with recycling, reuse, or disposal as waste.

Much of the literature on sustainability and metal cycles has been written by scientists and engineers, who take physical stocks and flows of materials — as well as environmental and social impacts measured in physical units — as the basic inputs to modeling and analysis. Efficiency in a scientific or engineering sense often is suggested as an ideal outcome (e.g., minimize particulate-matter emissions from coal-fired power plants, maximize the amount of metal extracted and recovered from a mineral deposit, maximize the recycling of electronic wastes).

Metal Sustainability: Global Challenges, Consequences, and Prospects, First Edition. Edited by Reed M. Izatt.

Economists, while starting with basic physical information on production and use of resources, emphasize in their models the way the behavior of producers and users influences production, use and prices. Efficiency in an economic sense reflects the optimal balance between the benefits and costs of resource extraction and use, with benefits and costs defined broadly to include the equivalent economic values (dollars) of environmental and other damages that scientists and engineers quantify in physical terms (e.g., parts per million of dissolved solids in water).

This chapter discusses economic perspectives on sustainability and metal life cycles. By and large, the perspectives I provide in this chapter represent the general consensus of the economics profession. There of course are differences of opinion among economists.

The chapter begins by reviewing the various meanings and approaches to sustainability, followed by introducing how economists approach the topic. It then discusses implications for mineral development and metal life cycles.

20.2 The Many Faces of Sustainability

Sustainability means different things to different people. *Environmental* sustainability gives priority to maintaining or enhancing the quality of the natural environment. *Social and cultural* sustainability emphasizes avoiding undesired negative social and cultural impacts associated with commercial activities. *Economic* sustainability focuses on sustaining improvements in human living standards. These faces of sustainability are, in effect, one-dimensional; they give priority to the natural environment, or to social and cultural dimensions of human activity, or to economic wellbeing.

Sustainable *development*, on the other hand, emphasizes the sustaining of development, which is inherently multidimensional and requires balancing the social, cultural, and economic dimensions of human activities and how they relate to the natural environment. Sustainable development represents economic growth that is consistent with social preferences for environmental quality and social/cultural development.

Cutting across all forms of sustainability and sustainable development are three issues. The first is the scale of focus, which can range from local to global. Local discussions of sustainability and mineral development often concentrate on the degree to which local communities benefit sufficiently from mining, given that they typically bear the brunt of social disruptions and environmental damages. Discussions of climate change and sustainability range from global to local — the observation that it is the global stock of carbon that influences global temperatures, the reality that impacts depend on location (coastline, desert, etc.). The second crosscutting issue is the carrying capacity of the natural environment and whether the scale and nature of human activity approach this carrying capacity. The lower the level of human activity, the greater the ability of the natural environment to absorb environmental damages and regenerate itself. The third issue is how to balance the interests of current and future generations. Many commercial activities, including mining, generate income today (a benefit) but may contribute to future environmental burdens or social disruptions (costs, broadly speaking, in the form of, for example, increased human sickness, reduced scenic vistas, or reduced species diversity). More specifically, is additional income today for a person living in poverty appropriate if the income-generating activity leads to environmental damage in the future? One needs a framework for evaluating such situations. Economists use a framework called discounting to balance benefits and costs that occur at different points in time; more on this framework in the next section of text.

20.3 Economic Concepts

The concepts of economic efficiency and equity, discounting, externalities, and capital are key to understanding how economists approach sustainability and the production and use of mineral resources and materials.

20.3.1 Economic Efficiency and Equity

Economic efficiency and equity are standards against which economists evaluate the desirability of outcomes. An outcome that is both economically efficient and equitable is "desirable." Stated slightly differently, economists consider efficiency and equity to be important goals.

Economic efficiency represents maximizing the net benefits of an activity. Think of it as optimization in a quantitative sense — for example, minimize the costs of achieving a certain level of metal production, or equivalently maximize revenues from metal production out of a given set of inputs and their associated costs (a mineral deposit, costs of labor, energy, equipment, and so on). In either case, net benefits are maximized. For a business, economic efficiency can be thought of as maximizing profits. For a household, it represents maximizing the utility or enjoyment the household experiences with how it uses its limited time and budget.[1]

Equity is fairness in outcomes — more specifically, fairness in the distribution of benefits and costs, as well as the allocation of net benefits, whether or not net benefits have been maximized. Whether a particular distribution is fair depends on one's definition of fairness. Important aspects of any definition of fairness in allocating net benefits of a project relate to questions such as:

- Who is eligible to share in the net benefits (workers, company owners, local communities, national governments, others)?
- What determines the relative priority among several eligible parties (contribution to the success of the project such as contribution of financial capital, labor, land; or need, such as when a non-mining region within a country has a higher poverty rate than the region of mining)?
- How should competing principles and criteria be reconciled (for example, how to balance the desire to distribute net benefits both to those who contributed directly to the success of the project and others in the region or nation who are poor or otherwise disadvantaged)?

Equity, and the questions above, also apply to the process by which public decisions are made. Who is eligible to participate in discussions about whether a new airport, convention center, shopping mall or mine is built?

[1] For more on economic efficiency, see any introductory textbook in economics, for example, Mankiw (2014).

Economists recognize that fairness is an important consideration. But they struggle with how to contribute to the debate about what is fair because it is largely an issue of moral philosophy.[2] In contrast, economic efficiency is more like an engineering problem or mathematical optimization.

20.3.2 Discounting

Economically efficient production and use of materials also requires that time be appropriately valued and incorporated into decision making. From the perspective of a mining company or community, producing one unit of mineral from a deposit today leaves one less unit available to be produced and used in the future. From the perspective of society, producing and using a resource brings benefits today but may lead to environmental damages to human health or ecosystems only in the future. More generally, discounting is perhaps the most important way economists balance the interests of current and future generations.

Economists account for time by discounting future benefits and costs to make them comparable to current benefits and costs. All current and future benefits and costs — financial values, as well as the broader external values described below — are identified and valued. The future benefits and costs are discounted by the social rate of time preference. Future benefits and costs are discounted, or reduced in value, for two reasons, both of which reflect the idea that a dollar of value received today is worth more than a dollar of value received in the future. First, humans are impatient and give higher value to the present than the future. Second, by waiting until the future to benefit from a resource we give up the opportunity to put that resource to productive use today. The opportunity for capital growth is lost. Both justifications are not without their critics, who argue that discounting is unethical or myopic and inherently unfair to future generations (Portney and Weyant 1999).

20.3.3 Externalities

Benefits and costs are defined broadly to include not only the obvious commercial revenues and financial costs of an activity but also the spillover effects of resource production and use. These spillover effects can be positive when, for example, a mine serves as a catalyst for broader economic development in a region. Spillover effects can be negative when, for example, there is damage to human health or ecosystems that is not considered in the decision to produce and use the mineral resource and material. These spillover effects are called externalities. They are external to the decisions to produce and use the mineral resources and materials.

Economically efficient production and use of materials require that the associated externalities be appropriately valued and incorporated into decision making. In some cases, values determined by a market can be used; for example, environmental damage to commercial fishing might be valued using the market value of fish not harvested and sold as a result of the environmental damage. In other cases, valuation is less straightforward because no direct market values exist. The value people place on maintaining wilderness might be estimated on the basis of surveys of how much people would be willing to pay to establish

[2] An excellent perspective on equity written by an economist is Young (1994).

a wilderness area. Or the value of damage to human health might be estimated on the basis of medical expenses and lost wages due to sickness. Or the external damage due to urban air pollution might be estimated by differences in real estate values between cities with clean and dirty air, after controlling for other factors that determine real estate values, such as quality of schools and medical care, proximity to cultural activities and other factors.

20.3.4 Capital

Finally, consider capital. In a narrow sense, capital is the stock of buildings, equipment, machinery and other man-made physical assets that firms use as inputs in the production of goods and services. In a broader sense, capital is the stock of all types of assets that create human wellbeing — encompassing, in addition to *man-made physical* capital, *human* capital, people's skills and talents; *social* capital, consisting of corporate, governmental, religious and other human institutions; and *natural* capital, representing air, land, water, flora, fauna, and nonrenewable energy and mineral resources. All four types of capital can be thought of, in some sense, as inputs to the creation of human wellbeing. They are complementary inputs because human wellbeing, including but not limited to benefits derived from purchased goods and services, requires all four types of capital in different combinations.

Sustaining human wellbeing over time, an economic version of sustainability, requires that the overall stock of capital remain intact over time. Capital in all its forms tends to wear out as it is used. To achieve sustainability, this degradation or depreciation must be offset with investments in new capital. To achieve sustainable development, the net effect of degradation and investment must be a sustainable increase in the stock of capital — the sum of the stocks of man-made physical, human, social, and natural capital.

Now consider the implications of these sustainability and economic concepts for mine development, national economies and metal life cycles.

20.4 Implications for Mine Development

20.4.1 Sustainability of Production

At an individual mine, mining is inherently unsustainable in a physical sense. A mine eventually will stop operating. It will cease production because it runs out of mineral reserves,[3] either physically running out of mineralized material or no longer making a profit as the quality of mineral resource declines. Nonetheless, mineral production at a mine usually is more sustainable than it appears at first. Rarely is the ultimate size of a mineral deposit known with certainty at the time mining begins. Defining a deposit requires an investment of time and funds. Rarely does it make commercial sense to fully define a deposit prior to mining. Instead, companies typically define a deposit sufficiently to justify mining at a scale that achieves most if not all of any scale economies. Then during mining, additional reserves are defined as a basis for later mining.

---

[3] Reserves are known to exist with a high degree of certainty (through drilling, sampling and other means) and are both technically and commercially feasible to mine under current market conditions. Thus they are a subset of all resources in the earth's crust.

Moreover, technological advances help make a mine more sustainable than otherwise. Technological improvements in production can be incremental or revolutionary.[4] Incremental improvements are process improvements to existing facilities that come from experience — for example, reducing the use of reagents in a froth flotation plant or improving the efficiency with which unwanted materials are separated from revenue-generating materials early in the production process. Revolutionary improvements are entirely new ways of producing a metal — for example, bacterial leaching as a means of recovering metals from ore. In either case, the effect is to enhance the technical and commercial feasibility of mining, often of a resource that otherwise would be infeasible to produce.

20.4.2 Sustainability of Wellbeing Originating from Mining

In an economic sense, however, mining and sustainability are compatible, under certain conditions. Sustainability requires (a) creating economic benefits through mineral exploration and mine development, (b) appropriately evaluating the environmental and sociocultural impacts of mining, both in deciding whether to mine and, if so, how to manage these impacts during operations and after a mine closes, and (c) sustaining the economic benefits from mineral resources after a mine closes. Economic efficiency is a guiding principle in all three areas.

Creating economic benefits requires that an otherwise dormant asset in the ground, a mineral deposit, be discovered, evaluated and developed. Mineral exploration and mine development can be thought of as creating natural capital, to use a concept introduced earlier in this chapter. Whether a private company develops a mineral deposit depends on a variety of factors that are shaped by the institutional and public policy framework of a region and nation. Some of this framework is common to all commercial activities — general rules and regulations, broad customs and ways of doing business, the quality of government and non-government institutions, the perceived stability of a region and other factors that define the overall business environment. Other parts of the institutional and public policy framework are specific to mineral development and mining — legislation and rules governing ownership of mineral resources, land access, security of tenure, fiscal rules, environmental and mine closure policies. Although public policy has multiple goals, an essential goal — from the perspective of economics — is economic efficiency.[5]

Appropriately evaluating and managing the environmental and socio-cultural impacts of mining, to an economist, means that the externalities associated with these impacts be included in deciding whether mining should occur and then in operating a mine. Otherwise, economic efficiency will not be achieved. Putting an economic value on external benefits and costs is not simple and straightforward, as noted in section 20.3.3. Nevertheless, even if it is not possible to put completely accurate values on external impacts that all interested parties agree to, establishing ledgers of full project benefits and costs, including externalities, allows direct comparison of external and private benefits and costs. This type of analysis is called benefit-cost analysis and is the subject of a large literature.[6]

---

[4] See National Research Council (2002).
[5] For more on institutions, public policies, and economic efficiency, see World Bank (2002, 2003, 2015a).
[6] For more on benefit-cost analysis, see Boardman *et al.* (2014).

Sustaining the economic benefits created from mining requires investment in assets that continue to generate wellbeing even after a mine closes. Think of this as economic efficiency over the long term. Use a portion of the proceeds from mining to invest in other, non-mining forms of capital (man-made physical, human, social, natural). The goal is to keep the overall capital stock intact — if not increase it — and as a result maintain and ideally enhance society's ability to generate wellbeing in the future. Investments to compensate for mineral depletion can take a variety of forms. Investing in education and health care enhances human capital. Investing in infrastructure — such as roads and airports, electric power systems, and public sanitation — enhances man-made physical capital. Another option is investing in other lines of business in which a company or region has a comparative advantage.

Finally, a major source of contention in mineral development relates to equity — the fairness, or not, of the distribution of benefits and costs among private companies, local communities, national governments and other interested parties. Many of the benefits go to companies, owners and workers (revenues, dividends, and salaries) and to national governments (tax and royalty revenues). Many of the costs arguably are disproportionately borne by local communities in the form of environmental damage and social disruption. Equity and fairness also apply to the process used to determine whether a proposed mine is developed and, if so, how. In most political jurisdictions, mining companies apply to government agencies for permissions and permits for mining. Government agencies review company applications. In theory, government agencies should act and make decisions based on what is in the best interest of society as a whole. Whether this happens in practice depends importantly on the degree to which and how other, non-governmental organizations can participate in the process of reviewing applications to mine. Government policies ultimately define roles for different parties. But a number of non-governmental and international organizations have established guidelines and frameworks for these issues, which are especially important in regions and nations with weak government institutions. Prominent examples include the World Bank's *Extractive Industries Source Book* (World Bank 2015a) and various publications and guidelines of the International Council on Mining and Metals (www.icmm.com).

20.5 Implications for Regional and National Mineral Development

The previous section took a mine as its object of study. Now consider regional and national economies and their efforts to put sustainability and mineral development into practice through sustainability funds and green accounting.

20.5.1 Sustainability Funds

Sustainability funds are investment funds that seek to convert depleting natural capital (minerals, oil) into other, more sustainable forms of capital. The goal is to make the mineral wealth permanent. Examples include the Alaska Permanent Fund in the United States,[7] the Alberta Heritage Trust Fund in Canada,[8] Norway's Government Pension

---

[7] www.apfc.org, accessed 10 July 2015.

[8] www.finance.alberta.ca/business/ahstf/, accessed 10 July 2015.

Fund Global[9] and more generally sovereign wealth funds that focus on saving and investment (rather than macroeconomic stabilization).

All sustainability funds face the following issues, as Hanneson (2001) describes:

- *How much to save and invest?* Answering this question requires identifying the investment goal. Is it to sustain or instead enhance the current level of wellbeing? The higher the goal, the higher the savings rate must be. The expected rate of return on the investment also is important: the higher the return, the lower the required saving rate.
- *In what?* One possibility is assets that generate returns either as income or through appreciation in value, such as stocks, bonds, real estate, and businesses. The specific sustainability funds identified above all invest in assets of this type; the income they generate is used for a variety of purposes including pensions and education or returned directly to citizens. Another possibility is to invest in assets with less visible, harder-to-quantify returns that, nevertheless, provide the social and physical infrastructure for assets and activities that generate income: education, health care, scientific and technical research and development, and systems for transportation, electricity, public sanitation and other forms of physical infrastructure. Governments normally undertake this type of investment.
- *Where?* In the mining community or region? Somewhere else in the nation? Wherever the investment returns are the greatest? The appropriate location of investment depends on the situation: the larger the economy, the more likely it is that there are productive investments at home. The lower the level of economic development, the greater the potential returns to investments in social infrastructure but the less developed financial institutions may be and the lower the absorption capacity for local investment. The more dependent an economy is on mineral production, the greater the urgency of domestic investment that will allow the economy to diversify.

20.5.2 Green Accounting

Green accounting expands the conventional national income and product accounts to include changes in the stock of natural capital — renewable and environmental assets (such as water, air, land, flora, fauna) and subsoil assets (crude oil, natural gas, coal and other minerals). The broad context for green accounting is the desire, articulated in section 20.4.2, to sustain capital stocks because the stock of capital determines the ability of society to create wellbeing.

The most prominent part of the conventional accounts is gross domestic product (GDP), the estimated value of all final goods and services produced by an economy. Once sold, this production also is income. So GDP represents both output and income, and GDP (or GDP per capita) usually is a starting point for discussions of a nation's economic wellbeing.

A simple example helps illustrate the deficiencies of the conventional accounts and why green accounting emerged. Imagine a simple island economy whose only asset (or capital) is a mineral resource. Mining is its only source of income. Last year its GDP, consisting only of mine output, was \$100. This year the value of its mine output increased to \$200, even after factoring out the effects of inflation. Has the economic wellbeing of the island's

9 www.nbim.no/en/the-fund/, accessed 10 July 2015.

citizens doubled? In the short term, wellbeing of course has doubled in the sense that citizens have doubled their incomes and purchasing power. In a longer-term sense, however, this interpretation may be misleading. It certainly would be misleading if doubling mine output also led to complete depletion of the mineral resource, the nation's only asset. There would be no asset left to generate wellbeing next year.

So GDP is an incomplete measure of wellbeing. It does not consider the degree to which a nation's stock of capital — its ability to generate wellbeing in the future — changed as a result of current production. GDP includes *additions* to the stock of man-made physical capital but excludes *subtractions* from this stock as it wears out (depreciation). Another measure, however, does include depreciation: net domestic product (NDP), which is GDP less depreciation of man-made physical capital. NDP is a better measure than GDP of longer-term, sustainable wellbeing. If net investment in man-made physical capital is zero, that is, if additions to the capital stock exactly offset depreciation, then an economy starts the next period with exactly the same level of man-made physical capital that it had the previous period. The economy has sustained its ability to generate goods and services the next period.

The GDP accounts ignore changes to the stocks of human, social, and natural capital. As noted at the beginning of this section, green accounting focuses on changes in stocks of natural capital. It starts with conventional NDP and adjusts it for additions to and subtractions from natural-capital stocks. The result can be thought of as "green NDP," which adjusts conventional GDP for changes in stocks of both man-made physical and natural capital. As such, green NDP provides a better indicators of sustainable wellbeing than conventional GDP, even though green NDP still ignores changes to stocks of human and social capital.

Green accounting still is very much exploratory. Green NDP has not supplanted GDP as an indicator of wellbeing. The United Nations has developed an international framework for environmental-economic accounting (United Nations 1993, 2003, and 2014). Most practical attempts at valuation have focused on valuing subsoil assets (oil, gas, coal, minerals), as they have market values. A number of national governments estimate and publish data on subsoil asset values (for example, Australian Bureau of Statistics 2015, Statistics Canada 2015). The United States does not have an ongoing program in this area. But its Bureau of Economic Analysis carried out an extensive preliminary effort in the 1990s (Bureau of Economic Analysis 1994a, 1994b) that was the subject of a U.S. National Research Council review (Nordhaus and Kokkelenberg 1999). The World Bank in recent years has published estimates of adjusted net savings that include changes to natural capital (World Bank 2015b).

Finally, it is worth going beyond the narrow confines of green accounting to note that exactly what is required of capital stocks over time to achieve sustainability is the subject of significant disagreement. One view, articulated earlier in section 20.4.2, is that sustainability requires only that the overall capital stock remain intact (known as *weak sustainability*). Depletion of a mineral resource or degradation of the environment is consistent with sustainability as long as offsetting investments in other forms of capital occur (e.g., education, a factory, pollution control technologies). This view emphasizes the substitutability among different forms of capital. The goal is not to sustain any particular type of asset; rather it is to sustain the wellbeing that results from different combinations of capital. This view emphasizes that different types and combinations of capital can create the same level of wellbeing.

The other view (known as *strong sustainability*) is that sustainability has two requirements: maintain both (a) the essential parts of the stock of natural capital and (b) the overall stock of capital. This view emphasizes the complementary nature of natural and other types of capital, and it gives special priority to sustaining natural capital. Without air, water, land, or other natural resources, we would not be able to grow the food and manufacture the product on which we depend. Wellbeing requires all four types of capital (man-made physical, human, social, natural).

This sharp distinction between weak and strong forms of sustainability — while helpful in focusing our attention on substitution possibilities — is simplistic. There are elements of truth in both views. Few would suggest that we can do completely without air or water, as long as we invest in people and factories. Few would suggest that we should cease mining altogether to maintain stocks of mineral resources.[10]

20.6 Implications for Metal Life Cycles, Material Efficiency, and the Circular Economy

Another conceptual lens through which to view metals and sustainability is the metal life cycle, as well as the related concepts of material efficiency and a circular economy.

A *metal life cycle*, as noted at the beginning of this chapter, starts with characterization of a mineral resource and then proceeds through mining, manufacturing, use, and ultimately re-use, recycling or disposal as waste. This idea is descriptive only — it makes no judgment about whether actual stocks and flows of metals are optimal from society's perspective. In contrast, *material efficiency* is the idea that we are not using earth resources and materials efficiently in a physical sense and that efficiency could be enhanced through deliberate efforts to design products that are (a) longer lasting, (b) increasingly modular and capable of being remanufactured, (c) made with components that can be reused, and (d) made with less material (Allwood *et al.* 2011) — think of these strategies as encompassing the first two of the popular trio of material strategies "reduce, reuse, recycle." The *circular economy* is the broader idea that society should dramatically reduce material wastes by aggressively pursuing all three of these strategies, reducing material needs, reusing products or at least components, and increasing recycling. The circular economy contrasts with a linear view of materials, which emphasizes, as the argument goes, using more primary raw materials, making more products, and disposing more manufacturing wastes and end-of-life products. A circular economy would close the loop through greater material efficiency and recycling.[11] The goal is

> to eradicate waste — not just from manufacturing processes, as lean management aspires to do, but systematically throughout the life cycles and uses of products and their components…the goal for durable components, such as metals and most plastics, is to reuse or upgrade them for other productive applications through as many cycles as possible (Zils 2014).

---

[10] See Pezzey and Toman (2005) for a richer discussion of weak and strong sustainability and their variants.

[11] See, for example, World Economic Forum (2014) or Webster (2015).

Material efficiency and the circular economy both start from the same concerns: first, we run the risk that metallic mineral resources will not be available in the future due to both the finite and nonrenewable nature of metals in the earth's crust and the expected dramatic increases in demand for materials as the world's population grows and its poorest people improve their living standards. Second, extensive wastes from mineral development and materials manufacturing create large environmental problems and damage to human health. Consider each of these concerns.

20.6.1 Nonrenewability of Mineral Resources and Metals

A physical view of nonrenewability and sustainability emphasizes "using up" the limited quantities of metals in the earth's crust. Common physical measures or indicators of the sustainability of mineral resources and metals are reserve/production (R/P) ratios, average crustal abundance of elements, and ore grade.

A *reserve/production (R/P) ratio* provides an indication of how long today's reserves will last if production rates remain the same over time. The ratio suggests the number of years until society "runs out" of a resource (Table 1). A simplistic and usually misleading interpretation of R/P ratios is: the smaller an R/P ratio, the more scarce and less sustainable the resource. A deeper interpretation recognizes the dynamic nature of the ratio's numerator (reserves) and denominator (production). Reserves change over time. Reserves decline as a result of ongoing mining, and they increase as a result of mineral exploration and development. Production changes over time, driven importantly by changes in metal demand. Macroeconomic recessions tend to reduce demand for most metals for the length of the recession. Over the longer term, demand for most metals has increased as population and human living standards also have increased, and production increases have followed. Using copper as an example of metals generally, there is no obvious upward or downward trend in the R/P ratio over the period 1950–2013, suggesting that reserves actually are more like inventory of canned goods in a supermarket than an indicator of long-term scarcity and unsustainability (Table 2).

Average crustal abundance ranges from greater than 50,000 parts per million (ppm) each for oxygen, silicon, aluminum, and iron to less than 0.0005 ppm for rhenium, iridium, rhodium, and osmium (Rudnick and Gao 2003). A simplistic and sometimes misleading interpretation of these figures on elemental abundance is that using more abundant elements such as iron and aluminum is inherently more sustainable than using less abundant elements such as rhenium and rhodium. A deeper understanding also considers the degree to which natural processes in the earth's crust lead to concentration (or not) of elements

Table 1 *World Reserve/Production Ratios for Selected Metals, 2014*[a]

Bauxite	120
Copper	37
Iron	60
Lead	16
Nickel	34
Zinc	17

[a]Calculated with data from U.S. Geological Survey (2015a).

Table 2 *World Reserve/Production Ratios for Copper, 1950–2013*[a]

1950	38
1960	39
1970	47
1980	68
1990	35
2000	26
2013	37

[a]Calculated with data on production from U.S. Geological Survey (2015b) and reserves from Bureau of Mines (1956, 1960, 1970, 1980 1990) and U.S. Geological Survey (2001, 2014).

Table 3 *Ore Grades in Australia, 1900–2009*[a]

	Copper	Gold	Lead	Nickel	Uranium
	(% copper)	(grams/tonne gold)	(% lead)	(% nickel)	(kilograms/tonne U_3O_8)
1900	2.77	24.04	13.55	na	na
1910	2.99	13.38	14.27	na	na
1920	3.89	14.22	13.49	na	na
1930	4.76	13.91	17.74	na	na
1940	0.84	6.33	12.63	na	na
1950	0.67	4.99	10.66	na	na
1960	1.97	4.83	9.94	na	1.85
1970	1.48	3.16	7.74	3.91	1.18
1980	1.90	3.37	6.24	2.06	3.38
1990	2.45	2.19	5.06	1.85	1.83
2000	1.56	1.97	4.78	1.21	1.10
2009	0.95	1.66	3.46	1.05	1.01

[a]Assembled with data from Mudd (2009); na = not available.

above their average concentrations. Some seemingly rare minerals, such as gold, are concentrated in the crust such that they are more available than suggested by their average abundance. Copper is only the 25th most abundant element on average (Rudnick and Gao 2003), and yet it is the 7th most-produced element (tonnage) due to its tendency to form discrete mineral assemblages and deposits [Skinner 1986]. The (rarely used) rare earths cerium and lanthanum are, on average, as abundant as copper and zinc, but they tend not to be concentrated significantly into discrete minerals and usually remain widely disseminated in the crust.

The third physical indicator is *ore grade*, the weight percent of an element in a unit of rock mined. Over time, ore grades generally have declined (see Table 3 with data from Australia), due presumably at least in part to the tendency to start with higher-grade and usually lower-cost deposits before calling on lower-grade and usually higher-cost deposits to satisfy our metal demands. Moving to lower-grade ores arguably is unsustainable in a somewhat different sense than with the R/P ratios and average crustal abundance. Rather than signaling running out of a metal, declining ore grades signal running out of affordable, low-cost metal that does not place excessive demands on environmental quality, as well as energy, water and other resource inputs. We defer to the next section of text our discussion of environmental damage and resource inputs. In this section, we limit discussion to interpreting the causes of declining ore grades and the implications for production costs.

Some of the decline in ore grades undoubtedly is the consequence of depleting higher-grade ores. Nevertheless, there may be another cause-and-effect relationship: declining ore grades also may be the consequence of technological improvements in mineral processing and extractive metallurgy that made it possible, in a commercial sense, to recover metals from ever lower-grade ores [Skinner 1986].

All three physical indicators — R/P ratios, average crustal abundance of elements, and ore grades — contain useful information about nonrenewability and sustainability. But by themselves they are incomplete and potentially misleading — providing an opening and starting point for an economic perspective.

Economists focus on two economic indicators of scarcity, which in turn influence inferences one might draw about sustainability. The first, *extraction cost,* reflects how much producers of metal have to give up to obtain the next unit of resource. We would expect costs to rise over time, given the tendency for production to move increasingly to lower-grade ores that are more difficult to process, deeper below the earth's surface and in more remote locations. The second indicator, a *metal's price,* reflects how much the users of metals have to give up to obtain the next unit of resource. We would expect metal prices to rise as metals become scarcer and users compete for the ever-shrinking stock of resources. To the extent that today's prices are forward looking, we can interpret longer-term price trends as signaling increased scarcity (or not). It is not unreasonable to believe that expectations about the future influence prices today; investors believing that future prices will be higher than current prices will buy additional metal today to sell in the future, leading to higher prices today than otherwise would occur. The historical evidence, as discussed by Tilton (2003), suggests that: (a) extraction costs generally have fallen due importantly to the cost-reducing effects of technological change and (b) long-term trends in metal prices over the past 130 years are more difficult to interpret but nevertheless do not suggest that mineral resources have become markedly more scarce. This is not to say that past trends should alleviate all concerns about the future; rather, simply, that we do not see long-term scarcity looming on the horizon.

An economic view of nonrenewability and sustainability also focuses on how people respond to physical and economic signals of increased scarcity and the threat of metal unavailability. There are three important responses. First, mineral exploration and mine development, pushing farther into the future the threat of material scarcity. Higher prices encourage explorers to search for undiscovered deposits and to further examine known mineralization or known deposits that previously were insufficiently attractive for further investigation. Higher prices encourage mine developers to expand existing operations or bring known but undeveloped deposits into production. These activities serve to reinvigorate reserves that otherwise would dwindle as a result of ongoing mining — illustrating the potentially misleading nature of R/P ratios. Recall the discussion above of R/P ratios and the data in Tables 1 and 2.

Second, substitution occurs in several forms, reducing the need for increasingly scarce materials. One form is element-for-element or material-for-material: e.g., in rare-earth permanent magnets, alter the ratios of various elements while achieving similar magnetic properties, or design a magnet with no rare earths at all. Another is resource saving in that it substitutes better technology for an element or material: e.g., reduce the amount of waste created in machining rare-earth magnets and as a result reduce the amount of material that needs to be purchased per magnet. Still another form of substitution takes place at the technology or system level: e.g., substitute away from rare-earth magnets by designing a wind turbine that does not use rare-earth magnets and instead has a gearbox or induction motor.

The third response is technological innovation. In a market economy, metal producers have an ongoing and continuing incentive to reduce discovery, development, extraction and processing costs — to increase profits. Innovation is one, very important means of reducing costs.[12] Manufacturers of engineered materials also have ongoing and continuing incentives to reduce their costs through innovations that facilitate all forms of substitution identified above. They can reduce costs by optimizing the use of particular elements in a material through advances in materials science and engineering. They can reduce the amount of material they need for a particular application through improvements in manufacturing efficiency and through increased recycling of manufacturing wastes. They can substitute away from expensive input materials by replacing one element with another. Manufacturers of engineered materials also have an incentive to design new materials with better functional properties.

20.6.2 Environmental Damages and Wastes

Declining ore grades, as noted above, also relate to sustainability in an environmental sense. In the face of declining ore grades, how significant are the environmental damages; how large are the energy, water and other resource requirements of metal extraction and recovery; and what is the likelihood of sustaining environmental quality and resource availability? Ore grade is inversely related to the amount of material that has to be blasted, transported and processed to extract and recover a metal. The lower the grade, the larger the amount of material movement (and vice versa). Although not an ironclad rule, one would expect greater movement of material to increase the risks of environmental damage and also increase the amount of energy, water, and other inputs to extract and recover metals; in other words, as ore grades fall, life-cycle impacts likely will rise. The perspectives of material efficiency and the circular economy focus clearly on reducing wastes and also on "closing the loop" — that is, pursuing sustainability by eliminating wastes and maximizing recycling.

An economic perspective on environmental damages and wastes starts with the concept of externalities, introduced earlier. An externality is an uncompensated spillover effect of an activity, such as mining or material production or use. To the extent that producers and users of materials do not consider in their decision making the environmental damages they cause, they impose external costs on society at large. The consequence is economically inefficient production and use of materials — that is, too much production and use from society's perspective. The basic form of an economic prescription for excessive environmental damage is to create mechanisms that force producers and users to recognize the external costs of their actions, known as internalizing what otherwise would be external costs. To internalize costs, economists typically emphasize changing economic incentives

[12] See National Research Council (2002), also mentioned in section 20.4.1.

through, for example, imposing taxes or fees on emissions, requiring transferable permits to be authorized to pollute, and subsidizing pollution control. Each mechanism has its advantages and disadvantages, which are beyond the scope of this paper. If properly implemented, using economic incentives for environmental protection achieves the same level of pollution control and environmental quality at lower costs than alternative regulations requiring specific pollution-control technologies or mandatory cuts in emissions from every polluter. Just as important, using economic incentives leads to more focus on technological innovation in pollution control than policies based on regulations of the sort identified above. The reason is that polluters have an ongoing and ever-present incentive to innovate to avoid paying a pollution tax or for a pollution permit, and innovation to pollute less is one way of avoiding these payments; in contrast, when faced with the requirement to install a specific piece of pollution-control equipment, a polluter will install the equipment and then focus on other (non-environmental) matters.[13]

20.7 What to Do?

What should we do to respond to the challenge set forth at the beginning of the paper, namely produce and use metals today in ways that are consistent with sustaining the natural environment, the social fabric of communities and societies, and human wellbeing?

An engineering approach emphasizes physical stocks and flows of materials and suggests pursuing efficiency in the technical sense — reduce wastes, reuse materials, and increase recycling to reduce environmental damages and escape the constraints imposed by the non-renewability of mineral resources and metals. Nothing in this approach is necessarily bad; more emphatically, all of these goals are, broadly speaking, good ideas.

But this approach does not clearly articulate how to determine specific goals (e.g., how much should we be willing to spend to increase recycling or reduce environmental damage?). Nor does it consider how to define appropriate roles and responsibilities for households, private companies, and governments in the pursuit of these goals. Is more recycling, less environmental damage, and less resource use always better — even if it requires giving up something else that we might value more highly? What activities is it reasonable to leave to the market? What are the essential roles for government?

An economic approach, as suggested throughout this chapter, provides a conceptual lens through which to answer these questions.

The economic approach that I and others suggest emphasizes putting into practice the concepts of economic efficiency and equity, discounting, externalities, and capital. This economic approach is not the cartoon version of economics, in which the market is always good and government activities always bad. Rather, it is an approach that seeks to focus government policies on those areas where private markets systematically fail. Moreover, it is an approach that recognizes that governments often do not have sufficient information to know which specific technological solutions are most efficient and thus should focus

[13] For more on economic approaches to environmental policy, including advantages and disadvantages of different forms of environmental policy, see Sterner and Coria (2011).

policies on providing the right incentives that will encourage private entities, which have more information, to make efficient decisions (Söderholm and Tilton 2012).

With respect to sustainability and mine development and the associated development of regions and nations, an economic approach:

- Recognizes that sustainability (indefinitely) of production is not a sensible goal, even though mines often have significantly longer lifetimes than initial reserves suggest, given the tendency to develop additional reserves during mining.
- Focuses instead on creating and sustaining the wellbeing that mining permits, in a manner that is economically efficient. This means developing public policies that facilitate mineral exploration and mine development, ensure that environmental and social externalities are considered in decisions about whether and how to mine, and sustain the economic benefits created from mining even after a mine closes through appropriate investment in other forms of capital (human, social, natural, man-made physical).

With respect to metal cycles, material efficiency, and the circular economy, an economic approach:

- Believes that, while there is no perfect indicator of resource non-renewability or scarcity, the economic indicators of extraction costs and prices provide valuable guides to the scarcity and thus sustainability of mineral resources and metals.
- Evaluates concerns about environmental damages and wastes as external costs, which must be valued (at least approximately) and incorporated into decision making and public policies.
- Prefers public policies that pursue economic efficiency by altering the incentives facing private decision makers, recognizing that governments often have insufficient information to directly choose the most efficient technological option.
- Prefers public policies that either fund or provide incentives for technological innovation. While not a panacea, innovation has the potential to improve both the technical and economic efficiencies of material production and use and also tends to be underprovided by the private sector alone, given the time and risks associated with research and the difficulty private entities have in fully capturing the benefits of pre-competitive research. That is, governments have an important role to play in the creation of knowledge.

Finally, rather than viewing engineering and economic approaches to these issues as competing or conflicting with one another, a more constructive view is that these approaches are important and complementary ways of thinking about and managing sustainability, mineral development and metal life cycles.

Acknowledgments

This work is supported in part by the Critical Materials Institute, an Energy Innovation Hub funded by the U.S. Department of Energy, Office of Energy Efficiency and Renewable Energy, Advanced Manufacturing Office.

References

Allwood, J.M., Ashby, M.F., Gutowski, T.G., & Worrell, E. (2011) Material efficiency: A white paper. *Resources, Conservation and Recycling* 55, 362–381.

Australian Bureau of Statistics (2015) *Australian Environment-Economic Accounts 2015*, no. 4655.0, Australian Bureau of Statistics, Canberra, [Online], Available: http://www.abs.gov.au/AUSSTATS/abs@.nsf/ProductsbyCatalogue/9EF05B385442E385CA257CAE000ED150?OpenDocument [10 July 2015].

Boardman, A., Greenberg, D., Vining, A., & Weimer, D. (2014) *Cost-Benefit Analysis: Concepts and Practice*, 4th edition, Prentice Hall, Upper Saddle River, New Jersey.

Bureau of Economic Analysis (1994a) Integrated economic and environmental satellite accounts. *Survey of Current Business*, April, pp. 33–49.

Bureau of Economic Analysis (1994b) Accounting for mineral resources. *Survey of Current Business*, April, pp. 50–72.

Bureau of Mines (1956) *Mineral Facts and Problems, 1956 Edition, Bulletin 556*, Washington, D.C., Government Printing Office.

Bureau of Mines (1960) *Mineral Facts and Problems, 1960 Edition, Bulletin 585*, Washington, D.C., Government Printing Office.

Bureau of Mines (1970) *Mineral Facts and Problems, 1970 Edition, Bulletin 650*, Washington, D.C., Government Printing Office.

Bureau of Mines (1980) *Mineral Facts and Problems, 1980 Edition, Bulletin 671*, Washington, D.C., Government Printing Office.

Bureau of Mines (1990) *Minerals Yearbook, Volume 1: Minerals and Metals*, Washington, D.C., Government Printing Office [Online], Available: http://digital.library.wisc.edu/1711.dl/EcoNatRes.MinYB1990v1 [22 July 2015).

Cox, P.A. (1989) *The Elements: Their Origin, Abundance, and Distribution*, Oxford University Press, Oxford.

Hanneson, R. (2001) *Investing for Sustainability: The Management of Mineral Wealth*, Kluwer Academic Publishers, Boston.

Izatt, R.M., Izatt, S.R., Bruening, R.L., Izatt, N.E., & Moyer, B.A. (2014) Challenges to achievement of metal sustainability in our high-tech society. *Chemical Society Reviews*, 43, 2451–2475.

Mankiw, N.G. (2014) *Principles of Economics*, 7th edition, South-Western College Publishers, Independence, Kentucky.

Mudd, G. (2009) *The Sustainability of Minining in Australia — Key Production Trends and Their Environmental Implications for the Future*, Research Report No. RR5, Department of Civil Engineering, Monash University and Mineral Policy Institute, Melbourne, Australia, [Online], Available: http://users.monash.edu.au/~gmudd/files/SustMining-Aust-Report-2009-Master.pdf [22 July 2015].

National Research Council (2002) *Evolutionary and Revolutionary Technologies for Mining*, National Academy Press, Washington, D.C.

Nordhaus, W.D. & Kokkelenberg, E.C. (eds) (1999) *Nature's Numbers: Expanding the National Economic Accounts to Include the Environment* National Academy Press, Washington, D.C.

Pezzey, J.C.V. & Toman, M.A. (2005) Sustainability and Its Economic Interpretations. In *Scarcity and Growth Revisited: Natural Resources and the Environment in the New Millennium* (eds R.D. Simpson, M.A. Toman, & R.U. Ayres), pp. 121–141. Resources for the Future, Washington, D.C.

Portney, P.R. & Weyant, J.P. (eds) (1999) *Discounting and Intergenerational Equity*, Resources for the Future, Washington, D.C.

Rudnick, R.G., & Gao, S. (2003) Composition of the Continental Crust. In *Treatise on Geochemistry*, volume 3 (eds H.D. Holland & Karl K. Turekian), pp. 1–64. Elsevier, Amsterdam.

Skinner, B.J. (1986) *Earth Resources*, 3rd edition, Prentice-Hall, Upper Saddle River, New Jersey.

Söderholm, P., & Tilton, J.E. (2012) Material efficiency: An economic perspective. *Resources, Conservation and Recycling* 61, 75–82.

Statistics Canada (2015) *Canada's Natural Resource Wealth, 2013*, Ottawa, [Online], Available: http://www.statcan.gc.ca/daily-quotidien/150326/dq150326b-eng.htm [10 July 2015].

Sterner, T., & Coria, J. (2011) *Policy Instruments for Environmental and Natural Resource Management*, RFF Press, Washington, D.C.

Tilton, J.E. (2003) *On Borrowed Time? Assessing the Threat of Mineral Depletion*, RFF Press, Washington, D.C.

United Nations (1993) *Handbook of National Accounting: Integrated Environmental and Economic Accounting*, Studies in Methods, Series F, No. 51, United Nations, New York, [Online], Available: unstats.un.org/unsd/publication/SeriesF/SeriesF_61E.pdf [23 July 2015].

United Nations (2003) *Handbook of National Accounting: Integrated Environmental and Economic Accounting 2003*, report of the United Nations, European Commission, International Monetary Fund, Organisation for Economic Co-operation and Development and the World Bank, United Nations, New York [Online], Available: unstats.un.org/unsd/envaccounting/seea2003.pdf [23 July 2015].

United Nations (2014) *System of Environmental-Economic Accounting 2012 — Central Framework*, report of the United Nations, European Commission, Food and Agriculture Organization of the United Nations, International Monetary Fund, Organisation for Economic Co-operation and Development and the World Bank, document symbol ST/ESA/STAT/Ser.F/109, United Nations, New York [Online], Available: unstats.un.org/unsd/envaccounting/seeaRev/SEEA_CF_Final_en.pdf [23 July 2015].

U.S. Geological Survey (2001) *Mineral Commodity Summaries 2001*, U.S. Geological Survey, Reston, Virginia, [Online], Available: minerals.usgs.gov [22 July 2015].

U.S. Geological Survey (2014) *Mineral Commodity Summaries 2014*, U.S. Geological Survey, Reston, Virginia, [Online], Available: minerals.usgs.gov [22 July 2015].

U.S. Geological Survey (2015a) *Mineral Commodity Summaries 2015*, U.S. Geological Survey, Reston, Virginia, [Online], Available: minerals.usgs.gov [17 July 2015].

U.S. Geological Survey (2015b) *Historical Statistics for Mineral and Metal Commodities in the United States*, U.S. Geological Survey, Reston, Virginia, [Online], Available: http://minerals.usgs.gov/minerals/pubs/historical-statistics/[22 July 2015].

Webster, K. (2015) *The Circular Economy: A Wealth of Flows*, Ellen MacArthur Foundation, Isle of Wight, United Kingdom.

World Bank (2002) *World Development Report 2002: Building Institutions for Markets*, World Bank, Washington, D.C., [Online], Available: http://www.worldbank.org/en/publication/wdr/wdr-archive [9 July 2015].

World Bank (2003) *World Development Report 2003: Sustainable Development in a Dynamic World*, World Bank, Washington, D.C., [Online], Available: http://www.worldbank.org/en/publication/wdr/wdr-archive [9 July 2015].

World Bank (2015a) *Extractive Industries Source Book*, World Bank, Washington, D.C., [Online], Available: http://www.eisourcebook.org/[9 July 2015].

World Bank (2015b) *World Development Indicators 2015*, World Bank, Washington, D.C., [Online], Available: https://openknowledge.worldbank.org/handle/10986/21634 [10 July 2015].

World Economic Forum (2014) *Towards the Circular Economy: Accelerating the Scale-Up Across Global Supply Chains*, World Economic Forum, Geneva, Switzerland, [Online], Available: http://www.weforum.org/reports/towards-circular-economy-accelerating-scale-across-global-supply-chains [14 July 2015].

Young, H.P. (1994) *Equity: In Theory and Practice*, Princeton University Press, Princeton, New Jersey.

Zils, M. (2014) *Moving Towards a Circular Economy*. McKinsey & Company, Munch [Online], Available: http://www.mckinsey.com/insights/manufacturing/moving_toward_a_circular_economy [14 July 2015).

21

Closing the Loop: Minerals Industry Issues

William J. Rankin and Nawshad Haque

CSIRO Mineral Resources, Clayton, VIC 3168, Australia

21.1 Introduction

Materials produced from minerals are essential for society and will be essential during the transition to sustainability. What a sustainable world might be like is open to speculation but it is hard to imagine any acceptable form of sustainability in which mineral-based materials would not have an important and ongoing role. In the transition to sustainability, it will be necessary progressively to close the materials cycle in ways that reduce the demand for virgin minerals and reduce the quantity of wastes. How this might be done is the focus of this chapter. We begin by introducing the closed loop and related concepts, then examine the strategies for closing the loop. We then review some of the tools for assessing strategies and illustrate their use with an example from steelmaking. Finally, we discuss the drivers and barriers to implementing closed loop strategies.

The well-known materials cycle shown in Figure 1 provides a useful reference point for our discussion, and we will refer to it throughout the paper. Figure 1 focuses on the materials and products of interest. However, it fails to show the waste that is produced at every stage of the materials cycle, not just at the end-of-life stage of a product.

Wastes are conventionally discharged to the environment or stored in purpose built facilities, and this approach provides the conventional definition of wastes as substances to be disposed of. However, a more useful definition is: wastes are substances produced as part of the materials cycle for which there are no present uses. This definition entertains the

Metal Sustainability: Global Challenges, Consequences, and Prospects, First Edition. Edited by Reed M. Izatt.

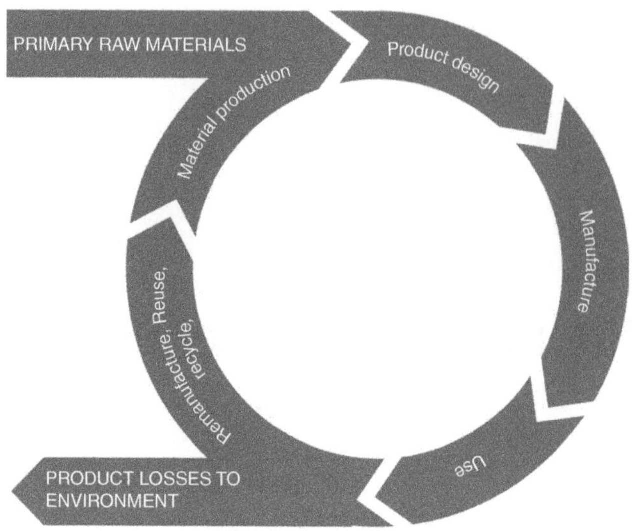

Figure 1 *The materials cycle*

possibility that wastes may, in future or under different circumstances, be useful. Hence, today wastes are often now referred to as co-products.

Our focus here is on the wastes produced during primary production of metals and mineral commodities — that is, the material production stage in Figure 1. However, as will become clearer, the materials production stage cannot be considered in isolation from the other stages of the materials cycle. Nor can the mining and mineral industry be considered in isolation from other industry sectors since they will become increasingly integrated as the focus on waste reduction (or closing the loop) continues to grow.

21.2 The Waste Hierarchy

The reduction and elimination of waste from all stages of the materials cycle are not new concepts and many strategies for waste minimisation have long been known. Our ancestors rarely wasted anything. Many products were remanufactured or reused and materials were recycled, often many times, before being ultimately discarded. Our present 'throw-away' culture largely resulted from the increasing availability of cheap energy and materials and the availability of cheap landfill. There has been a renewed focus on wastes only since the ever-increasing quantities of waste from energy and materials production and consumption began to have wide-scale environmental impacts, in the latter half of the 20th century.

The conventional waste hierarchy — reduce, reuse, recycle (the three Rs) — lists waste management strategies in decreasing order of desirability. Thus, the most desirable strategy is to reduce the quantity of materials and wastes associated with a product, and to consume less of the product. The next most desirable strategy is to reuse a product, and only as a last resort to recycle the materials comprising the product. The three Rs focus on the reduction or minimisation of waste rather than on their elimination and were developed with manufactured products, building and construction products, and domestic waste in mind. Their focus is on things that have had a useful life rather than on things or co-products that have not had a previous use, such as mining and processing wastes.

The focus on minimising waste started shifting towards eliminating wastes at source in the 1990s and, as often is the case, the European Union led the way. The European Union Council Directive 91/156/EEC of 1991 established the hierarchy: waste prevention; recovery; safe disposal. Importantly, it addressed things that have had a useful life and co-products that have had no previous use (such as mining wastes). After several revisions, Directive 2008/98/EC established the present hierarchy:

- prevention
- preparation for reuse
- recycling
- other recovery (e.g., energy recovery)
- disposal

Directive 2008/98/EC excludes wastes resulting from the prospecting, extraction, treatment and storage of mineral resources and the working of quarries, which had been included in earlier versions, since these were covered specifically by Directive 2006/21/EC, which follows a similar hierarchy.

21.3 Reducing and Eliminating Wastes

Mining and processing wastes are of two types: those produced directly in the production of a mineral or metal commodity (at the mine, smelter, etc.) and those produced indirectly (in the production of inputs used at the mine, smelter, etc.). Wastes may be in the form of solids, liquids or gases. The main direct and indirect wastes in primary production are summarised in Table 1. The total waste associated with the production of a metal or mineral commodity is the sum of the direct and indirect wastes.

The quantities of wastes produced in the materials production stage of Figure 1 are very large. This is illustrated in Figure 2, which shows the quantities of solid wastes produced directly in the production of some common commodities relative to the quantities produced of the commodities. Solids make up the largest quantity of primary production waste materials and can range from several times the mass of valuable element for abundant elements (e.g. iron and aluminium ores) to many thousands, and even millions, of times for some scarce elements. The total quantity of directly produced wastes decreases along the value chain from mining to manufacturing to recycling. The largest quantities of solid and liquid wastes are produced during mining and beneficiation; the major quantities of gaseous wastes are produced during high-temperature chemical processing, particularly smelting of metals and cement manufacture.

Wastes from mining and beneficiation have the largest potential environmental impact on land and water, and chemical processing wastes have the largest potential impact on the atmosphere. Some valuable final product is lost at each stage of value-adding and forms part of the waste from each stage. These losses are due to technological limitations, such as the need to leave some ore in mines as support material, incomplete liberation and separation during beneficiation, and bound or inaccessible minerals not amenable to leaching and chemical losses of metals in slags.

Table 1 Common wastes produced directly and indirectly as a result of mining and processing

Mining	Direct	waste rock; dust acid mine drainage CO_2 from fuel consumption
	Indirect	chemical wastes from explosives manufacture; fly ash from power generation CO_2 from power generation; CO_2 & other wastes from explosives manufacture
Beneficiation	Direct	tailings (including reagents) acid rock drainage (from tailings)
	Indirect	chemical wastes from reagent manufacture fly ash and CO_2 from power generation CO_2 & other wastes from reagent manufacture
Chemical transformation (leaching, smelting, electrowinning, etc)	Direct	leach residues; electro-refining residues slag; fumes from furnaces; spent refractories acid, alkaline & cyanide solutions from leach residues CO_2, SO_2, NO_x, etc from smelting
	Indirect	CO_2 & fly ash from power generation

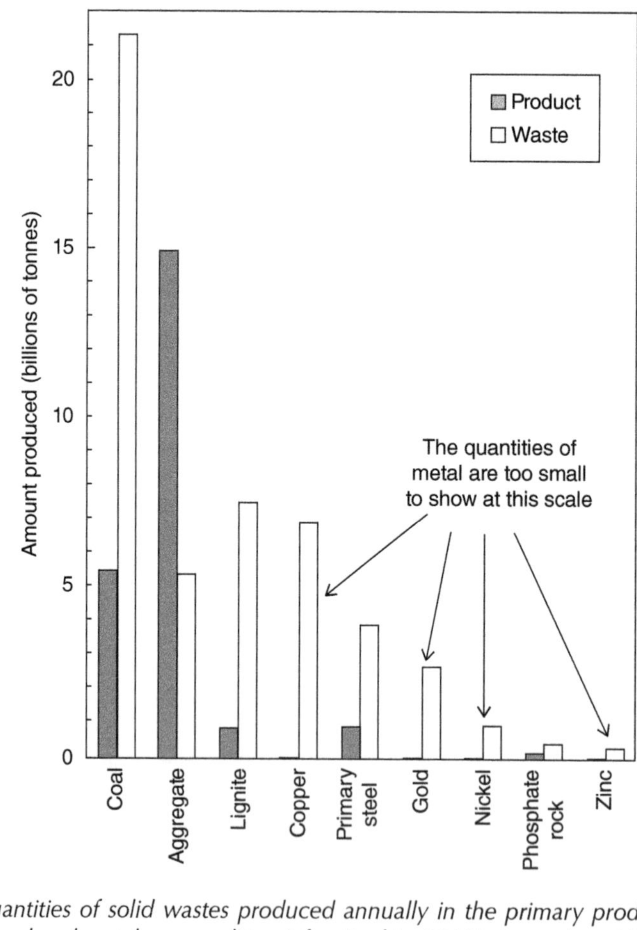

Figure 2 *Quantities of solid wastes produced annually in the primary production of some common mineral and metal commodities (after Rankin, 2011). Reproduced by permission of CSIRO*

The strategies for minimising and eliminating wastes in the production of mineral and metal commodities can be usefully grouped as follows:

- cleaner production
- use of waste as raw materials
- waste reduction through process re-engineering
- industrial ecology.

These are arranged in order of increasing capacity to minimise or eliminate wastes. This order also correlates with increasing degree of integration into the business of a company and the wider economy. Thus, cleaner production can be implemented at a single operation, whereas industrial ecology requires integration across companies and across industry sectors. These strategies are not mutually exclusive alternatives; there is considerable overlap between them and several strategies may be pursued in parallel.

Figure 3 illustrates the historical trend in approaches to addressing the environmental impact of wastes. Company behaviour has moved in recent decades from complying with regulations to being more aware of corporate social responsibility. In the next decades, corporate behaviour will need to move towards closing the loop to dramatically reduce the quantities of wastes. The drivers for change have moved from being almost exclusively profit to include regulations, stakeholders and increasingly to changing social values. In parallel, the materials cycle focus has shifted from a narrow view on products to include co-products. Increasingly, it will shift to the entire materials cycle and, ultimately, to the entire economy.

	Company behavior	Approach to the environment	Materials cycle focus	
Reactive ↓	**Pre-compliance** Driver: profits	• Foul and flee • Dilute and disperse	Products	↓
	Compliance Driver: regulations (1960s, 70s, 80s)	• Pollution control	Products and wastes	*Environmental performance*
	Corporate social responsibility Drivers: risk management; stakeholders; technology efficiency (1990s, 2000s, 2010s)	• Cleaner production • Process re-engineering	Products & by-products Entire materials cycle (stewardship)	*Technology* *Business integration*
Pro-active	**Closing the materials cycle** Drivers: new business opportunities; changing social values	• Wastes as raw materials • Closing-the-loop	Entire economy	↓

Figure 3 *The changing nature of industry's response to the environmental impact of wastes from mining and processing (after Rankin, 2011). Reproduced by permission of CSIRO*

21.3.1 Cleaner Production

Cleaner production, or pollution prevention, is the continuous application of an integrated preventive environmental strategy to processes, products, and services to increase overall efficiency and reduce risks to humans and the environment (UNEP, 2015). Cleaner production principles can be applied to the processes used in any industry, to the products themselves and to services. For production processes, cleaner production involves one or a combination of the following:

- conserving raw materials, water and energy
- eliminating toxic and dangerous raw materials
- reducing the quantity and toxicity of emissions and wastes at source during the production process

Waste is considered as a product with negative economic value. Each action to reduce consumption of raw materials and energy, and prevent or reduce generation of waste, can increase productivity and bring financial benefits to an enterprise.

Cleaner production differs from pollution control, which is an after-the-event, react-and-treat, end-of-pipe approach. Cleaner production attempts to anticipate and prevent. It aims to minimise or avoid practices such as waste treatment (including stabilisation, encapsulation and detoxification), waste dilution to comply with regulations (e.g., releasing contaminated water into rivers or streams during periods of high flow, blending arsenic residues with flotation tailings), and transferring hazardous or toxic substances from one medium to another (e.g., wet-scrubbing gases, then disposing of the contaminants as waste water).

Implementation of cleaner production requires a structured, holistic, common-sense approach using systems and people to both reduce environmental impact and improve the overall company performance.

21.3.2 Wastes as Co-products

Although processing of wastes could be considered as end-of-pipe pollution control and not particularly innovative, this is far from the truth. Many of the co-products from producing mineral and metal commodities that are now considered wastes contain much of value and technologies for extracting that value are often technically sophisticated.

There are several reasons why mining and processing waste should be considered as co-products:

- A mining and processing waste is essentially free (since it has already been mined) and is often in a form suitable for further processing (since it may have been crushed and ground).
- Use of a waste reduces the demand for new mined material since the product produced replaces product that would otherwise have to be mined.
- Use of a waste to make a saleable product reduces the volume of waste that must be stored or disposed of. This saves on storage and disposal costs and can lead to a reduction in the environmental impact. This is particularly advantageous for large-volume wastes such as beneficiation tailings.
- Producing a saleable product adds another source of income for a company.

Use of mining and processing wastes as raw materials for another process is not a new concept, and some wastes from mining and processing are already used to produce saleable products. Sulfur dioxide in smelter gases is routinely converted to sulfuric acid. Slags from iron and steel production are used as aggregate materials, as raw materials in cement manufacturing and as direct additions to concrete. However, mining and processing wastes are a largely untapped resource and there are many other potential applications.

A further advance would be to optimise processes so that all output streams are useful products, rather than optimising processes around the principal product. In this way, all outputs would be optimised (in terms of composition and morphology) to maximise their effectiveness as inputs for further value-adding or, as a last resort, for disposal.

While utilisation rather than disposal of mining and processing wastes is preferable, it must be recognised that not all, or even most, mining wastes can be used productively because the quantities of waste rock and overburden are so large that there are insufficient bulk applications, even in construction projects, to use significant quantities. Furthermore, mines are often located in remote and/or sparsely populated areas and the transportation of low-value construction products (sand, aggregate) to populated areas for infrastructure projects is uneconomic. Hence, the focus is necessarily on wastes from mines and processing operations close to populated areas and/or on higher-value products that can be transported economically over long distances.

Some examples to illustrate the possibilities include utilisation of red mud from Bayer processing of bauxite (Cooling, 2007; Jahanshahi et al., 2007), spent pot lining from aluminium smelting (Mansfield et al., 2002), fly ash from power generation, slags from smelting operations and the use of wastes in geopolymer concrete (Davidovits, 2008). There are many other examples in the literature.

21.3.3 Process Re-engineering

The strategy of waste reduction through re-engineering aims to minimise the quantity of waste produced or to produce a by-product in a form that can be used more readily. This involves process modification or even completely redesigning the flowsheet. There are three broad approaches:

- flowsheet simplification
- use of novel equipment
- use of novel processing conditions

Some examples of each are summarised in Table 2. Flowsheet simplification involves the removal or combination of stages to reduce the overall number of stages required to produce a mineral or metal commodity. This reduces the amount of transport, handling and physical processing of material and can potentially reduce the amount of chemical processing. This saves energy (CO_2 emissions) and reduces the quantities of other wastes. The use of novel reactors or other equipment involves utilising unique characteristics of a reactor or other item of equipment to do something that was previously difficult or impossible to do. The use of novel processing conditions involves utilising relatively standard reactors and flowsheet configurations with different reagents or processing conditions such as temperature, pressure or concentration. While one of these approaches often predominates, a major technological development usually combines aspects of two or all three.

Table 2 *Examples of processes that reduce waste production and/or energy consumption compared with conventional processing*

Flowsheet simplification	Use of novel equipment	Use of novel processing conditions
Finex (http://www.industry.siemens.com/verticals/metals-industry/en/metals/ironmaking/finex/pages/home.aspx); HIsmelt ironmaking (http://www.riotinto.com/ironore/hismelt-4724.aspx processes)	Ore sorting (Cutmore and Eberhardt, 2006)	Electrolytic production of iron using renewable energy (Sadoway, 2000; Cox and Fray, 2009; Xiao and Fray, 2010)
TiRo™ process for Ti metal production (Doblin and Wellwood, 2007; Doblin et al., 2011)	Underground and in-pit pre-concentration (Bamber, 2008) Castrip process for thin strip casting of steel (Sosinsky et al., 2008)	The top gas recycling blast furnace (ULCOS, 2015) Use of biomass as fuel and reductant in metallurgical processes (Birat and Borlee, 2008, Jahanshahi et al 2015)
	Dry granulation of slag, with heat recovery (Xie and Jahanshahi, 2008; Xie et al., 2010) Solar production of aluminium and other metals (Haueter et al., 1999; Murray, 1999, 2001; Halmann et al., 2007; Neelameggham, 2008)	

21.3.4 Closing the Loop

Figure 4(a) shows the conventional flow-through or open-loop production system we are most familiar with. A company takes in new materials and processes them using energy, and generates products and wastes. Both the products and wastes are external to the boundary of the system. They are considered externalities and their environmental impacts are borne largely by society as a whole. Some recycling may take place through recycling end-of-life products into the manufacturing system. Figure 4(b) shows a closed-loop production system. Note, however, it is not quite closed. New materials and energy still come from outside the system and certain wastes still leave it, but products and process wastes remain within the system. Responsibility for products and for process wastes, and for the impacts of their use, is borne within the system.

The unusable wastes which leave the closed-loop system are of three main types:

- wastes generated during the extraction of new materials (e.g., overburden and waste rock from mining);
- wastes that escape from the recycling loop (since some loss is inherent in recovering and recycling materials);
- wastes lost through the use of products (e.g., by being discarded to landfill or incinerated).

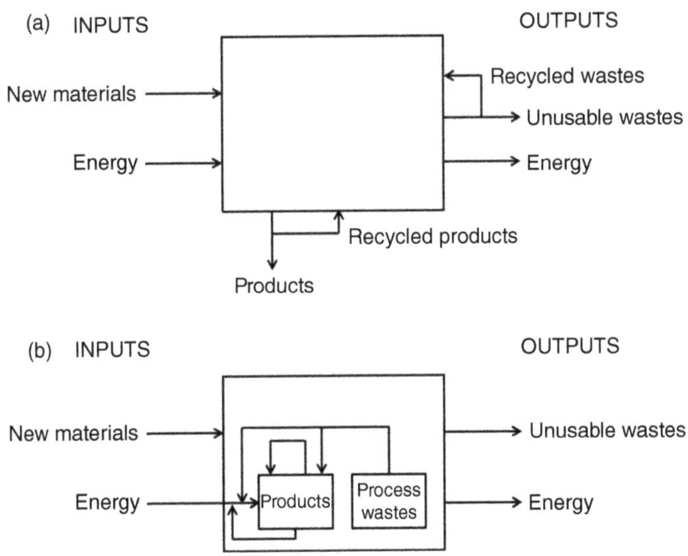

Figure 4 *Schematic representation of (a) open and (b) closed material flow systems (after Frosch, 1995). Reproduced by permission of CSIRO*

Closing the loop in practice is difficult. It means recognising that the industrial system consists of much more than separate stages of extraction, manufacture and disposal, and that the stages are linked across time, distance and economic sectors (Frosch, 1995). In many respects, a closed-loop system operates like a natural ecosystem. While ecosystems produce some wastes (substances that are not recycled), such as fossil fuels and limestone and phosphate deposits, they are largely self-contained and self-sustaining through the constant input of energy from the sun. In similar fashion, closed-loop systems, or industrial ecosystems, involve focusing less on the impacts of each industrial activity in isolation and more on the overall impact of all such activities. Thus, some companies in a system may be inefficient by conventional measures of environmental performance but the overall environmental performance of the group of companies may be superior because of the linkages. An example, discussed in detail in Section 21.4.1, is the use of waste biomass generated from adjacent forestry industries to substitute for coal and thereby reduce greenhouse gas emissions.

The evolution, or planned implementation, of a set of interrelated symbiotic links among groups of firms gives rise to an industrial ecosystem. When these symbiotic links occur within a relatively defined area the links are often referred to as *regional synergies* and the area as an *eco-industrial park*. At the other extreme, the links may occur over long distances to form a sort of virtual industrial ecosystem. There are many eco-industrial parks around the world and they are rapidly growing in number and complexity. Kalundborg (Denmark), Humberside (United Kingdom), Moerdijk and Rotterdam (the Netherlands) and Kwinana (Australia) are frequently cited examples (Kurip, 2007). Kalundborg is the location of a highly evolved eco-industrial park (Ehrenfeld and Gertler, 1997) and is probably the best known example of the implementation of industrial ecology principles. However, fear for the loss of flexibility or profit by individual industrial units hinders the development of eco-industrial parks.

21.3.5 Stewardship

An alternative way of viewing the materials cycle shown in Figure 1 is to consider it as consisting of three interrelated components (Figure 5):

- the resources from which materials are obtained
- the materials themselves
- the goods, products and infrastructure that contain materials

At any time there is a stock of each of these that may increase or decrease. The resource stock consists of all the things and substances from which materials are derived. These are present in or on the Earth; for example, trees, coal, oil, gas, rocks and minerals. The material stock consists of all the materials used to make goods or provide services. Materials are produced from the resource stock; for example, timber from trees, metals from ore minerals and plastics from oil. The product stock consists of the goods and services used by society. These may be in present use or be obsolete or disused products. The material stock is present mainly within products that are in use but it also includes primary and secondary materials and materials in obsolete and disused products.

The sustainability challenge posed by non-renewable resources can only be addressed through an integrated strategy for managing these three stocks. This is the concept of stewardship. In the context of the materials cycle, there are three types of stewardship to consider: resource stewardship, material stewardship and product stewardship.

Adopting stewardship strategies for each of the three forms of stock is a starting point for developing an integrated strategy. Such an integrated strategy is summarised in Table 3. There is already a degree of integration between the three forms that needs to be recognised and built on. Many primary metal smelters treat wastes of various kinds, including scrap metals and more complex materials, and combinations of materials (e.g. waste electrical and electronic equipment). Cement production is often highly integrated with other industry sectors; it utilises not only slag from the steel industry but a wide range of secondary materials, both as fuel and as raw material inputs.

Greater integration, both within the minerals industry and between the minerals industry and other industry sectors, is possible and the potential sustainability benefits are great. This can be achieved by using technologies and business strategies that link aspects of two or more stocks in complementary ways:

- Closed loop and industrial ecology principles can integrate industry sectors, commodities, products, wastes, etc., and are potentially the most powerful integrating strategies.
- Recycling and reuse. Mention has been made of the integration of primary smelters with recycling of metals. Cement kilns in urban areas have an invaluable role in consuming many secondary and waste materials (e.g. sawdust, spent solvents,) that would otherwise be sent to landfill.
- Vertically integrated mining and metals companies have scope for managing the use of primary metals and other commodities through the materials cycle and for optimising processes along the value-adding chain. Vertical integration was common among large minerals companies until recent decades, but the advantage has been lost by many through the business strategy of focusing on mining as the core business. As a result, many companies are now less able to respond to changing expectations, to move into areas of technological growth and to identify new business opportunities.

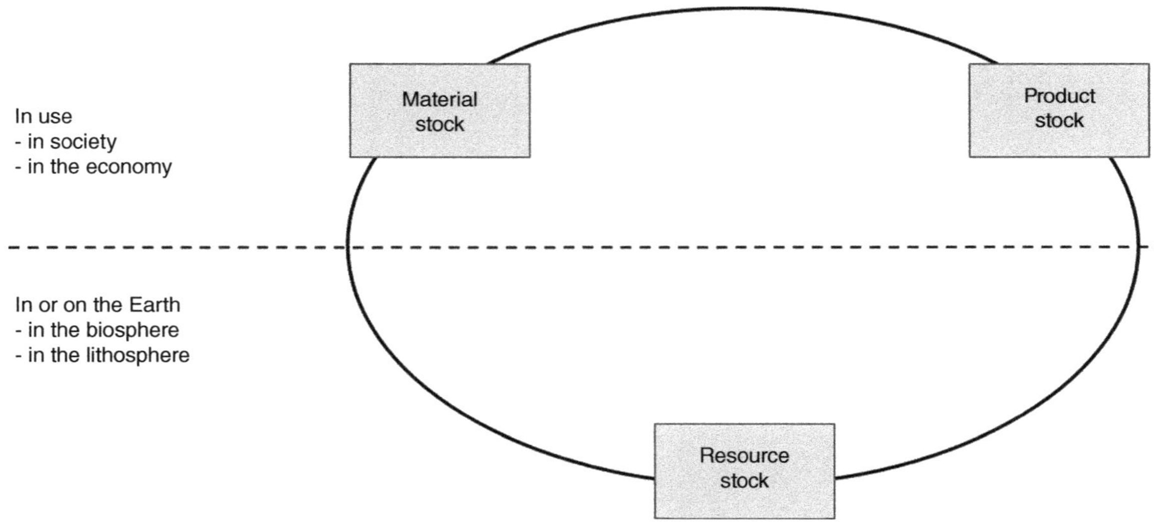

Figure 5 *The interrelationship of the resource stock, material stock and product stock. Reproduced by permission of CSIRO*

Table 3 An integrated stewardship strategy for minerals and metals

Resource stock	Material stock	Product stock
Objective Maximise resource use efficiency	Objective Maximise materials use efficiency	Objective Maximise end use efficiency
Strategies Mine the least quantity of material and extract the most value. Do maximum value-adding at the mine site. Disturb land and the environment to the least extent possible and restore it to its natural state. Minimise the direct and indirect usage of energy and water and the production of wastes through application of closed-loop principles. Prevent dispersion of toxic substances to prevent build-up in the biosphere. Close mines so they can be reopened and close them finally so that they will never become an environmental liability.	Strategies Use each material in applications where its characteristics make it the most appropriate material, bearing in mind its eco- and human-toxicity, relative scarcity and recyclability. Reuse and recycle materials in closed-loop systems that maintain their quality.	Strategies Design products to have the minimum materials required to achieve their function. Substitute services for goods where possible. Design products to maximise the scope for reuse, remanufacture and recycling. Provide systems for: collecting used products; decommissioning infrastructure; separating and pre-sorting materials before reusing or recycling Use landfill and incineration as a last resort and only for benign substances

- The integration of mining, milling and smelting/leaching by optimising the combined value chain rather than optimising each step independently is easier when the three stages are performed at the mine site. Wastes can be more easily managed and opportunities for new industries, based on extracting value from tailings, slags and other co-products, can be better realised. As with vertical integration, there has been a trend away from this in recent decades, with most large mining companies now preferring to do the minimum processing needed to produce a tradable commodity (usually a concentrate).
- The integration of mining, milling and smelting/leaching provides the opportunity to determine the best stage or stages in the value chain at which toxic elements should be removed or managed in order to prevent their dispersion into the environment. For example, arsenic is an impurity in most base-metal sulfide mineral deposits. Because specifications for traded concentrates require a relatively low upper limit for arsenic, arsenic usually has to be rejected at the concentrator. Rejection of arsenic-containing copper minerals (enargite, Cu_3AsS_4) in tailings, however, reduces the overall recovery of copper. A better approach might be to produce a higher arsenic concentrate, with a resulting higher copper recovery, and remove the arsenic as fume during smelting. This would have the added advantage of eliminating most of the arsenic in one highly concentrated stream, which could be stabilised and returned to the mine. This method would be possible only if the smelter and concentrator were co-located.

21.4 Tools for Closing the Loop

Because of the high degree of integration across the columns in Table 3, the identification of opportunities and development of approaches for particular companies, commodities and industry sectors requires tools and methodologies that cover the entire materials cycle, not just parts of it. These include techniques such as material flow analysis, life cycle assessment, the ecological footprint, the water footprint, hidden material flows and others.

Technology developers often claim that a technology is environmentally friendly and cheap. However, these claims need to be quantified and there are well-established techno-economic, environmental and social indicator sets that can be used to judge if a technology is superior to conventional or other alternative processing routes.

Techno-economic evaluation draws upon flowsheet development, mass and energy balances, sizing and scale-up of equipment and process cost engineering. Environmental evaluation can use conventional environmental impact assessment (EIS) methodologies for particular technologies for a geographical location. The environmental impact can be evaluated using internationally recognized tool such as life cycle assessment methodology following international standards (ISO, 2006).

21.4.1 A Case Study: Steelmaking Using Biomass

The use of charcoal derived from woody biomass can reduce carbon emissions from some metallurgical processes, in particular integrated steel making. CSIRO, Australia's national science agency, has investigated this area extensively over a number of years. One study evaluated the concept of a plant supplying charcoal to a steelworks to replace coal in pulverised fuel injection into the blast furnace. The evaluation included both a techno-economic evaluation and an evaluation of the environmental impact.

For this exercise, the conceptual flow sheet shown in Figure 6 was developed (Jahanshahi et al., 2015), and key items of equipment were identified. In the scenario evaluated, woody biomass in mixed shape and size (blocks of wood, small branches, woody stems) is transported from a nearby forest into storage. The biomass is conveyed from a supply bin to a hopper from which it is fed to a chipping unit to produce particles of generally uniform size within a specified acceptable range. Some classification and sorting of particles is required at this stage. Oversize chips are returned to the storage and supply bin for re-chipping; unwanted foreign material is screened out or collected at the bottom and discarded.

Generally, uniform-sized chips (a few mm to several cm long and wide) are acceptable for fixed bed drying. The wet chips at around 50% moisture content (dry basis) are conveyed into a bed dryer. Some cost provisions have been made for miscellaneous equipment such as a small log-handling unit (similar to front-end loaders), chainsaws and other saws for conversion of logs into smaller blocks of wood.

The operating temperature of the hot air over the bed is expected to be around 90 °C at 10% relative humidity. An infeed auger spreads the material across the width of the dryer. The hot air passes through the slowly advancing bed from below through perforated bottom plates. The nearly saturated moist exit gas is discharged by axial fans and is assumed to be at 72 °C and 96% relative humidity. Water may be condensed from the air and used elsewhere or in the process. The air may be recycled, after reheating, to the dryer depending on the configuration. The wet chips are dried to about 12% moisture

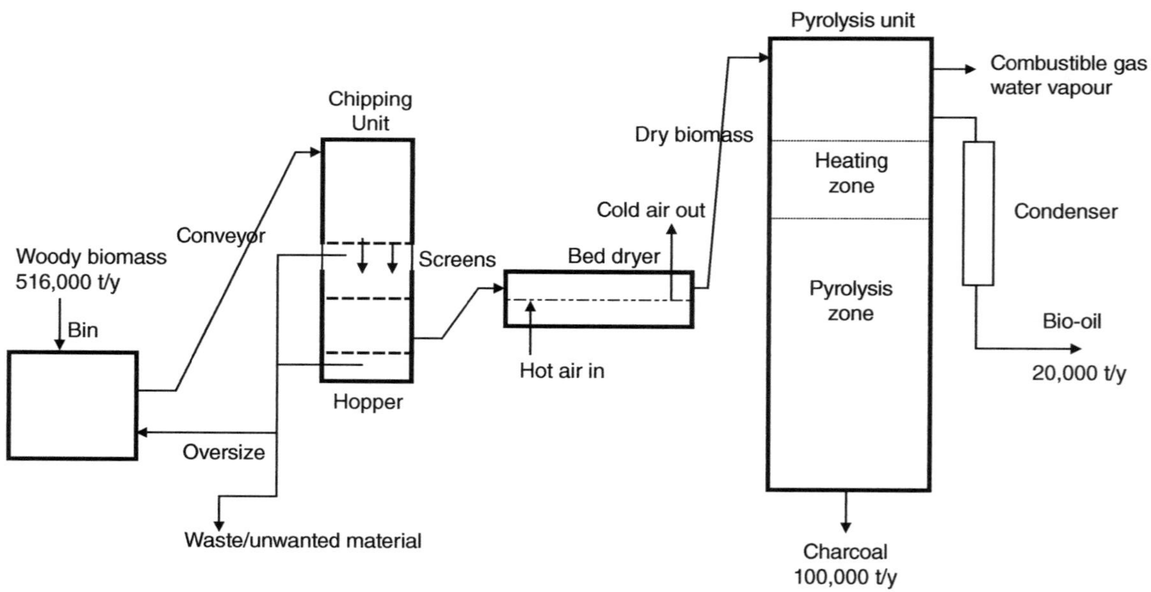

Figure 6 *Conceptual flowsheet of a charcoal plant for producing 100,000 tonnes per year*

(dry basis). It is assumed that the hot dry air for the dryer will be available as low-quality waste heat from the pyrolysis unit.

Pyrolysis is the thermal decomposition of biomass, typically in the range 200 to 500 °C, in an oxygen-free environment to produce solid char, volatiles (condensable hydrocarbons, or tar) and gases, the relative yields depending mostly on the rate of heating and on the final pyrolysis temperature. Charcoal is made by pyrolysis in purpose-built kilns, furnaces or retorts. For this exercise, a Lambiotte-type retort was assumed. The product yields expected were determined from small-scale slow pyrolysis tests on biomass conducted by CSIRO: ~0.3 t charcoal and 0.3 t of condensate per tonne of oven-dry biomass, with about 20% of the condensate being bio-oil.

The evaluation was based on the following assumptions:

- The charcoal plant is located at or near a sawmill and charcoal is transported to a steel plant about 300 km distant.
- The capacity of the charcoal plant is 100,000 tonnes of charcoal per year.
- Raw biomass is collected, chipped and dried close to the biomass source and transported an average of 150 km to the sawmill and charcoal plant.
- The average cost of wet biomass (at 50% moisture content) delivered to the charcoal plant is $60/t.
- The bio-oil product has a value of A$284/t (based on the March 2015 price of crude oil, $58/barrel, and assuming the energy content of bio-oil is 50% that of crude oil). Note that if the diesel price is used, the value of bio-oil will almost double.
- The net yield of bio-oil is 20,000 tonnes (333,000 t oven-dry biomass × 0.3 = 100,000 t condensate × 0.2).
- The 2015 price of standard Hunter Valley (New South Wales) PCI coal is A$131/t. Charcoal has a 50% greater value in-use, reflecting the predicted lower coke rates and improved hot metal Si, S and P contents (Jahanshahi *et al.*, 2015).

21.4.1.1 Economic Benefits

For the techno-economic evaluation, costs of equipment and materials were collected from various sources. The estimated total capital cost of the charcoal plant is A$111 million. The distribution of capital cost items is shown in Figure 7. By far the largest cost component is the pyrolysis unit.

Figure 8 shows the operating cost breakdown and the revenue streams. The total operating cost is $458/t charcoal. 45% of the operating cost was for the supply of raw materials and 16% for transport; maintenance and interest on capital, labour and utilities accounted for the remaining 39%.

Taking into account the carbon value in charcoal ($131/t charcoal, assuming the same as in PCI coal), the value-in-use ($65/t charcoal) and the value obtained from bio-oil ($57/t charcoal), the operating cost shortfall is $458 – $131 – $65 – $57 = $205 per tonne of charcoal.

A carbon tax of $47.4/t CO_2-e) would provide sufficient credits to cover this cost. A sensitivity analysis showed that for the same percentage change in the base case value, changes in the raw material costs had the greatest effect on the operating cost, followed by plant capital-related costs (maintenance and interest/depreciation) and raw material transport, with the labour cost having the least effect.

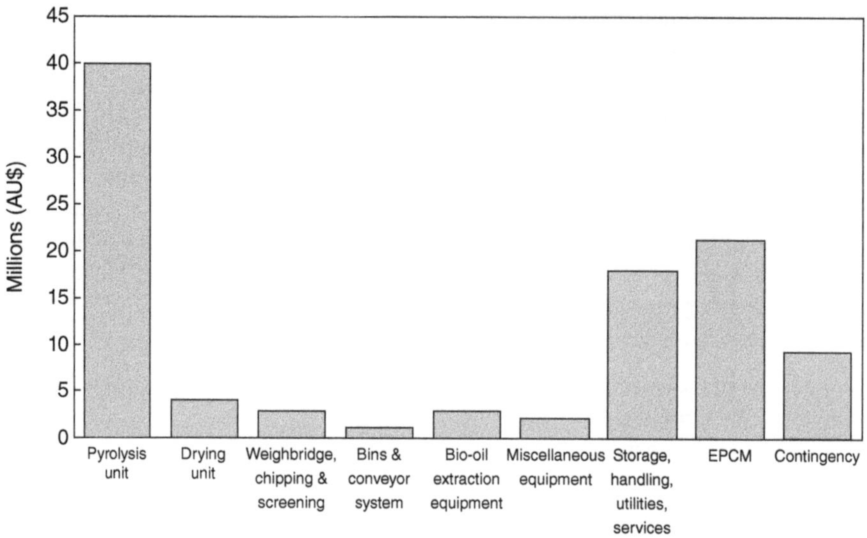

Figure 7 *Breakdown of estimated capital cost of the charcoal plant (total AU$111 million)*

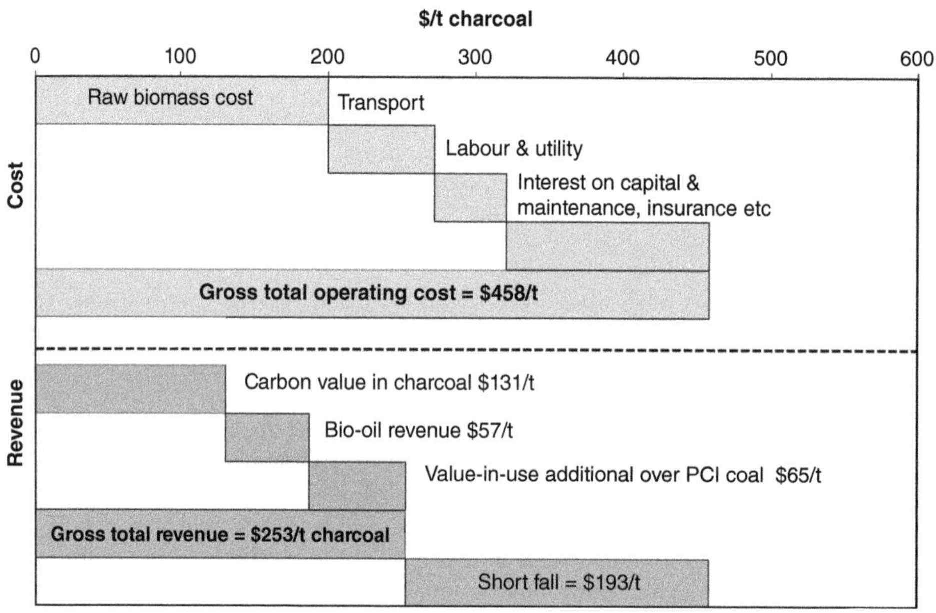

Figure 8 *Breakdown of the cost and revenue for the production of charcoal from biomass (per tonne charcoal basis)*

A more recent capital cost estimate using CSIRO's autogenous pyrolysis technology is A$53 million. This reduction is due to the 75% reduction in residence time for the CSIRO technology compared with a Lambiotte-type retort. The net cost of charcoal delivered to a steel plant then becomes A$264/t when by-product revenues are included. The value of bio-oil is \$134/t charcoal (a higher yield is expected) and the value of the electricity to the grid is \$16/t charcoal. The bio-oil yield in this recent estimate is 34,000t due to the assumption of higher condensate yield (52% compared with 33% in the previous estimate). If additional by-products such as the values of methanol and acetic acid (\$71/tonne) are included (with consideration of additional capital cost for tar processing), the net operating cost shortfall would be \$193/t. Application of a carbon tax of \$47.9/t CO2-e would yield revenue of \$193/t for credits from charcoal and by-products (e.g. bio-oil, electricity). This is similar to the carbon tax rate required in the previous analysis to cover the cost shortfall because, in the present case, credit for the excess electricity generation component is smaller.

Woody biomass can be used for a variety of applications that have environmental benefits. An estimate was made for determining maximum possible value-add per unit of oven-dry biomass. The results in Figure 9 show that pulp production is the highest-value use of oven-dry biomass. This is currently the main use of woody biomass. Electricity production is the least valuable use of biomass. Production of charcoal alone would not add much value since the yield of charcoal is low (~30% for oven-dry biomass). However, if a combination of products is made and credits for carbon emissions are accounted, then charcoal would be an attractive utilization of biomass. Although there are several assumptions in this analysis, and it is simplistic, it nevertheless provides aid in decision making regarding the use of biomass.

21.4.1.2 Environmental Benefits

The question now arises whether the biomass/charcoal option is both environmentally and economically sound. To answer this question, life-cycle assessment (LCA) methodology was used to estimate the greenhouse gas footprint of charcoal production from Mallee eucalypt biomass. This has been described in detail elsewhere by Norgate et al. (2012) and is summarized here.

The results, showing the contributions of the various stages to the greenhouse gas footprint of charcoal production, are given in Figure 10. The greenhouse gas footprint was found to be 105 kg CO_2-e/t charcoal, which reflects the fossil fuel used in its production (plantation management including fertilizer production and use, harvesting and transport). Since the charcoal plant is fuelled entirely by combustion of biomass by-products, any emissions associated with these stages are not fossil fuel-based and, coming from renewable biomass, do not contribute to carbon emissions.

Greenhouse gas credits for the bio-oil and combustible bio-gas by-products can significantly reduce the greenhouse gas impact: electricity is generated using the gas that replaces electricity generated from black coal, and bio-oil replaces diesel. The credits will normally outweigh the 105 kg/t charcoal, meaning the overall GWP is expected to be negative. However, it should be appreciated that the magnitude of the by-product credits depends on the by-product yields from the charcoal retort, which in turn are dependent on a number of factors, in particular the nature of the pyrolysis process (fast or slow) and the biomass feed composition. Slow pyrolysis tends to maximize the charcoal yield, with less gas and bio-oil by-products produced, thus by-product credits are less with slow pyrolysis.

21.4.1.3 Summary

Economic indicator sets (such as capital, operating costs, net present value, internal rate of return, payback period) and environmental scorecards (e.g. embodied energy, carbon and water footprints, waste and its impact) of new technologies or systems can assist industry and

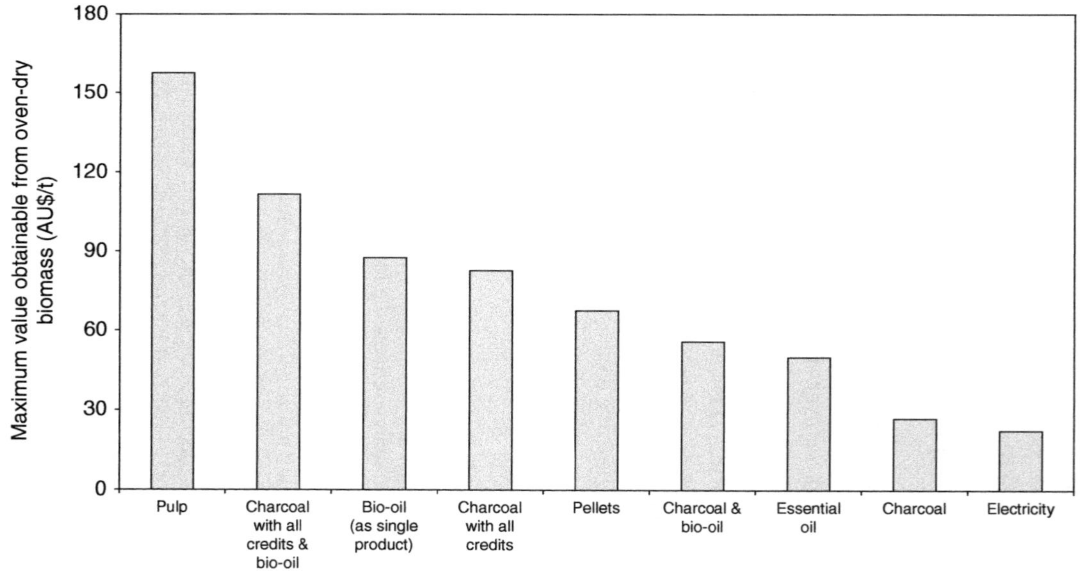

Figure 9 *Estimated value-added per unit of oven-dry biomass for various products*

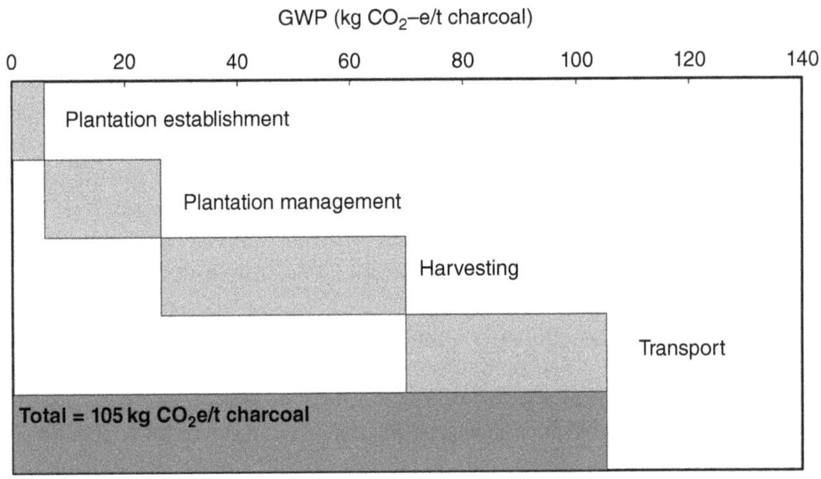

Figure 10 *Stage contributions to Global Warming Potential (GWP), or greenhouse gas footprint, of charcoal production (no by-product credits)*

stakeholders evaluate their potential. If a new technology that is environmentally beneficial is also economically viable, the chances are higher for it implementation. Thus, it is prudent to undertake process evaluation using robust methodology before proposing any technology for environmental solutions. Attempting to make multiple products from single raw material feeds often is an attractive proposition both economically and environmentally and may improve the chance of adoption. However, this frequently requires a different business model from that of the single-product company.

21.5 Closing the Loop: Barriers and Drivers

Very large financial investments are needed in the minerals sector for major changes in technologies. Established technologies have been refined over many years to be reliable and robust, and operations usually give financial returns long after the capital costs have been depreciated. The introduction of new technologies introduces production risks that can, and often do, prove very costly. Minerals companies are reluctant to introduce new technologies unless it can be done in an incremental way with minimum risk to production. Furthermore, the relatively low cost of disposal of mining and mineral processing wastes in most mineral resource-rich countries is a disincentive to do anything other than discard them.

The technical barriers are usually less important than financial and other barriers. Frequently, technical solutions are available or can be developed and implemented but are too costly, risky or difficult to implement for systemic, organisational or regulatory reasons. There are no shortages of ideas and possible approaches even for radically new technologies, many of which would lead to more efficient or cleaner processes. However, developing the right technology at the right time and in such a way that it can be introduced with minimum risk to production has proved a challenge for both technologists and business strategists in the minerals industry.

There is a large degree of entwinement between minerals companies and other industry sectors such as power generation, cement production, infrastructure (roads, rail, ports) and suppliers of reagents and other consumables. Technological changes in one area have implications that flow through the entire system. The co-production of multiple products (due to the complex nature of many mineral deposits), and the need to sell these to different markets with differing and changing demand cycles, adds another layer of complexity. The range of solid, liquid and gaseous wastes produced and the technologies for managing them is another complication. These combine to make mineral and metal production companies technologically complex. This constrains the changes that can be made easily, cheaply and with little risk.

Organisational barriers are primarily cultural. Companies often perceive themselves as in the business of making a particular commodity (alumina, iron ore, steel, aluminium, gold, etc.). All other materials created in making their product are seen as wastes to be disposed of as cheaply as possible. Changing the culture of a company so that it perceives the resource in its entirety, not just part of it, as its greatest asset is a challenge that no minerals company has yet tackled. Internal reward systems and the compartmentalisation of company activities discourage the interactions necessary to achieve the change.

The main issue in most jurisdictions is that regulations fail to promote closed-loop systems and may actually discourage or prevent it. Of particular concern are regulations relating to the use of wastes or co-products as substitutes for virgin materials, and the assignment of liabilities. In many countries, defining a material as a waste or secondary raw material has consequences for what uses are permitted, what administrative procedures apply to its transport, export and processing, and what costs will be incurred (Pongracz, 2002). In some jurisdictions, a company that sells material classified as waste remains liable for any damages that may result from its use, even if it has been reused several times before the damage occurs (Frosch, 1995).

Governments have a leading role in helping and encouraging companies to overcome market failure barriers. They can do this by creating an environment that encourages adoption of closed loop principles and that facilitates formation of industrial synergies, for example by:

- developing more appropriate regulations concerning wastes;
- entering into voluntary agreements with companies or industry associations on targets to achieve;
- applying market-based financial instruments such as tax concessions, taxes on emissions, emissions trading schemes and special purpose grants.

Historically, governments have responded to community expectations for better environmental outcomes through regulatory responses. This approach often prescribes conditions for resource access and use. Mining regulations often specify the maximum allowable level of pollution, minimum requirements for mine-site rehabilitation and the type of management processes that should be used to reduce environmental damage. However, in many situations the regulatory approach has failed to achieve the goals or has proved very expensive. Market-based instruments (MBIs) for environmental management are increasingly used for the management of natural resources and the environment. They are financial instruments that encourage behaviour through market signals rather than through explicit directives (Stavins, 2001). MBIs have been used successfully to control NOx and SOx emissions in the United States. They are appropriate where regulatory approaches have failed to stop ongoing degradation or where the cost is prohibitive. The focus in applying MBIs is on achieving outcomes through the self-interest of companies and individuals. MBIs have two potential financial advantages over more traditional instruments (Whitten et al., 2004). They allow different companies to make different adjustments in response to their unique business structures and opportunities; and they provide companies with an incentive to discover cheaper ways to achieve outcomes.

It is now generally accepted that a combination of increased regulation and use of market-based financial instruments (cap-and-trade schemes, carbon taxes) will drive stewardship. Already, extended producer responsibility, integrated product policy and other product-focused policies are being used to assign responsibility to companies along the materials cycle to manage environmental impacts and optimise resource recovery and recycling. There are currently local, regional and international schemes developing Environmental Product Declarations (EPDs) that document and communicate information about the life-cycle environmental impact of products based on LCA studies and product category rules (IEPD, 2015).

Schemes such as the REACH Directive (EC, 2007) in Europe seek to limit the impacts of toxic materials on human health and the environment. Regulatory controls on end-of-life vehicles and waste electrical and electronic equipment have been introduced in a number of countries. This pressures manufacturers and material suppliers to consider how materials selection will facilitate easier recycling. Directives originating in the European Union are becoming the standard for the development of legislation in other countries. The directives also have a broad impact on global product design, as manufacturers recognise that the trends being established in Europe are spreading around the world.

References

Bamber, A.S., B. Klein, R.C. and M.J. Scoble (2008) Integrated mining, processing and waste disposal systems for reduced energy and operating costs at Xstrata Nickel's Sudbury Operations, *Mining Technology*, vol. 117(3), pp. 142–153.

Birat, J.P. and J. Borlee (2008) ULCOS, the European steel industry's effort to find breakthrough technologies to cut its CO2 emissions significantly. In *Carbon Dioxide Reduction Metallurgy* (eds N.R. Neelameggham and R.G. Reddy), pp. 59, Minerals, Metals and Materials Society, Warrendale, PA, 2008.

Cooling, D. (2007) Improving the sustainability of residue management practices – Alcoa World Alumina Australia. In *Paste* (eds A Fourie and RJ Newell), pp. 3, Australian Centre for Geomechanics: Perth.

Cox, A. and D.J. Fray (2009) Electrolytic reduction of ferric oxide to yield iron and oxygen. In *Energy Technology Perspectives* (eds G. Reddy, C.K. Belt and E.E. Vidal), Minerals, Metals and Materials Society, Warrendale, PA, pp. 77.

Cutmore, N.G., and J.E. Eberhardt (2004) The future of ore sorting in sustainable processing. In *Green Processing 2006*, Australian Institute of Mining and Metallurgy: Melbourne, pp. 287–289.

Davidovits, J. (2008) *Geopolymer Chemistry and Applications*, 2nd ed., Geopolymer Institute, Saint-Quentin, France.

Doblin, C. and G.A. Wellwood (2007) TiRo™ — the development of a new process to produce titanium, 2008, in *CHEMECA: Academia and Industry Strengthening the Profession*, Curran Associates, Red Hook, New York, pp. 280–286.

Doblin, C., A. Chryss and A. Monch (2011) Titanium powder from the TiRO(TM) process. In *Powder Metallurgy of Titanium: Powder Processing, Consolidation and Metallurgy of Titanium,* Brisbane, Australia, 4–7 December 2011, pp. 95–100. Trans Tech Publications Inc, Switzerland.

EC (2007) *REACH in brief*, European Commission, Department Directorate General, October 2007; http://ec.europa.eu/environment/chemicals/reach/pdf/publications/2007_02_reach_in_brief.pdf; accessed 17/3/2015.

Ehrenfeld, J.R. and N. Gertler (1997) Industrial ecology in practice: the evolution of interdependence at Kallundborg. *Journal of Industrial Ecology*, vol. 1(1), 67–79.

Frosch, R.A. (1995) Industrial ecology: adapting technology for sustainable development. *Environment,* 1995, vol 37(10), 16–28.

Halmann, M., A. Frei and A. Steinfeld (2007) Carbothermal reduction of alumina: thermochemical equilibrium calculations and experimental investigation. *Energy*, vol. 32(12), 2420–2427.

Haueter, P., T. Seitz and A. Steinfeld (1999) Anew high-flux solar furnace for high-temperature thermochemical research. *Journal of Solar Energy Engineering – Transactions of the ASME*, vol. 121(1), 77–80.

IEPD (2015) International EPD System – Environmental Product Declarations http://www.environdec.com/; accessed 26/03/2015.

ISO (2006) Environmental management – Life cycle assessment – Requirements and guidelines, ISO 14044. International Standard Organisation, Geneva, Switzerland, 46 pp.

Jahanshahi, S., W.J. Bruckard and M.A. Somerville (2007) Towards zero waste and sustainable resource processing. In *International Conference on Processing and Disposal of Mineral and Industry Waste (PDMIW'07)*, pp.1, Falmouth, UK, 14–15 June 2007.

Jahanshahi, S., Mathieson, J.G., Somerville, M.A., Haque, N., Norgate, T.E., Deev, A., Pan, Y., Xie, D., Ridgeway, P., Zulli, P. (2015) Development of low-emission integrated steelmaking process. *Journal of Sustainable Metallurgy* 1(1):94–114.

Kurip, B. *Methodology for capturing environmental, social and economic implications of industrial symbiosis in heavy industrial areas*, PhD thesis, Curtin University of Technology, Perth, WA, December 2007; http://espace.library.curtin.edu.au/cgi-bin/espace.pdf?file=/2009/07/28/file_1/128365; accessed 17/3/2015.

Mansfield, K., G. Swayn and J. Harpley (2002) The spent pot lining treatment and fluoride recycling project. In *Green Processing 2002*, Australian Institute of Mining and Metallurgy, Melbourne, 2002, pp. 133–142.

Murray, J.P. (1999) Aluminium production using high-temerature solar process heat. *Solar Energy*, vol 66(2), 133–142.

Murray, J.P. (2001) Solar production of aluminum by direct reduction: preliminary results for two processes. *Journal of Solar Energy Engineering*, vol. 123, 125–132.

Neelameggham, N.R. (2008) Solar pyrometallurgy—An historic review. *JOM*, 60, 48–50.

Norgate T, Haque N, Somerville M, Jahanshahi S (2012) Biomass as a source of renewable carbon for iron and steelmaking. *ISIJ Intern* 52(8):1472–1481.

Pongracz, E. (2002) *Re-defining the concepts of waste and waste management: evolving the theory of waste management*, doctoral dissertation, University of Oulu, Department of Process and Environmental Engineering, Oulu, Finland.

Rankin, W.J. (2011) *Minerals Metals and Sustainability: Meeting Future Material Needs*, CRC Press/Balkema, Leiden, Netherlands.

Sadoway, D.R. (2000) Electrochemical processing in molten salts: from green metals extraction to lunar colonization, *Sixth International Conference on Molten Slags, Fluxes and Salts*, Stockholm and Helsinki, 12–17 June, 2000.

Sosinsky, D.J., P. Campbell, R. Mahapatra, W. Blejde and F. Fisher (2008) The CASTRIP® process – recent developments at Nucor steel's commercial strip casting plant. *Metallurgist*, 2008, vol. 52(11–12), 691–699.

Stavins, R.N. (2001) Experience with market-based environmental policy instruments. Discussion Paper 01–58, Resources for the Future, Washington, DC, November 2001

ULCOS (2015) http://www.ulcos.org/en/steel/home.php; accessed 17/3/2015.

UNEP (2015) Understanding cleaner production; www.unep.fr/scp/cp/understanding; accessed 17/3/15

Whitten, S., M. Van Bueren and D. Collins: An overview of market-based instruments and environmental policy in Australia. In *Market-based Tools for Environmental Management: Proceedings of the 6th Annual AARES National Symposium* (eds S. Whitten, M. Carter and G. Stoneham), 2004; www.ecosystemservicesproject.org/html/publications/docs/MBIs_overview.pdf

Xiao, Y.Y. and D.J. Fray (2010) Molten salts electrolysis for sustainable metals extraction and materials processing. In *Molten Salt Electrolysis: Theory, Types and Applications* (eds S. Kuai and J. Meng), Nova Science Publishers, New York, pp. 255–301.

Xie, D. and S. Jahanshahi (2008) Waste heat recovery from molten slags. In *4th International Congress on the Science and Technology of Steelmaking (ICS 2008)*, Gifu, Japan, October 2008, Iron and Steel Institute of Japan, pp. 674–677.

Xie, D., S. Jahanshahi and T. Norgate (2010) Dry granulation to provide a sustainable option for slag treatment. In *Sustainable Mining*, Australasian Institute for Mining and Metallurgy, Melbourne, pp. 22–28.

Index

abundance of elements (on Earth) 11
  average crustal abundance 477–8, 479
  see also critical/finite resources
accounting, green 474–6
acid leaching, backyard 69, 70
acid purification unit/APU (precipitation by acid streams at Asarco's Amarillo Copper Refinery) 403–6, 409–19
actinides 88, 295–316
  ionic liquids for production of 295–316
  lanthanide separations from 301
activated carbon, precious metal extraction 357
Administration Regulation for the Recovery and Treatment of Waste Electric and Electronic Products (China WEEE regulation) 135, 136
Administrative Measures on Pollution Control for Electronic Information Products (2006 - China) 114, 115, 155
Africa 101, 105–6
  dumpsites and scavenging 92–3
  e-waste quantities generated 62
  resources 91
  sub-Saharan 90, 101
  theft of scrap metal 101, 105
  West 93, 98, 99, 102, 105
air poles, platinum and platinum alloy catalysts for 350–2
air pollution (atmospheric pollution and introduction into air) 175–7
  biomarkers 211
  China, toxic (incl. heavy) metals 121, 175–7, 203–4
Ajaokuta Steel Company 90
Aladja, Delta Steel Company 90
alkali metals 88
alkaline earth elements 88
alloying elements 463
  for copper products 457
alumin(i)um (Al) 457–63
  e-waste 119–20
    China 125, 126, 129
  remelting 457–63
amalgamation technology, mercury, in informal gold mining 376
Amarillo, Asarco in (397–399, 402,406, 410, 421)
  history of 399–400
American Smelting and Refining Company *see* Asarco
AnaLig® (MRT-SPE) 272, 273, 279–87
AngloGold Ashanti 369, 386, 387
anions, coordinating, to make ionic liquids 299
anthropogenic sources (incl. urban mining) 3, 13–16, 94–5, 222–3
  e-waste 76–7
  non-essential metals 222–3
  platinum group metals *see* platinum group metals
antimony (Sb)
  in copper production, impurities 404, 405, 406, 407
  pollution in China 173–4, 175
  recycling in China 190–1
  sources and stocks 25
aqua regia for precious metal dissolution 356
aqueous environments *see* seawater; water
aqueous solutions of precious metal, chemistry 356
Arizona, Asarco's activities 399, 400, 424, 436–50

Metal Sustainability: Global Challenges, Consequences, and Prospects, First Edition. Edited by Reed M. Izatt.

Arizona Department of Environmental Quality (ADEQ) 436
arsenic (As) 221–43
as by-product of precious metal processing 370
as impurity 496
in copper production 399, 404–6, 410–14,
management 496
low-level exposures in humans 232–42
pollution/environmental contamination 221–43
China 172, 173, 174
ecosystem health and 242
sample analysis 279–80
recycling
China 187
in nickel carbonate recovery 409–13, 415
sources and stocks 25
speciation in food analysis 275
artisanal mining
mercury (AMM) 203, 204, 205, 212–15, 215
precious metals 375–6
Asarco (American Smelting and Refining Company) 397–452
Amarillo Copper Refinery 397–423, 425
exploration programs 427–33
history 399–400
Asia
dumpsites and scavenging 94
e-waste quantities generated 62
Asia-Pacific WEEE Database and Management Center 158–9
atmosphere, introduction into *see* air pollution
atmospheric slime leaching process in tellurium recovery 415–16
atomic absorption spectrometry (AAS) 273, 274, 277, 280
attitudes to waste, need for changes in 20
Australia
ore grades 478
steelmaking using biomass 497–503
autocatalysts 337–45
automotive/car catalytic converters 14, 15–16, 18, 105, 337–41, 342
demand 337–41
recycling systems 341–2, 383
automation in precious metal mining and ore movement 386, 388
automotive (car) catalytic converters 14, 15–16, 18, 105, 337–41, 342
Avalon Solar Project 442, 448–9

backyard acid leaching 69, 70
bacteria 253–70
in environment 253–4
metal interactions with 255–7
toxic metals and 257–65
barium (Ba), sources and stocks 25
base metals 88
investment 95
precious metal recovery from 388
platinum group metals 323, 362
rare earth metal recovery from processing of 323
recycling from e-waste 119–20
smelting *see* smelting
Basel Convention 53, 54, 103–4, 115, 116
batteries, lead-acid 189
beneficiation 95, 330
waste from 487, 488
beryllium (Be), sources and stocks 25
bifunctional ionic liquids 301–2
bioaccumulation of cadmium, case study 229–31
see also dynamic multi-pathway bioaccumulation model; persistence–bioaccumulation–inherent toxicity approach
bioavailability of toxic/non-essential metals 237–9
mercury in soil 208
biodiversity 254, 257, 265
biogeochemical cycles 222–3
bioleaching 42
Biological Opinion, Rock Creek project 430
biomass, steel production using 497–503
biometallurgy 42, 163
biomonitoring, non-essential metal releases 235
biosolids in land reclamation 436–8, 450
biosorption 42
biota (organisms)
detoxification by 229
mercury 200–1, 209–11
non-essential metal releases and 226–29
biotic ligand model (BLM) 227, 228, 230–1
bismuth (Bi) in copper production, impurities and their removal 399, 404–10, 413–14, 421–3

blast furnace
  iron, zinc recycling from dust 190
  lead, distribution of elements 457
BLM (biotic ligand model) 227, 228, 230–1
blood
  China and heavy metals 122, 214
  non-essential metals 239–40
Bokan–Dotson Ridge, Molecular Recognition Technology to separations of rare earth and platinum group metals from pregnant leach solution derived from 327–30
Botswana 92, 93, 101, 373
Brazil, dumpsites and scavenging 94
brominated flame retardants (BFRs) 58
  health effects 64
business, global 94–6
business-to-business (B2B) markets 1, 32, 39
business-to-consumer (B2C) markets 2, 32
by-products
  deleterious, precious metal extraction 370–1
  metals obtained as *see* companionality

Cabinet Mountains Wilderness Area 427, 429
cactus salvage nursery 442, 443
cadmium (Cd) 120, 221–43
  human exposures and health effects 64, 122, 232–42
    low-level exposures in humans 232–43
  pollution/environmental contamination 221–43
    case study 229–32
    China 172, 173
    ecosystem health and 242
  sources and stocks 25
calcines *see* tailings
Cameroun 98
  Camp Caiman Gold Exploration Project 427, 431–5
Canada
  cadmium environmental toxicology study 229–32
  Environmental Quality Guidelines on acceptable standards for metal contaminants in soil/water/sediment 259
  see also North America
capital and capital expenditure/costs (capex) 471
  charcoal production from biomass 500
  Molecular Recognition Technology and solvent extraction in separation and recovery of rare earth metals 330–1
car (automotive) catalytic converters 14, 15–16, 18, 105, 337–41, 342
carbon, activated, precious metal extraction 357
carbon dioxide, supercritical, in f-element isolation 304
carbonate fuel cell, molten 350
carbonates (precipitation method at Asarco) 409
casual waste workers, China 141–3
catalyst(s)
  automotive (catalytic converters) 14, 15–16, 18, 105, 342
  for fuel cells 349–55
  mercury as 188
  platinum as 336–7
cathode(s), highly durable platinum catalysts and platinum alloy catalysts for 350–1
cathode ray tube (CRT) 58, 64, 68, 75, 102, 119, 120, 121, 152, 155, 158
cation(s), coordinating, to make ionic liquids 299
cationic metals
  speciation and 223–5
  toxicology 226, 227
  in water 224
cattle and reclamation of copper mine tailings 436, 439
cell membrane damage in bacteria 264
cell phones *see* mobile phones
cement 494
  Portland, copper mine/smelter slag-based geopolymers as substitute for 448, 450
cementation (precipitation method at Asarco) 409, 417
  nickel 410–14
cerium (Ce), sources and stocks 25
cesium (Cs), sources and stocks 25
charcoal, steel production using 497–503
charities and e-waste re-use 34
chelating agents
  human excess non-essential metal exposure 235
  resins as, in elemental analyses 275–7, 278
chemical equilibrium approach (BLM model) 227, 228, 230–1
chemical fertilizers, China 175

chemical kinetics approach (dynamic multi-pathway bioaccumulation model; DYM-BAM) 228–9, 230–1
chemical thermodynamic analysis of element distribution in smelting and remelting processes 453, 454–6, 463
chimney flue gas, recycling in China from 189–91
China 112–19, 125–30, 134–220
  autocatalyst demand 341
  e-waste and its management (incl. recycling) 18–19, 112–19, 125–30, 134–50, 151–68, 191
    challenges and efforts in sustainability 125–30
    environmental issues 72, 113, 121
    health impact 72
    legislation and regulations 113–15, 135–7, 139, 154–7
    mechanism and practice 137–9
    metallurgical recovery 151–68
    sources 152–3
    strategies 153–4
  gold
    demand 337
    supply 336
  heavy metals *see* heavy metals
  pollution *see* pollution
  rare earth metals 10–12, 13
    inefficient production capacity 182
    supply to Japan 10–12
  recycling 185–93
    e-waste *see subheading above*
    reuse vs 32
  sustainability 185–93
    future 192–3
chitin, uranium recovery from seawater 308
chloride ions and precious metal aqueous solutions 356
chromium (Cr)
  e-waste 120
    health effects 64, 122
  pollution in China 172, 174
cinnabar/cinnabar ores 200, 201, 202, 204, 208
circuit boards 18, 58
  printed (PCBs; PWBs) 58, 68, 121, 123, 140, 141, 152, 347
circular economy 476–7
Circular Economy Promotion Law, China 113, 115, 125–6
clean chemistry *see* green and clean chemistry
cleaner production *see* prevention and control of pollution
Clēnera 442, 448
closed cycles 1, 2, 16
  lead-acid batteries 189
closing the loop (in production) 492–3, 494, 496–505
  barriers and drivers 503–5
  tools for 497–503
coal use (combustion/consumption), China 177, 189–90, 234
cobalt (Co), mobile phones and computers 15
cofactors of proteins 255–6, 262, 263, 264
collection of end-of-life wastes (incl. e-waste) 16–19
  as barrier to reuse 34
  China 135–9, 157
  developing countries 67–8
  formal sector *see* formal sector
  informal sector *see* informal sector
  see also scavenging
commensalism 254
comminuting 161–2
Commonwealth Scientific and Industrial Research Organisation (CSIRO) 497, 499, 500
companies and corporations
  behaviour 489
  organisational barriers to closing the loop 504
  precious metal mining, and their interactions with government 371–3
  social responsibility 371, 377, 489
companionality (production as by-product) 99
  indium 10
complex matrices, recovery/separations from
  analysis of trace metals after 277–9, 283, 288
  selective, rare earth and platinum group metals 317–32
computers (incl. PCs and laptops) 14, 15, 36, 56, 61, 120, 161
conditioning, end-of-life waste 16–17
consumer market and scrap metal 96–9
consumption, China 124, 125, 153
control of pollution *see* prevention and control of pollution
coordinating ions make ionic liquids 299

copper (Cu) 397–453
Amarillo Copper Refinery 397–423
Australia, ore grades 478
bacterial iron-sulfur clusters 262
China 125, 126, 128, 129
pollution 172
e-waste 119–20
mobile phones and computers 14, 15
industry 397–452
sustainability from perspective of 397–9, 424–26, 438
production 125, 400–21, 458
reserve/production ratios 478
smelting *see* smelting
tailings/calcines 401, 430
land reclamation 436–47
utilization 448, 450
co-products, waste as 490–1
corporations *see* companies
cost-effectiveness of Molecular Recognition Technology in metal separations 322
see also economic issues
countries (various)
e-waste quantities/generation 62, 110
regional areas *see* regional areas
sources and stocks of metals 25–7
see also developed countries; developing countries; national mineral development
crime
illegal mining of precious metals 375–6, 388
scrap metal theft 101, 105–6
critical/finite resources (of scarce metals) 4, 10, 39, 425
increasing demand 15, 24, 317
people's responses to scarcity 479–80
rare earth elements as 13, 24
raw materials (CRMs) 28–9
see also abundance; non-renewable resources; precious metals; specialty metals
crops, mercury 210–11
crust (Earth's) 24, 477
average crustal abundance 477–8, 479
crustal abundance, average 477–8, 479
CSIRO 497, 499, 500
cultural barriers (in organisations) to closing the loop 504
see also sociocultural sustainability and mining
Cupriavidus metallidurans 258, 259, 261
cutaneous exposure to non-essential metals 239
cyanobacteria 258

D4R concept (D4R =design for recycling, repair, refurbishment and reuse) 32, 35, 36
deep eutectic solvents 305
deep-shaft mining of precious metals 379
Delta Steel Company (Aladja) 90
demand 89–90
for precious metals 336–7, 364–6
autocatalysts 337–41
increasing 365–6
for scarce metals, increasing 15, 24, 317
scrap metal, WWII 87
Department of Health and Environmental Services, Montana 428
Department of Highways, Montana 428
Department of Natural Resources and Conservation, Montana 428
Department of State Lands, Montana 428
Department of Game and Fish, Montana 428
Department of Community Affairs, Montana 428
dermatological exposure to non-essential metals 239
detoxification 229–31
developed countries
e-waste flows to developing countries from to 67, 110, 111
e-waste reuse vs recycling 31–2
developing and economy-in-transition countries 52–108
e-waste 52–84, 102–5
environmental and health challenges 71
management challenges 65–8
reuse (vs recycling) 31, 32, 34, 36
solutions for present and future challenges 73–7
informal recycling 43, 44, 85–108
reuse 85–108
e-waste 31, 32, 34, 36
N,N-dioctyldiglycolamic acid (DODGAA) 303
Directives (EU) 29, 30–1, 154, 487
discounting 469, 470
dismantling/disassembly of e-waste in informal sector 140
China 140–5, 147, 159, 161
disposal of end-of-life products, magnitude of problem concerning 7–8

dissolution for precious metal extraction 356
DNA damage and repair in bacteria 263–4
Dodd–Frank Act (2010) 374
domestic appliances *see* household/domestic appliances
drinking water, non-essential metals in 238
  arsenic 239
    in China 173
dumps (waste) 92–6, 96
dust, China
  recycling from 190–1
  toxic metals 72, 121
dynamic multi-pathway bioaccumulation model (DYM-BAM) 228–9, 230–1
dysprosium (Dy), sources and stocks 26

e-waste *see* electronic and electrical equipment waste
eco-industrial parks 493
ecological friendliness with SuperLig® processes 322
economic issues (incl. cost; finance) 467–85
  closed-loop systems 499–501
  e-waste 65
    reuse 34
  formal recycling 43
  Molecular Recognition Technology in rare earth metal separations 322, 330–1
  precious metals
    cost-reduction by miniaturization 384
    production 369–70
  steelmaking using biomass 501–3
  sustainability 468–84
  wellbeing *see* wellbeing
  see also capital; funds; investment; prices; socioeconomic enterprises and development
ecosystem
  biodiversity and maintenance of 254
  contamination by non-essential metals 242
  land reclamation and 436, 438–9, 441
education as key aspect of global vision 20
efficiency
  economic 469–70
  material 476–7, 480, 482
electrical conductivity-based separation with e-waste 162–3
electrochemical recovery of actinides from spent nuclear fuel 306–7
electrodeposition 306, 307
electroless nickel plating 280
electrolysis, refining/purification by *see* electrorefining
electrometallurgical processing with e-waste 163
electronic and electrical equipment (EEE) 23–51, 151
  China, production 153
  ionic liquids in recovery of lanthanides from 307
electronic and electrical equipment waste/scrap (e-waste; WEEE) 2, 23–84, 109–68
  composition 56–8
  definitions 28, 52–3, 109–10
  developing countries *see* developing and economy-in-transition countries
  generation 61–3, 110, 112–13
    China 137, 156, 171, 188
    estimates of global generated quantities 61–3
  key aspects of global vision 20–1
  management (general aspects) 109–33
    challenges 65–8, 124
  problems of 63–5
    response to 29–31
  processing *see* processing
  recycling 29–31, 32, 34–46, 65, 68, 72, 74, 75, 76, 77, 78, 191, 345–9
    barriers to effective recycling 119–20
    China *see* China
    environmental impact in various impacts 72
    formal *see* formal sector
    informal *see* informal sector
    precious metals 345–9
  reuse 31–6
  scavenging 102–5
  typology/categories 53–4
electrorefining/electrolyte purification
  copper 402–9
  gold 359, 382
  silver 359
electrowinning
  copper 402, 410
  gold and silver 387
elements
  abundance *see* abundance
  analyzers (elemental analysis instrumentation) 272–3
  distribution in metal/slag/gas phases during metallurgical processing 453–66

employment *see* occupation
end-of-life products 2, 73–5, 102–5, 119
  autocatalysts, extraction and refining technologies 342–3
  magnitude of problem concerning disposal 7–8
  management (incl. recovery and recycling) 16–19, 74–5
  product design and 73–5
  separations 16–19
energy (power)
  consumption in China 181–2
  precious metals industry requirements 379–80
  renewable *see* renewable energy
Energy-Using Products Directive (EuP Directive 2005/32/EC) 31
environment 221–43, 472, 480–1
  Asarco and 426, 427, 428, 429, 430–3, 436,450
    copper refining and 402
  bacteria in 253–4
  company approaches to 489
  dynamics of metal flow in 90
  e-waste and 70, 71, 75
    China 72, 113, 121
    toxic components 64
    various countries 72
  economic perspective on damage to 480–1
  informal recycling and 45–6
  market-based instruments (MBIs) for management of 504–5
  non-essential metal releases into 221–43
  pollution *see* pollution
  precious metal impacts 373–4
  SuperLig® processes and 322
  sustainability 468, 472
  toxicology 226–32
Environmental Protection Agency (U.S.) acceptable standards for metal contaminants in soil/water/sediment 259
Environmental Quality Guidelines (Canada) on acceptable standards for metal contaminants in soil/water/sediment 259
enzymes and metals 255–6, 258, 263
equipment, novel, waste reduction via use of 491
equity 469–70, 473
erbium (Er), sources and stocks 26
Escherichia coli 258, 259, 262
Europe
  autocatalyst demand 337–41
  e-waste quantities generated 62
European Union (EU)
  acceptable standards for metal contaminants in soil/water/sediment 259
  Directives 29, 30–1, 154, 487
  European Commission's "Sustainable Consumption and Production and Sustainable Industrial Policy (SCP/SIP) Action Plan" 397–8
europium (Eu) 6
  sources and stocks 26
evaluation
  in precious metal recycling 381
  techno-economic 497, 499
exploitation spiral (with non-renewable resources) 385–6
exploration (and exploration programs) 424–427, 450, 472
  Asarco 427–35
export of scrap metal 99–102
externalities 319, 470–1, 480
extraction *see* mining; separation/extraction/purification

f-block elements *see* rare earth elements/metals
fairness (in outcomes) 469, 470, 473
feedstock collection *see* collection
ferrous metals (from high-tech sources), recycling 36–7
  see also iron
fertilizers, China 175
financial issues *see* economic issues; funds
finite resources *see* non-renewable resources
fish
  cadmium 231–2, 231, 233
  mercury 200, 209–10
five-year plan (China)
  11th 189
  12th 182, 183, 184, 186
flame retardants *see* brominated flame retardants
flow of materials *see* mobility
flowsheet simplification 491, 492
flue gas, recycling in China from 189–91
food
  arsenic speciation and analysis of 275
  non-essential metals in 239

formal sector (collection and recycling), e-waste 39–43, 135–9
  China 115–16, 117, 135–9
  informal sector interactions with 145–6
French Guiana, Camp Caiman Gold Exploration Project 427, 431–3
freshwater *see* water
fuel cells, catalysts for 347–55
Fukushima Daiichi nuclear disaster 287
functionalized materials
  ionic liquids 298, 300, 301–2, 303, 305
  in precious metal recycling 387–8
funds and funding 473–4
  e-waste in China 138–9
fungibility, precious metals 374

Gadani, ship graveyard 86
gadolinium (Gd), sources and stocks 26
gallium (Ga)
  bacterial nutrient assimilation and 264
  sources and stocks 26
gas phases during metallurgical recovery/processing, distribution of elements in 453–66
gastrointestinal tract, non-essential metals 233, 238–9
genotoxicity in bacteria 263–4
geopolitics *see* political concerns
geopolymers, copper mine/smelter slag-based 448, 450
germanium (Ge), sources and stocks 26
Ghana
  dumpsites and scavenging 93
  scrap metal
    export 102
    narrative of collector 103
global warming potential, steelmaking using biomass 501, 503
goals, appropriate adjustments in 20
gold (Au) 337
  Australia, ore grades 478
  demand 337, 364, 365
    increasing 365–6
  industry
    Camp Caiman Gold Exploration Project in French Guiana 431–3
    refining 359, 362–4, 381, 382
    structure 362
    sustainability and sustainable development 376–7
  mining 368, 369, 370–1, 374, 375, 376, 379
  mobile phones and computers 14, 15
  price increases and volatility 366, 367
  sources and stocks/reserves 25
    China 126
  supply 336, 365
  uses 364, 365
government
  market failure barriers and role of 504
  policy *see* policy
  precious metal mining companies and their interactions with 371–3
  see also political concerns
grades (ore) 478–9
  Australia 478
  precious metals
    decreasing grades 368
    high-grade materials 382–3
    lower-grade materials 382–3, 385–6, 387, 388
green accounting 474–6
green and clean chemistry 20, 271–94, 317–32
  definition 318
  informal recycling 19
greenhouse gas footprint, steelmaking using biomass 501, 503
gross domestic product (GDP) 474–5
groundwater
  naturally occurring non-essential metals 238–9
  pollution in China 171
Guangzhou SE-MBA New Plastics Technology Co., Ltd. 158
Guiyu 118, 121, 122, 191
Guizhou province
  antimony and its mining 175
  mercury mining 200, 201, 203, 204, 205, 210, 213, 215
    Lanmuchang 204, 205, 208, 210
    Wanshan 201, 203, 204, 205, 208, 209, 211, 212, 213, 214, 215, 216
    Wuchuan 203, 204, 205, 208, 209, 212, 213, 214, 215
    Xunyang 203, 208

hafnium (Hf), sources and stocks 26
hair, mercury 212–13, 215–16
Harmony Gold Mine 362–4

hazardous materials and components in e-waste 58, 64, 152
example items containing 56
screening plastics for 74–5
see also Restriction of Hazardous Substances in Electrical and Electronic Equipment
health issues (incl. burden on human body of exposures) 211–16, 232–42
China
e-waste 72, 122, 143
heavy metals 211–16
e-waste
China 72, 122, 143
toxic components 65, 68, 70, 71–2
Vietnam 72
informal recycling 45–6
non-essential metals at low-level exposures 232–42
platinum refining 373
see also pollution
heavy metals, China 169–99
burden on human body 122
example accidents (2008-2012) 172
pollution 170–85, 192
see also non-essential metals; toxic metals
Hecla Mining Company 427
high-value scrap, precious metals from 381
highly-selective resins for metal ion separations 273, 278
holmium (Ho), sources and stocks 26
Hong Kong, environmental impact of e-waste recycling 72
household/domestic appliances
China, e-waste generation 112
examples 55
human
exposures *see* health issues
microbiome 254, 265
wellbeing *see* wellbeing
hydrometallurgy 42, 70, 163, 164, 166, 304
precious metals 382
hydroxides (precipitation method at Asarco) 409–12,

IAMGOLD Corporation 433
IBC Advanced Technologies, Inc. (IBC) 273, 323, 328, 359
illegal mining of precious metals 375–6, 388
imidazolium ionic liquids 301
Imperial smelting process 457
import, China 125, 153
iron ore 127
impurities (in production and processing) and their removal 454, 456, 463
aluminum remelting 460
arsenic as *see* arsenic
copper production 401, 403–5, 410, 413, 422, 454, 456
incineration 8, 9
India
autocatalyst demand 341
environmental impact of e-waste recycling 72
gold
demand 337
supply 336
indium (In) 10
recycling from flue gas in China 190
sources and stocks 10, 26
inductively-coupled plasma mass spectrometry (ICP-MS) 273–5, 275, 277, 278, 279, 283
inductively-coupled plasma optical emission spectrometry (ICP-OES) 273–4, 275, 277, 278, 280, 283, 284, 285
industrial action, precious metals industry 374–5
industry 485–507
China
admittance threshold to limit projects with heavy metal emissions 183–4
future initiatives 193
processes for metal recovery from e-waste 166–7
copper *see* copper
liquid-liquid extraction (LLE) technology 304
precious metals *see* precious metals
informal sector (collection and recycling incl. e-waste) 18–19, 43–6, 139–46
China 116–19, 140–5
developing countries 43, 44, 85–108
formal sector interactions with 145–6
see also artisanal mining
information and communication technology, waste from (ICT; IT) 77
developing countries 63, 77
informal recycling 44
reuse 35, 36
integration 494–6
smelting 17–18, 19, 167, 168, 494, 496

internalizing costs 480–1
international initiatives (incl. cooperation)
  e-waste management in China 158–9
  precious metals and sustainability 378
investment 474
  developing countries 95–6
ion exchange (IX) resins 277, 283
  precious metals 357
ionic liquids for actinide and lanthanide production 295–316
iridium (Ir) 10
  selective recovery from complex matrices 323
  sources and stocks 10, 25
iron (Fe)
  blast furnace, zinc recycling from dust 190
  ore mining in China 127
  recycling (incl. e-waste) 119–20
    China 128
  see also ferrous metals
iron-sulfur clusters, bacterial 262
ISO-9000: 2000 400
ISO-9001:2008 33, 321, 400, 421
ISO-9002 400

Japan
  autocatalyst demand 337
  radionuclide analysis from Fukushima Daiichi nuclear disaster 287
  rare earth elements in
    rainwater (Tokyo) 279
    supply from China 10–12
  Tanaka Kikinzoku Kogyo K.K. 2, 18, 343, 352
jewelry, precious metal use 335, 336, 337, 365, 366
jobs *see* occupation

Kootenai National Forest (KNF) 428, 430
Krugersdorp landfill 92

labor relations, precious metals industry 374–5
land reclamation from tailings 436–42, 443–47, 450
landfill 7–8, 9
  Africa 92
  Asia 94
  Latin American countries 94
Lanmuchang Hg mining 204, 205, 208, 210
lanthanides 88, 295–316
  actinide separations from 301
  ionic liquids for production of 295–316
lanthanum (La), sources and stocks 26
laptops 14, 15, 36
Latin America
  dumpsites and scavenging 94
  e-waste quantities generated 62
law *see* legislation and regulations
leaching (process)
  backyard acid leaching 69, 70
  integrated 496
  in tellurium recovery 415–17
lead (Pb) 221–43
  Australia, ore grades 478
  China 126, 177
    industry (incl. smelting) 165–6, 184, 190
    recycling 189
  e-waste 119–20
    health effects 64, 122
  environmental releases and their effects 221–43
    ecosystem health 242
    humans and their health 232–42
  in high matrix plating solutions, analysis of trace levels 280–2
  recycling 102, 119, 120
    China 189
  smelting 165–6, 184, 190, 457, 459
lead-acid batteries 189
legislation and regulations 4
  Asarco's mining initiatives in Montana and 427–30,
  China 177–81
    e-waste 113–15, 135–7, 139, 154–7
    heavy metal pollution 177–81
  closing the loop and 504
  e-waste 29–31
    barriers to reuse 34
    China 113–15, 135–7, 139, 154–7
    developing countries, problems with 65, 68
    informal sector 45
  precious metals industry 371–3, 374
  scrap metal in South Africa 98
lichen as mercury pollution biomarkers 211
lifespans/life cycles 467, 468, 476
  e-waste 63, 67, 76

ligands
  biotic ligand model (BLM) 227, 228, 230–1
  metal-selective, precious metal recycling 387–8
lipid peroxidation in bacteria 264
liquid-liquid extractions (LLE)
  lanthanides and actinides 298–304, 308
  trace element analysis 273
liquid waste, precious metal extraction and refining 347–8
liver (fish), subcellular partitioning of cadmium 231
livestock and reclamation of copper mine tailings 439
local discussions of sustainability 468
  see also regional areas
lutetium (Lu), sources and stocks 26

magnetic separation with e-waste 162
Management Regulation on the Recycling and Treatment of Disposed Appliances and Electronics Products (MRRTDAEP) 156
manganese (Mn)
  and batteries 189
  and superoxide dismutase 264
manual sorting and dismantling e-waste, China 140–5
manufacturers, need for change in vision 21
manure and reclamation of copper mine tailings 439
market
  consumer, scrap metal and 96–9
  failure barriers, role of government 504
  market-based instruments (MBIs) for environmental management 504–5
marketing as barrier to e-waste re-use 34
material(s)
  cycle 485–6, 489, 494
  efficiency 476–7, 480, 482
  flow *see* mobility
  raw *see* raw materials
  stock, integration 496
    with resource stock and material stock 495
mechanization in precious metal mining and ore movement 386
media as key aspect of global vision 20
medical devices, examples 55
membrane (cell) damage in bacteria 264
mercury (Hg) 200–20, 221–43
  aquatic environments 225
  China 26
    environmental impact/pollution 174, 177, 202–11
    recycling 188
  ecosystem health and 242
  health effects (humans)
    e-waste 64
    non-essential metals and low-level exposures 232–5
  in precious metal industry
    amalgamation technology 376
    as by-product of precious metal processing 370
  sources and stocks 26
metabolism of non-essential metals 239
metal(s), definition 86–7
metalloenzymes and metalloproteins (metal-containing protein/enzymes) 255–6, 258, 263
metalloids 88
  metabolism 239
metallurgical recovery/processing
  China 151–68
  distribution of elements in metal/slag/gas phases during 453–66
  e-waste 42, 71, 151–68
  precious metals 382
MetaSep AnaLig® separation process 284
methylmercury (and other organomercury complexes) 203
  in biota 200–1, 209–11
  in calcines 203
  human exposure 211–13, 215–16
mutagentic effects 263
  in soil 205, 206
  in water 209, 298
microbiome and microbiota 265
  human 254, 265
β2-microglobulin, urinary 215
milling, integration with mining and smelting/leaching 496
mineral
  development (and mine) 426, 427, 428, 432, 471–6, 479, 482
  exploration *see* exploration
  integrated stewardship 496

mineralogy, precious metals 368–9
miniaturization (of devices), cost reduction by 384
minimum inhibitory concentrations (MICs), bacterial 258, 259
mining
  China 200–20
    heavy metal pollution of groundwater near to areas of, China 171
    mercury 200–20
  copper 400
  integration with milling and smelting/leaching 496
  as largest producer of solid waste 318
  mine and mineral development 426, 427, 428, 432, 471–6, 479, 482
  precious metals 362, 366, 369–80
    artisanal 375–6
    illegal 375–6, 388
    improvements 386–8
    lower-grade materials 385–6, 387, 388
    paradox of 'sustainable' mining 377
    safety 373
  raw materials *see* raw materials
  sustainability of wellbeing originating from 472–3
  sustainable, defining 425–7
  tailings *see* tailings
  urban/anthropogenic *see* anthropogenic sources
  waste from *see* waste
  wellbeing originating from *see* wellbeing
  see also ores
Mining and Reclamation Plan (MARP) 441–2
Mission Mine and Complex 399–401, 424, 436, 437, 438, 439, 441, 442, 448
mitigation measures, Rock Creek, Montana 430
mitigation measures, Camp Caiman Gold Exploration Project in French Guiana 431, 432
mobile phones/cell phones (incl. smart phones) 13, 14, 14–15, 29, 40, 57, 65, 77, 119, 345, 346
  gold content 384
mobility (incl. flow) of materials 86
  anthropogenic vs natural mobilization of non-essential metal releases 222
  China 112–13, 137
  e-waste from developed to developing countries 67, 110, 111
  metals 89–90
Molecular Recognition Technology (MRT) 13, 271–94, 317–32
  copper refining at Asarco 404–5, 406–9
  in green chemistry 271–94
  precious metals 359, 387
  in selective recovery of rare earth and platinum group metals from complex matrices 317–32
  for solid phase extraction (MRT-SPE) 272, 273, 279–87
molten carbonate fuel cell 350
Montana, Asarco in 427–31
  Department of Environmental Quality (MDEQ) 430
moss as mercury pollution biomarkers 211
mutagenicity in bacteria 263–4

national mineral development, implications 473–6
nationalism, sources, precious metals and 371–3
natural sources of non-essential metals 222–3
neodymium (Nd) 13
  sources and stocks 26
net domestic product (NDP) 475
nickel (Ni)
  Australia, ore grades 478
  electroless plating 280
Nigeria
  e-waste 57, 68, 72, 75, 75–6
    environmental impact 72
  scrap metal 96
    theft 105
Nippon PGM 343
nitric acid for precious metal dissolution 356
non-essential metals
  bacteria and 256–7
  releases into environment 221–43
  see also heavy metals; toxic metals
non-ferrous metals (from high-tech sources)
  China
    drain on production capacity 182
    industry 184
    recycling 185
  recycling 36–7, 89
    China 185–6
non-polar resins 278

non-renewable resources and metals 425, 477–8
economic view 477–8, 479
exploitation spiral 385–6
see also abundance; critical/finite resources
North America
autocatalyst demand 337, 341
e-waste quantities generated 62
see also United States
nuclear disaster, Fukushima Daiichi 287
nuclear fuel, spent, lanthanides and actinides from 296–7, 304, 306
nutrient assimilation in bacteria 264
nutritionally-essential metals 87–9

occupation (jobs; work; employment) in e-waste
China 70, 141–3
concerns over exposure (incl. health) 70, 143, 214, 215
see also labor relations
oligomeric ions 305
open cycles 2
operating expenses (opex), Molecular Recognition Technology and solvent extraction in separation and recovery of rare earth metals 330–1
Ordinance on Management of Prevention and Control of Pollution from Electronic and Information Products 135
ores 304–7
beneficiation *see* beneficiation
grades *see* grades
mining *see* mining
sources and stocks 25–7
organ(s), non-essential metals 238–9, 240
organics incorporated in copper mine tailings 437, 439
organisational barriers to closing the loop 504
organisms *see* biota
organomercury complexes *see* methylmercury
osmium (Os) 371
as by-product of precious metal processing 370
sources and stocks 25
oxidative stress in bacteria 262, 264
oxygen, bacteria and reactive species of 261–2

Pakistan, ship graveyard at Gadani 86
palladium (Pd)
demand 335, 364, 365
as autocatalyst 337–41
increasing 366
mobile phones and computers 14, 15
price increases and volatility 367
recycling 335
selective recovery from complex matrices 323
sources and stocks 25
supply 335, 365
uses 364, 365
pallets, plastic, Asarco 419–20
paradigm shift 20, 78
Patriot Act (2001) 374
PBT (persistence–bioaccumulation–inherent toxicity) approach with metal–organisms interactions in freshwater 226–7
perch, cadmium 231–2, 232, 233
periodic table (metals of) 86, 88, 94, 321
bacterial interactions with 256
recycling rates 100
persistence–bioaccumulation–inherent toxicity (PBT) approach with metal–organisms interactions in freshwater 226
Peru
dumpsites and scavenging 94
precious metal-related environmental incident 374
Philippines, San Mateo 92, 94
phosphor(s) 123–4
phosphoric acid fuel cells (PAFC) 340
photovoltaics (solar)
Asarco's Pima Country facility on disturbed lands 442, 448–9
high-tech waste from 36
phytoremediation 189
Pima County
land reclamation 436, 439
solar generating facility 442, 448–9
Regional Wastewater Reclamtion Department (Wastewater Department) 436, 450
placenta, heavy metal burden, China 122
plants, mercury 210–11
see also phytoremediation
plastics
in e-waste 152
screening for hazardous components in e-waste 74–5
plastic, at Asarco 419–20

plating solutions
high matrix, analysis of trace levels 280–2
precious metal extraction and refining 347–8
platinum (Pt) 10
demand 334, 336–7, 364, 365
as autocatalyst 337
increasing 366
uses 364, 365
for fuel cells 350–3
price increases and volatility 367
production cost increases 369–70
recycling 334, 387
refining, health risks 373
selective recovery from complex matrices 323
sources and stocks 10, 25
supply 334, 365
platinum group metals (PGM) 2, 9, 10
analysis of trace levels in recycled materials 282–3
industry
refining 381
structure 362
recycling 362, 387
refining 362, 363, 370, 381
selective recovery from complex matrices 317–32
urban mining 14–16
from catalytic converters 15–16, 18
platinum(Pt)/ruthenium(Ru) alloy catalysts, fuel cells 352–4
policy 20–1
e-waste management 71
China 113–15
heavy metal pollution in China 181
see also government
political concerns (incl. political stability and geopolitics) 10–12
China 24
precious metals industry 366, 371
pollution (environmental) 168–85, 481
China 121, 169–85
heavy and/or toxic metals 121, 169–85, 192, 202–11
laws and regulations 114, 177–81
point vs diffuse pollution 257
see also environment; health issues
polymer electrolyte membrane fuel cells (PEFC) 350, 350–1
polyvinyl chlorides (PVCs), China 188, 202
pond water, China, mercury 208–9
Portland cement, copper mine/smelter slag-based geopolymers as substitute for 448, 450
power *see* energy
praseodymium (Pr), sources and stocks 26
precious metals 1–22, 89, 123, 333–96
definition 333
demand *see* demand
fungibility 374
industry 361–96
current and emerging challenges 365–80
long-term view 388–9
responding to challenges 380–8
structure 362–4
investment 95–6
mining *see* mining
mobile phones and computers 14, 15
ore grades *see* grades
physicochemical characteristics 341
prices *see* prices
recycling 1–22, 37–9, 123, 337, 341–4, 345–9, 354, 365, 381–4
analysis of trace levels 282–5
as autocatalysts 341–2
global benefits 8–13
from high-tech sources 37–9, 341–4, 345–9, 354
from low-grade materials 382–3
specific metals 334, 335, 336
trends and efficiencies 383–4
refining *see* refining
sources and stocks 25
supply 334–6, 365
sustainable utilization 1–22
usage 364, 365
precipitation methods (for separation/purification)
at Asarco (copper from other metals - acid purification unit/APU) 404, 405–7, 409–19
precious metals 357–8
pre-processing, end-of-life waste 16–17
pressurized leaching process of anodic copper slimes in tellurium recovery 416–17
Prevention and Control of Environmental Pollution
by Solid Waste 154–5
by WEEE, Administrative Measures for the 114, 116, 155

prevention and control of pollution (and cleaner production) 490
by heavy metals in China 177–85
Prevention and Control of Pollution from Electronic and Information Products, Ordinance on Management of 135
prices 479–80
increasing/fluctuating/volatile 12–13
precious metals 366, 367, 388
responses to 479–80
precious metals 364
volatility 366, 367, 388
printed circuit (printed wiring) boards (PCBs; PWBs) 58, 68, 121, 123, 140, 141, 152, 347
private sector
e-waste reuse 35
scrap waste 97
processing
e-waste
Chinese companies 158, 159–60
metallurgical *see* metallurgical recovery/processing
for reuse 34
technologies for *see* technology
precious metals, safety 373
re-engineering of processes 491
see also separation
production (of metals)
actinides and lanthanides, ionic liquids for 295–316
as by-product *see* companionality
China 124, 125
drain on/inefficiencies in capacity 181–2
mercury 201
planning in future 193
cleaner *see* prevention and control of pollution
closing the loop *see* closing the loop
from complex matrices *see* complex matrices
copper 400–21, 428, 430
precious metals, increasing costs 369–70
reserve/production ratio 477, 478, 479
steel using biomass 497–503
stock (product stock) 496
with resource stock and material stock 495
sustainability of 471–2
waste *see* waste
see also by-products; co-products *and specific production processes*
Project Zerowin 35
proteins, cofactors of 255–6, 262, 263, 264
public perception of precious metals industry 371–2
purification *see* separation/extraction/purification
push and pull mechanisms 3
pyrolysis (biomass for steelmaking) 499, 501
pyrometallurgy 454
e-waste 42, 70, 163–4, 166
precious metals 382

quality management system at Amarillo copper refinery 421

radionuclide analysis using MRT Rad Disks 286–8
rainwater, analysis of trace levels of rare earth elements in 279
Ramsar List of Wetlands of International Importance 431
Rand Refinery 262, 364
rare earth elements/metals (REEs; f-elements) 6, 24, 123–4, 295–332
analysis of trace levels in rainwater 279
from China *see* China
as critical resources 13, 24
investment 96
ionic liquids for production of 295–316
recycling 123–4
selective recovery from complex matrices 317–32
urban mining 77
see also actinides; lanthanides
raw materials
critical (CRM) 28–9
mining 23
iron ore 127
Ray copper mine 399, 400, 401, 424, 439
reactive oxygen species and bacteria 261–2
rechargeable lamps 57
reclamation of land from tailings 436–47, 450
recovery (of resources)
from complex matrices *see* complex matrices
from e-waste 68
China 151–68
metallurgical *see* metallurgical recovery
technologies *see* technology
element distribution tendencies 456–63
from end-of-life waste 16–19, 74–5, 102–3
precious metals
improved 386–7
from lower-grade materials 385–6

tellurium, at Amarillo Copper Refinery 413–19
see also refining; separation
recycling 1–22, 453–66, 486, 494
for autocatalysts 341–2, 383
China *see* China
e-waste *see* electronic and electrical equipment waste
formal *see* formal sector
fuel cells 354–5
global benefits 8–13
informal *see* informal sector
key aspects of global vision 20
precious and specialty metals *see* platinum group metals; precious metals; specialty metals
sustainable mining and 426, 450
technology (incl. high-technology) products 3, 4, 10, 12, 36–46
reef-boring technology, gold 387
re-engineering of processes 491
refining
aluminum 459
copper, Amarillo 397–423
end-of-life waste (incl. e-waste) 17–18, 42, 163–7
China 163–7
of precious metals 342–3, 347–8
of precious metals 357–9, 362–4, 381–3, 387
end-of-life waste (incl. e-waste) 342–3, 347–8
platinum group metals 362, 363, 370, 381
three main activities 381
see also electrorefining
regional areas
of China (provinces)
prevention and control of pollution 184–5
waste copper and aluminum recycling 129
mineral development, implications 473–6
synergies 493
of World, autocatalyst demand 337–41
see also local discussions of sustainability
regulation *see* legislation and regulations
Regulation for Management of Recycling and Treatment of WEEE (2009 - China) 113, 114, 115–16, 116, 156
remelting processes 454, 464
aluminum 454, 457–63
renewable energy
on disturbed lands 442, 448
precious metal mining and 379
research in e-waste reuse 35–6
reserve/production ratio 477, 478, 479
resins for metal separation 275–9
resistance, bacterial
to antibiotics 265
to toxic metals 258–61
resources *see* sources and resources
respiratory system, non-essential metal uptake 237–9
responsible stewardship 6, 494–6
Restriction of Hazardous Substances (RoHS) in Electrical and Electronic Equipment 30–1, 115, 154
reuse 85–108, 486, 494
developing countries *see* developing countries
from high-tech wastes 31–6
revegetation of copper mine tailings 425, 432, 436–9, 441–2,, 446, 447
rhenium (Re)
selective recovery from complex matrices 323
sources and stocks 27
rhodium (Rh)
demand as autocatalyst 337–41
sources and stocks 25
rice, mercury 200, 210
rock, waste, Asarco and reclamation from deposition areas of 441–2
Rock Creek project 427, 429–30
"rod rinse" solution 420
Romania, precious metal-related environmental catastrophe 374
ROSE process 343
Rouyn-Noranda (northwestern Quebec), cadmium environmental toxicology study 229–32
Russia
palladium 335
platinum 334
ruthenium (Ru), sources and stocks 25
ruthenium (Ru)/platinum (Pt) alloy catalysts, fuel cells 352–4

safety issues
informal recycling 45–6
precious metal mining and processing 373
salts, molten, for metal ore processing 304–7
samarium (Sm), sources and stocks 27
San Mateo, Philippines 92, 94
San Xavier Tailings, land reclamation 441–2

saving rate 474
scandium (Sc), sources and stocks 27
scarce metals *see* critical/finite resources
scavenging 92–3, 94, 96–8
  e-waste 102–5
  on-site 92
schools, e-waste re-use by 34
science of metals 86–9
scrap metal 96–102
  Africa 93
  from electronic goods *see* electronic and electrical equipment waste
  export 99–102
  high-value, precious metals from 381
  theft 101, 105–6
  WWII and demand for 87
  see also waste
screening plastics for hazardous components in e-waste 74–5
seawater, uranium from 307–8
Second World War and scrap metal demand 87
sediment, standards acceptable for contamination 259
selective recovery of rare earth and platinum group metals from complex matrices 317–32
selenium (Se), sources and stocks 27
separation/extraction/purification
  for analysis 271–88
  by-products in *see* by-products
  from complex matrices *see* complex matrices
  copper electrolyte 404–9
  costs of (extraction cost) 479
  from end-of-life waste 16–19, 271
    for autocatalysts 342–3
    e-waste 42–3, 162–3
  precious metals 355–7, 381–2
    from e-waste 347–8
    improved technologies 386–7
  see also specific methods
sewage 174–5, 436
  organic matter recycled from (biosolids) in land reclamation 436–8, 450
ship graveyard, Pakistan 86
shredders and shredding 37
  e-waste 159
silver (Ag)
  bacteria affected by
    Ag nanoparticles in textiles 265
    iron-sulfur clusters of bacteria 262
  demand 364, 365
    increasing 365–6
  electrolytic refining 359
  industry
    refining 362, 363, 364, 381
    structure 362
  mobile phones and computers 15
  price increases and volatility 366, 367
  sources and stocks/reserves 25
    China 126
  uses 364, 365
Silver Bell copper mine 399, 400, 424, 439
Sims Metal Management 95
skin exposure to non-essential metals 239
SKS furnaces 164, 165, 166
slag (from smelting)
  from copper, utilization 448, 450
  distribution of elements in 453–66
slime leaching process in tellurium recovery 415–17
smart phones 29, 89, 119
smelting 42, 167, 168, 382–3, 453–7
  copper 164, 166, 183–4, 401, 404, 425, 456–7
    element distribution tendencies 456–7
    utilization of slag 448, 450
  integration 17–18, 19, 167, 168, 494, 496
  lead 165–6, 184, 190, 457, 459
  precious metal collection 383, 387, 388
  slag *see* slag
  waste treatment 188
  see also blast furnace
social responsibility, corporate 371, 377, 489
sociocultural sustainability and mining 468, 472
socioeconomic enterprises and development, e-waste in 75–6
  re-use 33–5
soil pollution/contamination (terrestrial introduction), toxic/heavy/non-essential metals 224–6
  China 72, 121, 174–5
    legislation 178–9
    mercury 205–8
  standards acceptable for 259
  see also land reclamation
solar power photovoltaics *see* photovoltaics
solid oxide fuel cell (SOFC) 349, 350
solid phase extraction (SPE) 273, 277, 286, 287, 288

ionic liquids in, actinides and lanthanides 303
Molecular Recognition Technology for (MRT-SPE) 272, 273, 279–87
rainwater rare earth elements 279
see also SuperLig®
solid waste
China
prevention and control of pollution by 154–5
recycling from 187–9
mining industry as largest producer of 318
precious metal extraction and refining 347
reducing and eliminating 487, 488
solutions
aqueous, of precious metal, chemistry 356
plating *see* plating solutions
solvent extraction
precious metals 358
rare earth metals 330–1
sorting of metals for recycling
formal sector 44–5
informal sector 39–40, 147
China 140–5, 147
sources and resources (metal) 24–8
abundance of, abundance
Africa 91
critical/limited *see* critical/finite resources
e-waste
China 152–3
as resource for socioeconomic development 75–6
non-renewable and finite *see* critical/finite resources; non-renewable resources
non-essential metals 222–3
precious metals and resource nationalism 371–3
stock, integration 496
with material stock and product stock 495
see also recovery
South Africa
gold industry 362–4, 369
illegal mining 376
informal recycling 95, 97, 98–9
South and Central America *see* Latin America
specialty metals 1–22
dumpsites and scavenging 92–3
recycling 1–22, 37–9
global benefits 8–13
from high-tech sources 37–9
sources and stocks 25–7
sustainable utilization 1–22
speciation, metal 223–5
arsenic, in food analysis 275
SuperLig® products from a wide range of species 321
sphalerite (ZnS), indium recovered (as by-product) 10
spillover effects 470
spiral of exploitation (with non-renewable resources) 385–6
stakeholders 4
standards
heavy metal pollution in China 179–81, 192
industry 183–4
reuse 32, 33
soil/water/sediment contamination 259
State Environmental Protection Administration of China (SEPA) 154, 155, 156
steel
production using biomass 497–503
recycling 119
stewardship
integrated 496
responsible 6, 494–6
stock *see* materials; production; sources and resources *and specific metal elements*
strikes (labor), precious metals industry 375
strontium (Sr)
radioactive, analysis using MRT Rad Disks 286–8
sources and stocks 27
subcellular partitioning of cadmium 231
sub-Saharan Africa 90, 101
substitution 479–80
with precious metals 384–5
Sudbury (Ontario), cadmium environmental toxicology study 229–32
sulfides and sulfates (precipitation method at Asarco) 409
sulfuric acid and nickel carbonate recovery 410
SuperLig® 319–31
bismuth removal at Amarillo Copper Refinery 406–7, 409
precious metals 359
superoxide dismutase 255, 264
supply
of e-waste for reuse 34
precious metals 334–6, 365
rare earth metals from China to Japan 10–12

supported liquid membranes (SLM), ionic liquid-based 303
surface water, China, mercury 208–9
sustainability (basic/general references) 376–8, 468
  definition 376, 424
    sustainable mining 425–7
  economic perspectives 468–84
  many faces of 468–9
  precious metals and 376–8
  week vs strong 475–6
"Sustainable Consumption and Production and Sustainable Industrial Policy (SCP/SIP) Action Plan" (European Commission) 397–8
sustainable development (basic/general references) 376–8, 426, 468

tailings (calcines) 401, 436–42, 448, 450
  copper *see* copper
  mercury 202, 204–5, 205, 208, 209
Taizhou 18, 48, 113, 118, 121, 141–3, 145, 191
takeback schemes, mobile phone 40, 41
Tanaka Kikinzoku Kogyo K.K. 2, 18, 343, 352
tantalum (Ta), sources and stocks 27
Tanzania, artisanal precious metal mining 375–6
target species in rare earth and platinum group metals from complex matrices, high-selectivity for 320–1
techno-economic evaluation 497, 499
technology (incl. high-technology)
  copper refinery 397–399, 402, 404, 410, 421
  for green chemistry trace element analysis 272–3
  for production/recovery/processing 70
    China 160, 187, 192–3
    innovation in, as response to scarcity and unavailability 480
    precious metals 355–9, 386–7
    state-of-the-art, lack 68
  products for 23–51
    alternative sources 10
    precious metals 380
    recycling 3, 4, 10, 12, 36–46
    reuse from waste 31–6
technosphere 9, 13, 89
tellurite and bacterial iron-sulfur clusters 262
tellurium (Te)
  recovery, at Asarco 413–19
  sources and stocks 27
terbium (Tb), sources and stocks 27
terrestrial introduction *see* soil
thallium (Tl), sources and stocks 27
theft, scrap metal 101, 105–6
thermodynamic analysis of element distribution in smelting and remelting processes 453, 454–6, 463
thiols (bacterial) 262
third-sector organizations and e-waste recycling 40
thrifting with precious metals 380, 385
thulium (Tm), sources and stocks 27
Tohono O'odham Nation 441, 448
Tokyo, analysis of trace levels of rare earth elements in rainwater 279
toxic metals (and toxic components) 57–8, 221–52, 257–65
  bacterial response to 257–65
  e-waste 64, 120–1
    example items 56
    recycling 120–1
  pollution/environmental contamination with 121, 170, 221–52
    air, China 121, 175–7, 203–4
  recycling and its benefits regarding 9
  see also heavy metals; non-essential metals
trace element analysis, green chemistry 272–3
transport e-waste, as barrier to reuse 34
triple bottom line concept 376–7, 426
Troy mine 427–9, 430
Tucson Electric Power (TEP) 442, 448
tungsten (W), sources and stocks 27

umbilical cord blood, heavy metal burden, China 122
Umicore 166–7
  integrated smelter 17–18, 167
United States (USA)
  copper industry 397–452
  Bureau of Indian Affairs 441
  Bureau of Land Management 428, 441
  Bureau of Mines 428
  Environmental Protection Agency 442, 448
  Forest Service (USFS) 428, 430–1
  land reclamation 436–42, 443–7
  material flow 86
  precious metals law and regulations 374
  recycling rates 89
  see also North America
upgrading 160, 161–3

uranium (U)
  Australia, ore grades 478
  from seawater 307–8
urban mining *see* anthropogenic sources
urine, heavy metal burden in China 214
use/utilization
  precious metals 336, 337, 364, 365
  waste 491
utilization, sustainable, precious and specialty metals 1–22

vertical integration 494

Wanshan Hg area 201, 203, 204, 205, 208, 209, 211, 212, 213, 214, 215, 216
waste(s) (from mining and production) 9, 480–1, 485–96
  Asarco's (from mining)
    reclamation of waste rock deposition areas 441–2
    reduction projects 419–20
  definition 485–6
  dumps 92–6, 96
  electronic and electrical equipment *see* electronic and electrical equipment waste; WEEE Directives
  hierarchy 486–7
  key aspects of global vision 20–1
  magnitude of problem concerning 7–8
  reduction/minimization/elimination 486–7, 487–96
    Asarco and 419–20
  reuse *see* reuse
  scavenging *see* scavenging
  solid *see* solid waste
  unusable, leaving closed-loop systems 492
  utilization 448, 450
  see also scrap
wastewater
  Arizona 432, 436, 450
  China 171, 173, 174–5, 178, 186–7
water (and freshwater/non-marine aqueous environments)
  heavy metal pollution 280
    analysis 280
    China 171–4, 208
  non-essential metal releases 221, 226
    organism–metal interactions 226
  precious metal mining and water use 380
  sample analysis of pollutants in 279–80
  standards acceptable for contamination 259
  see also aqueous solutions; drinking water; groundwater; rainwater; surface water; wetlands
WEEE (waste electronic and electrical equipment) Directives
  first (2002/96/EC) 29, 30, 53, 55
  new (2012/19/EU) 30, 31
WEEELABEX ("WEEE LABel of EXcellence") 33
wellbeing (economic and human) 471
  originating from mining 474–5
    sustainability of 472–3
West Africa 93, 98, 99, 102, 105
wetlands and Camp Caiman Gold Exploration Project in French Guiana 431
Wohlwill electrolysis/electrorefining 359, 382
woody biomass, steel production using 497–503
work *see* occupation
World War II and scrap metal demand 87
Wuchuan Hg mining 203, 204, 205, 208, 209, 212, 213, 214, 215

Xiangjiang River basin, heavy metal pollution 171, 173, 184–5, 192
Xunyang Hg mine 203, 208

yellow perch, cadmium 231–2, 232, 233
young people as key aspect of global vision 20
ytterbium (Yb), sources and stocks 27
yttrium (Y), sources and stocks 27

Zimbabwe, platinum from 334
Zinc (Zn)
  China 126
    industry 184
    pollution 172
    recycling 187, 188, 190
  smelting 457
zinc sulfide (sphalerite), indium recovered (as by-product) 10
zirconium (Zr), sources and stocks 27

www.ingramcontent.com/pod-product-compliance
Ingram Content Group UK Ltd.
Pitfield, Milton Keynes, MK11 3LW, UK
UKHW051034091026
15624UKWH00011B/430